Electric Machines

Offering a new perspective, this textbook demystifies the operation of electric machines by providing an integrated understanding of electromagnetic fields, electric circuits, numerical analysis, and computer programming. It presents fundamental concepts in a rigorous manner, emphasizing underlying physical modeling assumptions and limitations, and provides detailed explanations of how to implement the finite element method to explore these concepts using Python. It includes explanations of the conversion of concepts into algorithms, and algorithms into code, and examples building in complexity, from simple linear-motion electromagnets to rotating machines. Over 100 theoretical and computational end-of-chapter exercises test understanding, with solutions for instructors and downloadable Python code available online.

Ideal for graduates and senior undergraduates studying electric machines, electric machine design and control, and power electronic converters and power systems engineering, this textbook is also a solid reference for engineers interested in understanding, analyzing, and designing electric motors, generators, and transformers.

Dionysios Aliprantis is Professor of Electrical and Computer Engineering at Purdue University, West Lafayette, Indiana. He has developed and taught numerous undergraduate and graduate courses on electric machines, power electronics, and power systems. He is a Senior Member of the IEEE.

Oleg Wasynczuk is Professor of Electrical and Computer Engineering at Purdue University, West Lafayette, Indiana, and Chief Technical Officer of PC Krause and Associates, Inc. He is a Fellow of the IEEE.

"This book provides an innovative approach that is very relevant, particularly from the electric machine designer's perspective. After providing relevant background to vector and variational calculus, the finite element method and electromagnetic fields, their application to the analysis of electric machines at the design stage is illustrated by application to actual machines. The theoretical material presented is supplemented with numerous solved examples, building blocks of computer code, and exercise problem sets throughout the book to facilitate understanding for the reader. In this respect it is unique, as the common books on electric machines treat the material from the circuit analysis view rather than the basic field perspective as done here."

Om Malik, University of Calgary

"This text provides a valuable explanation of the theoretical and practical issues of solving for electric and magnetic fields in devices that perform energy conversion and position control. From fundamental Maxwell's equations to actual industrial-grade software technics of the finite element method including nonlinearities such as hysteresis, the text gives a path to understanding and implementation."

Peter Sauer, University of Illinois at Urbana-Champaign

Electric Machines

Theory and Analysis Using the Finite Element Method

DIONYSIOS ALIPRANTIS
Purdue University

OLEG WASYNCZUK
Purdue University

CAMBRIDGE
UNIVERSITY PRESS

University Printing House, Cambridge CB2 8BS, United Kingdom

One Liberty Plaza, 20th Floor, New York, NY 10006, USA

477 Williamstown Road, Port Melbourne, VIC 3207, Australia

314–321, 3rd Floor, Plot 3, Splendor Forum, Jasola District Centre, New Delhi – 110025, India

103 Penang Road, #05–06/07, Visioncrest Commercial, Singapore 238467

Cambridge University Press is part of the University of Cambridge.

It furthers the University's mission by disseminating knowledge in the pursuit of education, learning, and research at the highest international levels of excellence.

www.cambridge.org
Information on this title: www.cambridge.org/highereducation/isbn/9781108423748
DOI: 10.1017/9781108529280

First published 2023

Printed in the United Kingdom by TJ Books Ltd, Padstow, Cornwall, 2023

A catalogue record for this publication is available from the British Library.

ISBN 978-1-108-42374-8 Hardback

Additional resources for this publication at www.cambridge.org/Aliprantis_Wasynczuk

To Nadia, Alcmene, and Christopher (DA)

To my family (OW)

Contents

Preface *page* xi
Nomenclature xv

1 Review of Vector Calculus and Electromagnetic Fields 1
1.1 Overview of Maxwell's Equations 2
1.2 Vectors 3
1.2.1 Dot Product 4
1.2.2 Cross Product 8
1.2.3 Vectors in Python 11
1.2.4 Triple Products 13
1.3 Scalar and Vector Fields 15
1.4 Integrals of Scalar and Vector Fields 20
1.4.1 Line Integrals 20
1.4.2 Surface Integrals 23
1.4.3 Volume Integrals 26
1.5 Differential Operators on Scalar and Vector Fields 27
1.5.1 Gradient 27
1.5.2 Divergence 30
1.5.3 Curl 33
1.5.4 The Laplace Operator 41
1.5.5 Useful Identities 42
1.6 Integral Laws 47
1.6.1 The Divergence Theorem or Gauss' Law 47
1.6.2 Stokes' Theorem 52
1.6.3 Green's Identities 53
1.6.4 Green's Theorem in Two Dimensions 55
1.6.5 Helmholtz's Theorem 56
1.7 Case Studies on the Magnetostatic Field 63
1.7.1 Field from a Circular Loop and the Magnetic Dipole 64
1.7.2 Torque on a Magnetic Dipole 66
1.7.3 Field from a Distribution of Magnetic Dipoles 66
1.7.4 The Relationship between B and H in Materials 68
1.7.5 Two-Dimensional Problems 71
1.7.6 Axisymmetric Problems 78

1.7.7 An Elementary Three-Dimensional Problem Solved Using the Method of Moments 81
1.8 Summary 92
1.9 Further Reading 92
Problems 92
References 93

2 **Variational Form of Maxwell's Equations, Energy, and Force** 96
2.1 Calculus of Variations 97
2.1.1 The Euler–Lagrange Equation for a One-Dimensional Problem 98
2.1.2 The Euler–Lagrange Equation for a Two-Dimensional Problem 102
2.1.3 Problems in the Time Domain 104
2.2 Solving Maxwell's Equations with Variational Calculus 107
2.2.1 Functional for Poisson's Equation 107
2.2.2 The Electrostatic Problem 108
2.2.3 The Magnetostatic Problem 112
2.3 Energy Transfer Pathways 120
2.3.1 Conservation of Energy and Poynting's Theorem of Energy Flow 121
2.3.2 Energy Transfer through a Rotating Magnetic Field 122
2.3.3 Energy Balance in Conductors 140
2.3.4 The Coupling Field Energy 146
2.3.5 Core Loss 148
2.3.6 Circuit-Based Energy Flow Analysis 154
2.3.7 Mechanical Energy Transfer: Electromagnetic Force and Torque 162
2.4 Force and Torque Calculations from the Fields 178
2.4.1 Preliminary Modeling Assumptions 178
2.4.2 Torque Calculation Based on a Virtual Distortion of the Air Gap 181
2.4.3 Conservation of Electromagnetic Momentum: The Electromagnetic (Maxwell's) Stress Tensor 182
2.4.4 Force and Torque from Conservation of Energy 200
2.5 Summary 214
2.6 Further Reading 214
Problems 214
References 216

3 **The Finite Element Method** 219
3.1 Solution of Laplace's Equation Using the FEM 220
3.1.1 Functional Stationarity and the Weak Form 221
3.1.2 Single-Element Preliminary Calculations 221
3.1.3 Functional Contribution from a Single Element 223
3.1.4 Assembly of Global System Equation 225
3.2 Solution of Linear Poisson's Equation Using the FEM 230
3.2.1 Setting up the System of Equations 230
3.2.2 Functional Minimization and the Weak Form 234

3.2.3 Minimization of a Quadratic Functional 235
3.2.4 Handling of Boundary Conditions 237
3.3 Building Algorithms for Linear Magnetics 238
3.3.1 Data Structures 238
3.3.2 Building the Stiffness Matrix 242
3.3.3 A First FEM Solver 247
3.4 Calculation of Flux Linkage and Inductance 257
3.4.1 The Average MVP Method for Flux Linkage Calculations 257
3.4.2 Inductance Calculations Using the Flux Linkage 261
3.4.3 Converting Between Physical and Modeled Current Density 262
3.5 Solution of Nonlinear Poisson's Equation Using the FEM 273
3.5.1 Fixed-Point Method 275
3.5.2 Newton–Raphson Method 276
3.6 Building Algorithms for Nonlinear Magnetics 283
3.6.1 Modeling Nonlinear Materials 283
3.6.2 Building the Gradient and Hessian Matrix 288
3.6.3 FEA Program Implementation with Nonlinear Magnetics 291
3.7 Devices with Permanent Magnets 293
3.8 FEA of Axisymmetric Problems 294
3.8.1 Magnetically Linear Axisymmetric Problems 294
3.8.2 Nonlinear Axisymmetric Problems 296
3.8.3 Flux Linkage Calculation 296
3.9 Galerkin's Method 297
3.9.1 Distinguishing Features of Galerkin's Method 297
3.9.2 Energetic Interpretation of Galerkin's Method 300
3.9.3 Galerkin's Method for the Nonlinear Poisson Equation in Magnetic Devices 302
3.10 Summary 306
3.11 Further Reading 307
Problems 307
References 309

4 Electric Machine FEA Implementation Guidelines 311
4.1 Force and Torque Calculation 311
4.1.1 Coenergy Method 312
4.1.2 Maxwell Stress Tensor Methods 315
4.1.3 Virtual Displacement Method 321
4.1.4 An Air-Gap MVP-Based Method 330
4.2 Rotation without Remeshing 333
4.3 Wound-Rotor Synchronous Machines 336
4.3.1 Elementary Two-Phase Round-Rotor Synchronous Machine: Fundamentals of Distributed Windings 337
4.3.2 Elementary Three-Phase Salient-Rotor Synchronous Machine: Basics of Two-Reaction Theory 356

4.3.3 Analysis in *qd* Variables Using Park's Transformation 363
4.3.4 Derivation of a WRSM *qd* Equivalent Circuit from FEA 376
4.4 Permanent-Magnet Synchronous Machines 381
4.4.1 Elementary Surface-Mounted PMSM: Principles of Operation 381
4.4.2 Derivation of a PMSM *qd* Equivalent Circuit from FEA 387
4.4.3 Interior PMSM: Fundamentals of *qd*-Current Control 390
4.5 Switched Reluctance Machines 403
4.5.1 Magnetostatic Analysis 409
4.6 Exploiting Periodicity 413
4.6.1 Analysis of a Single Pole-Pair (Periodic Conditions) 414
4.6.2 Analysis of a Single Pole (Antiperiodic Conditions) 417
4.7 Summary 419
4.8 Further Reading 420
Problems 420
References 423

5 Problems in the Time Domain 425
5.1 Physics of Induced Currents 425
5.1.1 Electric Field in Two-Dimensional Problems 426
5.1.2 Motional EMF in Two-Dimensional Problems 427
5.2 Formulation of Two-Dimensional FEA in the Time Domain 427
5.3 Eddy Current Problems with Analytical Solutions 429
5.3.1 Eddy Currents in a Semi-Infinite Plate 429
5.3.2 Elliptical Model of Magnetic Hysteresis 435
5.3.3 The Complex Poynting Vector 437
5.3.4 The Semi-infinite Plate as a Conductor 444
5.3.5 Semi-infinite Plate with Twin Return Conductor 447
5.3.6 Circular Conductor 448
5.4 Phasor-Based FEA 449
5.4.1 Phasor-Based FEA in Lossy Materials 451
5.4.2 Calculating the *B*-Field Phasors and the Core Loss 452
5.5 Higher-Frequency Winding Model 463
5.6 Time-Stepping FEA 465
5.6.1 Time Stepping Using Newton's Method 466
5.6.2 Time Stepping with External Circuit (No Eddy Currents, No Motion) 468
5.6.3 Time Stepping with External Circuit and Eddy Currents (No Motion) 477
5.6.4 Time Stepping With Motion 478
5.7 Summary 481
5.8 Further Reading 481
Problems 481
References 482

Notes 483
Index 487

Preface

Ἀρχὴ γὰρ λέγεται μὲν ἥμισυ παντὸς
"For it is said that the beginning is the half of every work"
Plato[1]

This book is about devices that convert electrical to mechanical energy and vice versa. In the former case, we have an electric motor; in the latter case, an electric generator. Motors, generators, and transformers (which lack moving parts) are what we commonly call "electric machines." A tremendous number of electric machines have been invented over the last 200 years, so it would be futile to attempt a detailed analysis of each different type. Instead, we should provide engineers with an understanding of how common operating principles are obtained from classical laws of electromagnetism, regardless of rating, size, or application domain.

Hence, in this text everything is explained based on the electromagnetic field permeating a machine, which is governed by Maxwell's partial differential equations. So the field plays the leading role, but the Oscar for supporting role unquestionably goes to the finite element method (FEM) that computes the field. The FEM is a game changer because it allows us to take a look inside real machines with unparalleled resolution and accuracy. This textbook presents a fields-based theory of electric machines jointly with the FEM and its implementation using the Python programming language, in a bid to approach this subject in a pedagogical and, hopefully, more entertaining manner.

Electric Machines in the Twenty-First Century

The scientific area of electromechanical energy conversion is as vibrant as ever. Cost-effective, efficient, and reliable electric machines play a key role in addressing modern society's grand challenge of electrifying our energy and transportation infrastructures to combat climate change. Rotating generators still produce most of our electricity worldwide, while the penetration of semiconductor-based photovoltaic generation remains relatively low. In addition, the proliferation of new hydro, wind, and marine energy applications has increased demand for innovative generator designs. On the other hand, it is estimated that electric motors, such as pumps or fans in industrial, commercial, or residential settings, consume ultimately half of the electricity in the power grid. Stricter regulations concerning motor efficiencies are thus driving advances in motor designs and power electronics. In parallel with this unprecedented power-grid transformation, the transportation sector is undergoing rapid electrification. The commercial success of hybrid and all-electric vehicles, trains, marine vessels, drones, and next-generation aircraft hinges on continuous advancements in motor technology.

The design of electric machinery often begins with empirical rules of thumb or simple formulas that are derived after approximations. This approach is often adequate for obtaining "first-order" estimates of basic dimensions and parameters that come close to meeting nominal specifications. Nevertheless, novel machine designs and stringent performance requirements are challenging conventional wisdom and the validity of such methods. For more accuracy, industry practitioners rely on finite element analysis (FEA), which is the application of the FEM theory to a particular problem. Numerous sophisticated software packages have been developed that help us conduct FEA over multiple physical domains (e.g., electromagnetic, thermal, and mechanical). FEA is often embedded in an optimization loop that adjusts geometric and material parameters to maximize performance. However, FEA-based machine design is computationally demanding even with today's technology, but advances in optimization algorithms and computing hardware are changing this landscape very rapidly.

This text explains the operation of electric machines under a new light, bridging electromagnetic fields, electric circuits, numerical analysis, and computer programming. We do not shy away from presenting fundamental concepts in a rigorous fashion. In doing so, we emphasize underlying physical modeling assumptions and limitations. And all this is tightly integrated with nitty-gritty details of implementing the FEM using a modern language (Python) for programming on today's hardware. We have included examples covering several major classes of electric machines, which increase in complexity as new material is presented. From a simple linear-motion and constant-permeability electromagnet, we progress to devices with nonlinear ferromagnetic materials, and then introduce devices with rotational motion. We eventually reach the point where it becomes possible to analyze realistic permanent-magnet and wound-rotor synchronous machines, switched reluctance machines, and induction machines in the steady state and during transients. Each chapter includes a set of practice exercises, which are of a theoretical or programming nature.

Overview of Contents

The primary focus of this textbook is on the *electromagnetic* analysis of electric machines rather than their thermal or mechanical aspects. In **Chapter 1**, we recall relevant concepts from mathematics (vector calculus) and physics (classical electromagnetism). The reader may have encountered these during the course of an undergraduate degree. Nevertheless, they are reviewed to help progress smoothly throughout the material.

Chapter 1 also highlights basic Python syntax, which is the programming language that we use to implement the FEM.

We begin **Chapter 2** with a concise presentation of the calculus of variations. It has been our experience that this material is new to electrical engineering students. We prove that Maxwell's equations can be equivalently formulated as a variational problem, namely, the minimization of an energy-related functional, which inherently accounts for interface conditions due to material discontinuities. This approach, which forms the basis of the FEM, also provides useful physical insights into the principles of operation of electrical machinery. The second part of Chapter 2 is thus dedicated to an energy-oriented analysis of the magnetic field. These considerations naturally extend into theoretical results regarding the creation of electromagnetic force from the fields.

In **Chapter 3**, we introduce the FEM gently using elementary two-dimensional (2-D) problems involving Laplace and Poisson equations. We then adapt the method to the analysis of simple magnetic devices. We provide a thorough treatment of both magnetically linear and nonlinear problems, while emphasizing programming details. The chapter ends with further insights about Galerkin's method, setting forth the theoretical foundations of the FEM.

Practical FEA implementation guidelines for electric machines are offered next in **Chapter 4**. These include force and torque calculations, handling moving boundaries due to rotation without remeshing, and exploiting multi-pole symmetry by enforcing periodic boundary conditions. Detailed FEA examples are provided for wound-rotor synchronous machines, surface-mounted and interior permanent-magnet synchronous machines, and switched reluctance machines. These serve as a means to explain various important concepts, such as the magnetomotive force of distributed windings, the calculation of flux linkage and torque waveforms, Park's transformation in *qd* variables, and the derivation of equivalent circuits from FEA results.

Chapter 5 presents the analysis of time-varying fields with phasor-based FEA (under sinusoidal steady-state conditions) as well as time-stepping FEA (under arbitrary transients). This material begins with an overview of the physics of induced eddy currents. Finally, we provide details of coupling FEA models with external circuit equations, which is necessary in the analysis of squirrel-cage induction machines, transformer inrush currents, or the short-circuit transients of a synchronous generator.

The end-of-chapter problems (roughly 20 problems in each chapter) are a key pedagogical feature of our book and are tightly integrated with the material. The solutions manual accompanying the text provides answers to exercises that can be solved analytically, such as derivations and mathematical proofs. The programming exercises involve replicating results found in the main text; the reader is asked to write Python programs gluing together the various building blocks of code that are provided. We do not provide solutions to these exercises because there are many ways that such programs can be implemented, and we certainly do not wish to limit the creativity of the reader. Boxes are also sprinkled around the book, highlighting the most significant theoretical results and key equations.

Who Could Benefit by Reading this Book

This book is suitable for graduate-level students pursuing advanced engineering degrees with specialization in electric power systems, power electronics, or machines; and for practicing engineers who may be interested in understanding, analyzing, and designing electric motors, generators, and transformers. An undergraduate student at the senior level should have all the prerequisite knowledge to pick up this book, although our treatment is certainly more demanding than one typically encounters in introductory machine texts. The book contains a wealth of background information on vector calculus, variational calculus, and basic electromagnetism. Based on our experience with teaching this material, it is helpful to devote a few weeks at the beginning of a semester (or an independent study) to review these subjects before proceeding with the FEM and the analysis of electric machines.

The readers of this textbook will be fully capable of programming their own 2-D FEA program using Python, which is freely available and widely used by the scientific community. Apart from the personal satisfaction and sense of achievement gained from reaching this milestone demonstrating a solid grasp of the underlying concepts, such a tool could be useful for educational or research activities. Furthermore, a custom 2-D FEA program could feature the required flexibility for rapidly testing novel machine concepts that may be difficult to analyze using commercial software. This type of analysis (in contrast to a full-scale 3-D study) is lightweight and more than sufficient for obtaining accurate results over a broad range of applications. A reader will also gain a deeper understanding of what takes place "under the hood" of commercial FEA packages, and will be better informed regarding the limitations imposed by our imperfect modeling of the associated physical phenomena.

By trying to make the FEM more accessible, we aim to impart an appreciation for the numerical analysis of electromechanical devices, and perhaps to inspire further study of this subject. Our hope is that this book will serve for the years to come a new generation of engineers who are dedicated to changing our world for the better.

Nomenclature

This is a list of the main variables and mathematical operators encountered throughout this text. Ambiguity in notation may be resolved from context.

Greek-Letter Variables

α	Element basis function
α	Resistivity temperature coefficient
α	Scaling parameter of variation
$\alpha_p, \beta_p, \gamma_p, \delta_p$	Coefficients of p-pole-pair air-gap magnetic field
β	Generalized trapezoidal rule parameter
δ	Dirac delta function
δ	Skin depth
δ_{ij}	Kronecker delta
Δ	Area of triangle
ϵ	Permittivity
ζ	Radii ratio (inner over outer)
$\vec{\eta}, \eta$	Variation shape vector, scalar (test function)
θ	Angular or linear position
θ	Polar angle
$\hat{\boldsymbol{\theta}}$	Polar angle unit vector
θ_r	Rotor angle
λ	Flux linkage of winding
μ	Permeability
μ_0	Permeability of free space
μ_h	Hysteretic permeability
μ_r	Relative permeability
ν	Reluctivity
ξ	Relaxation factor
ρ	Charge density
ρ	Resistivity
ρ_m	Magnetic charge density
σ	Conductivity
$\boldsymbol{\sigma}$	Maxwell stress tensor
τ	Thickness

$\vec{\tau}$, τ	Torque vector, scalar value
$\tilde{\tau}$	Torque per unit depth
ϕ	Azimuthal angle
ϕ	Hat function
$\hat{\boldsymbol{\phi}}$	Azimuthal angle unit vector
ϕ_{pi}, ϕ_{po}	Phase angle of p-pole-pair air-gap magnetic potential at inner, outer boundary
φ	Scalar potential
Φ	Magnetic flux
$\tilde{\Phi}$	Magnetic flux phasor
χ_m	Magnetic susceptibility
ψ	Flux linkage of filament
ω	Electrical frequency
ω	Surface charge density
$\vec{\omega}$, ω	Angular velocity vector, scalar value
ω_m	Rotor angular velocity (mechanical)
ω_p	Angular velocity of p-pole-pair field (mechanical)
Ω	Region in space (the domain of a vector or scalar field)

Latin-Letter Variables

A	Cross-sectional area
$\mathbf{A}$	Array of magnetic potential values
$\vec{\mathbf{A}}$, A	Magnetic potential vector, magnitude
$\tilde{\mathbf{A}}$, $\tilde{A}$	Magnetic potential complex vector, phasor
$\tilde{A}$, $\tilde{\mathbf{A}}$	Modified magnetic potential function, array of values (for axisymmetric problems)
A_p	Function of p-pole-pair air-gap magnetic potential
A_{pi}, A_{po}	Amplitude of p-pole-pair air-gap magnetic potential at inner, outer boundary
$\mathbf{b}$	Source vector for linear FEA
$\vec{\mathbf{B}}$, B	Magnetic flux density vector, magnitude
$\tilde{\mathbf{B}}$, $\tilde{B}$	Magnetic flux density complex vector, phasor
B_{pr}, $B_{p\phi}$	Radial, tangential component of p-pole-pair air-gap B-field
B_{pri}, B_{pro}	Amplitude of radial component of p-pole-pair air-gap B-field at inner, outer boundary
D	Density
$\mathcal{D}$	Domain in 2-D or 3-D space
$\vec{\mathbf{D}}$, D	Electric displacement field vector, magnitude
e	Error
$\vec{\mathbf{E}}$, E	Electric field vector, magnitude
$\tilde{\mathbf{E}}$, $\tilde{E}$	Electric field complex vector, phasor
f	Frequency
$\vec{\mathbf{f}}$	Force density vector
F	Scalar field

$\vec{\mathbf{F}}$	Vector field
$\vec{\mathbf{F}}$, F	Force vector, magnitude
$\tilde{\mathbf{F}}$	Force per unit depth
$\mathcal{F}$	MMF vector
g	Acceleration of gravity
g	Air-gap width
$\mathbf{g}$	Gradient
h	Height
h	Time step
$\mathbf{H}$	Hessian matrix
$\vec{\mathbf{H}}$, H	Magnetic field vector, magnitude
$\tilde{\mathbf{H}}$, $\tilde{H}$	Magnetic field complex vector, phasor
$\mathcal{H}^1$	Hilbert space of functions with square integrable first derivatives
i or I	Current
$\tilde{i}$ or $\tilde{I}$	Current phasor
$\hat{\mathbf{i}}$	x-Axis unit vector of a Cartesian coordinate system
I	Functional
I	Moment of inertia
I_ν	Modified Bessel function of the first kind of order ν
j	The imaginary unit, $j = \sqrt{-1}$
$\hat{\mathbf{j}}$	y-Axis unit vector of a Cartesian coordinate system
$\mathbf{J}$	Jacobian matrix
$\vec{\mathbf{J}}$, J	Current density vector, magnitude
$\vec{\mathbf{J}}_m$, J_m	Magnetization current density vector, magnitude
k_e	Coefficient of eddy current loss
k_h	Coefficient of hysteresis loss
k_{pf}	Packing factor
k_{st}	Stacking factor
$\hat{\mathbf{k}}$	z-Axis unit vector of a Cartesian coordinate system
$\mathbf{K}$	Reference frame transformation matrix
$\vec{\mathbf{K}}$, K	Surface current density vector, magnitude
$\vec{\mathbf{K}}_m$, K_m	Magnetization surface current density vector, magnitude
ℓ	Length
L	Inductance
L	Lagrangian
$\vec{\mathbf{L}}$	Angular momentum
$\mathcal{L}^2$	Space of square integrable functions
m	Energy density of magnetic field
m	Mass
$\vec{\mathbf{m}}$	Magnetic dipole moment
$\vec{\mathbf{M}}$, M	Magnetization vector, magnitude
n	Number of slots
$\hat{\mathbf{n}}$	Unit normal vector
N	Number of turns

p	Basis function coefficient
p	Number of pole-pairs
p	Volumetric power
$\overline{p}$	Time-averaged volumetric power
$\vec{\mathbf{p}}$	Momentum per unit volume
P	Power
$\mathcal{P}$	Permeance
$\vec{\mathbf{P}}$	Momentum
P_{mp}	Mechanical power of p-pole-pair field
$P_{r\to s}$	Air-gap power flow from rotor to stator
q	Basis function coefficient
$\mathbf{q}$	Array of basis function coefficients
q_e	Electric charge of an electron
Q	Electric charge
Q	Reactive power
r	Basis function coefficient
r	Radial distance
r	Residual
$\mathbf{r}$	Array of basis function coefficients
$\mathbf{r}$, $\vec{\mathbf{r}}$	Displacement from the origin vector
$\hat{\mathbf{r}}$	Unit radial vector
$\dot{\mathbf{r}}$	Derivative of displacement vector $\mathbf{r}$ with respect to time
r_i	Inner radius
r_o	Outer radius
R	Radius of a circle through the middle of the air gap
R	Resistance
$\mathcal{R}$	Reluctance
$\hat{\mathbf{s}}$	Unit tangential vector
s_p	Slip of p-pole-pair field
S	Surface
$\mathbf{S}$	Complex Poynting vector
$\mathbf{S}$	Stiffness matrix
$\vec{\mathbf{S}}$	Poynting vector
t	Time
t	Step length
$\hat{\mathbf{t}}$	Unit tangential vector
$\vec{\mathbf{t}}$	Maxwell stress
T	Kinetic energy
$\mathbf{T}$	Matrix multiplying vector of time derivatives of magnetic potentials
v, V	Voltage
$\vec{\mathbf{v}}$, v	Velocity vector, magnitude
V	Potential energy
V	Region (volume) in 2-D or 3-D space
$\tilde{V}$	Voltage phasor

w	Solution guess
w	Width
w^*	True solution
W	Work or energy
W_{ag}	Energy stored in the air gap
W_c	Coenergy
W_f	Coupling field energy
X	Reactance
$\mathcal{X}^N$	N-dimensional trial space
Z	Impedance

Mathematical Operators

$\partial\Omega$	Boundary of region Ω
$\partial y/\partial x$	Partial derivative of y with respect to x
$\dot{x}$	Derivative of x with respect to time
$\lVert\vec{\mathbf{a}}\rVert$	Norm of vector
$\lVert w\rVert$	Norm of function
$\lVert w\rVert_E$	Energetic norm of function
$\mathbf{A}^\top$	Transpose of vector or matrix
$\overline{f}$	Time average of function
∇F	Gradient of scalar field
$\nabla\cdot\vec{\mathbf{F}}$	Divergence of vector field
$\nabla\times\vec{\mathbf{F}}$	Curl of vector field
$\nabla^2 F$	Laplacian of scalar field
$\nabla^2\vec{\mathbf{F}}$	Laplacian of vector field
$\mathbf{a}\cdot\mathbf{b}$	Dot product
$\vec{\mathbf{a}}\times\vec{\mathbf{b}}$	Cross product
δx	Small change in x
δw	Variation of solution
dx	Differential of x
da	Differential surface area
$d\mathbf{a}$	Differential surface area vector
$d\mathbf{r}$	Differential displacement vector
ds	Differential arc length
dv	Differential volume
dy/dx	Total derivative of y with respect to x
(u, v)	Inner product of functions
$(u, v)_E$	Energetic inner product of functions

1 Review of Vector Calculus and Electromagnetic Fields

Μηδείς ἀγεωμέτρητος εἰσίτω μοι τῇ θύρᾳ
"Let no one ignorant of geometry enter through this door"
Inscription on entrance of Plato's Academy[2]

In this chapter, we will be highlighting fundamental concepts of vector calculus, which is the branch of mathematics concerned with the differentiation and integration of vector fields. In doing so, we have assumed that the reader is already familiar with the main concepts of algebra, vectors, mechanics, and classical electromagnetism, most likely in the course of satisfying the requirements of an undergraduate collegiate degree. The material is self-contained to the extent possible, but it is not intended as a rigorous treatment. We will not be attempting to explain ideas such as: What is a derivative and an integral? What is force? What is energy? What is electricity? What is the electromagnetic field? ... Instead, we will emphasize the underlying mathematics (vector calculus) that enables us to *quantify* and *correlate* the physical variables in space.

Our exposition will progress from simple to more complicated ideas. To refresh our memory and support this material, we will be invoking many examples. Whenever possible, these will be based on electromagnetic field problems. The examples will also include relevant results that are necessary for a fields-based understanding and analysis of electric machines, while underpinning the mathematical formalism that we will adhere to for the remainder of this book.

As we set forth these wonderful concepts, we will not remain on a purely theoretical plane. Rather, we will immediately apply these ideas to perform relatively simple computational tasks using Python. These examples should be accessible by anyone with a basic knowledge of computer programming. We will thereby introduce Python (version 3) syntax and some key capabilities, and familiarize the reader with the programming language that we will use to code the finite element method (FEM) calculations. At the end of this chapter, we shall have a working knowledge of Python.

1.1 Overview of Maxwell's Equations

The equations that govern the electromagnetic field are known as **Maxwell's equations**.[3] They involve two vector variables, namely the **electric field** $\vec{\mathbf{E}}$ and the **magnetic flux density** $\vec{\mathbf{B}}$. The fields are created by the presence of a **charge density** ρ or a **current density** $\vec{\mathbf{J}}$ throughout space. In this first form, Maxwell's equations are applicable for problems in vacuum or free space.

Since we are primarily interested in the analysis of devices where the fields permeate matter, we need an alternate formulation of the equations, namely, a formulation in **macroscopic form**. In such form, the equations should capture the aggregate (i.e., space-averaged over many atoms) behavior of matter that is subject to an electromagnetic field. To this end, two additional fields are introduced. These are the **electric displacement field** $\vec{\mathbf{D}}$ and the **magnetic field** $\vec{\mathbf{H}}$, which help us model the effect of electric and magnetic polarization, respectively. We also introduce phenomenological, **constitutive laws**, such as

$$\vec{\mathbf{B}} = f(\vec{\mathbf{H}}) \quad \text{and} \quad \vec{\mathbf{E}} = g(\vec{\mathbf{D}}), \tag{1.1}$$

to describe the relationships of fields within materials.

The macroscopic variant of Maxwell's equations comes in two equivalent forms, namely, differential and integral. In **differential form**, the equations are

$$\text{Gauss'}^4 \text{ law for the electric field:} \quad \nabla \cdot \vec{\mathbf{D}} = \rho, \tag{1.2a}$$

$$\text{Gauss' law for the magnetic field:} \quad \nabla \cdot \vec{\mathbf{B}} = 0, \tag{1.2b}$$

$$\text{Faraday's}^5 \text{ law:} \quad \nabla \times \vec{\mathbf{E}} = -\frac{\partial \vec{\mathbf{B}}}{\partial t}, \tag{1.2c}$$

$$\text{Ampère's}^6 \text{ law:} \quad \nabla \times \vec{\mathbf{H}} = \vec{\mathbf{J}} + \frac{\partial \vec{\mathbf{D}}}{\partial t}. \tag{1.2d}$$

In **integral form**, the equations are

$$\text{Gauss' law for the electric field:} \quad \oiint_{\partial V} \vec{\mathbf{D}} \cdot d\mathbf{a} = \iiint_V \rho \, dv, \tag{1.3a}$$

$$\text{Gauss' law for the magnetic field:} \quad \oiint_{\partial V} \vec{\mathbf{B}} \cdot d\mathbf{a} = 0, \tag{1.3b}$$

$$\text{Faraday's law:} \quad \oint_{\partial S} \vec{\mathbf{E}} \cdot d\mathbf{r} = -\iint_S \frac{\partial \vec{\mathbf{B}}}{\partial t} \cdot d\mathbf{a}, \tag{1.3c}$$

$$\text{Ampère's law:} \quad \oint_{\partial S} \vec{\mathbf{H}} \cdot d\mathbf{r} = \iint_S \vec{\mathbf{J}} \cdot d\mathbf{a} + \iint_S \frac{\partial \vec{\mathbf{D}}}{\partial t} \cdot d\mathbf{a}. \tag{1.3d}$$

These are supplemented by an equation describing the **conservation of charge**:

$$\nabla \cdot \vec{\mathbf{J}} = -\frac{\partial \rho}{\partial t}. \tag{1.4}$$

When the sources are constant or changing "slowly" with time, we obtain simpler versions of Maxwell's equations by setting $\partial/\partial t = 0$:

$$\nabla \cdot \vec{\mathbf{D}} = \rho\,, \tag{1.5a}$$

$$\nabla \cdot \vec{\mathbf{B}} = 0\,, \tag{1.5b}$$

$$\nabla \times \vec{\mathbf{E}} = \mathbf{0}\,, \tag{1.5c}$$

$$\nabla \times \vec{\mathbf{H}} = \vec{\mathbf{J}}\,, \tag{1.5d}$$

$$\nabla \cdot \vec{\mathbf{J}} = 0\,. \tag{1.5e}$$

Clearly, this leads to a decoupling of the equations, so the electric and magnetic fields can be determined separately.

In practice, we encounter two types of problems, differentiated based on the main field source. If the field is created by a static or quasi-static distribution of charge, we obtain an **electrostatic problem**, where $\vec{\mathbf{B}} = \mathbf{0}$ or $\vec{\mathbf{B}} \approx \mathbf{0}$, and

$$\nabla \cdot \vec{\mathbf{D}} = \rho\,, \tag{1.6a}$$

$$\nabla \times \vec{\mathbf{E}} = \mathbf{0}\,. \tag{1.6b}$$

If the field is created by a static or quasi-static distribution of current, we obtain a **magnetostatic problem**, where $\vec{\mathbf{E}} = \mathbf{0}$ or $\vec{\mathbf{E}} \approx \mathbf{0}$, and

$$\nabla \cdot \vec{\mathbf{B}} = 0\,, \tag{1.7a}$$

$$\nabla \times \vec{\mathbf{H}} = \vec{\mathbf{J}}\,. \tag{1.7b}$$

All this may already seem overwhelming, but we hope that the material that follows will be helpful. In the subsequent sections, we will present background information on vector calculus that will enable us to comprehend what these formulas mean. We will start in §1.2 by defining vectors and mathematical operations on vectors, such as the dot and cross products. We will proceed in §1.3 by defining scalar and vector fields. In §1.4, we will define line, surface, and volume integrals, like the ones appearing in the integral form of Maxwell's equations. §1.5 will set forth differential operators on scalar and vector fields, whereas §1.6 will present various integral laws that are widely employed in vector calculus. We will conclude by putting a focus on the magnetic field in §1.7.

1.2 Vectors

Vector calculus entails the manipulation of **vectors** in three-dimensional (3-D) space, which are commonly used to describe physical quantities like force or velocity. These physical vectors are geometric objects that are associated with a magnitude and a direction. They are denoted using boldface notation and an arrow. For example, the magnetic flux density at a point (x, y, z) may be the vector

$$\vec{\mathbf{B}} = -0.5\,\hat{\mathbf{i}} + 1.2\,\hat{\mathbf{j}} + 0.5\,\hat{\mathbf{k}}\ \ \mathrm{T}, \tag{1.8}$$

where T stands for tesla,[7] which is the SI[8] unit for field density. Here, $\hat{\mathbf{i}}$, $\hat{\mathbf{j}}$, and $\hat{\mathbf{k}}$ are commonly used symbols for the three unit vectors of a Cartesian[9] coordinate system. Unit vectors will be denoted by hats throughout this text. In Equation (1.8), the vector was written as a vector sum of three components. A different way to express a vector is by listing three scalar coefficients in a given coordinate system as a tuple, e.g.

$$\mathbf{B} = (-0.5, 1.2, 0.5)\,, \tag{1.9}$$

or as a column vector

$$\mathbf{B} = \begin{bmatrix} -0.5 \\ 1.2 \\ 0.5 \end{bmatrix}. \tag{1.10}$$

Note that we did not use an arrow. The distinction is subtle, and often these two notations are used interchangeably. For instance, we may also write $\vec{\mathbf{B}} = (-0.5, 1.2, 0.5)$.

The **length** or **magnitude** or Euclidean[10] **norm** of a vector $\vec{\mathbf{v}}$ will be denoted as $||\vec{\mathbf{v}}||$. The norm is calculated using the Pythagorean[11] theorem. For instance, the magnitude of the above B-vector is

$$||\vec{\mathbf{B}}|| = \sqrt{(-0.5)^2 + 1.2^2 + 0.5^2} \approx 1.393 \text{ T}. \tag{1.11}$$

We may also define vectors in higher dimensions; however, these may not be representative of physical vector fields. These high-dimensional vectors will be denoted using a boldface symbol but without an arrow. For example, the solution of a linear system of equations with 100 unknowns, which is a column vector of dimension 100×1, may be denoted as $\mathbf{x}$. Strictly speaking, placing an arrow on top of a physical vector is not necessary since a boldface symbol should suffice. Nevertheless, this notational distinction is made for added clarity because in this book we will encounter both physical and higher-dimensional vectors stemming from the numerical solution of Maxwell's equations. The Euclidean norm for such vectors is denoted similarly by $||\mathbf{x}||$. If $\mathbf{x}$ has n components, then

$$||\mathbf{x}|| = \sqrt{x_1^2 + x_2^2 + \cdots + x_n^2}\,. \tag{1.12}$$

1.2.1 Dot Product

The **dot product** or **inner product** between two vectors $\mathbf{a} = (a_1, a_2, \ldots, a_n)$ and $\mathbf{b} = (b_1, b_2, \ldots, b_n)$ from the same n-dimensional space (in Cartesian coordinates) is defined algebraically as

$$\mathbf{a} \cdot \mathbf{b} = \mathbf{a}^\top \mathbf{b} = \sum_{i=1}^{n} a_i b_i = a_1 b_1 + a_2 b_2 + \cdots + a_n b_n\,. \tag{1.13}$$

The dot product returns a scalar, that is, $\mathbf{a} \cdot \mathbf{b} \in \mathbb{R}$. Note that the vector dimension n is arbitrary. Using this definition, we can see that

$$\mathbf{a} \cdot \mathbf{a} = ||\mathbf{a}||^2\,. \tag{1.14}$$

An equivalent geometric definition is

$$\vec{\mathbf{a}} \cdot \vec{\mathbf{b}} = \|\vec{\mathbf{a}}\|\,\|\vec{\mathbf{b}}\| \cos\theta\,, \tag{1.15}$$

where θ is the angle between the two vectors, such that $0 \leq \theta \leq 180°$ (see Figure 1.1). The latter definition is easier to visualize in two or three dimensions, hence the arrow notation. Note that

$$\mathbf{a} \cdot \mathbf{b} = \mathbf{b} \cdot \mathbf{a}\,. \tag{1.16}$$

This is called the **commutative** property. Another important property of the dot product is related to orthogonality: *Two vectors are* ***orthogonal*** *if and only if their dot product is zero.*

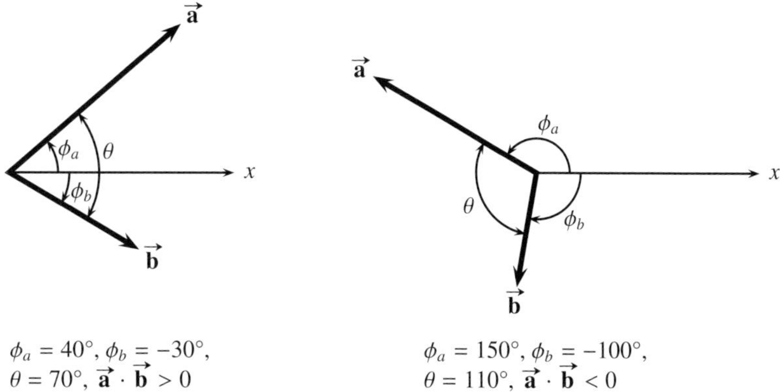

Figure 1.1 Dot product examples. Note that vector angles are positive when measured counterclockwise, and negative in the clockwise direction.

Example 1.1 *Equivalence of dot product definitions.*
Confirm that the algebraic and geometric definitions of the dot product are equivalent.

Proof For simplicity, we will solve this problem in two dimensions. (Even in 3-D space, without loss of generality we can always use a 2-D coordinate system that sits on the plane defined by the two vectors.) The trick is to express the vectors

$$\vec{\mathbf{a}} = a_1\hat{\mathbf{i}} + a_2\hat{\mathbf{j}}\,, \tag{1.17a}$$

$$\vec{\mathbf{b}} = b_1\hat{\mathbf{i}} + b_2\hat{\mathbf{j}} \tag{1.17b}$$

using their **polar coordinates**, that is, using their magnitudes and angles:

$$\mathbf{a} = (a_1, a_2) = (\|\vec{\mathbf{a}}\| \cos\phi_a, \|\vec{\mathbf{a}}\| \sin\phi_a)\,, \tag{1.18a}$$

$$\mathbf{b} = (b_1, b_2) = (\|\vec{\mathbf{b}}\| \cos\phi_b, \|\vec{\mathbf{b}}\| \sin\phi_b)\,. \tag{1.18b}$$

Let us limit the vector angles to the interval $(-180°, 180°]$. Without loss of generality, we will assume that $\phi_a \geq \phi_b$, as in Figure 1.1. Hence

$$\vec{\mathbf{a}} \cdot \vec{\mathbf{b}} = a_1 b_1 + a_2 b_2 \tag{1.19a}$$

$$= \|\vec{\mathbf{a}}\| \, \|\vec{\mathbf{b}}\| \, (\cos\phi_a \cos\phi_b + \sin\phi_a \sin\phi_b) \tag{1.19b}$$

$$= \|\vec{\mathbf{a}}\| \, \|\vec{\mathbf{b}}\| \cos(\phi_a - \phi_b)\,, \tag{1.19c}$$

where $\Delta\phi = \phi_a - \phi_b$ is in the interval $[0°, 360°)$. It suffices to show that $\cos\Delta\phi = \cos\theta$, where the angle between vectors $\theta \in [0°, 180°]$. There are two cases to consider: (i) if $0° \leq \Delta\phi \leq 180°$, then $\theta = \Delta\phi$; (ii) if $180° < \Delta\phi < 360°$, then $\theta = 360° - \Delta\phi$. In both cases, we obtain $\cos\Delta\phi = \cos\theta$. □

Example 1.2 *Invariance of dot product to axes rotation or reflection.*
Confirm that the algebraic definition of the dot product yields the same answer regardless of the choice of coordinate system.

Proof It is convenient to use 2-D Cartesian coordinate systems lying on the plane defined by the two vectors. We need to verify that the dot product remains the same in a **rotated** and possibly **reflected** coordinate system that maintains the length of the vectors. The more general case where the new coordinate system is such that all three components are present is left as an exercise for the reader. (The geometric dot product definition does not employ the coordinates of the vectors, implying that the result should indeed be independent of choice of axes. Nevertheless, this example is instructive because such coordinate system changes are fundamental to electric machine analysis.) In linear algebra, this is called a **change of basis** from one orthonormal basis to another, also termed an **orthogonal transformation**.

As depicted in Figure 1.2, a vector $\vec{\mathbf{a}}$ can be expressed in terms of the unit vectors of the two coordinate systems as

$$\vec{\mathbf{a}} = a_1 \hat{\mathbf{i}} + a_2 \hat{\mathbf{j}} = a_1' \hat{\mathbf{i}}' + a_2' \hat{\mathbf{j}}'\,. \tag{1.20}$$

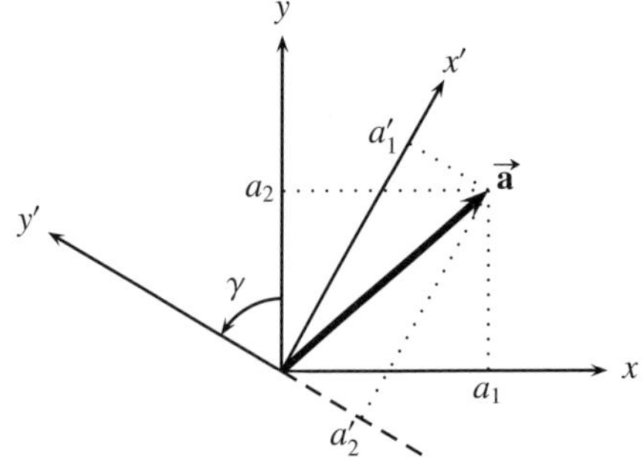

(a) Rotation of the axes by γ in the counterclockwise direction ($a_2' < 0$).

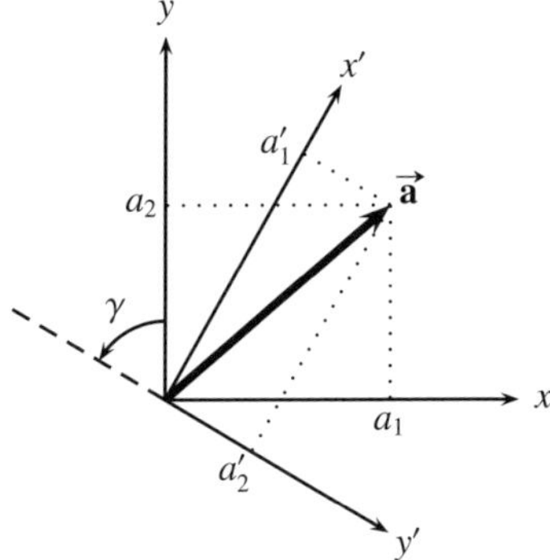

(b) Rotation by γ and subsequent reflection of the y'-axis across the x'-axis ($a_2' > 0$).

Figure 1.2 Cartesian coordinate system orthogonal transformation: from xy to $x'y'$.

By inspection, we can relate the unit vectors as

$$\hat{\mathbf{i}}' = \cos\gamma\,\hat{\mathbf{i}} + \sin\gamma\,\hat{\mathbf{j}}, \tag{1.21a}$$

$$\hat{\mathbf{j}}' = -\sin\gamma\,\hat{\mathbf{i}} + \cos\gamma\,\hat{\mathbf{j}} \tag{1.21b}$$

for a pure rotation, or

$$\hat{\mathbf{i}}' = \cos\gamma\,\hat{\mathbf{i}} + \sin\gamma\,\hat{\mathbf{j}}, \tag{1.22a}$$

$$\hat{\mathbf{j}}' = \sin\gamma\,\hat{\mathbf{i}} - \cos\gamma\,\hat{\mathbf{j}} \tag{1.22b}$$

for a rotation followed by a reflection of the y'-axis. Substitution of these equations in Equation (1.20) yields

$$\vec{\mathbf{a}} = a_1\hat{\mathbf{i}} + a_2\hat{\mathbf{j}} = (a_1' \cos\gamma \mp a_2' \sin\gamma)\,\hat{\mathbf{i}} + (a_1' \sin\gamma \pm a_2' \cos\gamma)\,\hat{\mathbf{j}}\,. \tag{1.23}$$

By equating coefficients, we obtain

$$\begin{bmatrix} a_1 \\ a_2 \end{bmatrix} = \begin{bmatrix} \cos\gamma & -\sin\gamma \\ \sin\gamma & \cos\gamma \end{bmatrix} \begin{bmatrix} a_1' \\ a_2' \end{bmatrix} \tag{1.24}$$

and

$$\begin{bmatrix} a_1 \\ a_2 \end{bmatrix} = \begin{bmatrix} \cos\gamma & \sin\gamma \\ \sin\gamma & -\cos\gamma \end{bmatrix} \begin{bmatrix} a_1' \\ a_2' \end{bmatrix}, \tag{1.25}$$

respectively, for the two cases. It is convenient to introduce a 2×2 **transformation matrix M** that defines the transformation from the original to the new coordinates, which is found by taking the inverse of the above matrices. For case (a), which involves only a rotation, we have

$$\begin{bmatrix} a_1' \\ a_2' \end{bmatrix} = \underbrace{\begin{bmatrix} \cos\gamma & \sin\gamma \\ -\sin\gamma & \cos\gamma \end{bmatrix}}_{\mathbf{M}_a} \begin{bmatrix} a_1 \\ a_2 \end{bmatrix}, \tag{1.26}$$

whereas for case (b), which involves rotation and reflection, we have

$$\begin{bmatrix} a_1' \\ a_2' \end{bmatrix} = \underbrace{\begin{bmatrix} \cos\gamma & \sin\gamma \\ \sin\gamma & -\cos\gamma \end{bmatrix}}_{\mathbf{M}_b} \begin{bmatrix} a_1 \\ a_2 \end{bmatrix}. \tag{1.27}$$

The components of a second vector $\vec{\mathbf{b}}$ are transformed in the same manner. Hence, the dot product using the new coordinates becomes

$$\vec{\mathbf{a}} \cdot \vec{\mathbf{b}} = a_1'b_1' + a_2'b_2' = \begin{bmatrix} a_1' & a_2' \end{bmatrix} \begin{bmatrix} b_1' \\ b_2' \end{bmatrix} \tag{1.28a}$$

$$= \left(\begin{bmatrix} a_1 & a_2 \end{bmatrix} \mathbf{M}^{\top}\right)\left(\mathbf{M}\begin{bmatrix} b_1 \\ b_2 \end{bmatrix}\right) \tag{1.28b}$$

$$= \begin{bmatrix} a_1 & a_2 \end{bmatrix} (\mathbf{M}^{\top}\mathbf{M}) \begin{bmatrix} b_1 \\ b_2 \end{bmatrix}, \tag{1.28c}$$

where $\mathbf{M}^{\top}$ is the **transpose** of **M**. It can be readily verified that $\mathbf{M}^{\top}\mathbf{M} = \mathbb{I}$, the identity matrix, for both cases. Hence, the dot product is invariant to orthogonal transformations.

□

1.2.2 Cross Product

We will define the **cross product** or **vector product** between two vectors $\vec{\mathbf{a}}$ and $\vec{\mathbf{b}}$ in 3-D space. The cross product yields a vector that satisfies the following conditions: (i) it is perpendicular to both $\vec{\mathbf{a}}$ and $\vec{\mathbf{b}}$; (ii) its direction is decided by the right-hand rule; and (iii) its magnitude is equal to the area of the parallelogram defined by $\vec{\mathbf{a}}$ and $\vec{\mathbf{b}}$ as its sides, as illustrated in Figure 1.3. We can write, therefore

$$\vec{\mathbf{a}} \times \vec{\mathbf{b}} = \|\vec{\mathbf{a}}\| \, \|\vec{\mathbf{b}}\| \sin\theta \, \hat{\mathbf{n}} \,. \tag{1.29}$$

Here, θ is the angle between the two vectors, so that $0 \leq \theta \leq 180°$. Also, $\hat{\mathbf{n}}$ is a unit normal vector that is perpendicular to the plane defined by $\vec{\mathbf{a}}$ and $\vec{\mathbf{b}}$, and points to a direction dictated by the right-hand rule. See Figure 1.4 for two examples. Note that

$$\vec{\mathbf{a}} \times \vec{\mathbf{b}} = -\vec{\mathbf{b}} \times \vec{\mathbf{a}} \,. \tag{1.30}$$

Hence, the cross product is **anti-commutative**. In contrast to the dot product, the magnitude of the cross product becomes zero when the vectors are parallel or antiparallel (i.e., when $\theta = 0$ or $\theta = 180°$, respectively). Examples of formulas from physics that employ the cross product are the torque, $\vec{\boldsymbol{\tau}} = \vec{\mathbf{r}} \times \vec{\mathbf{F}}$, and the Poynting[12] vector, $\vec{\mathbf{S}} = \vec{\mathbf{E}} \times \vec{\mathbf{H}}$.

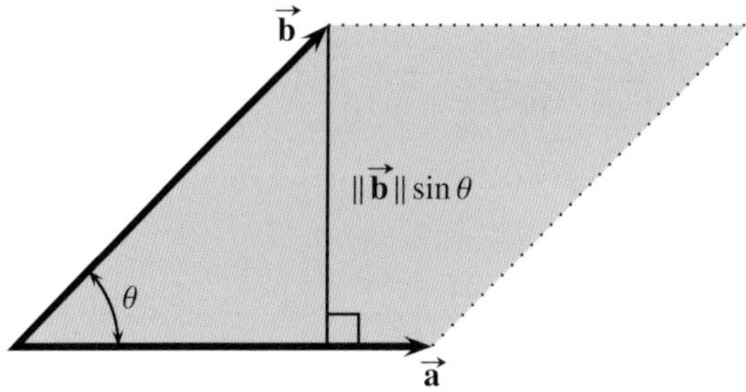

Figure 1.3 Geometric interpretation of the cross-product magnitude as the area of a parallelogram (base times height) formed by two vectors.

In Cartesian coordinates, the cross product can be evaluated as a determinant. If $\vec{\mathbf{a}} = (a_1, a_2, a_3)$ and $\vec{\mathbf{b}} = (b_1, b_2, b_3)$, then

$$\vec{\mathbf{a}} \times \vec{\mathbf{b}} = \begin{vmatrix} \hat{\mathbf{i}} & \hat{\mathbf{j}} & \hat{\mathbf{k}} \\ a_1 & a_2 & a_3 \\ b_1 & b_2 & b_3 \end{vmatrix} . \tag{1.31}$$

Using a cofactor expansion along the top row, this determinant evaluates to

$$\vec{\mathbf{a}} \times \vec{\mathbf{b}} = \begin{vmatrix} a_2 & a_3 \\ b_2 & b_3 \end{vmatrix} \hat{\mathbf{i}} - \begin{vmatrix} a_1 & a_3 \\ b_1 & b_3 \end{vmatrix} \hat{\mathbf{j}} + \begin{vmatrix} a_1 & a_2 \\ b_1 & b_2 \end{vmatrix} \hat{\mathbf{k}} \tag{1.32a}$$

$$= (a_2 b_3 - a_3 b_2)\,\hat{\mathbf{i}} + (a_3 b_1 - a_1 b_3)\,\hat{\mathbf{j}} + (a_1 b_2 - a_2 b_1)\,\hat{\mathbf{k}} \,. \tag{1.32b}$$

Strictly speaking, this is not a determinant of an actual matrix, but merely a mnemonic.

A special case of particular interest for us is the multiplication of vectors that sit on

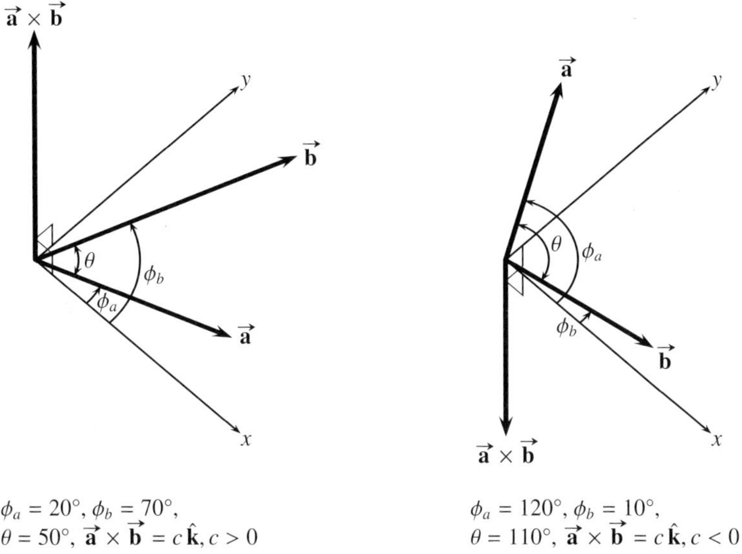

Figure 1.4 Cross-product examples. The vectors $\vec{\mathbf{a}}$ and $\vec{\mathbf{b}}$ are on the x–y plane.

the x–y plane, as shown in Figure 1.4. This means that $a_3 = b_3 = 0$, so the first two terms in the previous expression vanish, and we are left with

$$\vec{\mathbf{a}} \times \vec{\mathbf{b}} = (a_1 b_2 - a_2 b_1)\,\hat{\mathbf{k}}\,. \tag{1.33}$$

Example 1.3 *Cross product as a determinant.*

Confirm the validity of the formal determinant expression (1.31).

Proof Note that the standard basis vectors are defined such that

$$\hat{\mathbf{i}} \times \hat{\mathbf{j}} = \hat{\mathbf{k}}\,, \tag{1.34a}$$
$$\hat{\mathbf{j}} \times \hat{\mathbf{k}} = \hat{\mathbf{i}}\,, \tag{1.34b}$$
$$\hat{\mathbf{k}} \times \hat{\mathbf{i}} = \hat{\mathbf{j}}\,. \tag{1.34c}$$

This is the usual case for right-handed systems. Also, recall that the vector product is **distributive**, i.e.

$$\vec{\mathbf{a}} \times (\vec{\mathbf{b}} + \vec{\mathbf{c}}) = \vec{\mathbf{a}} \times \vec{\mathbf{b}} + \vec{\mathbf{a}} \times \vec{\mathbf{c}}\,. \tag{1.35}$$

Evaluating the cross product yields

$$\vec{\mathbf{a}} \times \vec{\mathbf{b}} = (a_1\hat{\mathbf{i}} + a_2\hat{\mathbf{j}} + a_3\hat{\mathbf{k}}) \times (b_1\hat{\mathbf{i}} + b_2\hat{\mathbf{j}} + b_3\hat{\mathbf{k}})\,. \tag{1.36}$$

Multiplying out the terms will lead to a sum of nine cross products. Three of these terms will be zero, since $\hat{\mathbf{i}} \times \hat{\mathbf{i}} = \hat{\mathbf{j}} \times \hat{\mathbf{j}} = \hat{\mathbf{k}} \times \hat{\mathbf{k}} = \vec{\mathbf{0}}$. The remaining six terms will contain cross products as in Equations (1.34a)–(1.34c), or in reverse order. Using the anti-commutative property (1.30) will lead to Equation (1.32b), which is the expansion of the determinant in Equation (1.31). □

Example 1.4 *The area of a triangle.*
Derive a formula for the area of a triangle based on the coordinates of its vertices.

Proof This result is invoked repeatedly in the FEM, so it is worthwhile taking a closer look. The proof hinges on the definition of the cross product. Consider the triangle ABC on the x–y plane, shown in Figure 1.5. The triangle vertices are located at (x_1, y_1), (x_2, y_2), and (x_3, y_3). In the ensuing analysis, it is necessary to assume that the vertices are ordered *counterclockwise*. Hence, if we define the vectors $\vec{\mathbf{c}}$ and $\vec{\mathbf{a}}$ as the sides AB and BC of the triangle, respectively, then $\vec{\mathbf{c}} \times \vec{\mathbf{a}}$ will be pointing along the positive z-axis. (To see this, you may translate $\vec{\mathbf{c}}$ mentally so that its origin coincides with the origin of $\vec{\mathbf{a}}$.) In turn, this implies that $\|\vec{\mathbf{c}} \times \vec{\mathbf{a}}\|$, which is the area of the parallelogram $ABCD$, equals the z-component of $\vec{\mathbf{c}} \times \vec{\mathbf{a}}$, as given by Equation (1.33). Denoting the triangle area by Δ, we have

$$\Delta = \frac{1}{2}\|\vec{\mathbf{c}} \times \vec{\mathbf{a}}\| = \frac{1}{2}(\vec{\mathbf{c}} \times \vec{\mathbf{a}}) \cdot \hat{\mathbf{k}} \tag{1.37a}$$

$$= \frac{1}{2}(c_1 a_2 - c_2 a_1) \tag{1.37b}$$

$$= \frac{1}{2}\left[(x_2 - x_1)(y_3 - y_2) - (y_2 - y_1)(x_3 - x_2)\right]. \tag{1.37c}$$

What if we had chosen another pair of sides? For instance, take $\vec{\mathbf{b}}$ to be the side CA. Then, $\vec{\mathbf{b}} \times \vec{\mathbf{c}}$ points towards the positive z-axis as well, and

$$\Delta = \frac{1}{2}\|\vec{\mathbf{b}} \times \vec{\mathbf{c}}\| = \frac{1}{2}(\vec{\mathbf{b}} \times \vec{\mathbf{c}}) \cdot \hat{\mathbf{k}} \tag{1.38a}$$

$$= \frac{1}{2}(b_1 c_2 - b_2 c_1) \tag{1.38b}$$

$$= \frac{1}{2}\left[(x_1 - x_3)(y_2 - y_1) - (y_1 - y_3)(x_2 - x_1)\right]. \tag{1.38c}$$

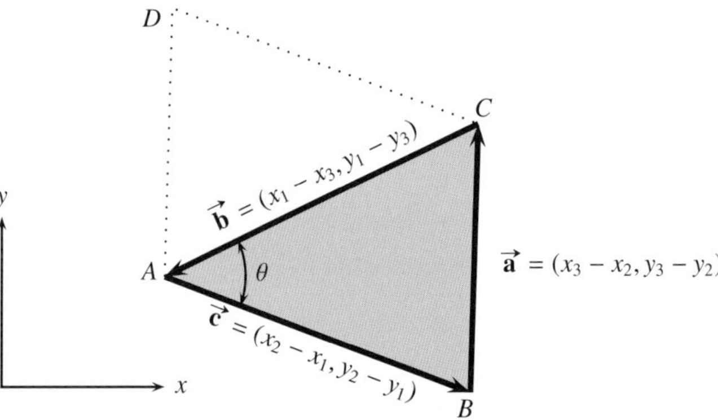

Figure 1.5 Calculating the area of a triangle on the x–y plane based on the cross product.

The last combination we could take is $\vec{\mathbf{a}} \times \vec{\mathbf{b}}$. This yields

$$\Delta = \frac{1}{2}\|\vec{\mathbf{a}} \times \vec{\mathbf{b}}\| = \frac{1}{2}(\vec{\mathbf{a}} \times \vec{\mathbf{b}}) \cdot \hat{\mathbf{k}} \tag{1.39a}$$

$$= \frac{1}{2}(a_1 b_2 - a_2 b_1) \tag{1.39b}$$

$$= \frac{1}{2}\left[(x_3 - x_2)(y_1 - y_3) - (y_3 - y_2)(x_1 - x_3)\right] . \tag{1.39c}$$

Note that these three formulas are similar in the sense that one leads to the other via cyclic permutation of the indices, i.e., replacing 1 by 2, 2 by 3, and 3 by 1. □

1.2.3 Vectors in Python

We proceed with a few examples that illustrate the application of the basic vector concepts we have presented using the Python programming language.

Example 1.5 *Vector operations in Python.*
In Python, calculations with vectors can be performed efficiently using the NumPy library. NumPy functions can operate on n-dimensional arrays, or **ndarray** objects. We present a simple Python program that calculates the length of the B-field vector introduced in Equation (1.8):

```
import numpy as np

# Define a vector as a NumPy array
B = np.array([-0.5, 1.2, 0.5])
print('B =',B)

# Calculate the norm of the vector
normB = np.linalg.norm(B)
print('||B|| =',normB)
```

Since this is our first encounter with a Python script, we note the following:

- To run this program, we may type each line successively in a Python interpreter, which can be started by typing `python` at the Unix shell or a Windows command prompt. Alternatively, these commands can be saved into a file that we can run by typing `python` <`filename.py`>.
- The line numbers are only shown for convenience when referring to certain parts of the code. They are not part of a Python program, unlike BASIC or FORTRAN.
- Single-line comments are preceded by a # character.

- NumPy commands are not loaded by default in Python. This is why we first **`import`** the NumPy library in line 1. All NumPy commands are then preceded by the `np` prefix, e.g., see the definition of the *B*-field in line 4. This naming convention is commonly followed in NumPy programs. In later examples, even though this command may not be always explicitly shown, you may rest assured that it is always there.
- The *B*-vector could have been defined more simply as `B = [-0.5, 1.2, 0.5]`. In Python this creates a **list** object, which is quite different from a NumPy ndarray. The program would still run without error, as NumPy would convert internally this list to an ndarray.
- Even though this is not shown in the program, it is noteworthy that *Python lists and arrays are indexed starting at zero*. To access the first element in the vector, we would type `B[0]`, for the second element, we would type `B[1]`, and so on. We can also access the last element as `B[-1]`, the second to last as `B[-2]`, etc. Note the use of brackets (rather than parentheses).
- In line 8, the `norm` command is defined inside the linear algebra (linalg) package of NumPy, so it is accessed by typing `np.linalg.norm`.
- The output of the script, which is generated by the two **`print`** commands, is

  ```
  B = [-0.5  1.2  0.5]
  ||B|| = 1.39283882772
  ```

  Without these commands, nothing would appear on the screen. In a Matlab program, the absence of a semicolon after a variable definition leads to the variable being displayed in the command window. However, as you can see from this example, semicolons are not used in Python to terminate commands.

Example 1.6 *Dot product using NumPy.*
Suppose we want to calculate the inner product of two vectors, $\mathbf{a} = (1, 2, 3, 4, 5)$ and $\mathbf{b} = (6, 7, 8, 9, 10)$. This can be achieved using the NumPy `dot` command. By running the following program, the reader can verify that $c = \mathbf{a} \cdot \mathbf{b} = 130$.

```
a = np.array([1, 2, 3, 4, 5])
b = np.array([6, 7, 8, 9, 10])
c = np.dot(a,b)
```

NumPy also provides a `cross` command, which can be invoked to evaluate cross products of vectors in $\mathbb{R}^2$ or $\mathbb{R}^3$.

Example 1.7 *Element-wise operations on vectors in Python.*
When programming the FEM, it is often required to operate element-wise on high-dimensional vectors. To this end, we may use various built-in NumPy functions, such as `sin` and `sqrt` (i.e., square root). For instance, suppose we have stored the x and y-components of a 2-D *B*-field at some points in our device in two vectors. Here is a Python program that calculates the magnitude of the *B*-field at each point. Arbitrary values have been set for the *B*-field at three points.

```
Bx = np.array([ 1, 0, 3])
By = np.array([-1, 2, 4])
normB = np.sqrt(np.square(Bx) + np.square(By))
print('||B|| =',normB)
```

The program returns

```
||B|| = [ 1.41421356  2.          5.        ]
```

The reader may readily verify that the mathematical operations were performed element-wise.

1.2.4 Triple Products

The following triple-product identities are useful for manipulating vector fields.

The **scalar triple product** involves the dot and cross products of three vectors, in the sense $\vec{\mathbf{a}} \cdot (\vec{\mathbf{b}} \times \vec{\mathbf{c}})$. Using elementary geometry, we can show that

$$\vec{\mathbf{a}} \cdot (\vec{\mathbf{b}} \times \vec{\mathbf{c}}) = \pm(\text{volume of parallelepiped}), \tag{1.40}$$

where the parallelepiped is the one defined by the three vectors as 3 (of its 12) edges, as shown in Figure 1.6. The parentheses may be dropped since there can be no ambiguity in the order of these operations, and we could write instead $\vec{\mathbf{a}} \cdot \vec{\mathbf{b}} \times \vec{\mathbf{c}}$. (We cannot dot-multiply $\vec{\mathbf{a}}$ and $\vec{\mathbf{b}}$ first because this yields a scalar that cannot be subsequently cross-multiplied with $\vec{\mathbf{c}}$.) It is easy to show that the scalar triple product is unaffected by circular permutation of its arguments:

$$\vec{\mathbf{a}} \cdot (\vec{\mathbf{b}} \times \vec{\mathbf{c}}) = \vec{\mathbf{b}} \cdot (\vec{\mathbf{c}} \times \vec{\mathbf{a}}) = \vec{\mathbf{c}} \cdot (\vec{\mathbf{a}} \times \vec{\mathbf{b}}). \tag{1.41}$$

This is left as an exercise for the reader.

The **vector triple product** is given by $\vec{\mathbf{a}} \times (\vec{\mathbf{b}} \times \vec{\mathbf{c}})$. It can be shown that

$$\vec{\mathbf{a}} \times (\vec{\mathbf{b}} \times \vec{\mathbf{c}}) = (\vec{\mathbf{a}} \cdot \vec{\mathbf{c}})\vec{\mathbf{b}} - (\vec{\mathbf{a}} \cdot \vec{\mathbf{b}})\vec{\mathbf{c}}. \tag{1.42}$$

First, we evaluate $\vec{\mathbf{b}} \times \vec{\mathbf{c}}$, which is a vector perpendicular to both $\vec{\mathbf{b}}$ and $\vec{\mathbf{c}}$. Let us

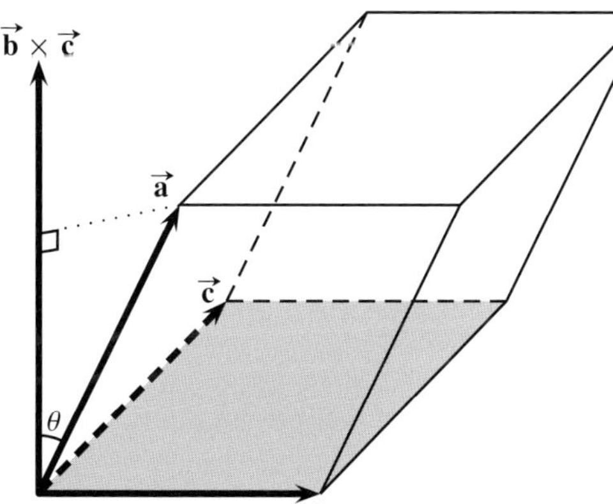

The parallelepiped volume is base times height. The gray-shaded base area is $||\vec{\mathbf{b}} \times \vec{\mathbf{c}}||$. The height is $||\vec{\mathbf{a}}|| \cdot |\cos\theta|$, where θ is the angle between $\vec{\mathbf{a}}$ and $(\vec{\mathbf{b}} \times \vec{\mathbf{c}})$, which could be greater than 90° (for instance, this would happen if $\vec{\mathbf{b}}$ and $\vec{\mathbf{c}}$ were reversed).

Figure 1.6 The scalar triple product as the volume of a parallelepiped.

denote this normal direction by $\hat{\mathbf{n}}$. Then, we cross multiply by $\vec{\mathbf{a}}$, which returns a vector that is perpendicular to $\hat{\mathbf{n}}$, i.e., it is on the plane defined by $\vec{\mathbf{b}}$ and $\vec{\mathbf{c}}$. Hence, the formula makes sense: it expresses the triple product as a linear combination of $\vec{\mathbf{b}}$ and $\vec{\mathbf{c}}$. A proof is provided in Example 1.10.

Example 1.8 *Scalar triple product as a determinant.*
Let $\vec{\mathbf{a}} = (a_1, a_2, a_3)$, $\vec{\mathbf{b}} = (b_1, b_2, b_3)$, $\vec{\mathbf{c}} = (c_1, c_2, c_3)$. Show that

$$\vec{\mathbf{a}} \cdot (\vec{\mathbf{b}} \times \vec{\mathbf{c}}) = \begin{vmatrix} a_1 & a_2 & a_3 \\ b_1 & b_2 & b_3 \\ c_1 & c_2 & c_3 \end{vmatrix} . \tag{1.43}$$

Proof Using the expansion of Equation (1.32b), we have

$$\vec{\mathbf{a}} \cdot (\vec{\mathbf{b}} \times \vec{\mathbf{c}}) = (a_1\,\hat{\mathbf{i}} + a_2\,\hat{\mathbf{j}} + a_3\,\hat{\mathbf{k}}) \cdot [(b_2c_3 - b_3c_2)\,\hat{\mathbf{i}} + (b_3c_1 - b_1c_3)\,\hat{\mathbf{j}} + (b_1c_2 - b_2c_1)\,\hat{\mathbf{k}}] \tag{1.44a}$$

$$= a_1(b_2c_3 - b_3c_2) + a_2(b_3c_1 - b_1c_3) + a_3(b_1c_2 - b_2c_1)\,. \tag{1.44b}$$

By inspection, we can confirm that this is the cofactor expansion of the determinant along its first row. □

Example 1.9 *Mechanical power from rotational motion.*
Suppose a force $\vec{\mathbf{F}}$ acts on a mass that is displaced by $\vec{\mathbf{r}}$ from the origin. The mass moves with velocity $\vec{\mathbf{v}} = d\vec{\mathbf{r}}/dt$. Find an expression for the mechanical power under rotational motion.

Solution
From physics, we know that the mechanical power exerted by the force is the dot product between force and velocity:

$$P_m = \frac{dW_m}{dt} = \vec{\mathbf{F}} \cdot \vec{\mathbf{v}}\,. \tag{1.45}$$

In other words, within a small time interval δt, where the body moves by $\delta\vec{\mathbf{r}} = \vec{\mathbf{v}}\delta t$, the force has done work equal to

$$\delta W_m = \vec{\mathbf{F}} \cdot \delta\vec{\mathbf{r}}\,. \tag{1.46}$$

Now suppose that the mass is rotating around an axis that passes through the origin so that its speed is given by

$$\vec{\mathbf{v}} = \vec{\omega} \times \vec{\mathbf{r}}\,, \tag{1.47}$$

where $\vec{\omega} = \omega\,\hat{\mathbf{a}}$ (rad/s) is an angular velocity vector. Here, $\hat{\mathbf{a}}$ is a unit vector that defines the axis of rotation. Hence, the mechanical power can be expressed as the scalar triple product

$$P_m = \vec{\mathbf{F}} \cdot (\vec{\omega} \times \vec{\mathbf{r}})\,. \tag{1.48}$$

Invoking the identity (1.41), we have

$$P_m = \vec{\omega} \cdot (\vec{\mathbf{r}} \times \vec{\mathbf{F}}), \tag{1.49}$$

or

$$P_m = \vec{\omega} \cdot \vec{\tau}, \tag{1.50}$$

where $\vec{\tau} = \vec{\mathbf{r}} \times \vec{\mathbf{F}}$ is the torque.

Example 1.10 *Vector triple-product expansion.*
Prove the identity (1.42).

Proof We have

$$\vec{\mathbf{a}} \times (\vec{\mathbf{b}} \times \vec{\mathbf{c}}) = \begin{vmatrix} \hat{\mathbf{i}} & \hat{\mathbf{j}} & \hat{\mathbf{k}} \\ a_1 & a_2 & a_3 \\ (b_2c_3 - b_3c_2) & (b_3c_1 - b_1c_3) & (b_1c_2 - b_2c_1) \end{vmatrix}. \tag{1.51}$$

Let us work with the $\hat{\mathbf{i}}$-component of the product:

$$a_2(b_1c_2 - b_2c_1) - a_3(b_3c_1 - b_1c_3) = (a_2c_2 + a_3c_3)b_1 - (a_2b_2 + a_3b_3)c_1 . \tag{1.52}$$

To bring it to the required form, we add and subtract $a_1b_1c_1$, which is incorporated into the existing b_1 and c_1 terms:

$$a_2(b_1c_2 - b_2c_1) - a_3(b_3c_1 - b_1c_3) = (a_1c_1 + a_2c_2 + a_3c_3)b_1 - (a_1b_1 + a_2b_2 + a_3b_3)c_1 \tag{1.53a}$$

$$= (\vec{\mathbf{a}} \cdot \vec{\mathbf{c}})b_1 - (\vec{\mathbf{a}} \cdot \vec{\mathbf{b}})c_1 . \tag{1.53b}$$

Manipulating in a similar manner the other two components, and then combining all three terms to form a vector, yields the desired result. □

1.3 Scalar and Vector Fields

We use the concept of a field to describe quantities throughout space, e.g., the effects of electric or magnetic sources. A **scalar field** is a function that assigns a real number value at each point of a domain Ω, where this could be the whole space or only a subset. On the other hand, a **vector field** is a function that assigns a vector at each point in Ω. Primarily, we are interested in fields in 3-D space that describe physical quantities of interest, such as magnetic fields.

In a more formal manner, we denote a scalar field as $F : \Omega \subseteq \mathbb{R}^3 \to \mathbb{R}$, whereas a vector field is denoted as $\vec{\mathbf{F}} : \Omega \subseteq \mathbb{R}^3 \to \mathbb{R}^3$. If $\mathbf{r}$ is a vector from the origin to an arbitrary point in Ω, we shall write $F(\mathbf{r})$ or $\vec{\mathbf{F}}(\mathbf{r})$ to obtain the scalar or vector field function value, respectively.

The vector $\mathbf{r}$ is still a vector in 3-D space, so we could have denoted it by $\vec{\mathbf{r}}$, and the vector field could be denoted as $\vec{\mathbf{F}}(\vec{\mathbf{r}})$. Invoking our poetic license, we will often choose to drop the arrow from $\vec{\mathbf{r}}$ so as to avoid cumbersome notation, especially when dealing with function arguments or differentials, such as the differential displacement $d\mathbf{r}$. However, arrows will always be used on top of electromagnetic vector fields or other physical quantities, such as the flux density $\vec{\mathbf{B}}$ or the force $\vec{\mathbf{F}}$.

Example 1.11 *Magnetic field of infinitely long, straight wire (part a: external field).* Determine the magnetic field that is created by an infinitely long, straight, and thin wire in free space that carries a constant current i, measured in A (amperes). Of course, infinitely long wires do not exist in practice, but they are useful from a pedagogical perspective. (This could also be a reasonable approximation of the field that exists very close to a wire, even if the wire is curved.)

Solution

From physics, recall that the magnetic flux density in this case is given by

$$B(r) = \frac{\mu_0 i}{2\pi r}, \tag{1.54}$$

where the physical constant

$$\mu_0 = 4\pi 10^{-7} \text{ H/m (henries}^{13} \text{ per meter)} \tag{1.55}$$

is the **magnetic permeability of free space**, and r is the distance (in m) from an arbitrary point of interest to the wire. In SI units, the B-field is measured in T (teslas). Another commonly encountered unit for the B-field is the gauss: 1 T = 10,000 gauss.

The above formula yields the magnitude of the vector field $\vec{\mathbf{B}} : \mathbb{R}^3 \to \mathbb{R}^3$ at any point in space (except on the wire itself) as a function of r. Clearly, to complete the definition of the vector field, we also need to define the direction of the vectors, which we will do descriptively (rather than algebraically): Each vector is perpendicular to both the wire and to the shortest line segment between the point of interest and the wire (whose length equals the distance r); its direction is determined by the right-hand rule.

Example 1.12 *Vector field visualization using Python.*

We can visualize vector fields using the built-in plotting capabilities of Python. For instance, let us consider the electric field in free space created by two point charges positioned at $\mathbf{r}_1 = (-0.5, 0, 0)$ and $\mathbf{r}_2 = (0.5, 0, 0)$, respectively, in m. The charge on the left is $Q_1 = +2$ C (coulomb[14]), and the charge on the right is $Q_2 = -1$ C.

The total electric field at an arbitrary point in space (displaced by $\mathbf{r}$ from the origin) is obtained by superposition:

$$\vec{\mathbf{E}}(\mathbf{r}) = \frac{1}{4\pi\epsilon_0}\left(Q_1 \frac{\mathbf{r}-\mathbf{r}_1}{\|\mathbf{r}-\mathbf{r}_1\|^3} + Q_2 \frac{\mathbf{r}-\mathbf{r}_2}{\|\mathbf{r}-\mathbf{r}_2\|^3}\right), \tag{1.56}$$

where the physical constant

$$\epsilon_0 \approx 8.854 \cdot 10^{-12} \text{ F/m (farads per meter)} \tag{1.57}$$

is the **permittivity of free space**. The E-field is measured in V/m (volts[15] per meter).

Below is a Python program that plots the vector field on the plane $z = 0$, where all vectors are horizontal (i.e., $E_z = 0$). Its output is shown in Figure 1.7, where two kinds of plots are illustrated, namely, a **quiver** plot and a **streamline** plot. The quiver plot consists of vectors at discrete user-specified points throughout the domain of interest. The streamline plot function uses the numerical data to produce a set of smooth lines, which have the property of being always tangent to the field. The streamlines originate from points that are selected internally by the plotting algorithm.

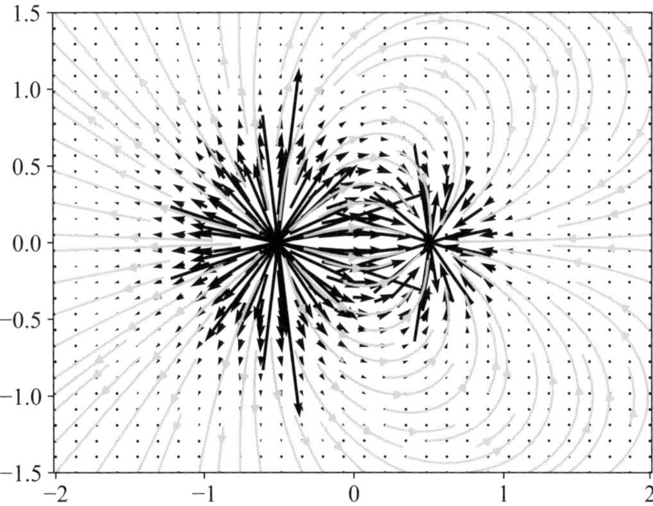

Figure 1.7 The electric field of two point charges. This figure was created by a superposition of a quiver and a streamline plot.

```
import numpy as np
import matplotlib.pyplot as plt

# Problem parameters
Q1 = 2
Q2 = -1
r1 = np.array([-0.5, 0, 0])
r2 = np.array([ 0.5, 0, 0])
z = 0 # we plot the projection of the E-field on this plane

# Calculate the electric field
nx = ny = 30
xv = np.linspace(-2, 2, nx)
yv = np.linspace(-1.5, 1.5, ny)
Ex = np.zeros((yv.size,xv.size))
Ey = np.zeros((yv.size,xv.size))

for i in range(xv.size):
    x = xv[i]
    for j in range(yv.size):
        y = yv[j]
        r = np.array([x, y, z])
        d1 = np.linalg.norm(r-r1)
```

```
        d2 = np.linalg.norm(r-r2)
25
        # We will not plot the field too close to the charges
        # because the quiver plot will be dominated by these
        # arrows. We define a minimum distance to exclude points
        # within a sphere of radius dmin around the charges.
30      dmin = 0.15
        if d1 < dmin or d2 < dmin:
            E1 = [0, 0, 0]
            E2 = [0, 0, 0]
        else:
35          E1 = Q1*(r-r1)/d1**3
            E2 = Q2*(r-r2)/d2**3
        E = E1 + E2
        Ex[j,i] = E[0]
        Ey[j,i] = E[1]
40
# Make the grid
xx, yy = np.meshgrid(xv, yv)

# Plot
45 fig = plt.figure()
ax = fig.gca()
ax.streamplot(xx, yy, Ex, Ey, color='0.75')
ax.quiver(xx, yy, Ex, Ey, width=.005, pivot = 'mid')
plt.axis('equal')
50 plt.show()
```

Let us dissect the script:

- NumPy is imported in line 1. Matplotlib Pyplot, which is a Matlab-like plotting library, is imported in line 2.
- Lines 5–9 contain self-explanatory problem-specific parameter definitions.
- Lines 12–16 contain array definitions. In Python, the chained assignment of line 12 is permitted, and saves us a bit of time defining the size of the arrays `xv` and `yv`. The arrays contain evenly spaced points created by a NumPy `linspace` command. (Here, the streamline plot function requires the grid points to be evenly spaced.) Note that it would have been incorrect to define `Ey = Ex` in line 16. In Python, this would create a single object in memory (at line 15), which both parameters would refer to, and would lead to a bug in the code.
- The double **`for`** loop in lines 18–39 calculates and stores a normalized electric field (we are ignoring the $4\pi\epsilon_0$ term since it does not affect this plot). In Python, the **`range`**`(x)` command creates a list of integers starting at zero and ending at `x-1`. Note the absence of `end` keywords to delimit the **`for`** loops. We need, however, to ensure proper indentation (typically, each level is indented by 4 spaces), otherwise the Python interpreter will complain.
- It is important to note the inverse order of indexing the electric field arrays: they are first indexed by `y` and then by `x`. This is how they were defined in lines 15 and 16 as well. This syntax is due to how the `meshgrid` function (line 42) works by default (similar to Matlab, although NumPy allows one to change this back to "normal" with an optional argument).

- The magnitude of $\vec{\mathbf{E}}$ close to the charges grows very fast and would dominate the plot (due to a $1/d^2$ term, d denoting distance); hence, we decide not to plot vectors in the vicinity of charges (see lines 26–33).
- Finally, plotting takes place in lines 44–50. First, a figure is created, and then a handle to its axes is obtained. The `quiver` and `streamplot` functions have similar syntax. The xy axes are drawn with equal scaling in line 49. The `show` command displays the plot.

For more information on the various differences between NumPy and Matlab, the reader is advised to consult online resources on this topic.

Example 1.13 *Scalar field visualization using Python.*

Let us plot contours (also called **isolines** or **level sets**) of the electrostatic potential based on the charges in the previous example. The potential is a scalar field measured in V, and is given by

$$\varphi(\mathbf{r}) = \frac{1}{4\pi\epsilon_0}\left(Q_1\frac{1}{\|\mathbf{r}-\mathbf{r}_1\|} + Q_2\frac{1}{\|\mathbf{r}-\mathbf{r}_2\|}\right). \tag{1.58}$$

The following Python code snippet illustrates the use of the built-in Matplotlib Pyplot `contour` function. The result is shown in Figure 1.8.

```
# Define contour values
V = [-12, -6, -2, -1, -0.5, -0.25, 0, \
     0.25, 0.5, 0.75, 1, 1.5, 2, 6, 12]

# Plot
fig = plt.figure()
```

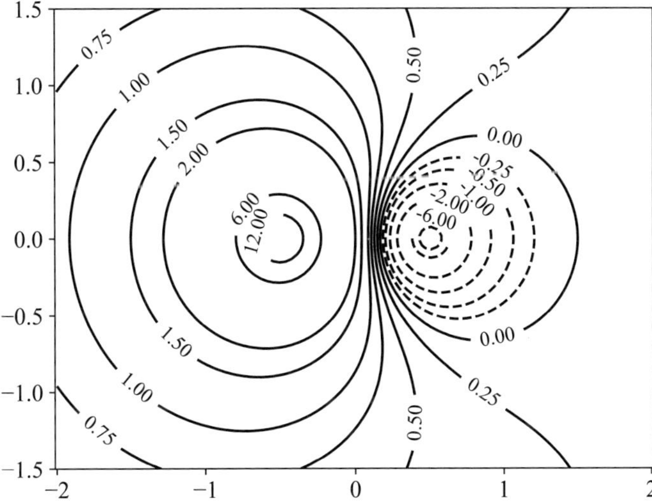

Figure 1.8 The electric potential of two point charges, generated by a contour plot. The potential values are normalized; the true potential (in V) is obtained after division by $4\pi\epsilon_0$.

```
ax = fig.gca()
CS = ax.contour(xx, yy, Phi, V, colors='k')
plt.axis('equal')
10 plt.clabel(CS, inline=1, fontsize=12, fmt = '%.2f', manual=True)
plt.show()
```

The code for computing the potential is not shown because it is similar to the previous example, and is performed over a regularly spaced grid. In line 2, we are defining which contour lines to plot. Note that the contours are not evenly distributed in this case. The contours look smooth because a relatively fine grid has been used. Also observe that negative potential values correspond to the dashed lines.

1.4 Integrals of Scalar and Vector Fields

We present several types of scalar and vector field integrals that we will encounter in this text, which are fundamental in understanding the properties of the magnetic field.

1.4.1 Line Integrals

The **line integral of a scalar field** $F : \Omega \to \mathbb{R}$ is defined over a piecewise smooth curve $C \subset \Omega$ (i.e., some curve within the space where the field exists) as

$$\int_C F(\mathbf{r})\, ds\,, \tag{1.59}$$

where ds represents a differential (infinitesimal) arc length along the curve. Note that the result does *not* depend on the direction of travel. For example, if $\rho(\mathbf{r})$ represents the density per unit length (kg/m) of a thin wire (which is not necessarily constant), its mass can be found as $\int_C \rho(\mathbf{r})\, ds$.

On the other hand, the **line integral of a vector field** $\vec{\mathbf{F}} : \Omega \to \mathbb{R}^3$ is defined over a piecewise smooth curve $C \subset \Omega$ as

$$\int_C \vec{\mathbf{F}}(\mathbf{r}) \cdot d\mathbf{r} = \int_C \vec{\mathbf{F}}(\mathbf{r}) \cdot \hat{\mathbf{t}}\, dr\,, \tag{1.60}$$

where $d\mathbf{r} = dr\,\hat{\mathbf{t}}$ is a differential displacement vector that points in the direction that we traverse the path, $\hat{\mathbf{t}}$ being a unit tangent vector. Note that the line integral is based on the dot product, so it accounts only for the component of $\vec{\mathbf{F}}$ that is tangent to the integration path. In a sense, we are measuring the degree to which $\vec{\mathbf{F}}$ is aligned with the integration path. As a result, the answer is a scalar. Also, the answer depends on the direction that we travel on the curve C; the result changes sign if we move in the opposite direction (since $d\mathbf{r}$ will become $-d\mathbf{r}$). Examples from physics are the work of a force, $W = \int_C \vec{\mathbf{F}}(\mathbf{r}) \cdot d\mathbf{r}$, and the electric field potential, $\varphi = -\int_C \vec{\mathbf{E}}(\mathbf{r}) \cdot d\mathbf{r}$.

A special case is when the path C is *closed*, that is, when it starts and ends at the same point. We call this kind of line integral a **circulation** of the vector field, and we denote it as

$$\oint_C \vec{\mathbf{F}}(\mathbf{r}) \cdot d\mathbf{r}\,. \tag{1.61}$$

Furthermore, if this happens to be zero for any path, we call the field **conservative**.

In general, to evaluate a line integral, we need to define a bijective parametrization of the curve, $\mathbf{r} : [a, b] \to C$, such that $\mathbf{r}(a)$ and $\mathbf{r}(b)$ are the two endpoints of C. The line integral of a scalar field is then

$$\int_C F(\mathbf{r})\, ds = \int_a^b F(\mathbf{r}(t))\, \|\mathbf{r}'(t)\|\, dt\,, \tag{1.62}$$

whereas the line integral of a vector field is

$$\int_C \vec{\mathbf{F}}(\mathbf{r}) \cdot d\mathbf{r} = \int_a^b \vec{\mathbf{F}}(\mathbf{r}(t)) \cdot \mathbf{r}'(t)\, dt\,. \tag{1.63}$$

Without getting too technical, we remind the reader that the line integral can be interpreted as a Riemann[16] sum, i.e., by splitting the curve into a finite number of segments, and then letting their width approach zero. The displacement between two adjacent points on the curve is

$$\delta\mathbf{r}(t) = \mathbf{r}(t + \delta t) - \mathbf{r}(t) \approx \mathbf{r}'(t)\, \delta t\,, \tag{1.64}$$

where we used a first-order approximation based on a Taylor[17] expansion. This also implies that the derivative $\mathbf{r}'(t)$ should be a vector that is tangent to the curve. The incremental arc length is

$$\delta s = \|\mathbf{r}'(t)\|\, \delta t \quad (\text{for } \delta t > 0)\,. \tag{1.65}$$

At the limit, the deltas become differentials (e.g., δs becomes ds), and the Riemann sums become the integrals (1.62) and (1.63).

An alternative notation for the line integral of a vector field, which is obtained by taking the dot product between $\vec{\mathbf{F}} = (F_x, F_y, F_z)$ and $d\mathbf{r} = (dx, dy, dz)$, is

$$\int_C \vec{\mathbf{F}}(\mathbf{r}) \cdot d\mathbf{r} = \int_C F_x\, dx + F_y\, dy + F_z\, dz\,. \tag{1.66}$$

The reader is cautioned that this should not be evaluated as three separate integrals. Rather, we should express the differential lengths based on the parametrization of the curve; for instance, we would substitute $dx = x'(t)\, dt$, etc.

If the curve is not smooth, that is, if there is a point, say $\mathbf{r}(c)$, where the curve changes direction abruptly, then the tangent directions right before and right after $\mathbf{r}(c)$ will be different. In this case, the line integral should be evaluated in two parts, as $\int_a^b = \int_a^c + \int_c^b$. This process may have to be repeated if the curve has a finite number of such discontinuities.

Example 1.14 *Line integral calculation.*
Consider the vector field

$$\vec{\mathbf{F}}(x,y) = -\operatorname{sgn}(y)F\,\hat{\mathbf{i}}, \tag{1.67}$$

where $F > 0$ is a constant. The field reverses direction across the x-axis. This situation is illustrated in Figure 1.9. Calculate the line integral of $\vec{\mathbf{F}}$ along the semicircular path C starting at $-90°$ and ending at $90°$ in the counterclockwise direction.

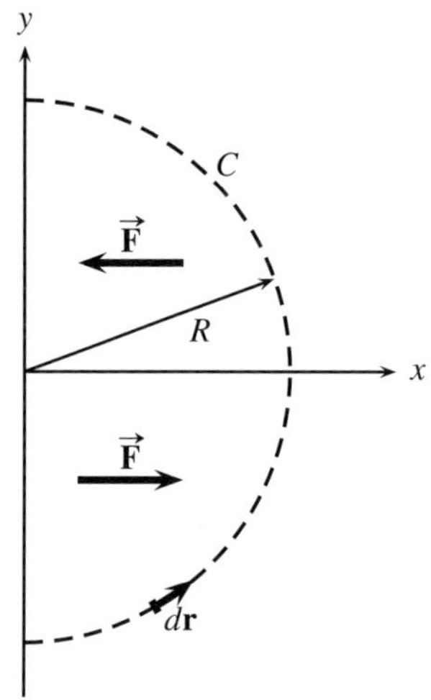

Figure 1.9 The geometry for the line integral of Example 1.14.

Solution
Note that $\vec{\mathbf{F}}(\mathbf{r}) \cdot d\mathbf{r} \geq 0$ throughout the path C, so we expect the answer to be positive. Let R be the radius of the semicircle. The integration path parametrization is

$$\mathbf{r}(t) = (x(t), y(t)) = (R\cos t, R\sin t)\,, \tag{1.68}$$

with parameter $t \in [-\pi/2, \pi/2]$ representing the angle with respect to the x-axis. To integrate, we need the derivative

$$\mathbf{r}'(t) = (-R\sin t, R\cos t)\,. \tag{1.69}$$

Hence

$$\int_C \vec{\mathbf{F}}(\mathbf{r}) \cdot d\mathbf{r} = \int_{-\pi/2}^{\pi/2} [-\operatorname{sgn}(y(t))F\,\hat{\mathbf{i}}] \cdot (-R\sin t\,\hat{\mathbf{i}} + R\cos t\,\hat{\mathbf{j}})\,dt \tag{1.70a}$$

$$= -FR\int_{-\pi/2}^{0} \sin t\,dt + FR\int_{0}^{\pi/2} \sin t\,dt \tag{1.70b}$$

$$= 2FR\,. \tag{1.70c}$$

1.4.2 Surface Integrals

The **surface integral of a scalar field** $F : \Omega \to \mathbb{R}$ is defined on a surface $S \subset \Omega$ as the double integral

$$\iint_S F(\mathbf{r})\, da\,. \tag{1.71}$$

To evaluate such an integral for a given surface, we need to parametrize the surface similarly to what we did for the line integral. A surface in $\mathbb{R}^3$ is parametrized with two variables; let us call them u and v. This means that we define a function $\mathbf{r} : U \times V \to S$ (so that $u \in U$ and $v \in V$), and we write $\mathbf{r}(u, v)$ for the function value at each point on the surface. The partial derivatives $\partial\mathbf{r}/\partial u$ and $\partial\mathbf{r}/\partial v$ yield vectors that are tangent to the surface at each point, pointing in the direction of increasing u and v, respectively.

We can approximate the surface integral as a Riemann sum of contributions from small surface elements:

$$\sum_k F(\mathbf{r}_k)\, \delta a_k\,, \tag{1.72}$$

where k is an element index, $F(\mathbf{r}_k)$ is the vector field value inside the element (since the elements are small, it can be assumed that F has approximately the same value within each element), and δa_k is the area of element k. Such a surface parametrization is depicted in Figure 1.10. Therefore, each surface element k resembles a parallelogram with sides

$$\mathbf{s}_k = \left.\frac{\partial\mathbf{r}}{\partial u}\right|_{\mathbf{r}=\mathbf{r}_k} \delta u_k \text{ and } \mathbf{t}_k = \left.\frac{\partial\mathbf{r}}{\partial v}\right|_{\mathbf{r}=\mathbf{r}_k} \delta v_k\,, \tag{1.73}$$

where the partial derivatives are evaluated inside the element k (e.g., $\mathbf{r}_k$ could be the parallelogram center). In general, elements do not have to be equal, so we kept the subscript k in δu_k and δv_k. The element area is found by taking a cross product:

$$\delta a_k = \|\mathbf{s}_k \times \mathbf{t}_k\| = \left\|\frac{\partial\mathbf{r}}{\partial u} \times \frac{\partial\mathbf{r}}{\partial v}\right\| \delta u_k\, \delta v_k\,, \tag{1.74}$$

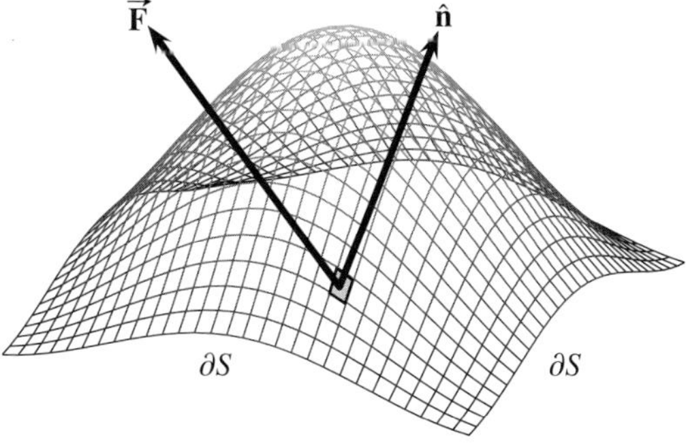

Figure 1.10 The integral of a vector field over the surface S is approximated by a Riemann sum of dot products.

assuming $\delta u_k > 0$ and $\delta v_k > 0$. Hence, the surface integral can be evaluated by

$$\iint_S F(\mathbf{r})\, da = \iint_{U\times V} F(\mathbf{r}(u,v)) \left\| \frac{\partial \mathbf{r}}{\partial u} \times \frac{\partial \mathbf{r}}{\partial v} \right\| du\, dv\,. \tag{1.75}$$

If the surface is piecewise smooth, e.g., a cube, we break the integral into pieces, one for each smooth part.

The **surface integral of a vector field** $\vec{\mathbf{F}} : \Omega \to \mathbb{R}^3$ is defined on a surface $S \subset \Omega$ as the double integral

$$\iint_S \vec{\mathbf{F}}(\mathbf{r}) \cdot d\mathbf{a} = \iint_S \vec{\mathbf{F}}(\mathbf{r}) \cdot \hat{\mathbf{n}}\, da\,, \tag{1.76}$$

where $d\mathbf{a} = da\,\hat{\mathbf{n}}$ is a differential area vector whose direction is determined by $\hat{\mathbf{n}}$, that is, a normal vector of unit length at each point $\mathbf{r}$ on the surface. (We could have denoted this as a function $\hat{\mathbf{n}}(\mathbf{r})$, but this would quickly become cumbersome.) The direction of $\hat{\mathbf{n}}$ is obtained from the assumed orientation of the surface.

This kind of surface integral can also be visualized as a Riemann sum of dot product-based contributions from small elements on the surface, as shown in Figure 1.10. Taking the dot product ensures that we account only for the component of $\vec{\mathbf{F}}$ that is normal to the surface. Therefore, this kind of vector field integral is essentially a surface integral of a scalar field! If the vector field has a physical meaning of flow density, which is often the case, the surface integral may be interpreted as the aggregate vector field flow or **flux** through the surface S. Examples from physics are the magnetic flux, $\Phi = \iint_S \vec{\mathbf{B}}(\mathbf{r}) \cdot d\mathbf{a}$, and the current, $i = \iint_S \vec{\mathbf{J}}(\mathbf{r}) \cdot d\mathbf{a}$.

The surface S shown in Figure 1.10 has a boundary that is denoted by ∂S. However, a surface can be *closed*, that is, it may not have a boundary itself, but it could be the boundary of a region in space (e.g., the surface that bounds a sphere). Then we use a special integral sign:

$$\oiint_S \vec{\mathbf{F}}(\mathbf{r}) \cdot d\mathbf{a}\,. \tag{1.77}$$

Using a parametrization of the surface, the surface integral can be evaluated using

$$\iint_S \vec{\mathbf{F}}(\mathbf{r}) \cdot d\mathbf{a} = \iint_{U\times V} \vec{\mathbf{F}}(\mathbf{r}(u,v)) \cdot \left(\frac{\partial \mathbf{r}}{\partial u} \times \frac{\partial \mathbf{r}}{\partial v} \right) du\, dv\,. \tag{1.78}$$

The partial derivatives yield vectors that are tangent to the surface at each point. Therefore, their cross product will be normal to the surface. Hence, to avoid the appearance of a minus sign, the parametrization needs to be defined so that this normal vector points in the desired direction.

Although the previous case is typically what one refers to as a "surface integral of a vector field," it is also possible to define a different kind of such an integral. In particular, we can take the surface integral

$$\vec{\mathbf{G}} = \iint_S \vec{\mathbf{F}}(\mathbf{r})\, da\,. \tag{1.79}$$

This should be interpreted as the component-wise integration of three scalar fields, which results in the vector $\vec{\mathbf{G}}$. For instance, the x-component of $\vec{\mathbf{G}}$ is

$$G_x = \iint_S F_x(\mathbf{r})\, da\,. \tag{1.80}$$

In any case, the type of integral that we are dealing with should be apparent from the context.

Example 1.15 *Surface integral calculation.*

Calculate a surface integral on a half-sphere of radius R (i.e., without integrating over its base), as shown in Figure 1.11. The surface is oriented such that the normal vector points outwards.

Solution

This surface may be described by the parametrization

$$\mathbf{r}(\theta,\phi) = R\sin\theta\cos\phi\,\hat{\mathbf{i}} + R\sin\theta\sin\phi\,\hat{\mathbf{j}} + R\cos\theta\,\hat{\mathbf{k}} = R\,\hat{\mathbf{r}}\,, \tag{1.81}$$

where $\theta \in U = [0,\pi/2]$ and $\phi \in V = (-\pi,\pi]$ are the polar and azimuthal angles of a spherical coordinate system, respectively, and $\hat{\mathbf{r}}$ is the unit radial vector. In this example, the unit normal vector of the integration surface is $\hat{\mathbf{n}} = \hat{\mathbf{n}}(\theta,\phi) = \hat{\mathbf{r}}$. Hence

$$\frac{\partial\mathbf{r}}{\partial\theta} = R\cos\theta\cos\phi\,\hat{\mathbf{i}} + R\cos\theta\sin\phi\,\hat{\mathbf{j}} - R\sin\theta\,\hat{\mathbf{k}} = R\,\hat{\boldsymbol{\theta}} \tag{1.82}$$

and

$$\frac{\partial\mathbf{r}}{\partial\phi} = -R\sin\theta\sin\phi\,\hat{\mathbf{i}} + R\sin\theta\cos\phi\,\hat{\mathbf{j}} = R\sin\theta\,\hat{\boldsymbol{\phi}}\,. \tag{1.83}$$

The tangential property may be readily verified since $\partial\mathbf{r}/\partial\theta\cdot\hat{\mathbf{n}} = \partial\mathbf{r}/\partial\phi\cdot\hat{\mathbf{n}} = 0$. Using

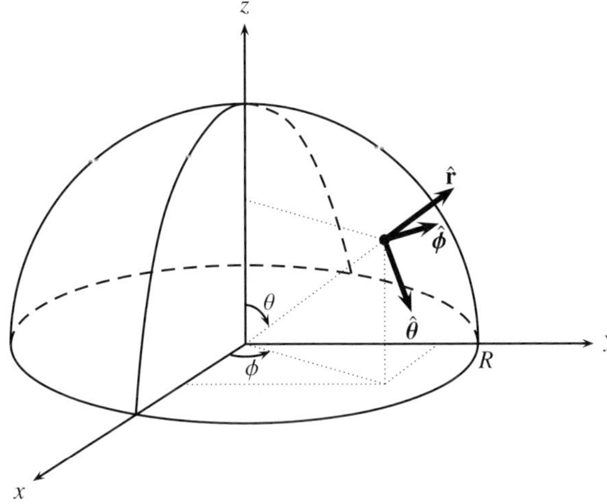

Figure 1.11 The geometry of a half-sphere displaying a spherical coordinate system.

a bit of elementary trigonometry, we obtain

$$\frac{\partial \mathbf{r}}{\partial \theta} \times \frac{\partial \mathbf{r}}{\partial \phi} = R^2(\sin^2\theta\cos\phi\,\hat{\mathbf{i}} + \sin^2\theta\sin\phi\,\hat{\mathbf{j}} + \sin\theta\cos\theta\,\hat{\mathbf{k}}) = R^2\sin\theta\,\hat{\mathbf{n}}\,. \tag{1.84}$$

Once the surface parametrization is complete, we can proceed with the evaluation of the integral. Suppose that $\vec{\mathbf{F}}(\mathbf{r}) = \hat{\mathbf{n}}$. We would then expect the integral to yield the surface area of the half-sphere (i.e., $2\pi R^2$). Let us verify this:

$$\begin{aligned}\iint_{U\times V} \vec{\mathbf{F}}(\mathbf{r}(\theta,\phi)) \cdot \left(\frac{\partial \mathbf{r}}{\partial \theta} \times \frac{\partial \mathbf{r}}{\partial \phi}\right) d\theta\, d\phi &= \iint_{U\times V} \hat{\mathbf{n}} \cdot (R^2 \sin\theta\,\hat{\mathbf{n}})\, d\theta\, d\phi \\ &= \int_{\phi=-\pi}^{\pi}\int_{\theta=0}^{\pi/2} R^2\sin\theta\, d\theta\, d\phi = 2\pi R^2\,. \end{aligned}\tag{1.85}$$

As a second example, consider the case where $\vec{\mathbf{F}} = F\,\hat{\mathbf{k}}$, $F > 0$. In this case, it is more convenient to work with the expansion in Cartesian coordinates:

$$\iint_{U\times V} \vec{\mathbf{F}}(\mathbf{r}(\theta,\phi)) \cdot \left(\frac{\partial \mathbf{r}}{\partial \theta} \times \frac{\partial \mathbf{r}}{\partial \phi}\right) d\theta\, d\phi \tag{1.86a}$$

$$= \iint_{U\times V} F\,\hat{\mathbf{k}} \cdot R^2(\sin^2\theta\cos\phi\,\hat{\mathbf{i}} + \sin^2\theta\sin\phi\,\hat{\mathbf{j}} + \sin\theta\cos\theta\,\hat{\mathbf{k}})\, d\theta\, d\phi \tag{1.86b}$$

$$= \int_{\phi=-\pi}^{\pi}\int_{\theta=0}^{\pi/2} FR^2 \sin\theta\cos\theta\, d\theta\, d\phi \tag{1.86c}$$

$$= \pi FR^2 \int_0^{\pi/2} \sin 2\theta\, d\theta \tag{1.86d}$$

$$= \pi FR^2\,. \tag{1.86e}$$

As an interesting observation, consider what would have happened if we had extended the integration to the base of the half-sphere, thus forming a closed surface. In this case, we would have obtained an additional term equal to

$$(\text{area of base})(-F) = -\pi R^2 F\,, \tag{1.87}$$

where the minus sign appears since the vector field $F\hat{\mathbf{k}}$ is entering the half-sphere from below, thereby opposing the outwards-pointing normal vector. Hence, the integral over the closed surface would have been $\oiint_S \vec{\mathbf{F}} \cdot d\mathbf{a} = 0$. This is not unexpected since the considered vector field has zero divergence, and we could have applied Gauss' law. More on this later in §1.6.

1.4.3 Volume Integrals

The **volume integral of a scalar field** $F : \Omega \to \mathbb{R}$ is defined on a volume $V \subset \Omega$ as the triple integral

$$\iiint_V F(\mathbf{r})\, dv\,, \tag{1.88}$$

where dv represents the differential volume. For instance, in Cartesian coordinates, $dv = dx\,dy\,dz$. In physics, this type of integral is often used when F represents some kind of density function, so integrating over a volume yields the total value of the quantity of interest, such as mass or energy. Again, the volume integral may be approximated by a Riemann sum over small volumes that cover the entire space V.

The **volume integral of a vector field** $\vec{\mathbf{F}} : \Omega \to \mathbb{R}^3$ is obtained by performing component-wise volume integration on $\vec{\mathbf{F}}$. It is denoted by

$$\vec{\mathbf{G}} = \iiint_V \vec{\mathbf{F}}(\mathbf{r})\, dv\,. \tag{1.89}$$

For instance, the x-component of $\vec{\mathbf{G}}$ is

$$G_x = \iiint_V F_x(\mathbf{r})\, dv\,, \tag{1.90}$$

so it is a volume integral of the scalar field F_x.

1.5 Differential Operators on Scalar and Vector Fields

In this section, we recall the various differential operators that we will encounter in this text while manipulating Maxwell's equations. Our approach here will be conceptual rather than numerical.

These operators will be associated with formulas that involve partial derivatives. For these to have meaning, it is required that the fields are differentiable (smooth) with respect to position. This will typically be the case within the volume of a single material whose properties change smoothly with position. However, discontinuities will be present across material boundaries. Since electric machines are constructed using materials of different properties, we need to be mindful of this fact.

1.5.1 Gradient

The **gradient** of a scalar field $F : \Omega \subset \mathbb{R}^3 \to \mathbb{R}$ is defined as

$$\nabla F(\mathbf{r}) = \frac{\partial F}{\partial x}\hat{\mathbf{i}} + \frac{\partial F}{\partial y}\hat{\mathbf{j}} + \frac{\partial F}{\partial z}\hat{\mathbf{k}}\,, \tag{1.91}$$

where the partial derivatives (assuming they exist) are evaluated at $\mathbf{r} = (x, y, z)$. If the scalar field is only defined in two dimensions, then we drop the third term in the above equation. For example, the electric field is defined as the (negative) gradient of the potential, $\vec{\mathbf{E}} = -\nabla\varphi$; see Examples 1.12 and 1.13.

The symbol ∇ is called the **nabla** or **del**. It plays the role of an *operator*, and can be defined as

$$\nabla = \frac{\partial}{\partial x}\hat{\mathbf{i}} + \frac{\partial}{\partial y}\hat{\mathbf{j}} + \frac{\partial}{\partial z}\hat{\mathbf{k}}\,. \tag{1.92}$$

The nabla operator is also used in the divergence and the curl, which will be introduced subsequently. Note that the gradient operates on a scalar function and returns a vector.

The expression (1.91) is valid only in Cartesian coordinates. In other coordinate systems (e.g., spherical or cylindrical) the expressions are different; these are not listed here, but they can be found in calculus texts or online.

For an interpretation of the gradient, consider two points, $P_1, P_2 \in \Omega$, displaced by $\mathbf{r}_1$ and $\mathbf{r}_2$ from the origin O, respectively, as shown in Figure 1.12. Let us fix P_1, allowing P_2 to move arbitrarily close to P_1, and denote by $\delta\mathbf{r} = \mathbf{r}_2 - \mathbf{r}_1$ their relative displacement. Suppose that $\delta\mathbf{r} = \delta x\,\hat{\mathbf{i}} + \delta y\,\hat{\mathbf{j}} + \delta z\,\hat{\mathbf{k}}$. Then, using a Taylor expansion of F at $\mathbf{r}_1$, the field changes by

$$\delta F = F(\mathbf{r}_2) - F(\mathbf{r}_1) \approx \frac{\partial F}{\partial x}\delta x + \frac{\partial F}{\partial y}\delta y + \frac{\partial F}{\partial z}\delta z\,, \tag{1.93}$$

as a first-order approximation, with partial derivatives evaluated at $\mathbf{r}_1$; hence,

$$\delta F \approx \nabla F(\mathbf{r}_1) \cdot \delta\mathbf{r}\,. \tag{1.94}$$

In the special case where P_1 is a local maximum or minimum point of F, the gradient is zero (i.e., all partial derivatives are zero). Otherwise, a nontrivial-level curve through P_1 exists (as shown in Figure 1.12). Suppose that both points are on this curve, $F(\mathbf{r}_1) = F(\mathbf{r}_2) = C$, so that $\delta F = 0$. As P_2 comes arbitrarily close to P_1, the displacement $\delta\mathbf{r}$ becomes tangent to the curve C. If $\hat{\mathbf{t}}$ is a unit tangent vector at $\mathbf{r}_1$, then $\mathbf{r}_2 = \mathbf{r}_1 + h\,\hat{\mathbf{t}}$. So, we have $\delta F = 0 = \nabla F(\mathbf{r}_1) \cdot h\,\hat{\mathbf{t}} + O(h^2)$, as $h \to 0$. Therefore, we can argue that *the gradient has to be normal to the level set* at P_1 since a small (but nonzero) step of length h in the tangent direction would lead to $\nabla F(\mathbf{r}_1) \cdot \hat{\mathbf{t}} \approx 0$. Figures 1.7 and 1.8 on pages 17 and 19, respectively, provide an illustration of this property using the electrostatic field as an example: the E-field is normal to the equipotential surfaces.

Furthermore, this line of thinking can lead us to the conclusion that *the gradient points at the direction where the field increases the most.* To see this, note that the dot product in Equation (1.94), $\delta F = \|\nabla F\|\,\|\delta\mathbf{r}\|\,\cos\theta$, becomes maximum when the angle θ between the displacement and the gradient is zero (maintaining a displacement of fixed length). In Figure 1.7, the electric field points towards the direction of maximum potential *decrease* due to the presence of the minus sign in its definition.

In general, the gradient is related to the **directional derivative** along the direction **v**,

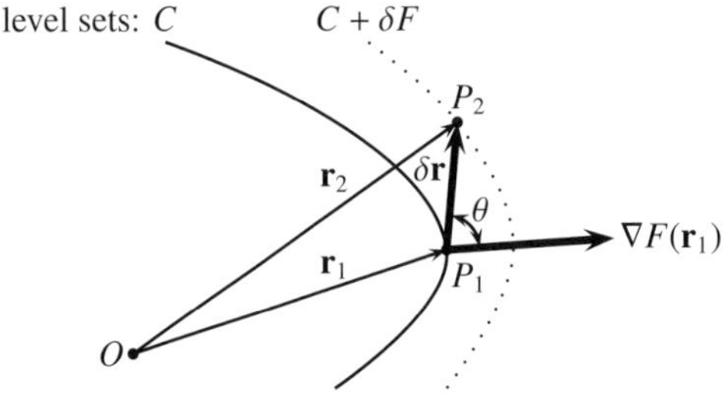

Figure 1.12 Geometrical interpretation of the gradient in two dimensions.

where it is customary to use a unit vector $\mathbf{v}$ (i.e., $\|\mathbf{v}\| = 1$). This is defined as

$$\nabla F_{\mathbf{v}}(\mathbf{r}) = \frac{d}{dh} F(\mathbf{r} + h\mathbf{v})\bigg|_{h=0} \tag{1.95a}$$

$$= \lim_{h\to 0} \frac{F(\mathbf{r} + h\mathbf{v}) - F(\mathbf{r})}{h} \tag{1.95b}$$

$$= \lim_{h\to 0} \frac{\nabla F(\mathbf{r}) \cdot (h\mathbf{v}) + (\text{high-order terms } O(h^2))}{h} \tag{1.95c}$$

$$= \nabla F(\mathbf{r}) \cdot \mathbf{v}\,. \tag{1.95d}$$

Example 1.16 *Gradient calculation.*
Calculate the gradient of the scalar field $F : \mathbb{R}^2 \to \mathbb{R}$ given by

$$F(x, y) = x^2 + y^2\,. \tag{1.96}$$

Solution
The level sets of F are circles centered at the origin. The gradient of F is

$$\nabla F(x, y) = \frac{\partial F}{\partial x}\hat{\mathbf{i}} + \frac{\partial F}{\partial y}\hat{\mathbf{j}} = 2x\,\hat{\mathbf{i}} + 2y\,\hat{\mathbf{j}}\,. \tag{1.97}$$

This vector points radially away from the origin, so it is perpendicular to the level sets, as expected. Note that $\nabla F = 0 \Rightarrow x = y = 0$. The gradient is zero at the origin, where F obtains its minimum value.

Example 1.17 *Conservative vector fields.*
Suppose that a vector field $\vec{\mathbf{F}} : \Omega \to \mathbb{R}^3$ is the gradient of a scalar field $\varphi : \Omega \to \mathbb{R}$, that is, $\vec{\mathbf{F}}(\mathbf{r}) = \nabla\varphi(\mathbf{r})$. For any (piecewise) smooth path C within Ω, which is parametrized by $t \in [a, b]$ so that $\mathbf{r}(t) = (x(t), y(t), z(t))$, we have

$$\int_C \vec{\mathbf{F}}(\mathbf{r}) \cdot d\mathbf{r} = \int_a^b \nabla\varphi(\mathbf{r}(t)) \cdot \mathbf{r}'(t)\,dt \tag{1.98a}$$

$$= \int_a^b \left(\frac{\partial\varphi}{\partial x}\frac{dx}{dt} + \frac{\partial\varphi}{\partial y}\frac{dy}{dt} + \frac{\partial\varphi}{\partial z}\frac{dz}{dt}\right) dt \tag{1.98b}$$

$$= \int_a^b \frac{d\varphi(\mathbf{r}(t))}{dt}\,dt \tag{1.98c}$$

$$= \varphi(\mathbf{r}(b)) - \varphi(\mathbf{r}(a))\,, \tag{1.98d}$$

where we used the chain rule for the derivative $d\varphi/dt$, and the fundamental theorem of calculus in the last step. This means that the line integral does not depend on the path but only on its two endpoints. The integral, in other words, is path independent. Also, if the two endpoints coincide so that the path is closed, the integral is zero. It can be shown that the converse is also true, i.e., if $\vec{\mathbf{F}}$ is path independent, then it is the gradient of some scalar field. This type of vector field is called **conservative**. For example, the electrostatic field is conservative since $\vec{\mathbf{E}} = -\nabla\varphi$ (a minus sign is added by convention).

1.5.2 Divergence

Often a vector field $\vec{\mathbf{F}} : \Omega \to \mathbb{R}^3$ has a physical meaning of flow density (i.e., flow per m^2), in which case we may be interested in calculating the total flow out of a given volume $V \in \Omega$ surrounding a point of interest P. This would be the case if there exist sources or sinks of the flow; for instance, this happens around point charges as shown in Figure 1.7 on page 17.

Hence, we define a scalar quantity called the **divergence** as the ratio

$$\nabla \cdot \vec{\mathbf{F}}(\mathbf{r}) = \lim_{|V| \to 0} \frac{\text{total outwards flow}}{|V|}, \tag{1.99}$$

where $|V| = \iiint_V dv$ denotes the volume size. Therefore, the divergence has units of flow per m^3. Examples from physics are the divergence of the electric displacement field, $\nabla \cdot \vec{\mathbf{D}} = \rho$, and the divergence of the magnetic field, $\nabla \cdot \vec{\mathbf{B}} = 0$. A vector field that has zero divergence everywhere, like the magnetic field, is called **incompressible** or **solenoidal**.

In Cartesian coordinates, when the vector field $\vec{\mathbf{F}} = (F_x, F_y, F_z)$ is differentiable, the divergence is given by the formula

$$\nabla \cdot \vec{\mathbf{F}}(\mathbf{r}) = \frac{\partial F_x}{\partial x} + \frac{\partial F_y}{\partial y} + \frac{\partial F_z}{\partial z}, \tag{1.100}$$

where the partial derivatives are evaluated at $\mathbf{r} = (x, y, z)$, that is, the displacement of P from the origin. For a 2-D field, we drop the last term. (In spherical or cylindrical coordinates, different formulas apply.) Note how the divergence can be obtained formally by taking the dot product between the nabla operator in Equation (1.92) and the vector field.

Proof Let us illustrate this concept using a small cube surrounding a center point $P = (x', y', z') = (a, a, a)/2$, as shown in Figure 1.13. For computational convenience, the

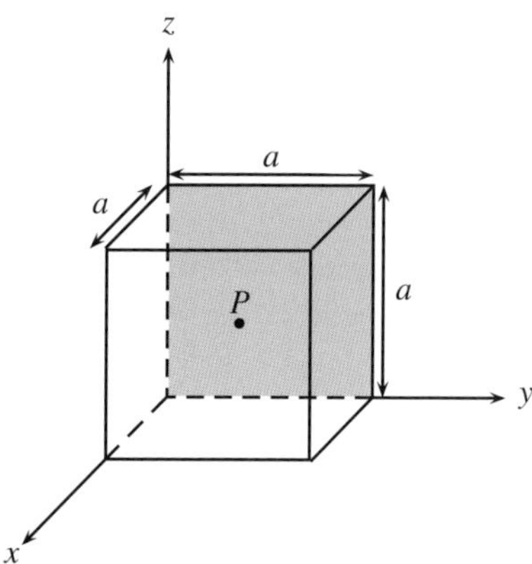

Figure 1.13 Illustration of the divergence as an outwards flow from a cube surrounding a point P at its center.

cube is oriented with the axes. We can use a surface integral to calculate

$$\text{total outwards flow} = \oiint_{\text{cube surface}} \vec{\mathbf{F}}(\mathbf{r}) \cdot d\mathbf{a} \,. \tag{1.101}$$

This integral can be evaluated by summing the contribution from the six sides. For instance, the contribution to the flow from the back side of the cube (which is shown in gray) is

$$\text{flow leaving from back} = \int_{y=0}^{a} \int_{z=0}^{a} \vec{\mathbf{F}}(x=0,y,z) \cdot (-\hat{\mathbf{i}})\, dy\, dz \,. \tag{1.102}$$

Note that only the x-component of the vector field will be needed in this calculation, where the outwards-pointing unit vector is $-\hat{\mathbf{i}}$. Using a first-order Taylor expansion of F_x, the vector field at the back side ($x = 0$) is related to its value at P by

$$F_x(x=0,y,z) = F_x(x',y',z') - \frac{a}{2}\frac{\partial F_x}{\partial x} + \left(y - \frac{a}{2}\right)\frac{\partial F_x}{\partial y} + \left(z - \frac{a}{2}\right)\frac{\partial F_x}{\partial z} \,, \tag{1.103}$$

with partial derivatives evaluated at P. Plugging this into Equation (1.102) yields

$$\text{flow leaving from back} = -\left(F_x(x',y',z') - \frac{a}{2}\frac{\partial F_x}{\partial x}\right) a^2 \,, \tag{1.104}$$

since the contributions from the last two terms involving the partials with respect to y and z vanish. (This is because $\int_0^a (\xi - a/2)\, d\xi = 0$.) Similarly, we can calculate

$$\text{flow leaving from front} = \left(F_x(x',y',z') + \frac{a}{2}\frac{\partial F_x}{\partial x}\right) a^2 \,. \tag{1.105}$$

Summing these two terms, we obtain

$$\text{flow leaving from front and back} = \frac{\partial F_x}{\partial x} a^3 = \frac{\partial F_x}{\partial x} |V| \,. \tag{1.106}$$

Similar expressions may be obtained for the other two sets of sides. Therefore, by allowing the cube to shrink arbitrarily close to the point P so that we can apply the definition (1.99), we obtain Equation (1.100). □

Example 1.18 *Divergence calculation.*

Calculate the divergence of the vector field $\vec{\mathbf{F}} : \mathbb{R}^3 \to \mathbb{R}^3$ given by

$$\vec{\mathbf{F}}(x,y,z) = 2xy\,\hat{\mathbf{i}} + (z^2 - y)\,\hat{\mathbf{j}} + (x - y)\hat{\mathbf{k}} \,. \tag{1.107}$$

Solution

The field is differentiable, and its divergence is

$$\nabla \cdot \vec{\mathbf{F}}(x,y,z) = \frac{\partial F_x}{\partial x} + \frac{\partial F_y}{\partial y} + \frac{\partial F_z}{\partial z} = 2y - 1 \,. \tag{1.108}$$

Example 1.19 *The interface between two magnetic materials (part a: normal components).*

Maxwell asserts that the magnetic flux density satisfies

$$\nabla \cdot \vec{\mathbf{B}}(\mathbf{r}) = 0 \tag{1.109}$$

everywhere in space. Now, as we may recall from physics, the magnetic field exhibits a discontinuity at the interface between two materials of different magnetic permeability. Hence, a technical difficulty arises, as we cannot apply the partial derivative formula (1.100) on the interface. However, the fundamental definition of Equation (1.99) still applies, where the total magnetic flux outflow is $\oiint_{\partial V} \vec{\mathbf{B}}(\mathbf{r}) \cdot d\mathbf{a}$. Since we can make the volume arbitrarily small, it is necessary that this outflow is zero (otherwise the ratio of outflow over volume would explode to an infinite value).

To be more specific, consider an integration surface that is centered around a point of interest P on the interface between the two materials. It is customary to use a **Gaussian pillbox**, which is a cylinder with short height. The cylinder bases, S_1 and S_2, are oriented parallel to the interface, as shown in Figure 1.14. For instance, we could parametrize the size of the pillbox by its radius $R > 0$ so that its base area $A = A(R) = \pi R^2$, and its height $h = h(R) = \kappa R$, with $0 < \kappa \ll 1$. As R goes to zero, the pillbox shrinks (its volume is $|V| = \kappa \pi R^3$), but always remains centered around the point of interest. Note that we are arbitrarily close to the point P, so the curvature of the surface (if any) can be neglected, and the interface resembles a plane. Hence, we can further assume that the magnetic field is constant in this tiny region, that is, we have constant $\vec{\mathbf{B}}_1$ in material 1 and constant $\vec{\mathbf{B}}_2$ in material 2.

The total outflow can be evaluated by adding the outflow from the side to the outflow from the bases. First, we note that the outflow from the side is zero. We can argue that this happens because the flux entering at any point (in either the upper or lower part) exits at the diametrically opposite point. This is readily shown by evaluating the dot product between $\vec{\mathbf{B}}_1$ or $\vec{\mathbf{B}}_2$ and the radially outwards-pointing unit vector for any two diametrical points on the cylinder side. Now, the unit normal vectors for the two bases

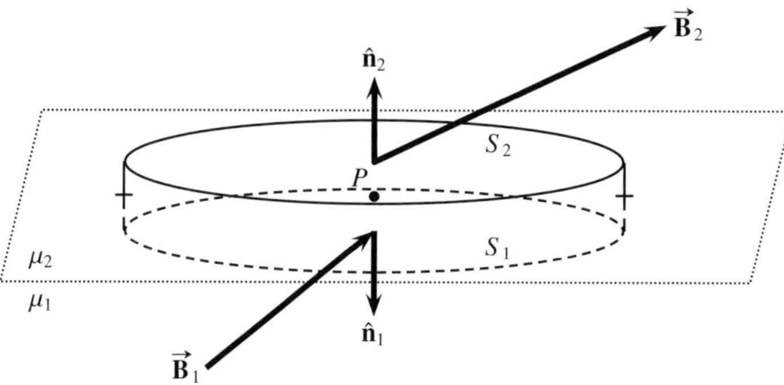

Figure 1.14 A Gaussian pillbox at the interface between two magnetic materials that have different magnetic permeability, μ_1 and μ_2. The bases straddle the interface. The point P is at the center of the cylinder.

of the pillbox satisfy $\hat{\mathbf{n}}_1 = -\hat{\mathbf{n}}_2$. Hence, the total outflow evaluates to

$$\iint_{S_1} \vec{\mathbf{B}}_1 \cdot \hat{\mathbf{n}}_1 \, da + \iint_{S_2} \vec{\mathbf{B}}_2 \cdot \hat{\mathbf{n}}_2 \, da = -(\vec{\mathbf{B}}_1 \cdot \hat{\mathbf{n}}_2)A + (\vec{\mathbf{B}}_2 \cdot \hat{\mathbf{n}}_2)A \,. \tag{1.110}$$

Application of the divergence definition (1.99) yields

$$\nabla \cdot \vec{\mathbf{B}}(\mathbf{r}) = \lim_{R\to 0} \frac{(-\vec{\mathbf{B}}_1 \cdot \hat{\mathbf{n}}_2 + \vec{\mathbf{B}}_2 \cdot \hat{\mathbf{n}}_2)\,\pi R^2}{\kappa \pi R^3} = \lim_{R\to 0} \frac{-\vec{\mathbf{B}}_1 \cdot \hat{\mathbf{n}}_2 + \vec{\mathbf{B}}_2 \cdot \hat{\mathbf{n}}_2}{\kappa R} \,. \tag{1.111}$$

Therefore, for this to equal zero, it is necessary that

$$(\vec{\mathbf{B}}_1 - \vec{\mathbf{B}}_2) \cdot \hat{\mathbf{n}}_2 = 0 \,. \tag{1.112}$$

In other words, *the normal component of the B-field must not change across the interface*. (The tangential components can be unequal. Their relationship will depend on the material permeabilities and the possible presence of a surface current.)

1.5.3 Curl

Another basic property of a vector field is its vorticity, that is, the degree to which it tends to whirl around at each point in space. This is quantified using the **curl** operator, denoted by "$\nabla\times$."

Suppose we wish to measure the vorticity of a vector field $\vec{\mathbf{F}} : \Omega \subset \mathbb{R}^3 \to \mathbb{R}^3$ around a point $P \in \Omega$, displaced by $\mathbf{r}$ from the origin. First, we define a rotation axis, which is associated with a unit vector $\hat{\mathbf{n}}$. Then, we take a small flat surface $S \in \Omega$ that is perpendicular to $\hat{\mathbf{n}}$, so that $P \in S$. The two sides of the surface S are called "positive" and "negative," and the normal vector $\hat{\mathbf{n}}$ exits from the positive side. The boundary of S is denoted by ∂S. The component of the curl along the rotation axis is defined as

$$(\nabla \times \vec{\mathbf{F}}(\mathbf{r})) \cdot \hat{\mathbf{n}} = \lim_{|S|\to 0} \frac{1}{|S|} \oint_{\partial S} \vec{\mathbf{F}}(\mathbf{r}) \cdot d\mathbf{r} \,, \tag{1.113}$$

where $|S| = \iint_S da$ denotes surface area. By convention, when evaluating the line integral, we walk along ∂S so that the positive side of S is on our left. This is the right-hand rule in this context. Note that this is an implicit definition of the curl vector, as it only provides one out of three components. We have to calculate the integral for two more (orthogonal) directions in space, to get the complete picture, i.e., the three components of the curl vector in 3-D space. For example, the curl appears in Maxwell's equation, $\nabla \times \vec{\mathbf{E}} = -\partial \vec{\mathbf{B}}/\partial t$. A vector field that has zero curl everywhere, like the electrostatic field, is called **irrotational**.

In Cartesian coordinates, the curl of a vector field $\vec{\mathbf{F}} = (F_x, F_y, F_z)$ is

$$\nabla \times \vec{\mathbf{F}}(\mathbf{r}) = \begin{vmatrix} \hat{\mathbf{i}} & \hat{\mathbf{j}} & \hat{\mathbf{k}} \\ \frac{\partial}{\partial x} & \frac{\partial}{\partial y} & \frac{\partial}{\partial z} \\ F_x & F_y & F_z \end{vmatrix} , \tag{1.114}$$

where the partial derivatives (assuming they exist) are evaluated at $\mathbf{r} = (x, y, z)$. This determinant does not correspond to a real matrix but is only a mnemonic; however, it explains why the curl is denoted formally as the cross product of the nabla with the vector field, using Equations (1.92) and (1.31) (see page 8).

In the special case of a 2-D field in x–y coordinates, we can embed the field in 3-D space so that $F_z = 0$ and $\partial/\partial z = 0$. This leads to

$$(\nabla \times \vec{\mathbf{F}})_{\text{2-D}} = \left(\frac{\partial F_y}{\partial x} - \frac{\partial F_x}{\partial y} \right) \hat{\mathbf{k}} . \tag{1.115}$$

For simplicity in 2-D problems, we may treat the 2-D curl as a scalar field. This will occur later on when analyzing cross-sections of electric machines.

Proof To derive Equation (1.114), consider the simple geometry shown in Figure 1.15, where a point $P = (x', y', z') = (0, a, a)/2$ is at the center of a small square $OABC$, which is aligned with the axes for convenience. The square defines the surface S that is perpendicular to the axis of rotation, which in this case is the x-axis. The line integral has to be split into four parts, one for each edge, so

$$\oint_{\partial S} \vec{\mathbf{F}}(\mathbf{r}) \cdot d\mathbf{r} = \underbrace{\int_0^a F_y(0, t, 0)\, dt}_{OA} + \underbrace{\int_0^a F_z(0, a, t)\, dt}_{AB} \\ \underbrace{- \int_0^a F_y(0, a-t, a)\, dt}_{BC} \underbrace{- \int_0^a F_z(0, 0, a-t)\, dt}_{CO} . \tag{1.116}$$

The minus signs appearing in the last two terms are due to how we are moving along those edges. The top edge BC, for instance, is parametrized as $\mathbf{r}(t) = \mathbf{r}_B - t\hat{\mathbf{j}}$, $t \in [0, a]$, so $d\mathbf{r} = -dt\,\hat{\mathbf{j}}$, and $\vec{\mathbf{F}} \cdot d\mathbf{r} = -F_y(0, a-t, a)\, dt$. To calculate the line integral, we take a first-order Taylor expansion of F_y around P:

$$F_y(x = 0, y = a - t, z = a) \approx F_y(x', y', z') + \left(\frac{a}{2} - t \right) \frac{\partial F_y}{\partial y} + \frac{a}{2} \frac{\partial F_y}{\partial z} , \tag{1.117}$$

with partial derivatives evaluated at P. Similar expansions are readily found for the other

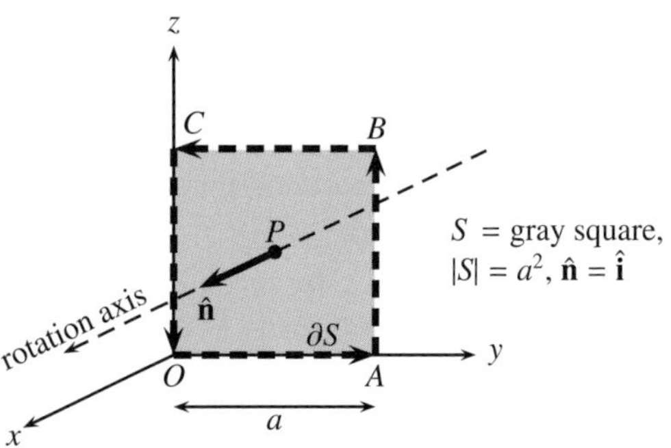

Figure 1.15 Illustration of the curl as a line integral along a square path around a point P at its center.

three edges. Integrating and adding all four terms yields

$$\oint_{\partial S} \vec{\mathbf{F}}(\mathbf{r}) \cdot d\mathbf{r} \approx \left(\frac{\partial F_z}{\partial y} - \frac{\partial F_y}{\partial z} \right) a^2 . \tag{1.118}$$

Hence, by applying the definition (1.113), the x-component of the curl is $\partial F_z/\partial y - \partial F_y/\partial z$, which agrees with Equation (1.114). For the remaining two components, we orient the surface accordingly, and repeat. □

Example 1.20 *Curl calculation.*
Calculate the curl of the vector field $\vec{\mathbf{F}} : \mathbb{R}^3 \to \mathbb{R}^3$ given by

$$\vec{\mathbf{F}}(x, y, z) = 2xy\,\hat{\mathbf{i}} + (z^2 - y)\,\hat{\mathbf{j}} + (x - y)\hat{\mathbf{k}} . \tag{1.119}$$

Solution
The field is differentiable, and its curl is

$$\nabla \times \vec{\mathbf{F}}(x, y, z) = (-1 - 2z)\,\hat{\mathbf{i}} + (-1)\,\hat{\mathbf{j}} + (-2x)\,\hat{\mathbf{k}} . \tag{1.120}$$

Example 1.21 *The curl in polar coordinates.*
Derive the curl operator in two dimensions (on a constant-z plane) using a special case of cylindrical coordinates called **polar coordinates**.

Proof The geometry of the calculation is depicted in Figure 1.16. An arbitrary point $P = (r, \phi)$ is surrounded by a curved box, whose sides are determined by lines of constant radius, $r \pm dr/2$, and constant angle, $\phi \pm d\phi/2$, where dr and $d\phi$ represent small displacements. The vector field is given in polar coordinates as $\vec{\mathbf{F}}(r, \phi) = F_r(r, \phi)\,\hat{\mathbf{r}} + F_\phi(r, \phi)\,\hat{\boldsymbol{\phi}}$.

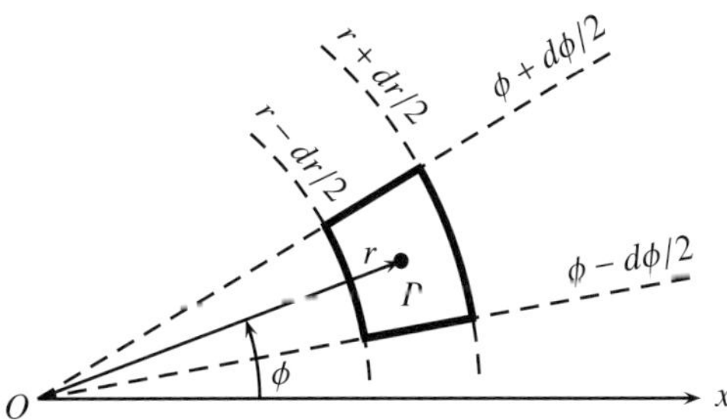

Figure 1.16 The calculation of the curl in polar coordinates.

We repeat the same steps as for the Cartesian coordinates, which begin by taking a first-order Taylor expansion of $\vec{\mathbf{F}}$ around P (in polar coordinates), and proceed with calculating the line integral along the four sides of the curved box. The algebra is similar, so its details will not be repeated here. The result is somewhat different, however:

$$\oint_{\partial S} \vec{\mathbf{F}}(\mathbf{r}) \cdot d\mathbf{r} \approx \left(F_\phi + r\frac{\partial F_\phi}{\partial r} - \frac{\partial F_r}{\partial \phi} \right) dr\, d\phi . \tag{1.121}$$

The answer depends on the value of F_ϕ, which is a new term that was not there in Cartesian coordinates (where only partial derivatives were present). This is more concisely written as

$$\oint_{\partial S} \vec{\mathbf{F}}(\mathbf{r}) \cdot d\mathbf{r} \approx \left(\frac{\partial(rF_\phi)}{\partial r} - \frac{\partial F_r}{\partial \phi} \right) dr\, d\phi \,. \tag{1.122}$$

Another difference is that we need to be careful when calculating the surface area, which is now found by subtracting the areas of the two sectors as follows:

$$|S| = \frac{1}{2}(d\phi)(r + dr/2)^2 - \frac{1}{2}(d\phi)(r - dr/2)^2 = r\, dr\, d\phi \,. \tag{1.123}$$

Finally, we divide the line integral by the surface area to obtain

$$(\nabla \times \vec{\mathbf{F}}) \cdot \hat{\mathbf{k}} = \frac{1}{r}\frac{\partial(rF_\phi)}{\partial r} - \frac{1}{r}\frac{\partial F_r}{\partial \phi} \,. \tag{1.124}$$

□

Example 1.22 *Magnetic field of infinitely long, straight wire (part b: internal and external field).*

The curl in cylindrical coordinates (from the previous example) can be used to calculate the magnetic field from an infinitely long, straight wire of cross-sectional area $A = \pi R^2$, carrying constant current i. This situation is conveniently analyzed with polar coordinates due to its symmetry.

Solution

Let us orient the z-axis with the current flow direction. From physics (by application of Ampère's law, and exploiting the cylindrical symmetry of this problem), we know that the magnetic field is given by

$$\vec{\mathbf{H}}(r) = \begin{cases} \dfrac{ir}{2\pi R^2}\hat{\boldsymbol{\phi}} & \text{for } r \le R \text{ (inside the conductor),} \\ \dfrac{i}{2\pi r}\hat{\boldsymbol{\phi}} & \text{for } r > R \text{ (outside the conductor).} \end{cases} \tag{1.125}$$

Inside the conductor, the field increases linearly from the center. Once outside, it drops as $1/r$. The SI unit for the H-field is A/m. Another commonly encountered unit is Oe (oersted[18]), with 1 Oe $\approx$ 79.58 A/m. In vacuum, the H and B-fields are related by $\vec{\mathbf{B}} = \mu_0 \vec{\mathbf{H}}$.

If we were to generate a quiver or streamline plot of the magnetic field, either inside or outside of the conductor, it would seem at first glance as if the magnetic field is rotating. However, by applying Equation (1.124), we obtain

$$(\nabla \times \vec{\mathbf{H}})_z = \begin{cases} \dfrac{1}{r}\dfrac{\partial}{\partial r}\left(\dfrac{ir^2}{2\pi R^2}\right) = \dfrac{i}{\pi R^2} & \text{inside the conductor,} \\ \dfrac{1}{r}\dfrac{\partial}{\partial r}\left(\dfrac{i}{2\pi}\right) = 0 & \text{outside the conductor.} \end{cases} \tag{1.126}$$

Similarly, for the electrostatic field that was depicted in Figure 1.7 on page 17, we know that $\nabla\times\vec{\mathbf{E}} = 0$. We conclude this discussion with the following observation: The bending of streamlines can be deceiving as it does not necessarily imply that the curl is nonzero.

Example 1.23 *The interface between magnetic materials (part b: tangential components).*
In magnetostatics, Ampère's law states that

$$\nabla \times \vec{\mathbf{H}}(\mathbf{r}) = \vec{\mathbf{J}}(\mathbf{r})\,, \tag{1.127}$$

relating the magnetic field to the (free) current density[a] at each point in space. Here, similar to Example 1.19, we are considering the interface between two materials. Our goal is to find a relationship between the tangential components of the H-field. Due to the discontinuity of the magnetic field across the interface, we cannot apply Equation (1.114), since it involves derivatives. However, we can always apply the fundamental definition of Equation (1.113).

The geometry is similar to that of Figure 1.14. Instead of a Gaussian pillbox, we are using a rectangular line integral path $ABCDEFA$ with P at its center, as shown in Figure 1.17. Again, we are very close to the point P, so the same assumptions are valid: the surface is locally flat, and the fields $\vec{\mathbf{H}}_1$ and $\vec{\mathbf{H}}_2$ are constant. Orienting the rectangle in the manner shown allows the calculation of the tangential component of the curl along the $\hat{\mathbf{t}}_\alpha$-axis. If the rectangle is rotated by 90°, we will obtain the curl along the $\hat{\mathbf{t}}_\beta$-axis. It is possible for these two components to be different, depending on the current components along the two tangential axes.

The contributions to the circulation of $\vec{\mathbf{H}}$ from the vertical sides are zero. For example,

[a] The magnetization inside magnetic materials, which represents magnetic dipole moment density due to phenomena at the atom level, is associated with so-called "bound" currents, $\vec{\mathbf{J}}_m = \nabla \times \vec{\mathbf{M}}$. In contrast, the free current density is associated with charges that are free to move within a conducting medium.

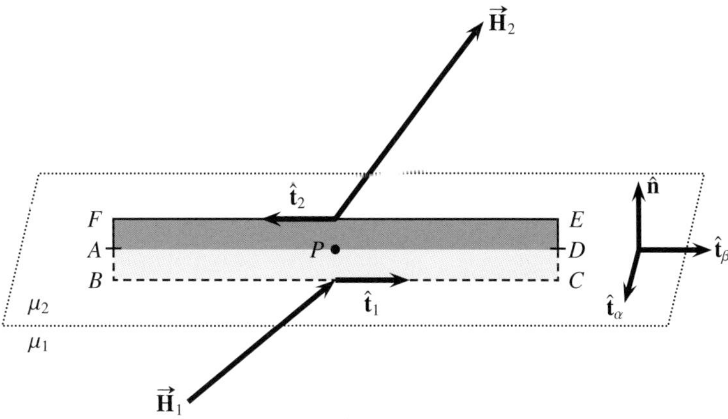

Figure 1.17 Calculation of the circulation of the magnetic field at the interface between two magnetic materials that have different magnetic permeability, μ_1 and μ_2. The point P is at the center of the integration path.

$\int_A^B \vec{\mathbf{H}}_1 \cdot d\mathbf{r} = -\int_C^D \vec{\mathbf{H}}_1 \cdot d\mathbf{r}$, and similarly for $\vec{\mathbf{H}}_2$. Therefore, the circulation is

$$\int_B^C \vec{\mathbf{H}}_1 \cdot \hat{\mathbf{t}}_1 \, dr + \int_E^F \vec{\mathbf{H}}_2 \cdot \hat{\mathbf{t}}_2 \, dr = (\vec{\mathbf{H}}_1 \cdot \hat{\mathbf{t}}_1 - \vec{\mathbf{H}}_2 \cdot \hat{\mathbf{t}}_1) w \,, \tag{1.128}$$

where w is the width of the rectangle. We also used the fact that $\hat{\mathbf{t}}_2 = -\hat{\mathbf{t}}_1 = -\hat{\mathbf{t}}_\beta$. Let us assume that the rectangle height is $h = \kappa w$, with $0 < \kappa \ll 1$. (This implies that a fixed aspect ratio is maintained as the rectangle shrinks.) Hence, using the definition of Equation (1.113), this tangential component of the curl becomes

$$[\nabla \times \vec{\mathbf{H}}(\mathbf{r})]_{t_\alpha} = \lim_{w \to 0} \frac{(\vec{\mathbf{H}}_1 \cdot \hat{\mathbf{t}}_\beta - \vec{\mathbf{H}}_2 \cdot \hat{\mathbf{t}}_\beta) w}{\kappa w^2} = \lim_{w \to 0} \frac{\vec{\mathbf{H}}_1 \cdot \hat{\mathbf{t}}_\beta - \vec{\mathbf{H}}_2 \cdot \hat{\mathbf{t}}_\beta}{\kappa w} \,. \tag{1.129}$$

What happens at the limit? The presence of a nonzero (finite) current density J (A/m^2) in either material is irrelevant because it does not affect the current through the point P, which is exactly on the interface between the two materials. For example, this occurs if the first material is a current-carrying copper conductor and the other is air. The limit should be zero. Therefore, in this case, it is necessary that $\vec{\mathbf{H}}_1 \cdot \hat{\mathbf{t}}_\beta = \vec{\mathbf{H}}_2 \cdot \hat{\mathbf{t}}_\beta$. Since the choice of axes is entirely arbitrary, this will hold in any direction, and we can conclude that *the tangential component of the H-field must be continuous across the boundary.*

However, in certain situations, we must assume the presence of a **surface current density**, i.e., an extremely thin layer of current flowing on the interface, and hence through the point P. It is customary to denote this by $\vec{\mathbf{K}}$ (measured in A/m). The orientation of the current is arbitrary, so

$$\vec{\mathbf{K}} = K_\alpha \hat{\mathbf{t}}_\alpha + K_\beta \hat{\mathbf{t}}_\beta \,. \tag{1.130}$$

Note that only the α-component contributes to current through the rectangular surface.

To model this phenomenon, we can think of a surface current density J_α flowing through the rectangle, such that $J_\alpha \to \infty$ as the rectangle height $h \to 0$, so that the product $J_\alpha h = K_\alpha$ remains constant. In other words, K_α represents $\hat{\mathbf{t}}_\alpha$-axis current per unit length along the $\hat{\mathbf{t}}_\beta$-axis. In this case, Ampère's law is interpreted as

$$[\nabla \times \vec{\mathbf{H}}(\mathbf{r})]_{t_\alpha} = \lim_{w \to 0} \frac{\vec{\mathbf{H}}_1 \cdot \hat{\mathbf{t}}_\beta - \vec{\mathbf{H}}_2 \cdot \hat{\mathbf{t}}_\beta}{\kappa w} = \lim_{w \to 0} \frac{(\vec{\mathbf{K}} \cdot \hat{\mathbf{t}}_\alpha) w}{\kappa w^2} \,, \tag{1.131}$$

where the current density is found as the total current, $K_\alpha w$, over the rectangle area. This implies that

$$(\vec{\mathbf{H}}_1 - \vec{\mathbf{H}}_2) \cdot \hat{\mathbf{t}}_\beta = \vec{\mathbf{K}} \cdot \hat{\mathbf{t}}_\alpha \,. \tag{1.132}$$

We can rewrite this expression as

$$(\vec{\mathbf{H}}_1 - \vec{\mathbf{H}}_2) \cdot (\hat{\mathbf{n}} \times \hat{\mathbf{t}}_\alpha) = \vec{\mathbf{K}} \cdot \hat{\mathbf{t}}_\alpha \,, \tag{1.133}$$

and then use the scalar triple-product identity (1.41) to obtain

$$(\vec{\mathbf{H}}_1 - \vec{\mathbf{H}}_2) \times \hat{\mathbf{n}} \cdot \hat{\mathbf{t}}_\alpha = \vec{\mathbf{K}} \cdot \hat{\mathbf{t}}_\alpha \,. \tag{1.134}$$

Since the choice of axes is arbitrary, we conclude that

$$(\vec{\mathbf{H}}_1 - \vec{\mathbf{H}}_2) \times \hat{\mathbf{n}} = \vec{\mathbf{K}} \,. \tag{1.135}$$

In other words, *in the presence of a surface current density, the tangential components of the H-field are no longer continuous across the interface; they now have a difference that is determined by the magnitude of the surface current.*

Example 1.24 *Invariance of operators to the selection of coordinate axes.*
The formulas for the three differential operators, namely Equation (1.91) for the gradient, Equation (1.100) for the divergence, and Equation (1.114) for the curl, were derived using convenient choices of coordinate axes. The question arises whether the result depends on this choice. Intuitively, we expect that it does not, given the more fundamental meanings of the operators, as expressed by Equations (1.94), (1.99), and (1.113); however, this is not an obvious fact. Here, we attempt a proof of this important, but often overlooked, property. If true, this will mean that the operator will yield the same vector in any orthogonal curvilinear coordinate system, such as spherical and cylindrical coordinates. Even though this exercise is somewhat extensive, it is intended to refresh some concepts from linear algebra.

Let $\hat{\mathbf{i}}$, $\hat{\mathbf{j}}$, $\hat{\mathbf{k}}$ be the mutually orthogonal unit vectors of the original coordinate system, and $\hat{\mathbf{i}}'$, $\hat{\mathbf{j}}'$, $\hat{\mathbf{k}}'$ be the new unit vectors, which are rotated with respect to the originals. The rotation can be arbitrary, but the order is important, as we assume that the cross products satisfy $\hat{\mathbf{i}}' \times \hat{\mathbf{j}}' = \hat{\mathbf{k}}'$, etc., as in Equations (1.34a)–(1.34c); that is, we need the coordinate systems to be right-handed. Now let $\mathbf{q}_1$, $\mathbf{q}_2$, $\mathbf{q}_3$ represent three 3×1 (column) arrays containing the components of the three new basis vectors in the original coordinate system. This means, for example, that $\hat{\mathbf{j}}' = q_{12}\hat{\mathbf{i}} + q_{22}\hat{\mathbf{j}} + q_{32}\hat{\mathbf{k}}$, where q_{ij} represents the ith component of q_j. These components can be arranged in a 3×3 matrix $\mathbf{Q} = [\mathbf{q}_1 \;\; \mathbf{q}_2 \;\; \mathbf{q}_3] = [q_{ij}]$. The elements q_{ij} are also called **directional cosines** of the new unit vectors with respect to the original. They are cosines because of the dot product; for instance, $\hat{\mathbf{j}}' \cdot \hat{\mathbf{k}} = q_{32} = \cos\alpha$, where α is the angle between $\hat{\mathbf{j}}'$ and $\hat{\mathbf{k}}$.

Since $\|\hat{\mathbf{j}}'\|^2 = \hat{\mathbf{j}}' \cdot \hat{\mathbf{j}}' = 1$, we have $\mathbf{q}_2^{\mathsf{T}}\mathbf{q}_2 = 1$. Also, the orthogonality of $\hat{\mathbf{j}}'$ with $\hat{\mathbf{i}}'$ and $\hat{\mathbf{k}}'$ yields $\mathbf{q}_2^{\mathsf{T}}\mathbf{q}_1 = 0$ and $\mathbf{q}_2^{\mathsf{T}}\mathbf{q}_3 = 0$, respectively. Similar expressions are obtained for the other basis vectors. Combined, these yield the matrix equation

$$\mathbf{Q}^{\mathsf{T}}\mathbf{Q} = \mathbb{I}\,, \tag{1.136}$$

which represents the **orthonormality** property of the coordinate axes.

An arbitrary vector $\vec{\mathbf{x}}$ can be expressed in the two coordinate systems as

$$\begin{aligned} \vec{\mathbf{x}} = x_1\hat{\mathbf{i}} + x_2\hat{\mathbf{j}} + x_3\hat{\mathbf{k}} &= x_1'\hat{\mathbf{i}}' + x_2'\hat{\mathbf{j}}' + x_3'\hat{\mathbf{k}}' && (1.137\text{a}) \\ &= x_1'(q_{11}\hat{\mathbf{i}} + q_{21}\hat{\mathbf{j}} + q_{31}\hat{\mathbf{k}}) \\ &\quad + x_2'(q_{12}\hat{\mathbf{i}} + q_{22}\hat{\mathbf{j}} + q_{32}\hat{\mathbf{k}}) \\ &\quad + x_3'(q_{13}\hat{\mathbf{i}} + q_{23}\hat{\mathbf{j}} + q_{33}\hat{\mathbf{k}})\,. && (1.137\text{b}) \end{aligned}$$

Gathering terms and equating coefficients yields

$$\mathbf{x} = \mathbf{Q}\mathbf{x}'\,, \tag{1.138}$$

where $\mathbf{x}$ and $\mathbf{x}'$ are 3×1 arrays containing the coefficients of the vector $\vec{\mathbf{x}}$ in the two coordinate systems. Taking the inverse using the orthonormality property ($\mathbf{Q}^{-1} = \mathbf{Q}^\top$):

$$\mathbf{x}' = \mathbf{Q}^\top \mathbf{x}. \tag{1.139}$$

The stage is now set for analyzing the three operators. Three distinct proofs will follow.

Proof For the **invariance of the gradient**, consider a scalar field A. Taking the partial derivative with respect to one of the new coordinates, x'_ℓ, $\ell \in \{1, 2, 3\}$:

$$\frac{\partial A}{\partial x'_\ell} = \sum_{k=1}^{3} \frac{\partial A}{\partial x_k} \frac{\partial x_k}{\partial x'_\ell} = \sum_{k=1}^{3} q_{k\ell} \frac{\partial A}{\partial x_k} = \mathbf{q}_\ell^\top \mathbf{G}, \tag{1.140}$$

where $\mathbf{G}$ is a 3×1 column vector containing the components of the gradient in the original coordinates, i.e., $\mathbf{G} = [\partial A/\partial x_k]$, $k \in \{1, 2, 3\}$. We have used Equation (1.138) to replace $\partial x_k/\partial x'_\ell$ by $q_{k\ell}$. We can also define $\mathbf{G}' = [\partial A/\partial x'_\ell]$, $\ell \in \{1, 2, 3\}$, as a column vector containing the new gradient components so that Equation (1.140) can be written more concisely as

$$\mathbf{G}' = \mathbf{Q}^\top \mathbf{G}. \tag{1.141}$$

In view of Equation (1.139), $\mathbf{G}$ and $\mathbf{G}'$ describe the same vector in space. □

Now suppose $\vec{\mathbf{A}}$ is a vector field. The components of $\vec{\mathbf{A}}$ in the two coordinate systems will be denoted by A_i and A'_i, respectively, where $i \in \{1, 2, 3\}$. Using the chain rule:

$$\frac{\partial A'_i}{\partial x'_\ell} = \sum_{k=1}^{3} \frac{\partial A'_i}{\partial x_k} \frac{\partial x_k}{\partial x'_\ell} = \sum_{k=1}^{3} \frac{\partial x_k}{\partial x'_\ell} \sum_{j=1}^{3} \frac{\partial A'_i}{\partial A_j} \frac{\partial A_j}{\partial x_k}. \tag{1.142}$$

From Equations (1.138) and (1.139), note that $\partial x_k/\partial x'_\ell = q_{k\ell}$ and $\partial A'_i/\partial A_j = q_{ji}$. Hence

$$\frac{\partial A'_i}{\partial x'_\ell} = \sum_{k=1}^{3} \sum_{j=1}^{3} q_{k\ell} q_{ji} \frac{\partial A_j}{\partial x_k}. \tag{1.143}$$

We introduce the **Jacobian**[19] **matrix** to express these terms concisely as

$$\mathbf{J}' = \mathbf{Q}^\top \mathbf{J} \mathbf{Q}. \tag{1.144}$$

(A Jacobian matrix contains all derivatives of a vector-valued function, $J_{jk} = \partial A_j/\partial x_k$.) Here, the Jacobian matrices are 3×3, with $\mathbf{J}' = [\partial A'_i/\partial x'_\ell]$ and $\mathbf{J} = [\partial A_j/\partial x_k]$.

Proof Our proof for the **invariance of the divergence** will make use of the **trace** operator, which is the sum of the elements on the main diagonal of a square matrix. The key idea is to realize that *the divergence equals the trace of the Jacobian* (by definition). Therefore, it suffices to show that $\mathrm{tr}(\mathbf{J}') = \mathrm{tr}(\mathbf{J})$. This can be readily shown by invoking the cyclic property of the trace, which says that *the trace of a product is invariant to cyclic permutations*. Hence

$$\mathrm{tr}(\mathbf{J}') = \mathrm{tr}(\mathbf{Q}^\top \mathbf{J} \mathbf{Q}) = \mathrm{tr}(\mathbf{J} \mathbf{Q} \mathbf{Q}^\top) = \mathrm{tr}(\mathbf{J}). \tag{1.145}$$

□

Proof For proving the **invariance of the curl**, we first note that it is constructed by differences of terms as in Equation (1.143). For example, the first component of $\vec{\mathbf{B}} = \nabla \times \vec{\mathbf{A}}$ in the new coordinate system is

$$B_1' = \frac{\partial A_3'}{\partial x_2'} - \frac{\partial A_2'}{\partial x_3'} = \sum_{k=1}^{3} \sum_{j=1}^{3} (q_{k2}q_{j3} - q_{k3}q_{j2}) \frac{\partial A_j}{\partial x_k} . \tag{1.146}$$

Now note that the terms in the above sum where $k = j$ will vanish, and that for $k \neq j$, the terms $q_{k2}q_{j3} - q_{k3}q_{j2}$ are actually the components of $\hat{\mathbf{j}}' \times \hat{\mathbf{k}}' = \hat{\mathbf{i}}'$:

$$\hat{\mathbf{j}}' \times \hat{\mathbf{k}}' = (q_{12}\hat{\mathbf{i}} + q_{22}\hat{\mathbf{j}} + q_{32}\hat{\mathbf{k}}) \times (q_{13}\hat{\mathbf{i}} + q_{23}\hat{\mathbf{j}} + q_{33}\hat{\mathbf{k}}) , \tag{1.147a}$$

$$\hat{\mathbf{i}}' = q_{11}\hat{\mathbf{i}} + q_{21}\hat{\mathbf{j}} + q_{31}\hat{\mathbf{k}} . \tag{1.147b}$$

So, let us expand the sum, which consists of six terms, as

$$\begin{aligned} B_1' = \frac{\partial A_3'}{\partial x_2'} - \frac{\partial A_2'}{\partial x_3'} = & \underbrace{(q_{12}q_{23} - q_{13}q_{22})}_{q_{31}} \frac{\partial A_2}{\partial x_1} + \underbrace{(q_{12}q_{33} - q_{13}q_{32})}_{-q_{21}} \frac{\partial A_3}{\partial x_1} + \\ & \underbrace{(q_{22}q_{13} - q_{23}q_{12})}_{-q_{31}} \frac{\partial A_1}{\partial x_2} + \underbrace{(q_{22}q_{33} - q_{23}q_{32})}_{q_{11}} \frac{\partial A_3}{\partial x_2} + \\ & \underbrace{(q_{32}q_{13} - q_{33}q_{12})}_{q_{21}} \frac{\partial A_1}{\partial x_3} + \underbrace{(q_{32}q_{23} - q_{33}q_{22})}_{-q_{11}} \frac{\partial A_2}{\partial x_3} , \end{aligned} \tag{1.148}$$

which leads to

$$\frac{\partial A_3'}{\partial x_2'} - \frac{\partial A_2'}{\partial x_3'} = q_{11} \left(\frac{\partial A_3}{\partial x_2} - \frac{\partial A_2}{\partial x_3} \right) + q_{21} \left(\frac{\partial A_1}{\partial x_3} - \frac{\partial A_3}{\partial x_1} \right) + q_{31} \left(\frac{\partial A_2}{\partial x_1} - \frac{\partial A_1}{\partial x_2} \right) , \tag{1.149}$$

or

$$B_1' = q_{11}B_1 + q_{21}B_2 + q_{31}B_3 = \mathbf{q}_1^\top \mathbf{B} , \tag{1.150}$$

where $\mathbf{B}$ is an array containing the coordinates of $\vec{\mathbf{B}}$ in the original system. By permutation of the indices, similar expressions are obtained for the second and third components of $\vec{\mathbf{B}} = \nabla \times \vec{\mathbf{A}}$ in the new coordinate system, i.e., $B_2' = \mathbf{q}_2^\top B$ and $B_3' = \mathbf{q}_3^\top B$. Hence, $\mathbf{B}' = \mathbf{Q}^\top \mathbf{B}$. □

1.5.4 The Laplace Operator

Consider a scalar field $F : \Omega \subset \mathbb{R}^3 \to \mathbb{R}$. First calculate its gradient (which yields a vector field), and then calculate the divergence of its gradient (which is a scalar field). This sequence of operations is called the **Laplace**[20] **operator** or **Laplacian**, and is denoted by $\nabla \cdot \nabla$, ∇^2, or Δ (although the latter notation may be confused with a change in a quantity, so we will not use it here). The Laplace operator appears in second-order partial differential equations, such as **Laplace's equation**, $\nabla^2 F = 0$, and **Poisson's**[21] **equation**, $\nabla^2 F = h$.

When the field is twice differentiable, it is straightforward to obtain the following expression in Cartesian coordinates using Equations (1.91) and (1.100):

$$\nabla^2 F(\mathbf{r}) = \frac{\partial^2 F}{\partial x^2} + \frac{\partial^2 F}{\partial y^2} + \frac{\partial^2 F}{\partial z^2}, \tag{1.151}$$

where partial derivatives are evaluated at $\mathbf{r} \in \Omega$. We drop the third term in 2-D fields.

Example 1.25 *The Laplacian of* $1/R$ *(basic calculations).*
Consider two points in space, $\mathbf{r} = (x, y, z)$ and $\mathbf{r}' = (x', y', z')$, and take their distance

$$R = \|\mathbf{r} - \mathbf{r}'\| = \sqrt{(x - x')^2 + (y - y')^2 + (z - z')^2}. \tag{1.152}$$

We can define two gradients, namely, $\nabla(1/R)$ with respect to (x, y, z), and $\nabla'(1/R)$ with respect to (x', y', z'), where

$$\nabla' = \frac{\partial}{\partial x'}\hat{\mathbf{i}} + \frac{\partial}{\partial y'}\hat{\mathbf{j}} + \frac{\partial}{\partial z'}\hat{\mathbf{k}}. \tag{1.153}$$

These are defined everywhere in space, except where the two points coincide, causing a singularity. For instance, we have

$$\nabla\left(\frac{1}{R}\right) = -\frac{1}{R^2}\nabla R = -\frac{1}{R^3}[(x - x')\,\hat{\mathbf{i}} + (y - y')\,\hat{\mathbf{j}} + (z - z')\,\hat{\mathbf{k}}] \tag{1.154a}$$

$$= -\frac{1}{R^3}(\mathbf{r} - \mathbf{r}') = -\frac{1}{R^2}\hat{\mathbf{n}}_{\mathbf{r}}, \tag{1.154b}$$

where $\hat{\mathbf{n}}_{\mathbf{r}}$ is a unit vector directed from $\mathbf{r}'$ to $\mathbf{r}$. Since $\nabla R = -\nabla' R = (\mathbf{r} - \mathbf{r}')/R$, it can be seen that

$$\nabla\left(\frac{1}{R}\right) = -\nabla'\left(\frac{1}{R}\right). \tag{1.155}$$

We may also take the two Laplacians with respect to (x, y, z) and (x', y', z'), and show by differentiating the above formula that

$$\nabla^2\left(\frac{1}{R}\right) = \nabla'^2\left(\frac{1}{R}\right) = 0. \tag{1.156}$$

This result should not surprise us, since the electric potential around a point charge, $\varphi = Q/4\pi\epsilon_0 R$, which also follows a $1/R$ relationship, has zero Laplacian everywhere in space (except on the charge itself).[b]

1.5.5 Useful Identities

We list several useful identities involving the differential operators, thus implicitly assuming that the fields are differentiable. We will be invoking these as we derive various results of interest. The proofs of these formulas involve elementary algebraic manipulations, and are left as exercises for the reader.

[b] This is because $-\nabla^2\varphi = -\nabla \cdot \nabla\varphi = \nabla \cdot \vec{\mathbf{E}} = \nabla \cdot \vec{\mathbf{D}}/\epsilon_0 = \rho/\epsilon_0$, ρ being the charge density. This is zero everywhere in space when only a point charge is present.

- The gradient of the product of two scalar fields is

$$\nabla(\varphi\psi) = \varphi\nabla\psi + \psi\nabla\varphi\,. \tag{1.157}$$

- The divergence of the product of a scalar field and a vector field is

$$\nabla\cdot(\varphi\vec{\mathbf{F}}) = \vec{\mathbf{F}}\cdot\nabla\varphi + \varphi\nabla\cdot\vec{\mathbf{F}}\,. \tag{1.158}$$

- The divergence of the cross product of two vector fields is

$$\nabla\cdot(\vec{\mathbf{F}}\times\vec{\mathbf{G}}) = \vec{\mathbf{G}}\cdot(\nabla\times\vec{\mathbf{F}}) - \vec{\mathbf{F}}\cdot(\nabla\times\vec{\mathbf{G}})\,. \tag{1.159}$$

- The curl of the product of a scalar field and a vector field is

$$\nabla\times(\varphi\vec{\mathbf{F}}) = \nabla\varphi\times\vec{\mathbf{F}} + \varphi\nabla\times\vec{\mathbf{F}}\,. \tag{1.160}$$

- The divergence of the curl is zero:

$$\nabla\cdot(\nabla\times\vec{\mathbf{F}}) = 0\,. \tag{1.161}$$

  This identity is fundamental for the analysis of electric machines. It allows us to express the magnetic field density, for which $\nabla\cdot\vec{\mathbf{B}} = 0$, as the curl of a **magnetic vector potential** (MVP), $\vec{\mathbf{B}} = \nabla\times\vec{\mathbf{A}}$.

- The curl of the gradient is zero:

$$\nabla\times(\nabla F) = \mathbf{0}\,. \tag{1.162}$$

  In an electrostatic problem, the electric field is the gradient of a scalar potential, $\vec{\mathbf{E}} = -\nabla\varphi$, and this formula yields $\nabla\times\vec{\mathbf{E}} = \mathbf{0}$.

- The curl of the curl is

$$\nabla\times(\nabla\times\vec{\mathbf{F}}) = \nabla(\nabla\cdot\vec{\mathbf{F}}) - \nabla^2\vec{\mathbf{F}}\,. \tag{1.163}$$

  Here, we have the gradient of the divergence minus a **vector Laplacian**, which in Cartesian coordinates is

$$\nabla^2\vec{\mathbf{F}} = (\nabla^2 F_x, \nabla^2 F_y, \nabla^2 F_z)\,. \tag{1.164}$$

  For example, consider Ampère's law in a magnetostatic problem, which states that

$$\nabla\times\vec{\mathbf{H}} = \vec{\mathbf{J}}\,, \tag{1.165}$$

  where $\vec{\mathbf{H}}$ is the magnetic field and $\vec{\mathbf{J}}$ is the free (i.e., not bound) current density. In free space and inside non-ferromagnetic materials (e.g., aluminum or copper conductors), where $\vec{\mathbf{B}} = \mu\vec{\mathbf{H}}$ with $\mu \approx \mu_0$, this formula yields $\nabla\times\vec{\mathbf{B}} = \mu_0\vec{\mathbf{J}}$. Since we can write $\vec{\mathbf{B}} = \nabla\times\vec{\mathbf{A}}$ (because $\nabla\cdot\vec{\mathbf{B}} = 0$), we can employ the curl-of-the-curl identity to obtain

$$\nabla\times(\nabla\times\vec{\mathbf{A}}) = \nabla(\nabla\cdot\vec{\mathbf{A}}) - \nabla^2\vec{\mathbf{A}} = \mu_0\vec{\mathbf{J}}\,, \tag{1.166}$$

  where $\vec{\mathbf{A}}$ is the MVP. It is possible to employ a **Coulomb** or **Lorenz**[22] **gauge** condition on the MVP, that is, to require that $\nabla\cdot\vec{\mathbf{A}} = 0$. This leads to a second-order differential equation in terms of the MVP:

$$-\nabla^2\vec{\mathbf{A}} = \mu_0\vec{\mathbf{J}}\,. \tag{1.167}$$

- The curl of the cross product of two vector fields is

$$\nabla \times (\vec{\mathbf{F}} \times \vec{\mathbf{G}}) = \vec{\mathbf{F}}(\nabla \cdot \vec{\mathbf{G}}) - \vec{\mathbf{G}}(\nabla \cdot \vec{\mathbf{F}}) + (\vec{\mathbf{G}} \cdot \nabla)\vec{\mathbf{F}} - (\vec{\mathbf{F}} \cdot \nabla)\vec{\mathbf{G}} . \tag{1.168}$$

The first two terms on the right-hand side are products of vectors ($\vec{\mathbf{F}}$ and $\vec{\mathbf{G}}$) with scalars (the divergence of $\vec{\mathbf{G}}$ and $\vec{\mathbf{F}}$, respectively). However, the third and fourth terms are based on a notation that we have not encountered before. For instance, writing $\vec{\mathbf{G}} \cdot \nabla$ is a mnemonic for an operator that expands as

$$\vec{\mathbf{G}} \cdot \nabla = G_x \frac{\partial}{\partial x} + G_y \frac{\partial}{\partial y} + G_z \frac{\partial}{\partial z} , \tag{1.169}$$

acting on each element of the operand (a vector field $\vec{\mathbf{F}} = (F_x, F_y, F_z)$) as follows:

$$(\vec{\mathbf{G}} \cdot \nabla)\vec{\mathbf{F}} = \begin{bmatrix} G_x \dfrac{\partial F_x}{\partial x} + G_y \dfrac{\partial F_x}{\partial y} + G_z \dfrac{\partial F_x}{\partial z} \\ G_x \dfrac{\partial F_y}{\partial x} + G_y \dfrac{\partial F_y}{\partial y} + G_z \dfrac{\partial F_y}{\partial z} \\ G_x \dfrac{\partial F_z}{\partial x} + G_y \dfrac{\partial F_z}{\partial y} + G_z \dfrac{\partial F_z}{\partial z} \end{bmatrix} . \tag{1.170}$$

- Consider a time-dependent scalar field $F(\mathbf{r}, t)$. The position $\mathbf{r}$ may also be a function of time, for example, if we are tracking the motion $\mathbf{r}(t)$ of a particle in space. In this case, we can calculate the **total derivative** of F with respect to time using the chain rule as

$$\frac{dF}{dt} = \frac{\partial F}{\partial t} + \frac{\partial F}{\partial x}\frac{dx}{dt} + \frac{\partial F}{\partial y}\frac{dy}{dt} + \frac{\partial F}{\partial z}\frac{dz}{dt} . \tag{1.171}$$

A more concise way of writing this is

$$\frac{dF}{dt} = \frac{\partial F}{\partial t} + \mathbf{v} \cdot \nabla F , \tag{1.172}$$

where

$$\mathbf{v} = \frac{d\mathbf{r}}{dt} \tag{1.173}$$

is the velocity. The total derivative is also called the **material derivative** or the **convective derivative**.

- Consider a time-dependent vector field $\vec{\mathbf{F}}(\mathbf{r}, t)$. Similar to the scalar case, we can compute the total derivative by operating on each component of $\vec{\mathbf{F}}$ separately. This leads to

$$\frac{d\vec{\mathbf{F}}}{dt} = \frac{\partial \vec{\mathbf{F}}}{\partial t} + (\mathbf{v} \cdot \nabla)\vec{\mathbf{F}} . \tag{1.174}$$

- The gradient of the dot product of two vector fields is

$$\nabla(\vec{\mathbf{F}} \cdot \vec{\mathbf{G}}) = (\vec{\mathbf{F}} \cdot \nabla)\vec{\mathbf{G}} + (\vec{\mathbf{G}} \cdot \nabla)\vec{\mathbf{F}} + \vec{\mathbf{F}} \times (\nabla \times \vec{\mathbf{G}}) + \vec{\mathbf{G}} \times (\nabla \times \vec{\mathbf{F}}) . \tag{1.175}$$

In the special case where $\vec{\mathbf{F}} = \vec{\mathbf{G}}$, we obtain

$$\frac{1}{2}\nabla(\vec{\mathbf{F}} \cdot \vec{\mathbf{F}}) = (\vec{\mathbf{F}} \cdot \nabla)\vec{\mathbf{F}} + \vec{\mathbf{F}} \times (\nabla \times \vec{\mathbf{F}}) . \tag{1.176}$$

Example 1.26 *The magnetic vector potential of an infinitely long wire.*
We revisit the case of the infinitely long and straight wire in free space, whose magnetic field was calculated in Example 1.22. The purpose of this example is to calculate the MVP for this configuration. The conductor cross-sectional area is a circle of radius R. A uniform current density $\vec{\mathbf{J}}(r) = (i/\pi R^2)\hat{\mathbf{k}}$ is assumed (for $r \leq R$). In this geometry, which has an infinite length, the variation along the z-axis is zero; hence, $\partial/\partial z = 0$ for every variable. Furthermore, this geometry has cylindrical symmetry, so $\partial/\partial\phi = 0$.

The MVP can be found either (i) from $\nabla \times \vec{\mathbf{A}} = \vec{\mathbf{B}} = \mu_0\vec{\mathbf{H}}$, or (ii) from $-\nabla^2\vec{\mathbf{A}} = \mu_0\vec{\mathbf{J}}$ with $\nabla \cdot \vec{\mathbf{A}} = 0$. The former will require solving a first-order differential equation, whereas the latter is a second-order differential equation. It appears that the former approach is simpler, so we proceed with this one first.

Approach (i). Recall the curl operator in cylindrical coordinates:

$$\nabla \times \vec{\mathbf{A}} = \left(\frac{1}{r}\frac{\partial A_z}{\partial\phi} - \frac{\partial A_\phi}{\partial z}\right)\hat{\mathbf{r}} + \left(\frac{\partial A_r}{\partial z} - \frac{\partial A_z}{\partial r}\right)\hat{\boldsymbol{\phi}} + \frac{1}{r}\left(\frac{\partial(rA_\phi)}{\partial r} - \frac{\partial A_r}{\partial\phi}\right)\hat{\mathbf{k}}. \tag{1.177}$$

Setting all $\partial/\partial z$ and $\partial/\partial\phi$ terms to zero yields

$$\nabla \times \vec{\mathbf{A}} = \left(-\frac{\partial A_z}{\partial r}\right)\hat{\boldsymbol{\phi}} + \frac{1}{r}\frac{\partial(rA_\phi)}{\partial r}\hat{\mathbf{k}}. \tag{1.178}$$

This should equal

$$\mu_0\vec{\mathbf{H}}(r) = \begin{cases} \dfrac{\mu_0 i}{2\pi}\dfrac{r}{R^2}\hat{\boldsymbol{\phi}} & \text{for } r \leq R \text{ (inside the conductor)}, \\[2ex] \dfrac{\mu_0 i}{2\pi}\dfrac{1}{r}\hat{\boldsymbol{\phi}} & \text{for } r > R \text{ (outside the conductor)}. \end{cases} \tag{1.179}$$

Equating coefficients, we obtain the following two differential equations:

$$-\frac{\partial A_z}{\partial r} = \mu_0 H_\phi(r), \tag{1.180}$$

$$\frac{1}{r}\frac{\partial(rA_\phi)}{\partial r} = \frac{1}{r}A_\phi + \frac{\partial A_\phi}{\partial r} = 0. \tag{1.181}$$

Equation (1.180) can be integrated:

$$A_z(r) = A_z(0) - \int_0^r \mu_0 H_\phi(\xi)\,d\xi = \begin{cases} A_z(0) - \dfrac{\mu_0 i}{4\pi}\left(\dfrac{r}{R}\right)^2 & \text{for } r \leq R, \\[2ex] A_z(R) - \dfrac{\mu_0 i}{2\pi}\ln\left(\dfrac{r}{R}\right) & \text{for } r > R. \end{cases} \tag{1.182}$$

Clearly, $A_z(0)$ is an arbitrary constant of integration. It is interesting to note that the potential value reduces indefinitely to minus infinity as we move away from the conductor due to the $\ln(r/R)$ term. Nevertheless, this does not lead to an infinite magnetic field, as taking the curl of $\vec{\mathbf{A}}$ (which involves differentiation of A_z with respect to r) yields a finite magnetic field that drops as $1/r$.

The solution of Equation (1.181) is

$$A_\phi(r) = \frac{C_\phi}{r}, \tag{1.183}$$

with C_ϕ a constant. Note that we have not obtained any information regarding A_r.

Approach (ii). Let us see if the second approach is more informative. Using the divergence and the vector Laplacian operators in cylindrical coordinates, we obtain

$$\nabla \cdot \vec{\mathbf{A}} = \frac{1}{r}\frac{\partial (rA_r)}{\partial r} = 0\,, \tag{1.184}$$

$$\nabla^2 \vec{\mathbf{A}} = \left(\nabla^2 A_r - \frac{A_r}{r^2}\right)\hat{\mathbf{r}} + \left(\nabla^2 A_\phi - \frac{A_\phi}{r^2}\right)\hat{\boldsymbol{\phi}} + \nabla^2 A_z\,\hat{\mathbf{k}} = -\mu_0 J\,\hat{\mathbf{k}}\,, \tag{1.185}$$

where we have set all $\partial/\partial z$ and $\partial/\partial\phi$ terms to zero. Note that the radial and tangential terms have the same form.

Equation (1.184) implies that

$$A_r(r) = \frac{C_r}{r}\,. \tag{1.186}$$

This is a new piece of information regarding A_r, which the previous approach did not provide (as we were not asking for the divergence to equal zero).

We proceed with the analysis of (1.185), equating components on both sides. The radial component of $\nabla^2\vec{\mathbf{A}}$ should vanish because the current does not have a radial component. To proceed, we need the Laplacian of a scalar field f in cylindrical coordinates, which is (without $\partial/\partial z$ and $\partial/\partial\phi$ terms)

$$\nabla^2 f = \frac{1}{r}\frac{\partial}{\partial r}\left(r\frac{\partial f}{\partial r}\right) = \frac{1}{r}\frac{\partial f}{\partial r} + \frac{\partial^2 f}{\partial r^2}\,. \tag{1.187}$$

Using Equation (1.186), we have

$$(\nabla^2\vec{\mathbf{A}})_r = \nabla^2 A_r - \frac{A_r}{r^2} = \frac{1}{r}\frac{\partial}{\partial r}\left(-r\frac{C_r}{r^2}\right) - \frac{C_r}{r^3} = 0\,. \tag{1.188}$$

Therefore, by virtue of the divergence being zero, the condition for the radial component is satisfied as an identity.

The tangential ϕ-component needs to satisfy the second-order differential equation

$$(\nabla^2\vec{\mathbf{A}})_\phi = \frac{1}{r}\frac{\partial A_\phi}{\partial r} + \frac{\partial^2 A_\phi}{\partial r^2} - \frac{A_\phi}{r^2} = 0\,. \tag{1.189}$$

We already know that a potential $A_\phi = C_\phi/r$ will satisfy this equation (as it did for A_r). However, this differential equation also admits $A_\phi = C_\phi r$ as a second solution. Hence, in general:

$$A_\phi(r) = \frac{C_{\phi 1}}{r} + C_{\phi 2}\,r\,. \tag{1.190}$$

Nevertheless, we can argue that the $C_{\phi 2}r$ term should be dropped because it yields a constant component of magnetic field in the z-direction, which is not present in this problem. (From Equation (1.178), this would be $2C_{\phi 2}$.) In the absence of any further information for A_r and A_ϕ, since the terms C_r/r and $C_{\phi 1}/r$ do not affect the magnetic field in any way, and to avoid the singularity at the origin, we can set $C_r = C_{\phi 1} = 0$. Therefore, $A_r(r) = A_\phi(r) = 0$.

Finally, for the z-component (which is the significant one here), we must solve the second-order differential equation

$$\nabla^2 A_z = \frac{1}{r}\frac{\partial}{\partial r}\left(r\frac{\partial A_z}{\partial r}\right) = \begin{cases} -\dfrac{\mu_0 i}{\pi R^2} & \text{for } r \leq R\,, \\ 0 & \text{for } r > R\,. \end{cases} \tag{1.191}$$

This is equivalent to

$$\frac{\partial}{\partial r}\left(r\frac{\partial A_z}{\partial r}\right) = \begin{cases} -\dfrac{\mu_0 i r}{\pi R^2} & \text{for } r \leq R\,, \\ 0 & \text{for } r > R\,, \end{cases} \tag{1.192}$$

leading to

$$r\frac{\partial A_z}{\partial r} = \begin{cases} C_{z1} - \dfrac{\mu_0 i}{2\pi}\left(\dfrac{r}{R}\right)^2 & \text{for } r \leq R\,, \\ C_{z2} & \text{for } r > R\,, \end{cases} \tag{1.193}$$

and finally to

$$A_z(r) = \begin{cases} C_{z3} + C_{z1}\ln r - \dfrac{\mu_0 i}{4\pi}\left(\dfrac{r}{R}\right)^2 & \text{for } r \leq R\,, \\ C_{z4} + C_{z2}\ln r & \text{for } r > R\,, \end{cases} \tag{1.194}$$

where C_{z1}–C_{z4} are constants of integration. We can set $C_{z1} = 0$ to avoid a singularity at the origin. We also require continuity of the potential and its derivative at $r = R$. It can be readily shown that we obtain an expression that is identical to Equation (1.182).

1.6 Integral Laws

Having defined the various differential operators, we can now proceed with the description of integral laws that are essential for our purposes. Again, we focus on the ideas rather than the technical details to the extent possible. The common thread among these laws is that they apply to integrations of differential operators. Hence, there is a similarity with the fundamental theorem of calculus, which states that $\int_a^b f'(x)\,dx = f(b)-f(a)$. Intuitively, we expect the result to depend on the field values along the integration boundary.

1.6.1 The Divergence Theorem or Gauss' Law

The **divergence theorem**, also known as **Gauss' law**, Gauss' theorem, or **Ostrogradsky's theorem**,[23] applies to a volume integral of the divergence of a vector field $\vec{\mathbf{F}} : \Omega \subset \mathbb{R}^3 \to \mathbb{R}^3$, which becomes a surface integral on the boundary ∂V of the volume $V \subset \Omega$:

$$\iiint_V \nabla \cdot \vec{\mathbf{F}}(\mathbf{r})\,dv = \oiint_{\partial V} \vec{\mathbf{F}}(\mathbf{r}) \cdot d\mathbf{a}\,. \tag{1.195}$$

This should not surprise us given the definition of the divergence (see Equation (1.99) on page 30), which is based on an infinitesimal volume and the associated surface integral.

As a Riemann sum, we can imagine the volume V being divided into a large number of very small cubes, where the flow coming out of one cube's side is entering the adjacent cube. Hence, all these terms cancel out, and only the flows on the outer surface ∂V are left. This boundary is also called a **Gaussian surface**.

An interesting corollary is obtained if we suppose that $\vec{\mathbf{F}} = \varphi\vec{\mathbf{C}}$, with $\varphi = \varphi(\mathbf{r})$ a scalar field, and $\vec{\mathbf{C}}$ a constant vector. Then, using Equation (1.158), we have

$$\iiint_V \nabla \cdot (\varphi\vec{\mathbf{C}})\, dv = \iiint_V \vec{\mathbf{C}} \cdot \nabla\varphi\, dv = \vec{\mathbf{C}} \cdot \iiint_V \nabla\varphi\, dv. \tag{1.196}$$

Applying the divergence theorem to the first term, we obtain

$$\vec{\mathbf{C}} \cdot \iiint_V \nabla\varphi\, dv = \oiint_{\partial V} \varphi\vec{\mathbf{C}} \cdot d\mathbf{a} = \vec{\mathbf{C}} \cdot \oiint_{\partial V} \varphi\,\hat{\mathbf{n}}\, da. \tag{1.197}$$

Because $\vec{\mathbf{C}}$ is arbitrary, we conclude that the volume integral of a gradient can be converted to a surface integral:

$$\iiint_V \nabla\varphi\, dv = \oiint_{\partial V} \varphi\,\hat{\mathbf{n}}\, da. \tag{1.198}$$

A second interesting corollary is obtained when we apply the divergence theorem to a vector that has three divergences as components. For example, let

$$\vec{\mathbf{G}} = \begin{bmatrix} \nabla \cdot \vec{\mathbf{F}}_1 \\ \nabla \cdot \vec{\mathbf{F}}_2 \\ \nabla \cdot \vec{\mathbf{F}}_3 \end{bmatrix}. \tag{1.199}$$

We can arrange the components of the three vectors in a 3×3 matrix, as follows:

$$\mathbf{F} = \begin{bmatrix} \vec{\mathbf{F}}_1 \\ \vec{\mathbf{F}}_2 \\ \vec{\mathbf{F}}_3 \end{bmatrix} = \begin{bmatrix} F_{1x} & F_{1y} & F_{1z} \\ F_{2x} & F_{2y} & F_{2z} \\ F_{3x} & F_{3y} & F_{3z} \end{bmatrix}, \tag{1.200}$$

which allows us to rewrite the previous equation more concisely as

$$\vec{\mathbf{G}} = \nabla \cdot \mathbf{F}, \tag{1.201}$$

as if the divergence operator acts on each row. The divergence theorem is now applied component-wise, leading to

$$\iiint_V \vec{\mathbf{G}}\, dv = \iiint_V \nabla \cdot \mathbf{F}\, dv = \oiint_{\partial V} \mathbf{F} \cdot \hat{\mathbf{n}}\, da. \tag{1.202}$$

Here, $\mathbf{F} \cdot \hat{\mathbf{n}}$ should be interpreted as three dot products, one per row of $\mathbf{F}$. This operation is similar to a matrix–vector multiplication.

Example 1.27 *The Laplacian of* $1/R$ *revisited (similarity to a Dirac*[24] *delta function).* Recall from Example 1.25 that $\nabla^2(1/R)$ has a peculiar property: it is everywhere equal to zero, *except* at $\mathbf{r} = \mathbf{r}'$, where a singularity occurs. In this example, we intend to show that the Laplacian of $1/R$ acts as a **Dirac delta function** in 3-D space. A Dirac delta function (or distribution) in $\mathbb{R}$ is zero everywhere, except where its argument equals zero, where it is undefined as it grows indefinitely towards ∞. However, this growth is such that the integral of the delta function satisfies

$$\int_{-\infty}^{\infty} \delta(x)\, dx = 1 \tag{1.203}$$

and

$$\int_{-\infty}^{\infty} f(x)\delta(x-a)\, dx = f(a), \tag{1.204}$$

for a continuous function $f : \mathbb{R} \to \mathbb{R}$.[c] Similarly, we will show that

$$-\frac{1}{4\pi} \iiint_V \nabla^2 \left(\frac{1}{\|\mathbf{r}-\mathbf{r}'\|} \right) dv = 1 \tag{1.205}$$

and

$$-\frac{1}{4\pi} \iiint_V F(\mathbf{r})\, \nabla^2 \left(\frac{1}{\|\mathbf{r}-\mathbf{r}'\|} \right) dv = F(\mathbf{r}'), \tag{1.206}$$

for a continuous scalar field $F : V \subset \mathbb{R}^3 \to \mathbb{R}$. Here, $\mathbf{r}$ is the variable of integration, whereas the point $\mathbf{r}'$ is fixed and assumed to be contained within the volume V. (Otherwise, the volume integrals become zero.)

Proof We will employ a modified distance

$$R_\epsilon = \sqrt{(x-x')^2 + (y-y')^2 + (z-z')^2 + \epsilon^2} = \sqrt{R^2 + \epsilon^2}, \tag{1.207}$$

where $\epsilon > 0$ is a parameter. Now the Laplacian $\nabla^2(1/R_\epsilon)$ is well-defined everywhere in space, even for $\mathbf{r} = \mathbf{r}'$. The idea is to let ϵ get arbitrarily close to zero, since $\lim_{\epsilon\to 0} R_\epsilon = R$, and to investigate what happens in this regime. We have

$$\nabla \left(\frac{1}{R_\epsilon} \right) = -\frac{1}{R_\epsilon^2} \nabla R_\epsilon = -\frac{1}{R_\epsilon^3} [(x-x')\,\hat{\mathbf{i}} + (y-y')\,\hat{\mathbf{j}} + (z-z')\,\hat{\mathbf{k}}] = -\frac{1}{R_\epsilon^3} (\mathbf{r}-\mathbf{r}'). \tag{1.208}$$

The reader can also verify that

$$\nabla^2 \left(\frac{1}{R_\epsilon} \right) = -\frac{3}{R_\epsilon^3} + \frac{3R^2}{R_\epsilon^5} = -\frac{3\epsilon^2}{R_\epsilon^5}, \tag{1.209}$$

which is plotted on a logarithmic scale in Figure 1.18 for various values of the parameter ϵ. This function obtains the value $-3/\epsilon^3$ for $R = 0$. Its resemblance to a one-dimensional (1-D) Dirac delta function is obvious. Note that as ϵ becomes smaller, the function value at the origin tends to infinity, but it also goes to zero much faster as we move away from the origin.

Let us use the divergence theorem to evaluate

$$\iiint_V \nabla^2 \left(\frac{1}{R_\epsilon} \right) dv = \oiint_{\partial V} \nabla \left(\frac{1}{R_\epsilon} \right) \cdot d\mathbf{a} = \oiint_{\partial V} -\frac{1}{R_\epsilon^3} (\mathbf{r}-\mathbf{r}') \cdot d\mathbf{a}, \tag{1.210}$$

[c] Mathematicians will argue that these integrals are not well-defined Riemann integrals, and that this is an abuse of notation. However, a more rigorous definition of the Dirac delta function is outside the scope of this book.

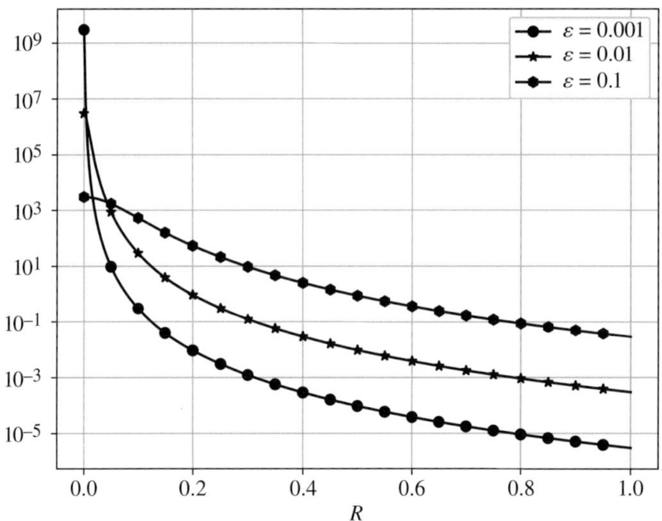

Figure 1.18 Parametric plots of the (negative) Laplacian of $1/R_\epsilon$, $f(R;\epsilon) = -\nabla^2(1/R_\epsilon)$.

where V is a sphere of radius R that is centered at the point $\mathbf{r}'$. Here, we are integrating the gradient (1.208) on the surface of the sphere ∂V. This implies that for all points $\mathbf{r}$ on ∂V, we have $\mathbf{r} - \mathbf{r}' = R\,\hat{\mathbf{n}}_\mathbf{r}$, where $\hat{\mathbf{n}}_\mathbf{r}$ is the outwards-pointing unit normal vector at point $\mathbf{r}$, which of course is parallel to $d\mathbf{a}$. Therefore

$$\iiint_V \nabla^2\left(\frac{1}{R_\epsilon}\right) dv = -\oiint_{\partial V} \left(R^2+\epsilon^2\right)^{-\frac{3}{2}} R\,da = -4\pi R^3 \left(R^2+\epsilon^2\right)^{-\frac{3}{2}}. \tag{1.211}$$

We now take the limit of the expression

$$\lim_{\epsilon\to 0} \iiint_V \nabla^2\left(\frac{1}{R_\epsilon}\right) dv = -4\pi\,, \tag{1.212}$$

which proves Equation (1.205).

An informal way to show the validity of Equation (1.206) is to consider a small sphere centered at $\mathbf{r}'$. We shrink the sphere as much as needed so that the approximation $F(\mathbf{r}) \approx F(\mathbf{r}')$ holds throughout the sphere. Then, we can pull this constant factor outside of the integral and use Equation (1.205) to get to the desired result.

Finally, it is straightforward to extend this result to vector fields. By applying Equation (1.206) to each component of a vector field $\vec{\mathbf{F}} : V \to \mathbb{R}^3$, we have

$$-\frac{1}{4\pi} \iiint_V \vec{\mathbf{F}}(\mathbf{r})\,\nabla^2\left(\frac{1}{\|\mathbf{r}-\mathbf{r}'\|}\right) dv = \vec{\mathbf{F}}(\mathbf{r}')\,. \tag{1.213}$$

This is a volume integral of a vector field obtained by performing the integration on a component-wise basis. For example:

$$[\vec{\mathbf{F}}(\mathbf{r}')]_x = \iiint_V [\vec{\mathbf{F}}(\mathbf{r})]_x\,\nabla^2\left(\frac{1}{\|\mathbf{r}-\mathbf{r}'\|}\right) dv\,. \tag{1.214}$$

□

Example 1.28 *Defining functions in Python.*

In this example, our intent is to explain how a **Python function** can be defined. We shall illustrate this using the generation of the plot of Figure 1.18 as a case study. To create this plot, a function was called repeatedly to obtain the values of the Laplacian, using the following Python code:

```
1  from math import pi
2  import numpy as np
3  import matplotlib.pyplot as plt
4
5  def Lapl_1oR_eps(R, eps):
6      Reps = np.sqrt(np.square(R) + eps**2)
7      y = np.divide(-3,np.power(Reps,3)) + \
8          np.divide(3*np.square(R),np.power(Reps,5))
9      return y
10
11 R = np.linspace(0,1,10000)
12 eps = [0.001, 0.01, 0.1]
13 markers = ['o','*','h']
14 k = 0
15 for e in eps:
16     plt.semilogy(R, -Lapl_1oR_eps(R = R,eps = e), \
17                  marker = markers[k], markevery = 500,
18                  color='k')
19     k += 1
20
21 plt.xlabel(r'$R$')
22 plt.legend([r'$\epsilon = ' + str(i) + '$' for i in eps])
23 plt.grid()
24 plt.show()
```

Let us look into this script further.

- The Laplacian function `Lapl_1oR_eps` is defined in lines 5–9, based on Equations (1.207) and (1.209). To define a function, we must use the keyword **`def`** that precedes the function name and the list of arguments, ending with a colon. Everything after this must be indented by the same number of space characters, which is typically 4 in Python. Note the line wrapping in line 7 using the backslash symbol.

- The function `Lapl_1oR_eps` has two inputs, namely, the radius R and the parameter ϵ (epsilon). Note that R can be a NumPy array, allowing all operations to be performed element-wise (see Example 1.7, page 12), using `np.sqrt`, `np.square`, `np.power`, and `np.divide`.

- Sending arrays as arguments to the function, rather than single elements, has the advantage of reducing the required number of function calls, and increases computational efficiency. In this example, the vector R has 10,000 evenly spaced elements in $[0, 1]$, as shown in line 11 (external to the function), which are generated using the `np.linspace` command. Finally, note that the function returns a value in line 9, using the keyword **`return`**.

- The syntax of line 15 is interesting. Here, this **`for`** loop takes every element contained **`in`** the list `eps`, names it `e`, and then does something with it. Of course, this is used for plotting a curve for each value of ϵ. The plot is performed with a `semilogy` command, which reminds us of the Matlab command with the same name. This generates a logarithmic y-axis, but the x-axis is linear.
- A closer look at the function call in line 16 reveals that we can specify function arguments by the at-first-glance awkward syntax `(R = R, eps = e)`. This is Python's way of making code simpler to read. It means that the function was defined with arguments named `R` and `eps`, as in line 5. Then, when the function is actually called, this syntax says that we assign the variable `R` (defined in line 11) and the variable `e` (from line 15) to the corresponding function arguments. It is all right for the variables to share the same name. Of course, this is not necessary, and we could have just called the function by `Lapl_1oR_eps(R, e)`.
- Plotting takes place in lines 21–24. Note the peculiar syntax for strings, e.g., `r'$R$'`. The `r` prefix makes this a **raw string**, and instructs the Python interpreter to treat backslash characters differently than the standard way (i.e., as escape sequences). Here, this is useful because they are used as special characters for typesetting mathematical symbols, which are placed inside a pair of dollar signs as in `$\epsilon$`, which produces a Greek letter ϵ on the plot.
- Line 22, which plots the legend, is interesting in two ways. First, it shows us how to create a string by **concatenation** of other strings. In Python, this can be simply achieved using the + operator. Note that the **`str`** command converts its numerical argument to a string. The second thing that is interesting is the overall syntax inside the `legend` command, whose argument is enclosed in brackets, `[...]`. This syntax generates a list of strings for the legend, and it is called a **list comprehension**. The list is created by iterating along the elements of the list `eps`, as signified by the latter part of the expression "**`for`** `i` **`in`** `eps`." In other words, a string is created for each element `i` of `eps`, and the result is a list of strings.

1.6.2 Stokes' Theorem

Stokes'[25] **theorem** (also known as the Kelvin[26]–Stokes theorem) applies to a surface integral of the curl of a vector field $\vec{\mathbf{F}} : \Omega \subset \mathbb{R}^3 \to \mathbb{R}^3$. The theorem states that

$$\iint_S \nabla \times \vec{\mathbf{F}}(\mathbf{r}) \cdot d\mathbf{a} = \oint_{\partial S} \vec{\mathbf{F}}(\mathbf{r}) \cdot d\mathbf{r}, \tag{1.215}$$

thus equating the surface integral of the curl on an open surface $S \subset \Omega$ to the circulation of the field along the surface boundary ∂S. The surface can be three-dimensional. The direction of the line integral is such that the positive side of the surface (i.e., the side where the normal vector is emanating from) is on our left as we walk down the path.

Intuitively, this theorem makes sense given the definition of the curl (see Equation (1.113) on page 33), which is based on an infinitesimal surface and the associated surface integral. If we subdivide the surface S into a large number of small patches, we can argue that contributions to the curl integral from adjacent sides cancel, except from those on the boundary.

1.6.3 Green's Identities

Green's[27] **identities** relate volume and surface integrals in 3-D space, so they are derived from Gauss' theorem. Suppose φ and ψ are twice-differentiable scalar fields defined in $\Omega \subset \mathbb{R}^3$. Applying the divergence theorem to the vector field $\varphi\nabla\psi$ (i.e., the product of a scalar with a gradient) over $V \subset \Omega$ yields

$$\iiint_V \nabla \cdot (\varphi\nabla\psi)\, dv = \oiint_{\partial V} \varphi\nabla\psi \cdot d\mathbf{a}\,. \tag{1.216}$$

Now, recalling the vector identity (1.158) that involves the divergence of the product of a scalar field and a vector field, we can rewrite this as

$$\iiint_V (\nabla\varphi \cdot \nabla\psi + \varphi\nabla^2\psi)\, dv = \oiint_{\partial V} \varphi\nabla\psi \cdot d\mathbf{a}\,. \tag{1.217}$$

This formula is known as **Green's first identity**.

An interesting corollary is obtained when we let $\nabla\psi = \vec{\mathbf{G}}$, while at the same time we think of φ as a component of another vector field $\vec{\mathbf{F}}$, e.g., $\varphi = F_x$. Then we obtain

$$\iiint_V (\nabla F_x \cdot \vec{\mathbf{G}} + F_x\nabla \cdot \vec{\mathbf{G}})\, dv = \oiint_{\partial V} F_x\vec{\mathbf{G}} \cdot d\mathbf{a}\,. \tag{1.218}$$

We can write this equation for the other two components of $\vec{\mathbf{F}}$, and if we collect all three equations in a vector, we obtain surface and volume integrals of vector fields. These are evaluated component-wise, as in Equations (1.79) and (1.89), and we can write the identity concisely as

$$\iiint_V [(\vec{\mathbf{G}} \cdot \nabla)\vec{\mathbf{F}} + (\nabla \cdot \vec{\mathbf{G}})\vec{\mathbf{F}}]\, dv = \oiint_{\partial V} \vec{\mathbf{F}}(\vec{\mathbf{G}} \cdot \hat{\mathbf{n}})\, da\,. \tag{1.219}$$

If we change the role of φ and ψ in Green's first identity, we obtain

$$\iiint_V (\nabla\varphi \cdot \nabla\psi + \psi\nabla^2\varphi)\, dv = \oiint_{\partial V} (\psi\nabla\varphi) \cdot d\mathbf{a}\,. \tag{1.220}$$

Subtracting the two formulas yields

$$\iiint_V (\varphi\nabla^2\psi - \psi\nabla^2\varphi)\, dv = \oiint_{\partial V} (\varphi\nabla\psi - \psi\nabla\varphi) \cdot d\mathbf{a}\,, \tag{1.221}$$

which is known as **Green's second identity**. This formula can be used to solve boundary-value problems in electromagnetic theory, as illustrated by the following example.

Example 1.29 *Integration of Poisson's equation.*

Integrate Poisson's equation for an electrostatic problem in a homogeneous medium of permitivity ϵ over a given volume V:

$$\nabla^2\varphi(\mathbf{r}) = -\frac{1}{\epsilon}\rho(\mathbf{r})\,. \tag{1.222}$$

A charge density ρ within V is the source of the potential φ.

Proof This differential equation can be solved by integration using Green's second identity, Equation (1.221). The key is to choose the function $\psi = 1/R$, where $R = \|\mathbf{r}-\mathbf{r}'\|$ is the distance between two points. Now Green's identity involves the Laplacian of $1/R$ on the left-hand side. Using the results of Examples 1.25 and 1.27, we get

$$-4\pi\varphi(\mathbf{r}') = \iiint_V \frac{\nabla^2\varphi}{R}\,dv - \oiint_{\partial V}\left(\frac{\varphi}{R^2}\hat{\mathbf{n}}_{\mathbf{r}} + \frac{\nabla\varphi}{R}\right)\cdot d\mathbf{a}\,, \tag{1.223}$$

where $\mathbf{r}'$ is any point within V, and $\hat{\mathbf{n}}_{\mathbf{r}}$ is a unit vector directed from $\mathbf{r}'$ to $\mathbf{r}$. Here, R is the distance between the point of observation at $\mathbf{r}'$ and the variable of integration at $\mathbf{r}$. The variable of integration moves either within the region of the sources (as the first integral indicates) or at the boundary of the space that we are considering (as the second integral indicates).

Since φ satisfies Poisson's equation, the previous equation may be rewritten as

$$\varphi(\mathbf{r}') = \frac{1}{4\pi\epsilon}\iiint_V \frac{\rho}{R}\,dv + \frac{1}{4\pi}\oiint_{\partial V}\left(\frac{\varphi}{R^2}\hat{\mathbf{n}}_{\mathbf{r}} + \frac{\nabla\varphi}{R}\right)\cdot d\mathbf{a}\,, \tag{1.224}$$

or

$$\varphi(\mathbf{r}') = \frac{1}{4\pi\epsilon}\iiint_V \frac{\rho}{R}\,dv + \frac{1}{4\pi}\oiint_{\partial V}\left[-\varphi\frac{\partial}{\partial n}\left(\frac{1}{R}\right) + \frac{1}{R}\frac{\partial\varphi}{\partial n}\right]da\,, \tag{1.225}$$

where $\partial/\partial n$ denotes the directional derivative along the outwards-pointing normal vector at each point on the surface ∂V. □

We have thus integrated Poisson's equation. For us, this result is mostly of theoretical interest, as we will not use this method to solve for the field inside electric machines. We can make, however, the following observations. The volume integral corresponds to the potential from charges within V, which could be zero if no charges are contained within the volume. The surface integral provides the contribution to the potential from charges outside V. This component of the potential would be the solution to a homogeneous boundary-value problem (i.e., the Laplace equation), where the potential or its normal derivative are specified on the boundary. Based on the type of boundary condition that is specified, we obtain Dirichlet[28] or Neumann[29] problems.

An in-depth discussion regarding the various types of boundary conditions is not appropriate at this point, but the special case where the entire boundary is an equipotential surface is interesting. Suppose that $\varphi(\mathbf{r}) = \varphi_0$, a constant, for all $\mathbf{r} \in \partial V$. The contribution to $\varphi(\mathbf{r}')$ from the first term in the surface integral is then

$$-\frac{1}{4\pi}\oiint_{\partial V}\varphi_0\frac{\partial}{\partial n}\left(\frac{1}{R}\right)da = -\frac{\varphi_0}{4\pi}\oiint_{\partial V}\nabla\left(\frac{1}{R}\right)\cdot d\mathbf{a} = -\frac{\varphi_0}{4\pi}\iiint_V \nabla^2\left(\frac{1}{R}\right)dv = \varphi_0\,, \tag{1.226}$$

using the divergence theorem and Equation (1.205). In other words, the potential of every point within V is increased by this amount.

1.6.4 Green's Theorem in Two Dimensions

Stokes' theorem in two dimensions is commonly known as **Green's theorem.**[d] Suppose we are on the x–y plane, integrating on a flat surface S the curl of a vector field that only has x and y-components. Let us define the 2-D vector field $\vec{\mathbf{F}} : \Omega \subset \mathbb{R}^2 \to \mathbb{R}^2$ as

$$\vec{\mathbf{F}}(x, y) = P(x, y)\,\hat{\mathbf{i}} + Q(x, y)\,\hat{\mathbf{j}}\,. \tag{1.227}$$

As we have noted earlier (see Equation (1.115) on page 34), the 2-D curl can be found by embedding the field in 3-D space:

$$\nabla \times \vec{\mathbf{F}} = \left(\frac{\partial Q}{\partial x} - \frac{\partial P}{\partial y} \right) \hat{\mathbf{k}}\,. \tag{1.228}$$

To apply Stokes' theorem, we note that the differential surface vector is oriented along the z-axis, i.e., $d\mathbf{a} = da\,\hat{\mathbf{k}}$. Hence, we obtain Green's theorem:

$$\iint_S \left(\frac{\partial Q}{\partial x} - \frac{\partial P}{\partial y} \right) da = \oint_{\partial S} P\,dx + Q\,dy\,, \tag{1.229}$$

where we wrote the line integral using the notation of Equation (1.66) on page 21.

Example 1.30 *Integration by parts in two dimensions.*
A useful corollary to Green's theorem is obtained by assuming that

$$P(x, y) = -\eta(x, y)g(x, y) \quad \text{and} \quad Q(x, y) = \eta(x, y)h(x, y)\,, \tag{1.230}$$

for appropriate functions η, g, and h. In this case, Green's theorem yields

$$\iint_S \left(\frac{\partial(\eta h)}{\partial x} + \frac{\partial(\eta g)}{\partial y} \right) da = \oint_{\partial S} -\eta g\,dx + \eta h\,dy\,. \tag{1.231}$$

Equivalently

$$\iint_S \left(h\frac{\partial \eta}{\partial x} + g\frac{\partial \eta}{\partial y} \right) da = \oint_{\partial S} -\eta g\,dx + \eta h\,dy - \iint_S \left(\eta\frac{\partial h}{\partial x} + \eta\frac{\partial g}{\partial y} \right) da\,. \tag{1.232}$$

This equation may be viewed as integration by parts in two dimensions. This may be easier to see if we define a 2-D vector field $\vec{\mathbf{v}} = (h, g)$. Then, the above formula can be written more concisely as

$$\iint_S \nabla\eta \cdot \vec{\mathbf{v}}\,da = \oint_{\partial S} \eta\vec{\mathbf{v}} \cdot \hat{\mathbf{n}}\,dr - \iint_S \eta\nabla \cdot \vec{\mathbf{v}}\,da\,, \tag{1.233}$$

where $\hat{\mathbf{n}}$ is an outwards-pointing unit normal vector. Here, the surface boundary line integral term was obtained by manipulating the previous term as follows:

$$\oint_{\partial S} -\eta g\,dx + \eta h\,dy = \oint_{\partial S} \eta(-g, h) \cdot (dx, dy) = \oint_{\partial S} \eta(h, g) \cdot (dy, -dx)\,. \tag{1.234}$$

[d] Green's theorem in two dimensions can be derived without prior knowledge of Stokes' theorem. These details are omitted here, but can be found in elementary calculus textbooks.

Note that (dx, dy) is a tangential vector along the curve, i.e., $(dx, dy) = dr\,\hat{\mathbf{t}}$, where $\hat{\mathbf{t}} = (dx, dy)/\sqrt{dx^2 + dy^2}$ and $dr = \sqrt{dx^2 + dy^2}$. On the other hand, $(dy, -dx)$ is a normal vector to the curve, i.e., $(dy, -dx) = dr\,\hat{\mathbf{n}}$, since $(dx, dy) \cdot (dy, -dx) = 0$. To convince ourselves that $\hat{\mathbf{n}}$ is outwards pointing, we can evaluate $\hat{\mathbf{n}} \times \hat{\mathbf{t}}$, which should be pointing towards the positive z-direction. Indeed, it can be readily shown that $\hat{\mathbf{n}} \times \hat{\mathbf{t}} = \hat{\mathbf{k}}$.

It is interesting to note that Equation (1.233) could have been obtained alternatively (and perhaps more directly) by applying Gauss' theorem to the surface integral of the divergence of $\eta\vec{\mathbf{v}}$. Here, it was obtained by Green's theorem, which was in turn derived from Stokes' theorem. Apparently, in two dimensions, Gauss' and Stokes' theorems are related to each other.

1.6.5 Helmholtz's Theorem

This chapter would be incomplete without a discussion of **Helmholtz's**[30] **theorem**, also known as the **fundamental theorem of vector calculus**. This theorem states that any vector field $\vec{\mathbf{F}} : \Omega \to \mathbb{R}^3$ can be expressed as the sum of an irrotational plus a solenoidal field. Hence, there exist two differentiable potential functions, namely, a *scalar potential* function φ and a *vector potential* function $\vec{\mathbf{A}}$, which satisfy

$$\vec{\mathbf{F}}(\mathbf{r}) = \underbrace{-\nabla\varphi(\mathbf{r})}_{\text{irrotational}} + \underbrace{\nabla \times \vec{\mathbf{A}}(\mathbf{r})}_{\text{solenoidal}} . \tag{1.235}$$

The minus sign in front of the gradient is used by convention. It can be shown that this decomposition is unique since the potential functions can be expressed in terms of the gradient and the curl of $\vec{\mathbf{F}}$, respectively.

Box 1.1 FEM fundamentals: Helmholtz's theorem (the fundamental theorem of vector calculus)

Any vector field $\vec{\mathbf{F}}$ can be decomposed as

$$\vec{\mathbf{F}}(\mathbf{r}) = -\nabla\varphi(\mathbf{r}) + \nabla \times \vec{\mathbf{A}}(\mathbf{r}) , \tag{1.236}$$

where

$$\varphi(\mathbf{r}) = \frac{1}{4\pi}\iiint_V [\nabla' \cdot \vec{\mathbf{F}}(\mathbf{r}')]\frac{1}{\|\mathbf{r} - \mathbf{r}'\|}\,dv' - \frac{1}{4\pi}\oiint_{\partial V} [\vec{\mathbf{F}}(\mathbf{r}') \cdot \hat{\mathbf{n}}]\frac{1}{\|\mathbf{r} - \mathbf{r}'\|}\,da' \tag{1.237}$$

and

$$\vec{\mathbf{A}}(\mathbf{r}) = \frac{1}{4\pi}\iiint_V [\nabla' \times \vec{\mathbf{F}}(\mathbf{r}')]\frac{1}{\|\mathbf{r} - \mathbf{r}'\|}\,dv' + \frac{1}{4\pi}\oiint_{\partial V} [\vec{\mathbf{F}}(\mathbf{r}') \times \hat{\mathbf{n}}]\frac{1}{\|\mathbf{r} - \mathbf{r}'\|}\,da' . \tag{1.238}$$

This result relates the vector field value at any point in space to its divergence and curl at all other points within a volume V, as well as its normal and tangential components at the boundary ∂V.

Proof The proof of Helmholtz's theorem is rather long but very instructive, as it combines many properties of vector fields that have been presented in this chapter. We begin by expressing the field at the arbitrary point $\mathbf{r}$ as

$$\vec{\mathbf{F}}(\mathbf{r}) = -\frac{1}{4\pi}\iiint_V \vec{\mathbf{F}}(\mathbf{r}')\,\nabla'^2\left(\frac{1}{\|\mathbf{r}-\mathbf{r}'\|}\right)dv', \tag{1.239}$$

which we encountered previously as Equation (1.213) in Example 1.27. A minor (purely notational) change from (1.213) is that here the primed quantities are the variables of integration. To visualize this integral, imagine a closed volume $V \subset \Omega$ within the domain of $\vec{\mathbf{F}}$, as shown in Figure 1.19. Here, the point $\mathbf{r}$ is fixed, and $\mathbf{r}'$ is moving around V.

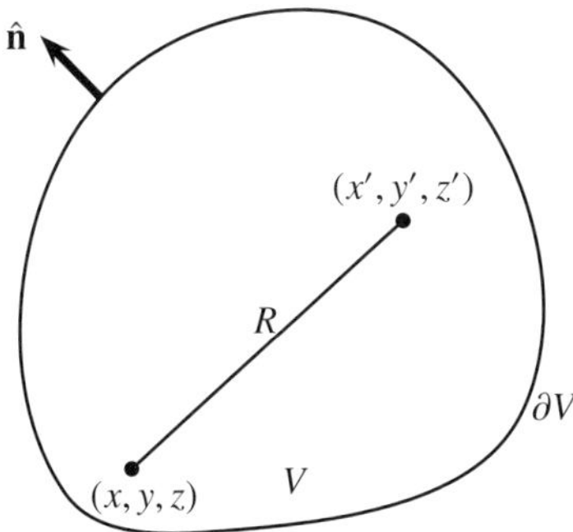

Figure 1.19 Illustration for the proof of Helmholtz's theorem.

In Equation (1.239), the derivatives in the Laplacian are with respect to (x', y', z'). The trick in this proof is to take derivatives with respect to (x, y, z), thus forming the Laplacian $\nabla^2(1/R)$, as discussed in Example 1.25. In fact, we have shown that $\nabla^2(1/R) = \nabla'^2(1/R)$.[e] Because of this, we can pull the Laplacian operator outside the integral (where it becomes a vector Laplacian) as it does not affect the variables of integration:

$$\vec{\mathbf{F}}(\mathbf{r}) = -\frac{1}{4\pi}\nabla^2\iiint_V \vec{\mathbf{F}}(\mathbf{r}')\,\frac{1}{\|\mathbf{r}-\mathbf{r}'\|}\,dv'. \tag{1.240}$$

Now, we invoke the identity (1.163):

$$\nabla\times(\nabla\times\vec{\mathbf{F}}) = \nabla(\nabla\cdot\vec{\mathbf{F}}) - \nabla^2\vec{\mathbf{F}}, \tag{1.241}$$

which allows us to rewrite the above as

$$\vec{\mathbf{F}}(\mathbf{r}) = \frac{1}{4\pi}\nabla\times\left(\nabla\times\iiint_V \vec{\mathbf{F}}(\mathbf{r}')\,\frac{1}{\|\mathbf{r}-\mathbf{r}'\|}\,dv'\right) - \frac{1}{4\pi}\nabla\left(\nabla\cdot\iiint_V \vec{\mathbf{F}}(\mathbf{r}')\,\frac{1}{\|\mathbf{r}-\mathbf{r}'\|}\,dv'\right). \tag{1.242}$$

We move the ∇ operators in the parentheses back inside the integrals, leading to

$$\vec{\mathbf{F}}(\mathbf{r}) = -\frac{1}{4\pi}\nabla\times\iiint_V \vec{\mathbf{F}}(\mathbf{r}')\times\nabla\frac{1}{\|\mathbf{r}-\mathbf{r}'\|}\,dv' - \frac{1}{4\pi}\nabla\iiint_V \vec{\mathbf{F}}(\mathbf{r}')\cdot\nabla\frac{1}{\|\mathbf{r}-\mathbf{r}'\|}\,dv', \tag{1.243}$$

[e] If one becomes uncomfortable with the singularity at $\mathbf{r} = \mathbf{r}'$, it is possible to replace $R = \|\mathbf{r}-\mathbf{r}'\|$ by R_ϵ, as we did in Example 1.27, and proceed with the proof using $\nabla'_\epsilon(1/R) = -\nabla_\epsilon(1/R)$ and $\nabla^2_\epsilon(1/R) = \nabla'^2_\epsilon(1/R)$.

where we used the identities (1.160) and (1.158) for products of scalar and vector fields, repeated here for convenience:

$$\nabla \times (\psi \vec{\mathbf{F}}) = \nabla\psi \times \vec{\mathbf{F}} + \psi \nabla \times \vec{\mathbf{F}}, \tag{1.244a}$$

$$\nabla \cdot (\psi \vec{\mathbf{F}}) = \vec{\mathbf{F}} \cdot \nabla\psi + \psi \nabla \cdot \vec{\mathbf{F}}. \tag{1.244b}$$

Note that $\nabla \times \vec{\mathbf{F}}(\mathbf{r}') = \mathbf{0}$ and $\nabla \cdot \vec{\mathbf{F}}(\mathbf{r}') = 0$, since the nabla operates on the unprimed coordinates. Now, we switch variables in the gradients using $\nabla(1/R) = -\nabla'(1/R)$ (see Example 1.25), and obtain

$$\vec{\mathbf{F}}(\mathbf{r}) = \frac{1}{4\pi}\nabla \times \underbrace{\iiint_V \vec{\mathbf{F}}(\mathbf{r}') \times \nabla' \frac{1}{\|\mathbf{r}-\mathbf{r}'\|}\, dv'}_{I_1} + \frac{1}{4\pi}\nabla \underbrace{\iiint_V \vec{\mathbf{F}}(\mathbf{r}') \cdot \nabla' \frac{1}{\|\mathbf{r}-\mathbf{r}'\|}\, dv'}_{I_2}. \tag{1.245}$$

Taking a moment to reflect, we see that the first term I_1 is a volume integral of a vector field obtained by the cross product between $\vec{\mathbf{F}}$ and a gradient, which results in the vector field $4\pi\vec{\mathbf{A}}(\mathbf{r})$ of Equation (1.235). The second term I_2 is the volume integral of a scalar field formed by a dot product between $\vec{\mathbf{F}}$ and a gradient, which equals the scalar field $-4\pi\varphi(\mathbf{r})$ of Equation (1.235). In other words, the previous expression is already in the form of Helmholtz's theorem, as it expresses the vector field as a sum of a solenoidal plus an irrotational component. In the next few steps, we proceed with algebraic manipulations that will yield alternative but more informative expressions for these components.

We proceed by again employing the identities (1.160) and (1.158), but now using the primed nablas. For convenience, let us treat each term separately. We have

$$I_1 = \iiint_V [\nabla' \times \vec{\mathbf{F}}(\mathbf{r}')]\frac{1}{\|\mathbf{r}-\mathbf{r}'\|}\, dv' - \iiint_V \nabla' \times \underbrace{\left(\vec{\mathbf{F}}(\mathbf{r}')\frac{1}{\|\mathbf{r}-\mathbf{r}'\|}\right)}_{\vec{\mathbf{f}}} dv'. \tag{1.246}$$

We momentarily shift the focus on the second term above, for which we need to prove the following result:

$$\iiint_V (\nabla' \times \vec{\mathbf{f}})\, dv' = -\oint\!\!\!\oint_{\partial V} \vec{\mathbf{f}} \times \hat{\mathbf{n}}\, da'. \tag{1.247}$$

To prove this identity, let $\vec{\mathbf{C}}$ be any constant vector so that

$$\iiint_V \nabla' \cdot (\vec{\mathbf{C}} \times \vec{\mathbf{f}})\, dv' = -\iiint_V \vec{\mathbf{C}} \cdot (\nabla' \times \vec{\mathbf{f}})\, dv' \tag{1.248a}$$

$$= \oint\!\!\!\oint_{\partial V} (\vec{\mathbf{C}} \times \vec{\mathbf{f}}) \cdot \hat{\mathbf{n}}\, da', \tag{1.248b}$$

where we used the identity (1.159) for the first equality, and the divergence theorem for the second. Using the scalar triple-product identity (1.41) on page 13, we have

$$(\vec{\mathbf{C}} \times \vec{\mathbf{f}}) \cdot \hat{\mathbf{n}} = \vec{\mathbf{C}} \cdot (\vec{\mathbf{f}} \times \hat{\mathbf{n}}). \tag{1.249}$$

Therefore

$$-\vec{\mathbf{C}} \cdot \iiint_V (\nabla' \times \vec{\mathbf{f}})\, dv' = \vec{\mathbf{C}} \cdot \oint\!\!\!\oint_{\partial V} \vec{\mathbf{f}} \times \hat{\mathbf{n}}\, da', \tag{1.250}$$

and since $\vec{\mathbf{C}}$ is arbitrary, the integrals must be equal. (By moving $\vec{\mathbf{C}}$ outside of the volume and surface integrals, we have effectively converted these from integrals of scalar fields to integrals of vector fields, which should be performed component-wise.) This result is a vector identity, which can be thought of as a form of Stokes' theorem. Hence, Equation (1.246) becomes

$$I_1 = \iiint_V [\nabla' \times \vec{\mathbf{F}}(\mathbf{r}')] \frac{1}{\|\mathbf{r} - \mathbf{r}'\|} dv' + \oiint_{\partial V} [\vec{\mathbf{F}}(\mathbf{r}') \times \hat{\mathbf{n}}] \frac{1}{\|\mathbf{r} - \mathbf{r}'\|} da' . \tag{1.251}$$

For the second term, we have

$$I_2 = \iiint_V \nabla' \cdot \left(\vec{\mathbf{F}}(\mathbf{r}') \frac{1}{\|\mathbf{r} - \mathbf{r}'\|} \right) dv' - \iiint_V [\nabla' \cdot \vec{\mathbf{F}}(\mathbf{r}')] \frac{1}{\|\mathbf{r} - \mathbf{r}'\|} dv' , \tag{1.252}$$

which becomes, after applying the divergence theorem:

$$I_2 = \oiint_{\partial V} \left(\vec{\mathbf{F}}(\mathbf{r}') \frac{1}{\|\mathbf{r} - \mathbf{r}'\|} \right) \cdot \hat{\mathbf{n}} \, da' - \iiint_V [\nabla' \cdot \vec{\mathbf{F}}(\mathbf{r}')] \frac{1}{\|\mathbf{r} - \mathbf{r}'\|} dv' . \tag{1.253}$$

This completes the proof of Helmholtz's theorem, which is summarized in Box 1.1. □

At this point, this is a purely mathematical result. Let us attempt to give it some context by considering physical vector fields. We let the integration volume V resemble a sphere whose radius tends to infinity, so we are including all sources that exist in the universe within the volume integrals. It is also assumed that the magnitude of the vector field drops as $\sim 1/R^2$; for instance, consider the electric field of a point charge that follows such a relationship. In this case, the surface integrals in the expressions for φ and $\vec{\mathbf{A}}$ will vanish. Hence, the potentials, and therefore the vector field itself, can be determined purely on the divergence and curl of the vector field throughout space, which are in turn associated with field *sources*.

Recall Maxwell's equations (1.2a)–(1.2d). These four equations, together with the constitutive laws in Equation (1.1), completely specify the divergence and curl of $\vec{\mathbf{E}}$ and $\vec{\mathbf{B}}$! Now we should be able to understand that this is not by accident. It naturally leads to the definition of potentials for the electromagnetic field and, hence, through Helmholtz's theorem to the field itself.

For instance, let us consider the electrostatic field in vacuum, where the role of $\vec{\mathbf{F}}$ is played by the electric field $\vec{\mathbf{E}}$, which is measured in V/m. In this case, physics dictates that the divergence of the electric field is associated with a volume charge density ρ C/m^3 (coulombs per cubic meter) that acts as the source of the field:

$$\nabla \cdot \vec{\mathbf{E}} = \frac{\rho}{\epsilon_0} . \tag{1.254}$$

Recall that the physical constant

$$\epsilon_0 \approx 8.854 \cdot 10^{-12} \text{ F/m (farads per meter)} \tag{1.255}$$

is the permittivity of free space. We also know that the curl of the electrostatic field is zero:

$$\nabla \times \vec{\mathbf{E}} = \mathbf{0} . \tag{1.256}$$

Using Helmholtz's theorem, we can deduce that the vector potential is zero, and thereby we express the electrostatic field as

$$\vec{\mathbf{E}}(\mathbf{r}) = -\nabla\varphi(\mathbf{r}), \tag{1.257}$$

where the scalar potential is measured in V, and is given by

$$\varphi(\mathbf{r}) = \frac{1}{4\pi\epsilon_0} \iiint_V \frac{\rho(\mathbf{r}')}{\|\mathbf{r} - \mathbf{r}'\|}\, dv'. \tag{1.258}$$

This integral provides the electrostatic potential at a point $\mathbf{r} \in V$. (See also Example 1.29, where the same result was obtained by integration of Poisson's equation.) The potential can be thought of as the superposition of contributions from all elementary charges $dq(\mathbf{r}') = \rho(\mathbf{r}')\, dv'$, with $\mathbf{r}' \in V$.

Now, consider the magnetostatic field in vacuum, where the role of $\vec{\mathbf{F}}$ is played by the magnetic field density $\vec{\mathbf{B}}$, which is measured in T. According to the fundamental postulates of electromagnetism, the divergence of the B-field is always zero:

$$\nabla \cdot \vec{\mathbf{B}} = 0, \tag{1.259}$$

whereas the curl satisfies Ampère's law:

$$\nabla \times \vec{\mathbf{B}} = \mu_0 \vec{\mathbf{J}}. \tag{1.260}$$

Here, the physical constant

$$\mu_0 = 4\pi 10^{-7}\ \text{H/m} \tag{1.261}$$

is the permeability of free space. The source of the field is the current density $\vec{\mathbf{J}}$, which is measured in A/m^2. Therefore, Helmholtz's theorem yields

$$\vec{\mathbf{B}}(\mathbf{r}) = \nabla \times \vec{\mathbf{A}}(\mathbf{r}), \tag{1.262}$$

with the MVP

$$\vec{\mathbf{A}}(\mathbf{r}) = \frac{\mu_0}{4\pi} \iiint_V \frac{\vec{\mathbf{J}}(\mathbf{r}')}{\|\mathbf{r} - \mathbf{r}'\|}\, dv', \tag{1.263}$$

which may be measured in T·m or Wb/m (webers[31] per meter). This volume integral provides the MVP at a point $\mathbf{r} \in V$, where $\mathbf{r}'$ is the variable of integration. Integration is performed component-wise for each of three separate scalar potentials. For instance, the x-component of the current, J_x, only affects the x-component of the MVP, A_x. The potential can be thought of as the superposition of contributions from all elementary current elements $\vec{\mathbf{J}}(\mathbf{r}')\, dv'$, with $\mathbf{r}' \in V$.

Example 1.31 *The law of Biot–Savart.*

By applying our toolbox of vector calculus to the MVP given by Equation (1.263), it is possible to derive an expression for the magnetic field created by an arbitrary current distribution in empty space. This result is commonly referred to as the law of Biot[32] and Savart,[33] based on experimental work they conducted in 1820.

Proof Taking the curl of Equation (1.263), we have

$$\vec{\mathbf{B}}(\mathbf{r}) = \nabla \times \frac{\mu_0}{4\pi} \iiint_V \frac{\vec{\mathbf{J}}(\mathbf{r}')}{\|\mathbf{r} - \mathbf{r}'\|} dv' = \frac{\mu_0}{4\pi} \iiint_V \nabla \times \left(\frac{\vec{\mathbf{J}}(\mathbf{r}')}{\|\mathbf{r} - \mathbf{r}'\|} \right) dv' , \tag{1.264}$$

where we were able to exchange the order of integration and differentiation because the curl operates on the coordinates at the point of observation $\mathbf{r} = (x, y, z)$, but does not affect the variables of integration $\mathbf{r}' = (x', y', z')$. Now we invoke the identity (1.160) to expand the integrand as

$$\nabla \times \left(\frac{\vec{\mathbf{J}}(\mathbf{r}')}{\|\mathbf{r} - \mathbf{r}'\|} \right) = \nabla \left(\frac{1}{\|\mathbf{r} - \mathbf{r}'\|} \right) \times \vec{\mathbf{J}}(\mathbf{r}') + \frac{1}{\|\mathbf{r} - \mathbf{r}'\|} \nabla \times \vec{\mathbf{J}}(\mathbf{r}') . \tag{1.265}$$

The second term on the right-hand side vanishes because the current is a function of the variables of integration only. Also, we employ Equation (1.155) to change the differentiation variables, thereby switching from ∇ to ∇'. This leads to

$$\vec{\mathbf{B}}(\mathbf{r}) = \frac{\mu_0}{4\pi} \iiint_V \vec{\mathbf{J}}(\mathbf{r}') \times \nabla' \left(\frac{1}{\|\mathbf{r} - \mathbf{r}'\|} \right) dv' , \tag{1.266}$$

and by performing the differentiation:

$$\vec{\mathbf{B}}(\mathbf{r}) = \frac{\mu_0}{4\pi} \iiint_V \vec{\mathbf{J}}(\mathbf{r}') \times \frac{\mathbf{r} - \mathbf{r}'}{\|\mathbf{r} - \mathbf{r}'\|^3} dv' . \tag{1.267}$$

If we define

$$\hat{\mathbf{a}}(\mathbf{r}, \mathbf{r}') = \frac{\mathbf{r} - \mathbf{r}'}{\|\mathbf{r} - \mathbf{r}'\|} \tag{1.268}$$

as the unit vector pointing from the current element at $\mathbf{r}'$ to the point of observation at $\mathbf{r}$, this expression may be rewritten as

$$\vec{\mathbf{B}}(\mathbf{r}) = \frac{\mu_0}{4\pi} \iiint_V \vec{\mathbf{J}}(\mathbf{r}') \times \frac{\hat{\mathbf{a}}(\mathbf{r}, \mathbf{r}')}{\|\mathbf{r} - \mathbf{r}'\|^2} dv' . \tag{1.269}$$

When the field source is current flowing inside a single filament, i.e., a conductor of a very small cross-sectional area, we can approximate

$$\vec{\mathbf{J}}(\mathbf{r}')\, dv' \approx i\, \hat{\mathbf{t}}(\mathbf{r}')\, dr' , \tag{1.270}$$

where i is the current in A, $\hat{\mathbf{t}}(\mathbf{r}')$ is a unit vector that is tangential to the filament at point $\mathbf{r}'$ and points to the direction of positive current flow, and dr' denotes an infinitesimal arc length along the filament. Hence, the volume integral becomes a line integral along the closed path C defined by the filament:

$$\vec{\mathbf{B}}(\mathbf{r}) = \frac{\mu_0 i}{4\pi} \oint_C \frac{\hat{\mathbf{t}}(\mathbf{r}') \times \hat{\mathbf{a}}(\mathbf{r}, \mathbf{r}')}{\|\mathbf{r} - \mathbf{r}'\|^2} dr' . \tag{1.271}$$

This should be interpreted as three scalar field line integrals of the form (1.59), one per component of the vector integrand. An illustration of this law is provided in Figure 1.20. We may thus think of the magnetic field as being created by contributions $d\vec{\mathbf{B}}$ from individual current elements (although it is impossible to physically isolate a single element). The contributions are perpendicular to a plane that contains both the current element and the line segment joining the element with the point of observation, and are oriented according to the right-hand rule. □

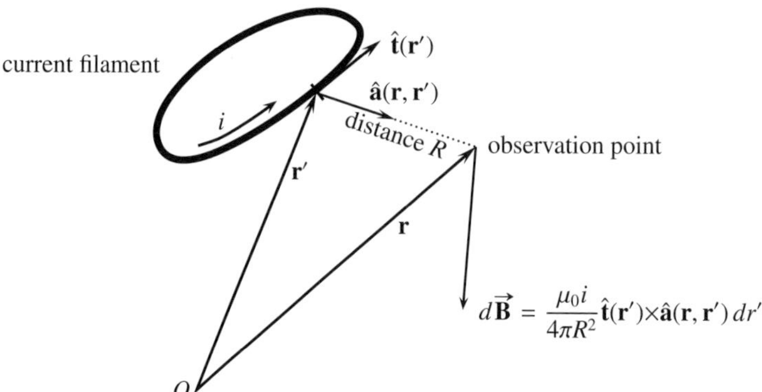

Figure 1.20 Illustration of the Biot–Savart law.

Interestingly, the vector potential is itself a vector field, hence Helmholtz's theorem applies. We have fixed its curl by setting it equal to $\vec{\mathbf{B}}$, but its divergence is still unspecified. This implies that there exist infinitely many vector potentials that will generate a given $\vec{\mathbf{B}}$. We often set the divergence to zero: $\nabla \cdot \vec{\mathbf{A}} = 0$. This particular choice is the Coulomb or Lorenz gauge that was mentioned earlier, and which was used to derive the Laplace equation (1.167) on page 43. As a matter of fact, it can be shown that the potential calculated using Equation (1.263), which was obtained by enclosing all current sources in the universe, satisfies $\nabla \cdot \vec{\mathbf{A}} = 0$. (See Problem 1.15.)

A more general choice for the divergence can be found by recalling Equation (1.162), which states that the curl of a gradient is zero. Hence, another feasible choice of vector potential, that is, one that still satisfies $\vec{\mathbf{B}} = \nabla \times \vec{\mathbf{A}}'$, is

$$\vec{\mathbf{A}}'(\mathbf{r}) = \vec{\mathbf{A}}(\mathbf{r}) + \nabla\psi(\mathbf{r}), \tag{1.272}$$

where $\vec{\mathbf{A}}$ is the Coulomb-gauge potential that has zero divergence, and ψ is an arbitrary scalar field. The divergence of $\vec{\mathbf{A}}'$ is, therefore, the Laplacian of ψ:

$$\nabla \cdot \vec{\mathbf{A}}'(\mathbf{r}) = \nabla^2\psi(\mathbf{r}). \tag{1.273}$$

Example 1.32 *The convergence of the potential integrals.*

We may be inclined to ask if the potential integrals of Equations (1.258) and (1.263) are well-defined, especially when the point of interest $\mathbf{r}$ enters regions in space where sources are present, since then the denominator of the integrand will become zero whenever $\mathbf{r} = \mathbf{r}'$. These are improper integrals, and some attention is required to understand them better.

Proof We will assume that the source term is bounded, which is typically the case in a real-world application. For instance, consider the x-component of Equation (1.263) for the MVP, for which we assume that $|J_x(\mathbf{r}')| < m$, for some $m > 0$ with units of A/m^2, and for all $\mathbf{r}'$. We surround the point of interest $\mathbf{r} = (x, y, z)$ by a small sphere of radius a, centered at $\mathbf{r}$, as shown in Figure 1.21.

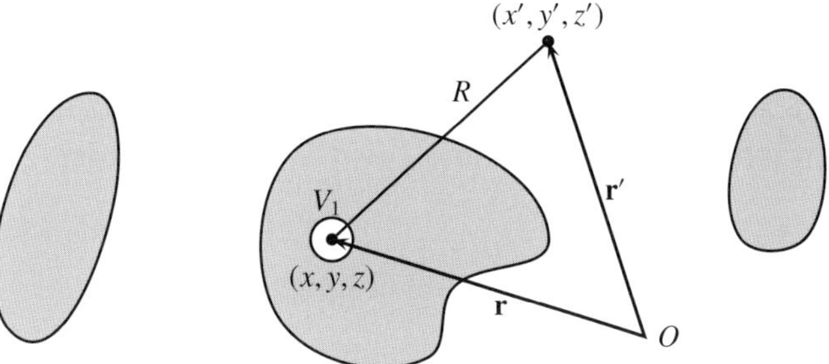

Figure 1.21 To calculate the potential when the point of interest (x, y, z) lies within a region where sources reside (shaded areas), we surround the point by a small sphere, whose volume V_1 is excluded from the integral. The variable of integration is represented by the point (x', y', z'), which moves all around space, and can enter the gray areas as well.

The volume contained within the sphere is denoted by V_1, and the volume external to the sphere is V_2, so the entire space is $V = V_1 + V_2$. The potential can be calculated by breaking up the integral of Equation (1.263) into two separate integrals over V_1 and V_2. The V_2 integral is not problematic, as the denominator $R = ||\mathbf{r} - \mathbf{r}'|| \geq a > 0$. For the V_1 integral, we can show that its contribution vanishes when the sphere shrinks around $\mathbf{r}$, by letting the radius $a \to 0$. In this simple spherical geometry, the differential volume can be expressed as $dv' = 4\pi R^2\, dR$. It follows that

$$\left| \iiint_{V_1} J_x(\mathbf{r}') \frac{1}{||\mathbf{r} - \mathbf{r}'||}\, dv' \right| < m \int_0^a \frac{1}{R} 4\pi R^2\, dR = 2m\pi a^2\,, \tag{1.274}$$

which vanishes as $a \to 0$. Hence, the integral is well-behaved, as the contribution from sources in the neighborhood of the point of interest is bounded by the above term. □

1.7 Case Studies on the Magnetostatic Field

In the previous sections, we presented fundamental concepts of vector calculus. We conclude this chapter with case studies that should enhance our understanding of how vector calculus facilitates the solution of magnetostatic field problems. This material will also highlight the physics that form the underlying foundation of electric machine analysis, with an emphasis on the role of the MVP.

Suppose that we are interested in calculating the magnetic field created by a (static) distribution of currents in space. Our previous discussion suggests two possible methods: (i) a direct method based on the law of Biot–Savart given by Equation (1.269); and

(ii) an indirect method where we first calculate the MVP via Equation (1.263), subsequently taking its curl to obtain the B-field. The potential can be seen as an intermediate quantity in this process. However, even in elementary geometries, an analytical evaluation of the integrals is not trivial due to the $R = ||\mathbf{r} - \mathbf{r}'||$ term in the denominator, which expands to

$$\frac{1}{R} = \frac{1}{\sqrt{||\mathbf{r}||^2 + ||\mathbf{r}'||^2 - 2\mathbf{r}\cdot\mathbf{r}'}} . \tag{1.275}$$

Practically, these sorts of calculations can be performed only in problems that involve special geometries, otherwise we must resort to numerical means.

1.7.1 Field from a Circular Loop and the Magnetic Dipole

We will calculate the magnetic field from a filamentary circular loop carrying a current i. To this end, we will use a modified version of Equation (1.263), integrating over the loop contour C:

$$\vec{\mathbf{A}}(\mathbf{r}) = \frac{\mu_0 i}{4\pi} \oint_C \frac{d\mathbf{r}'}{||\mathbf{r} - \mathbf{r}'||} . \tag{1.276}$$

Proof Let us denote the loop radius by a, and let us suppose that the loop is placed on the x–y plane, as shown in Figure 1.22. The integration path is parametrized by the angle $\phi' \in [0, 2\pi]$, so

$$\mathbf{r}' = (a\cos\phi', a\sin\phi', 0) \tag{1.277}$$

and

$$d\mathbf{r}' = (-a\sin\phi', a\cos\phi', 0)\, d\phi' . \tag{1.278}$$

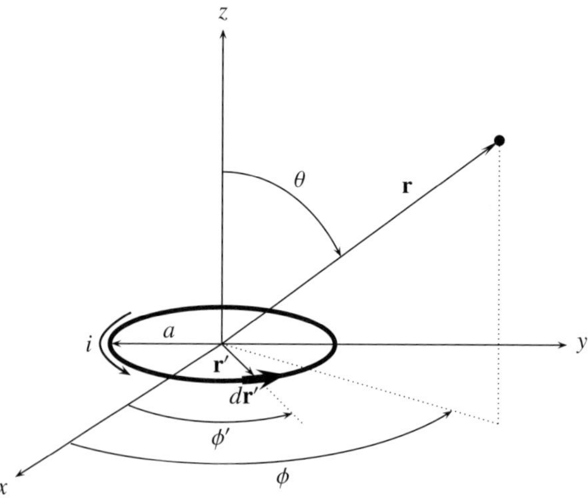

Figure 1.22 A circular current loop in a spherical coordinate system.

In particular, we will calculate the field far away from the loop itself, that is, at observation points where $\|\mathbf{r}\| \gg a$. We will first approximate the inverse distance between the point of observation and the current element by a first-order Taylor expansion:

$$\frac{1}{\|\mathbf{r}-\mathbf{r}'\|} \approx \frac{1}{r} + \frac{xx' + yy'}{r^3} \tag{1.279a}$$

$$= \frac{1}{r}\left(1 + \frac{ax}{r^2}\cos\phi' + \frac{ay}{r^2}\sin\phi'\right), \tag{1.279b}$$

where $r = \|\mathbf{r}\|$. Hence, Equation (1.276) yields

$$\vec{\mathbf{A}}(\mathbf{r}) = \frac{\mu_0 ia}{4\pi r}\int_0^{2\pi} (-\sin\phi'\,\hat{\mathbf{i}} + \cos\phi'\,\hat{\mathbf{j}})\left(1 + \frac{ax}{r^2}\cos\phi' + \frac{ay}{r^2}\sin\phi'\right) d\phi', \tag{1.280}$$

which evaluates to

$$\vec{\mathbf{A}}(\mathbf{r}) = \frac{\mu_0 i\pi a^2}{4\pi r^3}(-y\,\hat{\mathbf{i}} + x\,\hat{\mathbf{j}}), \tag{1.281}$$

or expressed in spherical coordinates:

$$\vec{\mathbf{A}}(r,\phi,\theta) = \frac{\mu_0 i\pi a^2}{4\pi r^2}\sin\theta\,\hat{\boldsymbol{\phi}}. \tag{1.282}$$

Due to symmetry about the z-axis, the MVP does not depend on ϕ. A few (logarithmically spaced) contour lines of $\vec{\mathbf{A}} \cdot \hat{\mathbf{i}}$ on the $x = 0$ plane are depicted in Figure 1.23. It should be noted that the plot is inaccurate in the region close to the origin because of the approximation we have made.

The flux density is found by taking the curl of the MVP. In spherical coordinates

$$\vec{\mathbf{B}}(\mathbf{r}) = \nabla \times \vec{\mathbf{A}}(\mathbf{r}) = \frac{1}{r\sin\theta}\frac{\partial}{\partial\theta}(A_\phi \sin\theta)\,\hat{\mathbf{r}} - \frac{1}{r}\frac{\partial}{\partial r}(rA_\phi)\,\hat{\boldsymbol{\theta}}, \tag{1.283}$$

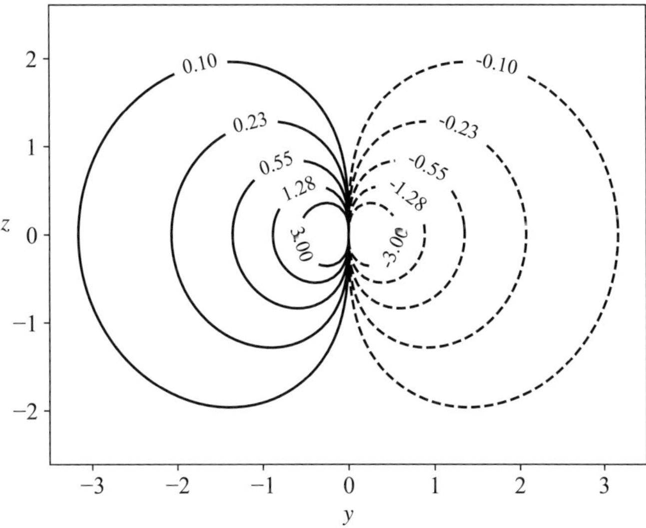

Figure 1.23 Contour plot of the MVP of a circular loop, normalized with respect to $(\mu_0 i\pi a^2)/(4\pi)$. The contour values are logarithmically spaced from ±0.1 to ±3.0.

which yields

$$\vec{\mathbf{B}}(\mathbf{r}) = \frac{\mu_0 i\pi a^2}{4\pi r^3}(2\cos\theta\,\hat{\mathbf{r}} + \sin\theta\,\hat{\boldsymbol{\theta}})\,. \tag{1.284}$$

□

If we let the loop radius a approach zero, while at the same time increasing the current so that $i\pi a^2$ maintains a finite value, we obtain what is called a **magnetic dipole**. The vector

$$\vec{\mathbf{m}} = i\pi a^2\,\hat{\mathbf{k}} = i\iint_S \hat{\mathbf{n}}\,da \tag{1.285}$$

is called the **magnetic dipole moment** of the circuit, measured in A·m^2. Here S is the loop surface (with $|S| = \pi a^2$), and $\hat{\mathbf{n}}$ is a unit normal vector directed so that the dipole moment points to the direction of an advancing screw that is rotated in the same sense as the current in the loop. Note that we can substitute the dipole moment magnitude $m = \|\vec{\mathbf{m}}\|$ in the equations for the MVP and the B-field to obtain equivalent expressions, e.g., $A_\phi = 10^{-7} m\sin\theta/r^2$.

1.7.2 Torque on a Magnetic Dipole

Suppose the dipole of Figure 1.22 is embedded in an external magnetic field $\vec{\mathbf{B}}_0$. The net force on the dipole is zero. This can be deduced from the Lorentz[34] force equation, which states that the force $d\vec{\mathbf{F}}$ on an infinitesimal current tube of axial length $d\mathbf{r}$ may be written as

$$d\vec{\mathbf{F}} = i\,d\mathbf{r}\times\vec{\mathbf{B}}_0\,. \tag{1.286}$$

This formula implies that force contributions from currents at diametrically opposed positions cancel each other out.

However, the net torque on the dipole will not be zero. The dipole torque may be calculated as

$$\vec{\tau} = \oint_C \mathbf{r}\times d\vec{\mathbf{F}}\,, \tag{1.287}$$

and it can be shown that

$$\vec{\tau} = \vec{\mathbf{m}}\times\vec{\mathbf{B}}_0\,. \tag{1.288}$$

The proof is left as an exercise. Note how the torque tends to align the dipole moment with the imposed magnetic field, as shown in Figure 1.24.

1.7.3 Field from a Distribution of Magnetic Dipoles

Let us assume the presence of a distribution of magnetic dipoles in a region of space V occupied by a material. We will denote by $\vec{\mathbf{M}}$ the magnetic dipole moment per unit volume:

$$\vec{\mathbf{M}}(\mathbf{r}) = \frac{d\vec{\mathbf{m}}}{dv}\ \text{A/m}\,, \tag{1.289}$$

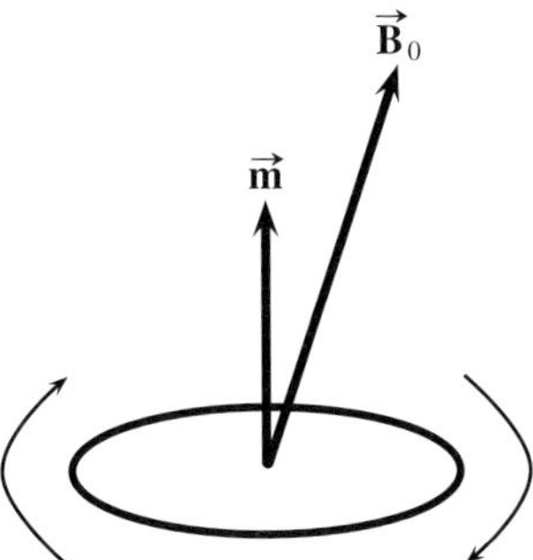

Figure 1.24 Torque on a magnetic dipole. If the dipole moment and the magnetic field, as shown, are on the page surface, then the developed torque is turning the dipole clockwise.

which means that within an infinitesimal volume dv centered at $\mathbf{r}$, there exists a dipole moment $d\vec{\mathbf{m}}(\mathbf{r})$. This quantity is called **magnetization** or **magnetic polarization**. The terminology stems from the fact that, macroscopically, materials subject to an external magnetic field exhibit a behavior that can be modeled as an alignment of atomic-level dipole moments. Our goal will be to show that an arbitrary magnetization $\vec{\mathbf{M}}(\mathbf{r})$ is equivalent to volumetric and surface magnetization currents, in the sense that both yield the same magnetic fields. To this end, we must compute the magnetic field from the magnetization.

Proof First, we note that the dipoles are distributed around space, so we must adapt our previous formula, where we assumed that the dipole was centered at the origin. Instead, we will calculate the MVP at point $\mathbf{r}$ created by a dipole at $\mathbf{r}'$. Using the definition of magnetic dipole moment, we can rewrite Equation (1.281) as

$$d\vec{\mathbf{A}}(\mathbf{r},\mathbf{r}') = \frac{\mu_0}{4\pi}\, d\vec{\mathbf{m}}(\mathbf{r}') \times \frac{\mathbf{r}-\mathbf{r}'}{\|\mathbf{r}-\mathbf{r}'\|^3}\,, \tag{1.290}$$

or (see Example 1.25)

$$d\vec{\mathbf{A}}(\mathbf{r},\mathbf{r}') = -\frac{\mu_0}{4\pi}\, d\vec{\mathbf{m}}(\mathbf{r}') \times \nabla\left(\frac{1}{R}\right), \tag{1.291}$$

where $R = \|\mathbf{r}-\mathbf{r}'\|$ is the distance between the observation point and the dipole. We have introduced the notation $d\vec{\mathbf{A}}$ for the MVP (far away, at a distance R) from an infinitesimal magnetic dipole $d\vec{\mathbf{m}}$. By virtue of the fact that matter at the atomic level is mostly vacuum, we may still use μ_0 as the permeability.

To capture the potential from all dipoles in space, we must integrate over $\mathbf{r}' \in V$:

$$\vec{\mathbf{A}}(\mathbf{r}) = \iiint_V d\vec{\mathbf{A}} = -\frac{\mu_0}{4\pi}\iiint_V \vec{\mathbf{M}}(\mathbf{r}') \times \nabla\left(\frac{1}{R}\right) dv'\,. \tag{1.292}$$

We can change $-\nabla(1/R)$ to $\nabla'(1/R)$ using Equation (1.155), thus switching the differentiation variables from the coordinates at the point of observation to those at the location of the dipoles. This yields

$$\vec{\mathbf{A}}(\mathbf{r}) = \frac{\mu_0}{4\pi} \iiint_V \vec{\mathbf{M}}(\mathbf{r}') \times \nabla' \left(\frac{1}{R} \right) dv' . \tag{1.293}$$

Now we invoke the identity (1.160), to obtain

$$\vec{\mathbf{A}}(\mathbf{r}) = \frac{\mu_0}{4\pi} \iiint_V \frac{\nabla' \times \vec{\mathbf{M}}(\mathbf{r}')}{R} dv' - \frac{\mu_0}{4\pi} \iiint_V \nabla' \times \frac{\vec{\mathbf{M}}(\mathbf{r}')}{R} dv' . \tag{1.294}$$

We can convert the second volume integral to a surface integral using the identity (1.247), which we derived within the proof of Helmholtz's theorem. Therefore:

$$\vec{\mathbf{A}}(\mathbf{r}) = \frac{\mu_0}{4\pi} \iiint_V \frac{\nabla' \times \vec{\mathbf{M}}(\mathbf{r}')}{R} dv' + \frac{\mu_0}{4\pi} \oiint_{\partial V} \frac{\vec{\mathbf{M}}(\mathbf{r}') \times \hat{\mathbf{n}}}{R} da' , \tag{1.295}$$

where $\hat{\mathbf{n}}$ is a unit normal vector that points outwards from the magnetized region.

Comparing Equations (1.295) and (1.263), it becomes clear that the field created by a magnetization distribution is equivalent to the following **magnetization current** sources:

$$\nabla' \times \vec{\mathbf{M}}(\mathbf{r}') = \vec{\mathbf{J}}_m(\mathbf{r}') , \tag{1.296a}$$

$$\vec{\mathbf{M}}(\mathbf{r}') \times \hat{\mathbf{n}} = \vec{\mathbf{K}}_m(\mathbf{r}') . \tag{1.296b}$$

These currents are also known as **bound currents** because they are created by the action of electrons that are bound (tied) to the atomic structure of a material, in contrast to the **free currents** that are caused by the flow of electrons within conductors. The first current source, $\vec{\mathbf{J}}_m$, is an equivalent current density distribution within the magnetized material volume. Note that this source is zero if the magnetization is uniform (i.e., when $\vec{\mathbf{M}}$ is the same within the material) because the curl vanishes in this case. It is also zero in a magnetically linear material, as we will explain in the next section. The second current source, $\vec{\mathbf{K}}_m$, is an equivalent surface current density, which appears at the interface between the magnetized material and its surroundings. (Recall that we have encountered a surface current density before in Example 1.23.) It should be noted that, usually, $\vec{\mathbf{K}}_m$ is the dominant magnetization current source, which is somewhat counterintuitive. □

1.7.4 The Relationship between B and H in Materials

The operation of electric machines is heavily dependent on ferromagnetic materials, such as iron, nickel, cobalt, and steel; and permanent magnets. Therefore, it is important to understand the modeling of free and bound currents, and how this is captured by the magnetic field variables B and H. For the purposes of our analysis, it should be understood that we are dealing with space-averaged fields, where the averaging takes place over many atoms, while still over a relatively small volume from a macroscopic viewpoint.

The **magnetic flux density** (also known as **magnetic induction**) is a vector field that satisfies

$$\nabla \cdot \vec{\mathbf{B}} = 0 \tag{1.297a}$$

and

$$\nabla \times \vec{\mathbf{B}} = \mu_0(\vec{\mathbf{J}} + \vec{\mathbf{J}}_m), \tag{1.297b}$$

which is a version of Ampère's law that has the total, free plus bound current on the right-hand side. We may also express the B-field as

$$\vec{\mathbf{B}} = \nabla \times \vec{\mathbf{A}}, \tag{1.298}$$

where the MVP $\vec{\mathbf{A}}$ is the superposition of the potentials due to free and bound currents. Physicists tell us that the B-field is a truly fundamental quantity because it causes the $q\vec{\mathbf{v}} \times \vec{\mathbf{B}}$ force in the Lorentz force law.

On the other hand, the **magnetic field intensity** (also known simply as the **magnetic field**) is a vector field that is created by the free currents only, i.e.

$$\nabla \times \vec{\mathbf{H}} = \vec{\mathbf{J}}, \tag{1.299}$$

which is a second version of Ampère's law. Note that we have ignored the displacement current $\partial\vec{\mathbf{D}}/\partial t$ that appears in the corresponding Maxwell equation (1.2d) because this term is usually insignificant in electric machine analysis.

In free space, B and H are related by

$$\vec{\mathbf{B}} = \mu_0\vec{\mathbf{H}}. \tag{1.300}$$

Within materials, however, their (macroscopic) relationship is

$$\vec{\mathbf{B}} = \mu_0(\vec{\mathbf{H}} + \vec{\mathbf{M}}). \tag{1.301}$$

In ferromagnetic materials, the magnitude of B is enhanced dramatically by the magnetization.

The magnetization can be expressed as a function of the magnetic field. In materials that get completely demagnetized when the external magnetic field (i.e., the field created by an external current excitation) is removed, we may write[f]

$$\vec{\mathbf{M}} = \chi_m\vec{\mathbf{H}}, \tag{1.302}$$

where χ_m is called the **magnetic susceptibility** of the material. Since M and H have the same units (A/m), χ_m is a dimensionless quantity. This leads to

$$\vec{\mathbf{B}} = \mu_0(1 + \chi_m)\vec{\mathbf{H}} = \mu_r\mu_0\vec{\mathbf{H}} = \mu\vec{\mathbf{H}}, \tag{1.303}$$

where

$$\mu = \mu_r\mu_0 \;\; \text{(H/m)} \tag{1.304}$$

is the **permeability** of the material, whereas the scalar, dimensionless quantity

$$\mu_r = 1 + \chi_m \tag{1.305}$$

is the **relative permeability** of the material. In ferromagnetic materials, χ_m and μ_r can

[f] This expression can capture both linear and nonlinear magnetization characteristics, the latter if we allow χ_m to be a function of the field.

obtain values on the order of 5000. If the material is magnetically linear, then χ_m is a constant, and we obtain

$$\vec{\mathbf{J}}_m = \nabla \times \vec{\mathbf{M}} = \chi_m \nabla \times \vec{\mathbf{H}} = \mathbf{0}\,, \tag{1.306}$$

assuming that there is no free current flowing within the material itself; this is typically the case, unless there are eddy currents present. This simple example reinforces our previous statement regarding the relative strengths of $\vec{\mathbf{J}}_m$ (typically weak) and $\vec{\mathbf{K}}_m$ (typically dominant).

In materials with a persistent magnetization (i.e., **permanent magnets**), the magnetization is

$$\vec{\mathbf{M}} = \chi_m \vec{\mathbf{H}} + \vec{\mathbf{M}}_0\,, \tag{1.307}$$

where $\vec{\mathbf{M}}_0 = \vec{\mathbf{M}}_0(\mathbf{r})$ is a constant component, in the sense that it does not depend on the value of the magnetic field over a broad range of external magnetic fields; however, the magnetization constant may depend on the location inside the permanent magnet in cases where the magnetization is not uniform. Therefore

$$\vec{\mathbf{B}} = \mu_r \mu_0 \vec{\mathbf{H}} + \mu_0 \vec{\mathbf{M}}_0\,. \tag{1.308}$$

Typically, permanent magnets have small relative permeability values, on the order of $\mu_r = 1.05$–1.2.

Interface Conditions

We need to pay special attention to the situation at the interface between a magnetized material and air (functionally equivalent to free space) or some other material. In the absence of magnetization, we have already examined what happens to the normal and tangential components of the fields in Examples 1.19 and 1.23 on pages 32 and 37, respectively. This background material is sufficient to understand the situation when magnetized materials are present. Now, the fields are determined by Equations (1.297a), (1.297b), and (1.299).

We infer the following.

- To satisfy $\nabla \cdot \vec{\mathbf{B}} = 0$, we obtained Equation (1.112), repeated here for convenience:

$$(\vec{\mathbf{B}}_1 - \vec{\mathbf{B}}_2) \cdot \hat{\mathbf{n}} = 0\,, \tag{1.309}$$

  with $\hat{\mathbf{n}}$ the unit normal vector. Hence, the normal component of $\vec{\mathbf{B}}$ is continuous across the interface. Of course, in view of Equations (1.303) or (1.308), the normal components of $\vec{\mathbf{H}}$ will differ because of unequal permeabilities.

- For example, consider the special case where flux crosses the interface between a ferromagnetic material and air at a normal angle (e.g., at the air gap of an electric machine). This result implies that the flux density $||\vec{\mathbf{B}}||$ will be the same on each side. Moreover, $||\vec{\mathbf{H}}||$ will be hundreds or thousands of times higher in the air than in the material.

- Another interesting case is the interface between a permanent magnet and surrounding air. Let us focus on the north pole, where flux is exiting the magnet. The previous equation tells us that

$$\mu_0 \vec{\mathbf{H}}_{\text{air}} \cdot \hat{\mathbf{n}} = (\mu_r \mu_0 \vec{\mathbf{H}}_{pm} + \mu_0 \vec{\mathbf{M}}_0) \cdot \hat{\mathbf{n}}\,, \tag{1.310}$$

where $\hat{\mathbf{n}}$ is an outwards-pointing vector, parallel to the magnetization. Equivalently

$$(\vec{\mathbf{H}}_{\text{air}} - \vec{\mathbf{H}}_{pm}) \cdot \hat{\mathbf{n}} = M_0 + \chi_m \vec{\mathbf{H}}_{pm} \cdot \hat{\mathbf{n}}\,. \tag{1.311}$$

This means that the divergence of the H-field is nonzero on the magnet surface. You could visualize this situation as if H-field flux lines are emanating from the magnet surface, pointing outwards and inwards on either side of the boundary. These H-field flux lines would then terminate at the south pole. In other words, inside the magnet, the H-field would be opposing the magnetization.

- In the absence of free surface current (the usual case in electric machine analysis), Equation (1.297b) implies that the tangential components of $\vec{\mathbf{B}}$ are constrained by

$$(\vec{\mathbf{B}}_1 - \vec{\mathbf{B}}_2) \times \hat{\mathbf{n}} = \mu_0(\vec{\mathbf{K}}_{m1} + \vec{\mathbf{K}}_{m2}) = \mu_0(\vec{\mathbf{M}}_1 - \vec{\mathbf{M}}_2) \times \hat{\mathbf{n}}\,, \tag{1.312}$$

with $\hat{\mathbf{n}}$ the unit normal vector from medium 1 to 2. Therefore, a discontinuity may exist in the tangential components of the B-field.

For example, if material 1 is a ferromagnetic material with magnetization $\vec{\mathbf{M}}_1$, and material 2 is air (where $\vec{\mathbf{M}}_2 = \mathbf{0}$), we can relate the tangential components by

$$B_{1t} = B_{2t} + \mu_0 M_{1t}\,. \tag{1.313}$$

This equation tells us that the magnetic flux density in the ferromagnetic material may be much higher than the flux density immediately outside in the air, especially if the flux is oriented mostly parallel to the interface.

- In the absence of free surface current (the usual case), Equation (1.299) implies that the tangential components of $\vec{\mathbf{H}}$ must be continuous.

- In general, at the interface between a ferromagnetic and a non-ferromagnetic material (e.g., air, copper, aluminum), a phenomenon called magnetic flux line **refraction** takes place. The reader is invited to work out the related Problem 1.14. It turns out that the flux lines change direction at the interface; they have a tendency to exit almost perpendicular to the interface right outside the ferromagnetic material, regardless of the direction they may approach the interface from within the material.

1.7.5 Two-Dimensional Problems

The example that follows introduces the special case of a 2-D problem. This type of problem arises when the current flows mostly in a straight line (i.e., when we can neglect the effects of wire bending). Mathematically, we would say that all interesting phenomena take place on an x–y plane, and there is nothing changing along the z-axis, or $\partial/\partial z = 0$. This is a common modeling assumption for electric machine analysis, and it is how we will formulate the FEM in subsequent chapters.

Example 1.33 *Numerical calculation of the magnetic field of a transmission line.* Let us consider two rectangular and perfectly straight copper conductors in free space, as shown in Figure 1.25, which form an elementary transmission line. The problem is to calculate the magnetic field in the rectangular region shown.

Solution

Let us assume that the conductors have an axial length $\ell = 10$ m. Current flows along the z-axis, so there will only be a z-component present for the MVP, which is obtained from

$$A_z(\mathbf{r}) = \frac{\mu_0}{4\pi} \iiint_V J_z(\mathbf{r}') \frac{1}{\|\mathbf{r} - \mathbf{r}'\|} \, dv' . \tag{1.314}$$

(The magnetization of copper is negligible.) As a primary objective, we are interested in calculating the magnetic field at the middle of this transmission line, that is, at $z = 0$. The impact of the conductor bending that takes place at the edges, i.e., at $z = \pm\ell/2$, is neglected due to the large distance. The conductor on the left carries current in the positive z-direction (coming out of the page), and the conductor on the right is the return path of the current. The current distribution is assumed to be uniform. However, the current density in the two conductors is unequal due to different cross-sectional areas. The area of the conductor on the left is 2 cm^2, and the area of the conductor on the right is 1 cm^2. Suppose the current is $i = 1$ A. Then $\vec{\mathbf{J}}_1 = 0.5\,\hat{\mathbf{k}}$ A/cm^2 and $\vec{\mathbf{J}}_2 = -1\,\hat{\mathbf{k}}$ A/cm^2. Hence, from Equation (1.314) we obtain

$$A_z(\mathbf{r}) = \frac{\mu_0 J_1}{4\pi} \iiint_{V_1} \frac{1}{\|\mathbf{r} - \mathbf{r}'\|} \, dv' - \frac{\mu_0 J_2}{4\pi} \iiint_{V_2} \frac{1}{\|\mathbf{r} - \mathbf{r}'\|} \, dv' , \tag{1.315}$$

where V_1 and V_2 are the volumes of the two conductors, $J_1 = 5000$ A/m^2 and $J_2 =$ 10,000 A/m^2.

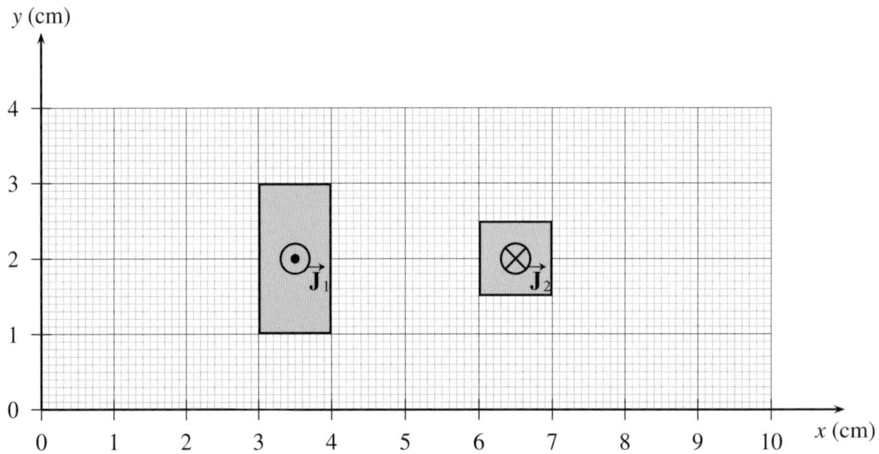

Figure 1.25 Conductor configuration (shaded areas) and discretization of space.

These integrals are difficult to evaluate analytically, so we proceed with a numerical calculation. We discretize the cross-section as shown in Figure 1.25, that is, using square elements with side $h = 1$ mm. Inside conductors, each square represents a conductor "filament" that extends from $-\ell/2$ to $\ell/2$ in the z-direction. We shall calculate the magnetic field at the center of each square element, located at $\mathbf{r}_k(z) = (x_k, y_k, z)$, $k \in \mathcal{K} = \{1, \ldots, 4000\}$. We will set $z = 0$ to obtain the field at the middle of the transmission line; however, we can also compute the field at an arbitrary axial position. (The symmetry about the $y = 2$ axis may be exploited to expedite the calculations, but we are not pursuing this approach here.)

We approximate the distance between the center of square k and a current source at $\mathbf{r}'$ as

$$\|\mathbf{r}_k(z) - \mathbf{r}'\| \approx \|\mathbf{r}_k(z) - \mathbf{r}'_j(z')\| = R_{kj}(z, z') \tag{1.316a}$$

$$= \sqrt{(x_k - x_j)^2 + (y_k - y_j)^2 + (z - z')^2}\,, \tag{1.316b}$$

where $\mathbf{r}' = \mathbf{r}'(x', y', z') \approx \mathbf{r}'_j(z') = (x_j, y_j, z')$. This represents the displacement from the origin to the center of the jth square that $\mathbf{r}'$ points to. Note that $\mathbf{r}'$ remains a function of the axial position z'. Because we have not discretized the geometry along the z-axis, both z and $z' \in [-\ell/2, \ell/2]$ are continuous variables.

Therefore, the integral for the MVP at square k is approximated by a sum accounting for the contribution from each current-carrying square filament j:

$$A_z(\mathbf{r}_k(z)) \approx \frac{\mu_0 J_1 h^2}{4\pi} \sum_{j \in \mathcal{J}_1} \int_{-\ell/2}^{\ell/2} \frac{1}{R_{kj}(z, z')}\, dz' - \frac{\mu_0 J_2 h^2}{4\pi} \sum_{j \in \mathcal{J}_2} \int_{-\ell/2}^{\ell/2} \frac{1}{R_{kj}(z, z')}\, dz'\,, \tag{1.317}$$

where $\mathcal{J}_1, \mathcal{J}_2 \subset \mathcal{K}$ are sets of indices for the two conductors. To obtain this expression, the triple integrals in Equation (1.315) were replaced by a summation of volume integrals with terms representing an integration within the volume V_j of the filament passing through square j. Furthermore, the volume integrals became single integrals by integrating out x and y, leading to the appearance of an h^2 factor. The integration is performed as follows:

$$\iiint_{V_j} \frac{1}{R_{kj}(z, z')}\, dv' = \int_{x_j - h/2}^{x_j + h/2} \int_{y_j - h/2}^{y_j + h/2} \int_{-\ell/2}^{\ell/2} \frac{1}{R_{kj}(z, z')}\, dx' dy' dz' \tag{1.318a}$$

$$= h^2 \int_{-\ell/2}^{\ell/2} \left[d_{kj}^2 + (z - z')^2\right]^{-1/2} dz' = h^2 \ln \frac{\left(\frac{\ell}{2} - z\right) + \sqrt{d_{kj}^2 + \left(\frac{\ell}{2} - z\right)^2}}{-\left(\frac{\ell}{2} + z\right) + \sqrt{d_{kj}^2 + \left(\frac{\ell}{2} + z\right)^2}} \tag{1.318b}$$

$$= h^2 \ln \frac{\left[\left(\frac{\ell}{2} - z\right) + \sqrt{d_{kj}^2 + \left(\frac{\ell}{2} - z\right)^2}\right]\left[\left(\frac{\ell}{2} + z\right) + \sqrt{d_{kj}^2 + \left(\frac{\ell}{2} + z\right)^2}\right]}{d_{kj}^2}\,, \tag{1.318c}$$

where we defined

$$d_{kj} = \sqrt{(x_k - x_j)^2 + (y_k - y_j)^2}\,. \tag{1.319}$$

For $z = 0$, we obtain

$$\iiint_{V_j} \frac{1}{R_{kj}(0,z')}\, dv' = 2h^2 \ln \frac{\frac{\ell}{2} + \sqrt{d_{kj}^2 + \left(\frac{\ell}{2}\right)^2}}{d_{kj}}. \tag{1.320}$$

Equation (1.320) is well-defined for $d_{kj} \neq 0$, but there is a problem when $k = j$ and $d_{kj} = 0$, that is, when calculating the potential contribution within a filament from its own current. To handle this special case, we will omit a small cube of the filament around $z = 0$, by not accounting for the contribution for $z \in [-h/2, h/2]$. This should not impact the result significantly according to the discussion in Example 1.32. Hence, when $k = j \in \mathcal{J}_1$ or $\mathcal{J}_2$, we have

$$\iiint_{V_j} \frac{1}{R_{jj}(0,z')}\, dv' = 2 \int_{x_j-h/2}^{x_j+h/2} \int_{y_j-h/2}^{y_j+h/2} \int_{h/2}^{\ell/2} \frac{1}{R_{jj}(0,z')}\, dx'dy'dz' \tag{1.321a}$$

$$= 2h^2 \int_{h/2}^{\ell/2} \frac{1}{z'}\, dz' = 2h^2 \ln \frac{\ell}{h}. \tag{1.321b}$$

At first glance, this result still appears to be problematic since the logarithm blows up as $h \to 0$; however, multiplication by h^2 saves the day because $\lim_{h\to 0} h^2 \ln \ell/h = 0$.

To summarize, we will use Equation (1.317) with $z = 0$, where

$$\int_{-\ell/2}^{\ell/2} \frac{1}{R_{kj}(0,z')}\, dz' = \begin{cases} 2\ln \dfrac{\frac{\ell}{2} + \sqrt{d_{kj}^2 + \left(\frac{\ell}{2}\right)^2}}{d_{kj}} & \text{for } k \neq j, \\ 2\ln \dfrac{\ell}{h} & \text{for } k = j. \end{cases} \tag{1.322}$$

We must evaluate this for each observation point of interest $k \in \mathcal{K}$, and for each filament $j \in \mathcal{J}_1, \mathcal{J}_2$. In our case, where we are interested in knowing the field throughout the rectangular domain, we have 4000 observation points and 300 filaments, leading to $4000 \times 300 = 1{,}200{,}000$ calculations. These lead to the result shown in Figure 1.26 for the MVP at $z = 0$, which was created using a contour plot as in Example 1.13. Note how the equipotentials resemble magnetic flux lines, and this is indeed the case. This is an important conclusion on its own, which is further discussed within Box 1.2.

To obtain the B-field, we can superimpose the contributions from each filament. To this end, we must calculate the partial derivatives $\partial A_z/\partial x$ and $\partial A_z/\partial y$ (at $z = 0$). The partial derivatives of Equation (1.317) with respect to x_k and y_k depend on those of Equation (1.322). The following result (for $k \neq j$) is key:

$$\frac{\partial}{\partial x_k}\left(2\ln \frac{\frac{\ell}{2} + \sqrt{d_{kj}^2 + \left(\frac{\ell}{2}\right)^2}}{d_{kj}}\right) = -\frac{\ell(x_k - x_j)}{d_{kj}^2 \sqrt{d_{kj}^2 + \left(\frac{\ell}{2}\right)^2}}, \tag{1.325}$$

which the reader should be able to verify using basic calculus. An identical formula can be derived for $\partial/\partial y_k$ of the term in the parentheses. Interestingly, the contribution to the B-field inside a filament due to its own current (for $k = j$) is zero.

Hence, the contribution to the B-field at square k from the filament passing through

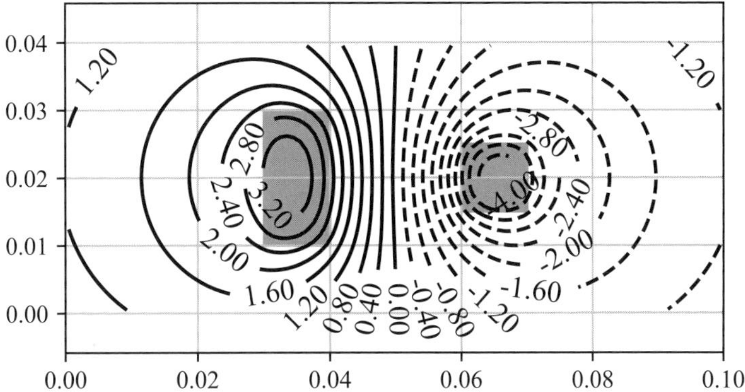

Figure 1.26 MVP contour plot. Numerical values are normalized with respect to $\mu_0/(4\pi)$.

> **Box 1.2** FEM fundamentals: Relationship between B-field and magnetic vector potential in 2-D problems
>
> A 2-D (x–y) magnetic field is obtained when all current sources and the MVP are parallel (to the z-axis). The magnetic flux density is
>
> $$\vec{\mathbf{B}} = \nabla \times \vec{\mathbf{A}} = \frac{\partial A_z}{\partial y}\hat{\mathbf{i}} - \frac{\partial A_z}{\partial x}\hat{\mathbf{j}} = -\hat{\mathbf{k}} \times \left(\frac{\partial A_z}{\partial x}\hat{\mathbf{i}} + \frac{\partial A_z}{\partial y}\hat{\mathbf{j}} + \frac{\partial A_z}{\partial z}\hat{\mathbf{k}} \right). \tag{1.323}$$
>
> In other words:
>
> $$\vec{\mathbf{B}} = \nabla A_z \times \hat{\mathbf{k}}. \tag{1.324}$$
>
> This implies:
>
> 1. The B-field lies on the x–y plane.
> 2. The B-field is perpendicular to the gradient of the MVP, ∇A_z.
> 3. Since ∇A_z is normal to the contours of A_z, $\vec{\mathbf{B}}$ is tangential to the contours.
> 4. Contours or equipotentials of A_z can be interpreted as magnetic flux lines.
> 5. If $\partial A_z/\partial z = 0$ (in infinite-length devices), then $\|\vec{\mathbf{B}}\| = \|\nabla A_z\|$.

square j carrying current i_j is

$$\vec{\mathbf{B}}_{kj} = \frac{\mu_0 i_j}{4\pi} \frac{\ell}{d_{kj}^2 \sqrt{d_{kj}^2 + \left(\frac{\ell}{2}\right)^2}} [-(y_k - y_j)\hat{\mathbf{i}} + (x_k - x_j)\hat{\mathbf{j}}]. \tag{1.326}$$

When $k = j$, we obtain $\vec{\mathbf{B}}_{jj} = \mathbf{0}$. Note that if $d_{kj} \ll \ell$, that is, if the line is very long, then

$$\vec{\mathbf{B}}_{kj} \approx \frac{\mu_0 i_j}{2\pi d_{kj}^2} [-(y_k - y_j)\hat{\mathbf{i}} + (x_k - x_j)\hat{\mathbf{j}}], \tag{1.327}$$

which agrees with the result from Example 1.22 on page 36 for the magnetic field of

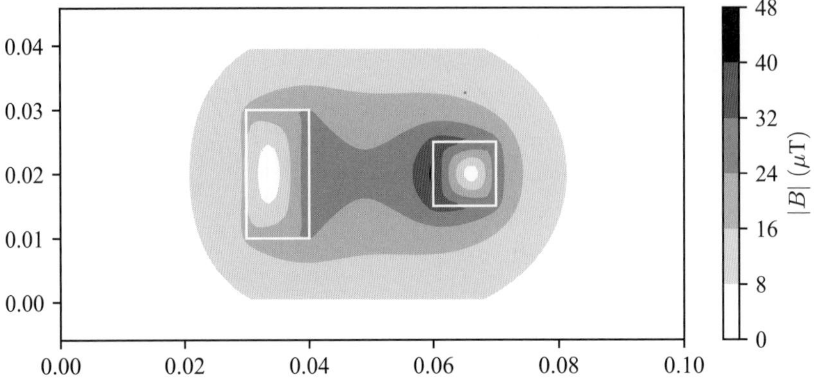

Figure 1.27 Magnitude of the B-field.

an infinitely long (round) wire. After these calculations are complete, it is possible to visualize the B-field magnitude on the $z = 0$ plane, which is plotted in Figure 1.27.

It is also instructive to examine the field over other planes for $z \neq 0$. The analysis reveals that the field does not change significantly for a wide range of z-values. In Figure 1.28, we plot the difference $\overline{A}_{z1}(z) - \overline{A}_{z2}(z)$, obtained by averaging the MVP over each conductor. This quantity stays more or less constant over most of the transmission line, and then drops rapidly as we approach the edges at $\pm\ell/2$.

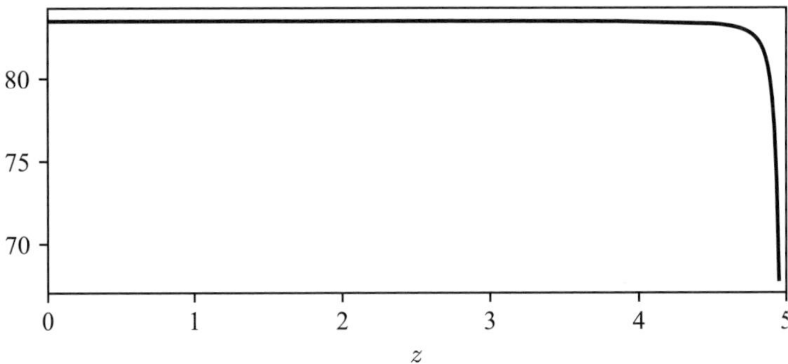

Figure 1.28 Difference of average MVPs between the two conductors with respect to axial position. The numerical values are normalized with respect to $\mu_0/(4\pi)$.

The previous example is representative of how electric machines are typically analyzed, in the sense that a 3-D geometry is reduced to a 2-D cross-section. A 2-D analysis is a compromise that trades off accuracy for substantial gains in computational efficiency. This approach is valid in the absence of geometric and field variation along the z-axis, which is a reasonable assumption for many electric machine designs. Hence, the remarks within Box 1.2 acquire a particular significance. However, the field calculations are not usually performed in the same manner as in the example. In the subsequent chapters, we will learn a different (better) way to perform the numerical analysis, namely, the FEM.

Another extremely useful fact for 2-D problems relates the flux to the MVP. Recall that the **magnetic flux** Φ through a surface S is measured in Wb and defined by the surface integral

$$\Phi = \iint_S \vec{\mathbf{B}} \cdot d\mathbf{a} \,. \tag{1.328}$$

Using the MVP and Stokes' theorem, we obtain

$$\Phi = \iint_S (\nabla \times \vec{\mathbf{A}}) \cdot d\mathbf{a} = \oint_{\partial S} \vec{\mathbf{A}} \cdot d\mathbf{r} \,. \tag{1.329}$$

Therefore, the flux is also the line integral of the MVP. We should evaluate the line integral by traversing the boundary ∂S as required by the presumed positive direction of flux, which defines the direction of the surface normal vector $\hat{\mathbf{n}}$.

In 2-D (x–y) problems, $\vec{\mathbf{A}} = A_z(x, y)\,\hat{\mathbf{k}}$, and $\vec{\mathbf{B}}$ lies on the x–y plane, as we have learned. In such problems, calculations of flux involve integration surfaces S with a rectangular boundary ∂S. The surface is oriented such that two of its sides are parallel to the z-axis, extending by ℓ in the axial direction, as shown in Figure 1.29. The MVP on these two sides is $\vec{\mathbf{A}}_1 = A_1\hat{\mathbf{k}}$ and $\vec{\mathbf{A}}_2 = A_2\hat{\mathbf{k}}$, respectively. The line integral of Equation (1.329) can be broken into four parts, one per side. We note that the sides that are parallel to the x–y plane are perpendicular to $\vec{\mathbf{A}}$, so they do not contribute to the integral. Hence, the flux is related to the difference of the MVP at the two sides, as highlighted in Box 1.3. The numerical values of the contours of A_z shown previously in Figure 1.26 should now obtain more physical sense. Subtracting any two of these values, as in Figure 1.28, yields the **flux per unit depth** between the corresponding points in space. The sign of the answer also determines the direction of flux.

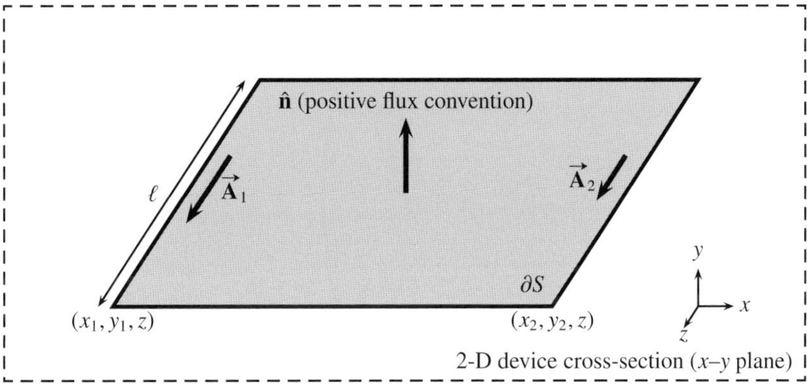

Figure 1.29 Calculation of magnetic flux in the 2-D case.

Box 1.3 FEM fundamentals: Relationship between magnetic flux and magnetic vector potential in 2-D problems

In 2-D problems, the magnetic flux through an arbitrary surface is found by a simple subtraction:

$$\Phi = \ell\,(A_1 - A_2)\,, \tag{1.330}$$

where A_1 and A_2 are the MVP values at two points on the x–y plane defining the sides of a rectangular boundary that extends by ℓ along the z-axis. The order of subtraction is determined by the presumed positive direction of flux.

We can conclude that:

1. The flux per unit depth (in Wb/m) equals the difference of the MVP at the two sides.
2. If we are standing on the x–y plane parallel to the z-axis, and we are facing in the direction of potential increase, then the flux is oriented from our left to our right.

1.7.6 Axisymmetric Problems

A special case of a "2-D" problem arises when the geometry, including the physical device layout as well as the current sources, exhibits **cylindrical symmetry**. These problems are called **axisymmetric**. For example, the elementary magnetic dipole and a solenoid exhibit such a symmetry. It is convenient to analyze such problems using cylindrical coordinates (r, ϕ, z), positioning the z-axis so that it coincides with the axis of symmetry. We would then impose the constraint that, due to symmetry, $\partial/\partial\phi = 0$. Hence, axisymmetric devices can be analyzed on any arbitrary constant-ϕ plane using (r, z) coordinates, as shown in Figure 1.30.

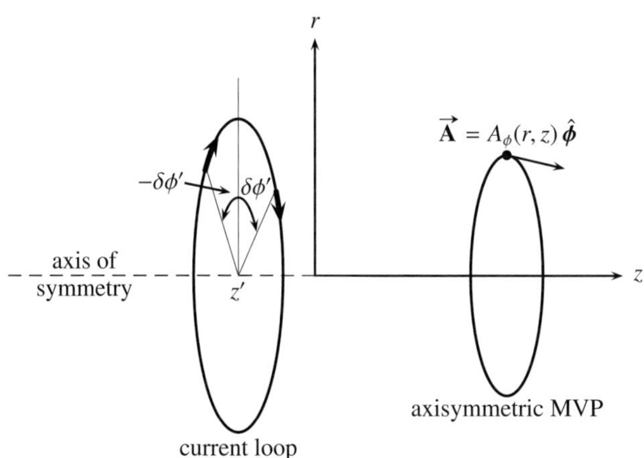

Figure 1.30 Axisymmetric problem analysis using cylindrical coordinates.

In axisymmetric problems, current sources are cylindrical, so the current density only has a ϕ-component whose magnitude does not depend on the angle ϕ:

$$\vec{\mathbf{J}}(\mathbf{r}) = J_\phi(r, z)\,\hat{\boldsymbol{\phi}}\,. \tag{1.331}$$

Note that this kind of symmetry is a property of free as well as bound current sources. Hence, the current can be decomposed to a distribution of circular filamentary loops around the axis of symmetry, like the one shown in Figure 1.30. Therefore, the total field is a superposition of magnetic dipole-like fields. This implies that the MVP has a ϕ-component only. This fact was shown in the analysis of the magnetic dipole in §1.7.1 (although the proof was for the field far from the dipole). There is a simpler way to see why this is so. Suppose that we are calculating the MVP at a point (r, ϕ, z) through an integral of current sources as in Equation (1.263). We can perform the integration conveniently by accounting for symmetric current elements from the circular filament at $(r', \phi \pm \delta\phi', z')$, letting $\delta\phi'$ vary from 0 to π; otherwise, r' and z' are arbitrary. Since the distances of any two symmetric sources from the point of observation are equal, vector addition of the MVP contributions from the current elements leads to a net contribution in the ϕ direction only:

$$\vec{\mathbf{A}}(\mathbf{r}) = A_\phi(r, z)\,\hat{\boldsymbol{\phi}}\,. \tag{1.332}$$

The magnetic field is obtained using the curl in cylindrical coordinates (setting $A_z = A_r = 0$):

$$\vec{\mathbf{B}} = -\frac{\partial A_\phi}{\partial z}\hat{\mathbf{r}} + \frac{1}{r}\frac{\partial (rA_\phi)}{\partial r}\hat{\mathbf{z}} = -\frac{\partial A_\phi}{\partial z}\hat{\mathbf{r}} + \left(\frac{A_\phi}{r} + \frac{\partial A_\phi}{\partial r}\right)\hat{\mathbf{z}}\,. \tag{1.333}$$

The apparent singularity at the $r = 0$ axis may be cause for concern, so let us carefully examine the implications for the two components of the B-field. First, the radial component must be zero along the $r = 0$ axis, otherwise a nonzero divergence would arise in violation of Maxwell's law. Hence

$$\left.\frac{\partial A_\phi}{\partial z}\right|_{r=0} = 0\,, \tag{1.334}$$

implying that A_ϕ is constant along the z-axis. Without loss of generality, we can impose the condition

$$A_\phi(0, z) = 0\,. \tag{1.335}$$

For the axial component, we revert to the fundamental definition of the curl, that is, Equation (1.113) on page 33. Taking the circulation of the MVP around a circular path of radius r centered on the z-axis, we have

$$B_z(0, z) = \lim_{r\to 0}\frac{A_\phi(r, z)\cdot(2\pi r)}{\pi r^2} = 2\lim_{r\to 0}\frac{A_\phi(r, z)}{r}\,. \tag{1.336}$$

For this limit to attain a finite value, A_ϕ must increase at least linearly in the vicinity of the axis. Hence, $A_\phi(r, z) \approx c(z)\cdot r$ for small r, where $c(z) \in \mathbb{R}$ is a function of axial position. Then $B_z(0, z) = 2c(z)$.

It is convenient (for reasons that will become clear when we learn more about the FEM) to introduce the function rA_ϕ. In cylindrical coordinates, we obtain the gradient

$$\nabla(rA_\phi) = \frac{\partial(rA_\phi)}{\partial r}\hat{\mathbf{r}} + \frac{\partial(rA_\phi)}{\partial z}\hat{\mathbf{z}}\,. \tag{1.337}$$

Also, the flux density can be expressed equivalently to Equation (1.333) as

$$\vec{\mathbf{B}} = \frac{1}{r}\left(-\frac{\partial(rA_\phi)}{\partial z}\hat{\mathbf{r}} + \frac{\partial(rA_\phi)}{\partial r}\hat{\mathbf{z}}\right). \tag{1.338}$$

Now, suppose we are interested in calculating the flux $\Phi(r, z)$ through a circular surface S of radius r centered on the axis of symmetry at an arbitrary value of z. The positive flux direction is taken to be the direction of the z-axis. Using the MVP, we have

$$\Phi(r, z) = \oint_{\partial S} \vec{\mathbf{A}} \cdot d\mathbf{r} = \int_0^{2\pi} A_\phi(r, z)\, r\, d\phi = 2\pi r A_\phi(r, z)\,. \tag{1.339}$$

This discussion leads us to the conclusions of Box 1.4.

Box 1.4 FEM fundamentals: Relationship between B-field, MVP, and flux in axisymmetric problems

An axisymmetric (r–z) magnetic field is obtained when the geometry of the problem has cylindrical symmetry ($\partial/\partial\phi = 0$), where all current sources and the MVP have a ϕ-independent ϕ-component. In this case, the magnetic flux density is

$$\vec{\mathbf{B}} = \nabla \times \vec{\mathbf{A}} = \frac{1}{r}\left(-\frac{\partial(rA_\phi)}{\partial z}\hat{\mathbf{r}} + \frac{\partial(rA_\phi)}{\partial r}\hat{\mathbf{z}}\right) = \frac{1}{r}\left(\frac{\partial(rA_\phi)}{\partial r}\hat{\mathbf{r}} + \frac{\partial(rA_\phi)}{\partial z}\hat{\mathbf{z}}\right) \times \hat{\boldsymbol{\phi}}\,. \tag{1.340}$$

In other words:

$$\vec{\mathbf{B}} = \frac{1}{r}\nabla(rA_\phi) \times \hat{\boldsymbol{\phi}}\,. \tag{1.341}$$

The flux through a circular surface of radius r centered on the symmetry axis at axial position z is

$$\Phi(r, z) = 2\pi r A_\phi(r, z)\,. \tag{1.342}$$

This implies:

1. The B-field does not have an azimuthal component ($B_\phi = 0$).
2. The B-field is perpendicular to the gradient $\nabla(rA_\phi)$.
3. Since $\nabla(rA_\phi)$ is normal to the contours of rA_ϕ, $\vec{\mathbf{B}}$ is tangential to these contours.
4. Contours or equipotentials of rA_ϕ can be interpreted as magnetic flux lines.
5. $\|\vec{\mathbf{B}}\| \neq \|\nabla A_\phi\|$, but $\|\vec{\mathbf{B}}\| = \|\nabla(rA_\phi)\|/r$.
6. The quantity rA_ϕ may be interpreted as flux per radian.
7. If $rA_\phi > 0$, then the flux is directed towards the z-axis.

1.7.7 An Elementary Three-Dimensional Problem Solved Using the Method of Moments

In the example that follows, we will calculate the magnetic field created by a cable that is embedded in a solid block of ferromagnetic material. We will attempt to solve this problem (numerically) in three dimensions. Instead of calculating the MVP, as we did in Example 1.33, we will calculate the B-field based on the law of Biot–Savart, using the so-called **method of moments** [1]. This method is sometimes used effectively in the analysis of electric machines, and is described herein for the sake of completeness. However, this section is not prerequisite for comprehending the material in the following chapters, and may safely be skipped. Nonetheless, the interested reader is encouraged to follow the example since it illustrates the application of several of the key concepts set forth earlier in this chapter.

In a ferromagnetic material, the magnetization is related to the magnetic field by

$$\vec{\mathbf{M}} = \chi_m \vec{\mathbf{H}} = \frac{\mu_r - 1}{\mu_r \mu_0} \vec{\mathbf{B}} . \tag{1.343}$$

As we have explained in §1.7.3, the magnetization can be represented by equivalent magnetization currents. Internally to a magnetically linear material, $\vec{\mathbf{J}}_m = \mathbf{0}$; however, nonzero surface magnetization currents are present, given by

$$\vec{\mathbf{K}}_m = \vec{\mathbf{M}} \times \hat{\mathbf{n}} = \frac{\mu_r - 1}{\mu_r \mu_0} \vec{\mathbf{B}} \times \hat{\mathbf{n}} , \tag{1.344}$$

where $\hat{\mathbf{n}}$ is a normal unit vector pointing outwards from the magnetized matter. The main objective of the exercise that follows is to demonstrate how a system of equations can be formulated to solve for $\vec{\mathbf{K}}_m$ over the entire surface of the magnetic material. Once the surface magnetization currents are known, we can determine the B-field everywhere in space using the law of Biot–Savart, i.e., Equations (1.269) and (1.271):

$$\vec{\mathbf{B}}(\mathbf{r}) = \frac{\mu_0 i}{4\pi} \oint_C \hat{\mathbf{t}}(\mathbf{r}') \times \frac{\mathbf{r} - \mathbf{r}'}{\|\mathbf{r} - \mathbf{r}'\|^3} \, dr' + \frac{\mu_0}{4\pi} \oiint_S \vec{\mathbf{K}}_m(\mathbf{r}') \times \frac{\mathbf{r} - \mathbf{r}'}{\|\mathbf{r} - \mathbf{r}'\|^3} \, da' . \tag{1.345}$$

The integration path C of the line integral follows the cable, which is modeled as a filament. The (closed) integration surface S is the boundary of the ferromagnetic material.

We note that Equation (1.345) is a continuous function of position $\mathbf{r}$, as long as the observation point does not coincide with a source, where a singularity occurs due to $\|\mathbf{r} - \mathbf{r}'\| \to 0$. This is problematic for us because we shall apply this formula very close to the material boundary, where surface magnetization currents are flowing. In our previous work (see §1.7.4), we have explained that there is a discontinuity in the tangential components of the B-field across a material interface (whereas the normal component is continuous). If material 1 is magnetized and material 2 is air, the difference in the tangential components is $B_{1t} - B_{2t} = \mu_0 M_{1t}$. Therefore, when $\mathbf{r}$ comes very close to a boundary, we will modify Equation (1.345) to account for the field that is created by the local surface magnetization current, which we will denote by $\vec{\mathbf{B}}_s(\mathbf{r})$:

$$\vec{\mathbf{B}}(\mathbf{r}) = \frac{\mu_0 i}{4\pi} \oint_C \hat{\mathbf{t}}(\mathbf{r}') \times \frac{\mathbf{r} - \mathbf{r}'}{\|\mathbf{r} - \mathbf{r}'\|^3} \, dr' + \frac{\mu_0}{4\pi} \iint_{S-\delta S} \vec{\mathbf{K}}_m(\mathbf{r}') \times \frac{\mathbf{r} - \mathbf{r}'}{\|\mathbf{r} - \mathbf{r}'\|^3} \, da' + \vec{\mathbf{B}}_s(\mathbf{r}) . \tag{1.346}$$

Note that we have excluded a small area δS surrounding $\mathbf{r}$ from the surface integral, which is now well-defined. In particular, if $\mathbf{r}$ points just inside the magnetized matter, then

$$\vec{\mathbf{B}}_s(\mathbf{r}) = \frac{\mu_0}{2}\vec{\mathbf{M}}_t(\mathbf{r}), \tag{1.347}$$

where

$$\vec{\mathbf{M}}_t = \vec{\mathbf{M}} - (\vec{\mathbf{M}} \cdot \hat{\mathbf{n}})\,\hat{\mathbf{n}} \tag{1.348}$$

denotes the tangential component of the magnetization. (If $\mathbf{r}$ points just outside the material, a change in the sign of $\vec{\mathbf{B}}_s$ occurs.)

Taking the cross product of Equation (1.346) with the outwards-pointing unit normal vector at the point of observation $\hat{\mathbf{n}}(\mathbf{r})$ (where $\mathbf{r}$ points right inside of the magnetized material surface), and in view of Equations (1.344), (1.347), and (1.348), we obtain

$$\begin{aligned}\frac{\mu_r}{\mu_r - 1}\vec{\mathbf{K}}_m(\mathbf{r}) &= \frac{i}{4\pi}\oint_C \left(\hat{\mathbf{t}}(\mathbf{r}') \times \frac{\mathbf{r} - \mathbf{r}'}{\|\mathbf{r} - \mathbf{r}'\|^3}\right) \times \hat{\mathbf{n}}(\mathbf{r})\, dr' + \\ &\frac{1}{4\pi}\iint_{S-\delta S}\left(\vec{\mathbf{K}}_m(\mathbf{r}') \times \frac{\mathbf{r} - \mathbf{r}'}{\|\mathbf{r} - \mathbf{r}'\|^3}\right) \times \hat{\mathbf{n}}(\mathbf{r})\, da' + \frac{1}{2}\underbrace{\vec{\mathbf{M}}(\mathbf{r}) \times \hat{\mathbf{n}}(\mathbf{r})}_{\vec{\mathbf{K}}_m}\,.\end{aligned} \tag{1.349}$$

Performing some elementary algebra and rearranging terms:

$$\begin{aligned}\vec{\mathbf{K}}_m(\mathbf{r}) - \frac{\mu_r - 1}{\mu_r + 1}\frac{1}{2\pi}\iint_{S-\delta S}\left(\vec{\mathbf{K}}_m(\mathbf{r}') \times \frac{\mathbf{r} - \mathbf{r}'}{\|\mathbf{r} - \mathbf{r}'\|^3}\right) \times \hat{\mathbf{n}}(\mathbf{r})\, da' \\ = \frac{\mu_r - 1}{\mu_r + 1}\frac{i}{2\pi}\oint_C \left(\hat{\mathbf{t}}(\mathbf{r}') \times \frac{\mathbf{r} - \mathbf{r}'}{\|\mathbf{r} - \mathbf{r}'\|^3}\right) \times \hat{\mathbf{n}}(\mathbf{r})\, dr'\,.\end{aligned} \tag{1.350}$$

This integral equation needs to be satisfied at all points right inside the surface of the magnetized matter. In mathematics, this is called a **Fredholm**[35] **equation**.

Example 1.34 *Numerical calculation of the magnetic field of a cable embedded in a magnetic material.*

A cable passes through a cube made of a magnetically linear material with relative permeability $\mu_r = 100$.[g] The cube has $\ell = 20$-cm long sides, it is centered at the origin, and is aligned with the axes of a Cartesian coordinate system, as shown in Figure 1.31. The cable carries current $i = 10$ A, and is inserted in the middle of 2-cm wide, square slots. (The cable is held in place by some non-magnetic filler material.) The slots are centered at $x = \pm 5$ cm and $y = 0$, and extend along the z-axis. At the two ends, the cable has a semicircular shape.

Solution

Our goal is to obtain an approximate numerical solution of Equation (1.350), that is, to determine the surface magnetization currents. To this end, we create a partition of the

[g] The accuracy of this method is known to deteriorate at high values of μ_r, therefore we are using a modest value in this example. An analysis of why this problem occurs is beyond the scope of this text.

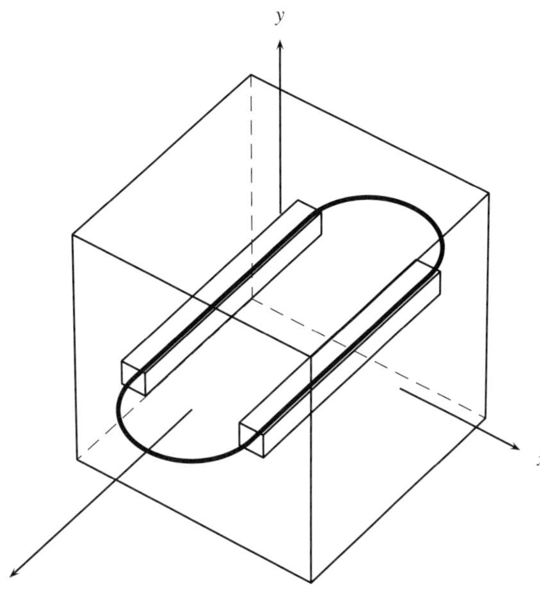

Figure 1.31 A cable that is inserted in a ferromagnetic cube. (The cube's outer surfaces are transparent so that the slots are visible.)

surface using non-overlapping rectangles, which is convenient for the geometry that we are considering. This partition, also known as a **mesh**, is shown in Figure 1.32. The tiles cover the surface in its entirety. With 100% hindsight (or after some trial and error), the rectangle dimensions are made smaller along the directions where we expect the solution to vary significantly.

We now assume a functional form of the solution. For simplicity, we assume that the surface magnetization current is constant in each rectangle. The general form of the solution we are seeking is thus

$$\vec{\mathbf{K}}_m(\mathbf{r}) = \sum_{n=1}^{N} \vec{\mathbf{K}}_m^n\, \alpha_n(\mathbf{r}) = \sum_{n=1}^{N} (K_x^n\, \hat{\mathbf{i}} + K_y^n\, \hat{\mathbf{j}} + K_z^n\, \hat{\mathbf{k}})\, \alpha_n(\mathbf{r})\,, \tag{1.351}$$

where N is the number of rectangular elements, and the **pulse basis function** is defined as

$$\alpha_n(\mathbf{r}) = \begin{cases} 1 & \text{if } \mathbf{r} \text{ points inside element } n, \\ 0 & \text{otherwise.} \end{cases} \tag{1.352}$$

In our case, because of the presumed orientation of all surfaces, only two components of $\vec{\mathbf{K}}_m$ can be nonzero in each rectangular element. For example, for an element with $\hat{\mathbf{n}} = \hat{\mathbf{k}}$, we only need to determine K_x^n and K_y^n. Therefore, our problem has $2N$ unknowns, corresponding to the two tangential magnetization currents in each element. The surface integral in Equation (1.350) can be written as a sum over all rectangular elements, leading to

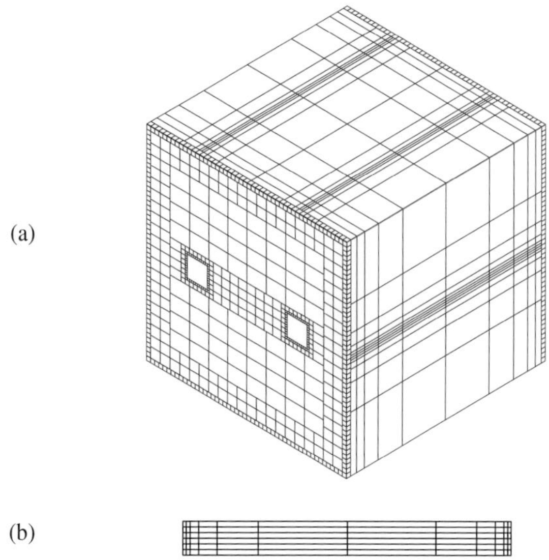

Figure 1.32 (a) Outer surface mesh and (b) slot surface mesh. All eight slot surfaces are meshed identically.

$$\vec{\mathbf{K}}_m(\mathbf{r}) - \frac{\mu_r - 1}{\mu_r + 1}\frac{1}{2\pi}\sum_{\substack{j=1\\ j\neq k}}^{N}\iint_{S_j}\left(\vec{\mathbf{K}}_m^j \times \frac{\mathbf{r}-\mathbf{r}'}{\|\mathbf{r}-\mathbf{r}'\|^3}\right)\times \hat{\mathbf{n}}^k\, da'$$
$$= \frac{\mu_r - 1}{\mu_r + 1}\frac{i}{2\pi}\oint_C\left(\hat{\mathbf{t}}(\mathbf{r}') \times \frac{\mathbf{r}-\mathbf{r}'}{\|\mathbf{r}-\mathbf{r}'\|^3}\right)\times \hat{\mathbf{n}}^k\, dr', \quad (1.353)$$

skipping the kth element that corresponds to the observation point $\mathbf{r}$ (where $\alpha_k(\mathbf{r}) = 1$). Here, j is the index for the source element, and S_j is its surface. Also, $\hat{\mathbf{n}}(\mathbf{r}) = \hat{\mathbf{n}}^k$ is the normal vector at the kth element, which is constant because the element is planar.

To proceed, we expand $[\vec{\mathbf{K}}_m^j \times (\mathbf{r} - \mathbf{r}')] \times \hat{\mathbf{n}}^k$, which is a vector triple product (see Example 1.10). Let $\vec{\mathbf{K}}_m^j = (K_x^j, K_y^j, K_z^j)$, $\mathbf{R} = \mathbf{r} - \mathbf{r}' = (R_x, R_y, R_z)$, and $\hat{\mathbf{n}}^k = (n_x^k, n_y^k, n_z^k)$. After elementary manipulations, we obtain

$$(\vec{\mathbf{K}}_m^j \times \mathbf{R}) \times \hat{\mathbf{n}}^k = [(K_y^j R_x - K_x^j R_y)\, n_y^k + (K_z^j R_x - K_x^j R_z)\, n_z^k]\,\hat{\mathbf{i}} +$$
$$[(K_z^j R_y - K_y^j R_z)\, n_z^k + (K_x^j R_y - K_y^j R_x)\, n_x^k]\,\hat{\mathbf{j}} +$$
$$[(K_x^j R_z - K_z^j R_x)\, n_x^k + (K_y^j R_z - K_z^j R_y)\, n_y^k]\,\hat{\mathbf{k}}. \quad (1.354)$$

For example, suppose that the source element j is aligned with the x–y plane so that $K_z^j = 0$. This element's impact on another x–y plane element k, where $\hat{\mathbf{n}}^k = (0, 0, \pm 1)$, is proportional to

$$(\vec{\mathbf{K}}_m^j \times \mathbf{R}) \times \hat{\mathbf{n}}^k|_{xy\to xy} = -K_x^j R_z n_z^k\,\hat{\mathbf{i}} - K_y^j R_z n_z^k\,\hat{\mathbf{j}}. \quad (1.355)$$

Note that elements on the same xy side of the cube do not affect each other, since $R_z = 0$. Similarly, the impact of an xy element on a yz element, where $\hat{\mathbf{n}}^k = (\pm 1, 0, 0)$, is proportional to

$$(\vec{\mathbf{K}}_m^j \times \mathbf{R}) \times \hat{\mathbf{n}}^k|_{xy \to yz} = (K_x^j R_y - K_y^j R_x)\, n_x^k\, \hat{\mathbf{j}} + K_x^j R_z n_x^k\, \hat{\mathbf{k}}\,, \tag{1.356}$$

and the impact of an xy element on a zx element, where $\hat{\mathbf{n}}^k = (0, \pm 1, 0)$, is proportional to

$$(\vec{\mathbf{K}}_m^j \times \mathbf{R}) \times \hat{\mathbf{n}}^k|_{xy \to zx} = (K_y^j R_x - K_x^j R_y)\, n_y^k\, \hat{\mathbf{i}} + K_y^j R_z n_y^k\, \hat{\mathbf{k}}\,. \tag{1.357}$$

An identical calculation, where $\vec{\mathbf{K}}_m^j$ is replaced by $\hat{\mathbf{t}}$, yields the impact of the filament current on the surface elements.

We will now invoke Equation (1.353) N times by pointing $\mathbf{r}$ within each element. In doing so, we are forming a linear system of $2N$ equations, $\mathbf{Ax} = \mathbf{b}$, for the $2N$ tangential components of $\vec{\mathbf{K}}_m$ all over the surface of the magnetized matter. Denoting the tangential components of the nth element as K_{t1}^n and K_{t2}^n, the vector of unknowns is

$$\mathbf{x}^\top = \begin{bmatrix} K_{t1}^1 & K_{t2}^1 & K_{t1}^2 & K_{t2}^2 & \dots & K_{t1}^N & K_{t2}^N \end{bmatrix}. \tag{1.358}$$

The mesh that we have created has $N = 2712$ square elements, and $\mathbf{A}$ is a 5424×5424 **influence matrix**. Both $\mathbf{A}$ and $\mathbf{b}$ can be assembled element-wise, looping over all elements. Implementation details are left as an exercise for the reader.

We may decide to satisfy Equation (1.353) at a certain point within each element, setting $\mathbf{r} = \mathbf{r}_k$, for $k = 1, \dots, N$. For instance, we could point $\mathbf{r}_k$ to the element centers. This approach is called a **point matching** or **collocation method**, and the centers would be called **collocation points**.

Alternatively, we will satisfy Equation (1.353) in a weighted average sense over each observation element. Arguably, this approach leads to better accuracy than the collocation method; however, this comes at greater computational effort. In particular, we will examine **Galerkin's**[36] **method**, named after the person who first developed this technique. The distinctive characteristic of Galerkin's method is the choice of weight functions: the integration of the equation involves the same weight functions as the basis functions of Equation (1.352) that we have assumed for the solution. (The FEM is also based on Galerkin's method.) Therefore, integrating both sides of Equation (1.353) as $\iint_{S_k} \alpha_k(\mathbf{r}) \vec{\mathbf{F}}(\mathbf{r})\, da$, we obtain

$$\vec{\mathbf{K}}_m^k - \frac{\mu_r - 1}{\mu_r + 1} \frac{1}{2\pi A_k} \sum_{\substack{j=1 \\ j \neq k}}^{N} \iint_{S_k} \iint_{S_j} \left(\vec{\mathbf{K}}_m^j \times \frac{\mathbf{r} - \mathbf{r}'}{\|\mathbf{r} - \mathbf{r}'\|^3} \right) \times \hat{\mathbf{n}}^k\, da'\, da$$

$$= \frac{\mu_r - 1}{\mu_r + 1} \frac{i}{2\pi A_k} \iint_{S_k} \oint_C \left(\hat{\mathbf{t}}(\mathbf{r}') \times \frac{\mathbf{r} - \mathbf{r}'}{\|\mathbf{r} - \mathbf{r}'\|^3} \right) \times \hat{\mathbf{n}}^k\, dr'\, da\,, \tag{1.359}$$

for $k = 1, \dots, N$, where $A_k = |S_k|$ is the surface area of element k. The calculation of the quadruple and triple integrals is not trivial. The use of rectangular elements facilitates this calculation, as the four integration limits are mutually independent.

Let us examine the $xy \to xy$ case first, meaning that some source element S_j lying on an x–y plane (with z constant) is affecting another xy observation element S_k. Hence, we must calculate the integral

$$\vec{\mathbf{I}}_{xy\to xy} = \iint_{S_k} \iint_{S_j} \left(\vec{\mathbf{K}}^j_m \times \frac{\mathbf{r} - \mathbf{r}'}{\|\mathbf{r} - \mathbf{r}'\|^3} \right) \times \hat{\mathbf{n}}^k \, dx' dy' dx \, dy \,. \tag{1.360}$$

Here, integration over the source rectangle S_j means that x' varies from x'_1 to x'_2, and that y' varies from y'_1 to y'_2, which are the coordinates of the rectangle vertices. Similarly, integration over the observation rectangle S_k means that x varies from x_1 to x_2, and that y varies from y_1 to y_2. Based on Equation (1.355):

$$\vec{\mathbf{I}}_{xy\to xy} = R_z n^k_z (-K^j_x \hat{\mathbf{i}} - K^j_y \hat{\mathbf{j}}) \iint_{S_k} \iint_{S_j} \frac{1}{R^3} \, dx' dy' dx \, dy \,, \tag{1.361}$$

with

$$R = \|\mathbf{R}\| = \sqrt{(x - x')^2 + (y - y')^2 + (z - z')^2} = \sqrt{R_x^2 + R_y^2 + R_z^2} \,. \tag{1.362}$$

It is convenient to integrate first with respect to x, and then with respect to x', which yields

$$\iint \iint \frac{1}{R^3} \, dx' dy' dx \, dy = - \iint \frac{R}{R^2 - R_x^2} \, dy' dy \,. \tag{1.363}$$

This result (as well as others that will follow in this example) is displayed as an indefinite integral, omitting constants of integration. This means that the antiderivative appearing as the integrand of the double integral is evaluated at the limits of the variables that are being integrated out. Although it may be possible to obtain an analytical expression of the double integral, we will instead compute it numerically. To this end, the integration library of SciPy (`scipy.integrate`) is a convenient tool that we have at our disposal. SciPy is Python's ecosystem of math, science, and engineering software.

The second case we examine is $xy \to yz$, for which we have

$$\vec{\mathbf{I}}_{xy\to yz} = \iint_{S_k} \iint_{S_j} \left(\vec{\mathbf{K}}^j_m \times \frac{\mathbf{r} - \mathbf{r}'}{\|\mathbf{r} - \mathbf{r}'\|^3} \right) \times \hat{\mathbf{n}}^k \, dx' dy' dy \, dz \,. \tag{1.364}$$

Based on Equation (1.356):

$$\vec{\mathbf{I}}_{xy\to yz} = n^k_x \iint_{S_k} \iint_{S_j} [(K^j_x R_y - K^j_y R_x) \hat{\mathbf{j}} + K^j_x R_z \hat{\mathbf{k}}] \frac{1}{R^3} \, dx' dy' dy \, dz \,. \tag{1.365}$$

This calculation involves three separate quadruple integrals, which can be reduced to double integrals as follows:

$$\iint \iint \frac{R_x}{R^3} \, dx' dy' dy \, dz = \iint \ln(R_z + R) \, dy' \, dy \,, \tag{1.366}$$

where we integrated first with respect to x' and then with respect to z;

$$\iint \iint \frac{R_y}{R^3} \, dx' dy' dy \, dz = - \iint \ln(R_z + R) \, dx' dy' \,, \tag{1.367}$$

where we integrated first with respect to y and then with respect to z; and

$$\iint \iint \frac{R_z}{R^3} \, dx' dy' dy \, dz = \iint \ln(R_x + R) \, dy' dy \,, \tag{1.368}$$

where we integrated first with respect to z and then with respect to x'. A minor difficulty arises for adjacent rectangles at the edges of the device. In this case, the antiderivatives of the double integration need to be evaluated at points lying on a common edge, where $R_x = R_z = 0$, thus leading to integrals of the form $\int \ln|y - y'|\, dy$. For adjacent elements, this is an improper integral of the second kind because $y - y'$ can attain a zero value within the integration interval, where the logarithm approaches infinity; nevertheless, this integral is convergent and thus the issue is resolved.

The third case we examine is $xy \to zx$, for which we have

$$\vec{\mathbf{I}}_{xy\to zx} = \iint_{S_k} \iint_{S_j} \left(\vec{\mathbf{K}}_m^j \times \frac{\mathbf{r} - \mathbf{r}'}{\|\mathbf{r} - \mathbf{r}'\|^3} \right) \times \hat{\mathbf{n}}^k \, dx'dy'dz\,dx\,. \tag{1.369}$$

Based on Equation (1.357):

$$\vec{\mathbf{I}}_{xy\to zx} = n_y^k \iint_{S_k} \iint_{S_j} [(K_y^j R_x - K_x^j R_y)\,\hat{\mathbf{i}} + K_y^j R_z\,\hat{\mathbf{k}}] \frac{1}{R^3}\, dx'dy'dz\,dx\,. \tag{1.370}$$

We perform a double integration on the three scalar integrals as follows:

$$\iint \iint \frac{R_x}{R^3}\, dx'dy'dz\,dx = -\iint \ln(R_z + R)\, dx'dy'\,, \tag{1.371}$$

where we integrated first with respect to x and then with respect to z;

$$\iint \iint \frac{R_y}{R^3}\, dx'dy'dz\,dx = \iint \ln(R_z + R)\, dx'dx\,, \tag{1.372}$$

where we integrated first with respect to y' and then with respect to z; and

$$\iint \iint \frac{R_z}{R^3}\, dx'dy'dz\,dx = \iint \ln(R_y + R)\, dx'dx\,, \tag{1.373}$$

where we integrated first with respect to z and then with respect to y'.

All other cases, e.g., $yz \to xy$, can be evaluated with the help of a coordinate transformation based on the three cases that we have worked out. The integrals are computed numerically in SciPy. It should be noted that the formation of the **A**-matrix can be time-consuming since we are computing the impact of each element on every other element of the domain.

Finally, we need to calculate the contribution of the filament current, which appears on the right-hand side of Equation (1.359) as an integral of the form

$$\vec{\mathbf{I}} = \iint_{S_k} \oint_C \left(\hat{\mathbf{t}}(\mathbf{r}') \times \frac{\mathbf{r} - \mathbf{r}'}{\|\mathbf{r} - \mathbf{r}'\|^3} \right) \times \hat{\mathbf{n}}^k \, dr'\, da\,. \tag{1.374}$$

First, let us focus on the two linear parts of the cable, extending from $z' = -\ell/2$ to $z' = \ell/2$ in the slots, where $\hat{\mathbf{t}} = \pm\hat{\mathbf{k}} = (0, 0, \pm 1)$. The integrals for xy, yz, and zx elements, respectively, are

$$\vec{\mathbf{I}}_{z\to xy} = t_z n_z^k \iint_{S_k} \int_{z'} (R_x\,\hat{\mathbf{i}} + R_y\,\hat{\mathbf{j}}) \frac{1}{R^3}\, dz'dx\,dy\,, \tag{1.375a}$$

$$\vec{\mathbf{I}}_{z\to yz} = -t_z R_x n_x^k\,\hat{\mathbf{k}} \iint_{S_k} \int_{z'} \frac{1}{R^3}\, dz'dy\,dz\,, \tag{1.375b}$$

$$\vec{\mathbf{I}}_{z\to zx} = -t_z R_y n_y^k \,\hat{\mathbf{k}} \iint_{S_k} \int_{z'} \frac{1}{R^3}\, dz'\, dz\, dx\,. \tag{1.375c}$$

Integrating the first integral with respect to the observation rectangle coordinates (integrating first with respect to x or y depending on whether R_x or R_y is in the numerator) yields a 1-D integral:

$$\vec{\mathbf{I}}_{z\to xy} = -t_z n_z^k \int_{z'} [\ln(R_y + R)\,\hat{\mathbf{i}} + \ln(R_x + R)\,\hat{\mathbf{j}}]\, dz'\,. \tag{1.376}$$

The second and third integrals are reduced to double integrals as follows:

$$\vec{\mathbf{I}}_{z\to yz} = -t_z R_x n_x^k \,\hat{\mathbf{k}} \int_z \int_{z'} \frac{R_y}{(R^2 - R_y^2)R}\, dz'\, dz\,, \tag{1.377a}$$

$$\vec{\mathbf{I}}_{z\to zx} = -t_z R_y n_y^k \,\hat{\mathbf{k}} \int_z \int_{z'} \frac{R_x}{(R^2 - R_x^2)R}\, dz'\, dz\,. \tag{1.377b}$$

These integrals can be evaluated numerically with SciPy.

Lastly, we add the contributions from the two cable ends. The calculation can be simplified by converting these two triple integrals into finite sums. First, let us approximate the semicircular filament (of radius $r_0 = 5$ cm) as a polygonal path with $N_e \gg 1$ segments. The line integral is thus approximated as a sum by pointing $\mathbf{r}'$ to the centers of the polygon edges, and multiplying by $\delta r' = 2r_0 \sin(\pi/2N_e)$. For example, the front side centers are positioned at

$$\mathbf{r}'_j = \frac{\ell}{2}\hat{\mathbf{k}} + r_0 \cos\frac{\pi}{2N_e}\,\hat{\mathbf{n}}_j, \text{ with } \hat{\mathbf{n}}_j = -\cos\phi_j\,\hat{\mathbf{i}} + \sin\phi_j\,\hat{\mathbf{k}}\,, \tag{1.378}$$

where

$$\phi_j = (2j-1)\frac{\pi}{2N_e}\,, \tag{1.379}$$

for $j = 1, \ldots, N_e$. Second, let us approximate the double integral over the kth observation element as the integrand value with $\mathbf{r}$ pointing at the element center $\mathbf{r}_k$ times the element area A_k. The impact on the kth element from the semicircular loop of the front side is found by approximating Equation (1.374) as

$$\vec{\mathbf{I}}_{\text{front end}\to k} \approx \sum_{j=1}^{N_e} \left(\hat{\mathbf{t}}_j \times \frac{\mathbf{r}_k - \mathbf{r}'_j}{\|\mathbf{r}_k - \mathbf{r}'_j\|^3} \right) \times \hat{\mathbf{n}}^k\, \delta r'\, A_k\,, \tag{1.380}$$

where

$$\hat{\mathbf{t}}_j = \sin\phi_j\,\hat{\mathbf{i}} + \cos\phi_j\,\hat{\mathbf{k}}\,. \tag{1.381}$$

At this point, we have worked out all the necessary mathematical details for assembling the linear system of equations for the surface magnetization currents. The calculated outer and inner surface current vectors are displayed as a quiver plot in Figure 1.33. There is an arrow for each mesh element. Arrow midpoints are centered at the elements, and their lengths are proportional to $\|\vec{\mathbf{K}}_m\|$.

The filament current is magnified by the surface current on the surrounding slot surfaces, which is generally flowing in the same direction, although a small deflection appears to take place close to the edges. We calculate an average z-axis magnetization

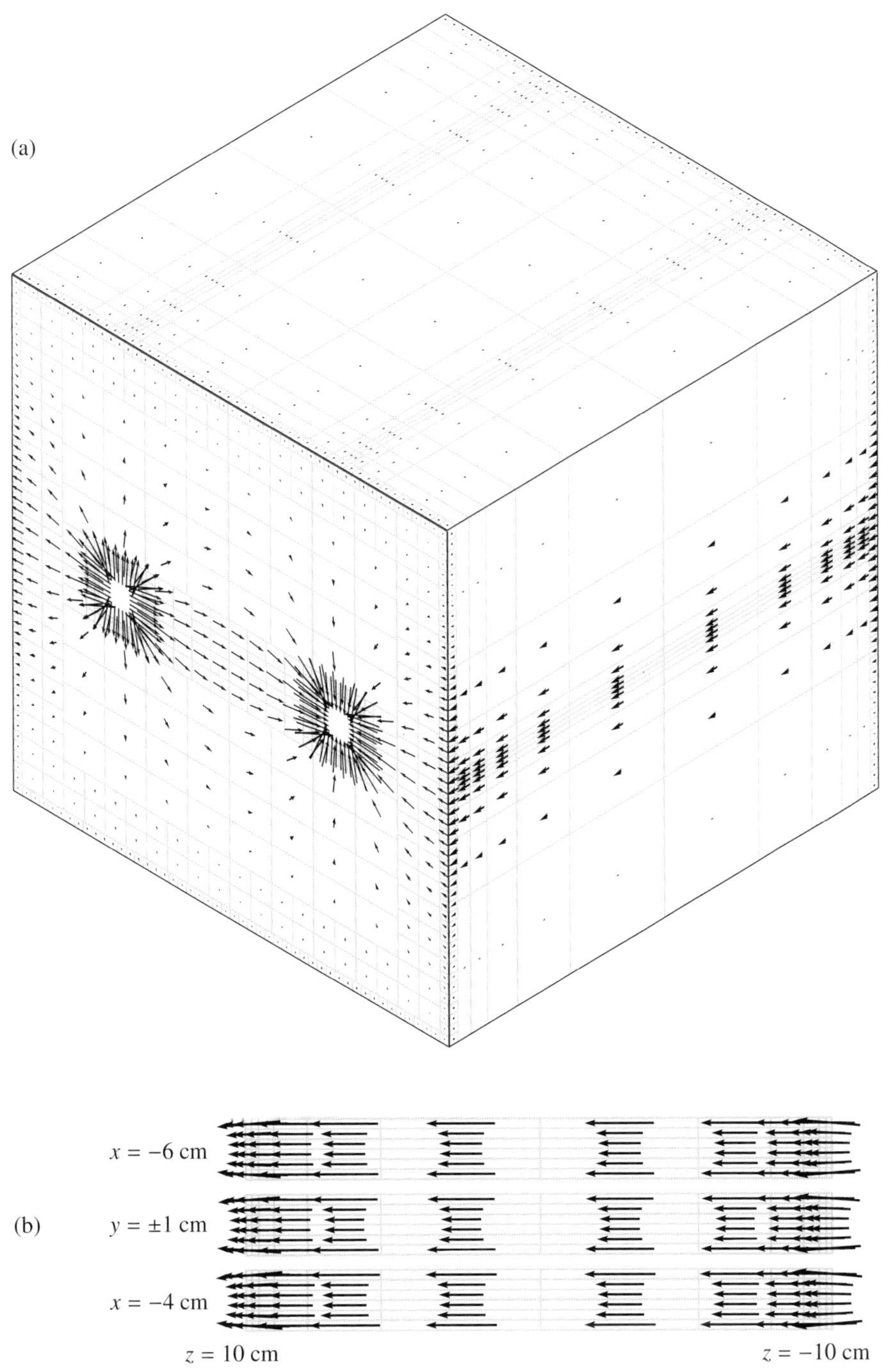

Figure 1.33 (a) Outer surface and (b) slot surface magnetization current density vectors. (Due to symmetry, only the slot centered at $x = -5$ cm is shown.)

current flowing on the slot surface equal to 893.5 A, which is many times larger than $i = 10$ A. It is interesting to note that the magnetization current is not uniform over the slot surface, tending to be stronger close to the slot edges (along the z-axis) and weaker at the middles. There is also a small reduction of the z-axis current magnitude as we move closer to the edges at $z = \pm\ell/2$. These effects are illustrated in Figure 1.34.

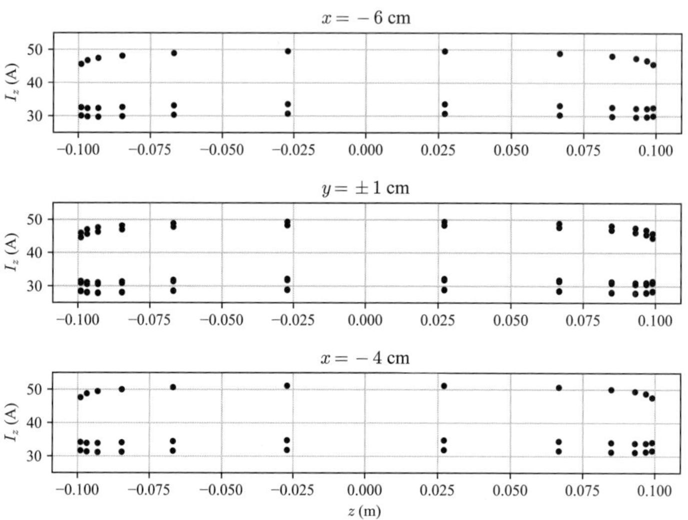

Figure 1.34 Slot surface currents. Each point corresponds to a mesh element centered at the corresponding z-value. Due to symmetry, certain elements have the same value.

This slot surface current is responsible for the high values of the B-field within the cube volume (due to the magnetization of matter). When slot surface currents reach the outer surface, a significant portion thereof travels through the surface towards the other slot, where it follows the return path of the filament inside the cube. A smaller portion of the slot surface current returns from the left and right sides, whereas an insignificant amount returns from the top and bottom. This distribution of surface currents can be visualized as forming additional current loops, whose combined effect is to contain the B-field within the cube.

After the system of equations is solved for the surface magnetization currents, we may calculate the B-field using Equation (1.345) as a finite sum over all current source elements. The calculations involve simpler single or double integrals. The B-field over a few representative cross-sections of the device is shown in Figure 1.35. The left column shows quiver plots of the xy-components, superimposed on filled contours of $|B_{xy}| = \|(B_x, B_y, 0)\|$. The plots on the right depict B_z. (Note that on the $z = 0$ plane, $B_z = 0$.) These plots were created by evaluating $\vec{\mathbf{B}}$ over a NumPy meshgrid of points at $x, y = [-11.5, -10.5, -9.5, \ldots, 11.5]$ cm. We observe that the field does not change noticeably over the axial length of the device, remaining similar to the solution at $z = 0$, although as we get closer to the edges a small but noticeable B_z-component is present.

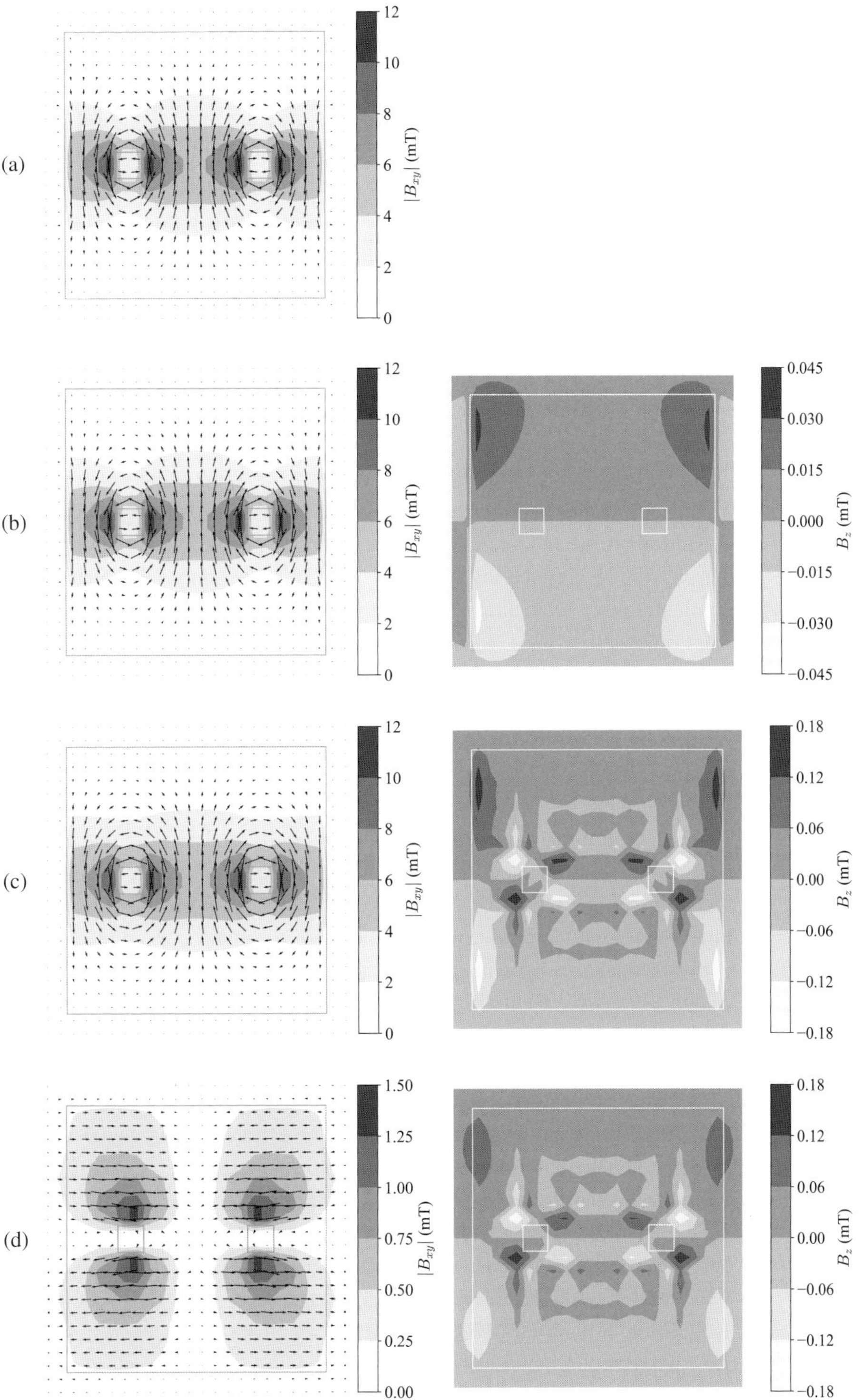

Figure 1.35 Plots of the B-field at (a) z = 0 cm, (b) z = 5 cm, (c) z = 9.5 cm, and (d) z = 10.5 cm.

1.8 Summary

In this chapter, we presented fundamental concepts of vectors and vector fields. Also, we introduced the Python programming language, and discussed several key results from electromagnetism. We have thus set the stage for the remainder of this text, which will rely on a fields-based approach for analyzing electric machines, and on Python for performing the numerical calculations on a computer.

In the next chapter, we will use these concepts to explain how Maxwell's partial differential equations can be solved by reformulating the problem as an optimization based on the principles of variational calculus.

1.9 Further Reading

The avid reader can choose among a variety of books for delving deeper into the details and proofs of the theorems of vector calculus. The treatments of this subject vary from the introductory and accessible (e.g., textbooks aimed at an audience of undergraduate physics or engineering students) [2–10] to the more advanced [11–15]. It is also common to find elementary treatments of vector calculus within textbooks dedicated to electromagnetism [16–27], which suggests the intimate relationship between these two scientific areas. For advanced treatments of electromagnetism at the graduate level, you may refer to classical textbooks such as [28–31].

There is no lack of resources for learning the Python language either. As one would expect, there are countless online resources and tutorials. There is also a variety of books for a broad audience from the beginner to the more seasoned programmer [32–37]. Lately, Python has been used extensively in modern data processing and machine learning applications [38–41]. Perhaps of more relevance to this text are books that focus on using Python for scientific computing, such as [42–46]. Basic Python is an interpreted language; however, it is possible to obtain compiled code with significant performance gains by using, e.g., Cython [47, 48].

Problems

1.1 Extend the proof of Example 1.2 to the general case of two 3-D Cartesian coordinate systems that are rotated with respect to each other.

1.2 Prove that the vector product is distributive.

1.3 Prove Equation (1.41).

1.4 Prove Equation (1.157).

1.5 Prove Equation (1.158).

1.6 Prove Equation (1.159).

1.7 Prove Equation (1.160).

1.8 Prove Equation (1.161).

1.9 Prove Equation (1.162).

1.10 Prove Equation (1.163).

1.11 Prove Equation (1.168).

1.12 Prove Equation (1.175).

1.13 Prove Equation (1.288).

1.14 *Refraction of magnetic flux lines.* Consider the interface between air and a steel material with relative permeability μ_r. The magnetic flux lines in the steel are approaching the boundary at an angle θ_1 with respect to the normal vector that points into the steel.

1. Derive a function for the angle of the magnetic flux lines as they exit into the air, $\theta_2 : \mathcal{D} \to \mathbb{R}$, where the domain $\mathcal{D} \subset \mathbb{R}^2$ represents pairs (θ_1, μ_r). The angle θ_2 is defined with respect to the normal vector that points into the air.
2. Plot this function over the domain $\mathcal{D} = [0, \pi/2) \times [100, 10{,}000]$. What conclusions can you draw?

1.15 Show that $\nabla \cdot \vec{\mathbf{A}} = 0$ for the MVP given by Equation (1.263). *Hint:* Use the fact that $\nabla \cdot \vec{\mathbf{J}} = 0$.

1.16 Write a program that replicates Figure 1.23.

1.17 Write a program that replicates the plots of Example 1.33.

1.18 Write a program that replicates the results of Example 1.34.

References

[1] R. Howard and S. Pekarek, "Two-dimensional Galerkin magnetostatic method of moments," *IEEE Trans. Magnetics*, vol. 53, no. 12, pp. 1–6, 2017.

[2] W. L. Briggs, L. Cochran, B. Gillett, and E. Schulz, *Calculus: Early Transcendentals*, 3rd ed. Boston: Pearson, 2018.

[3] J. J. Callahan, *Advanced Calculus: A Geometric View*. New York: Springer, 2010.

[4] S. J. Colley, *Vector Calculus*, 4th ed. Boston: Pearson, 2012.

[5] R. Larson and B. H. Edwards, *Multivariable Calculus*, 11th ed. Boston: Cengage Learning, 2016.

[6] J. E. Marsden and A. Tromba, *Vector Calculus*, 6th ed. New York: W. H. Freeman & Company, 2012.

[7] J. Rogawski, C. Adams, and R. Franzosa, *Calculus: Early Transcendentals*, 4th ed. New York: Macmillan Learning, 2019.

[8] H. M. Schey, *Div, Grad, Curl, and All That: An Informal Text on Vector Calculus*, 4th ed. New York: W. W. Norton & Company, 2004.

[9] J. Stewart, D. K. Clegg, and S. Watson, *Multivariable Calculus*, 9th ed. Boston: Cengage Learning, 2012.

[10] G. B. Thomas and R. L. Finney, *Calculus and Analytic Geometry*, 9th ed. Reading, MA: Addison-Wesley, 1996.

[11] S. Dineen, *Multivariate Calculus and Geometry*, 3rd ed. London: Springer, 2014.

[12] J. H. Hubbard and B. B. Hubbard, *Vector Calculus, Linear Algebra, and Differential Forms: A Unified Approach*, 5th ed. Ithaca, NY: Matrix Editions, 2015.

[13] L. H. Loomis and S. Sternberg, *Advanced Calculus*. Hackensack, NJ: World Scientific, 2014.

[14] J. R. Munkres, *Analysis on Manifolds*. Boca Raton, FL: CRC Press, 2018.
[15] M. Spivak, *Calculus on Manifolds: A Modern Approach to Classical Theorems of Advanced Calculus*. Boca Raton, FL: CRC Press, 2018.
[16] W. Hayt and J. Buck, *Engineering Electromagnetics*, 9th ed. New York: McGraw-Hill, 2019.
[17] D. J. Griffiths, *Introduction to Electrodynamics*, 4th ed. Cambridge: Cambridge University Press, 2017.
[18] O. D. Jefimenko, *Electricity and Magnetism*. New York: Appleton-Century-Crofts, 1966.
[19] W. K. H. Panofsky and M. Phillips, *Classical Electricity and Magnetism*, 2nd ed. Reading, MA: Addison-Wesley, 1962.
[20] D. T. Paris and F. K. Hurd, *Basic Electromagnetic Theory*. New York: McGraw-Hill, 1969.
[21] R. Plonsey and R. E. Collin, *Principles and Applications of Electromagnetic Fields*. New York: McGraw-Hill, 1961.
[22] E. M. Purcell and D. J. Morin, *Electricity and Magnetism*, 3rd ed. Cambridge: Cambridge University Press, 2013.
[23] M. N. O. Sadiku, *Elements of Electromagnetics*, 7th ed. Oxford: Oxford University Press, 2018.
[24] J. A. Stratton, *Electromagnetic Theory*. New York: McGraw-Hill, 1941.
[25] F. Ulaby and U. Ravaioli, *Fundamentals of Applied Electromagnetics*, 7th ed. Boston: Pearson, 2014.
[26] S. M. Wentworth, *Fundamentals of Electromagnetics with Engineering Applications*. Chichester, UK: Wiley, 2006.
[27] H. H. Woodson and J. R. Melcher, *Electromechanical Dynamics*. New York: John Wiley & Sons, 1968, vol. 1–3.
[28] J. D. Jackson, *Classical Electrodynamics*, 3rd ed. Hoboken, NJ: John Wiley & Sons, 1999.
[29] O. D. Kellogg, *Foundations of Potential Theory*. Berlin: Springer, 1929.
[30] L. D. Landau, L. P. Pitaevskii, and E. M. Lifshitz, *Electrodynamics of Continuous Media*, 2nd ed. Oxford: Butterworth-Heinemann, 1984.
[31] W. R. Smythe, *Static and Dynamic Electricity*, 2nd ed. New York: McGraw-Hill, 1950.
[32] D. Beazley and B. K. Jones, *Python Cookbook: Recipes for Mastering Python 3*, 3rd ed. Sebastopol, CA: O'Reilly, 2013.
[33] M. Lutz, *Learning Python*, 5th ed. Sebastopol, CA: O'Reilly, 2013.
[34] E. Matthes, *Python Crash Course: A Hands-On, Project-Based Introduction to Programming*, 2nd ed. San Francisco: No Starch Press, 2019.
[35] L. Ramalho, *Fluent Python: Clear, Concise, and Effective Programming*. Sebastopol, CA: O'Reilly, 2015.
[36] Z. A. Shaw, *Learn Python 3 the Hard Way: A Very Simple Introduction to the Terrifyingly Beautiful World of Computers and Code*. Boston: Addison-Wesley, 2017.
[37] A. Sweigart, *Automate the Boring Stuff with Python: Practical Programming for Total Beginners*, 2nd ed. San Francisco: No Starch Press, 2019.
[38] F. Chollet, *Deep Learning with Python*. Shelter Island, NY: Manning, 2017.
[39] A. Géron, *Hands-On Machine Learning with Scikit-Learn, Keras, and TensorFlow: Concepts, Tools, and Techniques to Build Intelligent Systems*, 2nd ed. Sebastopol, CA: O'Reilly, 2019.
[40] W. McKinney, *Python for Data Analysis: Data Wrangling with Pandas, NumPy, and IPython*, 2nd ed. Sebastopol, CA: O'Reilly, 2017.

[41] C. R. Severance, *Python for Everybody: Exploring Data Using Python 3*. CreateSpace Independent Publishing Platform, 2016.

[42] C. Führer, J. E. Solem, and O. Verdier, *Scientific Computing with Python 3*, 2nd ed. Birmingham: Packt, 2016.

[43] R. Johansson, *Numerical Python: Scientific Computing and Data Science Applications with Numpy, SciPy and Matplotlib*, 2nd ed. New York: Apress, 2018.

[44] H. P. Langtangen, *A Primer on Scientific Programming with Python*, 5th ed. Berlin: Springer, 2016.

[45] H. P. Langtangen, *Python Scripting for Computational Science*, 3rd ed. Berlin: Springer, 2008.

[46] H. K. Mehta, *Mastering Python Scientific Computing*. Birmingham: Packt, 2015.

[47] P. Herron, *Learning Cython Programming*, 2nd ed. Birmingham: Packt, 2016.

[48] K. Smith, *Cython: A Guide for Python Programmers*. Sebastopol, CA: O'Reilly, 2015.

2 Variational Form of Maxwell's Equations, Energy, and Force

Δός μοί ποῦ στῶ καὶ κινῶ τὴν γῆν

"Give me where to stand, and I can move the earth"

Archimedes[37]

This chapter begins with a concise discussion of variational calculus in §2.1. This mathematical theory is centered on the concept of a functional and, in particular, the case of a stationary functional. The discussion leads us to the derivation of the famous Euler–Lagrange equation. In §2.2, variational principles are applied to obtain a fundamental result for us, which states that Maxwell's equations for the electrostatic and magnetostatic problems can be derived by a variational calculus approach. This is important because it sets the stage for the finite element method (FEM), which will be introduced in the next chapter.

The intimate relationship of the functional to energy serves as a segue into the second half of this chapter, which is dedicated to the energy of the magnetic field. In §2.3, we turn our attention towards the various energy transfer pathways in electric machines. First, we take a close look at the electromagnetic field within the air gap of a rotating electric machine, and calculate the power transfer from the rotor to the stator using Poynting's theorem. Then, we study the energy balance within conductors, which are seen as entry points for exchanging energy between the magnetic field and the electrical system that is driving the current. This leads to the calculation of energy and coenergy using magnetic field variables. We also include a brief discussion of hysteresis and eddy current loss, which are important phenomena in every electric machine. We proceed with a circuit-based energy flow analysis, thereby connecting fields and circuit variables. Finally, this section explains how electromagnetic forces can be calculated from energy considerations in a conservative coupling field. The chapter concludes with §2.4, which is dedicated to fields-based approaches for calculating forces generated by the electromagnetic field. This forms the theoretical foundation of force/torque calculation from the finite element analysis (FEA) of an electric machine.

2.1 Calculus of Variations

The calculus of variations is a field in mathematics that deals with finding stationary points of functionals. We provide here a brief introduction to the subject, enough to help us proceed with our main goal of understanding the FEM.

A **functional** is a "function of a function," that is, it represents a mapping from a set of functions to a real number. Here, we will consider functionals in integral form, as in

$$I(w) = \int_a^b F(x, w, w')\, dx\,, \tag{2.1}$$

where $w = w(x)$, and the prime denotes differentiation with respect to x, i.e., $w' = dw(x)/dx$. The notation $I(w)$ means that the value of the functional depends on the function $w : [a, b] \to \mathbb{R}$. The function $F : \mathbb{R}^3 \to \mathbb{R}$ is determined from the problem at hand. It can have three arguments, namely, x, w, and w'; however, not all three need to be present. Since w and w' are functions of x, F can also be interpreted as a composite function of a single variable x, which is then integrated from a to b. We will assume that F and its derivatives are continuous functions.

The function w^* is a local maximizer of $I(w)$ if $I(w^*) \geq I(w)$ for all w in a neighborhood of w^*. Conversely, w^* is a local minimizer of $I(w)$ if $I(w^*) \leq I(w)$ for all w in a neighborhood of w^*.[a] In both cases, we say that the functional is **stationary** at w^*. The concept of a stationary functional is analogous to that of a stationary function. Here, we seek the function w^* that will make the first-order change in I with respect to the function vanish. We note that a stationary point may also correspond to a saddle point, so it does not necessarily have to be an extremum (i.e., a maximum or minimum).

Example 2.1 *Curve of shortest length.*

Consider the problem of finding a curve of shortest length connecting two points (x_1, y_1) and (x_2, y_2) on the x–y plane (with $x_1 < x_2$). Of course, we already know that the answer to this problem is a straight-line segment. We will illustrate how to formulate this elementary problem as the minimization of a functional.

Let $w(x)$ represent an arbitrary (smooth) curve that passes through the two points, so $w(x_1) = y_1$ and $w(x_2) = y_2$. The incremental arc length is $ds = \sqrt{dx^2 + dw^2}$. A functional representing the length of the curve can be expressed as

$$I(w) = \int ds = \int_{a=x_1}^{b=x_2} \sqrt{1 + w'(x)^2}\, dx\,. \tag{2.2}$$

Our goal is to identify the function $w^*(x)$ that minimizes this integral. In this example, the integrand F is a function of w' only.

---

[a] What is the neighborhood of a function? This concept may be defined based on the distance between two functions. The distance between functions is measured by an inner product in the function space, e.g.

$$\|w - w^*\| = \langle w - w^*, w - w^* \rangle^{1/2} = \sqrt{\int_a^b [w(x) - w^*(x)]^2\, dx}\,.$$

2.1.1 The Euler–Lagrange Equation for a One-Dimensional Problem

Example 2.1 describes an elementary problem of variational calculus. We will use this to explain the basic steps involved in the solution of a one-dimensional variational problem. This process, however, can be readily generalized to higher dimensions. At the end, we shall obtain the so-called Euler[38]–Lagrange[39] equation. We proceed as follows.

Step 1. Suppose the solution to the previous minimization problem (i.e., the function representing the straight-line segment) is the function $w^*(x)$. The main idea of variational calculus is to consider all functions that satisfy the boundary conditions of the problem at hand. These functions may be expressed as

$$w(x) = w^*(x) + \alpha\eta(x)\,, \tag{2.3}$$

where η is an arbitrary (smooth and differentiable) function with $\eta(a) = \eta(b) = 0$, and α is a scalar variable. The function η determines the shape of the deviation from w^*, whereas α is a scaling factor. The quantity $\alpha\eta$ is called a **variation** of w, and is denoted by δw.[b] Example variations of w^* are depicted in Figure 2.1.

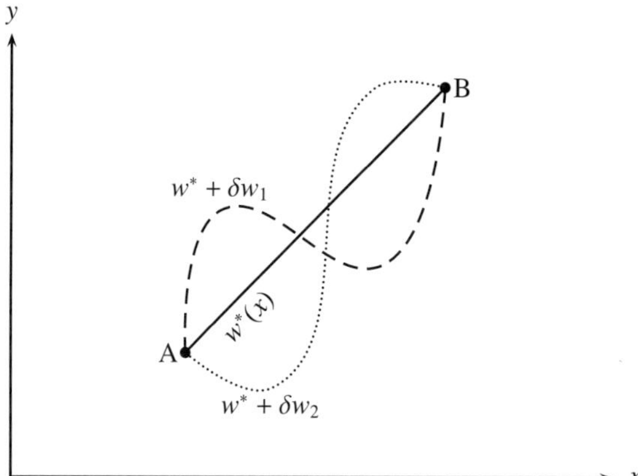

Figure 2.1 Two arbitrary variations of a function w^* representing the shortest path between two points A and B.

Step 2. Calculate the change in value of the functional from the optimum:

$$\delta I = I(w^* + \delta w) - I(w^*)\,. \tag{2.4}$$

We assume that the function F is analytic so that we can employ its first-order Taylor series expansion with respect to α. This change only affects w and its derivative, so

$$I(w^* + \delta w) = I(w^* + \alpha\eta) = \int_a^b F(x, w^* + \alpha\eta, w^{*\prime} + \alpha\eta')\,dx \tag{2.5a}$$

$$= \int_a^b \left[F + \alpha\eta\frac{\partial F}{\partial w} + \alpha\eta'\frac{\partial F}{\partial w'} + O(\alpha^2)\right] dx\,, \tag{2.5b}$$

[b] We use the lowercase Greek letter δ to denote small changes (variations) of quantities. This should not be confused with the d symbol, which denotes differentials.

where $O(\alpha^2)$ denotes higher-order terms. From our understanding of what a Taylor expansion is, it should be clear that F, $\partial F/\partial w$, and $\partial F/\partial w'$ are evaluated at the linearization points, $(x, w^*(x), w^{*\prime}(x))$, for $x \in [a, b]$. This yields

$$\delta I = \alpha \int_a^b \left[\eta \frac{\partial F}{\partial w} + \eta' \frac{\partial F}{\partial w'} \right] dx \; + \; O(\alpha^2) \, . \tag{2.6}$$

Step 3. Employ **integration by parts** on the second term of the integrand:

$$\delta I = \alpha \int_a^b \left[\eta \frac{\partial F}{\partial w} - \eta \frac{d}{dx} \left(\frac{\partial F}{\partial w'} \right) \right] dx \; + \; \alpha \eta \frac{\partial F}{\partial w'} \bigg|_a^b \; + \; O(\alpha^2) \, . \tag{2.7}$$

Since our choice of η satisfies $\eta(a) = \eta(b) = 0$, the term $\alpha\eta(\partial F/\partial w')\big|_a^b$ vanishes.

Step 4. Define a function $J : \mathbb{R} \to \mathbb{R}$ as

$$J(\alpha) = \delta I(\alpha; \eta) \, , \tag{2.8}$$

which yields the value of δI for a given choice of η function. Note that $J(0) = 0$. Now, because w^* minimizes the functional, the first derivative of J with respect to α should vanish at $\alpha = 0$. (In general, it suffices that w^* makes the functional stationary.) Therefore

$$\frac{dJ}{d\alpha} = \lim_{\alpha \to 0} \frac{\delta I}{\alpha} = \int_a^b \eta \left[\frac{\partial F}{\partial w} - \frac{d}{dx} \left(\frac{\partial F}{\partial w'} \right) \right] dx = 0 \, . \tag{2.9}$$

Step 5. Since our choice of η is arbitrary, we conclude that

$$\boxed{\frac{\partial F}{\partial w} - \frac{d}{dx} \left(\frac{\partial F}{\partial w'} \right) = 0 \, .} \tag{2.10}$$

This is called the **Euler–Lagrange equation**. It represents a second-order differential equation that the solution w^* should satisfy. Why second-order? Well, consider that $\partial F/\partial w' = G(x, w, w')$, another function of (x, w, w'). Then, from the second term of the Euler–Lagrange equation, the total derivative is

$$\frac{dG}{dx} = \frac{\partial G}{\partial x} + \frac{\partial G}{\partial w} \frac{dw}{dx} + \frac{\partial G}{\partial w'} \frac{dw'}{dx} \tag{2.11a}$$

$$= \frac{\partial G}{\partial x} + \frac{\partial G}{\partial w} w' + \frac{\partial G}{\partial w'} w'' \, , \tag{2.11b}$$

so there is a dependence on w'', thus leading to a second-order differential equation.

The argument that was used to obtain Equation (2.10) in Step 5 deserves some attention because it is central to the variational calculus line of thought. It is called the **fundamental lemma of calculus of variations**. We can prove it by contradiction. We assume that Equation (2.10) is *not* satisfied. This implies that there is (at least) one point $x_0 \in [a, b]$ for which

$$\frac{\partial F}{\partial w} - \frac{d}{dx} \left(\frac{\partial F}{\partial w'} \right) \bigg|_{x = x_0} = \kappa \neq 0 \, . \tag{2.12}$$

Also, due to the continuity of F and its derivatives, this will hold in a neighborhood of x_0. Since η is arbitrary, let us select an η that equals zero everywhere except in the vicinity of x_0. For instance, let us select an η that resembles a Dirac delta function centered around x_0, i.e., $\eta = \delta(x - x_0)$. Then, the integral in Equation (2.9) will evaluate to $\kappa \neq 0$, which is a contradiction.

Example 2.2 *Applying Euler–Lagrange to the shortest-curve problem.*
In the shortest-curve problem that was introduced in Example 2.1, the functional is based on the integral of

$$F(x, w, w') = \sqrt{1 + w'(x)^2}\,. \tag{2.13}$$

Here, we do not have an explicit dependence of F on w (or x for that matter, although this is not relevant for what follows). Applying the Euler–Lagrange equation (2.10) yields

$$\frac{d}{dx}\left(\frac{\partial F}{\partial w'}\right) = 0\,. \tag{2.14}$$

Naively, we could evaluate this to obtain a second-order differential equation, which would be somewhat involved. Instead, we note that this condition implies that $\partial F/\partial w'$ must be constant:

$$\frac{\partial F}{\partial w'} = \frac{w'(x)}{\sqrt{1 + w'(x)^2}} = C\,. \tag{2.15}$$

Squaring and rearranging terms yields

$$w'(x)^2 = \frac{C^2}{1 - C^2}\,. \tag{2.16}$$

Hence, the slope $w'(x)$ is constant, thus leading to a straight-line segment.

Example 2.3 *The brachistochrone problem.*
This famous problem was first posed by Johann Bernoulli[40] in 1696. Isaac Newton[41] was one of the few people who were able to provide a solution at that time. The name stems from the Greek word for "shortest time." It is one of the classical problems of variational calculus. In modern terms, we can formulate it as follows.

You are designing a roller coaster and are given x_a, x_b, $w_a = w(x_a)$, and $w_b = w(x_b)$. Here, x is the distance along the ground, and w is the height of the track. Assume $x_a < x_b$ and $w(x_a) > w(x_b)$. The mass of the car is m and the initial velocity at the top (x_a, w_a) is zero. Neglect friction and aerodynamic losses. The only force acting on the car is that of gravity. The objective is to maximize ride profits by *minimizing the time* needed to reach x_b (before hitting the brakes).

Solution
The situation is depicted in Figure 2.2. Let $v = ds/dt$ denote the velocity, which may be expressed as a function of the net distance s along the path, $v(s)$, with $v(0) = 0$. Here, ds is a differential length traveled over time dt. We use S to denote the total path length.

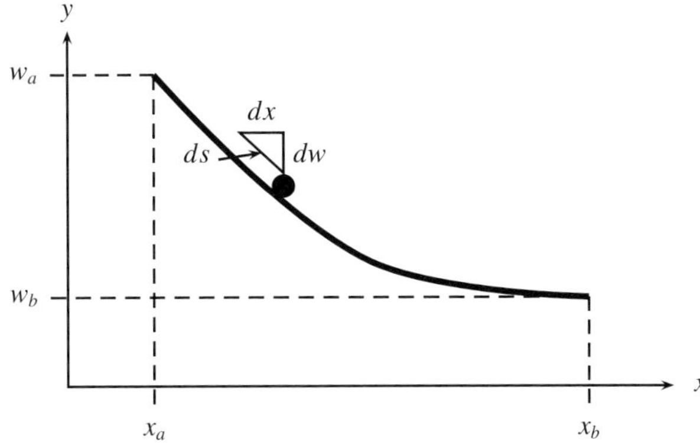

Figure 2.2 Illustration of the brachistochrone problem. Which curve leads to the shortest time of descent?

Since $(dt/ds) \cdot (ds/dt) = 1$, it follows that the total time can be found by[c]

$$I = \int dt = \int_0^S \frac{1}{v(s)}\,ds\,. \tag{2.17}$$

Similarly to Example 2.1, we express $ds = \sqrt{dx^2 + dw^2} = \sqrt{1 + w'(x)^2}\,dx$. Also, from the law of conservation of mechanical energy, we get

$$\frac{1}{2}mv^2 = mg(w_a - w)\,, \tag{2.18}$$

where g is the acceleration of gravity; hence

$$v = \sqrt{2g(w_a - w)}\,. \tag{2.19}$$

We have thus derived the functional

$$I(w) = \int_{x_a}^{x_b} \underbrace{\frac{1}{\sqrt{2g}}\sqrt{\frac{1 + w'^2}{w_a - w}}}_{F(x,w,w')}\,dx\,. \tag{2.20}$$

Here, the function F does not depend explicitly on x, but it does depend on both w and w'. Note that the solution will not depend on the mass. We may also omit $(2g)^{-1/2}$ from the functional since it is just a constant factor that will not affect its minimization.

The solution should satisfy the Euler–Lagrange equation (2.10), which implies that

$$\frac{\sqrt{1 + w'^2}}{2(w_a - w)^{\frac{3}{2}}} - \frac{d}{dx}\left[\frac{w'}{\sqrt{(1 + w'^2)(w_a - w)}}\right] = 0\,. \tag{2.21}$$

Without doing further algebra, it is obvious that this expression leads to a complicated second-order nonlinear differential equation that will be intractable analytically. However, since the functional does not depend directly on x, the **Beltrami identity**[42] applies,

[c] Note that this is an improper integral of the second kind because the velocity is zero at the initial point. However, it is expected to converge due to physical considerations.

which states that

$$F - w' \frac{\partial F}{\partial w'} = C \,, \tag{2.22}$$

with C a constant. Making use of this identity, we may obtain a simpler differential equation that is tractable analytically. The final solution has the shape of a *cycloid*. Details are left as an exercise for the reader (see Problem 2.1).

The fundamental idea of the preceding analysis was that finding a stationary point of the functional $I(w)$ *suffices* for satisfying the Euler–Lagrange equation. Of course, the opposite is also true since one can follow the steps backwards starting from the Euler–Lagrange equation. The forward direction is of particular interest to our purposes, as will be explained in §2.2. We proceed by extending the Euler–Lagrange theory to the 2-D case.

2.1.2 The Euler–Lagrange Equation for a Two-Dimensional Problem

In two dimensions (x, y), the function $w : \mathbb{R}^2 \to \mathbb{R}$ can be expressed as $w(x, y)$. The functional of interest is now a surface integral

$$I(w) = \iint_{\mathcal{D}} F(x, y, w, w_x, w_y)\, da \,, \tag{2.23}$$

where $\mathcal{D}$ is a 2-D domain of interest, and we define[d]

$$w_x = \frac{\partial w}{\partial x} \quad \text{and} \quad w_y = \frac{\partial w}{\partial y} \,. \tag{2.24}$$

The function $F : \mathbb{R}^5 \to \mathbb{R}$ has five arguments; however, not all need to be present. Since w, w_x, and w_y can be expressed as functions of x and y, F can also be interpreted as a composite function of x and y. To find a stationary point (extremum or saddle point) of this functional, we follow the same steps as previously.

Step 1. Suppose the solution to the problem is the function $w^*(x, y)$. We consider all functions that satisfy the boundary conditions of the problem at hand. These functions may be constructed by adding arbitrary variations to the solution:

$$w(x, y) = w^*(x, y) + \delta w(x, y) = w^*(x, y) + \alpha\eta(x, y) \,, \tag{2.25}$$

where η is an arbitrary (smooth and differentiable) function and α is a scalar variable. In general, we may divide the boundary of the domain, $\partial\mathcal{D}$, into two parts: first, there is a part $\partial\mathcal{D}_1$ where we shall specify the solution; and second, there is a part $\partial\mathcal{D}_2$ where the solution shall be unknown. This is depicted in Figure 2.3. (The boundary separation depends on the problem, so we leave it abstract for now.) We set $\eta(x, y) = 0$ for all $(x, y) \in \partial\mathcal{D}_1$, but let η be arbitrary on $\partial\mathcal{D}_2$. The function η determines the shape of the

[d] The notation w_x and w_y could also signify the x and y-components of a vector field w. In this context, where w is a scalar field, this does not make sense; so we are highjacking the notation for denoting the partial derivatives of w because we want to show them as explicit arguments of the function F.

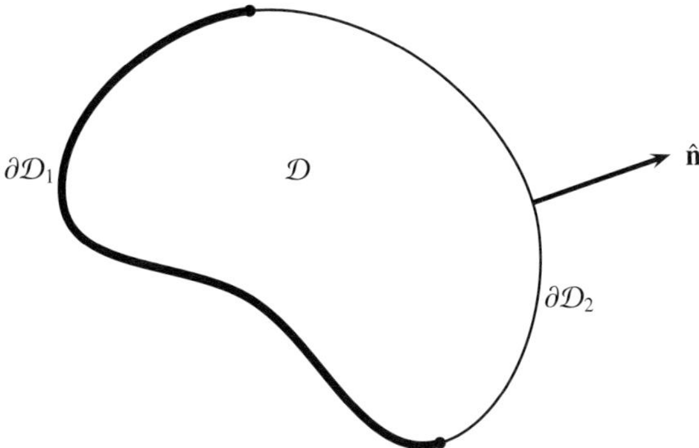

Figure 2.3 An abstract illustration of a 2-D domain with two types of boundary conditions. The boundary is separated into two parts by the dots. The solution is known on $\partial\mathcal{D}_1$ and unknown (free) on $\partial\mathcal{D}_2$. In practice, we may have to divide the boundary into more parts.

2-D deviation from w^*. It is clamped to zero on only that part of the boundary where the solution is known. You may visualize the shape of η as a sheet that can be shaken up and down while being held fixed at some part of its outer boundary.

Step 2. Calculate the change in the value of the functional from the optimum:

$$\delta I = I(w^* + \delta w) - I(w^*) . \tag{2.26}$$

Assuming that the function F is analytic, and taking a first-order Taylor series expansion with respect to α yields

$$\delta I = \alpha \iint_{\mathcal{D}} \left[\eta \frac{\partial F}{\partial w} + \frac{\partial \eta}{\partial x}\frac{\partial F}{\partial w_x} + \frac{\partial \eta}{\partial y}\frac{\partial F}{\partial w_y} \right] da \; + \; O(\alpha^2) . \tag{2.27}$$

Step 3. Employ **integration by parts** in two dimensions using Equation (1.232), which we repeat for convenience:

$$\iint_{\mathcal{D}} \left(\frac{\partial \eta}{\partial x} h + \frac{\partial \eta}{\partial y} g \right) da = - \iint_{\mathcal{D}} \left(\eta \frac{\partial h}{\partial x} + \eta \frac{\partial g}{\partial y} \right) da \; + \; \oint_{\partial\mathcal{D}} -\eta g \, dx + \eta h \, dy . \tag{2.28}$$

Here, this applies to the second and third terms of the integrand in Equation (2.27), with $h = \partial F / \partial w_x$ and $g = \partial F / \partial w_y$. Hence

$$\begin{aligned} \delta I = \alpha \iint_{\mathcal{D}} & \left[\eta \frac{\partial F}{\partial w} - \eta \frac{\partial}{\partial x}\left(\frac{\partial F}{\partial w_x} \right) - \eta \frac{\partial}{\partial y}\left(\frac{\partial F}{\partial w_y} \right) \right] da \\ & + \alpha \int_{\partial\mathcal{D}_2} \left(-\eta \frac{\partial F}{\partial w_y} dx + \eta \frac{\partial F}{\partial w_x} dy \right) \; + \; O(\alpha^2) . \end{aligned} \tag{2.29}$$

Since our choice of η satisfies $\eta = 0$ on $\partial\mathcal{D}_1$, that part of the line integral has vanished, so we are left with a line integral only on $\partial\mathcal{D}_2$.

Step 4. Define a function $J : \mathbb{R} \to \mathbb{R}$ as

$$J(\alpha) = \delta I(\alpha; \eta) , \tag{2.30}$$

which yields the value of δI for a given choice of η function. Note that $J(0) = 0$. Now, if the solution w^* yields a stationary point of the functional, it is necessary that

$$\frac{dJ}{d\alpha} = \lim_{\alpha\to 0}\frac{\delta I}{\alpha} = \iint_{\mathcal{D}} \eta\left[\frac{\partial F}{\partial w} - \frac{\partial}{\partial x}\left(\frac{\partial F}{\partial w_x}\right) - \frac{\partial}{\partial y}\left(\frac{\partial F}{\partial w_y}\right)\right] da + \int_{\partial\mathcal{D}_2} \eta\left(-\frac{\partial F}{\partial w_y}dx + \frac{\partial F}{\partial w_x}dy\right) = 0\,. \quad (2.31)$$

Step 5. Since our choice of η is arbitrary, the fundamental lemma of calculus of variations leads to the conclusion that the following two conditions must be satisfied by the solution w^*:

$$\frac{\partial F}{\partial w} - \frac{\partial}{\partial x}\left(\frac{\partial F}{\partial w_x}\right) - \frac{\partial}{\partial y}\left(\frac{\partial F}{\partial w_y}\right) = 0\,, \quad (2.32a)$$

$$\left(\frac{\partial F}{\partial w_x}, \frac{\partial F}{\partial w_y}\right)\cdot \hat{\mathbf{n}} = 0 \;\text{ on } \partial\mathcal{D}_2\,. \quad (2.32b)$$

Equation (2.32a) is the **Euler–Lagrange equation for the 2-D case**. Equation (2.32b) was obtained based on the interpretation of the line-integral term in Example 1.30. Here, $\hat{\mathbf{n}}$ is an outwards-pointing unit normal vector at the boundary, as shown in Figure 2.3. The significance of this result may not be immediately apparent, but it will become clear once we employ these conditions to solve the Poisson equation in §2.2.1.

2.1.3 Problems in the Time Domain

The previous examples involved problems where the independent variables are spatial coordinates (x, y). In this section, we illustrate applications of variational calculus in problems where the independent variable is time t. This formulation typically leads to the determination of the **equation of motion** of a system of particles. Such problems arise commonly in physics, forming the subject of Lagrangian mechanics. The following terminology applies: the functional is called the **action** S, and the function F is called the **Lagrangian** L.

We illustrate with an elementary example involving the motion of a single mass, whose position is denoted by $x(t)$. (Please note that, compared to the previous notation, x now plays the role of w, and t plays the role of x.) The **principle of stationary action** (also known as the principle of **least action**) requires that *the path of a particle is the one that makes the action stationary*, where (cf. Equation (2.1))

$$S(x) = \int_{t_1}^{t_2} L(t, x, \dot{x})\, dt\,. \quad (2.33)$$

The Lagrangian may have an explicit dependence on time, as well as on the position and velocity $\dot{x} = dx/dt$ of the particle. In mechanics, the Lagrangian is taken as the difference between the kinetic energy T and potential energy V:

$$L = T - V\,. \quad (2.34)$$

Consider the motion of a mass m that moves horizontally on a plane (without friction), and that is connected to the end of a spring of constant k. In this case, the Lagrangian is

$$L(x, \dot{x}) = \frac{1}{2}m\dot{x}^2 - \frac{1}{2}kx^2\,, \tag{2.35}$$

assuming that at $x = 0$ the spring is at rest. Note that there is no explicit dependence of L on time. A problem statement could be as follows: at time $t_1 = 0$, the mass is at rest at $x(0) = x_0 > 0$; then the mass is released, and we are interested in its equation of motion, that is, on the differential equation whose solution yields $x(t)$.

To solve this problem, we follow the steps of §2.1.1—only the notation is different. In Step 3, which involves the integration by parts, we observe that we also need to specify the boundary coundation $x(t_2)$. So, let us assume that $x(t_2)$ is known, so that $\eta(t_2) = 0$. We proceed to Step 4, and for stationarity we ask that $J'(\alpha) = 0$, which yields

$$\frac{dJ}{d\alpha} = \int_{t_1}^{t_2} \eta \left[\frac{\partial L}{\partial x} - \frac{d}{dt}\left(\frac{\partial L}{\partial \dot{x}}\right)\right] dt = 0\,. \tag{2.36}$$

Hence, the **Euler–Lagrange equation for the particle motion** becomes (cf. Equation (2.10))

$$\boxed{\frac{\partial L}{\partial x} - \frac{d}{dt}\left(\frac{\partial L}{\partial \dot{x}}\right) = 0\,.} \tag{2.37}$$

Using L from Equation (2.35), we obtain $\partial L/\partial x = -kx$, $\partial L/\partial \dot{x} = m\dot{x}$, and finally

$$-kx - m\ddot{x} = 0 \Rightarrow ma = -kx\,, \tag{2.38}$$

i.e., Newton's second law of motion (force = mass × acceleration) as it applies to this elementary example.

Although the previous example may seem trivial, this method is convenient for formulating the equations of motion of systems that involve many degrees of freedom (e.g., motion in 3-D space of a collection of particles). Also, it is powerful for modeling kinematic constraints that may be present (e.g., a motion that is restricted to lie on a given surface). We provide next an example that is more relevant to our topic.

Example 2.4 *The Lorentz force equation: a variational derivation.*

The force on a charged particle moving in an electromagnetic field is given by the Lorentz equation:

$$\vec{\mathbf{F}} = q(\vec{\mathbf{E}} + \vec{\mathbf{v}} \times \vec{\mathbf{B}})\,. \tag{2.39}$$

Here, q is the electric charge and $\mathbf{r}(t) = (x(t), y(t), z(t))$ is the position of the particle, leading to a particle velocity $\vec{\mathbf{v}} = \dot{\mathbf{r}} = (\dot{x}, \dot{y}, \dot{z})$. $\vec{\mathbf{E}}(\mathbf{r}(t), t)$ and $\vec{\mathbf{B}}(\mathbf{r}(t), t)$ are the electric and magnetic fields along the particle trajectory, respectively. These fields are measured by an observer in a stationary reference frame, and they are assumed to be imposed by exogenous electromagnetic sources. Hence, they are not affected by the tiny particle, also referred to as a **test charge**. The electromagnetic field is in turn related to the electric potential $\varphi(\mathbf{r}(t), t)$ and the magnetic vector potential $\vec{\mathbf{A}}(\mathbf{r}(t), t)$ by

$$\vec{\mathbf{E}} = -\nabla\varphi - \frac{\partial \vec{\mathbf{A}}}{\partial t}, \tag{2.40a}$$

$$\vec{\mathbf{B}} = \nabla \times \vec{\mathbf{A}}. \tag{2.40b}$$

Note that

$$\nabla \times \vec{\mathbf{E}} = -\nabla \times \frac{\partial \vec{\mathbf{A}}}{\partial t} = -\frac{\partial}{\partial t}(\nabla \times \vec{\mathbf{A}}) = -\frac{\partial}{\partial t}\vec{\mathbf{B}}, \tag{2.41}$$

because the curl of the gradient vanishes (see Equation (1.162)), leading to Maxwell's equation. The Lorentz force may be derived by a variational approach.

Proof The Lagrangian is

$$L(t, x, y, z, \dot{x}, \dot{y}, \dot{z}) = \underbrace{\tfrac{1}{2}m\,\dot{\mathbf{r}} \cdot \dot{\mathbf{r}}}_{T} - \underbrace{(q\varphi - q\vec{\mathbf{A}} \cdot \dot{\mathbf{r}})}_{V}, \tag{2.42}$$

where m is the particle mass. Note that the term $V = q\varphi - q\vec{\mathbf{A}} \cdot \dot{\mathbf{r}}$ is associated with the potential energy of the particle. Each of the three coordinates in this functional can be varied independently. Hence, we obtain three Euler–Lagrange equations:

$$\frac{\partial L}{\partial x} - \frac{d}{dt}\left(\frac{\partial L}{\partial \dot{x}}\right) = 0, \tag{2.43a}$$

$$\frac{\partial L}{\partial y} - \frac{d}{dt}\left(\frac{\partial L}{\partial \dot{y}}\right) = 0, \tag{2.43b}$$

$$\frac{\partial L}{\partial z} - \frac{d}{dt}\left(\frac{\partial L}{\partial \dot{z}}\right) = 0. \tag{2.43c}$$

Let us work with the first equation. Recall that $\vec{\mathbf{A}} = (A_x, A_y, A_z)$, where each of the three components is a function of (x, y, z, t). We have

$$\frac{\partial L}{\partial x} = -q\frac{\partial \varphi}{\partial x} + q\left(\frac{\partial A_x}{\partial x}\dot{x} + \frac{\partial A_y}{\partial x}\dot{y} + \frac{\partial A_z}{\partial x}\dot{z}\right) \tag{2.44}$$

and

$$\frac{\partial L}{\partial \dot{x}} = m\dot{x} + qA_x. \tag{2.45}$$

Next, we must evaluate $d/dt\,(\partial L/\partial \dot{x})$, where we must be careful in evaluating the total time derivative of $A_x(\mathbf{r}(t), t)$. This leads to

$$\begin{aligned} &-q\frac{\partial \varphi}{\partial x} + q\left(\frac{\partial A_x}{\partial x}\dot{x} + \frac{\partial A_y}{\partial x}\dot{y} + \frac{\partial A_z}{\partial x}\dot{z}\right) \\ &\qquad - m\ddot{x} - q\left(\frac{\partial A_x}{\partial x}\dot{x} + \frac{\partial A_x}{\partial y}\dot{y} + \frac{\partial A_x}{\partial z}\dot{z} + \frac{\partial A_x}{\partial t}\right) = 0. \end{aligned} \tag{2.46}$$

Rearranging terms, the x-component of force is

$$F_x = m\ddot{x} = -q\left(\frac{\partial \varphi}{\partial x} + \frac{\partial A_x}{\partial t}\right) + q\left[\left(\frac{\partial A_y}{\partial x} - \frac{\partial A_x}{\partial y}\right)\dot{y} + \left(\frac{\partial A_z}{\partial x} - \frac{\partial A_x}{\partial z}\right)\dot{z}\right]. \tag{2.47}$$

In view of Equations (2.40a) and (2.40b), we have

$$F_x = qE_x + q(B_z \dot{y} - B_y \dot{z}) \tag{2.48a}$$

$$= qE_x + q(\dot{\mathbf{r}} \times \vec{\mathbf{B}})_x . \tag{2.48b}$$

Using Equations (2.43b) and (2.43c), similar expressions can be obtained for the other two components of force, leading to Equation (2.39). □

2.2 Solving Maxwell's Equations with Variational Calculus

In the previous section, we introduced the basic concepts of the calculus of variations. We proceed by applying these to the problem at hand, that is, the solution of Maxwell's equations governing the electromagnetic field. The idea is to identify functionals whose stationary points are obtained by functions that satisfy Maxwell's equations. Of course, Maxwell's equations are given in the form of partial differential equations. But instead of solving these directly, the idea is to convert to an equivalent problem that will involve the minimization of a functional.

2.2.1 Functional for Poisson's Equation

Can a functional be identified so that its stationary point is obtained by a function that satisfies Poisson's equation, $\nabla^2 w = -h$? The answer is positive. Let us work in two dimensions for simplicity, keeping in mind that this result can be readily extended to three dimensions. This case is also relevant to us because we will analyze electric machines based on 2-D cross-sections.

With 20/20 hindsight, consider the functional

$$I(w) = \iint_{\mathcal{D}} \frac{1}{2} ||\nabla w||^2 dx\, dy - \iint_{\mathcal{D}} hw\, dx\, dy \tag{2.49a}$$

$$= \iint_{\mathcal{D}} \frac{1}{2}(w_x^2 + w_y^2)\, dx\, dy - \iint_{\mathcal{D}} hw\, dx\, dy . \tag{2.49b}$$

Using the notation of the previous section, where w_x and w_y denote partial derivatives of w with respect to the two axes:

$$F(x, y, w, w_x, w_y) = \frac{1}{2}(w_x^2 + w_y^2) - hw , \tag{2.50}$$

where $h = h(x, y)$ is a given source function. Now, let us assume that the functional is stationary, and apply the Euler–Lagrange equation (2.32a) and boundary condition (2.32b) to obtain

$$-h - \frac{\partial w_x}{\partial x} - \frac{\partial w_y}{\partial y} = 0 , \tag{2.51a}$$

$$(w_x, w_y) \cdot \hat{\mathbf{n}} = 0 \text{ on } \partial\mathcal{D}_2 . \tag{2.51b}$$

Indeed, the first equation implies that Poisson's equation is satisfied inside $\mathcal{D}$! The second equation tells us something about the solution on the part of the boundary where it is allowed to be free. It states that ∇w is perpendicular to the normal vector $\hat{\mathbf{n}}$, or equivalently that ∇w is tangential to the boundary. In other words, the solution only changes tangentially, but is constant in the normal direction. Alternatively, we may state this in terms of the directional derivative as a Neumann boundary condition:

$$(\nabla w)_n = \frac{\partial w}{\partial n} = 0 \text{ on } \partial\mathcal{D}_2 . \tag{2.51c}$$

Hence, you may imagine a thin layer close to the boundary (within the domain $\mathcal{D}$) where the gradient is tangential to the boundary. This further implies that the equipotentials of w (which are usually associated with flux lines) will approach the boundary $\partial\mathcal{D}_2$ at normal angles. This situation is depicted in Figure 2.4. Such boundary condition may arise along an axis of symmetry.

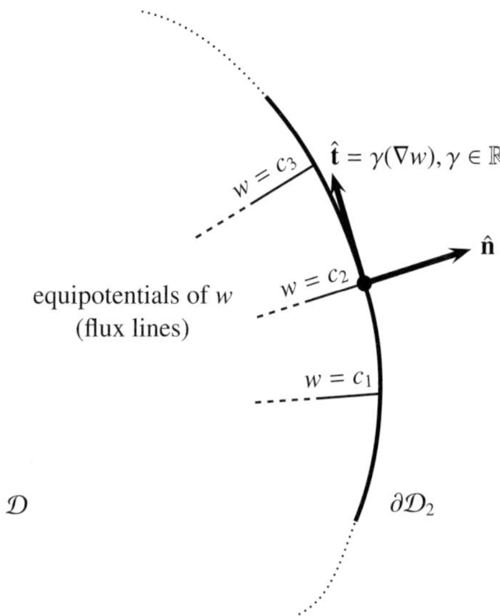

Figure 2.4 Poisson's equation coupled with a boundary condition where the solution is free to take arbitrary values leads to equipotential lines that are normal to the boundary. This illustration depicts three equipotential lines close to the boundary.

2.2.2 The Electrostatic Problem

In general, to solve an electrostatic problem, one needs to solve Poisson's equation for the electrostatic potential, $\nabla^2\varphi = -\rho/\epsilon$. However, this is a somewhat incomplete statement of the problem. A more precise description of the problem could be stated as follows.

Imagine a device that consists of an assembly of dielectric materials and conductors. For simplicity, assume we have just two different materials of constant permittivity ϵ_1

and ϵ_2, respectively, the former embedded within the latter. An arbitrary charge distribution within the materials, $\rho(\mathbf{r})$, is also given. For generality, one may also suppose that a given surface charge distribution $\omega(\mathbf{r})$ (i.e., a superthin layer of charge) is present on the interface between the dielectric materials. The conductors are positioned in various locations in space (possibly in both materials), and their potential $\varphi(\mathbf{r})$ is specified and constant, but possibly different within each conductor. (For instance, the potential on the conductors may be established by connecting to a battery or a direct current, d.c. generator.) Note that the electrostatic potential within electrical conductors is constant, and the internal electrical field is zero (since $\vec{\mathbf{E}} = -\nabla\varphi$). Hence, the conductor surfaces are associated with Dirichlet boundary conditions.

This abstract description of the "universe" is sketched in Figure 2.5. Regions in space where there is nonzero charge distribution (i.e., where $\rho(\mathbf{r}) \neq 0$ or $\omega(\mathbf{r}) \neq 0$) are lightly shaded. Conductors are represented by dark gray areas. Note that charge densities and potentials can be different. The region of material 2 is bounded externally by the surface S_2 and internally by the surface S_1, which also serves as the outer boundary of material 1. So, S_1 is the interface between the two materials. The materials also have internal boundaries with conductors, which will be collectively denoted as S_c. Since the potential is known within conductors, they may be considered holes in the domain.

According to Maxwell, the electrostatic field for this problem should satisfy the following set of equations:

$$\nabla \times \vec{\mathbf{E}} = \mathbf{0}, \tag{2.52a}$$

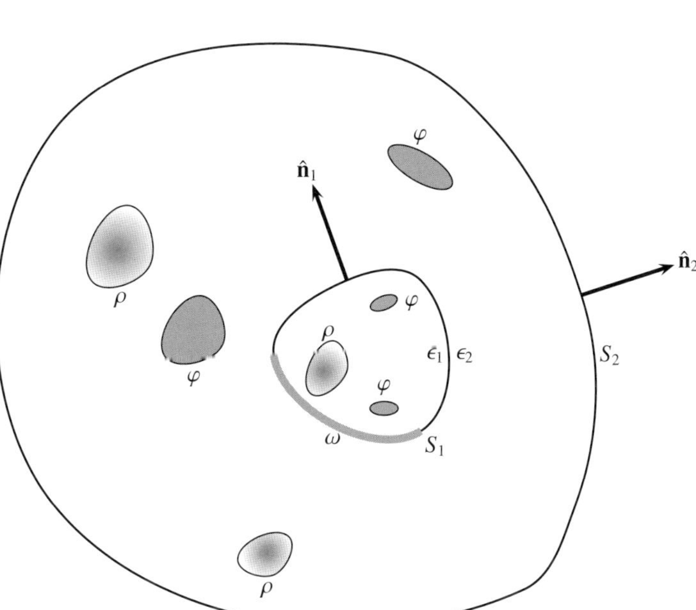

Figure 2.5 An abstract electrostatic problem consisting of two dielectric materials, charge distributions ρ and ω, and conductors with potential φ.

$$\nabla \cdot \vec{\mathbf{D}} = \rho\,, \tag{2.52b}$$

$$(\vec{\mathbf{E}}_2 - \vec{\mathbf{E}}_1) \times \hat{\mathbf{n}}_1 = \mathbf{0}\,, \tag{2.52c}$$

$$(\vec{\mathbf{D}}_2 - \vec{\mathbf{D}}_1) \cdot \hat{\mathbf{n}}_1 = \omega\,. \tag{2.52d}$$

The D-field is defined by

$$\vec{\mathbf{D}} = \epsilon \vec{\mathbf{E}}\,, \tag{2.53}$$

which is a constitutive law that describes the effect of polarization within the dielectric materials. In this context, the permittivity is a function of position, $\epsilon = \epsilon(\mathbf{r})$, in the sense that $\epsilon(\mathbf{r}) = \epsilon_1$, if $\mathbf{r}$ points to a location within material 1; and $\epsilon(\mathbf{r}) = \epsilon_2$, otherwise.

Equations (2.52a) and (2.52b) describe the variation of the fields within the materials. The fields are assumed to be differentiable functions of position. However, we expect the vector fields to experience a discontinuity across the interface, so the other two equations, (2.52c) and (2.52d), represent interface conditions for the tangential and normal components of the electric field, respectively, which should hold on S_1.[e]

Instead of solving for these vector fields directly, we will work with a scalar field φ, called the electrostatic potential. The potential must be differentiable within materials. This implies continuity of the potential within the two regions; however, we also want the potential to be *continuous across the material interface*. Furthermore, we need the potential to satisfy given boundary conditions on S_2 (either Dirichlet or Neumann) and S_c (Dirichlet). If we can find a potential that satisfies Poisson's equation:

$$\nabla^2 \varphi = -\frac{\rho}{\epsilon}\,, \tag{2.54}$$

within each material, then the E-field may be obtained by

$$\vec{\mathbf{E}} = -\nabla \varphi\,, \tag{2.55}$$

so that Equation (2.52a) is satisfied as an identity. It is also clear that Equation (2.52b) will be satisfied given the constitutive relation (2.53). Given our earlier discussion, we suspect that we should employ a function F similar to Equation (2.50) for the 2-D case.

The key result is highlighted in Box 2.1. The significance of this result lies in its simplicity and elegance. All of the pertinent Maxwell's equations (2.52a)–(2.52d) sprout from the stationarity of this functional, which notably only involves the scalar potential. The problem has been converted from a rather complicated formulation that involves solving a set of partial differential equations, to an optimization problem that is much more straightforward to set up.

Proof To prove the assertion of Box 2.1, we will follow the same steps that were outlined in §2.1.2. Essentially, we are re-deriving the Euler–Lagrange equation, while paying extra attention to the internal interface conditions. We begin by introducing a variation of the potential $\delta\varphi = \alpha\eta$ around the solution. The function η will be clamped

[e] It would have sufficed to require that Equations (2.52a) and (2.52b) hold throughout the entire space, since material interface conditions may be derived from these two fundamental equations. Our earlier discussion in Examples 1.19 and 1.23 for the magnetic field can be readily extended to electric fields. For clarity, we state the interface conditions explicitly as separate equations (2.52c) and (2.52d).

Box 2.1 Key finding: Solving the electrostatic problem by a variational approach

For an electrostatic problem, it suffices to make the following functional stationary:

$$I(\varphi) = \iiint_{\mathcal{D}} \frac{1}{2}\epsilon\|\nabla\varphi\|^2 dv - \iiint_{\mathcal{D}} \rho\varphi\, dv - \oiint_{S_1} \omega\varphi\, da\,. \tag{2.56}$$

Then Maxwell's equations will be satisfied, *including all interface conditions.*

to zero wherever the potential is known, that is, on (part of) the outer boundary and on the conductor surfaces. The key difference from our previous work is that we now have to be very careful in how we interpret the integration in Equation (2.56). Since the potential is not differentiable across the interface between materials, but only piecewise smooth, we must break the volume integral containing $\nabla\varphi$ into two parts, one for each material.

We calculate the variation of the functional δI, and require that

$$J'(\alpha) = \lim_{\alpha\to 0} \frac{\delta I}{\alpha} = 0\,. \tag{2.57}$$

This leads to

$$J'(\alpha) = \iiint_{\mathcal{D}} \left(\eta\frac{\partial F}{\partial\varphi} + \frac{\partial\eta}{\partial x}\frac{\partial F}{\partial\varphi_x} + \frac{\partial\eta}{\partial y}\frac{\partial F}{\partial\varphi_y} + \frac{\partial\eta}{\partial z}\frac{\partial F}{\partial\varphi_z}\right) dv + \oiint_{S_1} \eta\frac{\partial G}{\partial\varphi}\, da\,, \tag{2.58}$$

where

$$F(x, y, z, \varphi, \varphi_x, \varphi_y, \varphi_z) = \frac{1}{2}\epsilon(\varphi_x^2 + \varphi_y^2 + \varphi_z^2) - \rho\varphi \tag{2.59}$$

and

$$G(x, y, z, \varphi) = -\omega\varphi\,. \tag{2.60}$$

Note that, for example:

$$\frac{\partial F}{\partial\varphi_x} = \epsilon\varphi_x\,. \tag{2.61}$$

Hence, Equation (2.58) leads to

$$J'(\alpha) = \iiint_{\mathcal{D}} [-\eta\rho + \epsilon(\nabla\eta\cdot\nabla\varphi)]\, dv - \oiint_{S_1} \eta\omega\, da\,. \tag{2.62}$$

The next step in the process is integration by parts of the $\nabla\eta\cdot\nabla\varphi$ term. Since this is a 3-D problem, we must invoke Green's first identity (1.217):

$$\iiint_V (\nabla\varphi\cdot\nabla\psi + \varphi\nabla^2\psi)\, dv = \oiint_{\partial V} (\varphi\nabla\psi)\cdot d\mathbf{a}\,. \tag{2.63}$$

Here, the role of φ is played by η, and the role of ψ is played by $\epsilon\varphi$. Recall that Green's identity was obtained by application of Gauss' divergence theorem, which requires differentiable fields. Therefore, the volume integral over the entire domain $\mathcal{D} = \mathcal{D}_1 \cup \mathcal{D}_2$ must be evaluated as the sum of two separate integrals so that Green's identity can be employed. It is understood that the resulting surface integral terms are evaluated inside the corresponding boundaries, but arbitrarily close to the interface. With this in mind, we obtain

$$\begin{aligned} J'(\alpha) = \iiint_{\mathcal{D}_1} [-\eta(\rho + \epsilon_1 \nabla^2 \varphi)]\, dv + \oiint_{S_1} \eta \epsilon_1 (\nabla \varphi)_1 \cdot \hat{\mathbf{n}}_1 \, da \\ + \iiint_{\mathcal{D}_2} [-\eta(\rho + \epsilon_2 \nabla^2 \varphi)]\, dv - \oiint_{S_1} \eta \epsilon_2 (\nabla \varphi)_2 \cdot \hat{\mathbf{n}}_1 \, da \\ + \oiint_{S_2} \eta \epsilon_2 \nabla \varphi \cdot \hat{\mathbf{n}}_2 \, da - \iint_{S_1} \eta \omega \, da = 0 \, . \quad (2.64) \end{aligned}$$

Here, we have used the notation $(\nabla\varphi)_1$ and $(\nabla\varphi)_2$ to designate the gradients immediately inside and outside of the material interface S_1, respectively. We have also used the fact that the outwards-pointing vector from region 2 looking into region 1 is $-\hat{\mathbf{n}}_1$. Further, we note that the surface integral on S_2 practically only includes the part of that surface (if any) where we are leaving the potential free. Finally, we have not explicitly shown the surface integrals on conductor surfaces S_c; these vanish since $\eta = 0$ on those surfaces.

Because η is arbitrary, we conclude that stationarity of the functional implies the following:

- $\rho + \epsilon_1 \nabla^2 \varphi = 0$ inside $\mathcal{D}_1$;
- $\rho + \epsilon_2 \nabla^2 \varphi = 0$ inside $\mathcal{D}_2$;
- $[\epsilon_1(\nabla\varphi)_1 - \epsilon_2(\nabla\varphi)_2] \cdot \hat{\mathbf{n}}_1 - \omega = 0$ on S_1;
- $\nabla\varphi \cdot \hat{\mathbf{n}}_2 = 0$ on parts of S_2 where the potential is free.

In other words: (i) Poisson's equation is satisfied in each material; (ii) the interface condition for the normal component of the D-field is satisfied; and (iii) $\partial\varphi/\partial n = 0$ on the free outer boundary (implying a Neumann boundary condition).

An outstanding condition that remains to be verified is whether the solution will yield an E-field whose tangential component is continuous across the material interface. This did not arise from the stationarity of the integral. It follows, however, from the assumption of a continuous φ across the interface! The argument is quite simple. Continuity of φ leads to equality of the directional derivative along any tangential direction on both sides of the material interface. This in turn translates to equality of the tangential E-field components. □

2.2.3 The Magnetostatic Problem

The electrostatic problem that was presented in the previous section serves as an intermediate pedagogical step in our primary objective, which is the analysis of the magnetostatic problem governing the operation of electric machines. Strictly speaking, the problem cannot be classified as magnetostatic if fields are changing with respect to time, which is always the case in electric machines. However, if the rate of change is relatively slow, then common quasi-magnetostatic assumptions apply.

The analysis follows identical steps as in the electrostatic problem. The main difficulty is that the magnetic potential is a vector field $\vec{\mathbf{A}}$, which adds a layer of vector-calculus complexity to the algebra. The situation is depicted in Figure 2.6. We emphasize that this is only an abstraction, and that in practice we could have more than two materials (e.g., air, steel, copper) and many more current sources. The sources of the H-field are free (i.e., not bound) currents. These are represented by current densities $\vec{\mathbf{J}}$ (in A/m^2) and surface current densities $\vec{\mathbf{K}}$ (in A/m).

According to Maxwell, the magnetostatic field should satisfy the following set of equations:

$$\nabla \times \vec{\mathbf{H}} = \vec{\mathbf{J}}, \tag{2.65a}$$

$$\nabla \cdot \vec{\mathbf{B}} = 0, \tag{2.65b}$$

$$(\vec{\mathbf{H}}_1 - \vec{\mathbf{H}}_2) \times \hat{\mathbf{n}}_1 = \vec{\mathbf{K}}, \tag{2.65c}$$

$$(\vec{\mathbf{B}}_2 - \vec{\mathbf{B}}_1) \cdot \hat{\mathbf{n}}_1 = 0. \tag{2.65d}$$

Equations (2.65a) and (2.65b) describe the variation of the field within the materials, where the fields are assumed to be differentiable functions of position. The vector fields may experience a discontinuity across the interface; Equations (2.65c) and (2.65d) represent the interface conditions that should hold on S_1. It would have sufficed to require that Equations (2.65a) and (2.65b) hold throughout the entire space since material interface conditions may be derived from these two fundamental equations, based on our

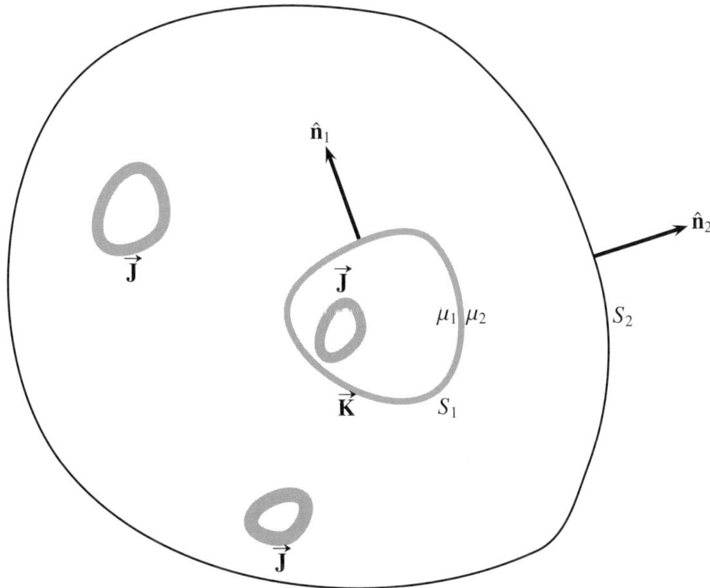

Figure 2.6 An abstract magnetostatic problem consisting of two materials and current distributions $\vec{\mathbf{J}}$ and $\vec{\mathbf{K}}$ (illustrated here as current loops).

earlier discussion in Examples 1.19 and 1.23. For clarity, we state the interface conditions explicitly as separate equations, (2.65c) and (2.65d).

The relationship between B and H in matter was discussed in §1.7.4. Magnetization in ferromagnetic materials is captured by a constitutive law such as

$$\vec{\mathbf{B}} = \mu \vec{\mathbf{H}} . \tag{2.66}$$

In permanent magnets:

$$\vec{\mathbf{B}} = \mu \vec{\mathbf{H}} + \mu_0 \vec{\mathbf{M}}_0 . \tag{2.67}$$

In this context, the permeability μ and the magnetization $\vec{\mathbf{M}}_0$ are functions of position. In the case of nonlinear materials, the permeability is also a function of the magnetic field. For instance, we could define $\mu = \mu(\mathbf{r}, \vec{\mathbf{B}})$, in the sense that

$$\mu(\mathbf{r}, \vec{\mathbf{B}}) = \begin{cases} \mu_1(\vec{\mathbf{B}}) & \text{if } \mathbf{r} \text{ points to a location within material 1,} \\ \mu_2(\vec{\mathbf{B}}) & \text{otherwise.} \end{cases} \tag{2.68}$$

Note that the expression (2.67) is more general since it can accommodate either a ferromagnetic material (with $\vec{\mathbf{M}}_0 = \mathbf{0}$, and $\mu = \mu_0(1 + \chi_m)$ that could be a function of $\vec{\mathbf{B}}$) or a permanent magnet with magnetization $\vec{\mathbf{M}}_0$ (and a typically constant $\mu \approx \mu_0$). By accepting the validity of such a law, it is implicitly assumed that magnetic hysteresis is not present. Otherwise, we would not be able to write a single-valued function relating $\vec{\mathbf{B}}$ and $\vec{\mathbf{H}}$ (for any given material that exists at point $\mathbf{r}$).

Instead of solving for the $\vec{\mathbf{B}}$ or $\vec{\mathbf{H}}$ vector fields directly, we will work with an intermediate vector field $\vec{\mathbf{A}}$, namely, the magnetic vector potential (MVP), which was introduced in Chapter 1. We require that $\vec{\mathbf{A}}$ is differentiable within the materials. Also, we want it to be *continuous across material interfaces*, and to satisfy given boundary conditions on S_2 (either Dirichlet or Neumann). For a given MVP field, we can use

$$\vec{\mathbf{B}} = \nabla \times \vec{\mathbf{A}} \tag{2.69}$$

to obtain the B-field so that Equation (2.65b) is satisfied as an identity. We can then use the constitutive law (2.67) to define a function

$$\vec{\mathbf{H}}(\mathbf{r}, \vec{\mathbf{B}}) = \mu(\mathbf{r}, \vec{\mathbf{B}})^{-1} \cdot (\vec{\mathbf{B}} - \mu_0 \vec{\mathbf{M}}_0(\mathbf{r})) . \tag{2.70}$$

The potential should be such that (2.65a) and the interface conditions are satisfied.

The key result is highlighted in Box 2.2. The functional of Equation (2.71) is slightly different from the ones we have encountered thus far because its argument is a vector (not scalar) field. The integrand now is a function of (x, y, z), the three components of $\vec{\mathbf{A}} = (A_x, A_y, A_z)$, and the partial derivatives of $\vec{\mathbf{A}}$ that are needed for $\vec{\mathbf{B}}$, which is the upper limit of the integral for m of Equation (2.72). The function m^f is a line integral of the vector field $\vec{\mathbf{H}}$ in the space of (B_x, B_y, B_z), evaluated using the constitutive law $\vec{\mathbf{H}}(\mathbf{r}, \vec{\mathbf{B}})$. The lowercase $\vec{\mathbf{b}}$ represents a dummy variable of integration. The explicit dependence on position is used to remind us of the presence of different materials in the device. An initial value m_0 and an arbitrary lower limit $\vec{\mathbf{B}}_0 = \vec{\mathbf{B}}_0(\mathbf{r})$ have been

[f] The reader may recall from physics that m carries a meaning of magnetic field energy density. We shall prove this fact later in this chapter.

Box 2.2 Key finding: Solving the magnetostatic problem by a variational approach

For a magnetostatic problem, it suffices to make the following functional stationary:

$$I(\vec{\mathbf{A}}) = \iiint_{\mathcal{D}} m(\mathbf{r}, \vec{\mathbf{B}}(\vec{\mathbf{A}}))\, dv - \iiint_{\mathcal{D}} \vec{\mathbf{J}} \cdot \vec{\mathbf{A}}\, dv - \oiint_{S_1} \vec{\mathbf{K}} \cdot \vec{\mathbf{A}}\, da\,, \tag{2.71}$$

where

$$m(\mathbf{r}, \vec{\mathbf{B}}(\vec{\mathbf{A}})) = m_0(\mathbf{r}) + \int_{\vec{\mathbf{B}}_0(\mathbf{r})}^{\vec{\mathbf{B}}(\vec{\mathbf{A}})} \vec{\mathbf{H}}(\mathbf{r}, \vec{\mathbf{b}}) \cdot d\vec{\mathbf{b}}\,. \tag{2.72}$$

Then Maxwell's equations will be satisfied, *including all interface conditions.*

introduced to help us model devices with permanent magnets. In general, we shall define m_0 and $\vec{\mathbf{B}}_0$ as corresponding to magnetic field $\vec{\mathbf{H}} = \mathbf{0}$. The $m(\mathbf{r}, \vec{\mathbf{B}}(\vec{\mathbf{A}}))$ integral returns a scalar value for each point in space since $\vec{\mathbf{A}} = \vec{\mathbf{A}}(\mathbf{r})$, which is then integrated over the entire domain in Equation (2.71). It should be emphasized that all of the pertinent Maxwell's equations (2.65a)–(2.65d) will stem from the stationarity of the functional (2.71), which only involves the vector potential. We have thus converted the problem from a set of partial differential equations to an optimization.

Proof To prove the assertion of Box 2.2, we will follow the same steps that were outlined in §2.1.2, and which were also followed in the analysis of the electrostatic problem in §2.2.2. We will not attempt to apply directly the Euler–Lagrange equation; however, we will follow an identical variational process.

We introduce a variation of the potential, $\delta\vec{\mathbf{A}} = \alpha\vec{\eta}$, around the solution $\vec{\mathbf{A}}^*$. The vector field $\vec{\eta}$ will be clamped to zero wherever the potential is known, that is, on (part of) the outer boundary. Since the potential is not differentiable at the material interface, we must break the volume integral term into two parts, one for each material; this is necessary to obtain a well-defined $\vec{\mathbf{H}}(\mathbf{r}, \vec{\mathbf{B}})$ with $\vec{\mathbf{B}} = \nabla \times \vec{\mathbf{A}}$. We denote $\vec{\mathbf{B}}^* = \vec{\mathbf{B}}(\vec{\mathbf{A}}^*) = \nabla \times \vec{\mathbf{A}}^*$, and $\delta\vec{\mathbf{B}} = \nabla \times (\delta\vec{\mathbf{A}}) = \alpha\nabla \times \vec{\eta}$. For simplicity, we denote $\vec{\mathbf{H}}(\mathbf{r}, \vec{\mathbf{B}}) = \vec{\mathbf{H}}(\vec{\mathbf{B}})$, so the explicit dependence on $\mathbf{r}$ is dropped, as with all other fields.

The variation of the functional is

$$\delta I(\alpha; \vec{\eta}) = I(\vec{\mathbf{A}}^* + \delta\vec{\mathbf{A}}) - I(\vec{\mathbf{A}}^*) \tag{2.73a}$$

$$= \iiint_{\mathcal{D}} \int_{\vec{\mathbf{B}}^*}^{\vec{\mathbf{B}}^*+\delta\vec{\mathbf{B}}} \vec{\mathbf{H}}(\vec{\mathbf{b}}) \cdot d\vec{\mathbf{b}}\, dv - \iiint_{\mathcal{D}} \vec{\mathbf{J}} \cdot \delta\vec{\mathbf{A}}\, dv - \oiint_{S_1} \vec{\mathbf{K}} \cdot \delta\vec{\mathbf{A}}\, da \tag{2.73b}$$

$$\approx \iiint_{\mathcal{D}} \vec{\mathbf{H}}(\vec{\mathbf{B}}^*) \cdot \delta\vec{\mathbf{B}}\, dv - \iiint_{\mathcal{D}} \vec{\mathbf{J}} \cdot \delta\vec{\mathbf{A}}\, dv - \oiint_{S_1} \vec{\mathbf{K}} \cdot \delta\vec{\mathbf{A}}\, da \tag{2.73c}$$

$$= \alpha\left\{\iiint_{\mathcal{D}} \vec{\mathbf{H}}(\vec{\mathbf{B}}^*) \cdot (\nabla \times \vec{\eta})\, dv - \iiint_{\mathcal{D}} \vec{\mathbf{J}} \cdot \vec{\eta}\, dv - \oiint_{S_1} \vec{\mathbf{K}} \cdot \vec{\eta}\, da\right\}, \tag{2.73d}$$

where the approximation in the third line is due to the first integral. The approximation is valid for small α, which is justified because we require that $\lim_{\alpha\to 0} \delta I/\alpha = 0$. Hence, the term in the curly brackets must vanish:

$$\iiint_{\mathcal{D}} \vec{\mathbf{H}}(\vec{\mathbf{B}}^*) \cdot (\nabla \times \vec{\eta})\, dv - \iiint_{\mathcal{D}} \vec{\mathbf{J}} \cdot \vec{\eta}\, dv - \oiint_{S_1} \vec{\mathbf{K}} \cdot \vec{\eta}\, da = 0\,. \tag{2.74}$$

Now we apply the vector calculus identity (1.159) to the first term:

$$\vec{\mathbf{H}} \cdot (\nabla \times \vec{\eta}) = \nabla \cdot (\vec{\eta} \times \vec{\mathbf{H}}) + \vec{\eta} \cdot (\nabla \times \vec{\mathbf{H}})\,, \tag{2.75}$$

where we use $\vec{\mathbf{H}} = \vec{\mathbf{H}}(\vec{\mathbf{B}}^*)$ to simplify notation. This term appears within an integral over the entire domain $\mathcal{D}$, which must be evaluated separately in each material. Furthermore, we can apply Gauss' theorem to the divergence component, thus leading to the appearance of $\vec{\eta} \times \vec{\mathbf{H}}$ surface integral terms. Hence, Equation (2.74) becomes

$$\begin{aligned} &\iiint_{\mathcal{D}_1} (\nabla \times \vec{\mathbf{H}} - \vec{\mathbf{J}}) \cdot \vec{\eta}\, dv + \iiint_{\mathcal{D}_2} (\nabla \times \vec{\mathbf{H}} - \vec{\mathbf{J}}) \cdot \vec{\eta}\, dv \\ &\quad + \oiint_{S_1} [(\vec{\eta} \times \vec{\mathbf{H}}_1) - (\vec{\eta} \times \vec{\mathbf{H}}_2)] \cdot \hat{\mathbf{n}}_1\, da - \oiint_{S_1} \vec{\mathbf{K}} \cdot \vec{\eta}\, da \\ &\quad + \oiint_{S_2} (\vec{\eta} \times \vec{\mathbf{H}}) \cdot \hat{\mathbf{n}}_2\, da = 0\,. \end{aligned} \tag{2.76}$$

Here, we have used subscripts 1 and 2 to signify that the H-fields are evaluated immediately inside the corresponding material at the boundary S_1. We have also used the fact that the outwards-pointing vector from region 2 looking into region 1 is $-\hat{\mathbf{n}}_1$. Further, we note that the surface integral on S_2 practically only includes the part of that surface (if any) where we are leaving the potential free.

To proceed, we invoke the scalar triple-product identity (1.41) to manipulate the surface integral terms as follows:

$$(\vec{\eta} \times \vec{\mathbf{H}}) \cdot \hat{\mathbf{n}} = (\vec{\mathbf{H}} \times \hat{\mathbf{n}}) \cdot \vec{\eta}\,. \tag{2.77}$$

Therefore, Equation (2.76) becomes

$$\begin{aligned} &\iiint_{\mathcal{D}_1} (\nabla \times \vec{\mathbf{H}} - \vec{\mathbf{J}}) \cdot \vec{\eta}\, dv + \iiint_{\mathcal{D}_2} (\nabla \times \vec{\mathbf{H}} - \vec{\mathbf{J}}) \cdot \vec{\eta}\, dv \\ &\quad + \oiint_{S_1} [(\vec{\mathbf{H}}_1 - \vec{\mathbf{H}}_2) \times \hat{\mathbf{n}}_1 - \vec{\mathbf{K}}] \cdot \vec{\eta}\, da + \oiint_{S_2} (\vec{\mathbf{H}} \times \hat{\mathbf{n}}_2) \cdot \vec{\eta}\, da = 0\,. \end{aligned} \tag{2.78}$$

Because $\vec{\eta}$ is arbitrary, we conclude that stationarity of the functional implies the following:

- $\nabla \times \vec{\mathbf{H}} = \vec{\mathbf{J}}$ inside $\mathcal{D}_1$;
- $\nabla \times \vec{\mathbf{H}} = \vec{\mathbf{J}}$ inside $\mathcal{D}_2$;
- $(\vec{\mathbf{H}}_1 - \vec{\mathbf{H}}_2) \times \hat{\mathbf{n}}_1 = \vec{\mathbf{K}}$ on S_1;
- $\vec{\mathbf{H}} \times \hat{\mathbf{n}}_2 = \mathbf{0}$ on parts of S_2 where the potential is free.

In other words: (i) Ampère's law is satisfied in each material; (ii) the interface condition for the tangential components of the H-field is satisfied on S_1; and (iii) the magnetic field is normal to the free outer boundary (similar to what is depicted in Figure 2.4 for Poisson's equation).

An outstanding condition that remains to be verified is whether the solution will yield a B-field whose normal component is continuous across the material interface. This did not arise from the stationarity of the integral. It follows, however, from the assumption of a continuous $\vec{\mathbf{A}}$ across the interface, similar to the electrostatic case! Again, the argument is quite simple. We need to show that the normal component of $\vec{\mathbf{B}} = \nabla \times \vec{\mathbf{A}}$ is equal on both sides of the material interface. The normal component of the curl is the circulation of the potential along a path that is tangent to the interface. Since $\vec{\mathbf{A}}$ is continuous, the same value will be obtained on either side of the interface. □

Example 2.5 *Isotropic nonlinear ferromagnetic materials.*
To model **isotropic** ferromagnetic materials with **nonlinear magnetization characteristics** (without hysteresis, i.e., **anhysteretic**), we express the B–H constitutive law (2.66) in the form

$$\vec{\mathbf{H}}(\vec{\mathbf{B}}) = \nu(B^2) \cdot \vec{\mathbf{B}}\,. \tag{2.79}$$

Here, the reciprocal of the permeability $\nu = 1/\mu$, called the **reluctivity**, is a scalar function of $B^2 = \|\vec{\mathbf{B}}\|^2$. Isotropy is a Greek word meaning that the behavior of the material is the same in any direction in space; hence, $\vec{\mathbf{H}}$ is parallel to $\vec{\mathbf{B}}$, and the permeability is only a function of the magnitude of the B-field. This general form of the constitutive law also captures the simpler case of magnetically linear materials (e.g., copper, air), where ν is a constant.

A typical anhysteretic B–H characteristic is sketched in Figure 2.7. More precisely, this curve represents the macroscopic behavior of a piece of ferromagnetic material, when subject to an external magnetic field. The material sample needs to be orders of magnitude larger than the individual magnetic Weiss[43] domains that are present in its microstructure, which typically are several microns wide.

As a first approximation, the B–H curve may be thought to consist of two linear segments at low and high magnetization, respectively, that are joined by a bent segment. Note that the shape is somewhat exaggerated (it is stretched out for illustration purposes) since a typical ferromagnetic material will have much steeper slope at the origin. Typical non-oriented electrical steel laminations exhibit an initial relative permeability on the order of $\mu_r = 1000$–5000. (At low magnetization, the B–H curve has a convex shape, which is attributed to Rayleigh's[44] law; this effect is not captured in the plot either.) At high values of H, the material becomes **saturated**, in the sense that it gradually reaches peak magnetization, and the B–H characteristic slope asymptotically drops to $dB/dH \to \mu_0$.

The shaded area represents the magnetic field energy density, $m(B) = \int_0^B H\,db$ (J/m^3), which we introduced as Equation (2.72) in the magnetostatic functional. When **r** points

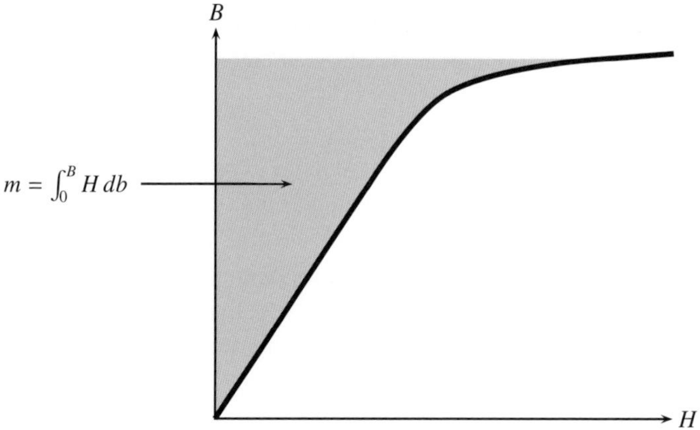

Figure 2.7 A typical anhysteretic B–H characteristic of a ferromagnetic material.

to regions in space occupied by the ferromagnetic material in question, Equation (2.72) becomes

$$m(\vec{\mathbf{B}}) = \int_{\mathbf{0}}^{\vec{\mathbf{B}}} \nu(b^2)\vec{\mathbf{b}} \cdot d\vec{\mathbf{b}} \tag{2.80a}$$

$$= \int_{\mathbf{0}}^{\vec{\mathbf{B}}} \nu(b^2)(b_x\,db_x + b_y\,db_y + b_z\,db_z) \tag{2.80b}$$

$$= \frac{1}{2}\int_0^{B^2} \nu(b^2)\,d(b^2)\,. \tag{2.80c}$$

Therefore, the line integral in the space of (B_x, B_y, B_z) has been converted into a single-variable integral with respect to B^2. This observation permits us to change the integration limits from vectors to scalars in the last line. It implies that the magnetization history of the material, in terms of the vector field variation over time, does not matter. The integral only depends on the final magnetization level, whereas the orientation of the B-field is irrelevant.

Example 2.6 *Permanent magnets.*
Permanent magnets (PMs) are modeled with the constitutive law (2.67), which can be expressed equivalently as

$$\vec{\mathbf{H}}(\vec{\mathbf{B}}) = \nu \cdot (\vec{\mathbf{B}} - \vec{\mathbf{B}}_r)\,, \tag{2.81}$$

where both $\vec{\mathbf{B}}_r = \mu_0\vec{\mathbf{M}}_0$ and the reluctivity ν are constant (i.e., independent of the magnetic field). Depending on the magnetization process, the permanent magnetization $\vec{\mathbf{M}}_0$ may be uniform inside the PM, or it may have a dependence on the position. Note that this vector equation allows $\vec{\mathbf{H}}$ and $\vec{\mathbf{B}}$ to be in different directions. The magnitude $B_r = \|\vec{\mathbf{B}}_r\|$ is called the **remanence**. It varies from ~0.4 T in hard ferrites (e.g., strontium ferrites, SrFe) to ~1.2 T in rare-earth magnets (e.g., samarium cobalt, SmCo, or neodymium iron boron, NdFeB). The remanence typically drops with increasing temperature (in a reversible manner, on the order of −0.1% per degree Celsius[45]), so this needs to be taken into account prior to an analysis.

The linear characteristic described by Equation (2.81) is valid for a broad range of operating conditions. However, PMs exhibit a **hysteresis loop**, as shown in Figure 2.8. (Such loops are typically measured and plotted for the special case where $\vec{\mathbf{H}}$, $\vec{\mathbf{B}}$, and $\vec{\mathbf{B}}_r$ are collinear.) Magnetic hysteresis will be discussed in more detail in §2.3.5, in the context of core loss in ferromagnetic materials. For the analysis of devices containing PMs, it is usually sufficient to assume that we remain on the linear part of this characteristic in the upper half plane. Otherwise, if for any reason the H-field exceeds the knee point of the loop, the curve drops rapidly, and we are in danger of demagnetizing the PM. Hence, we typically design devices such that this undesirable phenomenon is avoided. The **coercivity** H_c is where the B-field drops to zero. It can vary from ~250 kA/m in hard ferrites to greater than 1000 kA/m in rare-earth magnets. Note that the PM coercivity is tens of thousands of times higher than that of electrical steel.

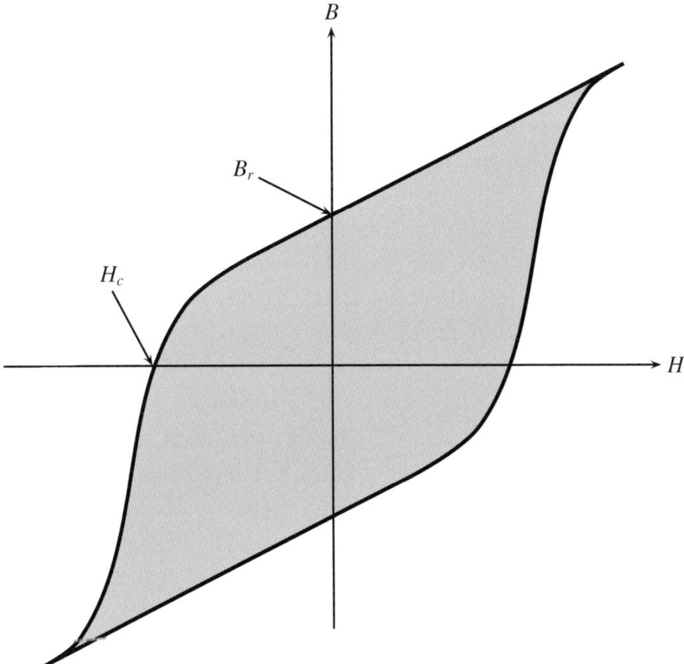

Figure 2.8 A typical B–H hysteresis loop of a permanent magnet.

The energy density in PMs can be evaluated as

$$m(\vec{\mathbf{B}}) = m_0 + \int_{\vec{\mathbf{B}}_r}^{\vec{\mathbf{B}}} \nu(\vec{\mathbf{b}} - \vec{\mathbf{B}}_r) \cdot d\vec{\mathbf{b}} \tag{2.82a}$$

$$= m_0 + \frac{\nu}{2}(B^2 + B_r^2) - \nu\vec{\mathbf{B}} \cdot \vec{\mathbf{B}}_r . \tag{2.82b}$$

The integration begins from the point where $\vec{\mathbf{H}} = \mathbf{0}$. The initial energy density $m_0 > 0$ is an unknown constant.

2.3 Energy Transfer Pathways

We have made the case that stationarity of the functional (2.71) leads to the solution of the magnetostatic problem. We have also alluded to the fact that the functional is related to energy, as we have mentioned that the first term of the functional is the energy "stored" in the magnetostatic field. In this section, we will explore the underlying energy transfer mechanisms further. Understanding these concepts lies at the core of electromechanical energy conversion theory.

How does energy transfer happen? Fundamentally, electric machines are devices that exchange energy with the outside world through electric circuits and one or more mechanically moving parts. Other significant mechanisms of energy exchange, such as heat flow, are also present. Modeling these is important for machine design, but we do not undertake it in earnest in this text, where we focus on understanding the electromagnetic phenomena.

The cornerstone of our analysis is the assumption of a **conservative magnetic field**. The field is often called a **coupling field** since it couples together the electrical and mechanical subsystems. Because of its conservative nature, the field can be visualized as a perfectly elastic sponge that may absorb an arbitrary amount of energy, which can be later returned intact to the outside world, as if the sponge has returned to its original state. This energy exchange can be visualized with a flow diagram such as the one in Figure 2.9. This figure depicts an instant in time where the device is operating as a motor because power is transferred from the two electrical subsystems (e.g., two voltage sources feeding power to two windings) to the mechanical side (in this case, there is only one moving component). At this particular instant in time, the coupling field is also contributing to the electromechanical energy conversion process, as can be seen by the amount of power that is being withdrawn.

Losses are incurred for every watt[46] that is injected to or absorbed from the coupling field due to ohmic[47] effects. Also, the net mechanical output is reduced by frictional loss that occurs when mechanical components are moving. Coupling fields are not lossless either, since magnetic materials exhibit hysteresis and are subject to eddy current flows. (These effects are not shown in the figure.) For FEA purposes, coupling field loss is typically ignored, mostly because it is notoriously challenging to model accurately and in a computationally efficient manner. Hence, the electromagnetic analysis is typically

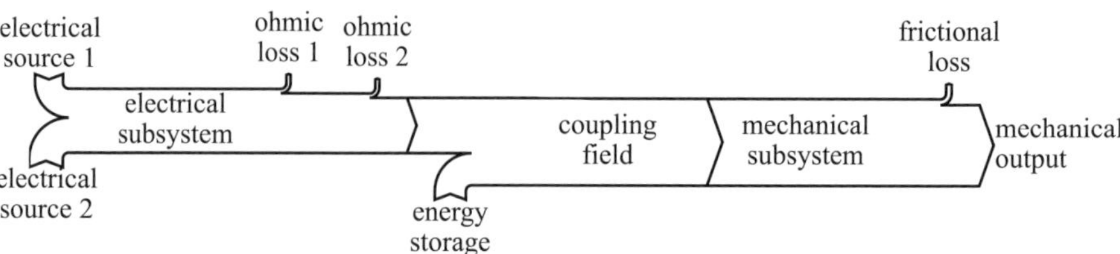

Figure 2.9 Conceptual diagram of power flow through a lossless magnetic coupling field interacting with two electrical subsystems and one mechanical subsystem.

conducted assuming lossless materials, which leads to a **conservative coupling field**. Hysteresis and eddy current loss can be accounted for at a post-processing stage.

We will now express mathematically the **energy balance** that we have conceptualized thus far. The assumed positive direction of energy flow is from the electrical side to the mechanical side, similar to what is depicted in Figure 2.9. Also, energy stored in the coupling field is assumed positive. (For the particular operating point depicted in the figure, the rate of change of coupling field energy storage would obtain a negative value.) Of course, this choice is entirely arbitrary, since energy flows can be either positive or negative (with the exception of losses, which are always non-negative). Hence, our model of the energy flow through a conservative coupling field is as follows:

$$\begin{pmatrix} \text{energy from} \\ \text{electric subsystem,} \\ W_E \end{pmatrix} = \begin{pmatrix} \text{ohmic loss,} \\ W_{E,\text{loss}} \end{pmatrix} + \begin{pmatrix} \text{magnetic energy to} \\ \text{coupling field,} \\ W_e \end{pmatrix}, \tag{2.83a}$$

$$\begin{pmatrix} \text{magnetic energy to} \\ \text{coupling field,} \\ W_e \end{pmatrix} = \begin{pmatrix} \text{energy stored in} \\ \text{coupling field,} \\ W_f \end{pmatrix} + \begin{pmatrix} \text{energy to} \\ \text{mechanical subsystem,} \\ W_m \end{pmatrix}, \tag{2.83b}$$

$$\begin{pmatrix} \text{energy to} \\ \text{mechanical subsystem,} \\ W_m \end{pmatrix} = \begin{pmatrix} \text{friction loss,} \\ W_{M,\text{loss}} \end{pmatrix} + \begin{pmatrix} \text{energy absorbed by} \\ \text{mechanical sources,} \\ W_M \end{pmatrix}. \tag{2.83c}$$

Combining these relationships, we obtain

$$W_E = W_{E,\text{loss}} + W_f + W_{M,\text{loss}} + W_M . \tag{2.84}$$

We will discuss these terms in more detail in the ensuing sections.

We will first analyze the transfer of energy through the air gap in a rotating electric machine. The air-gap field is only part of the coupling field that permeates the entire device. Nevertheless, the air gap is where most of the coupling field energy is stored, and always separates the moving parts from the stationary parts. Studying the properties of the electromagnetic field in the air gap is key in obtaining a deeper understanding of electric machine operation.

2.3.1 Conservation of Energy and Poynting's Theorem of Energy Flow

From the theory of electromagnetism, we may recall that the electromagnetic energy flow is quantified by the **Poynting vector**. The Poynting vector is defined as

$$\vec{\mathbf{S}} = \frac{1}{\mu_0} \vec{\mathbf{E}} \times \vec{\mathbf{B}} \quad \text{W/m}^2 . \tag{2.85}$$

This vector represents power flow surface density (i.e., energy per unit time and per unit area) crossing an infinitesimal area whose normal is oriented in the same direction as $\vec{\mathbf{S}}$.

This physical interpretation of $\vec{\mathbf{S}}$ is commonly obtained by using the divergence theorem together with Maxwell's equations. It is assumed that we are in free space, where

$\vec{\mathbf{H}} = \vec{\mathbf{B}}/\mu_0$. We start by applying the vector calculus identity (1.159):

$$\nabla \cdot \vec{\mathbf{S}} = \nabla \cdot (\vec{\mathbf{E}} \times \vec{\mathbf{H}}) = \vec{\mathbf{H}} \cdot (\nabla \times \vec{\mathbf{E}}) - \vec{\mathbf{E}} \cdot (\nabla \times \vec{\mathbf{H}}). \tag{2.86}$$

We can substitute the curls by recalling the respective Maxwell equations:

$$\nabla \cdot \vec{\mathbf{S}} = \vec{\mathbf{H}} \cdot \left(-\frac{\partial \vec{\mathbf{B}}}{\partial t}\right) - \vec{\mathbf{E}} \cdot \left(\vec{\mathbf{J}} + \frac{\partial \vec{\mathbf{D}}}{\partial t}\right), \tag{2.87}$$

which we rewrite as

$$\nabla \cdot \vec{\mathbf{S}} + \vec{\mathbf{E}} \cdot \vec{\mathbf{J}} = -\vec{\mathbf{H}} \cdot \frac{\partial \vec{\mathbf{B}}}{\partial t} - \vec{\mathbf{E}} \cdot \frac{\partial \vec{\mathbf{D}}}{\partial t}. \tag{2.88}$$

Note that $\vec{\mathbf{J}}$ represents the free current density (since it came from the curl of $\vec{\mathbf{H}}$). Now we integrate over the entire region $\mathcal{D}$ of the field, and apply the divergence theorem:

$$\oiint_{\partial \mathcal{D}} \vec{\mathbf{S}} \cdot d\mathbf{a} + \iiint_{\mathcal{D}} \vec{\mathbf{E}} \cdot \vec{\mathbf{J}}\, dv = -\iiint_{\mathcal{D}} \left(\vec{\mathbf{H}} \cdot \frac{\partial \vec{\mathbf{B}}}{\partial t} + \vec{\mathbf{E}} \cdot \frac{\partial \vec{\mathbf{D}}}{\partial t}\right) dv. \tag{2.89}$$

This is **Poynting's theorem**, derived in 1884. Equation (2.89) suggests a power balance similar to the one described previously but of broader applicability since it is not limited to a purely magnetic field. We will show in subsequent sections of this chapter that the term $\vec{\mathbf{E}} \cdot \vec{\mathbf{J}}$ is the ohmic heat loss density (W/m^3) within conductors, and that the term $\vec{\mathbf{H}} \cdot \partial \vec{\mathbf{B}}/\partial t$ can be associated with the rate of increase of magnetic energy density. Similarly, the term $\vec{\mathbf{E}} \cdot \partial \vec{\mathbf{D}}/\partial t$ is the rate of increase of electric energy density (we will not show this, but the proof can be found in physics texts). Hence, the Poynting vector $\vec{\mathbf{S}}$ may be interpreted as the electromagnetic energy flux per unit time and unit surface. Poynting's theorem thus provides a fields-based expression for the **conservation of energy** in an electromagnetic field: *The total electromagnetic power flowing outwards plus the ohmic loss (conversion of electromagnetic energy into heat) equals the rate of reduction of electromagnetic energy within the volume.*

Poynting's theorem was derived in free space. However, the physical significance of $\vec{\mathbf{S}}$ is maintained even when we are considering magnetized matter, and even when matter is allowed to move. In this case, the power balance should be modified by adding a term that corresponds to mechanical power. This will be the approach that we will take in §2.4.4, when we discuss the generation of electromagnetic force and torque based on the principle of conservation of energy.

2.3.2 Energy Transfer through a Rotating Magnetic Field

The operation of most alternating current (a.c.) electric machinery hinges on a **rotating magnetic field**. To understand this concept more deeply, we will take a close look at the magnetic field in the air gap of electric machines. We will then calculate the power that is transferred electromagnetically from stator to rotor (or vice versa) through the air gap using Poynting's theorem.

We will find that a rotating magnetic field is necessary to sustain a steady transfer of power from stator to rotor. For example, an a.c. squirrel-cage induction motor operates by absorbing electrical power from the stator terminals, also called the **armature**, which is converted (minus loss) to mechanical power at the rotor shaft. A synchronous generator functions in reverse, converting mechanical power input at the shaft to electric power output at the stator terminals. In any case, the armature consists of the coils that are directly in the electromechanical energy conversion path. Windings that generate magnetic fields without carrying a significant amount of power (other than what is needed to supply their ohmic loss) are called **field windings**. In contrast, (brushed) d.c. machines operate with a spatially stationary magnetic field. In d.c. machines, power cannot be transferred from stator to rotor. In a d.c. motor, electrical power is supplied to the rotor armature through brushes, and is converted "locally" to mechanical power at the shaft. So, the results of this section are mainly applicable to a.c. machines.

Since this is our first encounter with an a.c. rotating machine, we will describe it in some detail before proceeding with the analysis. Let us consider the elementary machine sketched in Figures 2.10 and 2.11. Like most rotating machines, this machine has a cylindrical shape and identical cross-section over its axial length ℓ.[g] Commonly, the **stator** is the outer part, which is restricted from moving as it is solidly affixed to the ground with a base. (There is always an outer frame between stator and base, which is not shown.) The **rotor** is the inner part, which can rotate freely around an axis through its center. The rotor is connected to an axial shaft that extends past the magnetically active part of the machine. The shaft is suspended at the two ends of the machine by bearings. Stator and rotor are concentric,[h] and are separated by a thin **air gap** (somewhat exaggerated in the figures for illustration purposes). The air-gap width, denoted by g, is the closest distance between rotor and stator. The mean radius of the air gap is R, based

[g] In real machines, the rotor is often skewed to mitigate undesired impacts of harmonic field components. Hence, the rotor cross-section could exhibit a small rotation from one end to the other. Alternatively, some machines are constructed with a number of discrete stacks. Nevertheless, the assumption of a constant cross-section is sufficient for our purposes.

[h] Real machines exhibit a slight eccentricity, which is ignored here.

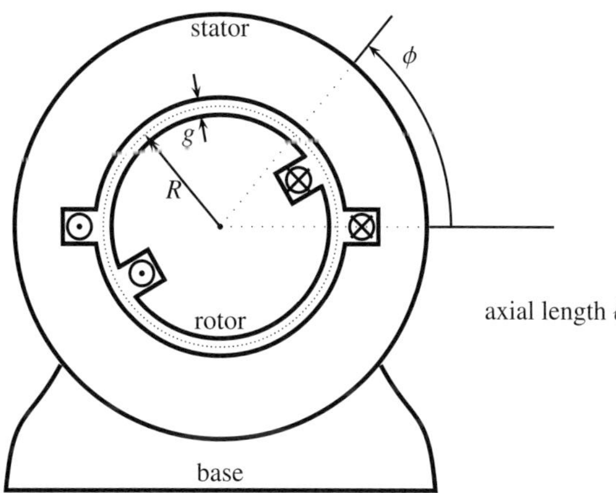

Figure 2.10 Cross-section of an elementary electric machine.

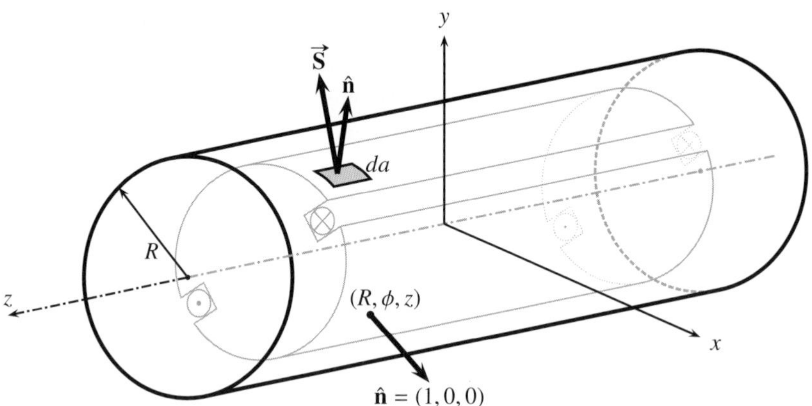

Figure 2.11 Electromagnetic power transfer calculation from rotor to stator using Poynting's theorem. The cylinder that encloses the rotor extends to the middle of the air gap in the radial direction. Vectors are expressed in a cylindrical coordinate system.

on averaging the radii of two circles that touch the rotor and stator surfaces (which are not necessarily circular due to slotting or other geometric features).

In some sense, a.c. electric machines are devices that are engineered to create a rotating magnetic field in the air gap. The exact manner in which this field is created depends on the placement of conductors and/or permanent magnets strategically inside the machine. However, the machine design is irrelevant at the moment. Our objective is to calculate the power transferred electromagnetically from rotor to stator (i.e., assumed positive in the radially outwards direction) through the air gap. According to Poynting's theorem, this may be obtained by a surface integral. We will employ the usual 2-D analysis assumptions ($\partial/\partial z = 0$, currents only along the z-axis, MVP $\vec{\mathbf{A}} = A\,\hat{\mathbf{k}}$, B-field on the r–ϕ plane). We will integrate over a cylinder that extends to the middle of the air gap, shown in Figure 2.11. We will ignore contributions to the integral from the cylinder bases.

The power from rotor to stator is

$$P_{r\to s} = \ell R \int_0^{2\pi} \vec{\mathbf{S}} \cdot \hat{\mathbf{n}}\, d\phi = \frac{\ell R}{\mu_0} \int_0^{2\pi} (\vec{\mathbf{E}} \times \vec{\mathbf{B}}) \cdot \hat{\mathbf{r}}\, d\phi\,. \tag{2.90}$$

The E-field is induced by a time-changing magnetic field:

$$\vec{\mathbf{E}} = -\frac{\partial \vec{\mathbf{A}}}{\partial t} = -\frac{\partial A}{\partial t}\hat{\mathbf{k}}\,. \tag{2.91}$$

Note that it has only a z-component, namely, E_z. The B-field is

$$\vec{\mathbf{B}} = \nabla \times \vec{\mathbf{A}} = \frac{1}{r}\frac{\partial A}{\partial \phi}\hat{\mathbf{r}} - \frac{\partial A}{\partial r}\hat{\boldsymbol{\phi}}\,. \tag{2.92}$$

The cross product

$$\vec{\mathbf{E}} \times \vec{\mathbf{B}} = -\frac{\partial A}{\partial t}\frac{\partial A}{\partial r}\hat{\mathbf{r}} - \frac{1}{r}\frac{\partial A}{\partial t}\frac{\partial A}{\partial \phi}\hat{\boldsymbol{\phi}} = -E_z B_\phi\, \hat{\mathbf{r}} - E_z B_r\, \hat{\boldsymbol{\phi}} \tag{2.93}$$

contains both radial and tangential components. The tangential component will be the largest one when the magnetic field is radially dominant, that is, $|B_r| > |B_\phi|$. This implies the presence of a tangential power flow, which does not contribute to energy flowing from rotor to stator. Using the radial component, we obtain the **power flow from rotor to stator** as

$$\boxed{P_{r\to s} = -\frac{\ell R}{\mu_0}\int_0^{2\pi} \frac{\partial A}{\partial t}\frac{\partial A}{\partial r}\, d\phi = -\frac{\ell R}{\mu_0}\int_0^{2\pi} E_z B_\phi \, d\phi \, .} \tag{2.94}$$

Interestingly, only the tangential component of the B-field ($B_\phi = -\partial A/\partial r$) enters into this formula. So, if either the fields are stationary (as in d.c. machines) or in the absence of B_ϕ (i.e., if the field is purely radial), then power transfer cannot take place. This is somewhat counterintuitive because it is often the case that introductory treatments of electric machines explain their operating principle based on a purely radial field in the air gap. To resolve this apparent paradox, and to appreciate what this formula is implying, we need a deeper understanding of the MVP in the air-gap region.

Analytical Solution of MVP in the Air Gap

Regardless of the shape of the stator and rotor surfaces surrounding the air gap, we can always identify an annular region of uniform width in the air gap. This annular region extends from an inner radius $r_i = R - g/2 + \epsilon$ to an outer radius $r_o = R + g/2 - \epsilon$, with $0 \le \epsilon < g/2$. Within the annular region, a magnetic vector potential under the Coulomb gauge ($\nabla \cdot \vec{\mathbf{A}} = 0$) satisfies the Laplace equation

$$\nabla^2 \vec{\mathbf{A}} = -\epsilon_0 \mu_0 \frac{\partial \vec{\mathbf{E}}}{\partial t} \approx \mathbf{0} \, . \tag{2.95}$$

The approximation is exact for static problems, and it should be reasonable for a broad range of frequencies within the quasi-magnetostatic regime.

The Laplacian for the z-component of the MVP in cylindrical coordinates is

$$\frac{1}{r}\frac{\partial}{\partial r}\left(r\frac{\partial A}{\partial r}\right) + \frac{1}{r^2}\frac{\partial^2 A}{\partial \phi^2} = 0 \, . \tag{2.96}$$

This partial differential equation can be solved using the method of separation of variables. The trick is to express the MVP as a product of two functions:

$$A(r,\phi) = f(r) g(\phi) \, . \tag{2.97}$$

The Laplacian becomes

$$\frac{1}{r} f' g + f'' g + \frac{1}{r^2} f g'' = 0 \, , \tag{2.98}$$

where derivatives of f and g are with respect to r and ϕ, respectively. Multiplying by r^2/fg:

$$r\frac{f'}{f} + r^2\frac{f''}{f} = -\frac{g''}{g} \, . \tag{2.99}$$

The left-hand side is a function of r only, whereas the right-hand side is a function of ϕ only. Therefore, there exists a number $\kappa \in \mathbb{R}$ so that

$$r\frac{f'}{f} + r^2\frac{f''}{f} = -\frac{g''}{g} = \kappa \,. \tag{2.100}$$

Solving the two ordinary differential equations, we obtain

$$f_p(r) = c_p r^{-p} + d_p r^p \,, \tag{2.101a}$$

$$g_p(\phi) = a_p \cos p\phi + b_p \sin p\phi \,, \tag{2.101b}$$

where the parameter $p > 0$ satisfies $p^2 = \kappa > 0$. Furthermore, $p \in \mathbb{Z}$ (i.e., it is an integer number) because g_p must be periodic every 2π.[i] We conclude that the general solution of the MVP in the air gap will be a superposition of terms of the form

$$A_p(r,\phi) = f_p(r) g_p(\phi) \,. \tag{2.102}$$

At any given r, this term represents a sinusoidal variation in the air gap with p positive and p negative peaks.

The coefficients a_p, b_p, c_p, d_p can be determined by solving a boundary value problem. We express the boundary conditions at $r = r_i$ and $r = r_o$ as Fourier[48] series

$$A(r_i,\phi) = \sum_{p=1}^{\infty} \alpha_p \cos p\phi + \beta_p \sin p\phi \,, \tag{2.103a}$$

$$A(r_o,\phi) = \sum_{p=1}^{\infty} \gamma_p \cos p\phi + \delta_p \sin p\phi \,, \tag{2.103b}$$

where α_p, β_p, γ_p, δ_p are given coefficients. The MVP can be written as the sum of two terms: the first term satisfies the boundary condition at $r = r_i$ and vanishes at $r = r_o$, while the second term achieves the opposite, that is, it satisfies the boundary condition at $r = r_o$ and vanishes at $r = r_i$. Of course, both terms must satisfy the Laplace equation, so they must be of the form (2.102). This leads to

$$A(r,\phi) = \sum_{p=1}^{\infty} \frac{1}{\zeta^p - \zeta^{-p}} \left[\left(\frac{r}{r_o}\right)^p - \left(\frac{r_o}{r}\right)^p \right] (\alpha_p \cos p\phi + \beta_p \sin p\phi)$$
$$+ \sum_{p=1}^{\infty} \frac{1}{\zeta^{-p} - \zeta^p} \left[\left(\frac{r}{r_i}\right)^p - \left(\frac{r_i}{r}\right)^p \right] (\gamma_p \cos p\phi + \delta_p \sin p\phi) \,, \tag{2.104}$$

where we defined the parameter

$$\zeta = \frac{r_i}{r_o} \,. \tag{2.105}$$

[i] The case $\kappa < 0$ is mathematically possible, but it is inconsistent with the physical problem at hand. It yields a solution $g_p(\phi) = a_p e^{p\phi} + b_p e^{-p\phi}$, where $\kappa = -p^2$. This solution is not periodic in the air gap, so it is rejected.
The special case $\kappa = 0$ leads to $g_0(\phi) = a_0 + b_0\phi$, $f_0(r) = c_0 + d_0 \ln r$. We can satisfy periodicity by setting $b_0 = 0$. However, f_0 is problematic because it represents a physical situation where the tangential component of the flux density in the air gap has a nonzero average value over 360°. Application of Ampère's law around any circular path within the annulus would preclude this from happening under normal operating conditions because the enclosed current (through the rotor cross-section) should be zero.

Interpretation of Air-Gap MVP Solution

We cannot quantify the air-gap MVP field without additional information about the boundary conditions. Here, let us focus on the most common **radial-flux** electric machine designs, where the air-gap magnetic field is oriented predominantly in the radial direction. To represent this operating principle, we will examine cases where the MVP boundary conditions are sinusoidal waveforms with peaks appearing at similar angles. Also, we will set the magnitudes of the waveforms to be approximately equal. A few illustrative examples are shown in Figures 2.12–2.14, which depict contour plots of the MVP. As explained in Box 1.2, MVP contours can be interpreted as magnetic flux lines.

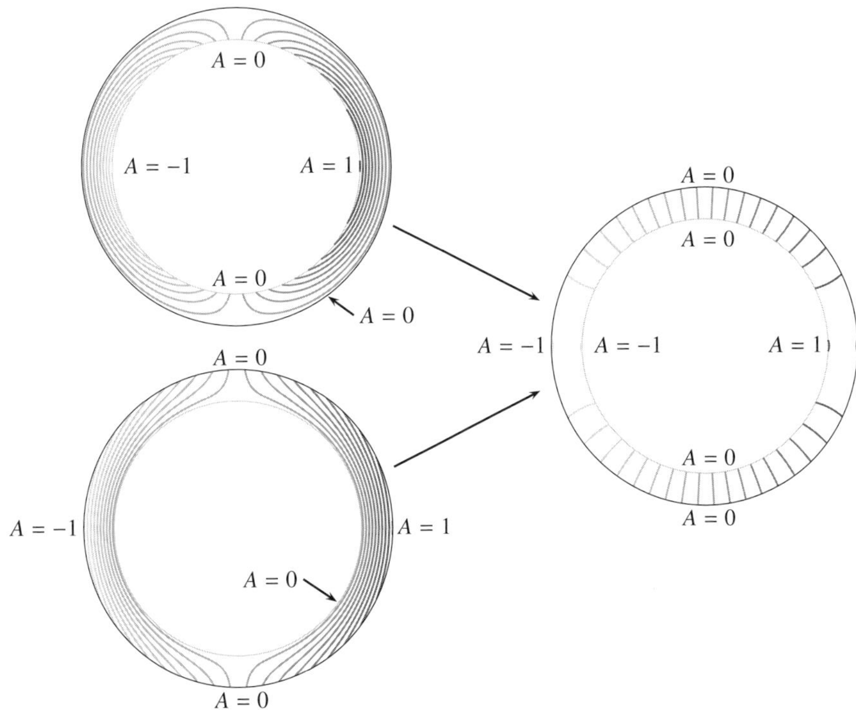

Figure 2.12 MVP contour plots in the air gap (case study 1).

In the first case study, we have set $p = 1$. Also, the MVP boundary waveforms are perfectly aligned, and the MVP peak value is 1 at both boundaries. On the left-hand side of Figure 2.12, we can observe the solutions to the two boundary value problems independently. On the top, we have the solution to the problem where the outer boundary potential is zero. On the bottom, we set the inner boundary to zero. On the right, we see the total solution obtained by the superposition of the two fields. We observe that the contours/flux lines connect points on the boundaries that have the same potential. For the plots on the left, the flux lines are between points on the same boundary. For the plot on the right, the flux lines connect the two boundaries. Interestingly, the flux density appears to be maximum where $A = 0$, and zero where $A = \pm 1$.

For the next two case studies, we set $p = 2$. We also increased the outer boundary

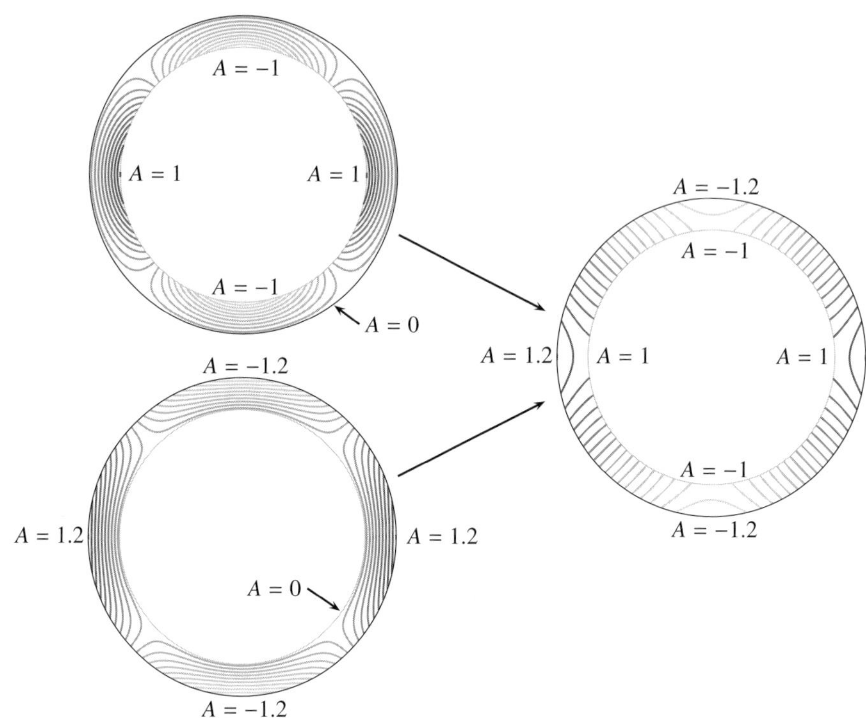

Figure 2.13 MVP contour plots in the air gap (case study 2).

peak from 1 to 1.2. The fields are depicted in Figures 2.13 and 2.14. This causes some of the flux lines originating from the outer boundary to never reach the inner boundary (since the potential there peaks at 1). Case study 3 is a variant where we added 10° to the angular offset of the outer boundary MVP. This has created an asymmetry in the flux lines, and an overall tendency for them to move in the counterclockwise direction. We have thus created a tangential component of flux density, which is a key ingredient for energy transfer!

We observe that the parameter p has the physical significance of number of magnetic **pole-pairs**, which can be visualized using the pattern of flux lines. For example, for $p = 1$, the MVP and flux density in the air gap repeat once per 360°. That is, for $p = 1$, there is a single north–south pole-pair. When $p = 2$, the waveforms repeat twice per 360°, so there are two north–south pole-pairs. And so on for $p = 3, 4, \ldots$. The designation of north and south poles is a matter of convention. Typically, *we designate as **north pole** a region in space where flux appears to be emerging outwards from a magnetic material (e.g., steel, permanent magnet) to its surroundings*. In contrast, ***south poles** are those regions where flux is entering a material from the outside*. For example, in Figure 2.12, the rotor has a north pole for $\phi = [-180, 0]°$ (i.e., on the bottom half, based on the definition of the angle ϕ shown in Figure 2.10), and a south pole for $\phi = [0, 180]°$

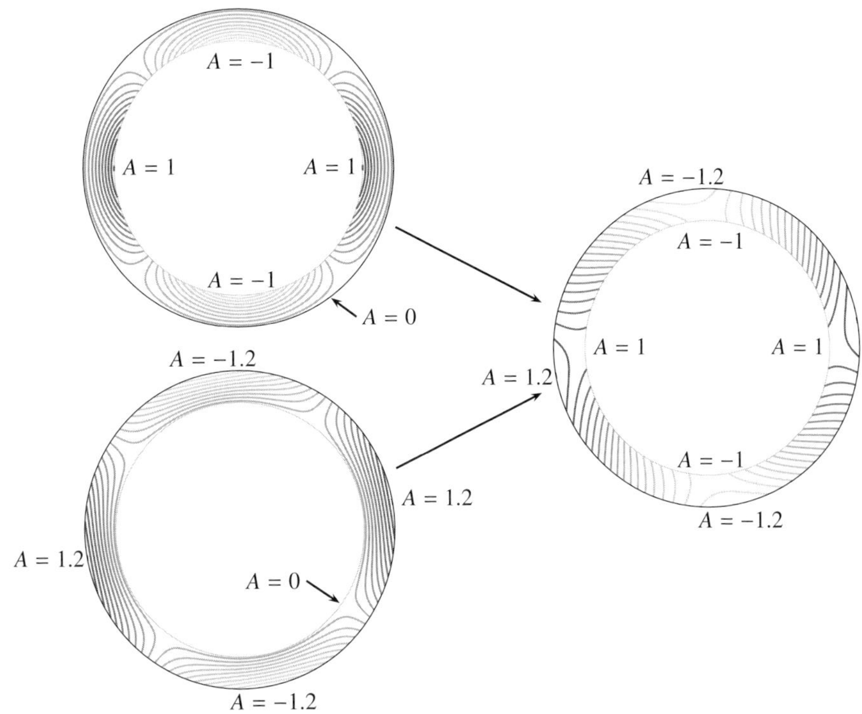

Figure 2.14 MVP contour plots in the air gap (case study 3).

(i.e., on the top half). On the stator, the poles are reversed: for instance, the north pole is at $\phi = [0, 180]°$, where flux is emerging.[j]

We can calculate the air-gap B-field by applying Equation (2.92) to the MVP expression given by Equation (2.104). Let us consider a single component of order p. Evaluating the partial derivatives, the radial component is

$$B_{pr}(r,\phi) = \frac{p}{r}\frac{1}{\zeta^{p}-\zeta^{-p}}\left[\left(\frac{r}{r_o}\right)^p - \left(\frac{r_o}{r}\right)^p\right](-\alpha_p \sin p\phi + \beta_p \cos p\phi)$$
$$+ \frac{p}{r}\frac{1}{\zeta^{-p}-\zeta^{p}}\left[\left(\frac{r}{r_i}\right)^p - \left(\frac{r_i}{r}\right)^p\right](-\gamma_p \sin p\phi + \delta_p \cos p\phi) \quad (2.106)$$

and the tangential component is

$$B_{p\phi}(r,\phi) = -\frac{p}{r}\frac{1}{\zeta^{p}-\zeta^{-p}}\left[\left(\frac{r}{r_o}\right)^p + \left(\frac{r_o}{r}\right)^p\right](\alpha_p \cos p\phi + \beta_p \sin p\phi)$$
$$- \frac{p}{r}\frac{1}{\zeta^{-p}-\zeta^{p}}\left[\left(\frac{r}{r_i}\right)^p + \left(\frac{r_i}{r}\right)^p\right](\gamma_p \cos p\phi + \delta_p \sin p\phi). \quad (2.107)$$

The radial flux density B_{pr} leads the MVP by 90 *electrical* degrees (which we will denote as 90° electrical) in space. The **electrical angles** are the arguments of the cosine

[j] The Earth's North Pole is actually a south magnetic pole since it attracts the north pole of compass magnets.

and sine terms in the expressions above, and they are defined as p times the actual **mechanical angle** ϕ. Therefore, the field undergoes one complete cycle within 360° electrical, or p cycles within a full mechanical revolution around the air gap. In other words, suppose the MVP is maximum at the following positions in the air gap (in terms of mechanical angles): $\phi_{\max} = \phi_0 + (2\pi/p)k$, $k \in \{0, 1, \ldots, p-1\}$, then the corresponding radial B-field is maximum at $\phi = \phi_{\max} - (\pi/2)/p$. The various waveforms that are depicted in Figure 2.15 for $p = 2$ (a four-pole field) correspond to the field of case study 3. The exact same condition is also illustrated as a quiver plot in the air-gap annulus in Figure 2.16, where rotor-side north/south poles are also indicated.

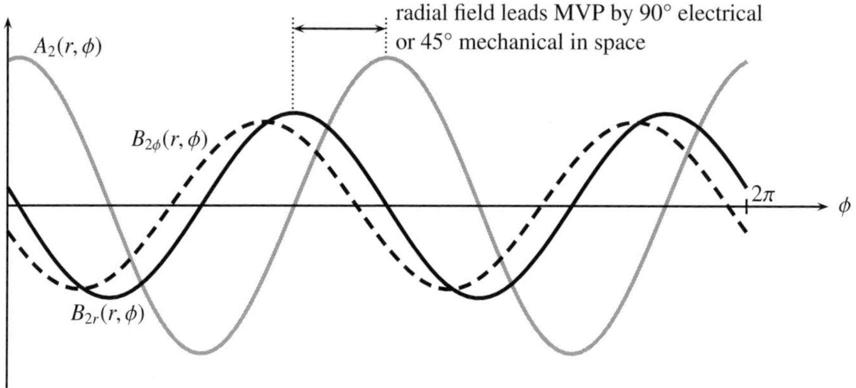

Figure 2.15 Four-pole MVP and magnetic flux density in the air gap (case study 3).

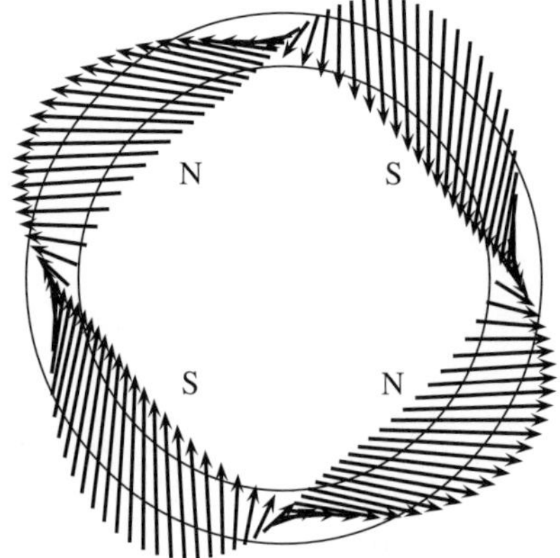

Figure 2.16 Quiver plot of a four-pole magnetic field in the air gap (case study 3). The arrows are representative of $\vec{\mathbf{B}}$ at various positions around the air gap (evaluated at $r = R$). They are drawn centered at R.

Electric machines are designed so that one of the pole-pair fields dominates over the rest. Ideally, there should be only one such component in the air gap. However, it is practically impossible to achieve a purely sinusoidal field, and multiple harmonics exist in the MVP and flux-density waveforms. The air-gap field thus consists of a superposition of an infinite number of pole-pair fields, but the dominant field designates the machine pole number. For instance, when we refer to a motor as a four-pole motor, what we are actually saying is that the four-pole field is the largest, and thus plays the central role in the operation of the motor.

Example 2.7 *Calculation of the magnetic energy stored in the air-gap annulus.*
Our objective here is to calculate the energy stored in the air-gap annulus. This result will be used in §2.4.2.

Solution
We integrate the volumetric energy density

$$W_{ag} = \frac{\ell \nu_0}{2} \int_{r=r_i}^{r_o} \int_{\phi=0}^{2\pi} B^2(r,\phi)\, r\, d\phi\, dr\,, \tag{2.108}$$

where

$$B^2(r,\phi) = B_r^2(r,\phi) + B_\phi^2(r,\phi)\,, \tag{2.109}$$

$$B_r(r,\phi) = \sum_{p=1}^{\infty} B_{pr}(r,\phi)\,,\ B_\phi(r,\phi) = \sum_{p=1}^{\infty} B_{p\phi}(r,\phi)\,. \tag{2.110}$$

The algebraic manipulations are tedious and are left as an exercise for the reader. While integrating over 2π with respect to ϕ, many sinusoidal terms will vanish. The result is

$$W_{ag} = \sum_{p=1}^{\infty} \frac{\pi \ell \nu_0 p}{\zeta^{-p} - \zeta^p} \left[\frac{\zeta^{-p} + \zeta^p}{2} (\alpha_p^2 + \beta_p^2 + \gamma_p^2 + \delta_p^2) - 2(\alpha_p \gamma_p + \beta_p \delta_p) \right] \tag{2.111a}$$

$$= \sum_{p=1}^{\infty} \frac{\pi \ell \nu_0 p}{\zeta^{-p} - \zeta^p} \left[\frac{\zeta^{-p} + \zeta^p}{2} (A_{pi}^2 + A_{po}^2) - 2A_{pi}A_{po} \cos(\phi_{po} - \phi_{pi}) \right]. \tag{2.111b}$$

The coefficients in the above expressions are related by

$$\alpha_p = A_{pi} \cos\phi_{pi},\ \beta_p = A_{pi} \sin\phi_{pi}\,, \tag{2.112a}$$

$$\gamma_p = A_{po} \cos\phi_{po},\ \delta_p = A_{po} \sin\phi_{po}\,. \tag{2.112b}$$

The parameters $A_{pi} > 0$, $A_{po} > 0$ represent the magnitudes of the boundary conditions at the inner and outer boundary, respectively. The angles ϕ_{pi} and ϕ_{po} determine the circumferential position of the boundary conditions. The air-gap energy consists of a sum of contributions from each separate pole-pair field. Fields with different pole-pair numbers do not interact as far as energy is concerned.

Rotating Magnetic Field

A.c. electric machines are engineered so that a continually rotating magnetic field is present in the air gap. Physically, this is created by the combined effects of stator and rotor currents in accordance with a variety of electric machine designs. Mathematically, rotation can be described uniformly, regardless of machine design, as an outcome of time-varying annulus boundary conditions, $\alpha_p(t)$, $\beta_p(t)$, $\gamma_p(t)$, $\delta_p(t)$. We define time-varying boundary MVP amplitudes and angular offsets through the following set of equations:

$$\alpha_p(t) = A_{pi}(t)\cos\phi_{pi}(t), \;\; \beta_p(t) = A_{pi}(t)\sin\phi_{pi}(t)\,, \tag{2.113a}$$

$$\gamma_p(t) = A_{po}(t)\cos\phi_{po}(t), \;\; \delta_p(t) = A_{po}(t)\sin\phi_{po}(t)\,. \tag{2.113b}$$

In particular, a p-pole-pair, **constant-amplitude**, **constant-speed**, rotating MVP is obtained by constant $A_{pi} > 0$, $A_{po} > 0$, and by linearly increasing angles

$$\phi_{pi}(t) = \omega t + \phi_{pi0}\,, \;\; \phi_{po}(t) = \omega t + \phi_{po0}\,. \tag{2.114}$$

The angles ϕ_{pi0} and ϕ_{po0} determine the circumferential position of the boundary conditions at time $t = 0$. The parameter ω (rad/s) describes how rapidly the MVP is changing with respect to time at any given point ϕ on the boundary and, consequently, within the entire air gap. In practice, ω is called the **electrical frequency** since it is the frequency of the current sources. This is a reasonable model for the steady-state behavior of the dominant pole-pair field in machines (by design), but it may not be as accurate for higher pole-pair fields.

The electrical frequency ω affects how fast the entire sinusoidal MVP waveform is moving around the air gap. We can write the MVP expression (2.104) as

$$\begin{aligned} A_p(r,\phi,t) = {} & \frac{1}{\zeta^p - \zeta^{-p}}\left[\left(\frac{r}{r_o}\right)^p - \left(\frac{r_o}{r}\right)^p\right] A_{pi}(t)\cos(p\phi - \phi_{pi}(t)) \\ & + \frac{1}{\zeta^{-p} - \zeta^p}\left[\left(\frac{r}{r_i}\right)^p - \left(\frac{r_i}{r}\right)^p\right] A_{po}(t)\cos(p\phi - \phi_{po}(t))\,, \end{aligned} \tag{2.115}$$

or more concisely

$$A_p(r,\phi,t) = \tilde{A}_{pi}(r,t)\cos(p\phi - \phi_{pi}(t)) + \tilde{A}_{po}(r,t)\cos(p\phi - \phi_{po}(t))\,. \tag{2.116}$$

We can readily verify the following properties (to be invoked in the ensuing analysis): $\tilde{A}_{pi}(r,t) > 0$, $\tilde{A}_{po}(r,t) > 0$, $\partial\tilde{A}_{pi}/\partial r < 0$, and $\partial\tilde{A}_{po}/\partial r > 0$. The waveform velocity can be found by setting the argument of either cosine to a constant value, say, zero. For instance, from the inner boundary cosine

$$\phi = \frac{1}{p}(\omega t + \phi_{pi0})\,. \tag{2.117}$$

Therefore, we conclude that the **rotating field mechanical angular velocity**, denoted as ω_p, is related to the electrical frequency and the number of pole-pairs by

$$\omega_p = \frac{d\phi}{dt} = \frac{\omega}{p}\,. \tag{2.118}$$

This is a well-known operating principle of rotating electric machinery: *The mechanical angular velocity of rotation (determined by the dominant pole-pair field) is reduced as we increase the number of magnetic poles for the same electrical frequency.*

Power Flow in the Air Gap

We are now ready to apply Poynting's theorem from Equation (2.94):

$$P_{r\to s} = -\frac{\ell r}{\mu_0}\int_0^{2\pi} \frac{\partial A}{\partial t}\frac{\partial A}{\partial r}\, d\phi\,, \tag{2.119}$$

to calculate the outwards power flow through a cylinder of arbitrary radius r in the air gap (i.e., $r_i \le r \le r_o$). We will use the concise Equation (2.116). We omit the dependence of the $\tilde{A}$ functions on (r, t) for the sake of notational brevity.

The total field is $A = \sum_p A_p$, so the expression for power will contain terms of products of p- and q-pole-pair fields:

$$\begin{aligned}\frac{\partial A_p}{\partial t}\frac{\partial A_q}{\partial r} = &\left[\frac{\partial \tilde{A}_{pi}}{\partial t}\cos(p\phi-\phi_{pi}) + \frac{\partial \tilde{A}_{po}}{\partial t}\cos(p\phi-\phi_{po})\right.\\ &\left. + \tilde{A}_{pi}\frac{d\phi_{pi}}{dt}\sin(p\phi-\phi_{pi}) + \tilde{A}_{po}\frac{d\phi_{po}}{dt}\sin(p\phi-\phi_{po})\right]\\ &\cdot\left[\frac{\partial \tilde{A}_{qi}}{\partial r}\cos(q\phi-\phi_{qi}) + \frac{\partial \tilde{A}_{qo}}{\partial r}\cos(q\phi-\phi_{qo})\right]. \end{aligned}\tag{2.120}$$

We transform these terms using trigonometric identities involving the sum and difference of the angles, such as

$$2\sin x\cos y = \sin(x+y) + \sin(x-y)\,. \tag{2.121}$$

After integrating over 2π with respect to ϕ, most terms that are sinusoidal functions of $(p \pm q)\phi$ will vanish. The only terms that remain correspond to $p = q$. Hence, we obtain

$$\begin{aligned}\frac{1}{\pi}\int_0^{2\pi}\frac{\partial A}{\partial t}\frac{\partial A}{\partial r}\,d\phi &= \frac{1}{\pi}\sum_{p=1}^{\infty}\int_0^{2\pi}\frac{\partial A_p}{\partial t}\frac{\partial A_p}{\partial r}\,d\phi\\ &= \sum_{p=1}^{\infty}\left(\frac{\partial \tilde{A}_{pi}}{\partial t}\frac{\partial \tilde{A}_{pi}}{\partial r} + \frac{\partial \tilde{A}_{po}}{\partial t}\frac{\partial \tilde{A}_{po}}{\partial r}\right)\\ &+ \sum_{p=1}^{\infty}\left(\frac{\partial \tilde{A}_{pi}}{\partial t}\frac{\partial \tilde{A}_{po}}{\partial r} + \frac{\partial \tilde{A}_{po}}{\partial t}\frac{\partial \tilde{A}_{pi}}{\partial r}\right)\cos(\phi_{po}-\phi_{pi})\\ &+ \sum_{p=1}^{\infty}\left(\tilde{A}_{pi}\frac{d\phi_{pi}}{dt}\frac{\partial \tilde{A}_{po}}{\partial r} - \tilde{A}_{po}\frac{d\phi_{po}}{dt}\frac{\partial \tilde{A}_{pi}}{\partial r}\right)\sin(\phi_{po}-\phi_{pi})\,. \end{aligned}\tag{2.122}$$

The above expression is quite general because it yields the power transfer for MVP boundary conditions that are time-varying in an arbitrary fashion. However, this also renders it rather difficult to interpret. One possible way to utilize the formula is by applying it twice at $r = r_i$ and $r = r_o$, thereby leading to a calculation of the outwards power transfer at the inner and outer boundaries, respectively. Subtracting the two answers should yield the rate of energy removal from the air gap. More precisely, employing an energy balance, we expect that

$$P_{r\to s}(r_o) - P_{r\to s}(r_i) = -\frac{dW_{ag}}{dt}\,, \tag{2.123}$$

where W_{ag} is given by Equation (2.111b). This proof is left as an exercise.

It is instructive to focus on the case of a single constant-amplitude, constant-speed rotating field. In this case, the $\partial\tilde{A}/\partial t$ terms vanish, and the boundary MVPs are rotating at the same speed, $\omega_p = p^{-1}\omega = p^{-1}d\phi_{pi}/dt = p^{-1}d\phi_{po}/dt$. This leads to

$$\int_0^{2\pi} \frac{\partial A_p}{\partial t}\frac{\partial A_p}{\partial r}\,d\phi = \pi\omega\left(\tilde{A}_{pi}\frac{\partial \tilde{A}_{po}}{\partial r} - \tilde{A}_{po}\frac{\partial \tilde{A}_{pi}}{\partial r}\right)\sin(\phi_{po0} - \phi_{pi0})\,. \tag{2.124}$$

Note that the term in the parentheses is always positive (based on properties of the coefficients of Equation (2.116)); therefore, the sign of this expression is determined by the $\sin(\phi_{po0} - \phi_{pi0})$ term.

We conclude that **the power transferred electromagnetically through the air gap by a constant-amplitude p-pole-pair rotating magnetic field** is

$$P_{r\to s,p} = \frac{\pi\ell r p\omega_p}{\mu_0}\left(\tilde{A}_{pi}\frac{\partial \tilde{A}_{po}}{\partial r} - \tilde{A}_{po}\frac{\partial \tilde{A}_{pi}}{\partial r}\right)\sin(\phi_{pi0} - \phi_{po0}) \tag{2.125a}$$

$$= \frac{2\pi\ell p^2\omega_p}{\mu_0(\zeta^{-p} - \zeta^p)}A_{pi}A_{po}\sin(\phi_{pi0} - \phi_{po0}) \tag{2.125b}$$

$$= \frac{2\pi\ell p^2\omega_p}{\mu_0(\zeta^{-p} - \zeta^p)}(\beta_p\gamma_p - \alpha_p\delta_p)\,. \tag{2.125c}$$

The second and third expressions are derived by algebraic manipulations (left as exercises for the reader). We observe that power is proportional to the angular velocity of field rotation, and is independent of the radius at which we are integrating (according to the second and third formulas). This makes intuitive sense because power should not be dissipated in air. The same amount of power that enters on one side exits on the other side, so that the total magnetic energy stored in the annulus remains constant. For generator operation (i.e., for positive power transferred from rotor to stator), the rotor-side MVP waveform must lead the stator-side MVP (i.e., $\phi_{pi0} > \phi_{po0}$), and vice versa for motor operation. Hence, conceptually, for generator operation the rotor-side north (south) poles are leading the opposing stator-side south (north) poles. Since opposite poles attract, it is as if the rotor field is "pulling" the stator field as it rotates, and thus mechanical power is transferred to the electrical side. For motor operation, the opposite is happening (e.g., as shown in Figure 2.16).

The power also depends on the MVP magnitude at the boundary. A more physically significant interpretation can be obtained based on an alternative expression that uses the flux density instead of the MVP. Let us denote by B_{pri} and B_{pro} the magnitudes of the radial components of flux density at the inner and outer boundaries, respectively. From Equation (2.106), these may be related to the respective MVP magnitudes by

$$B_{pri} = \frac{p}{r_i}A_{pi}\,,\ B_{pro} = \frac{p}{r_o}A_{po}\,. \tag{2.126}$$

Equivalently, the **air-gap power transfer based on flux density** is

$$P_{r\to s,p} = \frac{2\pi\ell\omega_p r_i r_o}{\mu_0(\zeta^{-p} - \zeta^p)}B_{pri}B_{pro}\sin(\phi_{pi0} - \phi_{po0})\,. \tag{2.127}$$

This formula provides a sense of the relationship between machine size and power. Let us consider the dominant pole-pair in the machine, which is responsible for the bulk of the power transfer. We can approximate $r_i \approx r_o \approx R$ and $B_{pri} \approx B_{pro}$. Hence, the power grows with the radius of the rotor squared, and is proportional to the axial length. In other words, power is proportional to the rotor volume, $V_r \approx \ell\pi R^2$. Power is also proportional to the square of the radial flux density in the air gap. Therefore, to increase power density (W/m^3), we need to increase the magnetic field as much as possible. Alternatively, power density may be increased by speeding up the field rotation, which is related to the speed of the rotor. For instance, the rotor velocity in a synchronous machine is the same as that of the dominant pole-pair field, ω_p.

Example 2.8 *Air-gap power calculation.*

Consider a four-pole machine with the following geometric parameters: $g = 1$ mm, $r_i = 50$ mm, $r_o = 51$ mm, $\ell = 100$ mm. The electrical angular frequency is $\omega = 120\pi$ rad/s (i.e., the electrical frequency is $f = 60$ Hz). These dimensions imply that we are dealing with a relatively small machine. Based on our best guess, this machine could be rated for roughly 1–10 kW. Calculate the power transferred by the dominant pole-pair field for $p = 2$.

Solution

According to Equation (2.118), the rotating field mechanical angular velocity is $\omega_p = 60\pi$ rad/s (or 1800 rpm, i.e., revolutions per minute, also commonly denoted as r/min or min^{-1}). Suppose that the radial flux density magnitudes are $B_{pri} = B_{pro}/\zeta = 1$ T. This choice leads to identical MVP peak values at the two boundaries, and thus to identical flux over each pole at both sides. (This condition may not be exactly what happens in practice, but it is a decent approximation.)

The air-gap power can be calculated from Equation (2.127):

$$P_{r\to s,p} = (2.97 \cdot 10^6) \cdot \sin(\phi_{pi0} - \phi_{po0})\ \text{W}\,. \tag{2.128}$$

The numerical value of the coefficient preceding the sine is enormous. This arises because we are dividing by μ_0 and again by $(\zeta^{-p} - \zeta^p)$, which are both very small numbers. Therefore, to obtain a power within the expected power range of this device, the only logical explanation is that the angular difference between the inner and outer boundary MVP cannot practically exceed a fraction of a degree. For instance, for a power transfer of -10 kW (motor operation), we obtain $\phi_{pi0} - \phi_{po0} \approx -0.19°$!

Waveforms of the radial and tangential components of flux density and Poynting's vector are provided in Figure 2.17. These quantities have been calculated at the middle of the air gap at some arbitrary instant in time. Due to the tiny angular difference between the MVP boundary conditions, the radial component of flux density dominates; however, a sizeable tangential component of flux density is also obtained. Poynting's vector has a dominant component in the tangential direction that points clockwise. Its useful component is the radial one, which is negative in this case, causing power transfer from stator to rotor.

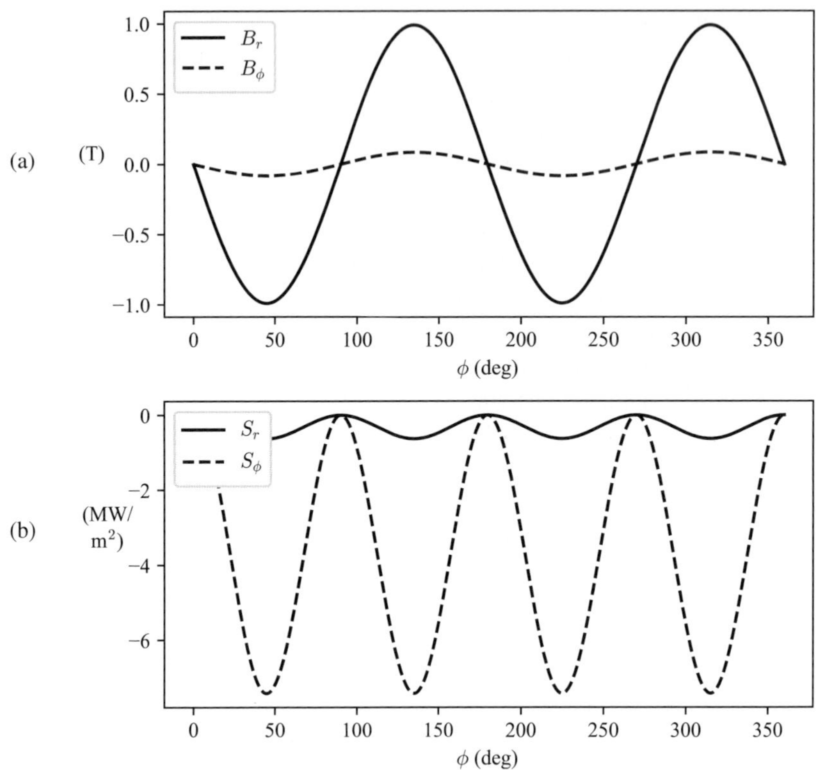

Figure 2.17 (a) Flux density and (b) Poynting's vector in the air gap.

Example 2.9 *Air-gap power based on a decomposition of the total field into stator and rotor components.*

The analysis thus far has been in terms of the *total* field in the air gap without consideration of how the field was created in the first place. Nevertheless, it is instructive to separate stator and rotor field components, that is, fields generated by current sources in the stator or rotor, respectively. This decomposition is valid in the case of linear magnetic materials, where the principle of superposition holds. In practice, when dealing with devices that are constructed with nonlinear materials, the total B-field cannot be calculated simply as the addition of two fields, since the permeability of the materials is not constant. However, for the sake of understanding a basic principle of operation, we can assume that the permeability of the rotor and stator materials (which can be thousands of times higher than that of air) is infinite.

Solution

We will work with a simplistic but effective machine model with two poles ($p = 1$). First, we will assume that the rotor and stator are perfectly round and isotropic, which means that there are no slots or other geometric features that could distort the field. This model could be representative of a squirrel-cage induction machine or a round-rotor synchronous machine. Therefore, conductors cannot be placed in slots, and the current sources are represented by equivalent surface current densities sitting on the rotor and stator surfaces. The air-gap region extends from just outside the rotor surface to just inside the stator surface. Since the permeability of the rotor and stator materials has been assumed infinite, it forces the flux lines to emerge from the steel in the radial direction. However, note that the presence of the rotor/stator surface current densities introduces a tangential component in the magnetic field (see Example 1.23 on page 37). Since the H-field is zero inside infinitely permeable steel (so that B attains a finite value), the tangential component of H right inside the air gap equals the adjacent surface current density. The stator field is perpendicular to the rotor surface, whereas the rotor field is perpendicular to the stator surface.

This situation is illustrated in Figure 2.18. The plots on the left depict (a) the stator and (b) the rotor field flux lines in the air gap. The flux lines inside the rotor and stator are irrelevant for this analysis, so they are not plotted. The dots and crosses indicate the polarity of the current densities. At this particular snapshot (say, at time $t = 0$), these can be described by the sinusoidal distributions

$$\vec{\mathbf{K}}_s = K_s \cos(\phi - 120°)\,\hat{\mathbf{k}}\,, \tag{2.129a}$$

$$\vec{\mathbf{K}}_r = K_r \cos\phi\,\hat{\mathbf{k}}\,, \tag{2.129b}$$

where $K_s, K_r > 0$ are unspecified constants. The current densities (as well as the corresponding magnetic fields) are rotating counterclockwise in synchronism with each other at constant angular velocity $\omega_p = \omega$, thus creating a rotating magnetic field in the air gap. The arrows indicate the principal directionality and magnitude of the two B-fields at the time instant of the snapshot. We are also identifying the location of the north and south poles of the two fields on the surface of the material where the respective current sources are located. In this case, we have defined the stator field to lead the rotor field by 120°, which translates to a time of $2\pi/3\omega$ seconds from the perspective of a stationary observer sitting on the stator surface. The fields are "stronger" close to their respective sources. The flux lines that do not cross the air gap are representative of a **leakage flux**, that is, a component of the flux that does not circulate around the entire magnetic circuit.

Since the field in this example comprises only a single pole-pair, let us drop the subscript p from the equations, and let us denote the Fourier coefficients of the two fields (see Equations (2.103a) and (2.103b)) as α_s, β_s, γ_s, δ_s and α_r, β_r, γ_r, δ_r. On the inner boundary, the MVP of the total field is

$$A(r_i, \phi) = (\alpha_s + \alpha_r)\cos\phi + (\beta_s + \beta_r)\sin\phi\,, \tag{2.130a}$$

and on the outer boundary, the MVP of the total field is

$$A(r_o, \phi) = (\gamma_s + \gamma_r)\cos\phi + (\delta_s + \delta_r)\sin\phi\,. \tag{2.130b}$$

To account for the effects of leakage, the Fourier coefficients are related by introducing scaling factors $c_s, c_r < 1$:

$$\alpha_s = c_s\gamma_s\,,\quad \beta_s = c_s\delta_s\,, \tag{2.131a}$$

$$\gamma_r = c_r\alpha_r\,,\quad \delta_r = c_r\beta_r\,. \tag{2.131b}$$

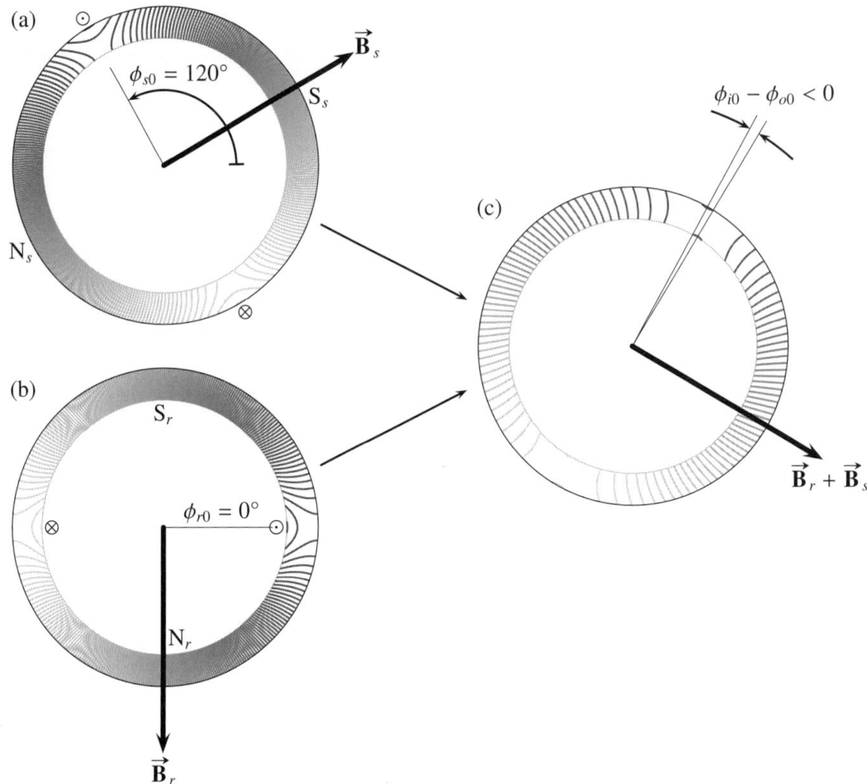

Figure 2.18 (a) Stator, (b) rotor, and (c) total field flux lines. The plots on the left include 150 flux lines in order to make the leakage flux lines clearly visible. The plot on the right uses 50 flux lines so that the tangential distortion of the field is visible.

Each individual field does not cause an angular displacement of the MVP waveform on the boundaries; that is, the inner and outer boundary MVP peaks are perfectly aligned with the winding dots over time. Hence, *each field acting on its own cannot transfer power across the air gap.*

The perpendicularity of the stator/rotor field to the rotor/stator surface means that the corresponding tangential component of the B-field vanishes. Hence, by setting $B_\phi = 0$ in Equation (2.107) for $r = r_i$ and $r = r_o$, we obtain the following relationships between the field coefficients:

$$c_s = c_r = \frac{\alpha_s}{\gamma_s} = \frac{\beta_s}{\delta_s} = \frac{\gamma_r}{\alpha_r} = \frac{\delta_r}{\beta_r} = \frac{2}{\zeta^p + \zeta^{-p}} . \tag{2.132}$$

Interestingly, this implies that the flux reduction factor is the same for both stator and rotor coils, $c = c_s = c_r$, with a value that is determined entirely by the air-gap dimensions and the pole-pair number. In this case study, we set $r_i = 2$ and $r_o = 2.5$, leading to $c = 0.976$. (This means that we lose 2.4% of the flux as leakage.)

Substituting these results in the power transfer equation (2.125c) and ignoring constant terms, yields

$$P_{r\to s} \sim \beta\gamma - \alpha\delta = (\beta_s + \beta_r)(\gamma_s + \gamma_r) - (\alpha_s + \alpha_r)(\delta_s + \delta_r) \tag{2.133a}$$

$$= (c\delta_s + \beta_r)(\gamma_s + c\alpha_r) - (c\gamma_s + \alpha_r)(\delta_s + c\beta_r) \tag{2.133b}$$

$$= (1 - c^2)(\beta_r\gamma_s - \alpha_r\delta_s)\,. \tag{2.133c}$$

It is helpful to rewrite this equation in terms of MVP magnitudes (A_{so}, A_{ri}) and peak angles ($\phi_s = \omega t + \phi_{s0}$, $\phi_r = \omega t + \phi_{r0}$) based on

$$\alpha_r = A_{ri}\cos\phi_r,\;\; \beta_r = A_{ri}\sin\phi_r\,, \tag{2.134a}$$

$$\gamma_s = A_{so}\cos\phi_s,\;\; \delta_s = A_{so}\sin\phi_s\,. \tag{2.134b}$$

Elementary trigonometric manipulations yield

$$P_{r\to s} \sim (1 - c^2)A_{ri}A_{so}\sin(\phi_{r0} - \phi_{s0})\,. \tag{2.135}$$

In contrast to what was happening when each field was acting alone, we see that *stator and rotor fields acting in unison can indeed lead to nonzero power transfer across the air gap*. The power depends sinusoidally on the angular displacement between the stator and rotor MVP waveforms. Although this result looks similar to Equation (2.125b), it should be noted that this is *not* the same formula. The previous equation was in terms of the *total* field, whereas the new equation is in terms of separate stator and rotor-field components. For example, to generate Figure 2.18, we have set $A_{so} = A_{ri} = 1$, $\phi_{s0} = 120°$, and $\phi_{r0} = 0°$. In terms of the total field, we obtain $A_i = A_o = 0.988$, $\phi_{i0} = 58.8°$, and $\phi_{o0} = 61.2°$. We can readily confirm that

$$A_iA_o\sin(\phi_{i0} - \phi_{o0}) = (1 - c^2)A_{ri}A_{so}\sin(\phi_{r0} - \phi_{s0}) = -0.0417\,. \tag{2.136}$$

We may also rewrite this in terms of the B-field radial components, B_{rr} and B_{sr}, based on Equation (2.126) with $p = 1$:

$$B_{rr} = \frac{1}{r_i}A_{ri}\,,\;\; B_{sr} = \frac{1}{r_o}A_{so}\,. \tag{2.137}$$

Note that B_{rr} is measured on the rotor surface, whereas B_{rs} is measured on the stator surface. Therefore

$$P_{r\to s} \sim (1 - c^2)r_ir_oB_{rr}B_{sr}\sin(\phi_{r0} - \phi_{s0})\,. \tag{2.138}$$

This result resembles Equation (2.127), but it is not the same formula because the former equation involves the total field. We are now better positioned to resolve the apparent paradox that was mentioned in our discussion of Equation (2.94) on page 125. The math is telling us that, under simplifying assumptions, we can express power in terms of the radial components of the stator/rotor fields and their relative angular displacement, with a coefficient that ultimately depends on the air-gap dimensions. However, even if the tangential field does not enter into the calculation, we should not be assuming that it is zero! Recall that we derived this result by applying Poynting's theorem, starting from a surface integral that involves a tangential B_ϕ, whose presence is essential for power transfer. We should understand that the two concepts are not mutually exclusive. Introductory machine treatments focus on the radial components because these are typically dominant and easier to estimate under simplifying assumptions. Low-level, physics-based analyses, which rely on a detailed solution of the air-gap field, may instead utilize the nonzero tangential B_ϕ component in a calculation of power (and torque) as a surface integral. A variety of such methods will be presented later in this text.

A geometric interpretation becomes possible if we define vectors for the two radial fields that are aligned with their principal directionality ($\vec{\mathbf{B}}_s$ and $\vec{\mathbf{B}}_r$ in Figure 2.18). We are assuming that the page is the x–y plane. To avoid any confusion, note that these vectors are not representative of the B-field at some point in space; rather, they are mathematical constructs of a different kind. The total field can be found as the vector sum of the stator and rotor fields, $\vec{\mathbf{B}} = \vec{\mathbf{B}}_s + \vec{\mathbf{B}}_r$. In this case, we find that $\vec{\mathbf{B}}$ is principally directed along $-30°$. Hence, we may conclude that **the power transfer may be visualized as a cross product between the rotating stator and rotor magnetic fields or between the total field and individual fields**:

$$P_{r\to s} \sim (\vec{\mathbf{B}}_s \times \vec{\mathbf{B}}_r) \cdot \hat{\mathbf{k}} \tag{2.139a}$$

$$= (\vec{\mathbf{B}} \times \vec{\mathbf{B}}_r) \cdot \hat{\mathbf{k}} \tag{2.139b}$$

$$= (\vec{\mathbf{B}}_s \times \vec{\mathbf{B}}) \cdot \hat{\mathbf{k}}\,. \tag{2.139c}$$

Since these cross products resemble the torque equation of the elementary magnetic dipole (1.288) on page 66, they appear commonly in introductory electric machine analyses for pedagogical reasons.

In this example, the cross product between the B-fields yields a vector that points along the negative z-axis, which means that power is transferred from stator to rotor. This could also be inferred by the positions of the north and south poles. It is as if the stator north/south poles are repelling the closest rotor north/south poles, which are ahead of them by 60° as they turn. In other words, it is as if the stator field is "pushing" the rotor field, thus leading to motor action.

2.3.3 Energy Balance in Conductors

The first of the energy balance equations, (2.83a) on page 121, encapsulates the effects of current flow in **conductors** forming electrical circuits. Conductors (e.g., magnet wire) are embedded in the device and interconnected appropriately to form **coils** or **windings**. Conductors may also come in the form of solid bars, commonly found in squirrel-cage induction motors. Eddy currents may also flow inside materials, which could be desirable (e.g., in an eddy current brake) or undesirable (e.g., in a transformer). The circuits may be moving in space, but we will assume that they remain rigid (i.e., their shape does not change). We will calculate the energy that is exchanged with the magnetic field through interaction with the electric current flowing in these circuits. In particular, the expressions that will be derived will be in terms of field quantities. At this point, details about the machine design and the placement of the conductors are not needed.

First, we recall some concepts from physics relating to current flow in conducting materials. Consider a conductor that carries current i. Within the conducting material of this circuit (e.g., copper or aluminum), imagine a tiny **filament** or current tube of cross-section da carrying an element of current di, as illustrated in Figure 2.19. For modeling purposes, the filament can be visualized as containing a stream of electrons flowing through the circuit; electrons do not escape the filament as they move through the circuit. The **law of conservation of charge** is what makes the current solenoidal, thereby making this macroscopic model of current flow realistic. Of course, this is only a simple illustration, whereas in reality a conductor may follow an arbitrary path within the device. For instance, a winding in an electric machine can have a complicated geometric pattern. Nevertheless, our analysis is valid irrespective of winding shape.

The shaded areas in Figure 2.19 represent coil terminals, where connections to external circuits are made. Let us call the terminal where current is entering the "positive terminal"; the other terminal thus becomes the "negative terminal." It is assumed that the terminals extend outside the magnetically active part of the device. Hence, $B \approx 0$ in the vicinity of the terminals; however, an electrostatic field exists in their vicinity. The terminals can be modeled as equipotential surfaces, where $\varphi = \varphi_A$ or $\varphi = \varphi_B$. This implies that the electric field $\vec{\mathbf{E}}$ is perpendicular to the terminal surfaces. We should keep in mind that electric machines may also contain internal conductors (e.g., the damper windings in synchronous generators), which are not connected to external circuits. The current filaments belonging to such conductors may not even come in the form of loops (e.g., the rotor bars of squirrel-cage motors).

Inside a metal conductor, the electrons move with an average **drift velocity** $\vec{\mathbf{v}}_d$, which is tangential to the filament. Due to collisions of electrons with the material lattice, energy is needed to maintain a constant flow of current. The power required for moving an average electron of charge q_e is $\vec{\mathbf{F}} \cdot \vec{\mathbf{v}}_d$, where $\vec{\mathbf{F}} = q_e\vec{\mathbf{E}}$ is the Lorentz force (see Example 2.4). The total power needed for a filamentary flow of current is found by summing over all electrons in the filament. Of course, it is impossible to calculate this

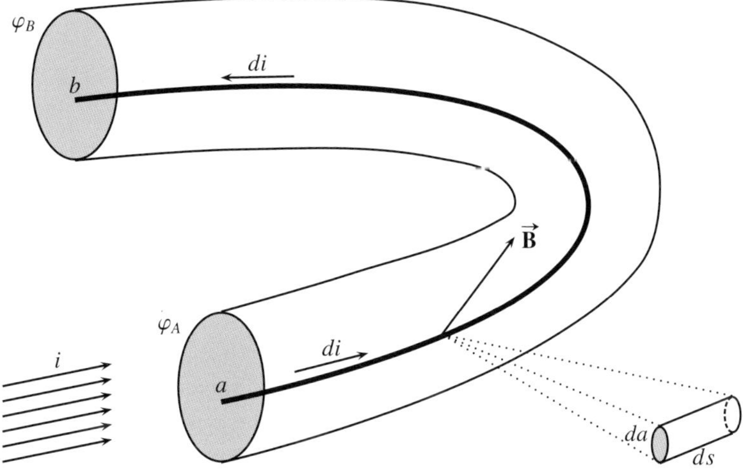

Figure 2.19 A filament or current tube inside a conductor.

sum literally, so it is conveniently converted to an integral by taking a macroscopic view. To this end, we introduce the **moving charge density** ρ_e, representing coulombs per cubic meter of electrons that are free to move within the conductor. Hence, over an entire filament occupying the volume $\mathcal{D}_f$, we need to expend power

$$P = \iiint_{\mathcal{D}_f} \rho_e \vec{\mathbf{E}} \cdot \vec{\mathbf{v}}_d \, dv \,. \tag{2.140}$$

The current density is

$$\vec{\mathbf{J}} = \rho_e \vec{\mathbf{v}}_d \,, \tag{2.141}$$

which is the flow of charge per conductor cross-sectional area per second. Therefore

$$P = \iiint_{\mathcal{D}_f} \vec{\mathbf{E}} \cdot \vec{\mathbf{J}} \, dv \,. \tag{2.142}$$

This formula reflects what is known as **Joule's**[49] **law**. The integrand $\vec{\mathbf{E}} \cdot \vec{\mathbf{J}}$ equals the watts per unit volume that are converted to heat from the steady electron flow, also known as **ohmic heating** or loss. It has been experimentally determined that the current density is related to the electric field by the constitutive law

$$\vec{\mathbf{J}} = \sigma \vec{\mathbf{E}} \,, \tag{2.143}$$

where σ is the electrical **conductivity** of the medium in S (siemens[50]) or mhos per meter. We can express the loss density of Joule's law in various forms, e.g.

$$\vec{\mathbf{E}} \cdot \vec{\mathbf{J}} = \sigma \vec{\mathbf{E}} \cdot \vec{\mathbf{E}} = \sigma E^2 = \vec{\mathbf{J}} \cdot \vec{\mathbf{J}} / \sigma = J^2 / \sigma \,. \tag{2.144}$$

The ohmic loss may be supplied by an external **electromotive force** (EMF), by the magnetic field itself, and/or by mechanical action (through the field). To understand this phenomenon better, consider the general expression for the electric field inside a filament that moves with velocity $\vec{\mathbf{v}}$, which is

$$\vec{\mathbf{E}} = -\nabla\varphi - \frac{\partial \vec{\mathbf{A}}}{\partial t} + \vec{\mathbf{v}} \times \vec{\mathbf{B}} \,. \tag{2.145}$$

Note that this differs from the previous expression (2.40a) in that it contains an extra $\vec{\mathbf{v}} \times \vec{\mathbf{B}}$ term. This is due to an important conceptual difference. Whereas Equation (2.40a) is the electric field in a stationary reference frame, Equation (2.145) is a non-relativistic expession of the electric field in a moving reference frame (that of the wire, where we must remain for this analysis). The presence of a $\nabla\varphi \neq 0$ term does *not* imply that the field is created by a charge distribution internal to the conductor; rather, this term is nonzero in the presence of surface charges on the conductor terminals and outer walls. The magnetic field variables $\vec{\mathbf{A}}$ and $\vec{\mathbf{B}} = \nabla \times \vec{\mathbf{A}}$ are representative of the field created by all current sources in the device.

If the device is excited by constant (d.c.) sources, and if the device does not contain moving parts, or even if it does but motion is constrained (i.e., $\vec{\mathbf{v}} = \mathbf{0}$), then we would expect that sooner or later a steady state will be obtained. This means that the fields at any location inside the device are not changing with time, i.e., $\partial \vec{\mathbf{A}} / \partial t = \mathbf{0}$. In this case, the electric field in conductors is just $\vec{\mathbf{E}} = -\nabla\varphi$. Hence, there is no energy exchange with the magnetic field ($W_e = 0$), and the energy supplied by the d.c. sources (W_E) is converted in its entirety to heat inside the conductors ($W_{E,\text{loss}}$).

In general, currents could be changing and/or conductor motion may be occurring. Hence, a nonzero $\partial\vec{\mathbf{A}}/\partial t$ and/or $\vec{\mathbf{v}}$ may be present in Equation (2.145), implying that the ohmic loss is supplied by energy coming from an external electrical source as well as from the magnetic field. Therefore, the energy balance of a current filament satisfies

$$\begin{aligned}\text{ohmic loss } [\vec{\mathbf{E}}\cdot\vec{\mathbf{J}}] &= \text{power from electrical source } [-\nabla\varphi\cdot\vec{\mathbf{J}}] \\ &\quad + \text{power from magnetic field } [(-\partial\vec{\mathbf{A}}/\partial t + \vec{\mathbf{v}}\times\vec{\mathbf{B}})\cdot\vec{\mathbf{J}}]. \quad (2.146)\end{aligned}$$

Whereas the left-hand side is always non-negative, the terms on the right-hand side can take positive or negative values, indicating the possibility of bi-directional energy flow between the electrical source(s) and the magnetic coupling field. Integrating over all filaments and over a period of time, we obtain Equation (2.83a) in the form

$$W_{E,\text{loss}} = W_E - W_e\,. \quad (2.147)$$

Example 2.10 *Power supplied to filament from an electrical source.*
Show that the power supplied to a filament from an external electrical source is given by the product of voltage times current.

Proof The power density (in W/m^3) supplied to a filament by the electrostatic (or slowly varying, quasi-electrostatic) field that is created by the action of an external electrical source is $-\nabla\varphi\cdot\vec{\mathbf{J}}$. Since the energy associated with a filament is infinitesimal, we will denote it by dW. In particular, filamentary energy variations over a small amount of time δt will be denoted as $\delta(dW)$. Hence, the amount of energy supplied to the filament by the electric source over δt, denoted as $\delta(dW_E)$, can be found by integrating over the entire filament volume:

$$\delta(dW_E) = \iiint_{\mathcal{D}_f} (-\nabla\varphi\cdot\vec{\mathbf{J}})\,dv\,\delta t\,. \quad (2.148)$$

Using Equation (1.158), we rewrite this as

$$\delta(dW_E) = \iiint_{\mathcal{D}_f} [\varphi\nabla\cdot\vec{\mathbf{J}} - \nabla\cdot(\varphi\vec{\mathbf{J}})]\,dv\,\delta t\,. \quad (2.149)$$

Charge may not accumulate inside conductors, so

$$\nabla\cdot\vec{\mathbf{J}} = -\frac{\partial\rho}{\partial t} = 0\,, \quad (2.150)$$

making the first term in the integrand vanish. The second term is transformed using the divergence theorem, leading to

$$\delta(dW_E) = -\oiint_{\partial\mathcal{D}_f} \varphi\vec{\mathbf{J}}\cdot d\mathbf{a}\,\delta t\,. \quad (2.151)$$

Since $\vec{\mathbf{J}} = di/da\,\hat{\mathbf{t}}$, where $\hat{\mathbf{t}}$ is the unit normal vector in the tangential direction of current flow, we obtain

$$\delta(dW_E) = di\,(\varphi_A - \varphi_B)\,\delta t = di\,V_{AB}\,\delta t\,. \tag{2.152}$$

Hence, the filament power equals the voltage across its endpoints times the current. □

Example 2.11 *Power exchange between filament and coupling field due to conductor motion in a constant magnetic field.*

Consider the component of power supplied to the filament from the coupling field due to the motional $\vec{\mathbf{v}} \times \vec{\mathbf{B}}$ term in Equation (2.146). Assume that the magnetic field is constant (so $\partial\vec{\mathbf{A}}/\partial t = \mathbf{0}$). Derive (i) the Lorentz force density and (ii) an expression for the energy exchanged between the filament and the magnetic field in terms of the flux linkage.

Solution

An infinitesimal amount of energy supplied over δt to the filament from the field is

$$-\delta(dW_e) = \iiint_{\mathcal{D}_f} (\vec{\mathbf{v}} \times \vec{\mathbf{B}}) \cdot \vec{\mathbf{J}}\,dv\,\delta t\,. \tag{2.153}$$

We can apply the scalar triple-product identity (1.41) to rewrite the integrand as

$$\vec{\mathbf{J}} \cdot (\vec{\mathbf{v}} \times \vec{\mathbf{B}}) = \vec{\mathbf{v}} \cdot (\vec{\mathbf{B}} \times \vec{\mathbf{J}}) = \vec{\mathbf{B}} \cdot (\vec{\mathbf{J}} \times \vec{\mathbf{v}})\,. \tag{2.154}$$

The middle term, which is in the familiar form of force times velocity, is thus recognized as mechanical power. By changing sign we obtain power flow *to* the coupling field. In the absence of energy being stored in the field (B is constant), this power must transfer to the mechanical side. Hence, the term

$$\vec{\mathbf{f}} = \vec{\mathbf{J}} \times \vec{\mathbf{B}} \tag{2.155}$$

represents the electromagnetic Lorentz force exerted by the field on a unit volume element of current. Of course, after integration over the entire volume, this also leads to a well-known formula for the force on a straight, current-carrying wire, $\vec{\mathbf{F}} = i\vec{\ell} \times \vec{\mathbf{B}}$.

To interpret the third term of the equality, it is instructive to draw a picture of a closed loop-forming filament in motion, e.g., see Figure 2.20. The displacement occurs within a small time δt, where the filament moves by $\delta\mathbf{r} = \vec{\mathbf{v}}\,\delta t$. Now take an infinitesimal element of the filament of length ds. Then

$$(\vec{\mathbf{J}} \times \vec{\mathbf{v}})\,dv = \left(\frac{di}{da}\,\hat{\mathbf{t}}\right) \times \left(\frac{\delta\mathbf{r}}{\delta t}\right)(ds\,da) = di\,\frac{d\mathbf{s} \times \delta\mathbf{r}}{\delta t}\,. \tag{2.156}$$

The cross product $d\mathbf{s}\times\delta\mathbf{r}$ has magnitude equal to the shaded area, and is directed parallel to the outwards-pointing normal vector $\hat{\mathbf{n}}$:

$$d\mathbf{s} \times \delta\mathbf{r} = da'\,\hat{\mathbf{n}}\,. \tag{2.157}$$

(We designate the shaded area by da', to differentiate it from the filament cross-sectional

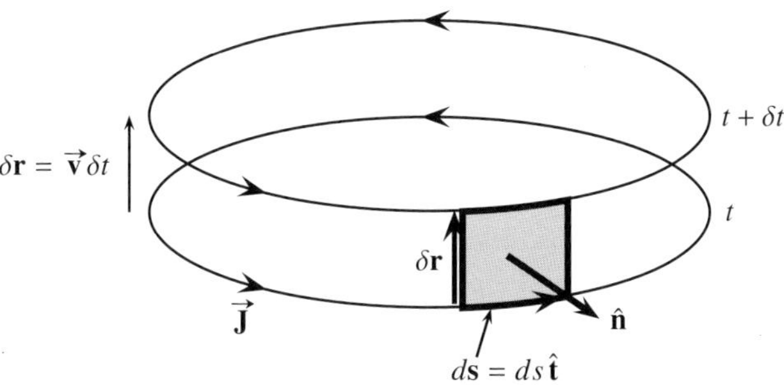

Figure 2.20 A filament in motion. For illustration purposes, the displacement is uniform throughout the filament, but this does not have to be necessarily the case. In other words, the velocity could be a function of position. For example, consider a (rigid) filament that rotates.

area da.) Hence, we can express the energy transferred to the filament within δt as the surface integral

$$-\delta(dW_e) = \iint_{\delta S} di\,\vec{\mathbf{B}} \cdot \hat{\mathbf{n}}\,da', \tag{2.158}$$

where δS denotes the surface of a "ribbon" defined by the two filament contours at times t and $t + \delta t$.

The reader may recognize that the term

$$\delta\psi = -\iint_{\delta S} \vec{\mathbf{B}} \cdot \hat{\mathbf{n}}\,da' \tag{2.159}$$

carries the physical significance of the incremental change in the flux linking the filament, or **flux linkage**, due to its motion. The minus sign is present because the normal vector $\hat{\mathbf{n}}$ is pointing in the opposite direction of positive flux through the filament, which is in turn determined by the direction of positive current. To see this, imagine an ashtray that is formed by the two filament contours, with a base defined by the first contour. The flux through the filament at time t can be found by integrating the B-field over this base. For the assumed direction of positive current, the flux should be positive if it crosses the base upwards. We can calculate the filament flux at time $t + \delta t$ by integrating the B-field over the ashtray base plus its sides (i.e., the ribbon δS) since we can select any open surface that has the filament as its boundary. If the magnetic field stays constant, the change in the filament flux is $\psi(t + \delta t) - \psi(t) = \delta\psi$ of Equation (2.159). Hence, we have arrived at the following expression for the incremental energy supplied from the electrical subsystem to the magnetic coupling field:

$$\delta(dW_e) = di\,\delta\psi\big|_{\vec{\mathbf{A}}\text{ constant}}, \tag{2.160}$$

where the change in flux $\delta\psi$ is due to the motion of the filament, but not due to a change of the magnetic potential.

2.3.4 The Coupling Field Energy

In this section, our main objective will be to calculate the energy that is stored in the coupling field. To this end, the fundamental assumption of a conservative coupling field plays a key role because it allows us to express the field energy as a single-valued function of the magnetic field within the device. In turn, the B-field depends on free and bound currents and on the position of the moving components. Therefore, in a device with a single mechanical degree of freedom θ (the usual case), the coupling field energy can be expressed as

$$W_f = W_f(\vec{\mathbf{J}}, \theta), \tag{2.161}$$

where θ could be representative of an angular or linear position. More precisely, W_f is the energy that we have transferred (i.e., stored) in the coupling field by introducing permanent magnets (if any), and then through the electrical and mechanical subsystems by energizing the coils from zero to a final current density $\vec{\mathbf{J}}$, and by moving to a position θ.

The field energy W_f can be calculated via a virtual assembly of the device as follows. We shall imagine that the device is constructed by bringing all its components from "infinity" to their actual locations in space, while maintaining all conductor currents to zero. If the device uses permanent magnets, we can imagine that these are placed at their respective locations in a totally demagnetized state. This assembly process does not affect the stored energy because in the absence of free or bound currents the magnetic field is null, and the electromagnetic force is zero. The moving components are then brought to their desired position θ, where they are locked (so that $d\theta = 0$ afterwards).

Subsequently, we magnetize the permanent magnets (if any are present) in situ, setting their persistent magnetization to the desired value $\vec{\mathbf{M}}_0$. This magnetization process is virtual, and does not need to be defined precisely. We have thus created a magnetic field $\vec{\mathbf{B}}_0'(\mathbf{r})$ permeating the device. At this stage, we should have a completely functional but electrically de-energized machine ($\vec{\mathbf{J}}_0 = \mathbf{0}$). The coupling field energy is unknown, but we can denote it as $W_{f0}(\theta) \geq 0$.[k] Note that the initial condition we have imposed here (i.e., that the free currents are zero) is not identical to the condition that we imposed when calculating the energy density in terms of the fields for the magnetostatic functional. (In §2.2.3, we defined the initial magnetic field $\vec{\mathbf{H}}_0 = \mathbf{0}$, and the initial flux density was $\vec{\mathbf{B}}_0 \neq \vec{\mathbf{B}}_0'$. Therefore, when permanent magnets are present, $W_{f0} \neq \iiint m_0 \, dv$.)

Finally, the currents are increased to the desired value $\vec{\mathbf{J}}$. The exact trajectory that is followed to change the currents is irrelevant due to the conservative field assumption. At the end of this process, an additional amount of energy, $W_f(\vec{\mathbf{J}}, \theta) - W_{f0}(\theta)$, has been transferred to the coupling field from the electrical sources through the conductor filaments. We can imagine that this process involves a stepwise change of currents (like a Riemann integral). Let us focus on an incremental change of the field energy through a single filament, $\delta(dW_f)$, taking place over a small time δt. Since motion is restricted, the change in mechanical energy is $\delta(dW_m) = 0$, and in view of Equation (2.83b)

[k] Part of this energy can be recovered through mechanical work done by the electromagnetic force created by the permanent-magnet field.

on page 121, the electrical energy is transferred in its entirety to the coupling field, $\delta(dW_e) = \delta(dW_f)$. Hence, using Equation (2.146):

$$\delta(dW_f) = \iiint_{\mathcal{D}_f} \vec{\mathbf{J}} \cdot \delta\vec{\mathbf{A}}\, dv\,. \tag{2.162}$$

Since all moving parts have been locked, $\vec{\mathbf{v}} = \mathbf{0}$, and the $\vec{\mathbf{v}} \times \vec{\mathbf{B}}$ term vanishes. Note that we have eliminated the minus sign appearing in $-\partial\vec{\mathbf{A}}/\partial t$, which implies that we have reversed the presumed direction of power flow so that $\delta(dW_f)$ is positive when energy is absorbed by the field.

We can now calculate the total energy transfer to the entire device by accounting for all filaments. Hence, **the incremental change of the coupling field energy** is

$$\boxed{\delta W_f = \iiint_{\mathcal{D}_c} \vec{\mathbf{J}} \cdot \delta\vec{\mathbf{A}}\, dv\,.} \tag{2.163}$$

Note that $\vec{\mathbf{J}} = \vec{\mathbf{J}}(\mathbf{r})$, signifying that conductors carry different currents, and even accounting for the possibility of an uneven current distribution within the same conductor. The integral is over the entire conducting region $\mathcal{D}_c$, but it can be extended over the entire device $\mathcal{D}$ since $\vec{\mathbf{J}} = \mathbf{0}$ outside conductors.

An important result can be obtained with a little vector calculus. (A similar manipulation was performed within the proof of the magnetostatic functional.) We replace $\vec{\mathbf{J}} = \nabla \times \vec{\mathbf{H}}$, and invoke the identity for the divergence of the cross product (1.159), which we rewrite as

$$(\nabla \times \vec{\mathbf{F}}) \cdot \vec{\mathbf{G}} = \nabla \cdot (\vec{\mathbf{F}} \times \vec{\mathbf{G}}) + \vec{\mathbf{F}} \cdot (\nabla \times \vec{\mathbf{G}})\,. \tag{2.164}$$

Hence

$$\delta W_f = \iiint_{\mathcal{D}} \nabla \cdot (\vec{\mathbf{H}} \times \delta\vec{\mathbf{A}})\, dv + \iiint_{\mathcal{D}} \vec{\mathbf{H}} \cdot (\nabla \times \delta\vec{\mathbf{A}})\, dv \tag{2.165a}$$

$$= \oiint_{\partial\mathcal{D}} (\vec{\mathbf{H}} \times \delta\vec{\mathbf{A}}) \cdot d\mathbf{a} + \iiint_{\mathcal{D}} \vec{\mathbf{H}} \cdot \delta\vec{\mathbf{B}}\, dv\,, \tag{2.165b}$$

where we made use of the divergence theorem. Finally, we make use of the fact that $\delta\vec{\mathbf{A}} = \mathbf{0}$ at the outer boundary of the device (implying no magnetic flux can escape this surface), thus obtaining an alternate expression for the **incremental change of coupling field energy based on the magnetic field throughout space**:

$$\boxed{\delta W_f = \iiint_{\mathcal{D}} \vec{\mathbf{H}} \cdot \delta\vec{\mathbf{B}}\, dv\,.} \tag{2.166}$$

To summarize, we have derived two expressions for the energy that is absorbed by the magnetic field from the electrical subsystem (under the condition that motion is restricted). Equation (2.163) depends on the current that flows in the conductors and the change of the magnetic potential in their vicinity. On the other hand, Equation (2.166) shows more clearly how this energy is distributed all over space through the action of the magnetic field. Although this expression has been derived as an integral, some people prefer to ascribe physical significance to the term $\vec{\mathbf{H}} \cdot \delta\vec{\mathbf{B}}$, as the change in the energy density locally at each point in space. Recall that we have previously encountered δW_f as the variation of the first term of the magnetostatic problem functional $\delta(\iiint m\,dv)$, where m was defined by Equations (2.71) and (2.72) on page 115.

Therefore, the total energy provided to the coupling field (starting from a demagnetized state, keeping the position fixed at θ) can be expressed as

$$W_f = \underbrace{\iiint_{\mathcal{D}} \left(m_0(\mathbf{r}) + \int_{\vec{\mathbf{B}}_0(\mathbf{r})}^{\vec{\mathbf{B}}_0'(\mathbf{r})} \vec{\mathbf{H}} \cdot d\vec{\mathbf{b}} \right) dv}_{W_{f0}} + \iiint_{\mathcal{D}} \int_{\vec{\mathbf{B}}_0'(\mathbf{r})}^{\vec{\mathbf{B}}(\mathbf{r})} \vec{\mathbf{H}} \cdot d\vec{\mathbf{b}}\, dv, \qquad (2.167)$$

where $\vec{\mathbf{B}}_0'(\mathbf{r})$ is the initial field due to the presence of permanent magnets (if any). It is implied that $\vec{\mathbf{H}}$ is given by a constitutive law as in Equation (2.70), based on the type of material present at point $\mathbf{r}$ in space when the moving member is positioned at θ. Since a unique solution for the field $\vec{\mathbf{B}} = \nabla \times \vec{\mathbf{A}}$ can be obtained from the stationarity of the functional for a given source distribution $\vec{\mathbf{J}}$, Equation (2.167) is essentially the same as (2.161).

It should be noted that the incremental energy expressions (2.163) and (2.166) do not rely on the assumption of a lossless material. They apply in general, and thus they may be used to calculate the energy that is injected into the field during any arbitrary transition. These expressions are, therefore, also valid for hysteretic materials. Nevertheless, whenever a single-valued function $\vec{\mathbf{H}} = \vec{\mathbf{H}}(\vec{\mathbf{B}})$ is representative of the material behavior, the path of integration is irrelevant, and the change in stored energy only depends on the initial and final magnetization (see Examples 2.5 and 2.6). This condition is reflected in Equation (2.167), which was written as a function of the magnetic field value only; we are thus implying that this integration is meaningful only for lossless materials and conservative fields.

Indeed, the definition of a function for the energy "stored" in the magnetic field, as in Equation (2.167), is only possible when the field is conservative. Such a function is meaningful because the field has the potential to return this amount of energy back to the outside world. Conversely, if the ferromagnetic material is lossy, then we cannot determine a single value for the energy stored based solely on the current magnetization state. In this case, the energy that is stored in the field depends on the trajectory followed during the magnetization process, and even on how fast the various physical quantities are changing.

2.3.5 Core Loss

Subjecting ferromagnetic and ferrimagnetic materials to alternating fields incurs energy loss. The case of the lossless isotropic nonlinear material that we described in Example 2.5 is only an idealization, which helps make the analysis tractable. This type of loss is typically referred to as **core loss** by electric machine analysts. Core loss stems from two different physical phenomena, namely, eddy current loss and hysteresis loss, which we shall describe briefly in the next sections. Material manufacturers always provide values for **total core loss** (i.e., eddy current plus hysteresis loss) in datasheets. (More precisely, they will typically guarantee that loss will not exceed a prescribed maximum value.) Usually, the loss is provided in terms of W/kg at certain operating frequencies.

Eddy Current Loss

Vortex-like currents, commonly known as **eddy currents** or **Foucault**[51] **currents**, are induced in conductive materials when subject to changing magnetic fields. The creation of these currents can be qualitatively explained by Maxwell's law for the electric field:

$$\nabla \times \vec{\mathbf{E}} = -\frac{\partial \vec{\mathbf{B}}}{\partial t}, \tag{2.168}$$

which implies that a time-varying magnetic field is responsible for the creation of an electric field in space with a nonzero curl. Hence, if this happens inside a conductive material, the creation of eddy currents is inevitable since $\vec{\mathbf{J}} = \sigma\vec{\mathbf{E}}$.

We shall conduct an analysis of this phenomenon within a long and thin piece of steel, namely, a **lamination**. Electric machines are commonly constructed by stacking together a number of steel laminations, whose thickness is in the approximate range 0.01–0.04 in (0.25–1.0 mm). The surface of the laminations is insulated to oppose the flow of eddy currents from one lamination to the next. The insulation is either via a naturally forming oxide on the lamination surface, or through a thin layer of organic or inorganic material coating, on the order of 0.2–5.0 μm thick. This helps reduce the magnitude of the eddy currents, as we shall see. Hence the associated ohmic losses are lowered, and the overall energy conversion efficiency of the device is improved.

Laminations are oriented by design so that the magnetic field travels within them along their long dimensions. Figure 2.21 depicts a lamination that is centered on the x–y plane. Here, for simplicity, let us suppose that the magnetic field enters perpendicularly to the thin side, so $\vec{\mathbf{B}} = B_x\,\hat{\mathbf{i}}$. Let us also assume that the B-field is uniform along the lamination y–z cross-section. The changing magnetic field is also a vector directed along the x-axis. Therefore, a circulating current will be induced within the lamination, which will travel parallel to its long sides, changing direction once it reaches close to the ends. For example, if the B-field is increasing in the direction shown in Figure 2.21, the induced eddy current will tend to oppose the change, thus circulating in the counterclockwise sense. The same eddy current will be present throughout the lamination width, i.e., in each y–z cross-section from $x = 0$ to $x = w$.

We shall study a single lamination subject to an externally imposed, uniform magnetic field. It is implicitly assumed that the surface insulation is perfect so that current does not flow between adjacent laminations (not shown). We also assume that the lamination height h is much larger than its thickness τ. This simplifies the analysis because we can neglect variations of the fields along the y-axis for most of the lamination height, as long as we stay relatively far from the edges. We can thus compute the electric field and current in any horizontal (x–z) cross-section. One limitation of our analysis here is that we are neglecting the fields that are created by the eddy currents themselves. This assumption will be relaxed later in §5.3.1.

These modeling assumptions imply that $\partial/\partial x = 0$ and $\partial/\partial y = 0$, and that the E-field

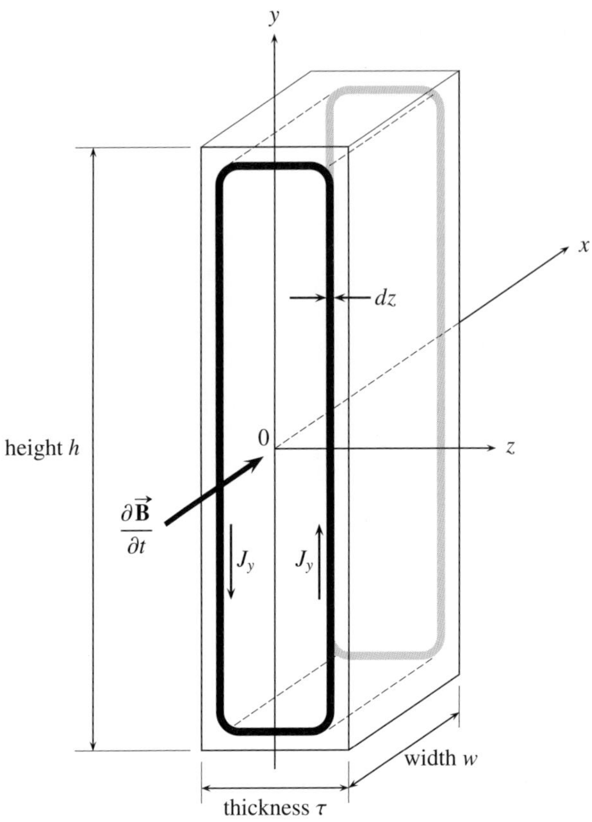

Figure 2.21 3-D view of a thin lamination (not drawn to scale) for eddy current analysis.

only has a y-component, $\vec{\mathbf{E}} = E_y\hat{\mathbf{j}}$. Hence

$$\nabla \times \vec{\mathbf{E}} = -\frac{\partial E_y}{\partial z}\hat{\mathbf{i}} = -\frac{\partial B_x}{\partial t}\hat{\mathbf{i}}, \tag{2.169}$$

or

$$\frac{\partial E_y}{\partial z} = \frac{\partial B_x}{\partial t}. \tag{2.170}$$

Under a uniform B-field, this implies that E_y experiences a constant rate of increase from $z = -\tau/2$ to $z = +\tau/2$. The solution is

$$E_y(z) = \frac{\partial B_x}{\partial t}z, \tag{2.171}$$

and the current density is

$$J_y(z) = \sigma\frac{\partial B_x}{\partial t}z, \tag{2.172}$$

where σ is the lamination conductivity. According to Equation (2.144), the instantaneous, local ohmic loss density (in W/m^3) is

$$\frac{J_y^2(z)}{\sigma} = \sigma\left(\frac{\partial B_x}{\partial t}z\right)^2 . \tag{2.173}$$

In other words, the total instantaneous loss (in W) in a current tube of thickness dz and width $2z$, as shown in Figure 2.21, is

$$2\sigma\left(\frac{\partial B_x}{\partial t}\right)^2 whz^2 dz , \tag{2.174}$$

ignoring the loss at the upper and lower edges. The total instantaneous loss (in W) can be found by integrating over the entire lamination volume:

$$P_{\text{eddy}} = \int_0^{\tau/2} 2\sigma\left(\frac{\partial B_x}{\partial t}\right)^2 whz^2 dz = \sigma\left(\frac{\partial B_x}{\partial t}\right)^2 \frac{\tau^3 wh}{12} . \tag{2.175}$$

Alternatively, we may calculate the instantaneous, volumetric average power loss (in W/m^3) by dividing the previous expression by the lamination volume τwh, which yields

$$p_{\text{eddy}} = \frac{P_{\text{eddy}}}{\tau wh} = \sigma\left(\frac{\partial B_x}{\partial t}\right)^2 \frac{\tau^2}{12} . \tag{2.176}$$

Now suppose that the lamination is subject to a purely sinusoidal excitation

$$B_x(t) = B_{\max} \cos \omega t . \tag{2.177}$$

Then, the instantaneous eddy current loss per unit volume is

$$p_{\text{eddy}}(t) = \frac{\sigma\omega^2 B_{\max}^2 \tau^2}{12} \sin^2 \omega t , \tag{2.178}$$

and the time-average eddy current loss per unit volume can be expressed as

$$\overline{p_{\text{eddy}}} = \frac{\sigma\pi^2 f^2 \tau^2 B_{\max}^2}{6} . \tag{2.179}$$

Example 2.12 *Eddy current loss estimation.*

Suppose we are considering using a 0.35-mm thick steel lamination that, in our application, will be subject to a 60-Hz sinusoidally varying magnetic field peaking at 1.7 T. The lamination material is a non-oriented electrical steel. The manufacturer provides the material resistivity as 54 μΩ·cm at 20°C, and its density as 7.65 g/cm^3. During operation, the material will be at 50°C. Estimate the eddy current loss at this operating point using Equation (2.179).

Solution

First, we may adjust the resistivity for temperature according to the linear law

$$\rho(T) = \rho_{20} \cdot [1 + \alpha(T - 20)] , \tag{2.180}$$

where ρ_{20} is the given value, and T is the operating temperature (in °C). Suppose the temperature coefficient of the material is $\alpha = 0.0047$ μΩ·cm/K. Hence

$$\rho(50) = 54 \cdot [1 + 0.0047 \cdot (50 - 20)] = 61.6 \ \mu\Omega\cdot\text{cm}\,, \tag{2.181}$$

which represents a 14% increase in resistivity from the nominal value. Therefore, the conductivity becomes

$$\sigma = \frac{1}{\rho} = \frac{1}{61.6 \cdot 10^{-8}} = 1.62 \cdot 10^{6} \ \text{S/m}\,. \tag{2.182}$$

The increase in temperature is associated with a small reduction in eddy current loss.

Equation (2.179) yields the time-average eddy current loss per unit volume:

$$\overline{p_{\text{eddy}}} = \frac{(1.62 \cdot 10^{6})(\pi^2)(60^2)(0.35^2 \cdot 10^{-6})(1.7^2)}{6} = 3.4 \ \text{kW/m}^3\,. \tag{2.183}$$

We may also calculate the time-average eddy current loss per unit mass, also known as the **specific eddy current loss**. To this end, we use the material density D.[l] Hence

$$\frac{\overline{p_{\text{eddy}}}}{D} = \frac{3.4 \cdot 10^3 \ \text{W/m}^3}{7.65 \cdot 10^3 \ \text{kg/m}^3} = 0.44 \ \text{W/kg} = 0.2 \ \text{W/lb}\,. \tag{2.184}$$

We note that Equation (2.179) should be applied with caution. It is not meant to be a highly precise formula for the calculation of eddy current loss. However, it provides insights into the various factors that affect this component. For instance, it reveals that the loss is expected to be proportional to the square of the lamination thickness, and (under purely sinusoidal excitation) the square of the operating frequency and the square of the B-field magnitude. These trends are useful to machine designers, who are tasked with selecting the best materials for their respective applications. More generally, the eddy current loss formula could be expressed as

$$\overline{p_{\text{eddy}}} = k_e f^2 \tau^2 B_{\max}^2\,, \tag{2.185}$$

where the coefficient k_e, which in general depends on operating conditions, may be obtained by performing experimental measurements on actual materials and devices, accounting for all sorts of non-idealities that have been ignored in the analysis.

Magnetic Hysteresis in Ferromagnetic Materials

Here, we will provide a brief description of the phenomenon of **magnetic hysteresis**, which is a Greek word meaning to "lag behind." In general, hysteresis is due to domain wall[m] motion and other effects at the atomic scale within ferromagnetic materials. Electric machine analysis relies on empirical loss models of this highly nonlinear phenomenon.

[l] Since the symbol ρ is used here for resistivity, we will denote the density as D. (D is also the symbol for the electric displacement field.)

[m] Domain walls form the interfaces between magnetic (Weiss) domains in magnetic materials.

Hysteresis is typically illustrated as in Figure 2.22, which depicts a **major loop**. This is the B–H trajectory followed while the material is subjected to a symmetric cyclic perturbation at a given frequency (the curve shape may depend on frequency). Here, as the magnetic field density and intensity are varied between $(H_1, B_1) = (-H_0, -B_0)$ and $(H_2, B_2) = (H_0, B_0)$, the magnetization of the material is such that we move along the path $CDEFC$ in a counterclockwise sense. In contrast, a **minor loop** is the trajectory that is obtained when the excitation is not symmetric around the origin.

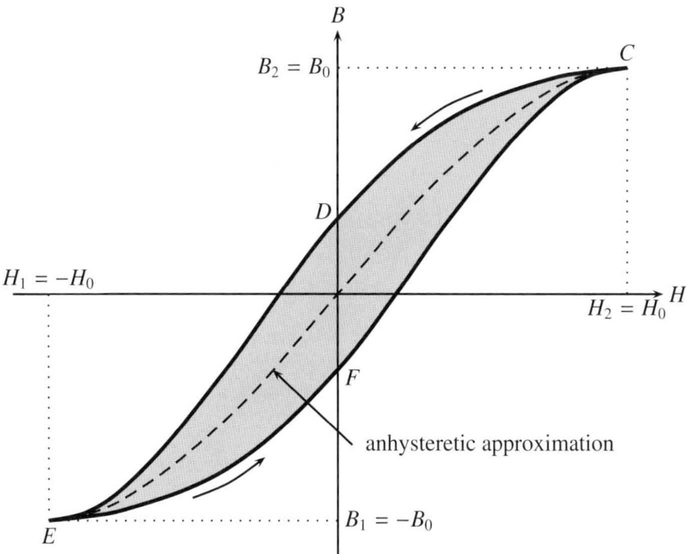

Figure 2.22 A typical hysteretic B–H characteristic of a ferromagnetic material.

The shaded area represents the net amount of energy (in terms of J/m^3) that is dissipated as heat over each cycle. To prove this, let us integrate the change in energy density as follows, introducing breakpoints based on the sign changes of H and dB:

$$\oint_{CDEFC} H\,dB = \int_{CD} H\,dB + \int_{DE} H\,dB + \int_{EF} H\,dB + \int_{FC} H\,dB \tag{2.186a}$$

$$= -(B_2DC) + (B_1DE) - (B_1FE) + (B_2FC) \tag{2.186b}$$

$$= [(B_2FC) - (B_2DC)] + [(B_1DE) - (B_1FE)] \tag{2.186c}$$

$$= (DFC) + (FDE) \tag{2.186d}$$

$$= \text{shaded area}\,. \tag{2.186e}$$

Here, we used $(\cdot)$ to denote positive-valued surface areas.

Under steady-state sinusoidal excitation conditions, a major loop is traversed continually with a frequency dictated by the source. Therefore, hysteresis loss in terms of power density (W/m^3) is directly proportional to frequency. In 1890, Steinmetz[52] observed experimentally that the hysteresis loss follows a remarkably simple mathematical law:

$$\overline{p_{\mathrm{hyst}}} = k_h f B_{\max}^{1.6}, \tag{2.187}$$

where k_h is a coefficient that Steinmetz referred to as "magnetic resistance" since it is a coefficient of conversion of magnetic energy into heat, similar to the electric resistance that can be thought of as a coefficient of conversion of electric energy into heat [1]. The exponent 1.6 was established empirically. However, the loop shape (and thus its area) may change with frequency. To capture this and other non-idealities, generalized Steinmetz formulas such as

$$\overline{p_{\mathrm{hyst}}} = k_h f^a B_{\max}^b \tag{2.188}$$

have been proposed. The exponents a and b are typically determined experimentally.

Figure 2.22 also depicts an anhysteretic approximation of the B–H characteristic, similar to the one shown in Example 2.5. This curve can be used for analysis purposes, e.g., for solving Maxwell's equations using the FEM, where the material is typically modeled as lossless. This is usually a decent approximation because modern electrical steels are manufactured for low core loss, and therefore exhibit relatively thin hysteresis loops.

2.3.6 Circuit-Based Energy Flow Analysis

At this point, it is useful to draw some connections between the fields-based results developed in the previous sections and circuit theory, which the reader may be more familiar with. We shall establish the relationships between the electromagnetic field quantities and circuit-level variables like the voltage and the flux linkage.

Let us consider a coil and filaments within. A closed path C can be formed by connecting the filament endpoints; starting from the positive terminal (where current is entering), we move towards the negative terminal in the direction of positive current flow inside the conductor, and we close the loop through an external path in air. This is depicted in Figure 2.23. (If the coil is moving, then this represents a snapshot of the coil position at an arbitrary instant in time.) We now evaluate the circulation of $\vec{\mathbf{E}}$ around this path:

$$\oint_C \vec{\mathbf{E}} \cdot d\mathbf{r} = \int_{a1b} \vec{\mathbf{E}} \cdot d\mathbf{r} + \int_{b2a} \vec{\mathbf{E}} \cdot d\mathbf{r}, \tag{2.189}$$

where $a1b$ is the path inside the filament, and $b2a$ is the path outside.

The left-hand-side integral will only contain the effect of nonconservative terms; that is, any contribution from an electrostatic field will vanish. The result is called the **induced voltage** V_{ind}. This voltage arises due to a changing magnetic field and/or motion of the coil with velocity $\vec{\mathbf{v}}(\mathbf{r})$ within a magnetic field (motional EMF). Using Equation (2.145), we have

$$V_{\mathrm{ind}} = \oint_C \vec{\mathbf{E}} \cdot d\mathbf{r} = \oint_C \left(-\frac{\partial \vec{\mathbf{A}}}{\partial t} + \vec{\mathbf{v}} \times \vec{\mathbf{B}} \right) \cdot d\mathbf{r}. \tag{2.190}$$

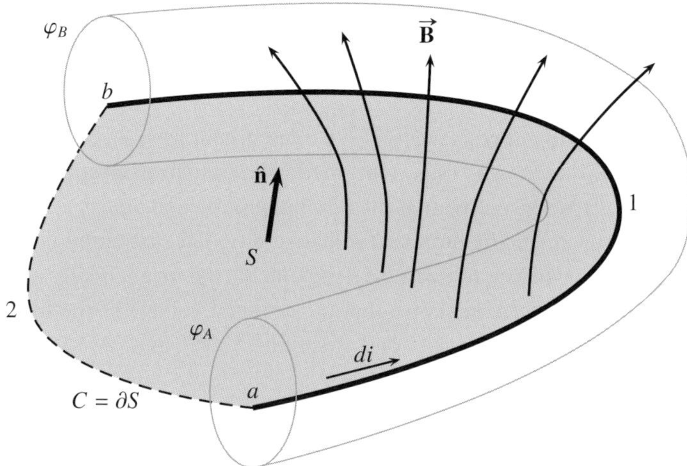

Figure 2.23 Magnetic flux linking a current filament.

The fact that $\vec{\mathbf{B}} = \nabla \times \vec{\mathbf{A}}$ implies that $\partial \vec{\mathbf{B}}/\partial t = \partial(\nabla \times \vec{\mathbf{A}})/\partial t = \nabla \times (\partial \vec{\mathbf{A}}/\partial t)$, by changing the order of differentiation. Invoking Stokes' theorem, we can convert the first term of the line integral into a surface integral, leading to

$$V_{\text{ind}} = -\iint_S \frac{\partial \vec{\mathbf{B}}}{\partial t} \cdot d\mathbf{a} + \oint_C \vec{\mathbf{v}} \times \vec{\mathbf{B}} \cdot d\mathbf{r}, \tag{2.191}$$

where S is any surface that is bounded by the contour $C = \partial S$. This surface may be hard to visualize in a real winding, which could have a complex (e.g., helical) shape. This is **Faraday's law**. It can be shown that the induced voltage equals the total derivative

$$V_{\text{ind}} = -\frac{d}{dt} \iint_S \vec{\mathbf{B}} \cdot d\mathbf{a} = -\frac{d\psi}{dt}, \tag{2.192}$$

where

$$\psi = \oint_C \vec{\mathbf{A}} \cdot d\mathbf{r} = \iint_S \vec{\mathbf{B}} \cdot d\mathbf{a} \tag{2.193}$$

is the **magnetic flux linked** by the filament. Taking the total derivative of flux implies that we are accounting for the effects of a changing magnetic potential with respect to time, as well as for the motion of the conductor in space.

The presence of the minus sign in the induced voltage equation is referred to as **Lenz's**[53] **law**. If the flux (which is positive in the direction shown in Figure 2.23, as determined by the right-hand rule applied to the line integral) is increasing ($d\psi/dt > 0$), then the induced voltage will be negative. This voltage tends to drive current out of terminal a and towards terminal b (through an external circuit). This current, in turn, creates a magnetic field that "opposes" the change in flux linkage.

Now, working with the right-hand side of Equation (2.189), we obtain

$$V_{\text{ind}} = \int_{a1b} \frac{\vec{\mathbf{J}}}{\sigma} \cdot d\mathbf{r} + (\varphi_B - \varphi_A) = R_f\, di - V_{ab}, \tag{2.194}$$

where

$$R_f = \int_{a1b} \frac{dr}{\sigma\, da} \tag{2.195}$$

is the filament **resistance**. (This integral will have a very large value due to division by da. However, it is then multiplied by di which is infinitesimal, so the voltage drop is finite.) We also note that the $b2a$ integral, which equals $-V_{ab}$, is valid only when there is no magnetic field present in its vicinity. The derivation is valid under the assumption that the winding terminals extend outside the magnetically active region of our device.

Hence, we have shown that the terminal voltage of the filament can be expressed as the familiar inductive circuit equation

$$V_{ab} = R_f\, di + \frac{d\psi}{dt}\,. \tag{2.196}$$

It follows that the incremental amount of energy supplied to the filament by the electric source, $\delta(dW_E)$, is

$$\delta(dW_E) = V_{ab}\, di\, \delta t = \quad R_f(di)^2\delta t \quad + \; di\, \delta\psi \tag{2.197a}$$

$$= \delta(dW_{E,\text{loss}}) \; + \; \delta(dW_f + dW_m)\,. \tag{2.197b}$$

The electrical energy is partially converted into ohmic loss (heat), whereas the remainder flows to the magnetic field. The coupling field stores some of this energy, $\delta(dW_f)$, while the rest is converted to mechanical energy, $\delta(dW_m)$. This energy balance equation is analogous to the fields-based expression (2.146).

Low-Frequency Winding Model

For circuit-level analysis, we are interested in the behavior of entire windings, not the individual filaments themselves. So, we rewrite the incremental change in field energy of Equation (2.163) (i.e., the amount of energy entering from the electrical side, keeping the position fixed) by performing the volume integral first along filamentary paths and then over conductor cross-sections S_c:

$$\delta W_f = \iint_{\tilde{\mathbf{r}}\in S_c} J(\tilde{\mathbf{r}}) \int_{a(\tilde{\mathbf{r}})}^{b(\tilde{\mathbf{r}})} \delta\vec{\mathbf{A}}(\mathbf{r}) \cdot d\mathbf{r}\, da = \iint_{\tilde{\mathbf{r}}\in S_c} J(\tilde{\mathbf{r}})\, \delta\psi(\tilde{\mathbf{r}})\, da\,. \tag{2.198}$$

Here, $\tilde{\mathbf{r}} \in S_c$ is a point on the conductor cross-section, so $a(\tilde{\mathbf{r}})$ and $b(\tilde{\mathbf{r}})$ are the filament start and end points, respectively. In turn, these points are sufficient to determine the change in filament flux linkage, $\delta\psi(\tilde{\mathbf{r}})$. This equation also reflects the fact that the current density $J(\tilde{\mathbf{r}})$ may depend on the position within the cross-section. The integration is over all conducting regions in the device.

It is important to note that $\psi(\tilde{\mathbf{r}})$ is not assumed to be constant throughout a conductor. The assumption of uniform flux linkage would have been valid for relatively thin conductor wires. For thicker wires, the validity of the assumption may deteriorate due to magnetic flux lines crossing the wire, so we leave it in this form for generality.

At high frequencies, it is known that the current tends to move closer to the conductor outer surface due to the skin effect (which will be presented in more detail in Chapter 5).

To avoid such complications, we proceed with a **low-frequency winding model**. Suppose that the device comprises K circuits (coils). In this case, it may be safe to assume that *the current density is constant over the cross-section of each wire*. The current in conductor k is $i_k = J_k A_k$, where $A_k = |S_k|$ is the conductor cross-sectional area. Then, Equation (2.198) yields

$$\delta W_f = \sum_{k=1}^{K} \frac{i_k}{A_k} \iint_{\tilde{\mathbf{r}} \in S_k} \delta\psi(\tilde{\mathbf{r}})\, da\,. \tag{2.199}$$

This process has led us to the following definition of the **winding flux linkage**:

$$\boxed{\lambda_k = \frac{1}{A_k} \iint_{\tilde{\mathbf{r}} \in S_k} \psi(\tilde{\mathbf{r}})\, da = \frac{1}{A_k} \iint_{\tilde{\mathbf{r}} \in S_k} \int_{a(\tilde{\mathbf{r}})}^{b(\tilde{\mathbf{r}})} \vec{\mathbf{A}}(\mathbf{r}) \cdot d\mathbf{r}\, da\,.} \tag{2.200}$$

In view of Equation (2.193), the winding flux linkage is defined as the average value of the filament fluxes over the coil.

We can thus derive a **winding voltage** equation by integrating Equation (2.196) over the terminal surface S_k of circuit k. This yields

$$V_k A_k = i_k \frac{\ell_k}{\sigma_k} + \iint_{S_k} \frac{d\psi}{dt}\, da\,, \tag{2.201}$$

assuming that filaments have equal length ℓ_k. Now the definition of the winding flux linkage comes in handy, as it allows us to write

$$V_k = R_k i_k + \frac{d\lambda_k}{dt}\,, \tag{2.202}$$

using the well-known definition for the (d.c.) resistance

$$R_k = \frac{\ell_k}{\sigma_k A_k}\,. \tag{2.203}$$

Using Equations (2.199) and (2.200), we can express the incremental change in field energy as

$$\delta W_f = \sum_{k=1}^{K} i_k\, \delta\lambda_k = \mathbf{i} \cdot \delta\boldsymbol{\lambda}\,. \tag{2.204}$$

The energy transferred to the magnetic field during a transition from magnetization state A to state B (restricting any mechanical movement) can be informally written as the integral

$$\Delta W_f = \int_A^B \mathbf{i} \cdot d\boldsymbol{\lambda}\,. \tag{2.205}$$

If the magnetic field is conservative (lossless), this integration will only depend on the endpoints, but not on the integration path. In this case, we can define the field energy similarly to Equation (2.167) as

$$W_f = W_{f0} + \int_{\boldsymbol{\lambda}_0}^{\boldsymbol{\lambda}} \mathbf{i} \cdot d\tilde{\boldsymbol{\lambda}}\,, \tag{2.206}$$

where W_{f0} is an initial permanent-magnet field energy component, and λ_0 is the initial flux linkage of the coils due to the presence of permanent magnets (if any) when the device is electrically de-energized ($\mathbf{i} = \mathbf{0}$).

The preceding discussion has led us to an interesting idea. When the device operates at relatively low frequencies, we can reduce the dimensionality of the problem. Whereas the original magnetostatic functional of Equation (2.71) has infinite degrees of freedom (i.e., the magnetic vector potential at each point in space, which yields the flux linking each filament), we now have obtained a finite-dimensional problem, commonly referred to as a **lumped-parameter model**. The imposed current density $\vec{\mathbf{J}}$, which acts as the source of the magnetic field (together with the permanent magnets), is replaced by K winding currents i_k. Moreover, the magnetic vector potential $\vec{\mathbf{A}}$ throughout space is replaced by K coil flux linkages λ_k. (The surface current density is assumed to be $\vec{\mathbf{K}} = \mathbf{0}$.) These thoughts lead to the result that is highlighted in Box 2.3, which connects the functional to the energy of a system of coils.

Box 2.3 Key finding: Relationship of magnetostatic functional to electric circuit variables, energy, and coenergy

In a low-frequency lumped-parameter device model, the magnetostatic functional of Equation (2.71) on page 115 has the same physical significance as the following function of winding flux linkages $g(\boldsymbol{\lambda})$:

$$I(\vec{\mathbf{A}}) = g(\boldsymbol{\lambda}) = W_{f0} + \int_{\boldsymbol{\lambda}_0}^{\boldsymbol{\lambda}} \mathbf{i} \cdot d\tilde{\boldsymbol{\lambda}} - \mathbf{i} \cdot \boldsymbol{\lambda} = W_f(\boldsymbol{\lambda}; \theta) - (i_1\lambda_1 + \cdots + i_K\lambda_K), \quad (2.207)$$

where $W_f(\boldsymbol{\lambda}; \theta)$ denotes the energy stored in the magnetic field when each winding has flux linkage $\boldsymbol{\lambda}$ (at mechanical position θ).

In circuit analysis, the **coenergy** is defined as

$$W_c = \mathbf{i} \cdot \boldsymbol{\lambda} - W_f . \quad (2.208)$$

Hence, the functional is also analogous to the negative coenergy:

$$I(\vec{\mathbf{A}}) = -W_c . \quad (2.209)$$

The concept of coenergy introduced in Box 2.3 is defined therein based on circuit quantities (flux linkages and currents). Coenergy may also be defined at a more fundamental level using field quantities. By analogy with Equation (2.208), **the coenergy of a conservative field** is defined by

$$\boxed{W_c = \iiint_{\mathcal{D}_c} \vec{\mathbf{J}} \cdot \vec{\mathbf{A}}\, dv - W_f .} \quad (2.210)$$

The coenergy is essentially a mathematical transformation, also known as a **Legendre**[54] **transform**. The physical significance of this concept is somewhat elusive.

We can derive a different formula for the coenergy starting from the fact that a small change of $\vec{\mathbf{J}} \cdot \vec{\mathbf{A}}$ can be expressed as

$$\delta(\vec{\mathbf{J}} \cdot \vec{\mathbf{A}}) = (\delta\vec{\mathbf{J}}) \cdot \vec{\mathbf{A}} + \vec{\mathbf{J}} \cdot (\delta\vec{\mathbf{A}}) . \tag{2.211}$$

Hence, we can write the incremental change in field energy (by restricting motion) using Equation (2.163) as

$$\delta W_f = \iiint_{\mathcal{D}_c} \vec{\mathbf{J}} \cdot \delta\vec{\mathbf{A}}\, dv = \iiint_{\mathcal{D}_c} \delta(\vec{\mathbf{J}} \cdot \vec{\mathbf{A}})\, dv - \iiint_{\mathcal{D}_c} \vec{\mathbf{A}} \cdot \delta\vec{\mathbf{J}}\, dv . \tag{2.212}$$

Therefore, in a conservative field, the total field energy (calculated by fixing the mechanical position θ) can be expressed as

$$W_f = W_{f0} + \iiint_{\mathcal{D}_c} \vec{\mathbf{J}} \cdot \vec{\mathbf{A}}\, dv - \iiint_{\mathcal{D}_c} \int_{\mathbf{0}}^{\vec{\mathbf{J}}(\mathbf{r})} \vec{\mathbf{A}} \cdot d\vec{\mathbf{j}}\, dv . \tag{2.213}$$

Hence, in view of the definition (2.210), the coenergy of a device can be calculated by

$$W_c = W_{c0} + \iiint_{\mathcal{D}_c} \int_{\mathbf{0}}^{\vec{\mathbf{J}}(\mathbf{r})} \vec{\mathbf{A}} \cdot d\vec{\mathbf{j}}\, dv , \tag{2.214}$$

where we defined $W_{c0} = -W_{f0}$ (cf. Equation (2.167)). Whereas the definition of field energy (2.167) hinges on a single-valued constitutive law $\vec{\mathbf{H}}(\vec{\mathbf{B}})$, the coenergy is based on a mapping $\vec{\mathbf{J}} \to \vec{\mathbf{A}}$ from the field source to the magnetic potential throughout space, which again is characteristic of a conservative field. An analogous expression for calculating coenergy using circuit quantities is

$$W_c = W_{c0} + \int_{\mathbf{0}}^{\mathbf{i}} \boldsymbol{\lambda} \cdot d\tilde{\mathbf{i}} . \tag{2.215}$$

The concepts of energy and coenergy are often illustrated as the two complementary areas under a λ–i curve, as in Figure 2.24. This visualization serves pedagogical reasons, but is not very useful for performing numerical calculations. Note that the $\lambda(i, \theta)$ curve can shift up or down depending on the mechanical position.

Example 2.13 *Magnetically linear circuits and inductance.*

In magnetically linear systems, the flux linkages of the coils are linearly related to free and bound currents, and the superposition principle can be applied. This can happen either because the materials that make up the device are magnetically linear (i.e., the permeability does not depend on the magnetic field), or because we are ignoring (for the sake of simplicity or with good reason) the nonlinearity of the B–H characteristic of ferromagnetic materials.

For instance, in a device with two coils and permanent magnets, we can write the system of flux linkage equations

$$\lambda_1 = L_{11} i_1 + L_{12} i_2 + \lambda_{m1} , \tag{2.216a}$$

$$\lambda_2 = L_{21} i_1 + L_{22} i_2 + \lambda_{m2} . \tag{2.216b}$$

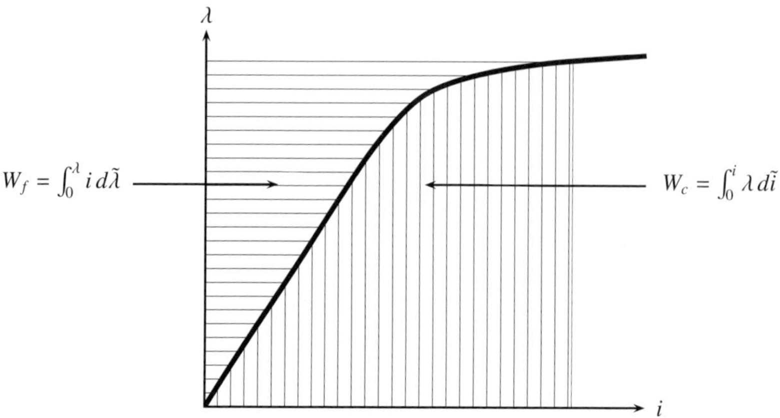

Figure 2.24 Energy and coenergy as areas under a λ–i curve.

Of course, the Ls in the above equations are the **self-** and **mutual inductances** of the coils, which are independent of the current values due to the assumption of magnetic linearity. However, the inductances can be functions of mechanical position. The λ_m terms are the flux linkages due to the permanent magnets; these are also functions of mechanical position. For a device with K coils and one mechanical degree of freedom, we can express these relationships as a matrix equation:

$$\boldsymbol{\lambda}(\mathbf{i}, \theta) = \mathbf{L}(\theta)\,\mathbf{i} + \boldsymbol{\lambda}_m(\theta)\,, \tag{2.217}$$

with $\mathbf{L}$ a square $K \times K$ **inductance matrix**, and where $\mathbf{i}$, $\boldsymbol{\lambda}$, and $\boldsymbol{\lambda}_m$ are $K \times 1$ column vectors.

The coupling field energy is obtained using Equation (2.206), which is written here in vector-matrix form:

$$W_f(\mathbf{i}, \theta) = W_{f0}(\theta) + \int_{\mathbf{0}}^{\mathbf{i}} \tilde{\mathbf{i}}^{\mathsf{T}}\,\mathbf{L}(\theta)\, d\tilde{\mathbf{i}}\,. \tag{2.218}$$

Note that the variable of integration has changed from flux linkage to current. The two-coil example is instructive. We have

$$W_f(\mathbf{i}, \theta) = W_{f0} + \int_{\mathbf{0}}^{\mathbf{i}} \begin{bmatrix} \tilde{i}_1 & \tilde{i}_2 \end{bmatrix} \begin{bmatrix} L_{11} & L_{12} \\ L_{21} & L_{22} \end{bmatrix} \begin{bmatrix} d\tilde{i}_1 \\ d\tilde{i}_2 \end{bmatrix} \tag{2.219a}$$

$$= W_{f0} + \int_{\mathbf{0}}^{\mathbf{i}} (L_{11}\tilde{i}_1 + L_{21}\tilde{i}_2)\, d\tilde{i}_1 + (L_{12}\tilde{i}_1 + L_{22}\tilde{i}_2)\, d\tilde{i}_2 \tag{2.219b}$$

$$= W_{f0} + \int_{\mathbf{0}}^{\mathbf{i}} f_1(\tilde{i}_1, \tilde{i}_2)\, d\tilde{i}_1 + f_2(\tilde{i}_1, \tilde{i}_2)\, d\tilde{i}_2\,. \tag{2.219c}$$

The dependence on θ is not shown for notational convenience. This is a line integral of the vector field $\mathbf{f} = (f_1, f_2)$ in the space of (i_1, i_2), where f has physical units of flux linkage.

Since the field is conservative, a mathematical theorem asserts that

$$\frac{\partial f_1}{\partial i_2} = \frac{\partial f_2}{\partial i_1} . \tag{2.220}$$

(Note that this condition is equivalent to $\nabla \times \mathbf{f} = \mathbf{0}$, which is also necessary for a conservative field since the circulation must be zero.) Substituting the expressions for f_1 and f_2, we obtain

$$L_{21} = L_{12} , \tag{2.221}$$

that is, the mutual inductances are necessarily equal, and $\mathbf{L}$ is symmetric. (This will be the case even in higher dimensions, i.e., when handling an arbitrary number of coils.) We note that this constitutes an alternative, energy-based proof of the **reciprocity theorem** for the mutual inductances.

Armed with this fact, let us work on calculating the line integral for the field energy. Since the field is conservative, we are free to choose any path joining the origin and the final set of currents $\mathbf{i}$. For convenience, let us integrate by ramping up i_1 first, keeping $i_2 = 0$ (so that $di_2 = 0$); and then ramp i_2 while maintaining i_1 at its final value (so that $di_1 = 0$). This trajectory yields

$$W_f(\mathbf{i}, \theta) = W_{f0} + \int_0^{i_1} (L_{11}\tilde{i}_1)\, d\tilde{i}_1 \ + \int_0^{i_2} (L_{12}i_1 + L_{22}\tilde{i}_2)\, d\tilde{i}_2 \tag{2.222a}$$

$$= W_{f0} + \left(\frac{1}{2}L_{11}i_1^2\right) + \left(L_{12}i_1 i_2 + \frac{1}{2}L_{22}i_2^2\right) \tag{2.222b}$$

$$= W_{f0}(\theta) + \frac{1}{2}\mathbf{i}^\top \mathbf{L}(\theta)\, \mathbf{i} \quad \text{(because } \mathbf{L} \text{ is symmetric)} \tag{2.222c}$$

$$= W_{f0}(\theta) + \frac{1}{2}\mathbf{i}^\top \boldsymbol{\lambda}(\mathbf{i}, \theta) - \frac{1}{2}\mathbf{i}^\top \boldsymbol{\lambda}_m(\theta) \quad \text{(in view of Equation (2.217))} . \tag{2.222d}$$

Therefore, in the magnetically linear case, the coupling field energy is a quadratic form of the currents. The inductance matrix $\mathbf{L}$ must be **positive definite** (and therefore, invertible) to guarantee that a strictly positive amount of energy is stored in the field. (A negative value does not make physical sense.) In devices without permanent magnets, we can express the field energy in terms of flux linkage:

$$W_f(\boldsymbol{\lambda}, \theta) = \frac{1}{2}\boldsymbol{\lambda}^\top \mathbf{L}^{-1}(\theta)\, \boldsymbol{\lambda} . \tag{2.223}$$

The coenergy is

$$W_c(\mathbf{i}, \theta) = \mathbf{i}^\top \boldsymbol{\lambda}(\mathbf{i}, \theta) - W_f(\mathbf{i}, \theta) \tag{2.224a}$$

$$= W_{c0}(\theta) + \frac{1}{2}\mathbf{i}^\top \boldsymbol{\lambda}(\mathbf{i}, \theta) + \frac{1}{2}\mathbf{i}^\top \boldsymbol{\lambda}_m(\theta) \tag{2.224b}$$

$$= W_{c0}(\theta) + \frac{1}{2}\mathbf{i}^\top \mathbf{L}(\theta)\, \mathbf{i} + \mathbf{i}^\top \boldsymbol{\lambda}_m(\theta) . \tag{2.224c}$$

We have thus shown that, in devices without permanent magnets, the field energy equals the coenergy, $W_f = W_c = \frac{1}{2}\mathbf{i}^\top \mathbf{L}(\theta)\, \mathbf{i}$. (However, note that this is not the case in magnetically nonlinear devices.)

Now that we have a more concrete expression for the field energy, let us revisit the observations made in Box 2.3. In view of Equation (2.207), the condition of a stationary functional translates to the requirement that

$$\frac{\partial g(\boldsymbol{\lambda})}{\partial \lambda_k} = 0, \quad k = 1, \ldots, K, \tag{2.225}$$

or that

$$\frac{\partial W_f(\boldsymbol{\lambda}; \theta)}{\partial \lambda_k} - i_k = 0, \quad k = 1, \ldots, K. \tag{2.226}$$

By inspection of Equation (2.223), we can readily verify that this condition holds in a magnetically linear circuit. (It will also hold in magnetically nonlinear circuits, as long as the coupling fields are conservative.)

2.3.7 Mechanical Energy Transfer: Electromagnetic Force and Torque

We will now discuss the process of mechanical energy exchange with the coupling field, as captured by W_m of Equation (2.83b), page 121. Of course, mechanical work requires motion (either displacement or rotation). The motion of a filament inside a constant magnetic field was the theme of Example 2.11, where it was found that this leads to the development of a force, and hence to electromagnetic-to-mechanical (or electromechanical, for short) energy conversion. It is thus tempting to conclude that the total electromagnetic force or torque can be obtained by summing up Lorentz force $\vec{\mathbf{J}} \times \vec{\mathbf{B}}$ contributions from all filaments in all conductors. However, this is not entirely correct.

As a matter of fact, in electric machines, typically the B-field is rather small within conductors! This happens because flux lines tend to bypass and flow around windings (under normal operating conditions), since they are embedded within ferromagnetic materials that have thousands of times higher permeability than copper or aluminum. Therefore, the $\vec{\mathbf{J}} \times \vec{\mathbf{B}}$ term is negligible within conductors, so the forces that are developed within them are insignificant. This leads to an apparent paradox. The confusion is further fomented by elementary "how motors work" machine treatments, where motor force is attributed to the action of current-carrying wires in a magnetic field.

The paradox is resolved when we realize that a magnetized ferromagnetic material is functionally equivalent to a current source, as shown in §1.7.3. So, the total electromagnetic force/torque is the outcome of the interaction of the magnetic field with *all* currents in the device, either within conductors (i.e., free currents) or within materials (i.e., bound currents), the latter actually contributing the major part in real machines. This also explains the creation of torque in machines without windings on the rotor, such as reluctance machines. We will delve into this further in §2.4.

For now, a more high-level view suffices. Again, we will rely on an energy balance argument, this time regarding the change in the coupling field energy. Any incremental change of magnetic potential $\delta\vec{\mathbf{A}}$ and position $\delta\mathbf{r}$ may be thought to occur in two stages: first, while motion is frozen, the coupling field absorbs an amount of energy from the electrical system δW_e through interaction with the current in the conductors; second, while keeping the magnetic state fixed, we let motion occur, thus allowing the coupling field to exchange energy δW_m with the mechanical subsystem. In a system with a single moving component, we may express the coupling field energy balance during this incremental change as

$$\delta W_f = \delta W_e - \delta W_m = \iiint_{\mathcal{D}_c} \vec{\mathbf{J}} \cdot \delta\vec{\mathbf{A}}\, dv \; - \; \vec{\mathbf{F}}_e \cdot \delta\mathbf{r}\,, \tag{2.227}$$

where $\vec{\mathbf{F}}_e$ is the total electromagnetic force or torque, depending on the physical significance of $\delta\mathbf{r}$. For example, $\delta\mathbf{r} = (\delta x, \delta y, \delta z)$ (in m) represents an arbitrary linear displacement of the moving member, in which case $\vec{\mathbf{F}}_e$ is the force vector (in N). Alternatively, $\delta\mathbf{r} = (\delta\theta_x, \delta\theta_y, \delta\theta_z)$ (in rad) may represent a small rotation along a certain axis in space, in which case $\vec{\mathbf{F}}_e$ is a torque vector (in N-m). The expression was written in this general form to allow for the case where the total force is not collinear with the displacement (e.g., due to some kinematic constraint). The sign convention that we have adopted implies that the direction of the force/torque vector is such that positive $\vec{\mathbf{F}}_e \cdot \delta\mathbf{r}$ represents a reduction in the coupling field energy. If a system has multiple moving components, the expression can be augmented by summing over all $\vec{\mathbf{F}}_{e,m} \cdot \delta\mathbf{r}_m$ terms, for $m = 1, 2, \ldots,$ representing the total work of the electromagnetic forces.

Therefore, assuming that a small movement occurs while keeping the magnetic potential over the conductors constant ($\delta\vec{\mathbf{A}} = \mathbf{0}$), thereby eliminating any energy exchange through the windings, we have

$$\delta W_f = -\vec{\mathbf{F}}_e \cdot \delta\mathbf{r}, \quad \text{with } \vec{\mathbf{A}} \text{ constant over conductors}\,. \tag{2.228}$$

Hence, by evaluating the partial derivative of the coupling field energy along any one of the coordinate system axes, denoted here by θ, we obtain the corresponding component of the **electromagnetic force or torque** vector:

$$\boxed{F_{e,\theta} = -\left.\frac{\partial W_f}{\partial \theta}\right|_{\vec{\mathbf{A}} \text{ constant over conductors}}} \,. \tag{2.229}$$

If a device has a single mechanical degree of freedom (e.g., a linear actuator or a common electric motor), it may thus be convenient to align the axes so that one of them coincides with the direction of motion. In this case, we can drop the subscript θ from the force since there is no ambiguity.

At first glance, the above expression seems like a result of a mostly theoretical nature. However, it forms the basis of a practical force/torque calculation method in FEA, namely, the virtual displacement method (see §4.1.3). Also, it leads directly through the definition of flux linkage (2.200) to a formula that is commonly used in (low-frequency) circuit-based analysis, namely

$$F_e = -\left.\frac{\partial W_f(\lambda, \theta)}{\partial \theta}\right|_{\lambda \text{ constant}} \,. \tag{2.230}$$

Identical arguments may be made using the coenergy. Combining Equation (2.227) with the definition (2.210), and using the mathematical identity (2.211), we obtain

$$\delta W_c = \iiint_{\mathcal{D}_c} \vec{\mathbf{A}} \cdot \delta \vec{\mathbf{J}}\, dv \; + \; \vec{\mathbf{F}}_e \cdot \delta \mathbf{r} \,. \tag{2.231}$$

Assuming that a small movement occurs by keeping the current constant ($\delta \vec{\mathbf{J}} = \mathbf{0}$):

$$\delta W_c = \vec{\mathbf{F}}_e \cdot \delta \mathbf{r}, \quad \text{with } \vec{\mathbf{J}} \text{ constant}\,. \tag{2.232}$$

Constraining the motion along the θ-axis, we obtain an alternative expression for the **electromagnetic force/torque in terms of the coenergy**:

$$\boxed{F_{e,\theta} = \left. \frac{\partial W_c}{\partial \theta} \right|_{\vec{\mathbf{J}} \text{ constant}} .} \tag{2.233}$$

In a circuit model (with a single mechanical degree of freedom), this is equivalent to

$$F_e = \left. \frac{\partial W_c(\mathbf{i}, \theta)}{\partial \theta} \right|_{\mathbf{i} \text{ constant}} . \tag{2.234}$$

Example 2.14 *Force in a magnetically linear device.*
Calculate the electromagnetic force or torque in a magnetically linear device.

Solution
It is convenient to apply the coenergy formula (2.234), where the coenergy is obtained from Equation (2.224c):

$$F_e = \frac{\partial W_{c0}}{\partial \theta} + \frac{1}{2} \mathbf{i}^\top \frac{\partial \mathbf{L}}{\partial \theta} \mathbf{i} + \mathbf{i}^\top \frac{\partial \boldsymbol{\lambda}_m}{\partial \theta} \,. \tag{2.235}$$

The first term of this expression yields a force that is due to the action of the permanent magnets alone. For example, this term would yield the force that attracts a magnet to a piece of iron, or the cogging torque in a de-energized permanent magnet motor. The other two terms represent forces developed by the interactions between coils and permanent magnets.

It could be argued that developing device models formulated in terms of currents as independent variables, rather than flux linkages, is more intuitive. (In layman's terms, it is easier to think of what happens when we first set currents in the coils and determine the flux linkages than the other way around, especially when multiple coils are mutually coupled.) When this is the case, force calculations based on the coenergy formula (2.234) are more direct. The following examples will illustrate the process of device modeling from first principles. This will be accomplished by proceeding along the conventional path:

$$\text{currents} \rightarrow \text{field} \rightarrow \text{flux} \rightarrow \text{inductance} \rightarrow \text{energy} \rightarrow \text{force}\,. \tag{2.236}$$

We will rely on various approximations since we do not yet possess the tools necessary to solve Maxwell's equations for the magnetic field. For a more accurate calculation, we will revisit these examples after we learn how to use the FEM.

Example 2.15 *Force developed by a horseshoe electromagnet.*

Consider a horseshoe-shaped (or U–I) electromagnet, as shown in Figure 2.25. An external source is providing voltage to the coil, which carries current i. The problem is to calculate the force on the piece of steel that we wish to pull. The steel is assumed to be magnetically linear, so the results of Example 2.14 apply.

The device parameters are:

- number of coil turns, $N = 20$;
- current in coil, $i = 10$ A;
- relative permeability of material in U, $\mu_{rU} = 5000$;
- relative permeability of material in I, $\mu_{rI} = 1000$;
- cross-sectional area of U, $A_U = 1\ \text{cm}^2$;
- cross-sectional area of I, $A_I = 0.5\ \text{cm}^2$;
- average radius of U, $R = 1.5$ cm;
- leg width of U, $w = 2$ cm;
- average flux path length in I, $h = 2R = 3$ cm;
- air-gap width, $x = 1$ mm.

Solution

These kinds of problems, involving relatively simple geometries, can be analyzed based on the "average" or "mean" flux path. This is the path of the "average flux line," which passes through the middle of the core, indicated by the dashed curve in the figure. Magnetic flux traverses the device centered on the average flux line. The flux density is assumed to be uniform over any cross-section around this mean flux path trajectory. Furthermore, for the sake of simplicity, all flux is assumed to be enclosed within the U–I core (except at those points where it jumps across the air gap). We are thus neglecting the small amount of flux that (always) leaks outside the material boundaries (e.g., crossing from the top to the bottom part of the U-core, bypassing the I). This should be a reasonably good approximation for devices with highly permeable ferromagnetic

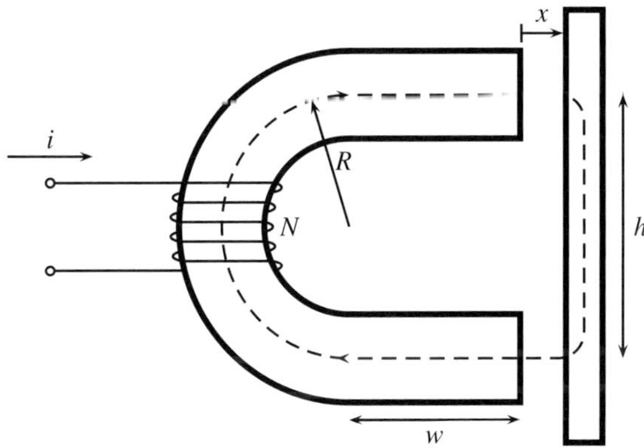

Figure 2.25 Geometry of an elementary horseshoe-shaped electromagnet. The dashed line represents the average flux line.

materials and small air gaps x along the average flux path. Sharp bends in the flux line are somewhat problematic in this model, but we will not let this trouble us. Such a simplified model should suffice for illustrating the force calculation method, although we recognize that the estimate may be rough.

First, we apply Ampère's law around the mean flux path ∂S that encloses a surface S, which may be taken to lie on the plane of the page:

$$\oint_{\partial S} \vec{\mathbf{H}} \cdot d\mathbf{r} = \iint_S \vec{\mathbf{J}} \cdot d\mathbf{a} = I_{\text{enclosed by contour } \partial S} = Ni\,. \tag{2.237}$$

The quantity that appears on the right-hand side of Ampère's law, Ni, is called the **magnetomotive force** (MMF), and is measured in At (ampere-turns). It is the total current enclosed by our integration path, crossing the integration surface at the N points where the coil intersects the page.

The circulation of the H-field is now broken into three parts: one for the U, one for the I, and one for the air gap (actually, this accounts for the top and bottom air-gap paths jointly). Each of these three parts will have potentially different B and H-fields. This leads to

$$H_U \underbrace{(2w + \pi R)}_{\ell_U} + H_g \underbrace{(2x)}_{\ell_g} + H_I \underbrace{h}_{\ell_I} = Ni\,, \tag{2.238}$$

where by introducing the three path lengths we may write more concisely

$$H_U \ell_U + H_g \ell_g + H_I \ell_I = Ni\,. \tag{2.239}$$

Recalling the constitutive laws, we obtain an equation involving the B-field:

$$\nu_U B_U \ell_U + \nu_g B_g \ell_g + \nu_I B_I \ell_I = Ni\,, \tag{2.240}$$

where $\nu_g = \mu_0^{-1}$, $\nu_U = (\mu_{rU}\mu_0)^{-1}$, and $\nu_I = (\mu_{rI}\mu_0)^{-1}$ are the reluctivities of the three different materials in the device.

Now we relate the flux density to the total flux via $\Phi = \iint B\,da = BA$, due to our uniform B-field assumption. Note that the magnetic flux is the same in each cross-section around the flux path. This leads to

$$\Phi \cdot \left(\frac{\nu_U \ell_U}{A_U} + \frac{\nu_g \ell_g}{A_g} + \frac{\nu_I \ell_I}{A_I} \right) = Ni\,. \tag{2.241}$$

Let's assume that $A_g = A_U$ for simplicity, although in reality a so-called fringing effect is present that tends to make flux expand.

The $\nu\ell/A$ terms have a special name: they are called **magnetic reluctances** and are designated by $\mathcal{R}$. These constants incorporate all the information about the device material and geometry. We could thus rewrite the previous expression as

$$\Phi \cdot (\mathcal{R}_U + \mathcal{R}_g + \mathcal{R}_I) = Ni\,. \tag{2.242}$$

This reminds us of Ohm's law, $iR = v$. Conceptually, the reluctance is a counterpart of the electric resistance $R = \rho\ell/A$, in the same manner that magnetic flux is an analogue to current, and the MMF is an analogue to the EMF in electric circuit analysis. In our device, we have three reluctances connected in series. The previous equation is called

Hopkinson's[55] **law**. We have thus formed a **magnetic circuit**, or a lumped-parameter representation of our magnetic device. Using the given device parameters, calculation of the reluctances is straightforward:

$$\mathcal{R}_U = 0.139 \cdot 10^6, \ \mathcal{R}_g = 15.915 \cdot 10^6, \ \mathcal{R}_I = 0.477 \cdot 10^6 . \tag{2.243}$$

An appropriate unit for the reluctance is the At/Wb. Note how the air-gap reluctance dominates the other two terms. This happens due to the low permeability of air compared to that of iron, even though the air-gap width is tiny.

Our end goal is to obtain inductance, energy, and then force. First, we get inductance by relating flux linkage and current:

$$L = \frac{\lambda}{i} = \frac{N\Phi}{i} = \frac{N^2}{\mathcal{R}_U + \mathcal{R}_g + \mathcal{R}_I} . \tag{2.244}$$

At the given distance x, this evaluates to $L = 24.2\ \mu\text{H}$. However, for force calculation, we need an expression of L as a function of x. This dependence comes from the air-gap reluctance, since

$$\mathcal{R}_g(x) = \frac{2\nu_g}{A_g} x . \tag{2.245}$$

Now, based on the previous example, we know that the field energy and coenergy are equal and given by

$$W_f = W_c = \frac{1}{2} L(x)\, i^2 = 1.21\ \text{mJ} . \tag{2.246}$$

The actual energy value is of little significance. What matters is the rate of change of energy, which yields force based on Equation (2.234). Here, the role of the generalized displacement θ is taken by the position of the moving piece x. Therefore

$$F_e = \frac{\partial W_c(i,x)}{\partial x} = \frac{1}{2}\frac{\partial L(x)}{\partial x} i^2 = -\frac{1}{2}\left(\frac{Ni}{\mathcal{R}_U + \mathcal{R}_g(x) + \mathcal{R}_I}\right)^2 \frac{\partial \mathcal{R}_g(x)}{\partial x} . \tag{2.247}$$

The force has a negative sign because it tends to pull the moving piece to the left, regardless of current polarity (i is squared in the equation). The numerical value of force obtained is $F_e = -1.16$ N.

Example 2.16 *Force developed by a cylindrical electromagnetic plunger.*
Electromagnetic plungers are linear motion actuators found in a variety of applications, such as relays, solenoid valves, and fuel injectors. Their principle of operation is identical to the horseshoe electromagnet that we analyzed in the previous example: activating the coil creates a force with a tendency to close the air gap. An interesting aspect of these devices is the fact that, by design, they exhibit **cylindrical symmetry**. In other words, these are **axisymmetric** devices. The basic principles of the magnetic field in axisymmetric problems have been presented in §1.7.6 on page 78.

Electromagnetic plungers come in various configurations, depending on the application; for instance, they can pull or push, and the plunger surface can be flat or it may have a conical shape in order to optimize the force profile with respect to mechanical position (e.g., to obtain high force for a short stroke or a weaker force but over an extended stroke length).

For the purposes of this example, let us consider the elementary device shown in Figure 2.26, where the plunger is a solid cylinder. The dashed horizontal line represents the axis of symmetry. Only the main magnetic structure is shown, and other structural and functional features (e.g., the spring or the housing) are omitted for simplicity. The plunger can move along the axial direction. We are tasked with computing the electromagnetic force.

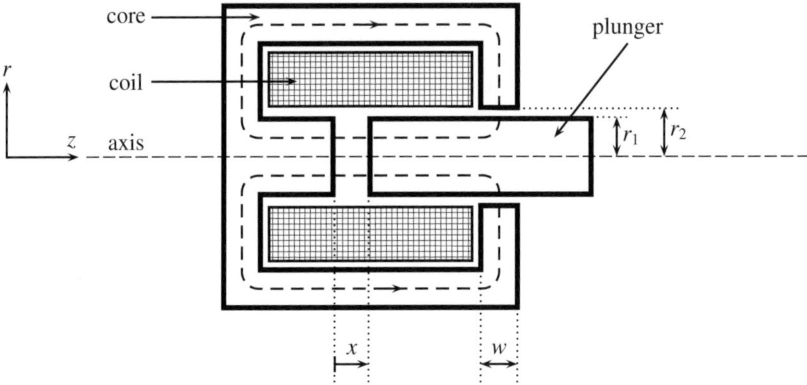

Figure 2.26 Geometry of an elementary axisymmetric electromagnetic plunger.

Solution

We proceed with the analysis of an elementary magnetic circuit. The solenoid has N turns and carries current i. For simplicity, let us assume that the permeability of the core and plunger material is very high so that the air-gap reluctances dominate. Also, let us neglect the presence of leakage flux inside the slot (although this could be significant). For argument's sake, suppose that positive current i in the coil creates a magnetic flux Φ that flows as indicated by the two dashed flux lines in the figure. Where flux exits the core towards the plunger, it moves radially inwards crossing a thin annular air gap of thickness $r_2 - r_1$. This area is occupied by a guiding tube made of a non-magnetic material (not shown). The cylindrical surface area at the core side is $A_2 = 2\pi w r_2$, whereas at the plunger side it is $A_1 = 2\pi w r_1$. Hence, if we assume a uniform flux density, Gauss' law implies the following relationship between the B-field at the two surfaces, B_1 and B_2:

$$\Phi = B_2 A_2 = B_1 A_1 \,. \tag{2.248}$$

Furthermore, at some arbitrary point inside the air gap at distance r from the axis, the B-field magnitude is

$$B(r) = B_1 \frac{r_1}{r} = B_2 \frac{r_2}{r} \,. \tag{2.249}$$

Now, suppose that the second air gap x is relatively small so that the flux moves axially through the entire plunger length. It then crosses to the core through a cylindrical air gap of base area A_a, so (ignoring fringing effects and assuming uniform flux density)

$$\Phi = B_a \pi r_1^2 = B_a A_a , \tag{2.250}$$

where B_a denotes the flux density in the axial direction.

Now, applying Ampère's law around the mean flux path yields

$$\nu_0 \int_{r_1}^{r_2} B(r)\, dr + \nu_0 B_a x = Ni , \tag{2.251}$$

or

$$\nu_0 B_1 r_1 \ln \frac{r_2}{r_1} + \nu_0 B_a x = Ni . \tag{2.252}$$

Hence, Hopkinson's law states

$$\Phi \cdot \Bigg(\underbrace{\frac{\nu_0 r_1 \ln \frac{r_2}{r_1}}{A_1}}_{\mathcal{R}_{12}} + \underbrace{\frac{\nu_0 x}{A_a}}_{\mathcal{R}_a} \Bigg) = Ni . \tag{2.253}$$

Finally, the electromagnetic force is

$$F_e = \frac{\partial W_c(i,x)}{\partial x} = \frac{1}{2} \frac{\partial L(x)}{\partial x} i^2 = -\frac{\nu_0}{2A_a} \Phi^2 = -\frac{\nu_0 A_a}{2} B_a^2 . \tag{2.254}$$

In the subsequent examples, we will analyze various types of rotating machines. The devices are elementary because they comprise the most rudimentary aspects of electric machines; however, they are not realistic machine designs as they cannot sustain continuous motion. Our goal will be to explain the creation of electromagnetic torque from a coupling field energy standpoint.

Example 2.17 *Torque developed by an elementary uniform air-gap machine.*

We will analyse a device resembling an elementary induction machine or a cylindrical-rotor synchronous machine, whose cross-section is depicted in Figure 2.27. The characteristic feature of this machine is its uniform air gap that has width g. This was also the subject of Example 2.9, where the current was assumed to be sinusoidally distributed on the surface of the stator and rotor. Instead of this ideal representation, here we will consider that the current is concentrated in slots.

Solution

We introduce a variable for the position of the rotor. This is the **rotor angle**, which is denoted by θ_r; when $\theta_r = 0$, the coils are aligned. We also need to define the angle of an arbitrary position inside the air gap with respect to the horizontal; this is denoted by ϕ (as usual). Both θ_r and ϕ are positive in the counterclockwise sense. We allow these angles to be unbounded; however, there is a clear physical significance to these

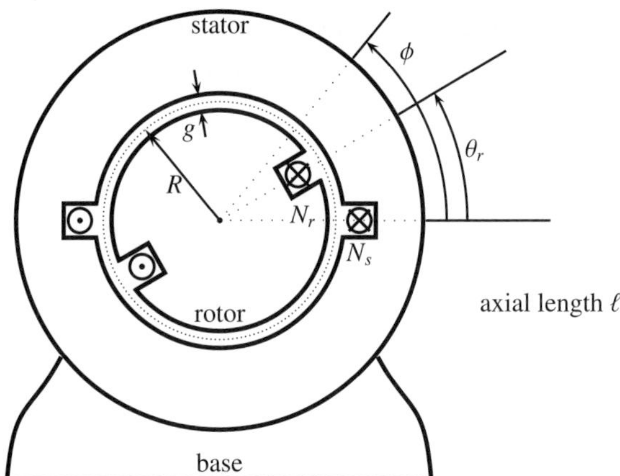

Figure 2.27 Cross-section of an elementary uniform air-gap electric machine.

variables that we should keep in mind. For instance, $\theta_r = 760°$ means that the rotor has made two revolutions and is physically at 40° from the horizontal.

The stator and rotor surfaces are circular, but they are **slotted** to allow for the placement of coils. This elementary machine has only two slots on the stator and two slots on the rotor. (Real machines have many more slots.) Windings are depicted by the circles in the slots. The crosses mean that current flows into the page, whereas the dots mean that the current returns flowing out of the page. Stator and rotor windings are formed by forming loops; the wire moves in the axial direction by ℓ, then as soon as it exits the slot it bends and re-enters the machine in the opposite direction at the other slot. These end-winding regions are not shown. The stator winding has N_s turns, whereas the rotor winding has N_r turns, that is, two possibly different numbers. The turns are not depicted separately, but the coil is tightly packed in the slot. The currents in the windings are i_s and i_r, respectively, at some instant in time. The currents may be supplied by an external circuit, or they may be generated endogenously through magnetic induction. Their exact nature does not matter for this example.

For simplicity, we shall assume that the stator and rotor are constructed of a material with very high permeability. This implies that the H-field within these parts of the device will be negligible. Hence, the coupling field energy density is zero in the stator and rotor, and nonzero only in the air-gap region. Due to the high permeability of stator and rotor and the small air-gap width, the magnetic flux can be assumed to flow radially in the air gap, exiting the iron surface perpendicularly (see §1.7.4). From our analysis of the air-gap field in §2.3.1, we know that this is not an exact representation of the field. However, it suffices for the purposes of this type of analysis.

Using polar coordinates, a radial magnetic field at an arbitrary location in the air gap is given by

$$\vec{\mathbf{H}}(r,\phi) = H(r,\phi)\,\hat{\mathbf{r}}\,. \tag{2.255}$$

Gauss' law requires that the magnetic flux crossing a circular arc between any two angles in the air gap is independent of the arc radius. Equivalently, the field should satisfy $\nabla \cdot \vec{\mathbf{B}} = \mu_0 \nabla \cdot \vec{\mathbf{H}} = 0$ in the air-gap region. To obtain this, we could model the magnetic field as

$$\vec{\mathbf{H}}(r,\phi) = H_R(\phi)\frac{R}{r}\,\hat{\mathbf{r}}\,, \tag{2.256}$$

where $H_R(\phi)$ is the field at the middle of the air gap. It can be readily shown using polar coordinates that this expression yields zero divergence. In addition, in the air gap we should satisfy $\nabla \times \vec{\mathbf{H}} = \mathbf{0}$. From Equation (2.256), the curl can be found in polar coordinates by Equation (1.124):

$$(\nabla \times \vec{\mathbf{H}}) \cdot \hat{\mathbf{k}} = -\frac{1}{r}\frac{\partial H_r}{\partial \phi} = -\frac{R}{r^2}\frac{\partial H_R(\phi)}{\partial \phi} = 0\,, \tag{2.257}$$

implying that $H_R(\phi)$ should be constant.

A crude sketch of how we would expect the magnetic flux lines to look is provided in Figure 2.28. (The flux lines will be found precisely once we learn how to implement the FEM.) As is common in machine analysis, we have superimposed arrows for the stator and rotor **magnetic axes**. In general, the axes are used to indicate the main directionality of the field source. The stator axis is fixed in space; the rotor axis is fixed on the rotor, so it turns together with the rotor. There is a similarity with the magnetic dipole moment (see §1.7.1 and 1.7.2). The torque on a magnetic dipole was found to be equal to $\vec{\mathbf{m}} \times \vec{\mathbf{B}}_0$, tending to align the dipole with the imposed external magnetic field. Our suspicion is that this machine behaves in a similar manner, that is, we expect that a torque will develop tending to align the rotor axis with the stator axis.

We are assuming that the device is magnetically linear, so we can excite each coil separately, and then we can superimpose the two fields to obtain the total field. We will use subscripts "s" and "r" to differentiate between the stator and rotor fields.

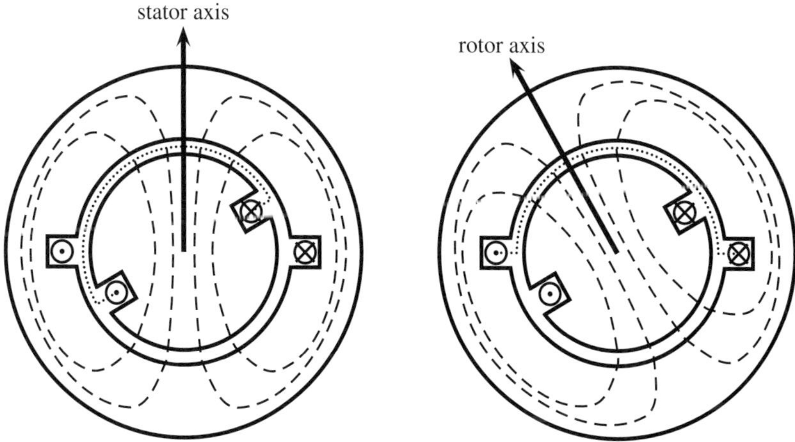

Figure 2.28 Flux lines in an elementary electric machine. Left: Created by stator. Right: Created by rotor. The dotted curves in the air gap represent integration surfaces for determining the flux passing through the two windings.

Let us first excite the stator. We apply Ampère's law following any one of the flux line paths shown on the left-hand side of Figure 2.28. Since the H-field is negligible within the magnetic materials, there is an **MMF drop** when we cross the air gap (twice for each flux line). In view of the uniformity of the air gap, the MMF drop is the same for any angle:

$$\int_{r_i}^{r_o} H_{Rs}(\phi)\frac{R}{r}\,dr = H_{Rs}(\phi)\underbrace{R\ln\frac{r_o}{r_i}}_{\tilde{g}} \approx H_{Rs}(\phi)g\,, \tag{2.258}$$

where $r_i = R - g/2$ and $r_o = R + g/2$ are the inner and outer radii of the air gap, respectively. The approximation

$$\tilde{g} = R\ln\frac{r_o}{r_i} \approx g \tag{2.259}$$

is valid when $g \ll R$. The situation is a bit uncertain right underneath the slots, at $\phi = 0$ and $\phi = 180°$. For lack of better information, let us assume that the H-field transitions like a step change. We have thus obtained a rectangular waveform

$$H_{Rs}(\phi) = \frac{N_s i_s}{2\tilde{g}} \cdot \begin{cases} +1 & \text{if } 0 < \phi < \pi\,, \\ -1 & \text{if } \pi < \phi < 2\pi\,. \end{cases} \tag{2.260}$$

(Strictly speaking, due to the unbounded range of ϕ, we should have added integer multiples of 2π to the angle conditions; however, we omit these for notational simplicity.)

Similarly, for the rotor field we have

$$H_{Rr}(\phi, \theta_r) = \frac{N_r i_r}{2\tilde{g}} \cdot \begin{cases} +1 & \text{if } 0 < \phi - \theta_r < \pi\,, \\ -1 & \text{if } \pi < \phi - \theta_r < 2\pi\,. \end{cases} \tag{2.261}$$

(Again, $\phi - \theta_r$ is unbounded, so what we are really referring to here is $\phi - \theta_r + 2k\pi$, $k \in \mathbb{Z}$.)

Next, we determine the inductances of the two coils. When we excite the stator, we obtain the stator winding self-inductance as

$$L_{ss} = \left.\frac{\lambda_s}{i_s}\right|_{i_r=0} = \left.\frac{N_s\Phi_s}{i_s}\right|_{i_r=0} = \frac{N_s}{i_s}\int_0^{\pi} \mu_0 H_{Rs}(\phi)\ell R\,d\phi = \frac{\mu_0 N_s^2 \ell R\pi}{2\tilde{g}}\,, \tag{2.262}$$

where the flux Φ_s is found by integrating the flux density $B_s = \mu_0 H_{Rs}$ over a half-cylinder passing through the middle of the air gap, as shown in Figure 2.28. Also, the rotor–stator mutual inductance is

$$L_{rs} = \left.\frac{\lambda_r}{i_s}\right|_{i_r=0} = \left.\frac{N_r\Phi_r}{i_s}\right|_{i_r=0} = \frac{N_r}{i_s}\int_{\theta_r}^{\theta_r+\pi} \mu_0 H_{Rs}(\phi)\ell R\,d\phi\,, \tag{2.263}$$

which yields

$$L_{rs} = \frac{\mu_0 N_s N_r \ell R}{2\tilde{g}} \cdot \begin{cases} (\pi - 2\theta_r) & \text{if } 0 < \theta_r < \pi, \\ (2\theta_r - 3\pi) & \text{if } \pi < \theta_r < 2\pi. \end{cases} \tag{2.264}$$

This is a periodic, triangular waveform oscillating between $\pm\mu_0 N_s N_r \ell R\pi/2\tilde{g}$ as the rotor turns (i.e., it repeats for θ_r values outside the interval $(0, 2\pi)$). Similarly, by exciting the rotor we obtain

$$L_{rr} = \left.\frac{\lambda_r}{i_r}\right|_{i_s=0} = \frac{\mu_0 N_r^2 \ell R \pi}{2\tilde{g}}\,. \tag{2.265}$$

The stator and rotor self-inductances are constants, not being affected by the rotation of the rotor. Note that we do not need to calculate L_{sr} because it is equal to L_{rs} due to the symmetry of the inductance matrix (see Example 2.13).

Knowledge of the inductances is sufficient for a calculation of the electromagnetic torque by Equation (2.234), with θ_r playing the role of the generic displacement θ:

$$\tau_e = \frac{1}{2}\begin{bmatrix} i_s & i_r \end{bmatrix} \frac{\partial}{\partial \theta_r}\left(\begin{bmatrix} L_{ss} & L_{sr}(\theta_r) \\ L_{sr}(\theta_r) & L_{rr} \end{bmatrix}\right)\begin{bmatrix} i_s \\ i_r \end{bmatrix} = i_s i_r \frac{\partial L_{sr}}{\partial \theta_r}\,, \tag{2.266}$$

which yields

$$\tau_e(\theta_r) = \frac{\mu_0 \ell R (N_s i_s)(N_r i_r)}{\tilde{g}} \cdot \begin{cases} -1 & \text{if } 0 < \theta_r < \pi, \\ +1 & \text{if } \pi < \theta_r < 2\pi. \end{cases} \tag{2.267}$$

As expected, the torque is negative for positive rotor angles ($\theta_r < 180°$), and positive for negative angles ($\theta_r > -180°$), thus always tending to align the rotor and stator axes. This equation provides some insight into basic machine design principles. It shows that the torque is proportional to the product of the two interacting magnetomotive forces, the machine axial length, and the mean air-gap radius, and is inversely proportional to the air-gap width.

Example 2.18 *Torque developed by an elementary reluctance machine.*

An elementary **reluctance machine** is shown in Figure 2.29. The stator of this type of machine is identical to the one from the previous example. The rotor is different since it does not have any windings; furthermore, the rotor shape is not circular, so the air gap is no longer uniform. High-permeability materials are used once more. We define the rotor angle θ_r so that it coincides with a rotor axis of symmetry where the air-gap width obtains its minimum value. (Note that the rotor has a second axis of symmetry, 90° away from θ_r, where the air-gap width is maximum.)

Solution

It is quite difficult to apply Ampère's law in this case, as we did in Example 2.17, because we cannot readily model and quantify the impact of the large air gap on the flux, unless we revert to additional modeling assumptions. To avoid these, we will approach this problem qualitatively. In this device, we observe that the inductance exhibits a periodicity every 180°. For the purposes of this example, we will model the stator inductance simply as

$$L_s(\theta_r) = L_A - L_B \cos 2\theta_r\,, \tag{2.268}$$

with coefficients $L_A > L_B > 0$ that we will not attempt to quantify further. Hence, it could be argued that our representation is based on the first two terms of a Fourier series expansion of $L_s(\theta_r)$, which is not necessarily sinusoidal; higher-order terms are

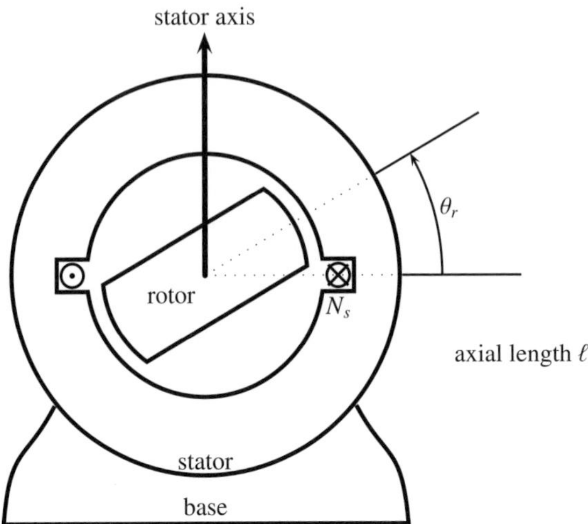

Figure 2.29 Cross-section of an elementary reluctance machine.

being ignored. (Electric machines are, by design, made to approach sinusoidal devices.) Our model guarantees that the inductance obtains a minimum value, $L_A - L_B$, at $\theta_r = 0, \pi, 2\pi, \ldots,$ where the rotor is horizontal. At these angles, the stator MMF needs to overcome the highest magnetic reluctance since the flux needs to cross the largest air gap. The inductance obtains a maximum value, $L_A + L_B$, at $\theta_r = \frac{\pi}{2}, \frac{3\pi}{2}, \frac{5\pi}{2}, \ldots$. These are positions where the rotor is vertical, in alignment with the stator axis.

The electromagnetic torque is given by Equation (2.234), which yields

$$\tau_e = i_s^2 L_B \sin 2\theta_r \,. \tag{2.269}$$

This type of torque, which is developed by the interaction of an external magnetic field with a piece of magnetizable matter in the absence of free current sources, is called **reluctance torque**. The name comes from the tendency of the field to rotate the moving piece to a point where the reluctance is minimum, or equivalently, where the inductance is maximum. At this point

$$\frac{\partial L_s}{\partial \theta_r} = \frac{\partial}{\partial \theta_r}\left(\frac{N_s^2}{\mathcal{R}_s}\right) = -\left(\frac{N_s}{\mathcal{R}_s}\right)^2 \frac{\partial \mathcal{R}_s}{\partial \theta_r} = 0\,, \tag{2.270}$$

so the torque is zero. In our device, these points are where the rotor is vertical.

These are **stable equilibrium points**, where small deviations create a torque that tends to bring the rotor back. Stable equilibria satisfy the condition

$$\left.\frac{\partial \tau_e}{\partial \theta_r}\right|_{\theta_r = \pi/2,\, 3\pi/2,\, \ldots} < 0\,. \tag{2.271}$$

If we expand the torque around a stable equilibrium using a first-order Taylor series, this condition guarantees that the torque will oppose the change in angle. It is interesting to observe that the electromagnetic torque will also be zero when the rotor is horizontal. However, these equilibria are unstable. If the rotor moves away from the horizontal ever so slightly, for whatever reason, the torque will tend to turn the rotor further away.

Example 2.19 *Torque developed by an elementary permanent-magnet machine.*

An elementary **permanent-magnet (PM) machine** is depicted in Figure 2.30. The stator is identical as in the previous examples. The rotor is cylindrical, but instead of a winding, we have now placed two permanent magnets on its surface, shown in gray. To accommodate the magnets, the air-gap width g has increased. This is only one out of many variants of PM machine designs; however, it suffices for our purpose, which is to illustrate the functional similarity with a wound-rotor machine, at least as far as torque creation is concerned.

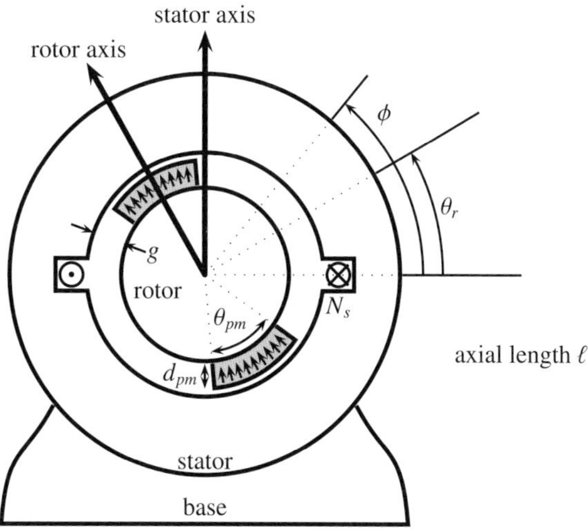

Figure 2.30 Cross-section of an elementary permanent-magnet electric machine.

Solution

The magnets are identical in shape, each spanning an angle θ_{pm}. They are attached to the surface with opposite polarity: the magnet that is centered at $\theta_r + \pi/2$ tends to create flux directed from the inside of the rotor to the air gap, whereas the magnet centered at $\theta_r - \pi/2$ tends to pull flux from the air gap into the rotor. In other words, the magnetization of the magnet at $\theta_r + \pi/2$ is pointing outwards, whereas that of the magnet at $\theta_r - \pi/2$ is pointing inwards. Alternatively, to describe this situation we would say that a **north rotor pole** appears at $\theta_r + \pi/2$, while a **south rotor pole** appears at $\theta_r - \pi/2$. The net magnetic field created by the two magnets acting in concert has the directionality of the **rotor axis**.

We employ identical simplifying assumptions as in the previous examples, namely, that stator and rotor are infinitely permeable, and that the magnetic field in the air gap is radial. Magnetic linearity implies that the stator winding flux linkage may be expressed as

$$\lambda_s = L_{ss} i_s + \lambda_{pm} \,, \tag{2.272}$$

where λ_{pm} is the flux linkage due to the permanent magnets. The magnetic field generated by the stator coil is given by Equation (2.260), and the stator self-inductance is given by Equation (2.262). To estimate the field generated by the magnets, we recall the results of §1.7.3, where it was shown that magnetized matter is equivalent to internal and surface current sources:

$$\vec{\mathbf{J}}_m = \nabla \times \vec{\mathbf{M}}, \tag{2.273a}$$

$$\vec{\mathbf{K}}_m = \vec{\mathbf{M}} \times \hat{\mathbf{n}}. \tag{2.273b}$$

It is convenient to assume that the magnet region magnetization is radial:

$$\vec{\mathbf{M}}(\phi, \theta_r) = M_0 \cdot \begin{cases} +\hat{\mathbf{r}} & \text{for } \phi \in (\theta_r + (\pi - \theta_{pm})/2,\ \theta_r + (\pi + \theta_{pm})/2), \\ -\hat{\mathbf{r}} & \text{for } \phi \in (\theta_r - (\pi + \theta_{pm})/2,\ \theta_r - (\pi - \theta_{pm})/2). \end{cases} \tag{2.274}$$

We will further assume that M_0 (A/m) is constant, so it is not affected by the stator field. This yields $\vec{\mathbf{J}}_m = \mathbf{0}$ in the interior of the magnets because $\nabla \times \hat{\mathbf{r}} = \mathbf{0}$. A nonzero surface current is present only on the magnet sides, where the outwards-pointing unit normal vector is $\hat{\mathbf{n}} = \pm\hat{\boldsymbol{\phi}}$:

$$\vec{\mathbf{K}}_m = M_0 \hat{\mathbf{k}} \cdot \begin{cases} -1 & \text{at } \phi = \theta_r + (\pi - \theta_{pm})/2, \\ +1 & \text{at } \phi = \theta_r + (\pi + \theta_{pm})/2, \\ -1 & \text{at } \phi = \theta_r - (\pi - \theta_{pm})/2, \\ +1 & \text{at } \phi = \theta_r - (\pi + \theta_{pm})/2. \end{cases} \tag{2.275}$$

These equivalent magnetization current sources are depicted in Figure 2.31, where magnets have been replaced by air. This is a reasonable approximation because the relative permeability of permanent magnets, such as samarium–cobalt or neodymium–iron–boron, is close to unity ($\mu_r \approx 1.05$). The magnitude of the equivalent current sources is $M_0 d_{pm}$, where $d_{pm} < g$ is the length of the magnet side.

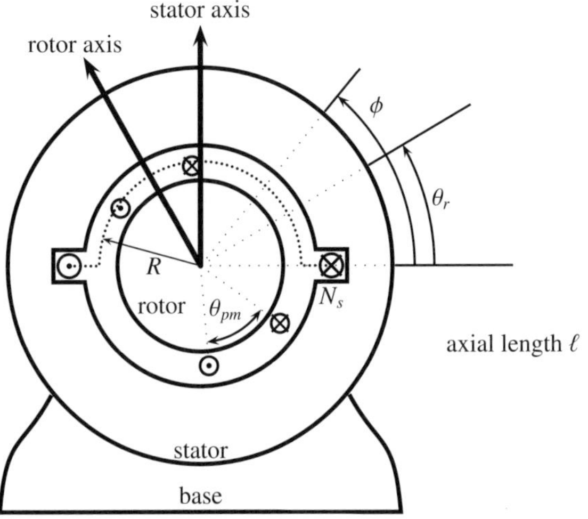

Figure 2.31 An elementary permanent-magnet electric machine with magnets replaced by equivalent current sources.

The substitution of the magnets by equivalent current sources permits us to apply Ampère's law for estimating the B-field using Equation (1.297b). The assumption of infinite permeability in stator and rotor implies that the MMF drop in these regions is zero. The MMF of the magnets is consumed entirely in the air gap. The magnet field has a radial component that varies as $1/r$, as discussed in Example 2.17. Its value at the middle of the air gap is

$$B_{Rr}(\phi,\theta_r) = \frac{\mu_0 M_0 d_{pm}}{\tilde{g}} \cdot \begin{cases} +1 & \text{if } (\pi-\theta_{pm})/2 < \phi-\theta_r < (\pi+\theta_{pm})/2\,, \\ -1 & \text{if } (-\pi-\theta_{pm})/2 < \phi-\theta_r < (-\pi+\theta_{pm})/2\,, \\ 0 & \text{otherwise}\,. \end{cases} \tag{2.276}$$

This resembles the expression of Equation (2.261); here, $2M_0 d_{pm}$ is the combined MMF of the two magnets. Note that the magnet field vanishes in between the magnets.

We can calculate the flux linked by the stator winding due to the permanent-magnet field as

$$\lambda_{pm}(\theta_r) = N_s \Phi_{pm}(\theta_r) = N_s \int_0^{\pi} B_{Rr}(\phi,\theta_r)\ell R\, d\phi\,. \tag{2.277}$$

At $\theta_r = 0$, where rotor and stator axes are aligned, the stator is linked to the maximum extent by the rotor flux. At this angle, evaluation of the integral yields

$$\lambda_{pm}(0) = \lambda_{pm,\max} = \frac{\mu_0 \ell R N_s M_0 d_{pm} \theta_{pm}}{\tilde{g}}\,. \tag{2.278}$$

As the rotor turns, the flux linked by the stator winding will oscillate between $\pm\lambda_{pm,\max}$. The variation of $\lambda_{pm}(\theta_r)$ has a trapezoidal shape, and is sketched in Figure 2.32.

The next step is to calculate the coupling field coenergy W_c and the electromagnetic torque τ_e using our prior results. Based on Equation (2.224c), we have

$$W_c(i_s,\theta_r) = W_{c0} + \frac{1}{2}L_{ss}i_s^2 + i_s\lambda_{pm}(\theta_r)\,, \tag{2.279}$$

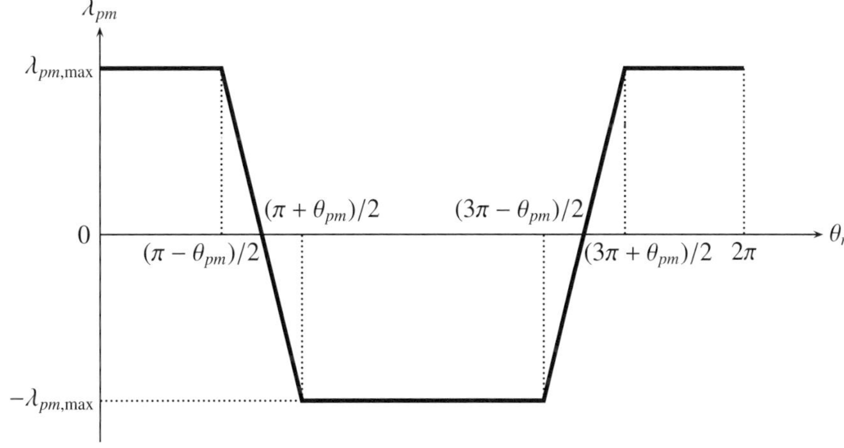

Figure 2.32 Variation of stator flux linkage from permanent magnets vs. rotor angle in an elementary machine.

where W_{c0} and L_{ss} are both constant (independent of θ_r) due to the cylindrical nature of the air gap in this example. Hence, from Equation (2.235), the electromagnetic torque is

$$\tau_e = \frac{\partial W_c}{\partial \theta_r} = i_s \frac{\partial \lambda_{pm}}{\partial \theta_r}. \tag{2.280}$$

By inspection of Figure 2.32, we obtain

$$\tau_e(\theta_r) = \frac{2 i_s \lambda_{pm,\max}}{\theta_{pm}} \cdot \begin{cases} -1 & \text{for } \theta_r \in ((\pi - \theta_{pm})/2,\ (\pi + \theta_{pm})/2), \\ 1 & \text{for } \theta_r \in ((3\pi - \theta_{pm})/2,\ (3\pi + \theta_{pm})/2), \\ 0 & \text{otherwise.} \end{cases} \tag{2.281}$$

Alternatively, we may rewrite this in a form that resembles Equation (2.267), involving the product of the two MMFs:

$$\tau_e(\theta_r) = \frac{\mu_0 \ell R (N_s i_s)(2 M_0 d_{pm})}{\tilde{g}} \cdot \begin{cases} -1 & \text{for } \theta_r \in ((\pi - \theta_{pm})/2,\ (\pi + \theta_{pm})/2), \\ 1 & \text{for } \theta_r \in ((3\pi - \theta_{pm})/2,\ (3\pi + \theta_{pm})/2), \\ 0 & \text{otherwise.} \end{cases} \tag{2.282}$$

The fact that torque is zero for certain angles is somewhat counterintuitive. This is an outcome of the presumed rectangular shape of the stator magnetic field. The Lorentz force equation can provide an alternative interpretation of the phenomenon. Within the zero-torque angular span, the entire magnet is at the same side of the stator coil. Hence, the forces $\vec{\mathbf{K}}_m \times \vec{\mathbf{B}}$ at the two sides of each magnet cancel. As soon as one side of the magnet crosses below a stator slot, the field is reversed, and a net torque is developed.

2.4 Force and Torque Calculations from the Fields

In this section, we will present three alternative methods for obtaining the total electromagnetic force or torque on a magnetized object, namely:

1. A method based on a virtual distortion of the air gap.
2. A method based on the law of conservation of momentum, commonly known as the Maxwell stress tensor method.
3. A method based on the law of conservation of energy, based on Poynting's power flow theorem.

The main difference from the approach in §2.3.7 is that these calculations will hinge on knowledge of the electromagnetic fields throughout space. The derivations will rely heavily on vector calculus, so the reader is warned that the material can be challenging. However, the end results are rewarding, elegant, and very significant.

2.4.1 Preliminary Modeling Assumptions

In the analysis that follows, we are assuming that our device has a single moving component that is **rigid**. In other words, as far as the calculations that follow are concerned, matter is not allowed to deform or compress. In reality, small deformations will take place due to the action of electromagnetic and inertial forces. Such effects are important to capture when studying vibrations and noise, but they are of minor significance for the main electromechanical energy conversion process.

A property of rigid-body motion is that it can be decomposed into the translation of an arbitrary point within the body followed by a rotation about this point. This is illustrated in Figure 2.33. The rigid body is surrounded by a stationary boundary in free space or air, depicted by the dashed line ∂V. The enclosed region, which contains the moving body, is denoted as V. We can think of the velocity as a vector field $\vec{\mathbf{v}} : V \to \mathbb{R}^3$. The velocity of any boundary point $\mathbf{r} \in \partial V$ is zero. The velocity field within the surrounding volume $V \setminus V_b$ can be arbitrary, as long as it transitions smoothly from zero at ∂V to the velocity of matter at ∂V_b. At any point within the body $\mathbf{r} \in V_b \subset V$, we have

$$\vec{\mathbf{v}}(\mathbf{r}) = \vec{\mathbf{v}}_0 + \vec{\omega} \times (\mathbf{r} - \mathbf{r}_0)\,, \tag{2.283}$$

where $\mathbf{r}_0$ and $\vec{\mathbf{v}}_0 = d\mathbf{r}_0/dt$ are the position and velocity of the center of rotation, and $\vec{\omega} = d\boldsymbol{\theta}/dt$ is the angular velocity vector. These three vectors are all independent of $\mathbf{r}$, thus we can show that the divergence of the velocity vector field vanishes for $\mathbf{r} \in V_b$:

$$\nabla \cdot \vec{\mathbf{v}} = \nabla \cdot (\vec{\omega} \times \mathbf{r}) = -\vec{\omega} \cdot (\nabla \times \mathbf{r}) = 0\,. \tag{2.284}$$

An electromagnetic force density $\vec{\mathbf{f}}$ is developed by the field acting on free currents and magnetized matter throughout the body, i.e., for $\mathbf{r} \in V_b$. Outside the body, in free space or air, the force is zero. We thus assume the presence of a vector field $\vec{\mathbf{f}} : V \to \mathbb{R}^3$. The work done by this force within a small time interval δt is

$$\delta W = \delta t \iiint_{V_b} \vec{\mathbf{v}} \cdot \vec{\mathbf{f}}\, dv = \delta t \iiint_{V} \vec{\mathbf{v}} \cdot \vec{\mathbf{f}}\, dv \tag{2.285a}$$

$$= \delta t \iiint_{V} \vec{\mathbf{v}}_0 \cdot \vec{\mathbf{f}}\, dv \;+\; \delta t \iiint_{V} [\vec{\omega} \times (\mathbf{r} - \mathbf{r}_0)] \cdot \vec{\mathbf{f}}\, dv \tag{2.285b}$$

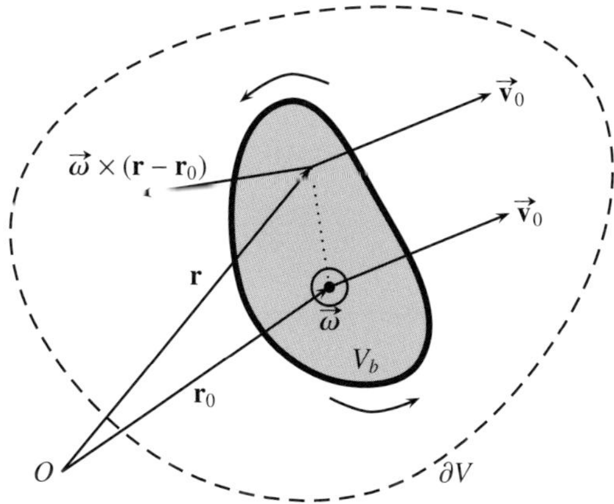

Figure 2.33 Rigid-body motion. Note $\vec{\omega}$ is a vector coming out of the page through the center of rotation, implying counterclockwise rotation.

$$= \delta t \vec{\mathbf{v}}_0 \cdot \iiint_V \vec{\mathbf{f}}\, dv + \delta t\, \vec{\omega} \cdot \iiint_V [(\mathbf{r} - \mathbf{r}_0) \times \vec{\mathbf{f}}\,]\, dv \qquad (2.285c)$$

$$= \delta \mathbf{r}_0 \cdot \vec{\mathbf{F}}_e + \delta \boldsymbol{\theta} \cdot \vec{\tau}_e , \qquad (2.285d)$$

where we introduced the **resultant electromagnetic force** $\vec{\mathbf{F}}_e$ and **torque** $\vec{\tau}_e$ as volume integrals. This equation shows that the work performed by the electromagnetic force is converted into two kinds of kinetic energy, namely, translational and angular. In general, the electromagnetic-to-mechanical energy conversion process may involve simultaneously both types of kinetic energy. Exotic machines that combine linear and angular motion have been designed. However, in the majority of cases we encounter devices of either the linear type ($\vec{\omega} = \mathbf{0}$), where motion is purely translational along a fixed direction (e.g., the electromagnets of Examples 2.15 and 2.16), or of the rotating type ($\vec{\mathbf{v}}_0 = \mathbf{0}$), where motion is purely rotational around a fixed axis (e.g., the elementary rotating machines of Examples 2.17–2.19). We will restrict our analysis to these two cases from now on.

Example 2.20 *Expansion of* $\nabla \times (\vec{\mathbf{v}} \times \vec{\mathbf{A}})$ *for rigid-body motion.*
It is instructive and useful for what follows to expand the term $\nabla \times (\vec{\mathbf{v}} \times \vec{\mathbf{A}})$ in the special case of rigid-body motion. Here, the vector field $\vec{\mathbf{A}}$ could represent any physical quantity, and $\vec{\mathbf{v}}$ is the velocity field.

Solution
According to the identity (1.168), and given that $\nabla \cdot \vec{\mathbf{v}} = 0$ for a rigid body, this yields

$$\nabla \times (\vec{\mathbf{v}} \times \vec{\mathbf{A}}) = \vec{\mathbf{v}}(\nabla \cdot \vec{\mathbf{A}}) + (\vec{\mathbf{A}} \cdot \nabla)\vec{\mathbf{v}} - (\vec{\mathbf{v}} \cdot \nabla)\vec{\mathbf{A}} . \qquad (2.286)$$

In the case of a purely translational motion, this further simplifies to

$$\nabla \times (\vec{\mathbf{v}} \times \vec{\mathbf{A}}) = \vec{\mathbf{v}}(\nabla \cdot \vec{\mathbf{A}}) - (\vec{\mathbf{v}} \cdot \nabla)\vec{\mathbf{A}} , \qquad (2.287)$$

because $\vec{\mathbf{v}}(\mathbf{r}) = \vec{\mathbf{v}}_0$ is independent of $\mathbf{r}$. When the motion has a rotational component, we have

$$(\vec{\mathbf{A}} \cdot \nabla)\vec{\mathbf{v}} = (\vec{\mathbf{A}} \cdot \nabla)\, \vec{\omega} \times \mathbf{r} = \vec{\omega} \times \vec{\mathbf{A}} , \qquad (2.288)$$

the proof of which is left as an exercise for the reader (see Problem 2.12). Hence, in general:

$$\nabla \times (\vec{\mathbf{v}} \times \vec{\mathbf{A}}) = \vec{\mathbf{v}}(\nabla \cdot \vec{\mathbf{A}}) + \vec{\omega} \times \vec{\mathbf{A}} - (\vec{\mathbf{v}} \cdot \nabla)\vec{\mathbf{A}} . \qquad (2.289)$$

The proofs that follow rely on results of vector calculus that involve differential operators and integral laws, which in turn require differentiable fields. To avoid the technicalities involved with abrupt changes in material properties and vector fields (such as the ones occurring at the interfaces between materials), we will consider that all transitions are smooth and differentiable. To this end, we will introduce a mathematical trick. For example, consider a hypothetical transition between two different materials, as shown in Figure 2.34. The trick is to replace sharp corners of step changes by smooth curves, thereby introducing a finite transition occurring in a thin region. The transition can be made arbitrarily small so that at the limit it will approach the original step change.

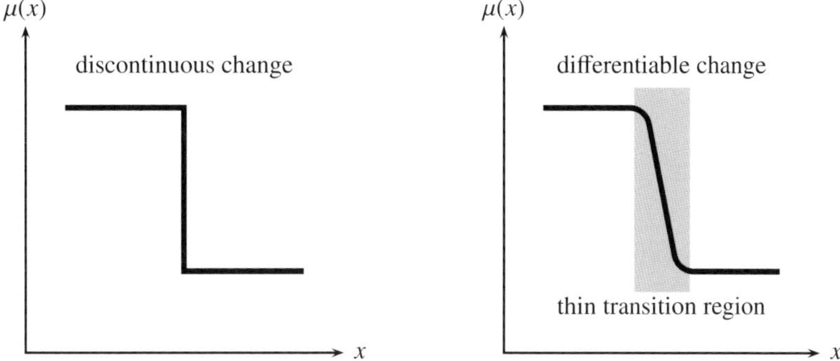

Figure 2.34 Representation of material interface as a smooth transition of permeability. A similar change is assumed for the underlying vector fields, e.g., the variation of the tangential component of the B-field or the normal component of the H-field.

2.4.2 Torque Calculation Based on a Virtual Distortion of the Air Gap

We are well positioned to derive an expression for the electromagnetic torque in a rotating electric machine. In Example 2.7 on page 131, we calculated the energy stored in the air gap of a machine based on the MVP boundary conditions as

$$W_{ag} = \sum_{p=1}^{\infty} \frac{\pi \nu_0 \ell p}{\zeta^{-p} - \zeta^{p}} \left[\frac{\zeta^{-p} + \zeta^{p}}{2} (A_{pi}^2 + A_{po}^2) - 2A_{pi}A_{po}\cos(\phi_{po} - \phi_{pi}) \right]. \tag{2.290}$$

This formula represents a sum of p-pole-pair field energies. Also, earlier in this chapter we found out that torque can be obtained as the partial derivative of the coupling field energy, arriving at Equation (2.229) on page 163:

$$\tau_e = -\left.\frac{\partial W_f}{\partial \theta}\right|_{\vec{\mathbf{A}}\text{ constant over conductors}}. \tag{2.291}$$

One way to implement this calculation is by representing a small rotation $\delta\theta$ as a **virtual distortion** of the air-gap annulus keeping the outer boundary fixed, but allowing the inside boundary (and everything that it encloses) to rotate as a rigid body. In doing so, we shall maintain the MVP constant on either side of the annulus. In particular, the MVP on the inner side will move in unison with each point in space; in other words, it is held constant in the reference frame of the rotor. Hence, the magnetic energy outside the annulus will not change. Therefore, the change of coupling-field energy in the entire device amounts to a change of the air-gap annulus energy only. According to the air-gap MVP expression (2.115) on page 132, a change in mechanical position reflects on the position of a p-pole-pair inner MVP component as $\delta\theta = \delta\phi_{pi}/p$. Hence

$$\tau_e = -\frac{\partial W_{ag}}{\partial \theta} = \sum_{p=1}^{\infty} \frac{2\pi \nu_0 \ell p^2}{\zeta^{-p} - \zeta^p} A_{pi} A_{po} \sin(\phi_{po} - \phi_{pi}) \,. \tag{2.292}$$

Note the similarity with the air-gap power transfer equation (2.125b) on page 134. More on this later.

Example 2.21 *Torque calculation in an electric machine.*

We revisit the motor of Example 2.8 on page 135. Now we will examine this device from the perspective of electromagnetic torque. The machine has geometric parameters $r_i = 50$ mm, $r_o = 51$ mm, $\ell = 100$ mm. Calculate the torque developed by the dominant pole-pair field ($p = 2$).

Solution

Let us recall that the radial flux density magnitudes are related to the MVP magnitudes by Equation (2.126):

$$B_{pri} = \frac{p}{r_i} A_{pi} \,, \; B_{pro} = \frac{p}{r_o} A_{po} \,. \tag{2.293}$$

Suppose that $B_{pri} = B_{pro}/\zeta = 1$ T. (This approximation leads to identical MVP peak values at the two boundaries, and thus to identical flux over each pole at both sides.) The electromagnetic torque can be calculated by an equivalent version of Equation (2.292):

$$\tau_e = \frac{2\pi \ell \nu_0 r_i r_o}{\zeta^{-p} - \zeta^p} B_{pri} B_{pro} \sin(\phi_{po} - \phi_{pi}) \,. \tag{2.294}$$

Substituting numerical values:

$$\tau_e = (15.8 \cdot 10^3) \cdot \sin(\phi_{po} - \phi_{pi}) \text{ Nm} \,. \tag{2.295}$$

As previously noted for power, the numerical value of the coefficient preceding the sine is enormous. (This arises because we are dividing by μ_0 and again by $(\zeta^{-p} - \zeta^p)$, which are both very small numbers.) Therefore, the angular difference between the inner and outer boundary MVP practically cannot exceed a fraction of a degree. For instance, for $\phi_{po} - \phi_{pi} = 0.1°$, we obtain $\tau_e = 27.5$ Nm.

2.4.3 Conservation of Electromagnetic Momentum: The Electromagnetic (Maxwell's) Stress Tensor

Our second approach for force/torque calculation will be based on the concept of **electromagnetic momentum**. Indeed, it is a lesser known fact that the electromagnetic field carries both energy *and* momentum. We will explain this concept for a magnetic device, that is, a device where the effect of free charge is insignificant ($\rho \approx 0$). We will represent magnetized matter by equivalent currents $\vec{\mathbf{J}}_m = \nabla \times \vec{\mathbf{M}}$, as explained in §1.7.3 and 1.7.4. Note that due to our assumption of smooth transitions, there is no need to introduce a surface magnetization current $\vec{\mathbf{K}}_m$. This is possible regardless of whether the material is magnetically linear or not. Our analysis should hold equally well for hysteretic materials. All we need is the magnetization throughout space at some instant in time.

In general, the fields may be time-varying, and Maxwell's equations are

$$\nabla \cdot \vec{\mathbf{E}} = 0\,, \qquad \nabla \times \vec{\mathbf{E}} = -\frac{\partial \vec{\mathbf{B}}}{\partial t}\,, \tag{2.296a}$$

$$\nabla \cdot \vec{\mathbf{B}} = 0\,, \qquad \nabla \times \vec{\mathbf{B}} = \mu_0(\vec{\mathbf{J}} + \vec{\mathbf{J}}_m) + \mu_0\epsilon_0\frac{\partial \vec{\mathbf{E}}}{\partial t}\,. \tag{2.296b}$$

Taking the cross product of the two curl equations with $\epsilon_0\vec{\mathbf{E}}$ and $\vec{\mathbf{B}}/\mu_0 = \nu_0\vec{\mathbf{B}}$, respectively, and adding yields

$$\epsilon_0(\nabla \times \vec{\mathbf{E}}) \times \vec{\mathbf{E}} + \nu_0(\nabla \times \vec{\mathbf{B}}) \times \vec{\mathbf{B}} = \epsilon_0\frac{\partial}{\partial t}(\vec{\mathbf{E}} \times \vec{\mathbf{B}}) + (\vec{\mathbf{J}} + \vec{\mathbf{J}}_m) \times \vec{\mathbf{B}}\,. \tag{2.297}$$

On the right-hand side, the term

$$\vec{\mathbf{E}} \times \vec{\mathbf{B}} = \mu_0\vec{\mathbf{S}} \tag{2.298}$$

is related to the Poynting vector that was introduced in §2.3.1, and

$$(\vec{\mathbf{J}} + \vec{\mathbf{J}}_m) \times \vec{\mathbf{B}} = \vec{\mathbf{f}}_L \tag{2.299}$$

is the Lorentz force density (in N/m^3) exerted on free and bound currents. We introduce the subscript "L" to differentiate this force density expression from one that will be obtained by a third method in the next subsection.

We invoke the vector calculus identity (1.176):

$$\frac{1}{2}\nabla(\vec{\mathbf{X}} \cdot \vec{\mathbf{X}}) = (\vec{\mathbf{X}} \cdot \nabla)\vec{\mathbf{X}} + \vec{\mathbf{X}} \times (\nabla \times \vec{\mathbf{X}})\,, \tag{2.300}$$

to replace the $(\nabla \times \vec{\mathbf{E}}) \times \vec{\mathbf{E}}$ and $(\nabla \times \vec{\mathbf{B}}) \times \vec{\mathbf{B}}$ terms. For example, for the B-field we have

$$(\nabla \times \vec{\mathbf{B}}) \times \vec{\mathbf{B}} = (\vec{\mathbf{B}} \cdot \nabla)\vec{\mathbf{B}} - \frac{1}{2}\nabla(\vec{\mathbf{B}} \cdot \vec{\mathbf{B}})\,. \tag{2.301}$$

Without affecting anything, we may add the zero vector $(\nabla \cdot \vec{\mathbf{B}})\vec{\mathbf{B}}$ to the right-hand side:

$$(\nabla \times \vec{\mathbf{B}}) \times \vec{\mathbf{B}} = (\vec{\mathbf{B}} \cdot \nabla)\vec{\mathbf{B}} + (\nabla \cdot \vec{\mathbf{B}})\vec{\mathbf{B}} - \frac{1}{2}\nabla(B^2)\,. \tag{2.302}$$

Expanding the first row and rearranging terms yields

$$[(\nabla \times \vec{\mathbf{B}}) \times \vec{\mathbf{B}}]_x = \left(B_x\frac{\partial B_x}{\partial x} + B_y\frac{\partial B_x}{\partial y} + B_z\frac{\partial B_x}{\partial z}\right) + \left(\frac{\partial B_x}{\partial x} + \frac{\partial B_y}{\partial y} + \frac{\partial B_z}{\partial z}\right)B_x - B\frac{\partial B}{\partial x} \tag{2.303a}$$

$$= \left(2B_x\frac{\partial B_x}{\partial x} - B\frac{\partial B}{\partial x}\right) + \left(B_y\frac{\partial B_x}{\partial y} + B_x\frac{\partial B_y}{\partial y}\right) + \left(B_z\frac{\partial B_x}{\partial z} + B_x\frac{\partial B_z}{\partial z}\right) \tag{2.303b}$$

$$= \nabla \cdot \left[B_x^2 - \tfrac{1}{2}B^2 \quad B_xB_y \quad B_xB_z\right]. \tag{2.303c}$$

Similar manipulations can be performed for the other two components of $\vec{\mathbf{B}}$ and for $\vec{\mathbf{E}}$.

We have thus shown that

$$\nu_0(\nabla \times \vec{\mathbf{B}}) \times \vec{\mathbf{B}} = \nabla \cdot \nu_0 \begin{bmatrix} B_x^2 - \frac{1}{2}B^2 & B_xB_y & B_xB_z \\ B_xB_y & B_y^2 - \frac{1}{2}B^2 & B_yB_z \\ B_xB_z & B_yB_z & B_z^2 - \frac{1}{2}B^2 \end{bmatrix} = \nabla \cdot \sigma^{(m)} \tag{2.304}$$

and

$$\epsilon_0(\nabla \times \vec{\mathbf{E}}) \times \vec{\mathbf{E}} = \nabla \cdot \epsilon_0 \begin{bmatrix} E_x^2 - \frac{1}{2}E^2 & E_xE_y & E_xE_z \\ E_xE_y & E_y^2 - \frac{1}{2}E^2 & E_yE_z \\ E_xE_z & E_yE_z & E_z^2 - \frac{1}{2}E^2 \end{bmatrix} = \nabla \cdot \sigma^{(e)} . \tag{2.305}$$

We introduced two 3×3 matrices (also known as **tensors**), $\sigma^{(m)}$ and $\sigma^{(e)}$, for the magnetic and electric fields, respectively.[n] Here, the divergence of a tensor is taken row-wise, and the result is a 3-D vector. The σ-matrices/tensors have elements

$$\sigma_{ij}^{(m)} = \nu_0(B_iB_j - \tfrac{1}{2}B^2\delta_{ij}) , \tag{2.306a}$$

$$\sigma_{ij}^{(e)} = \epsilon_0(E_iE_j - \tfrac{1}{2}E^2\delta_{ij}) , \tag{2.306b}$$

for $i, j \in \{1, 2, 3\}$, where δ_{ij} is the Kronecker[56] delta:

$$\delta_{ij} = \begin{cases} 0 & \text{if } i \neq j, \\ 1 & \text{if } i = j. \end{cases} \tag{2.307}$$

We define

$$\sigma = \sigma^{(m)} + \sigma^{(e)} . \tag{2.308}$$

The matrix σ has a special name in electromagnetism: it is called the **Maxwell stress tensor**.

In view of Equation (2.297), we conclude that

$$\nabla \cdot \sigma = \vec{\mathbf{f}}_L + \epsilon_0\mu_0 \frac{\partial \vec{\mathbf{S}}}{\partial t} . \tag{2.309}$$

In physics, the quantity

$$\vec{\mathbf{p}}_{\text{em}} = \epsilon_0\mu_0 \vec{\mathbf{S}} = \epsilon_0 \vec{\mathbf{E}} \times \vec{\mathbf{B}} = \frac{1}{c^2} \vec{\mathbf{S}} , \tag{2.310}$$

where c is the speed of light, is interpreted as **electromagnetic momentum per unit volume**. We may verify that this is indeed momentum density from the unit of $\vec{\mathbf{S}}/c^2$. Since we know that the Poynting vector represents power flow per unit area, the SI unit of $\vec{\mathbf{p}}_{\text{em}}$ is

$$1\ (\text{W/m}^2)/(\text{m/s})^2 = 1\ \text{J·s/m}^4 = 1\ (\text{Kg·(m/s)}^2)\text{·s/m}^4 = 1\ (\text{Kg·m/s})/\text{m}^3 , \tag{2.311}$$

[n] The electrical conductivity is also denoted by σ. These tensors have, of course, a different physical significance.

equal to a unit of mass times velocity (i.e., momentum) per unit volume. Therefore, we may write concisely the **divergence of the Maxwell stress tensor** as

$$\boxed{\nabla \cdot \sigma = \vec{\mathbf{f}}_L + \frac{\partial \vec{\mathbf{p}}_{\text{em}}}{\partial t}} \,. \tag{2.312}$$

Interestingly, the Poynting vector plays a role in both electromagnetic energy and momentum transfer.

Because of the $1/c^2$ factor, we suspect that the magnitude of the electromagnetic momentum density is small. We can estimate an upper bound by making some reasonable assumptions. Let us assume that $B < 2$ T throughout space, as dictated by ferromagnetic material properties. The electric fields should not exceed values that would cause breakdown of insulating materials and/or the ionization of air. From cable datasheets, we can obtain an approximate upper bound for the electric field within the insulation layer, say, $E_{\text{ins}} < 5 \cdot 10^6$ V/m. This leads to the following (conservative) upper bound for the electromagnetic momentum density in insulation, $p_{\text{em, ins}} < 8.85 \cdot 10^{-5}$ (kg·m/s)/m^3. We expect E_{ins} to be several orders of magnitude higher than the electric field that would exist within air, copper, aluminum, or steel, which actually make up most of the volume in a real device. For example, in a copper conductor, where $E = J/\sigma$, setting $J < 20$ A/mm^2 and $\sigma = 6 \cdot 10^7$ S/m yields the bounds $E_{\text{con}} < 0.33$ V/m and $p_{\text{em, con}} < 5.9 \cdot 10^{-12}$ (kg·m/s)/m^3. In laminations, we may use Equation (2.171) for a frequency of 100 Hz and a lamination width of 1 mm. This implies that $E_{\text{lam}} < 0.63$ V/m and $p_{\text{em, lam}} < 1.1 \cdot 10^{-11}$ (kg·m/s)/m^3. For all practical purposes, the electromagnetic momentum density is insignificant.

Linear Momentum and Resultant Force for Translational Motion

We will examine first the case of purely translational motion. From the second law of Newtonian mechanics, we recall that the sum of the forces acting on a body of mass m equals the rate of change of its (linear) momentum:

$$\vec{\mathbf{P}}_{\text{mech}} = m\vec{\mathbf{v}} \,. \tag{2.313}$$

In our case, we can imagine two forces exerted on the moving body, namely, the electromagnetic force $\vec{\mathbf{F}}_e$ plus a force of mechanical origin $\vec{\mathbf{F}}_m$. Hence

$$\vec{\mathbf{F}}_e + \vec{\mathbf{F}}_m = m\vec{\mathbf{a}} = \frac{d\vec{\mathbf{P}}_{\text{mech}}}{dt} \,, \tag{2.314}$$

where $\vec{\mathbf{a}}$ is the acceleration. The total (linear) electromagnetic momentum within a region of space that encloses the moving body is

$$\vec{\mathbf{P}}_{\text{em}} = \iiint_V \vec{\mathbf{p}}_{\text{em}} \, dv \,. \tag{2.315}$$

From Equation (2.312), the resultant electromagnetic force is

$$\vec{\mathbf{F}}_e = \iiint_V \vec{\mathbf{f}}_L \, dv = \iiint_V \nabla \cdot \sigma \, dv - \frac{d\vec{\mathbf{P}}_{\text{em}}}{dt} \,, \tag{2.316}$$

so we obtain

$$\frac{d}{dt}(\vec{\mathbf{P}}_{\text{mech}} + \vec{\mathbf{P}}_{\text{em}}) = \iiint_V \nabla \cdot \sigma \, dv \; + \; \vec{\mathbf{F}}_m \, . \tag{2.317}$$

To evaluate the volume integral, we invoke the divergence theorem corollary (1.202):

$$\iiint_V \nabla \cdot \sigma \, dv = \oiint_{\partial V} \sigma \cdot \hat{\mathbf{n}} \, da \, . \tag{2.318}$$

Alternatively, we can obtain the same result with a longer proof that shows the algebraic steps in more detail. We recall (2.302), which implies that

$$\nabla \cdot \sigma = \epsilon_0 \left[(\vec{\mathbf{E}} \cdot \nabla)\vec{\mathbf{E}} + (\nabla \cdot \vec{\mathbf{E}})\vec{\mathbf{E}} - \tfrac{1}{2}\nabla(E^2) \right] \\ + \nu_0 \left[(\vec{\mathbf{B}} \cdot \nabla)\vec{\mathbf{B}} + (\nabla \cdot \vec{\mathbf{B}})\vec{\mathbf{B}} - \tfrac{1}{2}\nabla(B^2) \right] . \tag{2.319}$$

The two terms in the brackets are similar, so we will illustrate the algebraic manipulations on the B-field term alone. To integrate, we will use corollary (1.219) of Green's first identity, which for our case signifies that

$$\iiint_V [(\vec{\mathbf{B}} \cdot \nabla)\vec{\mathbf{B}} + (\nabla \cdot \vec{\mathbf{B}})\vec{\mathbf{B}}] \, dv = \oiint_{\partial V} (\vec{\mathbf{B}} \cdot \hat{\mathbf{n}})\vec{\mathbf{B}} \, da \, . \tag{2.320}$$

Also, we use the divergence theorem corollary (1.198), which implies that

$$\iiint_V \nabla(B^2) \, dv = \oiint_{\partial V} B^2 \, \hat{\mathbf{n}} \, da \, . \tag{2.321}$$

Therefore, we have transformed volume integrals into surface integrals:

$$\iiint_V \nabla \cdot \sigma \, dv = \oiint_{\partial V} \underbrace{\epsilon_0 \left[(\vec{\mathbf{E}} \cdot \hat{\mathbf{n}})\vec{\mathbf{E}} - \tfrac{1}{2}E^2 \, \hat{\mathbf{n}} \right]}_{\vec{\mathbf{t}}_e} da \; + \; \oiint_{\partial V} \underbrace{\nu_0 \left[(\vec{\mathbf{B}} \cdot \hat{\mathbf{n}})\vec{\mathbf{B}} - \tfrac{1}{2}B^2 \, \hat{\mathbf{n}} \right]}_{\vec{\mathbf{t}}_m} da \, . \tag{2.322}$$

Let us work with the integrand of the second term on the right-hand side. Expressing $\vec{\mathbf{t}}_m$ in Cartesian coordinates:

$$\vec{\mathbf{t}}_m = \nu_0 \begin{bmatrix} (B_x n_x + B_y n_y + B_z n_z)B_x - \frac{1}{2}B^2 n_x \\ (B_x n_x + B_y n_y + B_z n_z)B_y - \frac{1}{2}B^2 n_y \\ (B_x n_x + B_y n_y + B_z n_z)B_z - \frac{1}{2}B^2 n_z \end{bmatrix} \tag{2.323a}$$

$$= \nu_0 \begin{bmatrix} (B_x^2 - \frac{1}{2}B^2)n_x + B_x B_y n_y + B_x B_z n_z \\ (B_y^2 - \frac{1}{2}B^2)n_y + B_x B_y n_x + B_y B_z n_z \\ (B_z^2 - \frac{1}{2}B^2)n_z + B_x B_z n_x + B_y B_z n_y \end{bmatrix} \tag{2.323b}$$

$$= \nu_0 \begin{bmatrix} B_x^2 - \frac{1}{2}B^2 & B_x B_y & B_x B_z \\ B_x B_y & B_y^2 - \frac{1}{2}B^2 & B_y B_z \\ B_x B_z & B_y B_z & B_z^2 - \frac{1}{2}B^2 \end{bmatrix} \begin{bmatrix} n_x \\ n_y \\ n_z \end{bmatrix} . \tag{2.323c}$$

An identical manipulation can be performed on $\vec{\mathbf{t}}_e$. Hence

$$\vec{\mathbf{t}}_e = \sigma^{(e)} \cdot \hat{\mathbf{n}} \; , \quad \vec{\mathbf{t}}_m = \sigma^{(m)} \cdot \hat{\mathbf{n}} \, , \tag{2.324}$$

and defining

$$\vec{\mathbf{t}} = \vec{\mathbf{t}}_e + \vec{\mathbf{t}}_m \, , \tag{2.325}$$

we obtain

$$\vec{\mathbf{t}} = \sigma \cdot \hat{\mathbf{n}} \,. \tag{2.326}$$

Summarizing these results:

$$\boxed{\frac{d}{dt}(\vec{\mathbf{P}}_{\text{mech}} + \vec{\mathbf{P}}_{\text{em}}) - \vec{\mathbf{F}}_m = \iiint_V \nabla \cdot \sigma \, dv = \oiint_{\partial V} \sigma \cdot \hat{\mathbf{n}} \, da = \oiint_{\partial V} \vec{\mathbf{t}} \, da \,.} \tag{2.327}$$

In physics, this equation is interpreted as a law of **conservation of momentum** for the combined mechanical and electromagnetic momentum of a system. Conservation of momentum is typically explained using an example with two masses that collide, or in general by hypothesizing a closed system that does not exchange matter with its surroundings and is not subject to external forces. In a real machine application, it may be difficult to conjure such a picture. One possible way to explain this equation is by thinking of it as a description of momentum transfer. In mechanics, momentum is transferred by the forces developed during the time when colliding masses make mechanical contact. In electromagnetics, momentum is transferred from a region of space to another without contact, through the action of the field. Since the region V encloses the moving part of the system, we may interpret the quantity $\vec{\mathbf{t}} = \sigma \cdot \hat{\mathbf{n}}$ as the **inward flow of momentum** per unit area and per unit time. Momentum is absorbed from the universe outside V. This typically consists of a component that is solidly affixed to the ground. In this case, due to the huge mass difference between the moving body and the earth, the momentum change of the external region is imperceptible. The inflow of momentum contributes to the increase of mechanical momentum of matter enclosed in V plus electromagnetic momentum of the field within V. As we discussed earlier, the electromagnetic momentum is insignificant, $\vec{\mathbf{P}}_{\text{em}} \approx \mathbf{0}$. Therefore, the left-hand side of Equation (2.327) essentially equals the resultant electromagnetic force.

When currents are stationary, the field is entirely magnetic since both the divergence and curl of the E-field are zero. However, even when quantities are changing with time, in magnetic devices the electromagnetic field is dominated by its magnetic component. So, $\|\vec{\mathbf{t}}_e\| \ll \|\vec{\mathbf{t}}_m\|$ and $\vec{\mathbf{t}} \approx \vec{\mathbf{t}}_m$. Unless otherwise specified, from now on we will assume that this approximation holds, so $\vec{\mathbf{t}}$ will denote its magnetic counterpart. We conclude that **the resultant force in quasi-magnetostatic problems** is

$$\boxed{\vec{\mathbf{F}}_e = \iiint_V \vec{\mathbf{f}}_L \, dv \approx \oiint_{\partial V} \vec{\mathbf{t}}_m \, da = \oiint_{\partial V} \nu_0 \left[(\vec{\mathbf{B}} \cdot \hat{\mathbf{n}})\vec{\mathbf{B}} - \tfrac{1}{2} B^2 \hat{\mathbf{n}} \right] da \,.} \tag{2.328}$$

This is a very important result! It implies that instead of calculating the force by summing Lorentz force contributions within the body, we can instead perform a surface integral of the electromagnetic B-field (only!) on any surface in free space that completely surrounds the body. The integration surface ∂V can be arbitrary because the resultant force consists of $\vec{\mathbf{f}}_L = (\vec{\mathbf{J}} + \vec{\mathbf{J}}_m) \times \vec{\mathbf{B}}$ contributions from within $V_b \subset V$. In other words, the answer does not depend on the particular shape of ∂V. Once the field solution has been obtained, e.g., through FEA, we can calculate the net force by focusing on the field around the moving body. We need not worry about calculating forces locally within the material.

In other words, the result implies the presence of a fictitious "tension" (force) that exists on the surface, with a **stress** (force per unit area) given by

$$\vec{\mathbf{t}}_m = \nu_0 \left[(\vec{\mathbf{B}} \cdot \hat{\mathbf{n}})\vec{\mathbf{B}} - \tfrac{1}{2} B^2\, \hat{\mathbf{n}} \right] . \tag{2.329}$$

We have termed this force fictitious because it is conceptually problematic to think of it as a real force since it is acting on free space (at the boundary), where no matter is present. This method of calculating the force as a surface integral is called the **Maxwell stress tensor (MST) method**.

Geometric Interpretation of the Maxwell Stress Tensor

The geometry of the MST method is illustrated in Figure 2.35. From Equation (2.329) we see that $\vec{\mathbf{t}}_m$ is a linear combination of $\vec{\mathbf{B}}$ and $\hat{\mathbf{n}}$, so the three vectors lie on the same plane. We will now show that $\vec{\mathbf{B}}$ bisects the angle between the surface normal $\hat{\mathbf{n}}$ and the stress vector $\vec{\mathbf{t}}_m$.

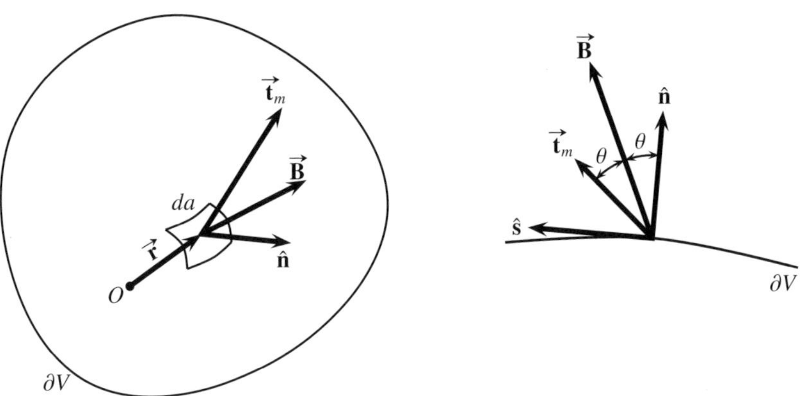

Figure 2.35 Illustration of the Maxwell stress tensor method as a surface integral. The elementary surface element da sits on the boundary ∂V of the volume that encloses the moving piece. On the right, we see a closer view of the three main vectors at an arbitrary point on the surface, such that all three are on the plane of the page.

Suppose $\vec{\mathbf{B}}$ forms an angle $\theta = \theta_B - \theta_n$ with the unit normal $\hat{\mathbf{n}}$, where angles are measured from the horizontal in a counterclockwise manner. The component of $\vec{\mathbf{t}}_m$ along the normal direction is

$$t_n = \vec{\mathbf{t}}_m \cdot \hat{\mathbf{n}} \tag{2.330a}$$
$$= \nu_0[(\vec{\mathbf{B}} \cdot \hat{\mathbf{n}})(\vec{\mathbf{B}} \cdot \hat{\mathbf{n}}) - \tfrac{1}{2}B^2] \tag{2.330b}$$
$$= \nu_0(B^2 \cos^2\theta - \tfrac{1}{2}B^2) \tag{2.330c}$$
$$= \tfrac{1}{2}\nu_0 B^2 \cos 2\theta . \tag{2.330d}$$

The component of $\vec{\mathbf{t}}_m$ along the tangential direction (here denoted by $\hat{\mathbf{s}}$, oriented at $\theta_s = \theta_n + \pi/2$) is

$$t_s = \vec{\mathbf{t}}_m \cdot \hat{\mathbf{s}} \tag{2.331a}$$

$$= \nu_0[(\vec{\mathbf{B}} \cdot \hat{\mathbf{n}})(\vec{\mathbf{B}} \cdot \hat{\mathbf{s}}) - \tfrac{1}{2}B^2\, \hat{\mathbf{n}} \cdot \hat{\mathbf{s}}] \tag{2.331b}$$

$$= \nu_0(B^2 \sin\theta \cos\theta - 0) \tag{2.331c}$$

$$= \tfrac{1}{2}\nu_0 B^2 \sin 2\theta\,. \tag{2.331d}$$

Hence, we conclude that

$$\|\vec{\mathbf{t}}_m\| = \tfrac{1}{2}\nu_0 B^2\,, \tag{2.332}$$

and that $\vec{\mathbf{t}}_m$ points at twice the angle that $\vec{\mathbf{B}}$ forms with the normal.

Therefore, if $\vec{\mathbf{B}}$ is collinear with the normal (either pointing outwards at 0° or pointing inwards at 180°), then at that point a tendency to pull the body towards the direction of the normal is added to the resultant force. If $\vec{\mathbf{B}}$ points tangentially at ±90°, then a tendency to push the body inwards is created. At any other angle, a "shear stress" is developed, generating a force in the tangential direction.

Since $\vec{\mathbf{B}}$ can be decomposed as

$$B_n = B\cos\theta \quad \text{and} \quad B_s = B\sin\theta\,, \tag{2.333}$$

we may also write

$$t_n = \nu_0\left(B_n^2 - \tfrac{1}{2}B^2\right) = \tfrac{1}{2}\nu_0(B_n^2 - B_s^2) \tag{2.334}$$

and

$$t_s = \nu_0 B_n B_s\,. \tag{2.335}$$

Example 2.22 *Force on an electromagnet: MST-based force calculation.*
Recall Example 2.15 that involved a horseshoe-shaped electromagnet, as depicted in Figure 2.36. Apply the MST method to calculate the force.

Solution

We enclose the moving piece by a boundary ∂V like the one shown by the dotted line in the figure. The two shaded areas represent the only regions where a nonzero B-field exists along the MST surface. Everywhere else the field is insignificant. For simplicity, let us assume that the field is horizontal, so there is only a normal component B_n, whereas the tangential component $B_s = 0$. At the top region (i.e., the horseshoe north pole), the field points in the direction opposite to the outwards-pointing normal vector; at the bottom region (i.e., the south pole), the field is directed in the same sense as the normal vector. In both cases, the stress $\vec{\mathbf{t}}$ points to the left, and tends to pull the piece closer to the U.

The total force is found by integration:

$$\vec{\mathbf{F}}_e = \oiint_{\partial V} \vec{\mathbf{t}}\, da = \left(\frac{B^2}{2\mu_0}\right)(2A_g)\,\hat{\mathbf{n}}\,, \tag{2.336}$$

where A_g is the cross-sectional area of the air-gap flux. In Example 2.15, where we used

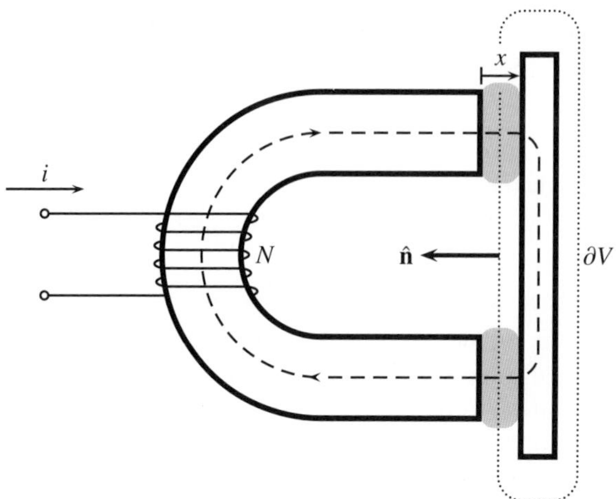

Figure 2.36 Application of the Maxwell stress tensor for calculating the force developed by a horseshoe-shaped electromagnet.

an energy-based approach, we had derived the expression (2.247) for force:

$$F_e = -\frac{1}{2}\left(\frac{Ni}{\mathcal{R}_U + \mathcal{R}_g + \mathcal{R}_I}\right)^2 \left(\frac{2\nu_g}{A_g}\right). \tag{2.337}$$

Using Equation (2.242), we rewrite this as

$$F_e = -\frac{\Phi^2}{\mu_0 A_g} = -\frac{B^2 A_g}{\mu_0}, \tag{2.338}$$

which is identical to the MST force of Equation (2.336)!

Angular Momentum and Torque from the Maxwell Stress Tensor for Rotational Motion

In the analysis of rotating electric machines, knowing the electromagnetic torque is of primary importance. We proceed similarly to how we treated linear motion. We recall from Newtonian mechanics the definition of **angular momentum** for a rigid body that rotates around a fixed axis:

$$\vec{\mathbf{L}}_{\text{mech}} = I\,\vec{\omega}, \tag{2.339}$$

where I denotes moment of inertia. Two torques are exerted on the moving body, namely, the electromagnetic torque $\vec{\tau}_e$ plus a torque of mechanical origin $\vec{\tau}_m$. Newton's second law states that

$$\vec{\tau}_e + \vec{\tau}_m = I\frac{d\vec{\omega}}{dt} = \frac{d\vec{\mathbf{L}}_{\text{mech}}}{dt}. \tag{2.340}$$

We define the total **angular electromagnetic momentum** within a region of space that encloses the moving body as

$$\vec{\mathbf{L}}_{\mathrm{em}} = \iiint_V \vec{\mathbf{r}} \times \vec{\mathbf{p}}_{\mathrm{em}}\, dv = \iiint_V \vec{\mathbf{r}} \times \frac{\vec{\mathbf{S}}}{c^2}\, dv\,, \tag{2.341}$$

where $\vec{\mathbf{r}} = (x, y, z)$ is the distance from the axis of rotation (taken here to be the origin of the coordinate system) to each elementary volume element. Using Equation (2.312), the resultant electromagnetic torque can be expressed as

$$\vec{\tau}_e = \iiint_V \vec{\mathbf{r}} \times \vec{\mathbf{f}}_L\, dv = \iiint_V \vec{\mathbf{r}} \times (\nabla \cdot \boldsymbol{\sigma})\, dv - \frac{d\vec{\mathbf{L}}_{\mathrm{em}}}{dt}\,. \tag{2.342}$$

We are thus collecting Lorentz-force contributions to torque from each point within V. Combining this with Equation (2.340), we obtain

$$\frac{d}{dt}(\vec{\mathbf{L}}_{\mathrm{mech}} + \vec{\mathbf{L}}_{\mathrm{em}}) - \vec{\tau}_m = \iiint_V \vec{\mathbf{r}} \times (\nabla \cdot \boldsymbol{\sigma})\, dv\,. \tag{2.343}$$

As we did for the linear motion case, we will neglect the angular electromagnetic momentum term. Therefore, this equation forms the basis for the calculation of electromagnetic torque.

Without loss of generality, let us assume that we are interested in calculating torque for an axis of rotation that is oriented along the z-axis of the coordinate system. In this case, we are only interested in integrating the z-component of $\vec{\mathbf{r}} \times (\nabla \cdot \boldsymbol{\sigma})$. Hence, by expanding the cross product, the z-component of torque is

$$\tau_{ez} = \iiint_V (x\, \nabla \cdot \boldsymbol{\sigma}_y - y\, \nabla \cdot \boldsymbol{\sigma}_x)\, dv\,. \tag{2.344}$$

Now, let us invoke the identity (1.158), which we will apply to each term of the integrand. First

$$x\, \nabla \cdot \boldsymbol{\sigma}_y = \nabla \cdot (x\boldsymbol{\sigma}_y) - \boldsymbol{\sigma}_y \cdot \nabla x\,, \tag{2.345}$$

and since the gradient $\nabla x = (1, 0, 0)$, we obtain

$$x\, \nabla \cdot \boldsymbol{\sigma}_y = \nabla \cdot (x\boldsymbol{\sigma}_y) - \sigma_{21}\,. \tag{2.346}$$

Similarly, one obtains

$$y\, \nabla \cdot \boldsymbol{\sigma}_x = \nabla \cdot (y\boldsymbol{\sigma}_x) - \sigma_{12}\,. \tag{2.347}$$

Subtracting the two expressions, and in view of the symmetry of the tensor $\boldsymbol{\sigma}$, the integrand becomes

$$x\, \nabla \cdot \boldsymbol{\sigma}_y - y\, \nabla \cdot \boldsymbol{\sigma}_x = \nabla \cdot (x\boldsymbol{\sigma}_y - y\boldsymbol{\sigma}_x)\,. \tag{2.348}$$

Substituting Equation (2.348) in (2.344), and applying the divergence theorem yields the surface integral

$$\tau_{ez} = \oiint_{\partial V} (x\boldsymbol{\sigma}_y - y\boldsymbol{\sigma}_x) \cdot \hat{\mathbf{n}}\, da\,. \tag{2.349}$$

This is the z-component of the cross product between $\vec{\mathbf{r}}$ and $\vec{\mathbf{t}}$, where $t_x = \sigma_x \cdot \hat{\mathbf{n}}$ and $t_y = \sigma_y \cdot \hat{\mathbf{n}}$:

$$\tau_{ez} = \oiint_{\partial V} (\vec{\mathbf{r}} \times \vec{\mathbf{t}})_z \, da \,. \tag{2.350}$$

Remarkably, this is how one would (perhaps naively) compute torque by assuming that it is created from the fictitious stress vector $\vec{\mathbf{t}}$ acting on the outer surface (in free space)!

Summarizing, we have shown that

$$\boxed{\vec{\tau}_e = \frac{d}{dt}(\vec{\mathbf{L}}_{\text{mech}} + \vec{\mathbf{L}}_{\text{em}}) - \vec{\tau}_m = \iiint_V \vec{\mathbf{r}} \times (\nabla \cdot \sigma) \, dv = \oiint_{\partial V} (\vec{\mathbf{r}} \times \vec{\mathbf{t}}) \, da \,.} \tag{2.351}$$

This equation may be interpreted as a law of **conservation of angular momentum**. The quantity $\vec{\mathbf{r}} \times \vec{\mathbf{t}} = \vec{\mathbf{r}} \times (\sigma \cdot \hat{\mathbf{n}})$ corresponds to an **inwards rate of transfer of angular momentum** per unit area through the boundary that encloses the moving body. This momentum is divided among angular momentum of matter and angular momentum of the electromagnetic field (which is typically negligible).

Torque on Cylindrical Surface Using the Maxwell Stress Tensor

To compute the torque developed by an electric machine using the MST method, we surround the rotor by a concentric, cylindrical surface of radius R and height h, as we did for the power transfer calculation in §2.3.1. It is common to choose a radius equal to the distance from the center of the rotor to the middle of the air gap, although this choice is not mandated by theory. The radius value is arbitrary, as long as the cylinder boundary lies in the air gap between rotor and stator. The cylinder height is such that the entire rotor, including the end turns, is enclosed. This choice of R and h ensures that we are accounting for all forces on the rotor. The coordinate system is oriented such that the z-axis is also the axis of rotation, as shown in Figure 2.37. The torque consists of two components, namely, the contribution from the curved part of the cylinder surface plus the contribution from the two bases.

The torque may be obtained by Equation (2.351). By expanding the stress in cylindrical coordinates, we have

$$(\vec{\mathbf{r}} \times \vec{\mathbf{t}})_z = [(r\hat{\mathbf{r}}) \times (t_r\hat{\mathbf{r}} + t_\phi\hat{\boldsymbol{\phi}} + t_z\hat{\mathbf{k}})]_z = rt_\phi \,. \tag{2.352}$$

The stress is given by Equation (2.329):

$$\vec{\mathbf{t}} = \nu_0[(\vec{\mathbf{B}} \cdot \hat{\mathbf{n}})\vec{\mathbf{B}} - \tfrac{1}{2}B^2\,\hat{\mathbf{n}}] \,, \tag{2.353}$$

and we note that it depends on the surface normal vector. At the bases, the normal vectors are $\hat{\mathbf{n}} = \pm\hat{\mathbf{k}}$, whereas on the curved part $\hat{\mathbf{n}} = \hat{\mathbf{r}}$. Therefore, at the bases

$$(\vec{\mathbf{r}} \times \vec{\mathbf{t}})_z = \pm\nu_0 r B_\phi B_z \,, \tag{2.354}$$

and at the circular surface

$$(\vec{\mathbf{r}} \times \vec{\mathbf{t}})_z = \nu_0 R B_\phi B_r \,. \tag{2.355}$$

We also substitute $da = r\,d\phi\,dr$ and $da = R\,d\phi\,dz$ for the differential area at the bases

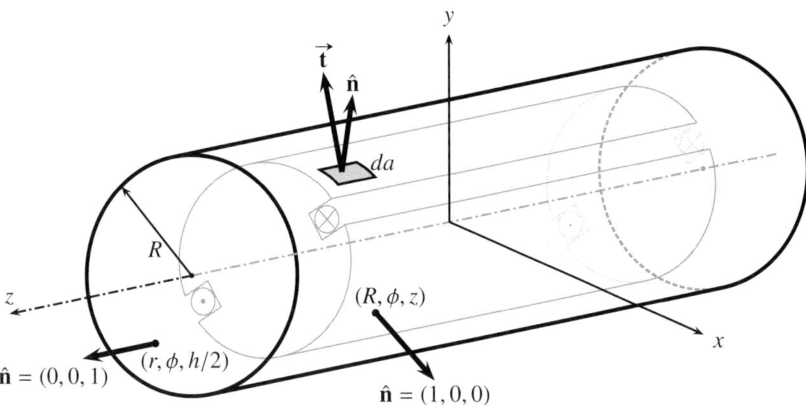

Figure 2.37 Torque calculation using the MST method on a cylindrical surface that is concentric with the rotor of an electric machine. Vectors are expressed in a cylindrical coordinate system.

and the curved part, respectively. This leads to a general expression for electromagnetic torque:

$$\tau_{ez} = \pm \int_{r=0}^{R} \int_{\phi=0}^{2\pi} \nu_0 r^2 B_\phi(r, \phi, \pm h/2) B_z(r, \phi, \pm h/2)\, d\phi\, dr + \int_{z=-h/2}^{h/2} \int_{\phi=0}^{2\pi} \nu_0 R^2 B_\phi(R, \phi, z) B_r(R, \phi, z)\, d\phi\, dz\,, \quad (2.356)$$

where the $\pm$ indicates that we actually have two terms, adding the one at $z = +h/2$ minus the one at $z = -h/2$.

It is often the case that (radial-flux) rotating electric machines are analyzed on the x–y plane using a 2-D approach (where $B_z = 0$). The **electromagnetic torque per unit axial length** is then

$$\tilde{\tau}_{ez} = \frac{R^2}{\mu_0} \int_0^{2\pi} B_\phi(R, \phi)\, B_r(R, \phi)\, d\phi\,. \quad (2.357)$$

The torque will be positive if it tends to accelerate the rotor towards the direction of increasing ϕ. In FEA, this formula has become almost synonymous with the MST method. There exist various numerical algorithms to implement this calculation. Several of these will be presented in Chapter 4.

From a theoretical perspective, this result indicates that the development of torque requires the presence of a tangential component in the B-field! A similar conclusion was reached in §2.3.1 regarding the power transfer across the air gap, e.g., see Equation (2.94). However, this is seemingly contradictory with the outcomes of Examples 2.17 and 2.19 that involved elementary, uniform air-gap machines, where we calculated torque as the partial derivative of the coenergy with respect to rotor angle under the reasonable assumption of a purely radial B-field. The same issue was also discussed in Example 2.9, where the power calculation was based on an analytical solution of the air-gap MVP field. We have resolved this apparent paradox by explaining that the field in fact has both radial and tangential components. The radial component appears in torque formulas that are commonly employed in analytical approaches where the dominant, radial field is estimated under simplifying assumptions. It could be argued that the energy-based approach is numerically more robust compared to the MST method. The volumetric energy density, $\nu_0 B^2 = \nu_0(B_r^2 + B_\phi^2)$, is approximately equal to $\nu_0 B_r^2$ when the radial component dominates. In contrast, the MST method is sensitive to the decomposition of the B-field in normal and tangential components along the integration path. Even when conducting FEA, we need to have adequate spatial resolution within the air gap to obtain an accurate estimate of the torque using the MST method.

MST Torque Based on the Air-Gap MVP Boundary Conditions

In §2.3.1 on page 121, we derived Equation (2.104) that yields the general expression for the MVP in the air gap of an electric machine, repeated here for convenience:

$$A(r,\phi) = \sum_{p=1}^{\infty} \frac{1}{\zeta^p - \zeta^{-p}} \left[\left(\frac{r}{r_o}\right)^p - \left(\frac{r_o}{r}\right)^p \right] (\alpha_p \cos p\phi + \beta_p \sin p\phi)$$
$$+ \sum_{p=1}^{\infty} \frac{1}{\zeta^{-p} - \zeta^p} \left[\left(\frac{r}{r_i}\right)^p - \left(\frac{r_i}{r}\right)^p \right] (\gamma_p \cos p\phi + \delta_p \sin p\phi). \quad (2.358)$$

The parameter $\zeta = r_i/r_o$, and the coefficients α_p, β_p, γ_p, δ_p correspond to the Fourier series expansion of the MVP boundary conditions, see Equations (2.103a) and (2.103b). The solution is a superposition of p-pole-pair magnetic fields. Based on this, we found the following expressions for the radial and tangential components of the flux density:

$$B_{pr}(r,\phi) = \frac{p}{r} \frac{1}{\zeta^p - \zeta^{-p}} \left[\left(\frac{r}{r_o}\right)^p - \left(\frac{r_o}{r}\right)^p \right] (-\alpha_p \sin p\phi + \beta_p \cos p\phi)$$
$$+ \frac{p}{r} \frac{1}{\zeta^{-p} - \zeta^p} \left[\left(\frac{r}{r_i}\right)^p - \left(\frac{r_i}{r}\right)^p \right] (-\gamma_p \sin p\phi + \delta_p \cos p\phi), \quad (2.359)$$

$$B_{p\phi}(r,\phi) = -\frac{p}{r} \frac{1}{\zeta^p - \zeta^{-p}} \left[\left(\frac{r}{r_o}\right)^p + \left(\frac{r_o}{r}\right)^p \right] (\alpha_p \cos p\phi + \beta_p \sin p\phi)$$
$$- \frac{p}{r} \frac{1}{\zeta^{-p} - \zeta^p} \left[\left(\frac{r}{r_i}\right)^p + \left(\frac{r_i}{r}\right)^p \right] (\gamma_p \cos p\phi + \delta_p \sin p\phi). \quad (2.360)$$

We now substitute the above expressions in Equation (2.357) and perform the integration. We omit the subscript "z" for notational simplicity. Algebraic manipulations are left as an exercise for the reader. We note that many terms will vanish after being integrated over 2π radians because of their sinusoidal nature. We obtain the following expressions for the **electromagnetic torque developed by a machine**:

$$\tau_e = \sum_{p=1}^{\infty} \tau_{ep} = \sum_{p=1}^{\infty} \frac{2\pi \ell p^2}{\mu_0(\zeta^{-p} - \zeta^p)} (\alpha_p \delta_p - \beta_p \gamma_p), \tag{2.361a}$$

$$= \sum_{p=1}^{\infty} \frac{2\pi \ell p^2}{\mu_0(\zeta^{-p} - \zeta^p)} A_{pi} A_{po} \sin(\phi_{po} - \phi_{pi}), \tag{2.361b}$$

$$= \sum_{p=1}^{\infty} \frac{2\pi \ell r_i r_o}{\mu_0(\zeta^{-p} - \zeta^p)} B_{pri} B_{pro} \sin(\phi_{po} - \phi_{pi}). \tag{2.361c}$$

The various coefficients that appear in the equations are related by

$$\alpha_p = A_{pi} \cos \phi_{pi}, \ \beta_p = A_{pi} \sin \phi_{pi}, \tag{2.362a}$$

$$\gamma_p = A_{po} \cos \phi_{po}, \ \delta_p = A_{po} \sin \phi_{po}, \tag{2.362b}$$

$$B_{pri} = \frac{p}{r_i} A_{pi}, \ B_{pro} = \frac{p}{r_o} A_{po}. \tag{2.362c}$$

The parameters $A_{pi} > 0$, $A_{po} > 0$ represent peak values of MVP boundary conditions at the inner and outer boundary, respectively. The angles ϕ_{pi} and ϕ_{po} determine the circumferential position of the boundary conditions at the arbitrary time instant of interest. The parameters B_{pri} and B_{pro} represent peak values of the radial component of flux density at the boundaries. It should be noted that these expressions are valid in general because they were not derived, e.g., under the assumption of a constant-speed, constant-amplitude rotating field. The electromagnetic torque is a sum of contributions from each pole-pair field. Fields of different pole-pair numbers do not interact as far as torque is concerned.

We can conclude that the MST approach led to an identical expression as Equation (2.292) on page 182, which was obtained by a virtual distortion of the air gap. *The two methods are thus theoretically equivalent.*

Relating MST Torque and Poynting's Power Flow

Equations (2.361a)–(2.361c), which were derived using the Maxwell stress tensor (reflecting the principle of conservation of electromagnetic angular momentum), bear a strong resemblance to the expressions for the air-gap power transfer, namely, Equations (2.125b), (2.125c), and (2.127) on page 134, which were derived from application of Poynting's theorem (reflecting the principle of conservation of energy). We will now tie the two approaches together, and in doing so, we will discover a basic principle of operation of a.c. electric machines. Our analysis will hold under the simplifying assumptions that we have made regarding **steady-state operation**, that is, considering a single dominant p-pole-pair field, which rotates at constant speed and has constant amplitude. We are implicitly assuming that under such steady-state conditions, the coupling field energy remains constant.

With a sign reversal of Equation (2.127), we obtain a Poynting-based expression for the stator-to-rotor air-gap power transfer by a p-pole-pair rotating magnetic field:

$$P_{s \to r,p} = \frac{2\pi \ell \omega_p r_i r_o}{\mu_0(\zeta^{-p} - \zeta^p)} B_{pri} B_{pro} \sin(\phi_{po0} - \phi_{pi0}) . \tag{2.363}$$

Here, recall that $\omega_p = \omega/p$ denotes the mechanical speed of rotation of the field in the air gap, where ω is the electrical frequency of the MVP at some arbitrary position in the air gap. The above formula resembles the terms of Equation (2.361c), so we may conclude that air-gap power and electromagnetic torque are related by

$$P_{s \to r,p} = \omega_p \tau_{ep} . \tag{2.364}$$

Now, suppose the machine rotor is mechanically spinning at ω_m rad/s. Then the rate of energy transfer to the mechanical subsystem is related to the electromagnetic torque by

$$P_{mp} = \omega_m \tau_{ep} . \tag{2.365}$$

Therefore, the mechanical power output is related to the air-gap power as follows:

$$P_{s \to r,p} = \omega_p \tau_{ep} = \omega_m \tau_{ep} + (\omega_p - \omega_m)\tau_{ep} = P_{mp} + \frac{\omega_p - \omega_m}{\omega_m} P_{mp} . \tag{2.366}$$

The last term in this equation depends on the difference between the angular velocities of the air-gap field and the rotor. Its magnitude depends on the type of machine and the operating point. In general, a.c. electric machines are classified as either **synchronous** or **asynchronous**. In synchronous machines, the rotor spins at exactly the same speed as the dominant pole-pair field ($\omega_m = \omega_p$). On the other hand, in asynchronous machines, the rotor is not synchronized with the field.

It is customary to define the **slip** of the p-pole-pair field as the relative difference between the angular velocity of the rotor and the rotating field:

$$\boxed{s_p = \frac{\omega_p - \omega_m}{\omega_p} = 1 - \frac{\omega_m}{\omega_p} .} \tag{2.367}$$

The slip is a way of expressing how fast the air-gap field "slips by" the rotor. If the rotor turns slower than the field, which is called **subsynchronous operation**, then $\omega_m < \omega_p$ and $s_p > 0$ (assuming that $\omega_p > 0$). It is also possible for the rotor to turn faster than the field, called **supersynchronous operation**, in which case $s_p < 0$. Often in electric machine analysis, while considering a single pole-pair field (the dominant one), we drop the subscript "p," denoting slip simply as s.

Introducing the slip in Equation (2.366), we can express the air-gap power as

$$P_{s \to r,p} = P_{mp} + \frac{s_p \omega_p}{\omega_m} P_{mp} = \frac{\omega_p}{\omega_m} P_{mp} , \tag{2.368}$$

or equivalently

$$\boxed{\begin{aligned} P_{mp} &= (1 - s_p) P_{s \to r,p} = P_{s \to r,p} - P_{\text{slip},p} , & (2.369\text{a}) \\ P_{\text{slip},p} &= s_p P_{s \to r,p} = \frac{s_p}{1 - s_p} P_{mp} . & (2.369\text{b}) \end{aligned}}$$

Here, we introduced the **slip power** as $P_{slip,p}$. This quantity equals the difference between the air-gap power and the mechanical power, and is proportional to the slip. The ratio of slip power over mechanical power is plotted as a function of the slip in Figure 2.38.

An energy balance argument can help us further understand the role of the slip power in the operation of a.c. electric machines. For the sake of argument, let us distinguish between two types of a.c. machine designs informally designated as: type (a), having a rotor that is isolated galvanically[57] from the outside world; type (b), having a rotor that contains windings that are connected to external circuits. Type (a) is representative of a **squirrel-cage induction machine** or of a **permanent-magnet synchronous machine**. Type (b) is representative of a **wound-rotor induction machine** or of a **wound-rotor synchronous machine**, which have rotor windings that are connected electrically to the outside via **slip rings**.

Therefore:

- In a machine of type (a), the slip power (if any) is consumed entirely as loss.
- In a machine of type (b), the slip power (if any) comprises loss *and* power that flows through the rotor towards the rotor electrical subsystem (captured by the rotor windings). Hence, this type of machine can have two armature windings, one on the stator and one on the rotor.
- In synchronous motors (of either type), where $P_{s\to r,p} = P_{mp} > 0$ and $s_p = 0$, the stator-to-rotor air-gap power is transferred in its entirety to the mechanical subsystem.
- In synchronous generators (of either type), where $P_{s\to r,p} = P_{mp} < 0$ and $s_p = 0$, the mechanical power is transferred in its entirety to the stator electrical subsystem through the air gap.
- In synchronous machines (either motors or generators), where $s_p = 0$, any amount of power that enters the rotor from an electrical subsystem (e.g., through the terminals of a rotor field winding) can only contribute to loss in the rotor. The rotor electrical subsystem cannot contribute to the mechanical power output of the machine. Conversely, it is not possible to transfer power from the shaft to the rotor-side electrical circuit.

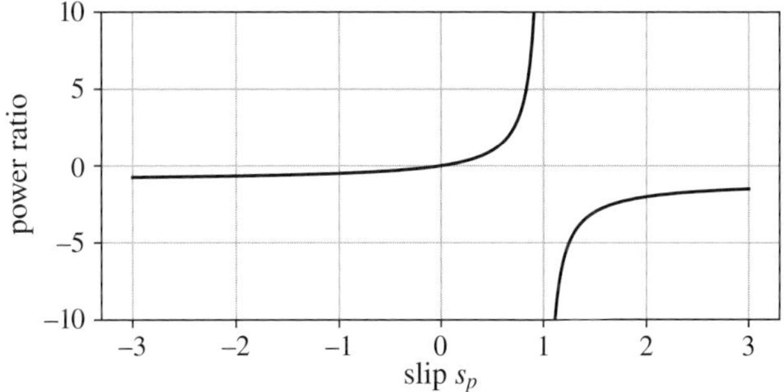

Figure 2.38 Ratio of slip power over mechanical power as a function of slip. Synchronous speed operation is obtained at $s_p = 0$. At $s_p = 1$, the rotor is not moving, so $P_{mp} = 0$.

- In asynchronous motors of type (a), where $P_{s\to r,p} > P_{mp} > 0$ and $s_p > 0$, the input stator electrical power is reduced by the slip power loss by the time it reaches the mechanical subsystem.
- In asynchronous generators of type (a), where $P_{mp} < P_{s\to r,p} < 0$ and $s_p < 0$, the input mechanical power is reduced by the slip power loss by the time it reaches the stator electrical subsystem.
- Slip values of a few percent (plus or minus) are common in squirrel-cage induction machines (motors or generators). This is by design, otherwise the slip power loss would be commensurate to the mechanical power, and the energy conversion efficiency would suffer.
- In asynchronous motors of type (b), we have $P_{s\to r,p} > 0$ and $P_{mp} > 0$, but the slip can be either positive or negative. For instance, if $s_p > 0$ (i.e., for subsynchronous operation), the slip power is positive, and this amount of power (minus loss) is absorbed by the rotor electrical subsystem.
- In asynchronous generators of type (b), we have $P_{s\to r,p} < 0$ and $P_{mp} < 0$, but the slip can be either positive or negative. For instance, if $s_p > 0$ (i.e., for subsynchronous operation), the slip power is negative, and this amount of power (plus loss) is supplied by the rotor electrical subsystem.
- Various electric machine topologies that take advantage of the slip power have been developed over the years, e.g., Kramer and Scherbius drives. A more modern example of a **slip power recovery drive** is that of a doubly fed induction generator (DFIG) wind turbine, which is described in the following example.

Lastly, let us make an observation regarding the mechanical power P_m, which holds more generally, even if the air-gap field contains multiple p-pole-pair fields. In a motor application, where power flows from stator to rotor, the quantity $P_m = \sum_p P_{mp}$ contains an amount of power that is wasted as heat due to frictional effects in the mechanical subsystem. This loss is commonly referred to as **friction and windage loss**, referring to losses that can be attributed to the friction of the bearings or slip rings, and to aerodynamic effects that oppose the rotor motion. Adhering to the notation that we introduced in Equation (2.83c), the power that is eventually reaching the shaft and is producing useful work can be expressed as $P_M = P_m - P_{M,\text{loss}}$.

Example 2.23 *Wind turbines with doubly fed induction generators.*
The topology of a wind turbine with a **doubly fed induction generator** (DFIG) is shown in Figure 2.39. This wind turbine employs a slip power recovery drive, which is one of the main reasons why it is widely used. According to our previous classification, this is a type (b) machine.

The DFIG rotor is connected to the power system via a back-to-back a.c.–d.c.–a.c. converter, which comprises two separate converters (allowing bi-directional power flow) coupled with a d.c. link. The role of the rotor-side converter, which operates based on

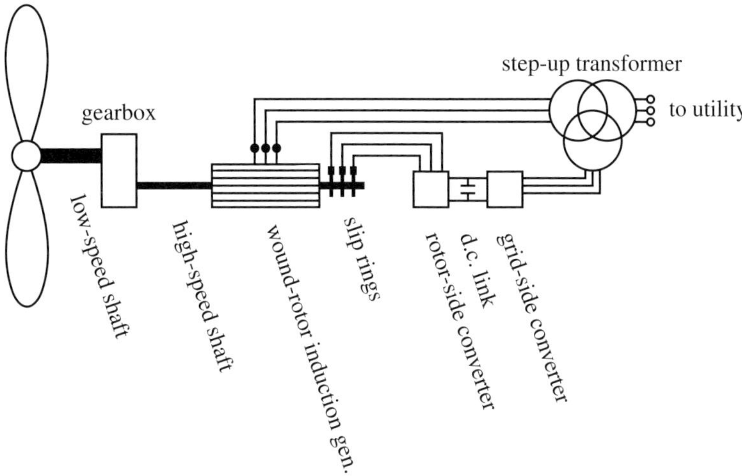

Figure 2.39 Topology of a DFIG wind generator.

an advanced motor drive control, is dual: (i) it controls the frequency of the currents that flow in the rotor windings so that synchronism is maintained between the stator and rotor magnetic fields at all times, which combine to create a single, dominant, p-pole-pair air-gap field; (ii) it controls the magnitude and phase of the currents that flow in the rotor windings, and indirectly by doing this it controls the real and reactive power that the wind turbine provides to the system. The grid-side converter has the role of transferring power to and from the rotor side, maintaining the d.c.-link capacitor at constant voltage. The step-up transformer is commonly a three-winding transformer, with two low-voltage windings for the stator and rotor, and a medium-voltage winding for connection to the wind farm collection system.

In the expressions that follow, we drop the subscript p from the variables because we are discussing the effects of the dominant pole-pair field. We maintain, however, the previous conventions for the assumed positive direction of power flow. Hence, the air-gap power satisfies $P_{s\to r} < 0$, since this machine is always generating electricity from the stator. The mechanical power is also negative, $P_m < 0$, representing power input due to the action of the wind that creates an aerodynamic force on the blades. Let us also assume that losses in stator and rotor are negligible. Therefore, the electrical power entering the stator terminals is $P_s = P_{s\to r} < 0$, whereas the electrical power entering the rotor terminals is $P_r = -P_{\text{slip}}$ (the negative sign appears because the slip power is assumed to be positive from stator to rotor). Adapting Equations (2.369a) and (2.369b), we have

$$P_m = (1 - s)P_s = P_s + P_r\,, \tag{2.370a}$$

$$P_r = -sP_s = -\frac{s}{1-s}P_m\,. \tag{2.370b}$$

We can draw the following conclusions:

- When $s > 0$ (subsynchronous operation), the rotor-side power has opposite sign to the stator-side power. So, for generator action ($P_s < 0$), the rotor windings are absorbing power from the power system (i.e., through the back-to-back converter).
- When $s < 0$ (supersynchronous operation), the rotor-side power has the same sign as the stator-side power. So, for generator action ($P_s < 0$), the rotor windings are supplying power to the power system.
- The slip power increases with $|s|$. In this type of turbine, we can *recover* this power using the power electronics and feed it back to the system. This permits efficient operation for relatively high values of slip.
- The power coming from the shaft splits between the stator and rotor sides. As the values of $|s|$ remain modest, the stator side carries the bulk of the power. Therefore, *the power electronics that are connected to the rotor side need to be rated for only the slip power!* The use of a *partially rated converter* leads to significant cost savings, and simplifies the design of the power electronics. It is in fact one of the main advantages that have led to the prevalence of this type of wind turbine. Typically, commercially available DFIG turbines operate within a ±30% slip range. Furthermore, variable-speed operation allows the turbine to capture more energy out of the wind stream.
- Something very interesting takes place for subsynchronous operation. The situation is easier to understand with a numerical example. Let us assume that the shaft power is $P_m = -1$ (units omitted), and that the slip is $s = 20\%$. Then the stator power will be $P_s = (-1)/(1 - 0.2) = -1.25$, which implies that the stator outputs more power than is generated from the mechanical side! This extra power comes from the rotor side, $P_r = -(0.2)(-1.25) = 0.25$. In other words, an amount of power equal to "0.25" circulates within the topology.
- For supersynchronous operation, the situation is as follows. Let us assume that the shaft power is $P_m = -1$ (units omitted), and that the slip is $s = -20\%$. Then the stator power will be $P_s = (-1)/(1+0.2) = -0.83$, and the rotor power is $P_r = -0.17$. In this case, the shaft power splits in two parts, with the bulk of the power flowing through the stator side. The two components of power meet again at the transformer, where they are combined and supplied to the power system.

2.4.4 Force and Torque from Conservation of Energy

Our third approach for force and torque calculation will be based on the law of conservation of energy. In §2.3.1, we obtained a mathematical expression representing the conservation of energy in the electromagnetic field, namely, Poynting's theorem. The main result that we had derived as Equation (2.89) is perfectly valid in its original form. However, the physical interpretation that was assigned to the various terms in the equation was based on the assumption of stationary matter. When motion occurs, some modifications are necessary to obtain the appropriate physical significance.

We shall now manipulate the equation to represent the following power balance within a volume that completely encloses the moving body:

$$\begin{pmatrix}\text{outflow of}\\ \text{E/M power}\end{pmatrix} + \begin{pmatrix}\text{ohmic}\\ \text{loss}\end{pmatrix} + \begin{pmatrix}\text{mechanical}\\ \text{power}\end{pmatrix} = \begin{pmatrix}\text{rate of reduction of}\\ \text{E/M energy stored}\end{pmatrix}. \tag{2.371}$$

For the signs to be consistent, mechanical work is exerted by a force created endogenously by the field. As shown in Figure 2.33 on page 179, the moving body V_b is completely enclosed by a surface ∂V that is fixed in space and time, and that leaves some finite gap to avoid kinematically obstructing the body's motion. Hence, the first term represents the electromagnetic power flowing outwards through this surface, being thus absorbed by the surrounding electromagnetic field. The other two terms on the left-hand side are also energy outflows, but of a different kind.

From the preceding discussion in this chapter, we can identify the proper expressions for the terms in Equation (2.371):

$$\begin{pmatrix}\text{outflow of}\\ \text{E/M power}\end{pmatrix} = \oiint_{\partial V} \vec{\mathbf{S}} \cdot d\mathbf{a}\,, \tag{2.372a}$$

$$\begin{pmatrix}\text{ohmic}\\ \text{loss}\end{pmatrix} = \iiint_V (\vec{\mathbf{E}} + \vec{\mathbf{v}} \times \vec{\mathbf{B}}) \cdot \vec{\mathbf{J}}\, dv\,, \tag{2.372b}$$

$$\begin{pmatrix}\text{mechanical}\\ \text{power}\end{pmatrix} = \iiint_V \vec{\mathbf{f}}_E \cdot \vec{\mathbf{v}}\, dv\,, \tag{2.372c}$$

$$\begin{pmatrix}\text{rate of reduction of}\\ \text{E/M energy stored}\end{pmatrix} = -\iiint_V \frac{d}{dt}\left\{ m_0(\mathbf{r}) + \int_{\vec{\mathbf{B}}_0(\mathbf{r})}^{\vec{\mathbf{B}}(\mathbf{r})} \vec{\mathbf{H}}(\vec{\mathbf{b}}) \cdot d\vec{\mathbf{b}} \right\} dv\,. \tag{2.372d}$$

In these equations, all variables are vector fields defined in a region V surrounding the moving body that is tightly enclosed in a (moving) region V_b. In the ohmic loss equation, $\vec{\mathbf{E}} = -\nabla\varphi - \partial\vec{\mathbf{A}}/\partial t$ is the electric field seen by a stationary observer; hence, $\vec{\mathbf{E}} + \vec{\mathbf{v}} \times \vec{\mathbf{B}}$ is the electric field seen by the electrons inside moving conductors. A free current $\vec{\mathbf{J}}$ can only exist within the body in V_b, so this vector field is necessarily zero outside V_b; this allows us to obtain the ohmic loss by integrating over the entire space V. A similar argument regarding the force $\vec{\mathbf{f}}_E$ can be made for extending the integration region of mechanical power from V_b to V. Here, we introduce the subscript "E" to differentiate the energy-based force density[o] from the Lorentz force density $\vec{\mathbf{f}}_L$.

The equation for the rate of reduction of stored energy requires careful interpretation, in particular regarding the meaning of the derivative d/dt of the energy density. We are referring to the rate of change of energy density at a point in space displaced by $\mathbf{r}$ from the origin. Here, $\mathbf{r}$ is just the variable of integration of the volume integral, and is independent of time. This local change of energy density may occur due to (i) a variation of the B-field at $\mathbf{r}$ with respect to time, and/or (ii) a change of material occupying point $\mathbf{r}$

[o] This is not the "official" name for this force density. We just use it here to differentiate it from the Lorentz force density, which is unambiguously known as such. The designation stems from the way this force is derived here, i.e., based on an energy balance.

in space, via a corresponding change of the *B–H* characteristic and associated initial magnetization energy (if any). More precisely, we could have written the upper limit as $\vec{\mathbf{B}}(\mathbf{r}, t)$, the lower limit as $\vec{\mathbf{B}}_0(\mathbf{r}, t)$, the integrand as $\vec{\mathbf{H}}(\mathbf{r}, t, \vec{\mathbf{b}})$, and the initial energy density as $m_0(\mathbf{r}, t)$. Here, it is convenient to define the initial conditions $\vec{\mathbf{B}}_0$ and m_0 such that $\vec{\mathbf{H}}(\vec{\mathbf{B}}_0) = \mathbf{0}$. We will suppose that materials are lossless, and thus can be modeled by a constitutive law of the form[p]

$$\vec{\mathbf{B}} = \mu\vec{\mathbf{H}} + \mu_0\vec{\mathbf{M}}_0 = \mu\vec{\mathbf{H}} + \vec{\mathbf{B}}_0 . \tag{2.373}$$

When a permanent magnet starts replacing a non-magnetic material that is originally occupying the space at point $\mathbf{r}$, we have $dm_0/dt > 0$. However, at other points on the magnet boundary the opposite scenario is developing, that is, magnet is being replaced by non-magnetic material. Hence, integrated over the entire volume, $\iiint_V dm_0/dt\, dv = 0$.

The Leibniz[58] integral rule comes in handy. It applies to the calculation as follows (omitting functional dependency on $\mathbf{r}$ and t for notational simplicity):

$$\frac{d}{dt}\int_{\vec{\mathbf{B}}_0}^{\vec{\mathbf{B}}} \vec{\mathbf{H}}(\vec{\mathbf{b}})\cdot d\vec{\mathbf{b}} = \vec{\mathbf{H}}(\vec{\mathbf{B}})\cdot\frac{\partial\vec{\mathbf{B}}}{\partial t} - \underbrace{\vec{\mathbf{H}}(\vec{\mathbf{B}}_0)}_{=\mathbf{0}}\cdot\frac{\partial\vec{\mathbf{B}}_0}{\partial t} + \int_{\vec{\mathbf{B}}_0}^{\vec{\mathbf{B}}} \frac{\partial\vec{\mathbf{H}}(\vec{\mathbf{b}})}{\partial t}\cdot d\vec{\mathbf{b}} . \tag{2.374}$$

The first term on the right-hand side represents power being stored due to a varying magnetic field; the second term is always zero by virtue of the definition of $\vec{\mathbf{B}}_0$; the third term should be interpreted as an integration of the rate of change of the *B–H* characteristic under constant limits. An illustration of the incremental change of stored magnetic energy is provided in Figure 2.40. The sketch captures an incremental change during the transition from a material of low permeability to one with higher permeability. The shape of the *B–H* curve suggests that a ferromagnetic material ($B_0 = 0$) is replacing the material that was previously at this location. The magnitude of change is exaggerated for illustration purposes. We will now switch the variable of integration from *B* to *H* as follows:

$$\frac{d}{dt}\int_{\vec{\mathbf{B}}_0}^{\vec{\mathbf{B}}} \vec{\mathbf{H}}(\vec{\mathbf{b}})\cdot d\vec{\mathbf{b}} = \vec{\mathbf{H}}(\vec{\mathbf{B}})\cdot\frac{\partial\vec{\mathbf{B}}}{\partial t} - \int_{\mathbf{0}}^{\vec{\mathbf{H}}} \frac{\partial\vec{\mathbf{B}}(\vec{\mathbf{h}})}{\partial t}\cdot d\vec{\mathbf{h}} , \tag{2.375}$$

where we also had to introduce a minus sign. The validity of this algebraic manipulation can be verified by inspection of Figure 2.40. Of course, strictly speaking, the upper limit is $\vec{\mathbf{H}}(\mathbf{r}, t)$, and the integrand is a function $\vec{\mathbf{B}}(\mathbf{r}, t, \vec{\mathbf{h}})$.

Now we can compare the original Poynting equation (2.89) (assuming $\partial\vec{\mathbf{D}}/\partial t = \mathbf{0}$ for a magnetostatic problem) with this newly derived version:

[p] The assumption of lossless materials implies that this calculation is more restrictive than the one based on the Lorentz force density, where we had argued that it is valid even for nonlinear, possibly hysteretic, materials.

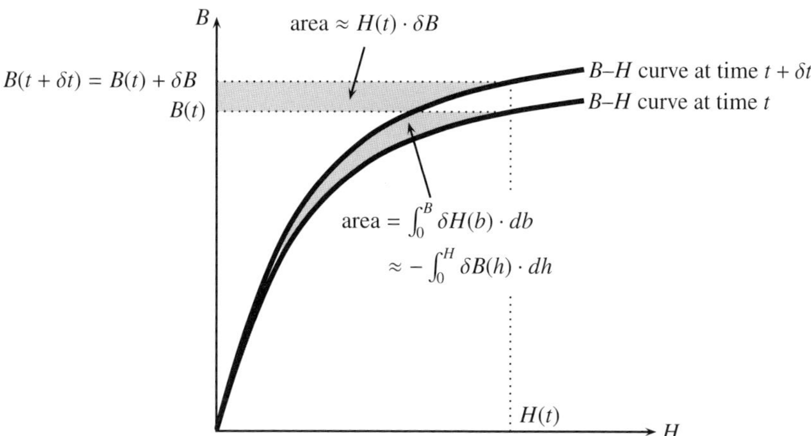

Figure 2.40 Incremental change of magnetic energy caused by combined field and material variation (shaded area). The quantity δH is the horizontal distance between the two curves (in this case, negative), whereas δB is their vertical distance (in this case, positive).

$$\text{original:}\quad \oiint_{\partial V} \vec{\mathbf{S}} \cdot d\mathbf{a} + \iiint_V \vec{\mathbf{E}} \cdot \vec{\mathbf{J}}\, dv = -\iiint_V \vec{\mathbf{H}}(\vec{\mathbf{B}}) \cdot \frac{\partial \vec{\mathbf{B}}}{\partial t}\, dv\,, \tag{2.376a}$$

$$\begin{aligned}\text{modified:}\quad &\oiint_{\partial V} \vec{\mathbf{S}} \cdot d\mathbf{a} + \iiint_V (\vec{\mathbf{E}} + \vec{\mathbf{v}} \times \vec{\mathbf{B}}) \cdot \vec{\mathbf{J}}\, dv + \iiint_V \vec{\mathbf{f}}_E \cdot \vec{\mathbf{v}}\, dv \\ &= -\iiint_V \vec{\mathbf{H}}(\vec{\mathbf{B}}) \cdot \frac{\partial \vec{\mathbf{B}}}{\partial t}\, dv + \iiint_V \int_{\mathbf{0}}^{\vec{\mathbf{H}}} \frac{\partial \vec{\mathbf{B}}(\vec{\mathbf{h}})}{\partial t} \cdot d\vec{\mathbf{h}}\, dv - \underbrace{\iiint_V \frac{dm_0}{dt}\, dv}_{0}\,.\end{aligned} \tag{2.376b}$$

Subtracting the two equations yields

$$\iiint_V \vec{\mathbf{f}}_E \cdot \vec{\mathbf{v}}\, dv = -\iiint_V (\vec{\mathbf{v}} \times \vec{\mathbf{B}}) \cdot \vec{\mathbf{J}}\, dv + \iiint_V \int_{\mathbf{0}}^{\vec{\mathbf{H}}} \frac{\partial \vec{\mathbf{B}}(\vec{\mathbf{h}})}{\partial t} \cdot d\vec{\mathbf{h}}\, dv\,, \tag{2.377}$$

and using the triple-scalar-product identity (1.41) for the middle term:

$$\iiint_V \vec{\mathbf{f}}_E \cdot \vec{\mathbf{v}}\, dv - \iiint_V (\vec{\mathbf{J}} \times \vec{\mathbf{B}}) \cdot \vec{\mathbf{v}}\, dv + \iiint_V \int_0^{\vec{\mathbf{H}}} \frac{\partial \vec{\mathbf{B}}(\vec{\mathbf{h}})}{\partial t} \cdot d\vec{\mathbf{h}}\, dv\,. \tag{2.378}$$

We will manipulate the last term, in an attempt to transform the integrand to the form $\vec{\mathbf{X}} \cdot \vec{\mathbf{v}}$. We have

$$\begin{aligned}\iiint_V \int_{\mathbf{0}}^{\vec{\mathbf{H}}} \frac{\partial \vec{\mathbf{B}}(\vec{\mathbf{h}})}{\partial t} \cdot d\vec{\mathbf{h}}\, dv &= \iiint_V \int_{\mathbf{0}}^{\vec{\mathbf{H}}} \left(\frac{\partial \mu}{\partial t} \vec{\mathbf{h}} + \mu_0 \frac{\partial \vec{\mathbf{M}}_0}{\partial t} \right) \cdot d\vec{\mathbf{h}}\, dv \qquad &(2.379a) \\ &= \iiint_V \left(\int_{\mathbf{0}}^{\vec{\mathbf{H}}} \frac{\partial \mu}{\partial t} \vec{\mathbf{h}} \cdot d\vec{\mathbf{h}} + \mu_0 \frac{\partial \vec{\mathbf{M}}_0}{\partial t} \cdot \vec{\mathbf{H}} \right) dv\,. \qquad &(2.379b)\end{aligned}$$

To evaluate the rate of change of $\mu = \mu(\mathbf{r}, t, \vec{\mathbf{h}})$ and $\vec{\mathbf{M}}_0 = \vec{\mathbf{M}}_0(\mathbf{r}, t, \vec{\mathbf{H}})$, consider that the material present at point $\mathbf{r}$ at time t is the same material that was at point $\mathbf{r} - \delta t \vec{\mathbf{v}}$ at time $t - \delta t$. Furthermore, if rotation is occurring, $\vec{\mathbf{M}}_0$ has also turned by a small angle, $\delta t\,\vec{\omega}$. So, within a small time interval δt, the function μ has changed by

$$\delta\mu = \mu(\mathbf{r} - \delta t\,\vec{\mathbf{v}}) - \mu(\mathbf{r}) \approx -\delta t\,\vec{\mathbf{v}} \cdot \nabla\mu\,, \tag{2.380}$$

using a first-order Taylor expansion, where the gradient is evaluated at $\mathbf{r}$. The change of permanent magnetization within the same time interval is found by displacing and then rotating the magnetization of the incoming material:

$$\begin{aligned}\delta\vec{\mathbf{M}}_0 &= \vec{\mathbf{M}}_0(\mathbf{r} - \delta t\,\vec{\mathbf{v}}) + \delta t\,\vec{\omega} \times \vec{\mathbf{M}}_0(\mathbf{r} - \delta t\,\vec{\mathbf{v}}) - \vec{\mathbf{M}}_0(\mathbf{r}) \\ &\approx \delta t[-(\vec{\mathbf{v}} \cdot \nabla)\vec{\mathbf{M}}_0 + \vec{\omega} \times \vec{\mathbf{M}}_0]\,,\end{aligned} \tag{2.381}$$

where we dropped higher-order terms from the Taylor expansion. Letting $\delta t \to 0$ so that derivatives are formed, and substituting these results back into Equation (2.378):

$$\begin{aligned}\iiint_V \vec{\mathbf{f}}_E \cdot \vec{\mathbf{v}}\, dv &= \iiint_V (\vec{\mathbf{J}} \times \vec{\mathbf{B}}) \cdot \vec{\mathbf{v}}\, dv \\ &- \iiint_V \left(\int_{\mathbf{0}}^{\vec{\mathbf{H}}} (\vec{\mathbf{v}} \cdot \nabla\mu)\vec{\mathbf{h}} \cdot d\vec{\mathbf{h}} + \mu_0(\vec{\mathbf{v}} \cdot \nabla)\vec{\mathbf{M}}_0 \cdot \vec{\mathbf{H}} - \mu_0\vec{\omega} \times \vec{\mathbf{M}}_0 \cdot \vec{\mathbf{H}} \right) dv\,.\end{aligned} \tag{2.382}$$

In view of Equation (2.289) on page 180, we rewrite the last two terms as follows:[q]

$$\begin{aligned}&-\mu_0 \iiint_V [(\vec{\mathbf{v}} \cdot \nabla)\vec{\mathbf{M}}_0 - \vec{\omega} \times \vec{\mathbf{M}}_0] \cdot \vec{\mathbf{H}}\, dv \\ &\qquad = \mu_0 \iiint_V [\nabla \times (\vec{\mathbf{v}} \times \vec{\mathbf{M}}_0) - \vec{\mathbf{v}}(\nabla \cdot \vec{\mathbf{M}}_0)] \cdot \vec{\mathbf{H}}\, dv\,.\end{aligned} \tag{2.383}$$

The first term inside the integrand can be changed using the vector calculus identity (1.159):

$$\nabla \times (\vec{\mathbf{v}} \times \vec{\mathbf{M}}_0) \cdot \vec{\mathbf{H}} = \nabla \cdot [(\vec{\mathbf{v}} \times \vec{\mathbf{M}}_0) \times \vec{\mathbf{H}}] + (\vec{\mathbf{v}} \times \vec{\mathbf{M}}_0) \cdot \underbrace{(\nabla \times \vec{\mathbf{H}})}_{\vec{\mathbf{J}}}\,. \tag{2.384}$$

Since we are integrating over V, we may apply the divergence theorem to the term $\nabla \cdot [(\vec{\mathbf{v}} \times \vec{\mathbf{M}}_0) \times \vec{\mathbf{H}}]$. The resulting surface integral will vanish because both $\vec{\mathbf{v}}$ and $\vec{\mathbf{M}}_0$ are zero at the outer boundary. So

$$\iiint_V \nabla \times (\vec{\mathbf{v}} \times \vec{\mathbf{M}}_0) \cdot \vec{\mathbf{H}}\, dv = \iiint_V (\vec{\mathbf{v}} \times \vec{\mathbf{M}}_0) \cdot \vec{\mathbf{J}}\, dv = \iiint_V \vec{\mathbf{v}} \cdot (\vec{\mathbf{M}}_0 \times \vec{\mathbf{J}})\, dv\,. \tag{2.385}$$

Collecting all these results back into Equation (2.382), we have

$$\begin{aligned}\iiint_V \vec{\mathbf{f}}_E \cdot \vec{\mathbf{v}}\, dv &= \iiint_V (\vec{\mathbf{J}} \times \vec{\mathbf{B}}) \cdot \vec{\mathbf{v}}\, dv - \iiint_V \vec{\mathbf{v}} \cdot \int_{\mathbf{0}}^{\vec{\mathbf{H}}} \nabla\mu\,\vec{\mathbf{h}} \cdot d\vec{\mathbf{h}}\, dv \\ &+ \iiint_V \vec{\mathbf{v}} \cdot (\mu_0\vec{\mathbf{M}}_0 \times \vec{\mathbf{J}})\, dv - \iiint_V \vec{\mathbf{v}}(\nabla \cdot \mu_0\vec{\mathbf{M}}_0) \cdot \vec{\mathbf{H}}\, dv\,.\end{aligned} \tag{2.386}$$

The most difficult part of the proof is now complete. We have set the stage to extract an expression for the force density.

[q] Strictly speaking, the result that we are invoking holds inside V_b, where we can guarantee that $\nabla \cdot \vec{\mathbf{v}} = 0$. However, $\vec{\mathbf{M}}_0 = \mathbf{0}$ outside V_b since that space is empty, so we can extend the integration to V.

Force Density from Conservation of Energy

In the case of translational motion (e.g., for a linear actuator), the velocity throughout the body is uniform ($\vec{\mathbf{v}}(\mathbf{r}) = \vec{\mathbf{v}}_0$). It suffices to ensure that the volume integrals on both sides of Equation (2.386) are equal for arbitrary $\vec{\mathbf{v}}_0$. We proceed with the understanding that the force density cannot be uniquely determined, since we may add any term that integrates to zero over V_b. For example, suppose that we add the gradient of an arbitrary (but differentiable) scalar function ψ to the estimated force density, such that $\psi = 0$ on ∂V_b; we also ask that $\psi = 0$ outside the body as well (there should be no force in empty space). Integrating the corresponding power component over V:

$$\iiint_V \nabla\psi \cdot \vec{\mathbf{v}}_0 \, dv = \iiint_{V_b} \nabla\psi \cdot \vec{\mathbf{v}}_0 \, dv = \iiint_{V_b} \nabla(\psi \vec{\mathbf{v}}_0) \, dv = \oint\!\!\oint_{\partial V_b} \psi \vec{\mathbf{v}}_0 \, d\mathbf{a} = 0 \,, \tag{2.387}$$

where we used the vector identity (1.158) and $\nabla \cdot \vec{\mathbf{v}}_0 = 0$ for the second equality, and the divergence theorem for the third equality.

Therefore, an expression for **the force density $\vec{\mathbf{f}}_E$ as a function of the magnetic field, the free current, and material properties** at point $\mathbf{r}$ and time t is

$$\boxed{\vec{\mathbf{f}}_E = \vec{\mathbf{J}} \times (\vec{\mathbf{B}} - \mu_0 \vec{\mathbf{M}}_0) - \mu_0(\nabla \cdot \vec{\mathbf{M}}_0)\vec{\mathbf{H}} - \int_{\mathbf{0}}^{\vec{\mathbf{H}}} \nabla\mu \vec{\mathbf{h}} \cdot d\vec{\mathbf{h}} \,.} \tag{2.388}$$

The three terms on the right-hand side can be interpreted as follows:

1. The first term yields the familiar Lorentz $\vec{\mathbf{J}} \times \vec{\mathbf{B}}$ force density on free currents. Within conductors, the formula applies with $\vec{\mathbf{M}}_0 = \mathbf{0}$. Also, it yields forces on ferromagnetic materials or permanent magnets in the presence of eddy currents.
2. The second term yields a force on permanent magnets. The force is collinear with $\vec{\mathbf{H}}$, and its magnitude is scaled by $\nabla \cdot \vec{\mathbf{M}}_0$. Historically, from the time of Poisson who hypothesized the presence of "magnetic matter," this term has been associated with a **magnetic charge density**:

   $$\rho_m = -\nabla \cdot \vec{\mathbf{M}}_0 \,. \tag{2.389}$$

   If the magnetization is uniform within a magnet, this force will only appear on its outer boundary. It will be more prominent at the north and south poles, where the divergence of the magnetization is significant, as it transitions to that of the surrounding material (where $M_0 = 0$).
3. The third term yields force due to a varying permeability. The math is a bit awkward, as at first glance it seems we are multiplying a vector ($\nabla\mu$) with another vector ($\vec{\mathbf{h}}$). But if we study the derivation carefully, we will realize that this term originated as $(\vec{\mathbf{v}} \cdot \nabla\mu)\vec{\mathbf{h}}$. So, it should be interpreted component-wise; e.g., $\int (\nabla\mu)_x \vec{\mathbf{h}} \cdot d\vec{\mathbf{h}}$. This term is nonzero only at the interface between materials. For example, considering an interface between air and steel, $-\nabla\mu$ is a vector that points from the steel to the air; hence, it is as if a force is developed that tends to peel off the outer layer from the steel. This term is responsible for the development of reluctance torque in electric machines. In a magnetically linear material, this force density becomes $-(\nabla\mu)H^2/2$.

It should be noted that the Lorentz force density, $\vec{\mathbf{f}}_L = (\vec{\mathbf{J}} + \vec{\mathbf{J}}_m) \times \vec{\mathbf{B}}$, is *different* from the energy-based force density $\vec{\mathbf{f}}_E$ given by Equation (2.388). Only their volume integrals should be equal so that an identical resultant force is obtained:

$$\vec{\mathbf{F}}_e = \iiint_V \vec{\mathbf{f}}_L \, dv = \iiint_V \vec{\mathbf{f}}_E \, dv \,. \tag{2.390}$$

(The fact that they are, indeed, equal will be proven shortly.) For instance, Lorentz forces are developed on the sides of magnets where magnetization currents $\vec{\mathbf{M}} \times \hat{\mathbf{n}}$ are present, whereas this method predicts forces on the north/south poles where $\nabla \cdot \vec{\mathbf{M}}_0$ is nonzero. Also, in ferromagnetic materials, a Lorentz force may be present within the material itself if $\nabla \times \vec{\mathbf{M}}$ is nonzero, whereas this method restricts the reluctance force on the boundary because the gradient of the permeability function vanishes within a material.

Torque from Conservation of Energy

In the case of purely rotational motion, the velocity throughout the body is $\vec{\mathbf{v}}(\mathbf{r}) = \vec{\omega} \times \mathbf{r}$, where we assumed that the center of rotation is at the origin. Substituting this in Equation (2.386), and using the scalar triple product to convert all terms to the form $\vec{\omega} \cdot \vec{\mathbf{X}}$, yields

$$\iiint_V \vec{\omega} \cdot (\mathbf{r} \times \vec{\mathbf{f}}_E) \, dv = \iiint_V \vec{\omega} \cdot \{\mathbf{r} \times [\vec{\mathbf{J}} \times (\vec{\mathbf{B}} - \mu_0 \vec{\mathbf{M}}_0)]\} \, dv$$
$$- \iiint_V \vec{\omega} \cdot [\mathbf{r} \times (\nabla \cdot \mu_0 \vec{\mathbf{M}}_0) \cdot \vec{\mathbf{H}}] \, dv - \iiint_V \vec{\omega} \cdot \left(\mathbf{r} \times \int_{\mathbf{0}}^{\vec{\mathbf{H}}} \nabla \mu \, \vec{\mathbf{h}} \cdot d\vec{\mathbf{h}} \right) dv \,. \tag{2.391}$$

Therefore, the electromagnetic torque is

$$\vec{\tau}_e = \iiint_V (\mathbf{r} \times \vec{\mathbf{f}}_E) \, dv \,, \tag{2.392}$$

where $\vec{\mathbf{f}}_E$ is given by Equation (2.388).

Equivalence with the Lorentz Force Density

We will show that the energy-based force density yields the same resultant force and torque as the one obtained by employing the Lorentz force density. Our approach will be to show equivalence with the surface integral of the MST method, see Equations (2.328) and (2.351).

Proof Equation (2.388) may be written as a function of the H-field only:

$$\vec{\mathbf{f}}_E = (\nabla \times \vec{\mathbf{H}}) \times (\mu \vec{\mathbf{H}}) + (\nabla \cdot (\mu \vec{\mathbf{H}}))\vec{\mathbf{H}} - \int_{\mathbf{0}}^{\vec{\mathbf{H}}} \nabla \mu \, \vec{\mathbf{h}} \cdot d\vec{\mathbf{h}} \,. \tag{2.393}$$

This was obtained by replacing $\vec{\mathbf{J}} = \nabla \times \vec{\mathbf{H}}$. We also used the constitutive law (2.373) and Maxwell's law, $\nabla \cdot \vec{\mathbf{B}} = \nabla \cdot (\mu \vec{\mathbf{H}}) + \mu_0 \nabla \cdot \vec{\mathbf{M}}_0 = 0$. We manipulate the first term on the right-hand side by invoking the vector calculus identity (1.176):

$$\frac{1}{2} \nabla (\vec{\mathbf{H}} \cdot \vec{\mathbf{H}}) = \frac{1}{2} \nabla (H^2) = (\vec{\mathbf{H}} \cdot \nabla)\vec{\mathbf{H}} + \vec{\mathbf{H}} \times (\nabla \times \vec{\mathbf{H}}) \,, \tag{2.394}$$

and then multiplying both sides by the scalar field μ. This leads to

$$(\nabla \times \vec{\mathbf{H}}) \times (\mu \vec{\mathbf{H}}) = (\mu \vec{\mathbf{H}} \cdot \nabla)\vec{\mathbf{H}} - \frac{\mu}{2}\nabla(H^2)\,. \tag{2.395}$$

Now we turn our attention to the integral. As was mentioned earlier, this should be interpreted as a component-wise integration. Using the x-component as an example, we employ the Leibniz integral rule to obtain

$$\int_0^{\vec{\mathbf{H}}} \frac{\partial \mu}{\partial x} \vec{\mathbf{h}} \cdot d\vec{\mathbf{h}} = \frac{\partial}{\partial x}\left(\int_0^{\vec{\mathbf{H}}} \mu \vec{\mathbf{h}} \cdot d\vec{\mathbf{h}}\right) - \mu \vec{\mathbf{H}} \cdot \frac{\partial \vec{\mathbf{H}}}{\partial x} \tag{2.396a}$$

$$= \frac{\partial}{\partial x}\left(\int_0^{\vec{\mathbf{H}}} \mu \vec{\mathbf{h}} \cdot d\vec{\mathbf{h}}\right) - \frac{\mu}{2}\frac{\partial (H^2)}{\partial x}\,. \tag{2.396b}$$

Stacking these three components to form a vector, we can write

$$\int_0^{\vec{\mathbf{H}}} \nabla\mu \vec{\mathbf{h}} \cdot d\vec{\mathbf{h}} = \nabla\left(\int_0^{\vec{\mathbf{H}}} \mu \vec{\mathbf{h}} \cdot d\vec{\mathbf{h}}\right) - \frac{\mu}{2}\nabla(H^2)\,. \tag{2.397}$$

Combining Equations (2.393), (2.395), and (2.397), we obtain

$$\vec{\mathbf{f}}_E = (\mu \vec{\mathbf{H}} \cdot \nabla)\vec{\mathbf{H}} + (\nabla \cdot (\mu \vec{\mathbf{H}}))\vec{\mathbf{H}} - \nabla\left(\int_0^{\vec{\mathbf{H}}} \mu \vec{\mathbf{h}} \cdot d\vec{\mathbf{h}}\right). \tag{2.398}$$

It can be readily shown that this equals the divergence of a tensor $\boldsymbol{\sigma}_E$:

$$\vec{\mathbf{f}}_E = \nabla \cdot \underbrace{\begin{bmatrix} \mu H_x^2 - \int_0^{\vec{\mathbf{H}}} \mu \vec{\mathbf{h}} \cdot d\vec{\mathbf{h}} & \mu H_x H_y & \mu H_x H_z \\ \mu H_x H_y & \mu H_y^2 - \int_0^{\vec{\mathbf{H}}} \mu \vec{\mathbf{h}} \cdot d\vec{\mathbf{h}} & \mu H_y H_z \\ \mu H_x H_z & \mu H_y H_z & \mu H_z^2 - \int_0^{\vec{\mathbf{H}}} \mu \vec{\mathbf{h}} \cdot d\vec{\mathbf{h}} \end{bmatrix}}_{\sigma_E}. \tag{2.399}$$

There is a similarity with the Maxwell stress tensor we derived earlier, see Equation (2.304), but the two tensors (and the force densities) are different. The Maxwell stress tensor depends on the B-field and the permeability of vacuum. On the other hand, the tensor for the energy-based force density depends on the H-field and the nonlinear permeability of the material that occupies point $\mathbf{r}$ in space.

The resultant force is $\vec{\mathbf{F}}_e = \iiint_V \vec{\mathbf{f}}_E\, dv$. For the first two terms on the right-hand side of Equation (2.398), we invoke the vector calculus identity (1.219):

$$\iiint_V [(\mu \vec{\mathbf{H}} \cdot \nabla)\vec{\mathbf{H}} + (\nabla \cdot (\mu \vec{\mathbf{H}}))\vec{\mathbf{H}}]\, dv = \oiint_{\partial V} \vec{\mathbf{H}}(\mu \vec{\mathbf{H}} \cdot \hat{\mathbf{n}})\, da\,. \tag{2.400}$$

For the third term, we invoke the identity (1.198):

$$\iiint_V \nabla\phi\, dv = \oiint_{\partial V} \phi\, \hat{\mathbf{n}}\, da\,, \tag{2.401}$$

setting $\phi = \int_0^{\vec{\mathbf{H}}} \mu \vec{\mathbf{h}} \cdot d\vec{\mathbf{h}}$. Therefore, we have obtained the resultant force as a surface integral:

$$\vec{\mathbf{F}}_e = \oiint_{\partial V} \left[\vec{\mathbf{H}}(\mu \vec{\mathbf{H}} \cdot \hat{\mathbf{n}}) - \left(\int_0^{\vec{\mathbf{H}}} \mu \vec{\mathbf{h}} \cdot d\vec{\mathbf{h}}\right)\hat{\mathbf{n}}\right] da\,. \tag{2.402}$$

Finally, we realize that integrating at the boundary means that we are in air ($\mu = \mu_0$), so the equation becomes

$$\vec{\mathbf{F}}_e = \mu_0 \oiint_{\partial V} \left[\vec{\mathbf{H}}(\vec{\mathbf{H}} \cdot \hat{\mathbf{n}}) - \tfrac{1}{2}H^2\,\hat{\mathbf{n}}\right] da = \nu_0 \oiint_{\partial V} \left[\vec{\mathbf{B}}(\vec{\mathbf{B}} \cdot \hat{\mathbf{n}}) - \tfrac{1}{2}B^2\hat{\mathbf{n}}\right] da\,. \tag{2.403}$$

This surface integral implies the presence of a surface stress that in fact is identical to $\vec{\mathbf{t}}_m$ of Equation (2.329), which we had previously derived for the MST method starting from the Lorentz force equation. We conclude that the two force densities, $\vec{\mathbf{f}}_L$ and $\vec{\mathbf{f}}_E$, yield identical answers as far as the resultant force is concerned.

Since $\vec{\mathbf{f}}_E = \nabla \cdot \boldsymbol{\sigma}_E$, electromagnetic torque is

$$\vec{\tau}_e = \iiint_V (\mathbf{r} \times \vec{\mathbf{f}}_E)\, dv = \iiint_V [(\mathbf{r} \times (\nabla \cdot \boldsymbol{\sigma}_E)]\, dv\,. \tag{2.404}$$

In exactly the same way as we did for the Maxwell stress tensor, this integral can be transformed into a surface integral based on the surface stress $\vec{\mathbf{t}}_m$, that is

$$\vec{\tau}_e = \iiint_V [\mathbf{r} \times (\nabla \cdot \boldsymbol{\sigma}_E)]\, dv = \oiint_{\partial V} (\mathbf{r} \times \vec{\mathbf{t}}_m)\, da\,. \tag{2.405}$$

Therefore, we may also conclude that the two force densities, $\vec{\mathbf{f}}_L$ and $\vec{\mathbf{f}}_E$, yield identical answers as far as the resultant torque is concerned. □

Example 2.24 *Torque on a permanent magnet.*

Consider a rectangular permanent magnet (shaped as a parallelepiped) in free space, which is free to spin around its center. The magnet has width w along the x-axis, height h along the y-axis, and length ℓ along the z-axis. The axis of rotation is the z-axis, as shown by the dots in the cross-section of Figure 2.41. A constant permanent magnetization, $\vec{\mathbf{M}}_0 = M_0\,\hat{\mathbf{j}}$, has been established within the magnet. The permeability of the magnet material is μ_0.[r] An external, uniform magnetic field $\vec{\mathbf{B}}_{\text{ext}}$ is present throughout space. The problem is to calculate the z-axis torque on the magnet.

Solution

The PM creates a magnetic field that, from a distance, resembles that of a magnetic dipole, see Figure 1.23. The PM field is superimposed on the external magnetic field throughout space, so the total field is

$$\vec{\mathbf{B}} = \vec{\mathbf{B}}_{\text{ext}} + \vec{\mathbf{B}}_{pm}\,. \tag{2.406}$$

If an accurate expression for the PM field were available, we could apply the Maxwell stress tensor method on some surface surrounding the magnet. However, lacking such an analytical formula, we shall instead calculate torque by accounting for the forces that are developed via interaction with magnetic charges or equivalent magnetization currents.

As we have seen, there are two ways to calculate the torque, namely: (A) based on a

[r] In general, magnets have permeability $\mu \neq \mu_0$. This simplification permits us to ignore an additional reluctance force component at the magnet surface, which would make this example more involved.

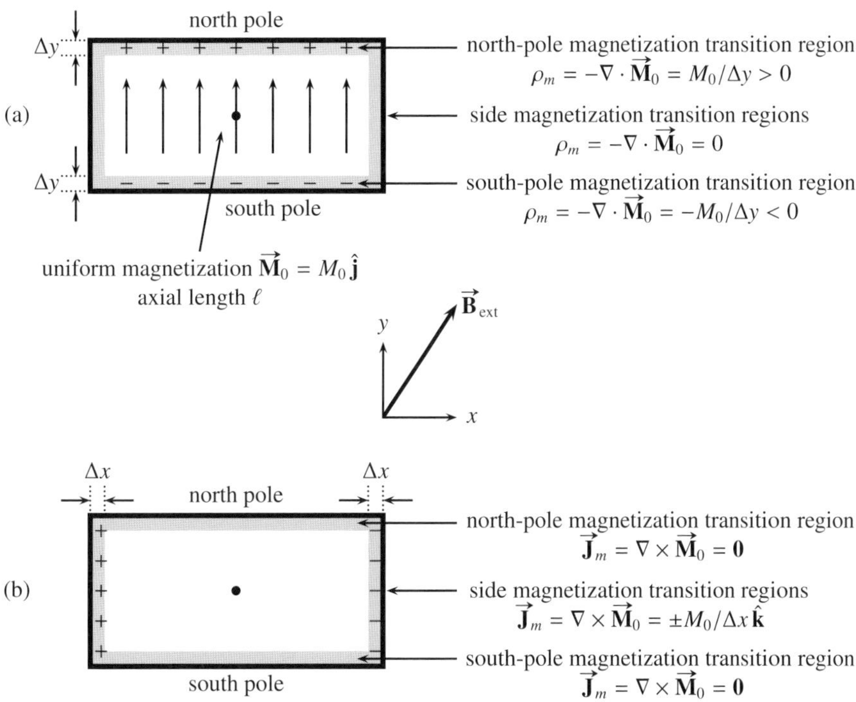

Figure 2.41 Calculation of force distribution on permanent magnet: (a) based on magnetic charge; (b) based on equivalent magnetization currents.

distribution of magnetic charge ρ_m that interacts with the H-field per Equation (2.388), as shown in Figure 2.41(a); or (B) based on a distribution of equivalent magnetization current $\vec{\mathbf{J}}_m$ that interacts with the B-field to create a Lorentz force given by Equation (2.299), as in Figure 2.41(b). We introduce a thin layer at the magnet's outer surface, where a smooth transition between the internal magnetization and that of its surroundings (zero) occurs. We are modeling this transition as a linear increase/decrease, like the one depicted in Figure 2.34, which leads to the magnetic charge and current expressions of Figure 2.41. Magnetic charge is found only on the north and south pole faces of the magnet, where $\rho_m = -\nabla \cdot \vec{\mathbf{M}}_0 \neq 0$. Equivalent magnetization current $\vec{\mathbf{J}}_m = \nabla \times \vec{\mathbf{M}}_0$ is present on the other four sides of the magnet, and is zero on the poles. (Only two sides are shown in the cross-section of Figure 2.41. The other two sides at $z = \pm\ell/2$ will carry magnetization current oriented along the $\pm x$-axis.) The two models are equivalent in the sense that they create the same B-field, and should also lead to the same force and torque on the magnet, as we explained in the preceding sections.

To proceed, we need to know the magnetic field inside the magnet, and in particular we need to know the field within the transition regions, where forces are developed.

Method (A). For convenience, let us take the origin to be at the center of the PM. The torque component of interest may be expressed as

$$\tau_{ez} = \iiint_V \mu_0 \rho_m (xH_y - yH_x)\, dv \tag{2.407a}$$

$$= \mu_0 \int_{x=-w/2}^{w/2} \int_{y=h/2-\Delta y}^{h/2} \int_{z=-\ell/2}^{\ell/2} \frac{M_0}{\Delta y} (xH_y - yH_x)\, dz\, dy\, dx$$
$$- \mu_0 \int_{x=-w/2}^{w/2} \int_{y=-h/2}^{-h/2+\Delta y} \int_{z=-\ell/2}^{\ell/2} \frac{M_0}{\Delta y} (xH_y - yH_x)\, dz\, dy\, dx\,. \tag{2.407b}$$

Here, $H_x = H_x(x, y, z)$ and $H_y = H_y(x, y, z)$. The volume integral is broken up into two terms representing contributions from the north and south poles, respectively.

According to Equation (2.373), inside the magnet the H-field satisfies

$$\vec{\mathbf{B}} = \mu_0 \vec{\mathbf{H}} + \mu_0 \vec{\mathbf{M}}_0\,. \tag{2.408}$$

Since

$$\nabla \cdot \vec{\mathbf{B}} = \nabla \cdot (\mu_0 \vec{\mathbf{H}} + \mu_0 \vec{\mathbf{M}}_0) = 0\,, \tag{2.409}$$

we have

$$\nabla \cdot \vec{\mathbf{H}} = -\nabla \cdot \vec{\mathbf{M}}_0 = \rho_m\,. \tag{2.410}$$

Also, in the absence of free currents in this problem, $\nabla \times \vec{\mathbf{H}} = \mathbf{0}$. Therefore, we may express the magnetic field as the negative gradient of a **scalar magnetic potential** function φ_m:

$$\vec{\mathbf{H}} = -\nabla \varphi_m\,. \tag{2.411}$$

Taking the divergence we obtain a Poisson equation for the magnetic scalar potential:

$$\nabla \cdot \vec{\mathbf{H}} = -\nabla^2 \varphi_m = \rho_m\,. \tag{2.412}$$

Based on our earlier work regarding the divergence and curl at material interfaces, see Examples 1.19 and 1.23, we observe the following regarding the normal and tangential components of the H-field at the north and south poles: (i) the normal component H_y changes by $\rho_m \Delta y = \pm M_0$; and (ii) the tangential component H_x is continuous across the interface.

The H-field is essentially identical to the electric field of a parallel plate capacitor! This well-known electrostatic problem (obtained by solving $\nabla \cdot \vec{\mathbf{D}} = \rho$) is mathematically equivalent to the magnetostatic problem, with the magnetic charge now playing the role of electric charge. Solving Poisson's partial differential equation in three dimensions is not trivial, even in this elementary example consisting of the permanent magnet sandwiched between two superthin layers of magnetic charge. Nevertheless, we know that the H-field emanates from the positive magnetic charge at the north pole, and terminates at the negative magnetic charge at the south pole. Inside the magnet, if $h \ll w$ and $h \ll \ell$, the lines of H are mostly parallel and are pointing downwards, but outwards fringing may occur at the vicinity of the sides.

With this picture of the PM H-field in mind, we can derive additional relationships that arise from PM symmetry considerations. Symmetry with respect to the $x = 0$ plane implies that

$$H_{pm,x}(x, y, z) = -H_{pm,x}(-x, y, z)\,, \tag{2.413a}$$
$$H_{pm,y}(x, y, z) = H_{pm,y}(-x, y, z)\,, \tag{2.413b}$$

$$H_{pm,z}(x,y,z) = H_{pm,z}(-x,y,z)\,. \tag{2.413c}$$

Symmetry with respect to the $y = 0$ plane implies that

$$H_{pm,x}(x,y,z) = -H_{pm,x}(x,-y,z)\,, \tag{2.414a}$$

$$H_{pm,y}(x,y,z) = H_{pm,y}(x,-y,z)\,, \tag{2.414b}$$

$$H_{pm,z}(x,y,z) = -H_{pm,z}(x,-y,z)\,. \tag{2.414c}$$

Symmetry with respect to the $z = 0$ plane implies that

$$H_{pm,x}(x,y,z) = H_{pm,x}(x,y,-z)\,, \tag{2.415a}$$

$$H_{pm,y}(x,y,z) = H_{pm,y}(x,y,-z)\,, \tag{2.415b}$$

$$H_{pm,z}(x,y,z) = -H_{pm,z}(x,y,-z)\,. \tag{2.415c}$$

These properties of the PM field simplify the torque expression (2.407b) because integrals of odd functions can be identified if we slice the device intelligently. The simplified integral is

$$\tau_{ez} = \mu_0 \int_{x=-w/2}^{w/2} \int_{y=h/2-\Delta y}^{h/2} \int_{z=-\ell/2}^{\ell/2} \frac{M_0}{\Delta y}(-yH_{\text{ext},x})\,dz\,dy\,dx$$
$$- \mu_0 \int_{x=-w/2}^{w/2} \int_{y=-h/2}^{-h/2+\Delta y} \int_{z=-\ell/2}^{\ell/2} \frac{M_0}{\Delta y}(-yH_{\text{ext},x})\,dz\,dy\,dx\,. \tag{2.416}$$

The result depends only on the external field, as expected, because the endogenous PM field should not create any torque. Evaluating this integral is straightforward, and yields

$$\tau_{ez} = -\mu_0 wh\ell M_0 H_{\text{ext},x}\,. \tag{2.417}$$

The minus sign means that if the external field is directed as shown in Figure 2.41, then the developed electromagnetic torque tends to rotate the magnet clockwise, trying to align the magnetization with the field. This is also the result that we would obtain in a more direct fashion by multiplying the PM volume ($wh\ell$) with the dipole torque density $\vec{\mathbf{M}}_0 \times \vec{\mathbf{B}}_{\text{ext}}$.

Method (B). In the second modeling approach, the effect of permanent magnetization is represented by equivalent magnetization currents appearing at the PM sides. Note that the H-field in this case will be *different* from the H-field in the previous method. Here, the H-field is obtained by $\vec{\mathbf{B}} = \mu_0 \vec{\mathbf{H}}$, where $\nabla \times \vec{\mathbf{B}} = \mu_0 \vec{\mathbf{J}}_m$.

The Lorentz forces that are developed at the two sides at $x = \mp w/2$ are

$$\vec{\mathbf{f}}_L = \vec{\mathbf{J}}_m \times \vec{\mathbf{B}} = \pm\frac{M_0}{\Delta x}\hat{\mathbf{k}} \times \vec{\mathbf{B}} = \pm\frac{M_0}{\Delta x}(-B_y\hat{\mathbf{i}} + B_x\hat{\mathbf{j}})\,. \tag{2.418}$$

Lorentz forces are also developed at the other two sides, at $z = \pm\ell/2$:

$$\vec{\mathbf{f}}_L = \vec{\mathbf{J}}_m \times \vec{\mathbf{B}} = \pm\frac{M_0}{\Delta z}\hat{\mathbf{i}} \times \vec{\mathbf{B}} = \pm\frac{M_0}{\Delta z}(-B_z\hat{\mathbf{j}} + B_y\hat{\mathbf{k}})\,. \tag{2.419}$$

The z-axis torque component is

$$\tau_{ez} = \iiint_V (xf_{Ly} - yf_{Lx})\,dv \tag{2.420a}$$

$$
\begin{aligned}
&= \int_{x=-w/2}^{-w/2+\Delta x} \int_{y=-h/2}^{h/2} \int_{z=-\ell/2}^{\ell/2} \frac{M_0}{\Delta x}(xB_x + yB_y)\,dz\,dy\,dx \\
&- \int_{x=w/2-\Delta x}^{w/2} \int_{y=-h/2}^{h/2} \int_{z=-\ell/2}^{\ell/2} \frac{M_0}{\Delta x}(xB_x + yB_y)\,dz\,dy\,dx \\
&+ \int_{x=-w/2}^{w/2} \int_{y=-h/2}^{h/2} \int_{z=\ell/2-\Delta z}^{\ell/2} \frac{M_0}{\Delta z}(-xB_z)\,dz\,dy\,dx \\
&- \int_{x=-w/2}^{w/2} \int_{y=-h/2}^{h/2} \int_{z=-\ell/2}^{-\ell/2+\Delta z} \frac{M_0}{\Delta z}(-xB_z)\,dz\,dy\,dx\,. \qquad (2.420b)
\end{aligned}
$$

Invoking symmetry, the torque simplifies to

$$
\begin{aligned}
\tau_{ez} = &\int_{x=-w/2}^{-w/2+\Delta x} \int_{y=-h/2}^{h/2} \int_{z=-\ell/2}^{\ell/2} \frac{M_0}{\Delta x}(xB_{\text{ext},x})\,dz\,dy\,dx \\
&- \int_{x=w/2-\Delta x}^{w/2} \int_{y=-h/2}^{h/2} \int_{z=-\ell/2}^{\ell/2} \frac{M_0}{\Delta x}(xB_{\text{ext},x})\,dz\,dy\,dx\,. \qquad (2.421)
\end{aligned}
$$

We observe that contributions from the front and back sides have vanished. The torque expression evaluates to

$$
\tau_{ez} = -wh\ell M_0 B_{\text{ext},x}\,. \qquad (2.422)
$$

Method (B) yields the same answer as Method (A).

Force calculation using surface distributions of charge and current. The presence of the superthin transition layers at the PM sides is a theoretical artifact, conveniently leading to differentiable functions for the vector calculus proofs. Alternatively, we may use *surface* charge and magnetization current density distributions. These are obtained at the limit, e.g., if $\Delta x \to 0$. The main question that arises in this approach is this: Since the field is changing at the interface, which value should we use in the formula for force? The preceding analysis suggests that we should use the *average* value of the field at the two sides of the interface.

Method (A). This method involves a magnetic charge density. Consider a point on the north pole of the magnet at arbitrary x and z. Converting to a surface density of magnetic charge means that the integral in Equation (2.407b) with respect to y is replaced by its value at the limit $\Delta y \to 0$:

$$
\lim_{\Delta y \to 0} \left\{ \int_{y=h/2-\Delta y}^{h/2} \frac{M_0}{\Delta y}(xH_y - yH_x)\,dy \right\} = M_0 \left(x\frac{H_y^+ + H_y^-}{2} - \frac{h}{2}H_x \right). \qquad (2.423)
$$

At the north pole, the normal components are $H_y^+ = H_y(x, h/2, z)$ and $H_y^- = H_y(x, h/2 - \Delta y, z)$. Since $\nabla \cdot \vec{\mathbf{H}} = \rho_m$, these components are related by

$$
H_y^+ - H_y^- = M_0 = \text{surface charge density}\,. \qquad (2.424)
$$

Also, because $\nabla \times \vec{\mathbf{H}} = \mathbf{0}$, the tangential component of the H-field is continuous across the interface, so its average value is $H_x = H_x(x, h/2, z)$. Therefore, the north-pole surface force density is

$$\vec{\mathbf{f}}_{E,s} = \mu_0 M_0 \vec{\mathbf{H}}_{\text{ave}} \,. \tag{2.425}$$

The torque integral in Equation (2.407b) has thus been converted from a volume integral to two surface integrals over each pole, of the form

$$\iint_S \mathbf{r} \times \vec{\mathbf{f}}_{E,s}\, da = \iint_S \mathbf{r} \times (\mu_0 M_0 \vec{\mathbf{H}}_{\text{ave}})\, da \,, \tag{2.426}$$

i.e.

$$\begin{aligned}\tau_{ez} = \mu_0 M_0 \int_{x=-w/2}^{w/2} \int_{z=-\ell/2}^{\ell/2} \left[x \frac{H_y^+ + H_y^-}{2} - \frac{h}{2} H_x \right]_{y=h/2} dz\, dx \\ - \mu_0 M_0 \int_{x=-w/2}^{w/2} \int_{z=-\ell/2}^{\ell/2} \left[x \frac{H_y^+ + H_y^-}{2} - \frac{h}{2} H_x \right]_{y=-h/2} dz\, dx \,.\end{aligned} \tag{2.427}$$

Invoking symmetry, all PM field terms will vanish, and we will obtain the same answer as before.

Method (B). This method is based on an equivalent surface density of magnetization current $\vec{\mathbf{K}}_m$. Take the $x = -w/2$ side as an example where the outwards normal is $\hat{\mathbf{n}} = -\hat{\mathbf{i}}$. Considering a point at arbitrary y and z, we have

$$\vec{\mathbf{K}}_m = \vec{\mathbf{M}}_0 \times \hat{\mathbf{n}} = M_0\, \hat{\mathbf{k}} = \Delta x\, \vec{\mathbf{J}}_m \,. \tag{2.428}$$

Converting to a surface density of magnetization current means that the integral in Equation (2.420b) with respect to x is replaced by its value at the limit $\Delta x \to 0$:

$$\lim_{\Delta x \to 0} \left\{ \int_{x=-w/2}^{-w/2+\Delta x} \frac{M_0}{\Delta x} (x B_x + y B_y)\, dx \right\} = M_0 \left(-\frac{w}{2} B_x + y \frac{B_y^+ + B_y^-}{2} \right) . \tag{2.429}$$

The tangential components at this PM boundary are defined as $B_y^+ = B_y(-w/2, y, z)$ and $B_y^- = B_y(-w/2 + \Delta x, y, z)$. Because $\nabla \times \vec{\mathbf{B}} = \mu_0 \vec{\mathbf{J}}_m$, these components are related by

$$-B_y^+ + B_y^- = \mu_0 M_0 = \mu_0 \cdot (\text{surface current density}) \,. \tag{2.430}$$

Also, the normal component of the B-field is continuous across the interface because $\nabla \cdot \vec{\mathbf{B}} = 0$, so its average value is $B_x = B_x(-w/2, y, z)$. Therefore, the corresponding surface force density is

$$\vec{\mathbf{f}}_{L,s} = \vec{\mathbf{K}}_m \times \vec{\mathbf{B}}_{\text{ave}} = M_0\, \hat{\mathbf{k}} \times \vec{\mathbf{B}}_{\text{ave}} = M_0 \left(-\frac{B_y^+ + B_y^-}{2}\, \hat{\mathbf{i}} + B_x \hat{\mathbf{j}} \right) . \tag{2.431}$$

The torque integral in Equation (2.420b) has been converted from a volume integral to four surface integrals over the PM sides of the form

$$\iint_S \mathbf{r} \times \vec{\mathbf{f}}_{L,s}\, da = \iint_S \mathbf{r} \times (\vec{\mathbf{K}}_m \times \vec{\mathbf{B}}_{\text{ave}})\, da \,. \tag{2.432}$$

This calculation yields the same answer as before.

2.5 Summary

This chapter started with a presentation of the main concepts of variational calculus. We then proceeded with a variational proof of the equivalence between Maxwell's equations for the electrostatic and magnetostatic problems and the stationarity of a functional. The inherent relationship of the functional to energy led us to an analysis of the main energy transfer pathways and loss mechanisms in electric machines. The properties of a conservative coupling field were highlighted. We also explained how circuit quantities can be obtained from the fields. Finally, we derived from basic physics various fundamental fields-based expressions for the electromagnetic force/torque developed by an electric machine.

Our approach in this chapter was of a theoretical nature—no Python! However, the concepts that were presented are fundamentals that are necessary for understanding the operation of motors based on field solutions. In the next chapter, we will set forth the finite element method, which will allow us to solve our first real-world problem.

2.6 Further Reading

The calculus of variations is a classical subject that finds broad application in mechanics and optics, as well as in modern control theory, engineering, and finite element methods. This subject is fascinating because it is not purely a computational method, but transcends the boundaries of mathematics and represents basic laws of physics, such as the principle of least action in mechanics or Fermat's[59] principle for the propagation of light. A broad bibliography is available to the interested reader [2–13].

Electric machine basics can be reviewed in undergraduate-level texts such as [14–25]. These books are representative of modern pedagogical approaches to the theory of electric machines. It is also interesting to examine how the subject has evolved over the years, by reading classical books on a.c./d.c. machines and electromechanical energy conversion dating back to the early 1900s [26–37]. To the reader who may be inclined to learn more about the intricacies of machine design and control, including that of permanent magnet and switched reluctance motors, we suggest [38–53]. References related to various methods that have been proposed for the calculation of the electromagnetic force in machines are listed at the end of Chapter 4.

Problems

2.1 Derive an analytical solution for the curve of the brachistochrone problem (see Example 2.3) using the Beltrami identity (2.22) as a starting point.

2.2 A numerical solution to the roller coaster problem (see Example 2.3) may be obtained by discretizing space. Suppose we subdivide $[x_a, x_b]$ into $N - 1$ equally spaced subintervals (it is not necessary to select equally spaced intervals, but let's keep it simple). Within each interval, assume $w(x)$ is approximated as a straight line connecting

(x_k, w_k) and (x_{k+1}, w_{k+1}), for $k = 1, \dots, N-1$, so that $x_1 = x_a$ and $x_N = x_b$. The unknown values of w_k, $k = 2, \dots, N-1$ are to be determined. (In general, these w_k will differ from the actual solution $w(x_k)$, but we would like them to be close.) To this end, convert the integral of Equation (2.20) to a finite sum

$$I(w) \approx \sum_{k=1}^{N-1} I_k(w_k, w_{k+1}), \tag{2.433}$$

where I_k is the approximate contribution to I from the kth element. Express I_k for a prototypical element. Simplify to the extent possible. How do we find the w_k that minimize this function (discussion question—do not actually implement yet)? *Hint:* This is an optimization problem. Try to formulate it as such, i.e., define an objective function, decision variables, and constraints. Identify a software tool that can be used to solve it.

2.3 Based on the previous problem, implement a numerical solution to the brachistochrone problem using boundary points of your choice. Compare with the theoretical curve. Explore what happens for different choices of the discretization parameter N.

2.4 Annotate Figures 2.12–2.14 with arrows indicating the direction of the magnetic field.

2.5 Recreate the flux line plots of Figures 2.12–2.14. Experiment with different pole-pairs and MVP boundary conditions.

2.6 Recreate the quiver plot of Figure 2.16. Experiment with different pole-pairs and MVP boundary conditions.

2.7 Derive Equations (2.111a) and (2.111b).

2.8 Derive Equation (2.123).

2.9 Derive Equations (2.125b) and (2.125c).

2.10 Recreate the plots of Example 2.8.

2.11 Recreate the results of Example 2.9.

2.12 Prove Equation (2.288).

2.13 Consider a copper conductor with a 1 mm^2 cross-sectional area. The conductor is carrying 10 A of current. Calculate the drift velocity of the electrons.

2.14 Prove that Faraday's law (2.192) is obtained from Equation (2.191), accounting for the possibility of a moving coil.

2.15 Consider Example 2.17. Sketch the variation of $W_f(\theta_r)$ and $\tau_e(\theta_r)$ for θ_r in the range $(-2\pi, 2\pi)$.

2.16 Consider Example 2.17. Derive the torque using a different approach, based on calculating the coupling field energy directly from the stator and rotor magnetic fields.

2.17 Consider Example 2.17 and introduce a slight variation where the magnetic field is no longer a rectangular function of position in the air gap, but is sinusoidal. (For example, take a Fourier series representation of the rectangular waveform, and use only the fundamental component.) Derive expressions for the inductances and the electromagnetic torque.

2.18 Consider Example 2.17, which depicts an elementary two-pole machine. Modify the stator and rotor by adding slots and coils to obtain a four- and a six-pole machine.

2.19 Consider Example 2.18. Sketch the variation of $W_f(\theta_r)$ and $\tau_e(\theta_r)$ for θ_r in the range $(-2\pi, 2\pi)$.

2.20 Similar to what happens to the resultant force based on $\vec{\mathbf{f}}_E$, show that the electromagnetic torque of Equation (2.392) is not affected by adding a force density component that is the gradient of an arbitrary scalar field that vanishes on the boundary of the moving body.

2.21 Consider Example 2.24. Sketch the PM H-field inside and outside the magnet based on a magnetic charge surface density, and confirm the symmetry relations (2.413a)–(2.415c). Also sketch the PM B-field based on an equivalent magnetization current surface density. Finally, go through the mathematical derivations for both methods.

References

[1] C. P. Steinmetz, "On the law of hysteresis," *Trans. of AIEE*, vol. IX, no. 1, pp. 3–64, Jan. 1892.

[2] I. M. Gelfand and S. V. Fomin, *Calculus of Variations*. Englewood Cliffs, NJ: Prentice-Hall, 1963.

[3] M. Giaquinta and S. Hildenbrandt, *Calculus of Variations*. Berlin: Springer, 2004, vol. 1–2.

[4] P. Hammond, *Energy Methods in Electromagnetism*. Oxford: Clarendon Press, 1981.

[5] H. Kielhöfer, *Calculus of Variations: An Introduction to the One-Dimensional Theory with Examples and Exercises*, ser. Universitext. Cham, Switzerland: Springer International, 2018.

[6] C. Lanczos, *The Variational Principles of Mechanics*, 4th ed. New York: Dover, 1970.

[7] G. Leitmann, *The Calculus of Variations and Optimal Control: An Introduction*. New York: Springer, 1981.

[8] M. Levi, *Classical Mechanics with Calculus of Variations and Optimal Control: An Intuitive Introduction*. Providence, RI: American Mathematical Society, 2014.

[9] D. Liberzon, *Calculus of Variations and Optimal Control Theory: A Concise Introduction*. Princeton, NJ: Princeton University Press, 2012.

[10] M. Mesterton-Gibbons, *A Primer on the Calculus of Variations and Optimal Control Theory*. Providence, RI: American Mathematical Society, 2009.

[11] H. Sagan, *Introduction to the Calculus of Variations*. New York: Dover, 1969.

[12] B. van Brunt, *The Calculus of Variations*, ser. Universitext. New York: Springer, 2004.

[13] R. Weinstock, *Calculus of Variations with Applications to Physics & Engineering*. New York: Dover, 1974.

[14] S. Chapman, *Electric Machinery Fundamentals*, 5th ed. New York: McGraw-Hill, 2012.

[15] T. Gönen, *Electrical Machines with Matlab*, 2nd ed. Boca Raton, FL: CRC Press, 2012.

[16] C. A. Gross, *Electric Machines*. Boca Raton, FL: CRC Press, 2006.

[17] P. Krause, O. Wasynczuk, S. D. Pekarek, and T. O'Connell, *Electromechanical Motion Devices: Rotating Magnetic Field-Based Analysis with Online Animations*, 3rd ed. Hoboken, NJ: Wiley-IEEE Press, 2020.

[18] J. Melkebeek, *Electrical Machines and Drives*. Cham, Switzerland: Springer International, 2018.

[19] N. Mohan, *Electric Machines and Drives: A First Course*. Hoboken, NJ: Wiley, 2012.

[20] M. A. Pai, *Power Circuits & Electromechanics*. Champaign, IL: Stipes, 2012.

[21] R. Ramshaw and R. G. van Heeswijk, *Energy Conversion: Electric Motors and Generators*. Philadelphia: Saunders, 1990.

[22] M. S. Sarma, *Electric Machines: Steady-State Theory and Dynamic Performance*, 2nd ed. Boston: Cengage Learning, 1997.

[23] P. C. Sen, *Principles of Electric Machines and Power Electronics*, 3rd ed. Hoboken, NJ: Wiley, 2013.

[24] S. D. Umans, *Fitzgerald & Kingsley's Electric Machinery*, 7th ed. New York: McGraw-Hill, 2014.

[25] T. Wildi, *Electrical Machines, Drives, and Power Systems*, 6th ed. Boston: Pearson, 2006.

[26] C. B. Gray, *Electrical Machines and Drive Systems*. Harlow, UK: Longman Scientific & Technical, 1989.

[27] C. V. Jones, *The Unified Theory of Electrical Machines*. London: Butterworths, 1967.

[28] M. Kostenko and L. Piotrovsky, *Electrical Machines*. Moscow: Mir Publishers, 1974, vol. 1-2.

[29] P. Krause, O. Wasynczuk, S. Sudhoff, and S. Pekarek, *Analysis of Electric Machinery and Drive Systems*, 3rd ed. Hoboken, NJ: Wiley-IEEE Press, 2013.

[30] Y. H. Ku, *Electric Energy Conversion*. New York: Ronald Press Company, 1959.

[31] A. S. Langsdorf, *Theory of Alternating-Current Machinery*, 2nd ed. New York: McGraw-Hill, 1955.

[32] A. S. Langsdorf, *Principles of Direct Current Machines*, 3rd ed. New York: McGraw-Hill, 1923.

[33] W. V. Lyon, *Transient Analysis of Alternating-Current Machinery*. Cambridge: Technology Press of MIT and John Wiley & Sons, 1954.

[34] S. Seely, *Electromechanical Energy Conversion*. New York: McGraw-Hill, 1962.

[35] G. R. Slemon and A. Straughen, *Electric Machines*. Reading, MA: Addison-Wesley, 1980.

[36] J. G. Tarboux, *Alternating-Current Machinery*. Scranton, PA: International Textbook Company, 1947.

[37] D. C. White and H. H. Woodson, *Electromechanical Energy Conversion*. New York: John Wiley & Sons, 1959.

[38] P. L. Alger, *Induction Machines: Their Behavior and Uses*, 2nd ed. Basel, Switzerland: Gordon and Breach, 1970.

[39] B. Bilgin, J. W. Jiang, and A. Emadi, *Switched Reluctance Motor Drives: Fundamentals to Applications*. Boca Raton, FL: CRC Press, 2019.

[40] A. Dymkov, *Transformer Design*. Moscow: Mir Publishers, 1975.

[41] D. C. Hanselman, *Brushless Permanent Magnet Motor Design*, 2nd ed. Lebanon, OH: Magna Physics Publishing, 2006.

[42] J. F. Gieras, *Permanent Magnet Motor Technology: Design and Applications*, 3rd ed. Boca Raton, FL: CRC Press, 2009.

[43] T. Kenjo and S. Nagamori, *Permanent-Magnet and Brushless DC Motors*. Oxford: Clarendon Press, 1986.

[44] R. Krishnan, *Permanent Magnet Synchronous and Brushless DC Motor Drives*. Boca Raton, FL: CRC Press, 2010.

[45] R. Krishnan, *Switched Reluctance Motor Drives: Modeling, Simulation, Analysis, Design, and Applications*. Boca Raton, FL: CRC Press, 2017.

[46] T. A. Lipo, *Introduction to AC Machine Design*. Hoboken, NJ: Wiley-IEEE Press, 2017.

[47] T. J. E. Miller, *Brushless Permanent Magnet and Reluctance Motor Drives*. Oxford: Oxford University Press, 1989.

[48] T. J. E. Miller, *Switched Reluctance Motors and their Control*. Oxford: Magna Physics Publishing and Clarendon Press, 1993.

[49] T. J. E. Miller and J. R. Hendershot, *Design of Brushless Permanent-Magnet Machines*. Venice, FL: Motor Design Books, 2010.

[50] S. A. Nasar, I. Boldea, and L. E. Unnewehr, *Permanent Magnet, Reluctance, and Self-Synchronous Motors*. Boca Raton, FL: CRC Press, 1993.

[51] J. Pyrhönen, T. Jokinen, and V. Hrabovcová, *Design of Rotating Electrical Machines*, 2nd ed. Chichester, UK: Wiley, 2014.

[52] S. D. Sudhoff, *Power Magnetic Devices: A Multi-Objective Design Approach*, 2nd ed. Hoboken, NJ: Wiley-IEEE Press, 2021.

[53] P. Vas, *Sensorless Vector and Direct Torque Control*. Oxford: Oxford University Press, 1998.

3 The Finite Element Method

Πάντα κατ' αριθμόν γίγνονται
"Everything happens according to numbers"
Pythagoras[60]

In the preceding chapter, we discussed theoretical concepts that are fundamental for a deeper understanding of fields-based electric machine analysis. The objective of this chapter is to set forth the finite element method (FEM) for computing numerical approximations of the fields. A rigorous mathematical treatment of the FEM can be quite challenging. Instead, here we will take a more pragmatic approach, where various aspects of the method will be revealed gradually as we progress.

In §3.1, we will introduce the FEM by learning how to solve the elementary Laplace's equation (without ascribing physical significance to the problem). We will learn the basics of meshing a 2-D domain using triangular elements, and we will explain how this leads to a linear system of equations for the field. Then, in §3.2 we will extend the theory to the linear Poisson's equation, which includes a source term. This will permit us to conduct finite element analysis (FEA)[a] of electromagnetic devices with magnetically linear materials. §3.3 will focus on coding the linear FEM using Python, including post-processing steps that allow us to visualize the solution. §3.4 is dedicated to the topic of extracting flux linkage and inductance from an FEA solution.

In the second part of this chapter, starting with §3.5, we will describe how to incorporate in the FEA ferromagnetic materials with nonlinear *B–H* characteristics, and how to solve the problem (which is no longer linear) using iterative methods, with a focus on the Newton–Raphson method. §3.6 will set forth important computer implementation details for the nonlinear problem. §3.7 will deal with the modeling of permanent magnets. §3.8 will explain how to handle devices with cylindrical symmetry. Finally, in §3.9, we will take a more abstract view in a bid to present important mathematical concepts related to Galerkin's method.

[a] FEM refers to the method itself. FEA refers to the application of the FEM to solve a particular problem.

3.1 Solution of Laplace's Equation Using the FEM

For our first introduction to the FEM, suppose we need to find the scalar field $w(x, y)$ that satisfies Laplace's equation:

$$\nabla^2 w = 0 \text{ inside } \mathcal{D} \subset \mathbb{R}^2, \tag{3.1}$$

with Dirichlet boundary constraint (see Figure 2.3 on page 103)

$$w(x, y) = g(x, y) \text{ on } \partial\mathcal{D}_1 \tag{3.2}$$

and Neumann constraint

$$\frac{\partial w}{\partial n} = 0 \text{ on } \partial\mathcal{D}_2. \tag{3.3}$$

Here, w could have the physical meaning of an electrostatic or magnetic potential. As explained in §2.2.1, the solution to this problem may be obtained with a variational approach by finding the function $w(x, y)$ that makes the following functional stationary:

$$I(w) = \frac{1}{2} \iint_{\mathcal{D}} \|\nabla w\|^2 dx\, dy. \tag{3.4}$$

In the FEM, instead of seeking a solution of the original problem in the space of smooth, twice-differentiable functions (since the Laplacian requires two derivatives), we will seek an approximate solution that lies in the space of continuous, piecewise linear functions. To define the breakpoints of the piecewise linear solution, we partition the domain $\mathcal{D}$ into a finite number of non-overlapping sub-domains, as illustrated in Figure 3.1. This is called a **mesh**. The sub-domains may be of many possible forms; however, we will consider only triangular elements.

Within each triangular element of the domain, we will assume that $w(x, y)$ is a linear interpolate of its (yet to be determined) vertex values. Hence, since adjacent triangles share common vertex values, continuity of the solution across triangle boundaries is guaranteed. If the dimensions of the triangles forming the mesh are small, we expect that our approximate solution will be close to the real solution. Estimating the error of

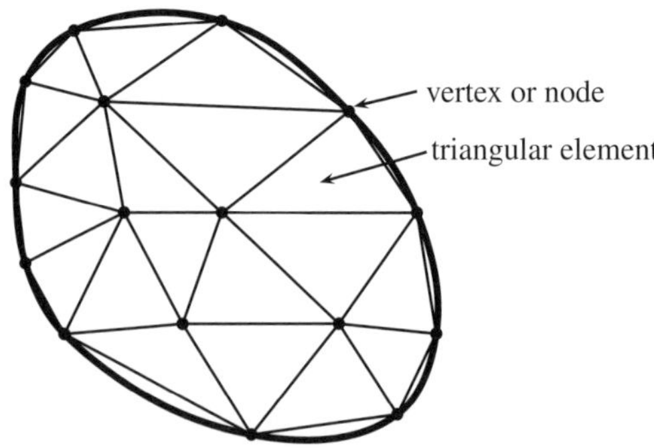

Figure 3.1 A 2-D domain that has been meshed with triangular elements.

the FEA solution is an important topic in numerical analysis (see references at the end of this chapter), but it is outside the scope of this text.

3.1.1 Functional Stationarity and the Weak Form

Let us recall briefly what happens when we ask that the functional is stationary. Because of Equation (2.27) in Step 2 of our variational process (see page 103), we know that stationarity of the functional (3.4) implies that the solution w^* satisfies

$$\iint_{\mathcal{D}} \left(\frac{\partial \eta}{\partial x}\frac{\partial w^*}{\partial x} + \frac{\partial \eta}{\partial y}\frac{\partial w^*}{\partial y} \right) dx\,dy = \iint_{\mathcal{D}} \nabla\eta \cdot \nabla w^*\, dx\,dy = 0\,, \tag{3.5}$$

where η defines the shape of the variation, which can be arbitrary. This equation is equivalent to the Euler–Lagrange equation (i.e., the Laplace equation in this case); what separates the two is only an integration by parts. In mathematics, this is called the **weak form** of the Laplace equation.

Now let us evaluate the functional for any admissible function $w = w^* + \alpha\eta$:

$$I(w^* + \alpha\eta) = \frac{1}{2}\iint_{\mathcal{D}} \nabla(w^* + \alpha\eta)\cdot\nabla(w^* + \alpha\eta)\, dx\,dy \tag{3.6a}$$

$$= \frac{1}{2}\iint_{\mathcal{D}} \nabla w^*\cdot\nabla w^*\, dx\,dy + \alpha\iint_{\mathcal{D}} \nabla\eta\cdot\nabla w^*\, dx\,dy$$

$$+ \frac{\alpha^2}{2}\iint_{\mathcal{D}} \nabla\eta\cdot\nabla\eta\, dx\,dy \tag{3.6b}$$

$$= I(w^*) + \frac{\alpha^2}{2}\iint_{\mathcal{D}} \|\nabla\eta\|^2\, dx\,dy\,, \tag{3.6c}$$

where we eliminated the term that vanishes due to the weak form. Therefore, we conclude that stationarity of the Laplace equation functional leads to a minimization problem. This is because we are looking for the solution w^* that satisfies

$$I(w^*) \leq I(w)\,, \tag{3.7}$$

for any function w that is twice differentiable (as required by the Laplace equation) and satisfies the prescribed Dirichlet boundary condition.

The FEM variational process operates on the set of piecewise linear functions on the given mesh. These functions cannot be solutions of the Laplace equation because they are not twice differentiable (since the first derivative is discontinuous). Nevertheless, there is no problem in having piecewise linear functions in $I(w)$, which only requires the existence of a gradient. So, *the FEM problem is to determine a piecewise linear function that minimizes the functional against any other piecewise linear function satisfying the boundary conditions.*

3.1.2 Single-Element Preliminary Calculations

Consider an individual triangle, as shown in Figure 3.2. Inside the triangular element, the solution takes the form

$$w(x, y) = a + bx + cy\,. \tag{3.8}$$

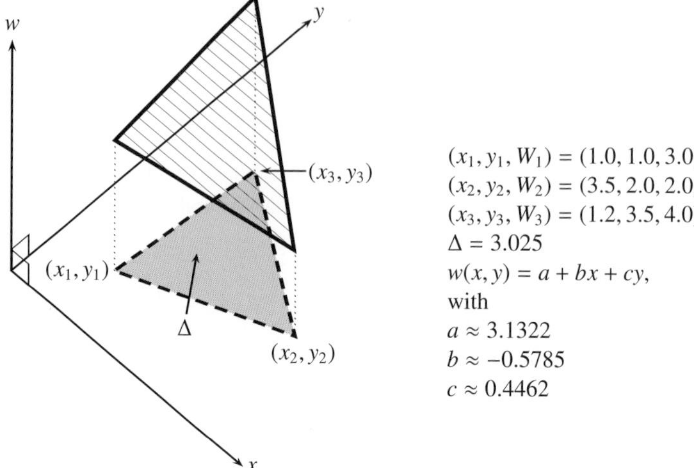

Figure 3.2 Example of a linear triangular element.

This is called a **linear element**. The coefficients a, b, and c can be related to the vertex potentials W_1, W_2, and W_3 by solving the linear system of equations

$$W_1 = a + bx_1 + cy_1\,, \tag{3.9a}$$
$$W_2 = a + bx_2 + cy_2\,, \tag{3.9b}$$
$$W_3 = a + bx_3 + cy_3\,. \tag{3.9c}$$

This yields

$$a = \frac{1}{2\Delta}[(x_2y_3 - x_3y_2)W_1 + (x_3y_1 - x_1y_3)W_2 + (x_1y_2 - x_2y_1)W_3]\,, \tag{3.10a}$$
$$b = \frac{1}{2\Delta}[(y_2 - y_3)W_1 + (y_3 - y_1)W_2 + (y_1 - y_2)W_3]\,, \tag{3.10b}$$
$$c = \frac{1}{2\Delta}[(x_3 - x_2)W_1 + (x_1 - x_3)W_2 + (x_2 - x_1)W_3]\,, \tag{3.10c}$$

where (see Example 1.4 and Equation (1.38c) on page 10)

$$\Delta = \frac{1}{2}[(x_2 - x_1)(y_3 - y_1) - (y_2 - y_1)(x_3 - x_1)] = \pm(\text{triangle area})\,. \tag{3.10d}$$

We recall that Δ = (triangle area) > 0, if the nodes are ordered in a counterclockwise sense (as in Figure 3.2); otherwise, $\Delta = -$(triangle area).

Substituting Equations (3.10a)–(3.10c) into (3.8):

$$\begin{aligned} w(x, y) &= \frac{1}{2\Delta}[(x_2y_3 - x_3y_2)W_1 + (x_3y_1 - x_1y_3)W_2 + (x_1y_2 - x_2y_1)W_3] \\ &+ \frac{1}{2\Delta}[(y_2 - y_3)W_1 + (y_3 - y_1)W_2 + (y_1 - y_2)W_3]x \\ &+ \frac{1}{2\Delta}[(x_3 - x_2)W_1 + (x_1 - x_3)W_2 + (x_2 - x_1)W_3]y\,. \end{aligned} \tag{3.11}$$

Rearranging:

$$\begin{aligned} w(x,y) &= W_1 \frac{1}{2\Delta}[(x_2y_3 - x_3y_2) + (y_2 - y_3)x + (x_3 - x_2)y] \\ &+ W_2 \frac{1}{2\Delta}[(x_3y_1 - x_1y_3) + (y_3 - y_1)x + (x_1 - x_3)y] \\ &+ W_3 \frac{1}{2\Delta}[(x_1y_2 - x_2y_1) + (y_1 - y_2)x + (x_2 - x_1)y]\,. \end{aligned} \tag{3.12}$$

More compactly, we can write the solution as the linear combination

$$w(x,y) = \sum_{i=1}^{3} W_i\, \alpha_i(x,y)\,, \tag{3.13}$$

where the α_i are called **element basis functions**. The basis functions may be expressed in the form

$$\boxed{\alpha_i(x,y) = \frac{1}{2\Delta}(p_i + q_i x + r_i y)} \tag{3.14}$$

where, for instance:

$$p_1 = x_2y_3 - x_3y_2\,, \tag{3.15a}$$

$$q_1 = y_2 - y_3\,, \tag{3.15b}$$

$$r_1 = x_3 - x_2\,. \tag{3.15c}$$

Expressions for p_2, q_2, r_2, p_3, q_3, and r_3 are obtained by inspection of Equation (3.12), or from Equations (3.15a)–(3.15c) by cyclic permutation of the subscripts, e.g.

$$p_2 = x_3y_1 - x_1y_3\,, \tag{3.16a}$$

$$q_2 = y_3 - y_1\,, \tag{3.16b}$$

$$r_2 = x_1 - x_3\,. \tag{3.16c}$$

It can be shown that the basis functions possess the property

$$\alpha_i(x_j, y_j) = \delta_{ij} = \begin{cases} 1 & \text{if } i = j, \\ 0 & \text{if } i \neq j, \end{cases} \tag{3.17}$$

using the Kronecker delta. As shown in Figure 3.3, the basis functions are pyramids with a height of 1 on top of the corresponding vertex.

3.1.3 Functional Contribution from a Single Element

Let us now return to the functional we want to minimize using continuous piecewise linear functions of the form (3.13). With the domain $\mathcal{D}$ broken up into triangular elements:

$$I(w) = \frac{1}{2}\iint_{\mathcal{D}} \|\nabla w\|^2 dx\,dy = \frac{1}{2}\sum_{k=1}^{N_\Delta} \iint_{\Delta_k} \|\nabla w\|^2 dx\,dy\,, \tag{3.18}$$

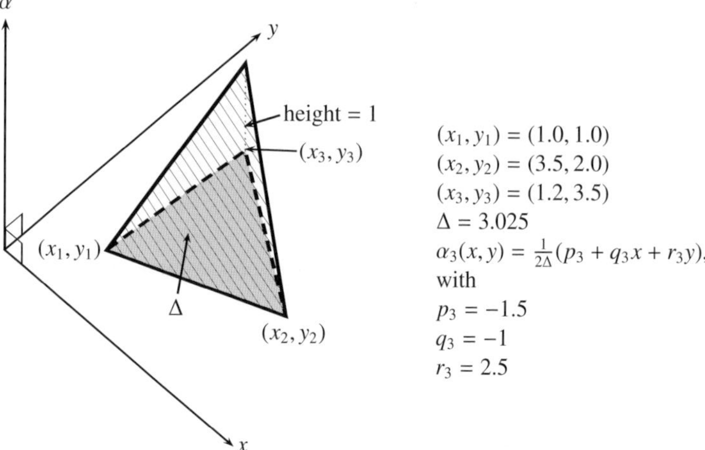

Figure 3.3 Example of a basis function of a linear triangular element.

where N_Δ represents the number of triangular elements in the domain, and Δ_k is the kth triangular element. Note that (with a slight abuse of notation) we are using the same symbol to denote the element itself as well as its surface area. Let us denote the arbitrary linear function in the k-th triangle as

$$w_k(x, y, \mathbf{W}^k) = \sum_{i=1}^{3} W_i^k \, \alpha_i^k(x, y) \,, \tag{3.19}$$

where we have added a subscript to the function w, and a superscript to the coefficients W_i and the basis functions α_i, to denote that they are associated with the kth triangle. In general, we will use a subscript for scalar quantities, such as scalar fields or parameters like the triangle area. Otherwise, a superscript will be used for vector or matrix quantities, or for functions that depend on the element vertex number, such as the basis functions; in these cases, the subscript will be reserved for the vertex index.

The contribution by triangle k to the functional (3.18) is

$$I_k(w_k) = \frac{1}{2} \iint_{\Delta_k} \|\nabla w_k\|^2 dx\, dy = \frac{1}{2} \iint_{\Delta_k} \nabla w_k \cdot \nabla w_k \, dx\, dy \,. \tag{3.20}$$

Substituting Equation (3.19) into (3.20):

$$I_k(w_k) = \frac{1}{2} \iint_{\Delta_k} \left(\sum_{i=1}^{3} W_i^k \, \nabla \alpha_i^k \right) \cdot \left(\sum_{j=1}^{3} W_j^k \, \nabla \alpha_j^k \right) dx\, dy \,. \tag{3.21}$$

Equivalently

$$I_k(w_k) = \frac{1}{2} \sum_{i=1}^{3} \sum_{j=1}^{3} W_i^k W_j^k \iint_{\Delta_k} \nabla \alpha_i^k \cdot \nabla \alpha_j^k \, dx\, dy \,. \tag{3.22}$$

Let us define

$$S_{ij}^k = \iint_{\Delta_k} \nabla\alpha_i^k \cdot \nabla\alpha_j^k \, dx\, dy. \tag{3.23}$$

Having thus integrated out the spatial coordinates, Equation (3.22) can be expressed in matrix form as a function of only the vertex potentials:

$$I_k(\mathbf{W}^k) = \frac{1}{2}\begin{bmatrix} W_1^k & W_2^k & W_3^k \end{bmatrix}\begin{bmatrix} S_{11}^k & S_{12}^k & S_{13}^k \\ S_{21}^k & S_{22}^k & S_{23}^k \\ S_{31}^k & S_{32}^k & S_{33}^k \end{bmatrix}\begin{bmatrix} W_1^k \\ W_2^k \\ W_3^k \end{bmatrix}, \tag{3.24}$$

or

$$I_k(\mathbf{W}^k) = \frac{1}{2}(\mathbf{W}^k)^\top \mathbf{S}^k\, \mathbf{W}^k. \tag{3.25}$$

Here, $\mathbf{S}^k$ is called the **element stiffness matrix** of the kth element.

An expression for S_{ij}^k may be obtained by substituting Equation (3.14) into (3.23). In particular:

$$S_{ij}^k = \left(\frac{1}{2\Delta_k}\right)^2 \iint_{\Delta_k} (q_i^k\,\hat{\mathbf{x}} + r_i^k\,\hat{\mathbf{y}}) \cdot (q_j^k\,\hat{\mathbf{x}} + r_j^k\,\hat{\mathbf{y}})\, dx\, dy, \tag{3.26}$$

where $\hat{\mathbf{x}}$ and $\hat{\mathbf{y}}$ are unit vectors, and where we added superscripts to q and r to associate them with triangle k. (In Chapter 1, we denoted the unit vectors as $\hat{\mathbf{i}}$ and $\hat{\mathbf{j}}$, respectively; however, now i and j are node indices, so we changed the notation to avoid confusion.) Evaluating the given integral, we have

$$S_{ij}^k = \frac{1}{4\Delta_k}(q_i^k q_j^k + r_i^k r_j^k). \tag{3.27}$$

More concisely, we can express **the element stiffness matrix** as

$$\boxed{\mathbf{S}^k = \frac{1}{4\Delta_k}\left[\mathbf{q}^k(\mathbf{q}^k)^\top + \mathbf{r}^k(\mathbf{r}^k)^\top\right],} \tag{3.28}$$

where we defined $\mathbf{q}^k$ and $\mathbf{r}^k$ as column vectors containing all q and r terms, respectively. Note that $\mathbf{S}^k$ is symmetric. Also, since $q_1 + q_2 + q_3 = r_1 + r_2 + r_3 = 0$, the matrix is not of full rank (adding its rows or columns will give zero). It is, however, a positive semi-definite matrix.

3.1.4 Assembly of Global System Equation

Writing Equation (3.25) for each of the triangles and summing the results, we can express Equation (3.18) as

$$I(\mathbf{W}_{\text{dis}}) = \frac{1}{2}\sum_{k=1}^{N_\Delta}(\mathbf{W}^k)^\top \mathbf{S}^k \mathbf{W}^k, \tag{3.29}$$

where

$$\mathbf{W}_{\text{dis}} = \begin{bmatrix}(\mathbf{W}^1)^\top & (\mathbf{W}^2)^\top & \dots & (\mathbf{W}^{N_\Delta})^\top\end{bmatrix}^\top \tag{3.30}$$

represents a vector of vertex potentials numbered in accordance with a **disjoint**, local numbering system. For example, consider a domain $\mathcal{D}$ with two triangular elements as shown in Figure 3.4. The disjoint numbering scheme is on the right side.

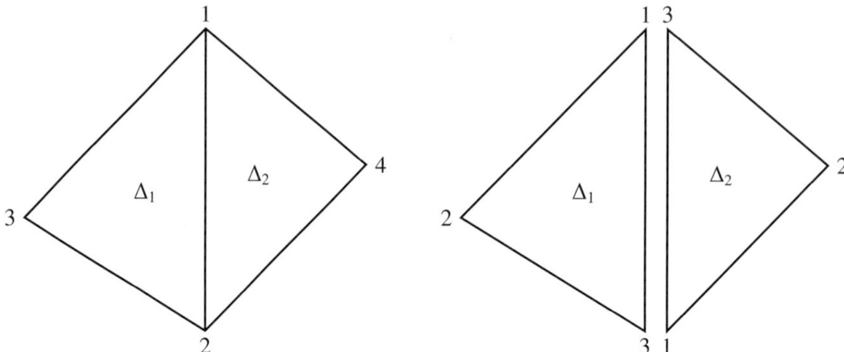

Figure 3.4 Global and disjoint (local) numbering systems.

Now we face the problem that a vertex potential of one triangular element may also represent the vertex potential of an adjacent triangular element. On the left side of Figure 3.4, where nodes are labeled in a global sense from 1 to 4, vertices 1 and 2 are shared by both triangles. However, our current expression for $I(\mathbf{W}_{\text{dis}})$ is based on a local or disjoint numbering system.

For example, for the two-element system we have

$$I(\mathbf{W}_{\text{dis}}) = \frac{1}{2}\begin{bmatrix} W_1^1 & W_2^1 & W_3^1 & W_1^2 & W_2^2 & W_3^2 \end{bmatrix} \cdot \begin{bmatrix} S_{11}^1 & S_{12}^1 & S_{13}^1 & 0 & 0 & 0 \\ S_{21}^1 & S_{22}^1 & S_{23}^1 & 0 & 0 & 0 \\ S_{31}^1 & S_{32}^1 & S_{33}^1 & 0 & 0 & 0 \\ 0 & 0 & 0 & S_{11}^2 & S_{12}^2 & S_{13}^2 \\ 0 & 0 & 0 & S_{21}^2 & S_{22}^2 & S_{23}^2 \\ 0 & 0 & 0 & S_{31}^2 & S_{32}^2 & S_{33}^2 \end{bmatrix} \begin{bmatrix} W_1^1 \\ W_2^1 \\ W_3^1 \\ W_1^2 \\ W_2^2 \\ W_3^2 \end{bmatrix}, \tag{3.31}$$

or

$$I(\mathbf{W}_{\text{dis}}) = \frac{1}{2}\mathbf{W}_{\text{dis}}^{\top}\mathbf{S}_{\text{dis}}\mathbf{W}_{\text{dis}}\,. \tag{3.32}$$

To obtain a similar expression in the global numbering system, we relate the disjoint and global potentials as

$$\mathbf{W}_{\text{dis}} = \mathbf{C}\mathbf{W}_{\text{con}}\,, \tag{3.33}$$

where $\mathbf{C}$ is the so-called **connection matrix**, and $\mathbf{W}_{\text{con}}$ is a vector of vertex potentials in the connected (global) numbering scheme. For the given two-element system, Equation (3.33) implies

$$\begin{bmatrix} W_1^1 \\ W_2^1 \\ W_3^1 \\ \\ W_1^2 \\ W_2^2 \\ W_3^2 \end{bmatrix} = \begin{bmatrix} 1 & 0 & 0 & 0 \\ 0 & 0 & 1 & 0 \\ 0 & 1 & 0 & 0 \\ \\ 0 & 1 & 0 & 0 \\ 0 & 0 & 0 & 1 \\ 1 & 0 & 0 & 0 \end{bmatrix} \begin{bmatrix} W_1 \\ W_2 \\ W_3 \\ W_4 \end{bmatrix}. \tag{3.34}$$

Substituting Equation (3.33) into (3.32):

$$I(\mathbf{W}_{\mathrm{con}}) = \frac{1}{2}(\mathbf{C}\mathbf{W}_{\mathrm{con}})^{\top}\mathbf{S}_{\mathrm{dis}}\,(\mathbf{C}\mathbf{W}_{\mathrm{con}}) \tag{3.35a}$$

$$= \frac{1}{2}\mathbf{W}_{\mathrm{con}}^{\top}(\mathbf{C}^{\top}\mathbf{S}_{\mathrm{dis}}\mathbf{C})\mathbf{W}_{\mathrm{con}} \tag{3.35b}$$

$$= \frac{1}{2}\mathbf{W}_{\mathrm{con}}^{\top}\mathbf{S}_{\mathrm{con}}\mathbf{W}_{\mathrm{con}}\,, \tag{3.35c}$$

where

$$\mathbf{S}_{\mathrm{con}} = \mathbf{C}^{\top}\mathbf{S}_{\mathrm{dis}}\mathbf{C} \tag{3.36}$$

is the **global stiffness matrix** of the connected system. If we evaluate $\mathbf{S}_{\mathrm{con}}$ for the given two-element system (see Problem 3.2), we would find

$$\mathbf{S}_{\mathrm{con}} = \begin{bmatrix} S_{11}^1 + S_{33}^2 & S_{13}^1 + S_{31}^2 & S_{12}^1 & S_{32}^2 \\ S_{31}^1 + S_{13}^2 & S_{33}^1 + S_{11}^2 & S_{32}^1 & S_{12}^2 \\ S_{21}^1 & S_{23}^1 & S_{22}^1 & 0 \\ S_{23}^2 & S_{21}^2 & 0 & S_{22}^2 \end{bmatrix}, \tag{3.37}$$

where all S_{ij}^k elements in the matrix are from the element stiffness matrices.

Referring to Figure 3.4, we find that a pattern has emerged. In particular, the global stiffness matrix elements are given by

$$S_{\mathrm{con},ii} = \sum_{k=1}^{N_\Delta} S_{ll}^k\,, \tag{3.38a}$$

$$S_{\mathrm{con},ij} = \sum_{k=1}^{N_\Delta} S_{lm}^k\,. \tag{3.38b}$$

In Equation (3.38a), the summation is taken only for those triangles that contain node i as a vertex, and in particular, a term is added only if node l of triangle k ($l \in \{1,2,3\}$, disjoint system) is coincident with node i (connected system). In Equation (3.38b), the summation is taken only for those triangles that contain both nodes i and j, and in particular, a term is added if nodes l and m of triangle k ($l, m \in \{1,2,3\}$, disjoint system) are coincident with nodes i and j (connected system), respectively.

If we substitute Equation (3.27) into (3.38a) and (3.38b), we see that the global stiffness matrix elements have the following form:

$$S_{\mathrm{con},ii} = \sum_{k=1}^{N_\Delta} \frac{1}{4\Delta_k}\left[(q_l^k)^2 + (r_l^k)^2\right], \tag{3.39a}$$

$$S_{\text{con},ij} = \sum_{k=1}^{N_\Delta} \frac{1}{4\Delta_k}(q_l^k q_m^k + r_l^k r_m^k). \tag{3.39b}$$

In other words, the global stiffness matrix is assembled by adding elements of the 3 × 3 $\mathbf{S}^k$ sub-matrices at row and column indices determined by the association between the local (disjoint) and global numbering schemes. Hence, $\mathbf{S}_{\text{con}}$ is also a symmetric and rank-deficient matrix.

Summarizing, the approximate solution of Laplace's equation in the given domain $\mathcal{D}$ is obtained by finding $\mathbf{W}_{\text{con}}$ that minimizes

$$I(\mathbf{W}_{\text{con}}) = \frac{1}{2}\mathbf{W}_{\text{con}}^{\mathsf{T}}\mathbf{S}_{\text{con}}\mathbf{W}_{\text{con}}, \tag{3.40}$$

where Equations (3.39a) and (3.39b) define the elements of $\mathbf{S}_{\text{con}}$, according to our prior discussion of which terms are added to the sums. We have yet to discuss how to find the minimizing $\mathbf{W}_{\text{con}}$, and how to handle boundary conditions. We will postpone discussion of these topics until §3.2.4.

Example 3.1 *Assembly of the stiffness matrix.*

Consider the uniform mesh of the rectangular domain shown in Figure 3.5. The domain is partitioned into 24 identical triangles, and there are 20 nodes. The triangles are isosceles and right-angled, with base = height = 1 and Δ = 1/2. Assemble the stiffness matrix.

Solution

For the two types of triangles shown, we can compute

$$\mathbf{q}^A = \begin{bmatrix}-1 & 1 & 0\end{bmatrix}^{\mathsf{T}}, \quad \mathbf{r}^A = \begin{bmatrix}-1 & 0 & 1\end{bmatrix}^{\mathsf{T}}, \tag{3.41a}$$

$$\mathbf{q}^B = \begin{bmatrix}1 & -1 & 0\end{bmatrix}^{\mathsf{T}}, \quad \mathbf{r}^B = \begin{bmatrix}1 & 0 & -1\end{bmatrix}^{\mathsf{T}}. \tag{3.41b}$$

Hence, using Equation (3.27) we obtain the corresponding element stiffness matrices:

$$\mathbf{S}^A = \mathbf{S}^B = \frac{1}{2}\begin{bmatrix} 2 & -1 & -1 \\ -1 & 1 & 0 \\ -1 & 0 & 1 \end{bmatrix}. \tag{3.42}$$

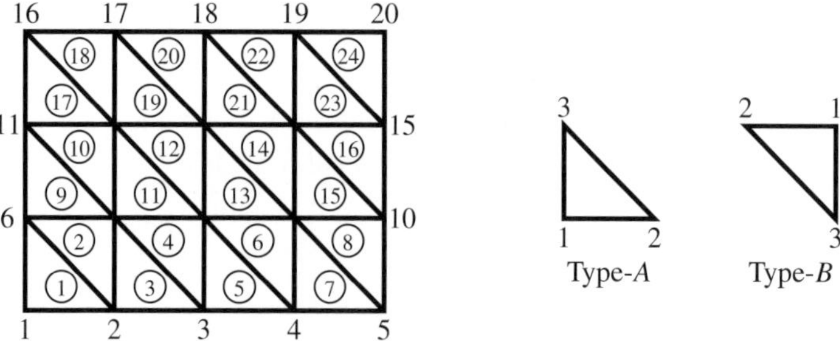

Figure 3.5 Mesh of a rectangular domain. The local numbering scheme is depicted on the right.

To assemble the global stiffness matrix we proceed by adding the contributions of each element in an iterative fashion, starting from a 20×20 zero matrix. The three nodes of triangle 1 are associated with global nodes 1, 2, and 6; its nine elements are added to the corresponding rows and columns of the global matrix, which becomes

$$\begin{bmatrix} \mathbf{1.0} & \mathbf{-0.5} & 0 & 0 & 0 & \mathbf{-0.5} & \dots & \text{(cols. 7–20 are zero)} \\ \mathbf{-0.5} & \mathbf{0.5} & 0 & 0 & 0 & \mathbf{0.0} & & \\ 0 & 0 & 0 & 0 & 0 & 0 & & \\ 0 & 0 & 0 & 0 & 0 & 0 & & \\ 0 & 0 & 0 & 0 & 0 & 0 & & \\ \mathbf{-0.5} & \mathbf{0.0} & 0 & 0 & 0 & \mathbf{0.5} & & \\ \vdots & \text{(rows 7–20 are zero)} & & & & & & \end{bmatrix}. \tag{3.43}$$

The matrix elements that were modified are shown in boldface. Note that zero elements are explicitly shown (as 0.0). These happen to be zero in this case since we have elements with sides that align with the coordinate axes.

The vertices of triangle 2 are associated with global nodes 7, 6, and 2. Adding its elements to the global matrix yields

$$\begin{bmatrix} 1.0 & -0.5 & 0 & 0 & 0 & -0.5 & 0 & \dots & \text{(cols. 8–20 are zero)} \\ -0.5 & \mathbf{1.0} & 0 & 0 & 0 & \mathbf{0.0} & \mathbf{-0.5} & & \\ 0 & 0 & 0 & 0 & 0 & 0 & 0 & & \\ 0 & 0 & 0 & 0 & 0 & 0 & 0 & & \\ 0 & 0 & 0 & 0 & 0 & 0 & 0 & & \\ -0.5 & \mathbf{0.0} & 0 & 0 & 0 & \mathbf{1.0} & \mathbf{-0.5} & & \\ 0 & \mathbf{-0.5} & 0 & 0 & 0 & \mathbf{-0.5} & \mathbf{1.0} & & \\ \vdots & \text{(rows 8–20 are zero)} & & & & & & & \end{bmatrix}. \tag{3.44}$$

The vertices of triangle 3 are associated with global nodes 2, 3, and 7. Adding its elements to the global matrix yields

$$\begin{bmatrix} 1.0 & -0.5 & 0 & 0 & 0 & -0.5 & 0 & \dots & \text{(cols. 8–20 are zero)} \\ -0.5 & \mathbf{2.0} & \mathbf{-0.5} & 0 & 0 & 0.0 & \mathbf{-1.0} & & \\ 0 & \mathbf{-0.5} & \mathbf{0.5} & 0 & 0 & 0 & \mathbf{0.0} & & \\ 0 & 0 & 0 & 0 & 0 & 0 & 0 & & \\ 0 & 0 & 0 & 0 & 0 & 0 & 0 & & \\ -0.5 & 0.0 & 0 & 0 & 0 & 1.0 & -0.5 & & \\ 0 & \mathbf{-1.0} & \mathbf{0.0} & 0 & 0 & -0.5 & \mathbf{1.5} & & \\ \vdots & \text{(rows 8–20 are zero)} & & & & & & & \end{bmatrix}. \tag{3.45}$$

And so on, until contributions from all 24 triangles are accounted for. This is left as an exercise for the reader. Note how the diagonal contains positive entries, that the matrix is symmetric, and that the rows and columns add to zero.

It should also be noted that the final matrix will be **sparse**. This term describes matrices that have a few nonzero values, whereas typically the majority of their elements is zero. For example, row 5 will have nonzero elements only at those locations where there is a neighboring node, i.e., at columns 4, 5, 9, and 10. This fact has practical implications on memory storage and computational efficiency.

3.2 Solution of Linear Poisson's Equation Using the FEM

We have discussed how electric machine analysis is often formulated as a two-dimensional problem over a cross-section, assuming that $\partial/\partial z = 0$, as if the device has "infinite length" along the axial z direction. Also, current sources are present only in the z-axis (i.e., so-called end effects are neglected as in Example 1.33), and thus the MVP $\vec{\mathbf{A}}$ only has a z component. For notational simplicity, we will denote these functions as $J = J(x, y)$ and $A = A(x, y)$, respectively (omitting an implicit z subscript).

We recall the key result of Box 1.2 on page 75: in a 2-D problem

$$\vec{\mathbf{B}} = \nabla \times \vec{\mathbf{A}} = \frac{\partial A}{\partial y}\hat{\mathbf{i}} - \frac{\partial A}{\partial x}\hat{\mathbf{j}} . \tag{3.46}$$

Hence, in view of the 2-D curl formula (1.115):

$$(\nabla \times \vec{\mathbf{H}})_z = \frac{\partial H_y}{\partial x} - \frac{\partial H_x}{\partial y} \tag{3.47a}$$

$$= \frac{\partial}{\partial x}(\nu B_y) - \frac{\partial}{\partial y}(\nu B_x) , \tag{3.47b}$$

and using Ampère's law:

$$\frac{\partial}{\partial x}\left(\nu\frac{\partial A}{\partial x}\right) + \frac{\partial}{\partial y}\left(\nu\frac{\partial A}{\partial y}\right) = -J . \tag{3.48}$$

Here, we have implicitly assumed the absence of any permanent magnetization since we wrote $\vec{\mathbf{H}} = \nu\vec{\mathbf{B}}$. The magnetic reluctivity ν will depend on the unknown solution itself if magnetic saturation is present, so we may not always assume it is constant. This is called a **nonlinear Poisson equation**. It is the partial differential equation that we must satisfy in a 2-D magnetostatic problem (in conjunction with the appropriate interface and boundary conditions).

For the purposes of this section, where we are still presenting key concepts of the FEM, let us assume for simplicity that the materials are magnetically linear. This means that $\nu = \mu^{-1}$ is constant within each sub-region of the device, as defined by material boundaries. Hence, we will explain first how to solve the *linear* Poisson equation

$$\frac{\partial^2 A}{\partial x^2} + \frac{\partial^2 A}{\partial y^2} = -\mu J . \tag{3.49}$$

The main difference from the previous section, where we worked on the Laplace equation, is the nonzero source term, which is present only in current-carrying regions.

3.2.1 Setting up the System of Equations

In the previous chapter, we explained that the magnetostatic problem can be solved by making an energy-related functional stationary (see §2.2.1 and §2.2.3). The functional (2.71) involves the magnetic energy density m. In a linear isotropic material of reluctivity ν (see Example 2.5), we have

$$m(B) = \frac{1}{2}\nu B^2 . \tag{3.50}$$

In 2-D problems, the B-field magnitude is equal to the magnitude of the gradient of the magnetic potential; this follows directly from Equation (3.46):

$$B = \|\vec{\mathbf{B}}\| = \|\nabla A\| . \tag{3.51}$$

Therefore, the corresponding functional is

$$I(A) = \frac{1}{2}\iint_{\mathcal{D}} \nu(x,y)\,\|\nabla A\|^2\,dx\,dy - \iint_{\mathcal{D}} JA\,dx\,dy , \tag{3.52}$$

where $\nu(x, y)$ shows the explicit dependence of reluctivity on position, i.e., the type of material in each location of the device. Note that the physical unit of this expression is J/m because the original volume integral has become a surface integral.

We assume that the solution is piecewise linear:

$$A_k(x, y, \mathbf{A}^k) = \sum_{i=1}^{3} A_i^k\,\alpha_i^k(x, y) . \tag{3.53}$$

We break up the functional integral into a sum of element contributions:

$$I(A) = \sum_{k=1}^{N_\Delta} I_k(A_k) = \sum_{k=1}^{N_\Delta} \frac{1}{2} \iint\limits_{\Delta_k} \nu_k\|\nabla A_k\|^2\,dx\,dy - \sum_{k=1}^{N_\Delta} \iint\limits_{\Delta_k} JA_k\,dx\,dy \tag{3.54a}$$

$$= \sum_{k=1}^{N_\Delta} I_{k,1}(A_k) \; - \; \sum_{k=1}^{N_\Delta} I_{k,2}(A_k) . \tag{3.54b}$$

The mesh should be such that material boundaries coincide with element edges; hence, there is a single reluctivity ν_k within each element. (Curved material boundaries are approximated by linear segments.) The terms $I_{k,1}$ are identical to those from the previous section, so we can immediately assert that

$$I_{k,1}(\mathbf{A}^k) = \frac{\nu_k}{2}(\mathbf{A}^k)^\top \mathbf{S}^k \mathbf{A}^k . \tag{3.55}$$

Compared to Equation (3.25), here we are also multiplying by the element reluctivity ν_k.

Now we focus on the $I_{k,2}$ terms that stem from the current sources. We will assume that the current density is constant within each triangle, $J(x, y) = J_k$. This is a mild assumption (see Problem 3.5 for a different approach) that allows us to evaluate the integral simply as

$$I_{k,2}(A_k) = J_k \iint\limits_{\Delta_k} A_k(x, y)\,dx\,dy \tag{3.56a}$$

$$= J_k \iint\limits_{\Delta_k} \sum_{i=1}^{3} A_i^k\,\alpha_i^k(x, y)\,dx\,dy \tag{3.56b}$$

$$= J_k \sum_{i=1}^{3} A_i^k \iint\limits_{\Delta_k} \alpha_i^k(x, y)\,dx\,dy . \tag{3.56c}$$

Substituting Equation (3.14) into (3.56c) and evaluating (see Problem 3.3), we obtain

$$I_{k,2}(\mathbf{A}^k) = J_k \sum_{i=1}^{3} A_i^k \frac{\Delta_k}{3} = \frac{J_k \Delta_k}{3} \sum_{i=1}^{3} A_i^k . \tag{3.57}$$

Because J_k represents the current density within triangle k, it follows that $J_k \Delta_k$ represents the total current in triangle k. For convenience, let us denote $J_k \Delta_k$ as i_k. Then

$$I_{k,2}(\mathbf{A}^k) = \frac{i_k}{3} \sum_{i=1}^{3} A_i \tag{3.58a}$$

$$= \frac{i_k}{3} \begin{bmatrix} A_1^k & A_2^k & A_3^k \end{bmatrix} \begin{bmatrix} 1 \\ 1 \\ 1 \end{bmatrix} \tag{3.58b}$$

$$= (\mathbf{A}^k)^\top \mathbf{b}^k , \tag{3.58c}$$

where we have defined

$$\mathbf{b}^k = \frac{i_k}{3} \begin{bmatrix} 1 \\ 1 \\ 1 \end{bmatrix} . \tag{3.59}$$

Therefore, the contribution to the functional (3.54a) from triangle k is the quadratic form

$$I_k(\mathbf{A}^k) = \frac{\nu_k}{2} (\mathbf{A}^k)^\top \mathbf{S}^k \mathbf{A}^k - (\mathbf{A}^k)^\top \mathbf{b}^k . \tag{3.60}$$

If we sum over all triangles, then in our disjoint system

$$I(\mathbf{A}_{\text{dis}}) = \frac{1}{2} \sum_{k=1}^{N_\Delta} \nu_k (\mathbf{A}^k)^\top \mathbf{S}^k \mathbf{A}^k - \sum_{k=1}^{N_\Delta} (\mathbf{A}^k)^\top \mathbf{b}^k , \tag{3.61}$$

with

$$\mathbf{A}_{\text{dis}} = \left[(\mathbf{A}^1)^\top \quad (\mathbf{A}^2)^\top \quad \ldots \quad (\mathbf{A}^{N_\Delta})^\top \right]^\top . \tag{3.62}$$

For example, let us apply this to the two-element system of Figure 3.4:

$$I(\mathbf{A}_{\text{dis}}) = \frac{1}{2} \left[(\mathbf{A}^1)^\top \; (\mathbf{A}^2)^\top \right] \begin{bmatrix} \nu_1 \mathbf{S}^1 & \mathbf{0} \\ \mathbf{0} & \nu_2 \mathbf{S}^2 \end{bmatrix} \begin{bmatrix} \mathbf{A}^1 \\ \mathbf{A}^2 \end{bmatrix} - \left[(\mathbf{A}^1)^\top \; (\mathbf{A}^2)^\top \right] \begin{bmatrix} \mathbf{b}^1 \\ \mathbf{b}^2 \end{bmatrix} \tag{3.63a}$$

$$= \frac{1}{2} \mathbf{A}_{\text{dis}}^\top \mathbf{S}_{\text{dis}} \mathbf{A}_{\text{dis}} - \mathbf{A}_{\text{dis}}^\top \mathbf{b}_{\text{dis}} . \tag{3.63b}$$

Relating the disjoint and connected systems as in Equation (3.33):

$$I(\mathbf{A}_{\text{con}}) = \frac{1}{2} (\mathbf{C}\mathbf{A}_{\text{con}})^\top \mathbf{S}_{\text{dis}} (\mathbf{C}\mathbf{A}_{\text{con}}) - (\mathbf{C}\mathbf{A}_{\text{con}})^\top \mathbf{b}_{\text{dis}} \tag{3.64a}$$

$$= \frac{1}{2} (\mathbf{A}_{\text{con}})^\top (\mathbf{C}^\top \mathbf{S}_{\text{dis}} \mathbf{C}) \mathbf{A}_{\text{con}} - (\mathbf{A}_{\text{con}})^\top (\mathbf{C}^\top \mathbf{b}_{\text{dis}}) \tag{3.64b}$$

$$= \frac{1}{2} (\mathbf{A}_{\text{con}})^\top \mathbf{S}_{\text{con}} \mathbf{A}_{\text{con}} - (\mathbf{A}_{\text{con}})^\top \mathbf{b}_{\text{con}} . \tag{3.64c}$$

It follows from our previous work that the global stiffness matrix elements are assembled from individual element contributions as in

$$S_{\mathrm{con},ii} = \sum_{k=1}^{N_\Delta} \frac{\nu_k}{4\Delta_k}\left[(q_l^k)^2 + (r_l^k)^2\right], \tag{3.65a}$$

$$S_{\mathrm{con},ij} = \sum_{k=1}^{N_\Delta} \frac{\nu_k}{4\Delta_k}(q_l^k q_m^k + r_l^k r_m^k). \tag{3.65b}$$

In Equation (3.65a), the summation is taken only for those triangles that contain node i as a vertex, and in particular, a term is added only if node l of triangle k ($l \in \{1,2,3\}$, disjoint system) is coincident with node i (connected system). In Equation (3.65b), the summation is taken only for those triangles that contain both nodes i and j, and in particular, a term is added if nodes l and m of triangle k ($l, m \in \{1,2,3\}$, disjoint system) are coincident with nodes i and j (connected system), respectively.

To establish the general form of $\mathbf{b}_{\mathrm{con}}$, let us evaluate the general form for the two-element system of Figure 3.4. Here

$$\mathbf{b}_{\mathrm{con}} = \mathbf{C}^\top \mathbf{b}_{\mathrm{dis}} \tag{3.66a}$$

$$= \begin{bmatrix} 1 & 0 & 0 & 0 & 0 & 1 \\ 0 & 0 & 1 & 1 & 0 & 0 \\ 0 & 1 & 0 & 0 & 0 & 0 \\ 0 & 0 & 0 & 0 & 1 & 0 \end{bmatrix} \begin{bmatrix} b^1 \\ b^1 \\ b^1 \\ b^2 \\ b^2 \\ b^2 \end{bmatrix} \tag{3.66b}$$

$$= \begin{bmatrix} b^1 + b^2 \\ b^1 + b^2 \\ b^1 \\ b^2 \end{bmatrix}, \tag{3.66c}$$

where according to Equation (3.59), $b^1 = i_1/3$ and $b^2 = i_2/3$. Apparently, the elements of $\mathbf{b}_{\mathrm{con}}$ are assembled by

$$b_{\mathrm{con},i} = \sum_{k=1}^{N_\Delta} b^k = \sum_{k=1}^{N_\Delta} \frac{i_k}{3}, \tag{3.67}$$

where the summation is taken only if triangle k contains node i as a vertex.

Summarizing, the approximate solution to the linear Poisson's equation (3.49) is obtained by finding $\mathbf{A}_{\mathrm{con}}$ that minimizes

$$I(\mathbf{A}_{\mathrm{con}}) = \frac{1}{2}\mathbf{A}_{\mathrm{con}}^\top \mathbf{S}_{\mathrm{con}} \mathbf{A}_{\mathrm{con}} - \mathbf{A}_{\mathrm{con}}^\top \mathbf{b}_{\mathrm{con}}, \tag{3.68}$$

where Equations (3.65a) and (3.65b) define the elements of $\mathbf{S}_{\mathrm{con}}$, and Equation (3.67) defines the elements of $\mathbf{b}_{\mathrm{con}}$. From now on, we will drop the subscript "con" from these terms for notational convenience. It will be always implied that we are working under a global numbering scheme with the global stiffness matrix and current source vector.

Example 3.2 *Assembly of the current vector.*

Consider the elementary rectangular domain mesh shown in Figure 3.6. We have introduced a current-carrying conductor in the shaded square. Suppose the current is $i = 1$ A in each triangle (the current flows out of the page, in the z-direction). What is $\mathbf{b}$?

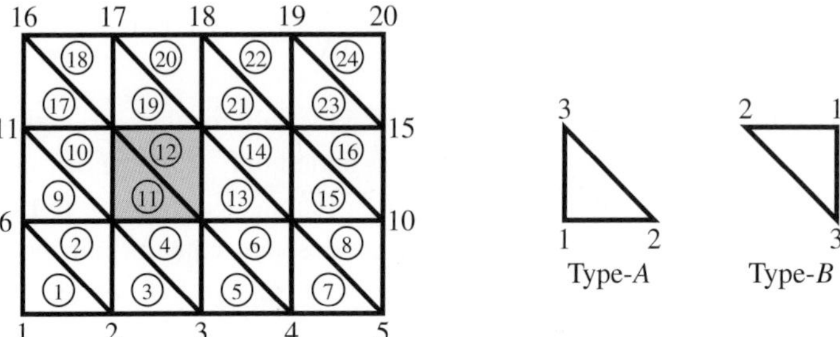

Figure 3.6 A rectangular domain mesh with a current source.

Solution

We proceed on an element-wise basis starting from a 20×1 zero column vector. Triangles 1–10 do not contribute anything. Triangle 11 adds $1/3$ to vector elements 7, 8, and 12. Note that the local node numbering is irrelevant because the same value is added to all three elements. Triangle 12 adds $1/3$ to vector elements 13, 12, and 8. Triangles 13–24 do not contribute anything further to the vector. Hence

$$\mathbf{b} = [0\ \ 0\ \ 0\ \ 0\ \ 0\ \ 0\ \ 1/3\ \ 2/3\ \ 0\ \ 0\ \ \ldots \\ 0\ \ 2/3\ \ 1/3\ \ 0\ \ 0\ \ 0\ \ 0\ \ 0\ \ 0\ \ 0]^{\mathsf{T}}\,. \tag{3.69}$$

3.2.2 Functional Minimization and the Weak Form

How do we know that we are minimizing the functional? Similar to how we addressed this question for the Laplace equation, we refer to Equation (2.27) in Step 2 of the variational process (see page 103). We can assert that stationarity of the functional (3.52) implies that the solution A^* satisfies

$$\iint_{\mathcal{D}} \left(-\eta J + \nu \frac{\partial \eta}{\partial x}\frac{\partial A^*}{\partial x} + \nu \frac{\partial \eta}{\partial y}\frac{\partial A^*}{\partial y} \right) dx\, dy = 0\,, \tag{3.70}$$

or

$$\iint_{\mathcal{D}} \nu \nabla \eta \cdot \nabla A^*\, dx\, dy = \iint_{\mathcal{D}} \eta J\, dx\, dy\,, \tag{3.71}$$

for arbitrary η. This is the **weak form** of the Poisson equation.

Evaluating the functional for any admissible function $A = A^* + \alpha\eta$ yields

$$I(A^* + \alpha\eta) = \frac{1}{2} \iint_{\mathcal{D}} \nu \nabla(A^* + \alpha\eta) \cdot \nabla(A^* + \alpha\eta)\, dx\, dy - \iint_{\mathcal{D}} (A^* + \alpha\eta) J\, dx\, dy \tag{3.72a}$$

$$= \frac{1}{2}\iint_{\mathcal{D}} \nu\nabla A^* \cdot \nabla A^* \,dx\,dy - \iint_{\mathcal{D}} A^* J \,dx\,dy$$
$$+ \alpha\left(\iint_{\mathcal{D}} \nu\nabla\eta \cdot \nabla A^* \,dx\,dy - \iint_{\mathcal{D}} \eta J \,dx\,dy\right) + \frac{\alpha^2}{2}\iint_{\mathcal{D}} \nu\nabla\eta \cdot \nabla\eta \,dx\,dy \tag{3.72b}$$

$$= I(A^*) + \frac{\alpha^2}{2}\iint_{\mathcal{D}} \nu\|\nabla\eta\|^2 \,dx\,dy\,, \tag{3.72c}$$

where we eliminated the term in parentheses that vanishes by virtue of Equation (3.71). Therefore, we conclude that stationarity of the linear Poisson equation functional is in fact a minimization (since $\nu > 0$ due to physical considerations). This is because

$$I(A^*) \leq I(A) \tag{3.73}$$

for any admissible A, that is, any function that makes the functional well-defined and satisfies the boundary conditions. As in the case of the Laplace equation, *the Poisson equation FEM problem is to determine a piecewise linear function that minimizes the functional against any other piecewise linear function satisfying the boundary conditions.*

3.2.3 Minimization of a Quadratic Functional

We have explained that the numerical solution of Poisson's equation using the FEM has the form of a piecewise linear function $A(x, y)$. It suffices to identify a vector of node potentials $\mathbf{A}$ minimizing the quadratic form

$$I(\mathbf{A}) = \frac{1}{2}\mathbf{A}^\top\mathbf{S}\mathbf{A} - \mathbf{A}^\top\mathbf{b}\,, \tag{3.74}$$

where $\mathbf{A}$, $\mathbf{S}$, and $\mathbf{b}$ are for the connected system. The minimization of the functional arising from a Laplace equation is just a special case of Equation (3.74), where initially $\mathbf{b} = \mathbf{0}$ (in the absence of any sources); eventually, however, this becomes nonzero due to the boundary conditions, so we obtain a nonzero solution.

For illustration purposes, suppose our system only has two nodes. Of course, this assumption is physically meaningless since we cannot define a triangular element with just two nodes. Nevertheless, it helps to illustrate the algebraic steps. The extension to higher dimensions is straightforward. Expanding the quadratic form (3.74), we obtain

$$I(A_1, A_2) = \frac{1}{2}S_{11}A_1^2 + S_{12}A_1A_2 + \frac{1}{2}S_{22}A_2^2 - A_1b_1 - A_2b_2\,, \tag{3.75}$$

since symmetry of the stiffness matrix implies that $S_{12} = S_{21}$. If both A_1 and A_2 are free variables, then the necessary conditions for minimization of I are

$$0 = \frac{\partial I}{\partial A_1} = S_{11}A_1 + S_{12}A_2 - b_1\,, \tag{3.76a}$$
$$0 = \frac{\partial I}{\partial A_2} = S_{12}A_1 + S_{22}A_2 - b_2\,. \tag{3.76b}$$

In matrix form, even for higher dimensions, we may write these conditions as

$$\mathbf{0} = \mathbf{S}\mathbf{A} - \mathbf{b}\,. \tag{3.77}$$

Therefore, we have shown that minimization of the quadratic form (3.74) boils down to solving a linear system of equations:

$$\mathbf{S}\mathbf{A} = \mathbf{b}\,. \tag{3.78}$$

But is there a unique solution to this problem? We have explained that $\mathbf{S}$ is a rank-deficient matrix. This means that it is not invertible, and therefore this linear system will have an infinite number of solutions. Let us consider what happens to the functional when we add an arbitrary constant A_0 to a vector of magnetic potentials $\mathbf{A}$:

$$\mathbf{A} + \mathbf{A}_0 = \mathbf{A} + \begin{bmatrix} 1 \\ \vdots \\ 1 \end{bmatrix} A_0\,. \tag{3.79}$$

This is possible if we are not imposing any Dirichlet boundary conditions, thereby allowing the solution to "float." Note that

$$\mathbf{S}\,(\mathbf{A} + \mathbf{A}_0) = \mathbf{S}\mathbf{A} \quad \text{and} \quad (\mathbf{A} + \mathbf{A}_0)^\top\,\mathbf{S} = \mathbf{A}^\top\mathbf{S}\,, \tag{3.80}$$

since the columns and rows of the symmetric matrix $\mathbf{S}$ add to zero. Hence

$$I(\mathbf{A} + \mathbf{A}_0) = \frac{1}{2}\,(\mathbf{A} + \mathbf{A}_0)^\top\,\mathbf{S}\,(\mathbf{A} + \mathbf{A}_0) - (\mathbf{A} + \mathbf{A}_0)^\top\,\mathbf{b} \tag{3.81a}$$

$$= I(\mathbf{A}) - \mathbf{A}_0^\top\mathbf{b}\,. \tag{3.81b}$$

We distinguish two cases based on the sum of the right-hand-side terms, $\sum b_i$. First, we examine the case where

$$\sum b_i \neq 0\,. \tag{3.82}$$

This occurs when we are modeling a 2-D domain where the currents do not sum to zero; for example, the case of a single infinitely long wire (with a return path at infinity), which we studied in Example 1.26. In this case, we can make the term $\mathbf{A}_0^\top\mathbf{b}$ arbitrarily low by adjusting the value of A_0. Therefore, the problem is ill-posed since it is not possible to find a solution that minimizes the functional. Also, the linear system (3.78) is inconsistent since adding all equations leads to $0 = \sum b_i \neq 0$.

Second, we examine the case where

$$\sum b_i = 0\,, \tag{3.83}$$

which occurs when the net current in the cross-section of our 2-D device sums to zero. Now $\mathbf{A}_0^\top\mathbf{b}$ vanishes, and the value of the functional is not affected. Hence, if $\mathbf{A}^*$ is a minimizing vector, then $\mathbf{A}^* + \mathbf{A}_0$ is also a minimizing vector. So, the problem does not have a unique solution.

Such complications are avoided once we impose Dirichlet boundary conditions, thus fixing the potential of some points in space. Then we would expect to obtain a unique solution. The details of setting boundary conditions will be presented next.

3.2.4 Handling of Boundary Conditions

Suppose we are given the functional

$$I(\mathbf{A}) = \frac{1}{2}\mathbf{A}^{\top}\mathbf{S}\mathbf{A} - \mathbf{A}^{\top}\mathbf{b}, \tag{3.84}$$

which is to be minimized subject to the constraint that some potentials are specified (corresponding to boundary nodes). How is this done?

Consider an elementary three-node system wherein

$$\begin{aligned} I(A_1, A_2, A_3) = \frac{1}{2}S_{11}A_1^2 + \frac{1}{2}S_{22}A_2^2 + \frac{1}{2}S_{33}A_3^2 \\ + S_{12}A_1A_2 + S_{23}A_2A_3 + S_{13}A_1A_3 - A_1b_1 - A_2b_2 - A_3b_3 . \end{aligned} \tag{3.85}$$

If the potential A_2 is fixed (i.e., $A_2 = A_{2f}$), then to find A_1 and A_3 that minimize Equation (3.85) we set

$$0 = \frac{\partial I}{\partial A_1} = S_{11}A_1 + S_{12}A_{2f} + S_{13}A_3 - b_1 , \tag{3.86a}$$

$$0 = \frac{\partial I}{\partial A_3} = S_{13}A_1 + S_{23}A_{2f} + S_{33}A_3 - b_3 , \tag{3.86b}$$

which gives us two equations with two unknowns. We can augment these with the identity $0 = A_2 - A_{2f}$, arriving at the matrix form

$$\begin{bmatrix} 0 \\ 0 \\ 0 \end{bmatrix} = \begin{bmatrix} S_{11} & S_{12} & S_{13} \\ 0 & 1 & 0 \\ S_{13} & S_{23} & S_{33} \end{bmatrix} \begin{bmatrix} A_1 \\ A_2 \\ A_3 \end{bmatrix} - \begin{bmatrix} b_1 \\ A_{2f} \\ b_3 \end{bmatrix} . \tag{3.87}$$

Equivalently

$$\begin{bmatrix} S_{11} & S_{12} & S_{13} \\ 0 & 1 & 0 \\ S_{13} & S_{23} & S_{33} \end{bmatrix} \begin{bmatrix} A_1 \\ A_2 \\ A_3 \end{bmatrix} = \begin{bmatrix} b_1 \\ A_{2f} \\ b_3 \end{bmatrix} , \tag{3.88}$$

which can be solved to obtain the potentials. Symmetry may be restored to the matrix for a more efficient numerical solution,[b]

$$\begin{bmatrix} S_{11} & 0 & S_{13} \\ 0 & 1 & 0 \\ S_{13} & 0 & S_{33} \end{bmatrix} \begin{bmatrix} A_1 \\ A_2 \\ A_3 \end{bmatrix} = \begin{bmatrix} b_1 - S_{12}A_{2f} \\ A_{2f} \\ b_3 - S_{23}A_{2f} \end{bmatrix} . \tag{3.89}$$

Generalizing, to introduce a number of fixed nodes, the algorithm in Box 3.1 may be used to modify our original linear system of equations. It is often the case that the boundary nodes are assigned a potential of zero. In this case, Step 2a of the process is unnecessary, and can be skipped to save a tiny amount of computational time.

[b] Symmetry is typically exploited by numerical linear algebra software packages.

Box 3.1 Programming the FEM: Algorithm for enforcing boundary conditions

1. Form **S** and **b** in accordance with the original procedure.
2. Iterate over all Dirichlet nodes j ($A_j = A_{jf}$):
   a. Subtract $S_{ij}A_{jf}$ from the ith element of **b**, for $i = 1, \ldots, N$. In other words, subtract $\mathbf{S}_j A_{jf}$ from **b**, where $\mathbf{S}_j$ is the jth column of **S**.
   b. Set the jth column of **S** to zero.
3. Iterate (a second time) over all Dirichlet nodes j ($A_j = A_{jf}$):
   a. Set the jth element of **b** to A_{jf}.
   b. Set the jth row of **S** to zero.
   c. Set $S_{jj} = 1$.

3.3 Building Algorithms for Linear Magnetics

In this section, we present details about the algorithms that can be used to establish **S** and **b** in a computer. First, we explain how the mesh data is stored and loaded from a computer using Python code. Then, we describe the building procedure for the element and global stiffness matrices, which set the stage for our first FEA program.

3.3.1 Data Structures

We will assume that the input data is structured as shown in Tables 3.1 and 3.2, for nodes and triangles, respectively. This information is typically the output of a meshing algorithm that triangulates (i.e., subdivides into triangles) the domain, and it is saved as two files that the FEA program "reads" at the beginning. The domain mesh consists of a total of N nodes and K triangles.

Table 3.1 Structure of input data related to nodes

node	x coord.	y coord.	identifier
1	x_1	y_1	integer
2	x_2	y_2	integer
$\vdots$	$\vdots$	$\vdots$	$\vdots$
N	x_N	y_N	integer

Table 3.1 has four columns: its first column contains a node index; its second and third columns contain floating-point numbers representing the (x, y) coordinates of the vertices; its fourth column contains integer identifiers. The node identifiers may be used, for example, to indicate that a particular node is at the boundary. The identifier convention is decided by the programmer. For instance, we may assign an identifier of 1 to all

Table 3.2 Structure of input data related to triangular elements

triangle	l	m	n	identifier
1	index to 1st node	index to 2nd node	index to 3rd node	integer
2	index to 1st node	index to 2nd node	index to 3rd node	integer
⋮	⋮	⋮	⋮	⋮
K	index to 1st node	index to 2nd node	index to 3rd node	integer

outer boundary nodes so that the program can assign the proper boundary conditions, whereas all other nodes could be assigned an identifier of 0.

Table 3.2 has five columns: its first column contains a triangle index; its second, third, and fourth columns contain integer numbers that relate the three vertices (l, m, n) to the nodes listed in Table 3.1; its fifth column contains integer identifiers. These are typically used to indicate the material. For instance, we could assign a value of 0 if the triangle is in air, 1 if the triangle is in steel, 2 if the triangle is in copper, etc.

Example 3.3 *Node and triangle data for a simple rectangular domain.*

Recall the elementary domain discussed in Example 3.1, where we have $N = 20$ vertices and $K = 24$ triangles. How would we store information defining the geometry of this problem in the computer?

Solution

Based on the previous discussion, the node and triangle data structures would be as shown in Tables 3.3 and 3.4.

Table 3.3 Input data related to nodes

node	x coord.	y coord.	identifier
1	0.0	0.0	1
2	1.0	0.0	1
3	2.0	0.0	1
4	3.0	0.0	1
5	4.0	0.0	1
6	0.0	1.0	1
7	1.0	1.0	0
8	2.0	1.0	0
⋮	⋮	⋮	⋮
20	4.0	3.0	1

Loading the Data from Files

In this section, we will present a way to load the data files from a drive using Python. The piece of code below defines the function `load_mesh` that reads two text files and

Table 3.4 Input data related to triangular elements

triangle	l	m	n	identifier
1	1	2	6	0
2	7	6	2	0
3	2	3	7	0
4	8	7	3	0
⋮	⋮	⋮	⋮	⋮
24	20	19	15	0

outputs two NumPy arrays that contain node and triangle information, thus completely defining the mesh.

```python
1  def load_mesh(fname, iter_no=1):
2      """
3      Loads & returns nodes and triangles (as NumPy arrays)
4      from files generated by Triangle. Also returns number of
5      nodes and triangles.
6      """
7  
8      nodes_fname = fname + '.{}.node'.format(iter_no)
9      ele_fname = fname + '.{}.ele'.format(iter_no)
10 
11     # load nodes
12     with open(nodes_fname,'r') as f:
13         line = f.readline().rstrip().split()
14         n_nodes = int(line[0])
15         n_attr = int(line[2])
16         n_bm = int(line[3])
17         N = n_nodes*(3 + n_attr + n_bm)
18     nodes = my_loadtxt(nodes_fname, N=N, delimiter=None,\
19                        skiprows=1, rowlength=3+n_attr+n_bm)
20 
21     # load triangles
22     with open(ele_fname,'r') as f:
23         line = f.readline().rstrip().split()
24         n_tri = int(line[0])
25         N = n_tri*5
26     triangles = my_loadtxt(ele_fname, N=N, delimiter=None,\
27                            dtype=int, skiprows=1, rowlength=5)
28 
29     return n_nodes, nodes, n_tri, triangles
```

The two input files are generated by Triangle, which is a very capable open-source meshing program for 2-D geometries written by J. R. Shewchuk.[61] All the examples in this text are meshed using Triangle, so it is worthwhile to take a closer look. After meshing the geometry, Triangle generates a .node and a .ele file. Also, it prepends a number to the extension depending on the meshing iteration; this is 1 by default. (See lines 8–9 for usage.) The first line of these files contains useful information about the number and properties of nodes and elements, respectively.

In particular, the syntax of .node files is as follows, where items enclosed by <·> are required, and items enclosed by [·] are optional:

- First line: <# of vertices> <dimension (must be 2)> <# of attributes> <# of boundary markers (0 or 1)>
- Remaining lines: <vertex #> <x> <y> [attributes] [boundary marker]

The attributes are floating-point numbers that can be used optionally to specify physical quantities for each node. This feature is not used here; hence, we should set the number of attributes as zero on the first line. However, the nodes will have boundary markers, corresponding to the identifier column of the data structure in Table 3.1.

With this in mind, the content of lines 11–17 should be clear. The nodes file is opened for reading in line 12 using the `open(nodes_fname,'r')` command, preceded by the **`with`** keyword. This syntax is good Python practice because it has the advantage that the file is automatically closed after its block of code finishes, even if an exception (i.e., an unexpected error) is raised on the way. Afterwards, the first line is read with the `readline` command, all trailing whitespace is removed with the `rstrip` command, and the line is split into a list of words with the `split` command. Then, we determine the number of nodes, attributes, and boundary markers through string-to-integer conversions. The variable `N` in line 17 represents the number of elements that the `nodes` array will contain, since it will have `n_nodes` rows and `3 + n_attr + n_bm` columns as in Table 3.1. (Even though the number of node attributes is zero, the code is written in a general way.)

For .ele files, the syntax is

- First line: <# of triangles> <nodes per triangle> <# of attributes>
- Remaining lines: <triangle #> <node> <node> <node> ... [attributes]

The attributes can be floating-point numbers. Here, they are simply integers that represent the material identifiers of Table 3.2. The dots signify that Triangle may generate more than three nodes per triangle. This happens when quadratic (i.e., not linear) elements are employed, but this is not the case here. This explains the content of lines 21–25. The variable `N` in line 25 represents the number of elements of the `triangles` array, which will have `n_tri` rows and five columns as in Table 3.2.

The `nodes` and `triangles` variables are obtained from two `my_loadtxt` function calls in lines 18–19 and 26–27, respectively. This function is described next.

```python
import numpy as np

def my_loadtxt(filename, N, rowlength, delimiter=',',
                skiprows=0, dtype=float):
    """
    Simple CSV file reader.
    """

    def iter_func():
        with open(filename, 'r') as infile:
            for _ in range(skiprows):
                next(infile)
            for line in infile:
```

```
14                 if line[0] == '#':
15                     next(infile)
16                 else:
17                     line = line.rstrip().split(delimiter)
18                     for item in line:
19                         yield dtype(item)
20 
21     data = np.fromiter(iter_func(), dtype=dtype, count=N).\
22                 reshape((-1, rowlength))
23     return data
```

The first thing that we encounter inside this function is another definition. In particular, `iter_func` is a local method that achieves the following:

- It opens the file for reading.
- It skips a predefined number of header rows. This is achieved using the **`next`** keyword, which returns the next line in the file.
- It reads the remaining lines, skipping those that begin with a "#" symbol, which are comments.
- Each line is split into words using a specified delimiter, so the output is a list of words. By default, the delimiter is a comma, so the default behavior is to read a comma-separated values (CSV) file. Nevertheless, the function is called from `load_mesh` using the delimiter `None`. This instructs the `split` method to treat blocks of consecutive whitespace as the separator.
- The **`yield`** keyword means that `iter_func` is not a function in the conventional sense but a Python generator. Simply speaking, this type of object executes once every time it is called, then it hibernates until is called again. So, when called, `iter_func` just returns (yields) the next word in the file, but not the entire file. The advantage is lower memory usage since we do not need to load the entire file in memory.
- In line 21, the variable `data` is created using the NumPy `fromiter` method, which accepts an iterator object (in this case, the generator that we defined above), a data type, and a number of elements to read. The optional `count` argument is specified to improve performance because it allows memory to be preallocated for holding the output array. At this point, `data` is a 1-D NumPy array, so in line 22 it is `reshape`'d as a 2-D array of the appropriate size.

3.3.2 Building the Stiffness Matrix

Once the mesh has been loaded, we can proceed with the calculation of all element stiffness matrices $\mathbf{S}^k$ using Equation (3.27), and then construct the global stiffness matrix $\mathbf{S}$. The "trick" behind the procedure that assembles the $\mathbf{S}$ matrix is that its elements are not calculated one at a time. Instead, the triangle list is scanned and, for each triangle, the appropriate elements are updated. The following Python examples illustrate how these tasks can be implemented in the computer.

We note that for all practical purposes, the $\mathbf{S}$ matrix needs to be stored as a sparse matrix. If we attempt to store the $\mathbf{S}$ matrix as a dense matrix, we may quickly run out of memory, needlessly storing all its zero elements. For instance, say we have a mesh that has $N = 20{,}000$ nodes. The stiffness matrix is $N \times N$, so we would need to store 400 million floating-point numbers (most of them zero). If we preallocate memory to store this dense matrix, using 8 bytes for each double-precision number, we would need about 3 GB of RAM. In contrast, the memory requirements for storing a sparse matrix are modest since we are only keeping track of its nonzero elements. For example, if each row has on average $n_{nz} = 8$ nonzero elements, then we would need to store about $n_{nz}N = 160{,}000$ doubles and $2n_{nz}N = 320{,}000$ integers (row/column indices for each value). At 8 bytes per double, and 2 bytes per unsigned integer, we only need about 2 MB of RAM. This is all done behind the scenes, but we need to be aware of its consequences. For instance, one limitation is that adding nonzero elements to a sparse matrix can be computationally costly.

Building the Element Stiffness Matrix

The element stiffness matrix can be assembled with `calculate_pqrDeltaSk`. In this implementation, apart from $\mathbf{S}^k$, we return the basis function coefficients $\mathbf{p}^k$, $\mathbf{q}^k$, and $\mathbf{r}^k$ of Equation (3.14). These variables are not used directly in the FEM solver, but they could be useful for determining the x and y-components of the B-field (or for debugging purposes). Also, the function returns the triangle areas Δ_k, which are needed to calculate various region areas (e.g., coil areas).

```
import numpy as np

def calculate_pqrDeltaSk(n_tri, nodes, triangles):
    """
    Calculates areas (delta) and p,q,r for all triangles.
    Also calculates the element stiffness matrix
    Sk = (q*q.T + r*r.T)/(4 * delta)
    for all triangles. For each element, we save the vector
    (exploiting symmetry):
    Sk = [s11 s12 s13 s22 s23 s33]
    """

    delta = np.zeros(n_tri)
    p = np.zeros((n_tri,3))
    q = np.zeros((n_tri,3))
    r = np.zeros((n_tri,3))
    Sk = np.zeros((n_tri,6))

    for k in range(n_tri):
        l = triangles[k,1]-1
        m = triangles[k,2]-1
        n = triangles[k,3]-1

        x1 = nodes[l,1]
        y1 = nodes[l,2]
        x2 = nodes[m,1]
        y2 = nodes[m,2]
        x3 = nodes[n,1]
        y3 = nodes[n,2]

```

```
        p[k,0] = x2*y3 - x3*y2
        q[k,0] = q1 = y2 - y3
        r[k,0] = r1 = x3 - x2
        p[k,1] = x3*y1 - x1*y3
        q[k,1] = q2 = y3 - y1
        r[k,1] = r2 = x1 - x3
        p[k,2] = x1*y2 - x2*y1
        q[k,2] = q3 = -q[k,0] - q[k,1]
        r[k,2] = r3 = -r[k,0] - r[k,1]

        # here we rely on the fact that Triangle generates
        # triangles with nodes listed in ccw order (so that
        # the area will turn out positive)
        delta[k] = dlt = 0.5 * (r3*q2 - q3*r2)

        Sk[k,0] = (q1*q1 + r1*r1)/4/dlt
        Sk[k,1] = (q1*q2 + r1*r2)/4/dlt
        Sk[k,2] = (q1*q3 + r1*r3)/4/dlt
        Sk[k,3] = (q2*q2 + r2*r2)/4/dlt
        Sk[k,4] = (q2*q3 + r2*r3)/4/dlt
        Sk[k,5] = (q3*q3 + r3*r3)/4/dlt

    return p, q, r, delta, Sk
```

Building the Global Stiffness Matrix

The global stiffness matrix is assembled using the element-wise building algorithm that has been described earlier. The following is an example of how this can be coded as a function `build_S_matrix`. The function needs (i) the number of nodes, (ii) the triangles data structure (from this, it only uses the node indices l, m, n), (iii) the element stiffness matrices, and (iv) a vector of triangle reluctivities needed to evaluate Equations (3.65a) and (3.65b), one per triangle. We are importing the `sparse` package from SciPy, to help us define a sparse stiffness matrix.

```
import scipy.sparse as sps
import numpy as np

def build_S_matrix(n_nodes, triangles, Sk, nu):
    """
    Builds global sparse (symmetric) S-matrix.
    """

    n_tri = nu.size
    row = np.zeros(9*n_tri, dtype = int)
    col = np.zeros(9*n_tri, dtype = int)
    s = np.zeros(9*n_tri)
    count = 0

    for k in range(n_tri):
        gamma = nu[k]

        l = triangles[k,1]-1
        m = triangles[k,2]-1
        n = triangles[k,3]-1
```

```
        row[count] = l
        col[count] = l
        s[count] = gamma * Sk[k,0]
25      count += 1

        row[count] = l
        col[count] = m
        tmp = gamma * Sk[k,1]
30      s[count] = tmp
        count += 1
        row[count] = m
        col[count] = l
        s[count] = tmp
35      count += 1

        row[count] = l
        col[count] = n
        tmp = gamma * Sk[k,2]
40      s[count] = tmp
        count += 1
        row[count] = n
        col[count] = l
        s[count] = tmp
45      count += 1

        row[count] = m
        col[count] = m
        s[count] = gamma * Sk[k,3]
50      count += 1

        row[count] = n
        col[count] = m
        tmp = gamma * Sk[k,4]
55      s[count] = tmp
        count += 1
        row[count] = m
        col[count] = n
        s[count] = tmp
60      count += 1

        row[count] = n
        col[count] = n
        s[count] = gamma * Sk[k,5]
65      count += 1

    S = sps.coo_matrix((s, (row,col)), \
                    shape=(n_nodes,n_nodes)).tocsc()
    return S
```

Within the **for** loop of rows 15–65, nine (row, column, value) triplets are stored for each element. The global stiffness matrix is obtained with the command in line 67, where a sparse matrix in COOrdinate format is created. The COO matrix is immediately converted to compressed sparse column (CSC) format, which adds the values of all duplicate (row, column) entries in the COO matrix. This is it; we do not need to write any extra code! This matrix assembly procedure is very common in FEA throughout all engineering disciplines.

Building the Source Vector

The procedure to assemble the source vector **b** is conceptually similar. One way that it could be implemented is to augment the `for` loop of `build_S_matrix` according to the following code snippet.

```
def build_S_matrix(n_nodes, triangles, Sk, nu, J, delta):
    # other initialization commands
    ...
    b = np.zeros(n_nodes)

    for k in range(n_tri):
        # previous S-matrix code is here
        ...

        t = J[k]*delta[k]/3
        b[l] += t
        b[m] += t
        b[n] += t

    S = ...
    return S, b
```

The previous `build_S_matrix` routine has been slightly modified in the following ways. (i) An array `J` is passed as an extra argument to `build_S_matrix`, containing the current density of each triangle in the mesh. Typically, this would be a vector of mostly zeros, except where conductors are present. However, since this is a vector (not a matrix), it is not really worthwhile storing the information in sparse format. (ii) The array `delta` of triangle areas is an additional argument. Recall that this has been calculated earlier, together with the element stiffness matrices, in `calculate_pqrDeltaSk`.

Enforcing Boundary Conditions

We now discuss how to enforce Dirichlet boundary conditions, which was described algorithmically in §3.2.4 (see Box 3.1). Here, we set the boundary potential to zero, so there is no need to modify the source vector (Step 2a of the algorithm).

First, we need the indices of the vertices that lie on the boundary of the domain, where we will apply a zero potential. For instance, this array of Dirichlet nodes may be obtained by the following command:

```
ind_Diri = np.int_(nodes[nodes[:,3] == 1,0])
```

This command returns the index of those elements from the first column of the nodes array whose identifier on the fourth column equals 1. The numerical data, which are originally floating-point numbers, are recast as integers (`np.int_` is the default integer data type in NumPy) so that they can be used to index arrays. The nodes are numbered starting from 1 in the data files; however, in Python we must make sure to use zero-based indexing. Hence, we should reference the corresponding elements by `ind_Diri-1`.

Now, we may enforce boundary conditions as follows:

```
for col in ind_Diri-1:
    S.data[S.indptr[col]:S.indptr[col+1]] = 0
S = S.tocsr()
```

```
for row in ind_Diri-1:
    S.data[S.indptr[row]:S.indptr[row+1]] = 0
    S[row,row] = 1
```

Recall that the stiffness matrix is originally in CSC format, as indicated in line 68 of `build_S_matrix`. This means that its data is saved one column at a time, so it can be readily accessed by column. In particular, `indptr` is an attribute of a SciPy sparse matrix, such that `indptr[col]` is the index of the first element of column `col` in the `data` array. In other words, `indptr[col]:indptr[col+1]` yields a list of indices for the elements of `data` corresponding to column `col`. After we set the corresponding column elements to zero, we convert the matrix to compressed sparse row (CSR) format, which rearranges the data by row, and repeat the process for the appropriate rows. Finally, we set the corresponding elements from the diagonal to 1.

Note that the conversion of elements to zero does not imply their removal from the sparse matrix. Doing this would be unnecessarily time-consuming. Rather, we choose to keep these zero elements, using up only a tiny amount of memory that is still allocated for their storage in order to gain in performance.

We should also make sure that the corresponding elements of the **b** array are set to zero (Step 3a of the algorithm). Usually, this is not coded because current sources are not adjacent to the boundary, in which case these values are already zero.

3.3.3 A First FEM Solver

We have thus set the stage for implementing our first FEM solver. The main program structure is outlined in Box 3.2.

Box 3.2 Programming the FEM: Main program structure (linear magnetics)

1. Define and mesh the geometry (call `Triangle`).
2. Read node and triangle data files (call `load_mesh`).
3. Build element stiffness matrices (call `calculate_pqrDeltaSk`).
4. Assign material properties (constant magnetic reluctivity) to triangles.
5. Build global stiffness matrix and current source vector (call `build_S_matrix`).
6. Enforce boundary conditions (update **S**).
7. Solve **SA** = **b** for the (free) node potentials.
8. Calculate quantities of interest and generate plots by post-processing **A**.

With the exception of Steps 1 (meshing) and 8 (post-processing), we have discussed these algorithms in detail. Writing this program is left as an exercise for the reader. The solution of the linear system in Step 7 can be obtained by invoking a solver from the sparse linear algebra library of Scipy:

```
import scipy.sparse.linalg as spsl
A = spsl.spsolve(S,b)
```

It should be noted that the algorithms for a sparse matrix solve are nontrivial, and could involve thousands of lines of code. However, we do not need to understand all these mathematical details in depth. Python comes with a variety of state-of-the-art numerical routines, written by experts in these fields, which are updated on a regular basis. A detailed exposition of the underlying numerical algorithms is outside the scope of this text.

One could question the reason why we need to invoke such a sophisticated numerical solver. Can we not just find the solution with a command such as `A = inv(S)*b`? This is not recommended because it implies that the inverse of the sparse matrix **S** is calculated first, and then multiplied by **b**. However, in general, the inverse of a sparse matrix is a full matrix, so this approach will require a tremendous amount of memory, and is computationally inefficient compared to more advanced sparse solvers.

Example 3.4 *Square-core inductor.*

The purpose of this rather extensive example is to help the reader develop (and debug) a first FEA program. In the previous pages, we have set forth programming details regarding how to store and read a mesh from a file, how to formulate the matrices for a sparse linear system of equations, and how to solve the system. We will now walk through all remaining steps, which involve the practical issues of (i) defining the geometry, (ii) generating a mesh, (iii) visualizing the mesh and the solution, and (iv) performing an elementary sanity check.

Problem definition. We will analyze the elementary square-core inductor shown in Figure 3.7. Our analysis will be two-dimensional. The relative permeability of the (magnetically linear, solid steel) core is $\mu_r = 500$. The inductor is formed by a copper coil that has $N = 20$ turns. The direction of current flow is arbitrary (it will not affect the result in any meaningful way), so let us assume that current enters the page on the left, and exits on the right.

The modeling of the coil merits some further discussion. In practice, the winding process may not lead to an ideal coil, in the sense that the coil turns may not be perfectly layered. The two gray parallelepipeds represent the boundaries of the two coil sides, which enclose all conductors. In reality, these areas are not filled entirely by copper. There are mainly two reasons for this: (i) the insulating outer layer of the wires; and (ii) the geometry of the wire itself, and the constraints that it imposes on packing the region. For example, it is impossible to completely fill a region with round conductors; there will always be some space between their circular cross-sections.

It is thus impractical to attempt a precise modeling of the placement of the coil conductors due to the inherent stochasticity, and because this would impose an unnecessary numerical burden due to the high number of elements that will be needed to capture these fine geometric details. Therefore, it is customary and usually sufficient to model coils by assigning a uniform current density J throughout a region in space that con-

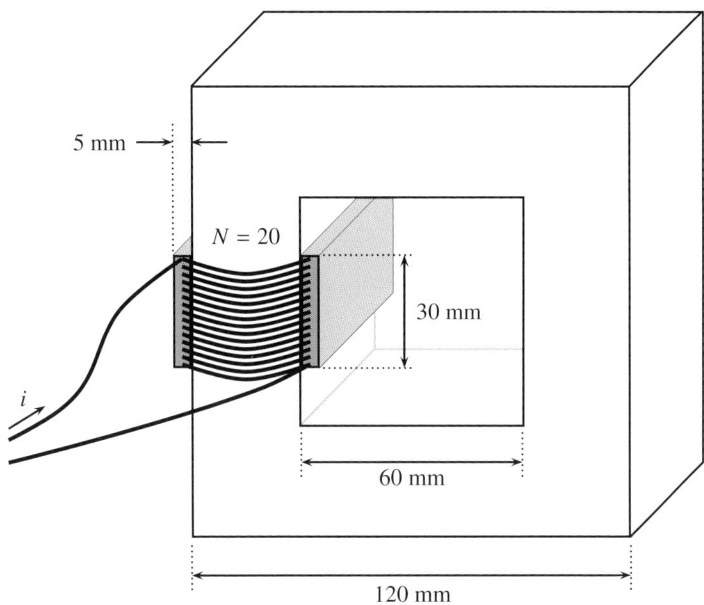

Figure 3.7 Square-core inductor geometry.

tains all conductors. We will call this the effective **coil area** A_c. Here, this may be calculated from the given dimensions as $A_c = 5 \times 30 = 150$ mm^2. Suppose the current is $i = 15$ A, d.c. Then, in the program we will assign a value of

$$J = \frac{Ni}{A_c} = 2 \cdot 10^6 \text{ A/m}^2 \tag{3.90}$$

to each triangle in the coil region, which will be used in the assembly of the source vector **b**. Of course, the triangles on the left will be assigned a value of -2 MA/m^2.

Meshing using Triangle. First, we will explain how to run the program Triangle to generate a mesh. Triangle will read a .poly file describing the geometry in the form of a **planar straight-line graph** (PSLG). This file contains a list of vertices, followed by a list of line segments each connecting two of the vertices. The segments are used to delineate material regions. In addition, the file may optionally contain information about holes (internal areas where a mesh is not needed) and device regions.

For this example, we create an ASCII file named `square_core_ind.poly`:

```
# vertices
20 2 0 1
1   0     0     1
2   200   0     1
3   200   -200  1
4   0     -200  1
5   40    -40   0
6   160   -40   0
7   160   -160  0
8   40    -160  0
```

```
9   40      -115    0
10  35      -115    0
11  35      -85     0
12  40      -85     0
13  70      -70     0
14  130     -70     0
15  130     -130    0
16  70      -130    0
17  70      -115    0
18  75      -115    0
19  75      -85     0
20  70      -85     0
# segments
22 0
1   1   2
2   2   3
3   3   4
4   4   1
5   5   6
6   6   7
7   7   8
8   8   9
9   9   10
10  10  11
11  11  12
12  12  5
13  12  9
14  13  14
15  14  15
16  15  16
17  16  17
18  17  18
19  18  19
20  19  20
21  20  17
22  20  13
# holes
0
# regional attributes
5
1   5   -5    3
2   100 -100  3
3   50  -50   1
4   39  -100  21
5   71  -100  22
```

We observe that the vertex syntax is identical to what has been described for the .node files earlier:

- First line: <# of vertices> <dimension (must be 2)> <# of attributes> <# of boundary markers (0 or 1)>
- Remaining lines: <vertex #> <x> <y> [attributes] [boundary marker]

The first four vertices (and associated segments) define a 200-mm × 200-mm bounding box (the edges of our universe for the purposes of this study), with its upper-left point as the origin of the coordinate system. These vertices are assigned the boundary marker 1, which is what Triangle assigns to boundary nodes and edges. The remaining 16 vertices define the inductor geometry, and are assigned the boundary marker 0. Here, all dimensions are given in millimeters for convenience. To avoid unit conversion bugs in the code, it is good practice to convert these coordinates to SI units (meters) as soon as they are loaded in memory.

The syntax for segments, holes, and regional attributes is as follows:

- One line: <# of segments> <# of boundary markers (0 or 1)>
- Following lines: <segment #> <endpoint> <endpoint> [boundary marker]
- One line: <# of holes>
- Following lines: <hole #> <x> <y>
- Optional line: <# of regional attributes and/or area constraints>
- Optional following lines: <region #> <x > <y> <attribute> <maximum area>

In this example, there are 22 segments. We are not setting the optional boundary marker for these. In this case, Triangle finds all nodes and edges at the boundary, and assigns to them a boundary marker 1. In general, when defining a geometry, it is good practice to avoid segments that intersect. Intersection points should be identified and explicitly entered as separate nodes. If we let Triangle determine intersection points, we may run into unforeseen numerical issues caused by floating-point errors.

Next, we specify that there are no holes present (line 48). Finally, we are specifying the properties of five regions using the attribute convention: 3 = air, 1 = steel, 21 = left side of coil, 22 = right side of coil. The coordinates of a single point inside each region need to be provided, as done in lines 51–55. Triangle will then propagate this attribute to all elements within the region. For this to succeed, it is important to define the segments carefully so that closed boundaries for each region are properly formed. (If we forget a segment, then the regional attribute will "leak" over to the adjacent region.)

It is certainly possible to mesh the geometry by manually running Triangle from the command line. Alternatively, we could run Triangle from a Python script. One way to achieve this is as follows:

```
1 from subprocess import check_call
2
3 fname = 'square_core_ind'
4 maxsize = 25.0
5
6 buf = '-pqDAa{:.2f}'.format(maxsize)
7 check_call(['./triangle',buf,fname+'.poly'])
```

We are importing the `check_call` function from the `subprocess` library, to issue a system command for running Triangle. It is assumed that the Triangle executable is in the same directory as the Python script. Triangle is typically called with several command line switches. In this case, the switches that we are specifying in the string `buf` in line 6 are indicating that: (i) we are reading a PSLG file (-p); (ii) we are asking for a quality mesh generation where elements have a minimum angle of 20° (-q); (iii) all triangles are Delaunay (-D); (iv) a regional attribute is assigned to each triangle (-A); and (v) a maximum triangle area constraint (-a25.00) is imposed so that no triangle will have an area larger than 25 mm^2. The reader should consult the documentation of Triangle for more information about these and other possible options for calling Triangle.

The generated mesh is shown in Figure 3.8. The triangulation algorithm maintains all vertices specified in the PSLG .poly file, as a close inspection reveals, but it also refines the geometry. The goal is to obtain a so-called **Delaunay**[62] **triangulation**, so extra nodes are added in the domain, called **Steiner**[63] **points**. The original PSLG segments are split into smaller edges as well. The definition of a Delaunay triangulation is somewhat esoteric, as it requires some knowledge of Euclidean geometry: "*A Delaunay triangulation of a vertex set has the property that no vertex in the vertex set falls in the interior of the circumcircle (circle that passes through all three vertices) of any triangle in the triangulation.*" Simply speaking, this property implies that thin, sliver-like triangles are avoided, as this is generally considered to be undesirable from a numerical standpoint. Here, the mesh has 1,314 vertices and 2,498 triangles. One easy way to control these numbers is by adjusting the maximum triangle size. The reader may experiment with the command-line switches, observing the differences of the triangulations that are obtained from the same PSLG input file.

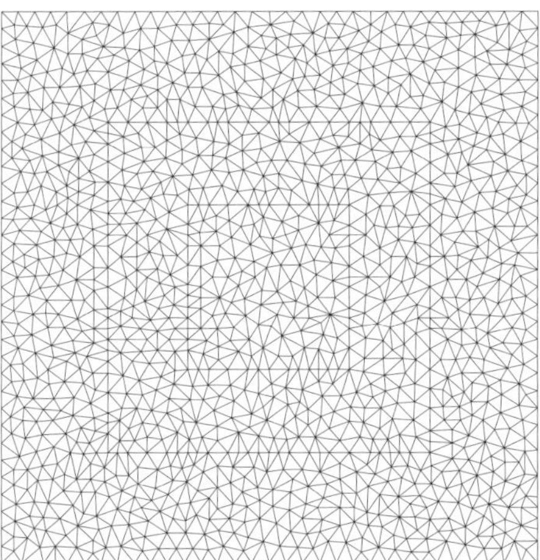

Figure 3.8 Square-core inductor mesh.

Plotting the mesh. After the meshing is complete, it is usually prudent to inspect the mesh for errors. The mesh may be plotted using the following function:

```
import matplotlib.pyplot as plt

def plot_mesh():
    C = np.ones(n_tri)
    C[ind_st-1] = 0.8
    C[ind_coil_1-1] = 0.6
    C[ind_coil_2-1] = 0.6
    plt.tripcolor(triangulation,C,edgecolors='0.9',\
                             cmap=plt.cm.gray,vmin=0.0,vmax=1.0)
```

We are plotting the triangulation using a gray colormap, where the variable C sets the color of each triangle. A value of 1.0 corresponds to white, and 0.0 corresponds to black. For instance, the edge color is 0.9, so they appear as light gray line segments (see Figure 3.9). The variables ind_st, ind_coil_1, and ind_coil_2 contain indices to nodes in steel and the two coils, respectively, similar to the ind_Diri array that was described earlier. Note that plot_mesh does not have any explicit argument. The variables are accessed from the global namespace, where they should be already defined.

The mesh can be conveniently plotted using the Matplotlib Pyplot tripcolor function, which accepts a Triangulation object as its main argument. A Triangulation object contains information about an unstructured triangular grid (i.e., its vertices and triangles), and can be created using the tri library of Matplotlib as follows:

```
from matplotlib.tri import Triangulation

def make_triangulation(nodes, triangles):
    x = nodes[:,1]
    y = nodes[:,2]
    tr = triangles[:,1:4]-1
    return Triangulation(x,y,tr)

triangulation = make_triangulation(nodes, triangles)
```

Plotting the solution. We proceed with Steps 2–7 of the main program. Writing this program is left as an exercise for the reader (see Problem 3.8). After the system has been solved, we can add code to help us visualize the results (Step 8).

It is straightforward to plot the flux lines (equipotentials of A) with one of Python's built-in functions, namely, Matplotlib Pyplot tricontour:

```
def plot_flux_lines(Nlines = 10):
    plt.tricontour(triangulation,A,Nlines,colors='k')

plot_mesh()
plot_flux_lines()
```

The result is depicted in Figure 3.9. Here, we called tricontour mostly with default options, but this function has many optional keywords that can be used to change how the plot looks, and allows the user to specify which contour values should be plotted. We observe that each contour is formed by joining linear segments in each triangle, which is expected since we are using linear elements. Smoother flux lines may be obtained by further reducing the triangle size.

Moreover, we see that most of the magnetism "flows" through the steel material. Increasing the number of flux lines is possible through the Nlines parameter. By doing so, the density of flux lines in the steel will increase, and more lines will appear crossing into the air from the steel, like the one shown on the left side. These lines are representative of magnetic flux that escapes from the **main flux path** through the core, which is termed **leakage flux**.

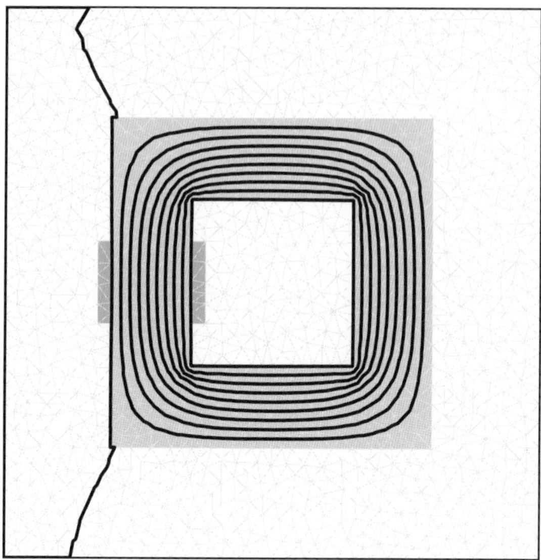

Figure 3.9 Square-core inductor flux lines.

It is interesting to note that:

- The flux lines do not cross the outer boundary, as expected due to the enforced boundary conditions. (This is not a physically accurate result. It is adequate for the purposes of this preliminary study because most of the flux goes through the core. However, if the precise calculation of leakage flux had been one of our objectives, we should consider moving the boundaries further away from the core.)

- The flux lines in the upper and lower halves are not completely symmetric (most apparent in the leakage flux line). However, in theory, the solution should have a horizontal axis of symmetry through the middle of the core. This issue may be attributed to the absence of symmetry or structure in the mesh. To exploit symmetry, we can reduce computational time by modeling only the top part of the core (see Problem 3.10). Adding an axis of symmetry requires some simple modifications in the PSLG, namely, deleting all bottom-half nodes and segments, and terminating the upper half by adding axial vertices and segments. Although the axis of symmetry would now be the lower boundary of our domain, we would not assign a boundary marker of 1; rather, we would let the potential be free along this line. In theory, this Neumann boundary condition leads to flux lines that cross the boundary at right angles, as explained in §2.2.1; however, the numerical solution may not be perfectly perpendicular at the boundary.

It is also possible to plot the B-field (flux density) magnitude over the domain. This kind of plot can help us visualize areas of the device that may be saturated. Recall that in triangle k, the magnetic vector potential has a linear dependence on position:

$$A_k(x, y, \mathbf{A}^k) = \sum_{i=1}^{3} A_i^k \alpha_i^k(x, y), \tag{3.91}$$

which implies that

$$\nabla A_k(x, y, \mathbf{A}^k) = \begin{bmatrix} (\nabla A_k)_x \\ (\nabla A_k)_y \end{bmatrix} = \frac{1}{2\Delta_k} \begin{bmatrix} (\mathbf{q}^k)^\top \\ (\mathbf{r}^k)^\top \end{bmatrix} \mathbf{A}^k. \tag{3.92}$$

We recall Equation (3.46):

$$\vec{\mathbf{B}} = \nabla \times \vec{\mathbf{A}} = \frac{\partial A}{\partial y}\hat{\mathbf{i}} - \frac{\partial A}{\partial x}\hat{\mathbf{j}}, \tag{3.93}$$

which yields the **element flux density as a function of the MVP values**:

$$\boxed{\vec{\mathbf{B}}^k = \frac{1}{2\Delta_k} \begin{bmatrix} (\mathbf{r}^k)^\top \\ -(\mathbf{q}^k)^\top \end{bmatrix} \mathbf{A}^k.} \tag{3.94}$$

Hence, the B-field is constant in each element. Furthermore, its magnitude satisfies

$$B_k^2 = \|\nabla A_k\|^2 = \frac{1}{(2\Delta_k)^2}(\mathbf{A}^k)^\top \begin{bmatrix} \mathbf{q}^k & \mathbf{r}^k \end{bmatrix} \begin{bmatrix} (\mathbf{q}^k)^\top \\ (\mathbf{r}^k)^\top \end{bmatrix} \mathbf{A}^k. \tag{3.95}$$

In view of Equation (3.28), the **magnitude of the B-field is related quadratically to the MVP values through the element stiffness matrix**:

$$\boxed{B_k^2 = \frac{1}{\Delta_k}(\mathbf{A}^k)^\top \mathbf{S}^k \mathbf{A}^k.} \tag{3.96}$$

We can calculate the B-field magnitude in each triangle using the element stiffness matrices and areas that have been obtained from `calculate_pqrDeltaSk`. This is plotted in Figure 3.10, which was generated using Matplotlib Pyplot `tripcolor`. A colorbar has been added to the plot to facilitate its interpretation. Note how the flux density is higher at the inner corners of the core, where also the flux lines are closer together.

Performing a sanity check. It is always a good idea to perform sanity checks after modifications to FEA code are done and/or new results are obtained. Debugging an FEA program can be time-consuming since inconspicuous, hard-to-find errors may be present. Usually, inspecting and evaluating the results can reveal that something is wrong. In this example, we will examine the variation of the A-field (in SI units) along the horizontal axis of symmetry through the middle of the core, $y = -100$ mm, as shown in Figure 3.11. (Generating this figure is left as an exercise for the reader.) The vertical dashed lines represent the boundaries of the core and coils. But what is the interpretation of this plot? And how does it help us check the validity of the result?

The A-field plot may be interpreted with the help of Box 1.3 on page 78, where it was explained that the difference of A between any two points equals flux per unit depth. Here, the potential starts from the value zero at $x = 0$, satisfying the imposed boundary condition. Then it drops slightly until $x = 40$ mm, where the steel core begins. This drop indicates a small amount of leakage flux moving upwards. Inside the core, from

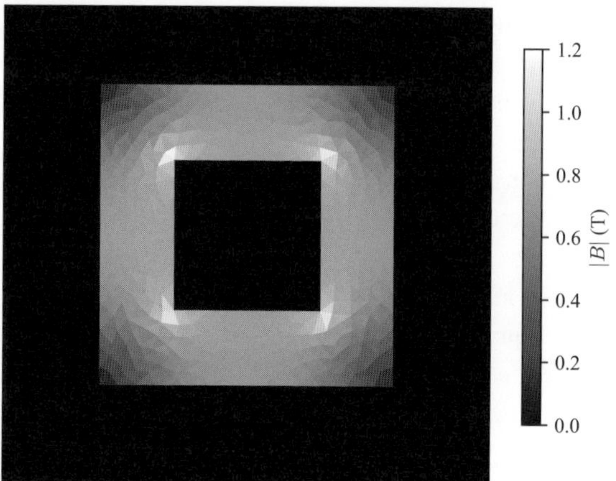

Figure 3.10 Square-core inductor B-field magnitude.

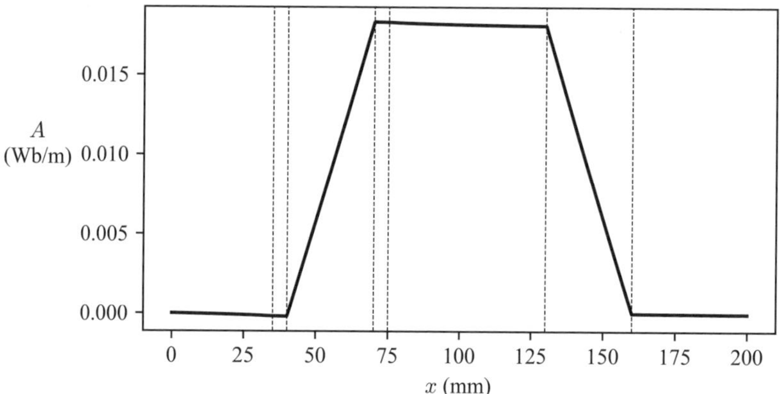

Figure 3.11 Square-core inductor A-field variation along a horizontal axis of symmetry.

$x = 40$ to $x = 70$ mm, the potential increases at a higher rate, indicating a relatively large value of flux crossing in the downwards direction. In the internal core "window," we observe that the potential exhibits a slight drop, which means that a small amount of leakage flux crosses in the upwards direction. From 130 to 160 mm, we are again inside the core, and the main flux is in the upwards direction. Once we exit the core, the potential remains almost constant, indicating an insignificant amount of flux in this region. The direction of flux in all regions is consistent with the right-hand rule.

We can also perform a quantitative sanity check. To this end, we will conduct a simple magnetic circuit analysis, as we did in Example 2.15 on page 165. We invoke Ampère's law in its integral form:

$$\oint_{\partial S} \vec{\mathbf{H}} \cdot d\mathbf{r} = \iint_S \vec{\mathbf{J}} \cdot d\mathbf{a}\,. \tag{3.97}$$

It is convenient to define a square integration path ∂S through the middle of the core, which is traversed counterclockwise (in the same sense as the magnetic field). This path bounds a square surface S, which is crossed perpendicularly by the N conductors on the right side of the coil. For simplicity, and in view of the fact that each side of the core has the same width (so that the flux density is approximately the same throughout the core), we can assume that the H-field has a constant magnitude and is parallel to each side of the path. As can be seen from Figure 3.9, this is a decent approximation close to the midpoints of the core sides, but not as good close to the corners, where the flux lines are bending and become concentrated towards the inner corners. Physically speaking, we are assuming that the flux density is uniform throughout the core, that somehow magically it changes direction without bending, and that no leakage flux exists. This crude approximation is sufficient for our purposes. Applying Ampère's law, we obtain

$$4\ell H = JA_c = Ni \Rightarrow H = \frac{Ni}{4\ell} = \frac{300}{0.36} = 833.33 \text{ A/m}, \tag{3.98}$$

because the length of each side of the path is $\ell = 90$ mm. Hence, a crude estimate of the B-field is

$$B = \mu_r \mu_0 H = (500)(4\pi 10^{-7})(833.33) \approx 0.52 \text{ T}. \tag{3.99}$$

Is this consistent with Figure 3.11? From the quasi-linear increase and decrease of the potential in the core between 0.0 and 0.018 Wb/m, we deduce that the flux density in the core is more or less uniform along the horizontal axis, and that the flux per unit depth is

$$A_1 - A_2 \approx 0.018 \text{ Wb/m}. \tag{3.100}$$

The flux density is found by dividing this result by the width of the core:

$$B \approx \frac{0.018}{0.03} = 0.6 \text{ T}, \tag{3.101}$$

which is fairly close to the estimated value of Equation (3.99), thus enhancing our confidence in the validity of the FEA solution.

3.4 Calculation of Flux Linkage and Inductance

It is often the case that a fields-based solution must be interfaced with an external electric circuit. As explained in §2.3.6, we can extract macroscopic quantities such as the flux linkage and the inductance from the field. We remain in the low-frequency regime, where the current is uniformly distributed over the wire cross-section.

3.4.1 The Average MVP Method for Flux Linkage Calculations

We have defined the flux of a filament $C = \partial S$ in Equation (2.193) on page 155 as

$$\psi = \oint_C \vec{\mathbf{A}}(\mathbf{r}) \cdot d\mathbf{r} = \int_a^b \vec{\mathbf{A}}(\mathbf{r}) \cdot d\mathbf{r}, \tag{3.102}$$

where the orientation of the closed path C is the same as that of positive current flow, which in turn is related to the assumed direction of positive magnetic flux through surface S. A related open path integral is obtained when a and b are filament endpoints outside the magnetically active region. In particular, point a is the positive terminal, which is defined as the terminal where positive current enters the coil. This designation leads to positive flux from a positive current, and thus to a positive value for the self-inductance of the coil. The calculation is valid regardless of whether the coil current is positive, negative, or zero. It will yield the total flux through the filament, generated by all current sources in the universe.

Based on the representation of a conductor by infinitely many filaments in parallel, we further defined a winding flux linkage in Equation (2.200) as the average of all filament fluxes over the wire cross-section S:

$$\lambda = \frac{1}{A_w} \iint_{\tilde{\mathbf{r}} \in S} \psi(\tilde{\mathbf{r}})\, da = \frac{1}{A_w} \iint_{\tilde{\mathbf{r}} \in S} \int_{a(\tilde{\mathbf{r}})}^{b(\tilde{\mathbf{r}})} \vec{\mathbf{A}}(\mathbf{r}) \cdot d\mathbf{r}\, da\,, \tag{3.103}$$

where $A_w = |S|$ is the wire cross-sectional area (determined by its gauge).[c] Here, the variable of integration $\tilde{\mathbf{r}}$ points to an initial location on the positive terminal cross-section S, which subsequently defines the filament endpoints, $a(\tilde{\mathbf{r}}) = \tilde{\mathbf{r}}$ and $b(\tilde{\mathbf{r}})$, in the positive and negative coil terminals, respectively.

For a practical application of this formula, let us consider a 2-D problem with the coil shown in Figure 3.12. This particular coil is formed by winding $N = 23$ turns of a round magnet wire in two rectangular slots. The turns are designated as $t_1^{\pm}, t_2^{\pm}, \ldots, t_N^{\pm}$, based on how positive current density aligns with the z-axis. Here, we have a case where the coil is wound in such a way so that the positive terminal (where current enters) is associated with a negative current density (based on the orientation of the z-axis, which points out of the page). Current i enters on the left side at t_1^-, where it spreads evenly over the conductor cross-sectional area S (gray circle on the left-hand side), then travels into the page and returns at t_1^+, loops around and enters at t_2^-, and so on, finally exiting at t_N^+ on the right. The conductor consists of many filaments connected in parallel, each carrying an infinitesimal amount of current, di.

Consider a filament that starts at point $a(\tilde{\mathbf{r}}) = \tilde{\mathbf{r}} = \mathbf{r}_1^- + \delta\tilde{\mathbf{r}}$, displaced by $\delta\tilde{\mathbf{r}}$ from the center of t_1^-, which in turn is displaced by $\mathbf{r}_1^-$ from the origin. The dots in the figure signify all the points where the chosen filament intersects the cross-section. (In practice, the filament may not be at exactly the same spot relative to the center in each turn, in contrast to how it has been drawn, due to conductor turning and bending.) The filament ends at some location within t_N^+, for instance, at $b(\tilde{\mathbf{r}}) = \mathbf{r}_N^+ + \delta\tilde{\mathbf{r}} = (\mathbf{r}_N^+ - \mathbf{r}_1^-) + \tilde{\mathbf{r}}$.

In a 2-D problem of axial length ℓ, we can simplify Equation (3.103) by evaluating the line integral over the entire filament as a sum of line integrals over each consecutive turn. We introduce the notation $A_n^+(\tilde{\mathbf{r}})$ to denote the magnetic potential at certain locations within the filament that starts at $\tilde{\mathbf{r}}$. In particular, the superscript and subscript indicate that we are within a filament segment that corresponds to the positive side of turn n.

[c] Even though they both share the same symbol, the wire area A_w is not related to the MVP $\vec{\mathbf{A}}$.

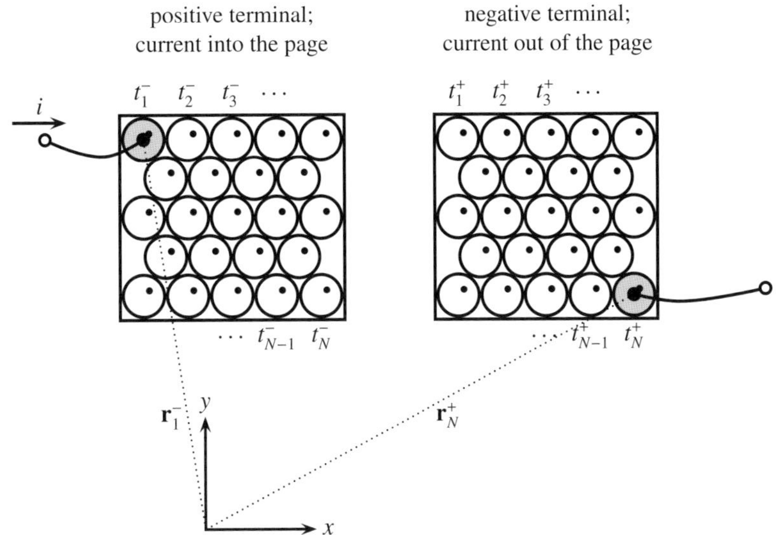

Figure 3.12 Cross-section of a coil formed by placing round magnet wire in a rectangular slot.

These potentials are found on the dots at the right-hand side of the coil in Figure 3.12. Similarly, $A_n^-(\tilde{\mathbf{r}})$ denotes the potential of a filament in the negative side of turn n.

Let us integrate $\int \vec{\mathbf{A}}(\mathbf{r}) \cdot d\mathbf{r}$ along the filament that originates at $\tilde{\mathbf{r}}$, considering only the linear segment inside turn n at the positive coil side. The position $\mathbf{r}$ moves parallel to the z-axis, say from $\mathbf{r}_{an}^+(\tilde{\mathbf{r}})$ at $z = -\ell$ to $\mathbf{r}_{bn}^+(\tilde{\mathbf{r}})$ at $z = 0$. We parametrize this path as $\mathbf{r}(\xi; \tilde{\mathbf{r}}) = \mathbf{r}_{an}^+(\tilde{\mathbf{r}}) + \xi \ell \,\hat{\mathbf{k}}$, with $\xi \in [0, 1]$, so that $\mathbf{r}_{bn}^+(\tilde{\mathbf{r}}) = \mathbf{r}_{an}^+(\tilde{\mathbf{r}}) + \ell \,\hat{\mathbf{k}}$. Hence, $d\mathbf{r}/d\xi = \ell \,\hat{\mathbf{k}}$, and

$$\int_{\mathbf{r}_{an}^+(\tilde{\mathbf{r}})}^{\mathbf{r}_{bn}^+(\tilde{\mathbf{r}})} \vec{\mathbf{A}}(\mathbf{r}) \cdot d\mathbf{r} = \int_{\xi=0}^{\xi=1} [A_n^+(\tilde{\mathbf{r}}) \,\hat{\mathbf{k}}] \cdot [\ell \,\hat{\mathbf{k}}] \, d\xi = \ell \, A_n^+(\tilde{\mathbf{r}}) \,. \tag{3.104}$$

A similar path parametrization is defined along the negative coil side, from $\mathbf{r}_{an}^-(\tilde{\mathbf{r}})$ at $z = 0$ to $\mathbf{r}_{bn}^-(\tilde{\mathbf{r}})$ at $z = -\ell$. We have $\mathbf{r}(\xi; \tilde{\mathbf{r}}) = \mathbf{r}_{an}^-(\tilde{\mathbf{r}}) - \xi \ell \,\hat{\mathbf{k}}$, with $\xi \in [0, 1]$, and $\mathbf{r}_{bn}^-(\tilde{\mathbf{r}}) = \mathbf{r}_{an}^-(\tilde{\mathbf{r}}) - \ell \,\hat{\mathbf{k}}$. Now, $d\mathbf{r} = -\ell \,\hat{\mathbf{k}} \, d\xi$. We have thus obtained the following expression for the coil flux linkage:

$$\lambda = \frac{\ell}{A_w} \iint_{\tilde{\mathbf{r}} \in S} \left\{ \sum_{n=1}^{N} [A_n^+(\tilde{\mathbf{r}}) - A_n^-(\tilde{\mathbf{r}})] \right\} da \,. \tag{3.105}$$

The potentials can be summed in no particular order without affecting the result of the integration. In other words, we do not need to know the exact placement of each turn! Also, the integral is not affected if we first integrate over the cross-section of each turn, and then add all turn contributions. This means that we do not need to know the exact position of the filament in the conductor either! Hence, the above flux linkage integral becomes

$$\lambda = \frac{\ell}{A_w} \left\{ \iint_{\mathbf{r} \in S^+} A(\mathbf{r}) \, da - \iint_{\mathbf{r} \in S^-} A(\mathbf{r}) \, da \right\} , \tag{3.106}$$

where S^+ and S^- represent the union of all positive and negative conductor surfaces, respectively.

As we explained in Example 3.4, low-frequency FEM models typically assume a uniform current density within the entire slot. The shape and arrangement details of the coils are ignored. For evaluating Equation (3.106) based on an FEA solution, the integration areas S^+ and S^- are different from their physical counterparts; they are now representative of the entire cross-sections of the slots, which are assumed to be 100% filled with conductive material. The wire area A_w is no longer significant (since there are no distinct wires visible in an FEM mesh). Hence, A_w is replaced by the equivalent quantity A_c/N, where $A_c = |S^+| = |S^-|$ is the total slot cross-sectional area. Therefore, Equation (3.106) becomes

$$\lambda = \frac{\ell N}{A_c}\left\{\iint_{\mathbf{r}\in S^+} A(\mathbf{r})\,da - \iint_{\mathbf{r}\in S^-} A(\mathbf{r})\,da\right\}. \tag{3.107}$$

These integrals are converted to sums over the K^+ and K^- triangles covering the two coil regions. The surface integral of $A(\mathbf{r})$—a linear function—over each triangle is quite easy to obtain (see Problem 3.3). Hence

$$\boxed{\lambda = \frac{\ell N}{A_c}\left\{\sum_{k=1}^{K^+}\frac{A_1^k + A_2^k + A_3^k}{3}\Delta_k - \sum_{k=1}^{K^-}\frac{A_1^k + A_2^k + A_3^k}{3}\Delta_k\right\}.} \tag{3.108}$$

We are therefore subtracting the average magnetic vector potential (MVP) between the two sides of the coil, which are designated as positive or negative based on the directionality of positive coil current with respect to the z-axis. This method for calculating the flux linkage may be called the **average MVP method**.

The average MVP method applies in general, and it even holds for nonlinear magnetics. In particular, the method is not restricted to two-sided coils like the one shown earlier in the square-core inductor example. This type of coil is called a **concentrated** winding because ideally the same flux Φ passes through, or links, all of its turns. Neglecting the variation of magnetic potential within the conductor region, Equation (3.108) leads to the elementary relationship, $\lambda = N\Phi$. Such a macroscopic view of the coil does not accurately capture the variation of flux density within the coil conductors, which is attributed to a varying MVP within the coil region. Indeed, a close inspection of Figure 3.11 shows that the MVP varies slightly within the two coil regions. In this case, Equation (3.108) indicates that the flux linkage can be interpreted as the product of turns times the spatial average of the flux over the coil region:

$$\lambda = N\overline{\Phi}. \tag{3.109}$$

A different type of winding is commonly encountered in electric machines, where windings can form more complex patterns by virtue of being **distributed** in many slots. Equation (3.108) is still valid, as it was derived without any particular consideration of the winding shape, based on the generic line integral $\oint \vec{\mathbf{A}} \cdot d\mathbf{r}$ over a filament length. To apply this formula, we just need to identify all positively and negatively oriented coil sides and corresponding triangles, whose average potentials will be added or subtracted, respectively, to yield the flux linkage.

3.4.2 Inductance Calculations Using the Flux Linkage

Inductance calculations are often needed, especially when modeling magnetically linear devices. (See Example 2.13.) Clearly, the self-inductance of a single coil is

$$L = \frac{\lambda}{i} . \tag{3.110}$$

This ratio will be constant for a linear circuit, regardless of the value of the current.

This procedure may be generalized for self- and mutual-inductance calculations when $K > 1$ current sources are present. We may be asked to calculate the device inductance matrix, which relates flux linkages and currents as $\boldsymbol{\lambda} = \mathbf{Li}$:

$$\begin{bmatrix} \lambda_1 \\ \lambda_2 \\ \vdots \\ \lambda_K \end{bmatrix} = \begin{bmatrix} L_{11} & L_{12} & \dots & L_{1K} \\ L_{21} & L_{22} & \dots & L_{2K} \\ \vdots & \vdots & & \vdots \\ L_{K1} & L_{K2} & \dots & L_{KK} \end{bmatrix} \begin{bmatrix} i_1 \\ i_2 \\ \vdots \\ i_K \end{bmatrix} . \tag{3.111}$$

To obtain this matrix, we conduct K FEA studies. In each consecutive simulation, we excite the kth coil in the device with a nominal 1-A current, leaving all other coils open-circuited. After solving for the MVP, we calculate the flux linkage of all K coils by applying Equation (3.108). Exciting the kth coil leads to the calculation of the kth column of the inductance matrix. In theory, the obtained matrix should be symmetric. However, we should be cognizant of the fact that due to numerical error, it may not turn out perfectly symmetric.

When a single coil, say the kth, is excited in a magnetically linear device, the field energy is

$$W_f = \frac{1}{2} L_{kk} i_k^2 . \tag{3.112}$$

This leads to an alternative formula for the self-inductance. The field energy can be obtained directly from an FEA solution as

$$W_f = \frac{\ell}{2} \mathbf{A}^\top \mathbf{SA} , \tag{3.113}$$

where ℓ is the device depth and $\mathbf{S}$ is the original global stiffness matrix, that is, before we introduce any modifications required to enforce boundary conditions. Hence

$$L_{kk} = \frac{2W_f}{i_k^2} = \frac{\ell}{i_k^2} \mathbf{A}^\top \mathbf{SA} . \tag{3.114}$$

In general, when more than one coil is excited, then $\frac{\ell}{2}\mathbf{A}^\top\mathbf{SA}$ should equal $\frac{1}{2}\mathbf{i}^\top\mathbf{Li}$. This energy-based calculation serves as a sanity check for debugging the average MVP method. It should be performed only when the device is modeled as magnetically linear. Once we have confidence that our flux linkage calculation is correct, we can switch to a magnetically nonlinear model (which will be described in a subsequent section). The exact same flux linkage calculation function should work in a nonlinear device; however, the coupling field energy is no longer given by Equation (3.113) or by $\frac{1}{2}\mathbf{i}^\top\mathbf{Li}$.

3.4.3 Converting Between Physical and Modeled Current Density

The current density in a physical wire of cross-sectional area A_w that carries current i is

$$J_w = \frac{i}{A_w}. \tag{3.115}$$

In the FEA, we assign a (different) virtual current density

$$J = \frac{Ni}{A_c}. \tag{3.116}$$

The fundamental property of this current density scaling is that it does not affect the total number of ampere-turns, which remains invariant: $Ni = JA_c = NJ_wA_w$.

To understand better how the two current densities are related, consider that the total physical conductor area is only a fraction of the coil area:

$$NA_w < A_c. \tag{3.117}$$

A **packing factor** or **slot fill factor** is defined as the ratio

$$k_{pf} = \frac{NA_w}{A_c} < 1, \tag{3.118}$$

and is usually specified as a design parameter for any device. In our case, the round shape of the wires imposes an upper bound on k_{pf} (that can be calculated as an interesting geometrical problem for various configurations), so a complete fill of the slot with conductor material is not possible. The packing factor is further reduced by the presence of a thin layer of insulating material on the surface of the wires. So, even if a square magnet wire had been used, and even if it had been neatly and tightly arranged in layers within the slot, the packing factor would still be less than unity. To summarize, the FEM current density is a scaled version of the actual wire current density:

$$J = k_{pf} \cdot J_w. \tag{3.119}$$

The conversion between physical and modeled current density should be accounted for while calculating ohmic loss in an FEA program. From Equation (2.144) on page 142, we can express the total ohmic loss from all N wires in a slot as

$$P_{\text{loss}} = \frac{J_w^2}{\sigma} NA_w\ell, \tag{3.120}$$

where ℓ is the device depth and σ is the conductivity of the conductors. Using the above definitions, we can obtain an equivalent expression in terms of the FEM current density:

$$P_{\text{loss}} = \frac{J^2}{k_{pf}\sigma} A_c\ell. \tag{3.121}$$

This implies that in an FEA program calculation based on J and the total coil area, we should effectively reduce the conductivity of the material by the packing factor.

Example 3.5 *Single-phase transformer—mutually coupled coils.*

Suppose we add a second coil on the square core of Example 3.4, thus creating an elementary **single-phase transformer**. The two coils are **mutually coupled** by the magnetic field. The cross-section of this device is depicted in Figure 3.13. The device has length (into the page) $\ell = 30$ mm. The primary side has $N_1 = 20$ turns, and the secondary side has $N_2 = 10$ turns. The current polarities have been selected such that the two fluxes will oppose each other: positive current in the primary coil creates magnetic flux moving in a counterclockwise sense, whereas positive current in the secondary moves flux in the clockwise sense. Let us analyze the device assuming linearity in the magnetic circuit. The goal is to obtain the inductance matrix.

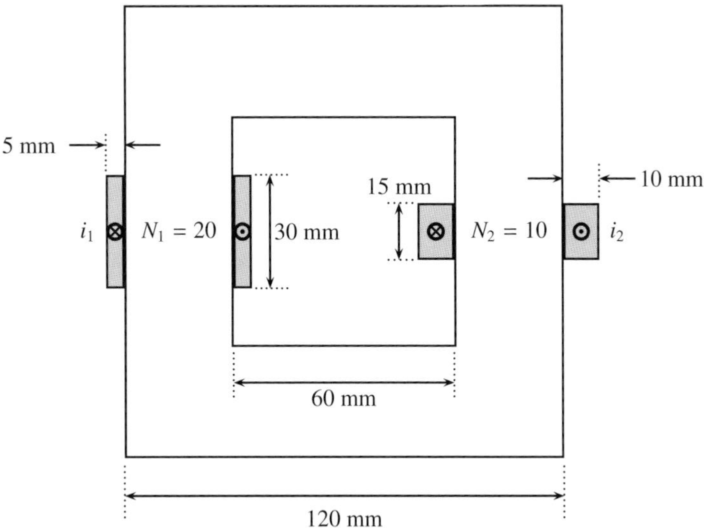

Figure 3.13 Square-core transformer cross-section.

Solution

We will conduct two studies: first, we will set $i_1 = 1$ A, $i_2 = 0$, and second, we will set $i_1 = 0$, $i_2 = 1$ A. In preparation for this, we should write a Python function that calculates the flux linkage in a coil, implementing the average MVP method of Equation (3.108). This is left as an exercise for the reader. After each FEM problem is solved, we call this function to obtain the flux linkages of each coil, thus forming directly the two columns of the inductance matrix. In this case, we obtain

$$\mathbf{L} = \begin{bmatrix} 0.75044 & -0.36568 \\ -0.36568 & 0.18801 \end{bmatrix} \text{mH}. \tag{3.122}$$

This result was calculated with a Delaunay mesh generated using Triangle, consisting of approximately 93,000 elements with an imposed maximum area constraint of 1 mm^2. The negative signs of the off-diagonal terms are indicative of the opposing mutual flux linkages. The number of decimal digits shown is probably more than enough. In practice, two or three significant digits would suffice. However, we report these additional digits here for the purposes of establishing the numerical symmetry of the matrix.

Knowing the inductance matrix is sufficient for analyzing any magnetically linear device during transient or steady-state conditions. Additional FEA studies are not required; the information that is crystallized in the inductance coefficients is all that is needed. Of course, we also need the resistance of the coils, which can be placed in a diagonal matrix **R**. In this simple case, where **L** is constant, we can model the system behavior by a system of ordinary differential equations (ODEs):

$$\mathbf{v} = \mathbf{R}\mathbf{i} + \mathbf{L}\frac{d}{dt}\mathbf{i}, \tag{3.123}$$

where **v** is a vector of terminal voltages. To proceed, we need more information about the terminal voltages (including how these may depend on the currents). Then it is trivial to solve these equations numerically in the computer using an ODE solver.

Example 3.6 *The T-equivalent circuit of two mutually coupled coils.*
The **equivalent circuit** of a single-phase transformer is broadly used by electrical engineers. This two-port network, which resembles the letter T as seen in Figure 3.14, captures the essential electrical and magnetic properties of two coupled coils that are mutually coupled through a lossy magnetic medium. The circuit (minus the core loss element R_c) is equivalent to the expressions that one would obtain by combining Kirchhoff's voltage law for the two coils with the corresponding flux linkage equations; in other words, it is equivalent to Equation (3.123). Nevertheless, it is valuable in its own right because it provides insight about transformer operation, and it permits us to analyze more complex electric circuits containing transformers as components. The derivation of the T-equivalent circuit can be found in many introductory texts on electric power systems. We will show how it can be obtained by algebraic manipulations that involve a simple rearrangement of the terms in the flux linkage equations.

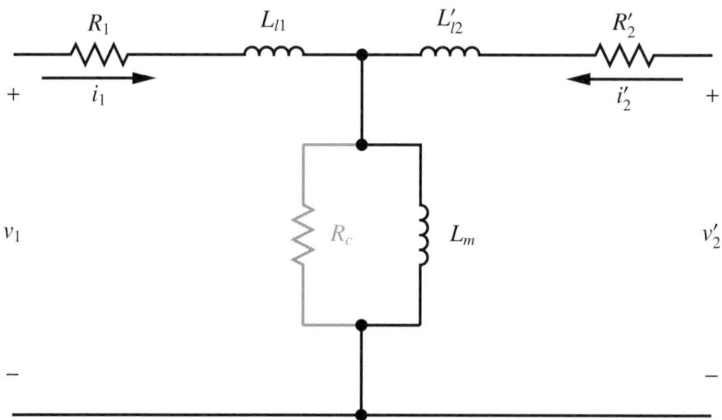

Figure 3.14 The T-equivalent transformer circuit.

Solution

We first note that Figure 3.14 depicts a case where, due to the choice of positive current directions on the equivalent circuit, the effect of the two currents is additive in the vertical branch. From a physical perspective, this implies (as we will see later on) that the magnetic fluxes created by positive current in each coil are in the same direction. However, this was not the case of Example 3.5, where the two fluxes were opposing each other. To maintain the circuit in this "canonical" form (purely for the purposes of the analysis right below), let us simply reverse the presumed positive polarity of coil 2 from the previous example. Effectively, this changes the sign of the off-diagonal terms of Equation (3.122), which are now both positive.

In the equivalent circuit, all variables and parameters have been expressed by "reflecting" or "referring" the electrical quantities of the secondary to the primary side through a mathematical transformation that uses the **turns ratio** N_1/N_2. The transformed quantities are denoted with primes. The equivalent-circuit secondary current (i'_2), voltage (v'_2), and flux linkage (λ'_2) are scaled versions of their actual physical values:

$$i'_2 = \frac{N_2}{N_1} i_2, \; v'_2 = \frac{N_1}{N_2} v_2, \text{ and } \lambda'_2 = \frac{N_1}{N_2} \lambda_2 . \tag{3.124}$$

Furthermore, the primed resistance and inductance are related to their actual physical values by

$$R'_2 = \left(\frac{N_1}{N_2}\right)^2 R_2 \quad \text{and} \quad L'_{l2} = \left(\frac{N_1}{N_2}\right)^2 L_{l2} . \tag{3.125}$$

A reverse transformation can be applied to reflect all primary quantities to the secondary. This choice depends on the problem at hand, which determines which side of the transformer is of interest.

Armed with these definitions, we can rewrite the flux linkage equations as

$$\lambda_1 = L_{11} i_1 + L_{12} i_2 = L_{11} i_1 + L'_{12} i'_2 , \tag{3.126a}$$

$$\lambda'_2 = \frac{N_1}{N_2} (L_{12} i_1 + L_{22} i_2) = L'_{12} i_1 + L'_{22} i'_2 , \tag{3.126b}$$

where we introduced

$$L'_{12} - \frac{N_1}{N_2} L_{12} \quad \text{and} \quad L'_{22} = \left(\frac{N_1}{N_2}\right)^2 L_{22} . \tag{3.127}$$

Our goal is to make a common $(i_1 + i'_2)$ term appear, as suggested by the vertical branch of the equivalent circuit. This is achieved by adding and subtracting a term in each equation, leading to

$$\lambda_1 = (L_{11} - L'_{12})\, i_1 + L'_{12} (i_1 + i'_2) , \tag{3.128a}$$

$$\lambda'_2 = (L'_{22} - L'_{12})\, i'_2 + L'_{12} (i_1 + i'_2) , \tag{3.128b}$$

or

$$\lambda_1 = L_{l1} i_1 + L_m (i_1 + i'_2) , \tag{3.129a}$$

$$\lambda'_2 = L'_{l2} i'_2 + L_m (i_1 + i'_2) , \tag{3.129b}$$

where

$$L_{l1} = L_{11} - L'_{12},\ L'_{l2} = L'_{22} - L'_{12},\ \text{and}\ L_m = L'_{12} \tag{3.130}$$

are the three inductances that appear on the equivalent circuit. The inductance of the vertical branch, L_m, is called the **magnetizing inductance**. The other two inductances, L_{l1} and L'_{l2}, are called **leakage inductances**.

This is sufficient to extract the numerical values of the inductive elements of the equivalent circuit. Once we obtain an inductance matrix, e.g., the one given in Equation (3.122), we can use Equation (3.130) and the definitions of the primed quantities. In our case, this leads to the numerical values

$$L_{l1} = 19.08\ \mu\text{H} \quad \text{and} \quad L'_{l2} = 10.34\ \mu\text{H}\,. \tag{3.131}$$

For a deeper, more physical understanding, one must look at the fields. The main feature of the T-equivalent circuit is that it represents a conceptual separation of the total flux of each coil in two components, namely a **leakage** plus a **magnetizing flux**:

$$\Phi_1 = \Phi_{l1} + \Phi_m \quad \text{and} \quad \Phi_2 = \Phi_{l2} + \Phi_m\,. \tag{3.132}$$

Such a separation is possible because of Gauss' law for the magnetic flux, which leads to the solenoidal nature of the B-field. This law applies to the flux through each individual filament turn, and by extension to the entire coil on a spatial average basis. The magnetizing flux Φ_m is defined as the component of flux that is proportional to

$$N_1 i_1 + N_2 i_2 = N_1(i_1 + i'_2)\,. \tag{3.133}$$

This quantity is the **total magnetomotive force** (MMF), measured in At (ampere-turns). (See also Example 2.15 on page 165, where the MMF was introduced in the context of a single coil.) The total MMF appears on the right-hand side of Ampère's law:

$$\oint_{\partial S} \vec{\mathbf{H}} \cdot d\mathbf{r} = \iint_S \vec{\mathbf{J}} \cdot d\mathbf{a} = I_{\text{enclosed by contour } \partial S} = N_1 i_1 + N_2 i_2\,, \tag{3.134}$$

if the path ∂S is such that it encloses the total current of both coils. All the paths that circle through the core are such paths.[d] Of course, we may also imagine a path that only encloses the current of a single coil; then, the circulation of the H-field would equal the MMF of that particular coil.

Figure 3.15 depicts a condition where the total MMF is zero, or equivalently, where the magnetizing branch current $i_1 + i'_2 = 0$. This is characteristic of what happens when two equal and opposing magnetic poles are brought together, causing flux lines to bend away from each other. It seems as if the flux lines are leaking outside the core. In other words, leakage flux consists of flux lines passing through one of the coils without crossing through, or "linking," the other coil. In the equivalent circuit, the corresponding flux linkages are represented by current flowing through the leakage inductances, L_{l1} and L'_{l2}.

[d] Note that a counterclockwise closed path of integration ∂S defines a normal vector to the surface S coming out of the page. Hence, given how we originally defined the positive polarities of currents in Figure 3.13, the total MMF would be $N_i i_1 - N_2 i_2$. In this example, we have reversed the polarity of coil 2.

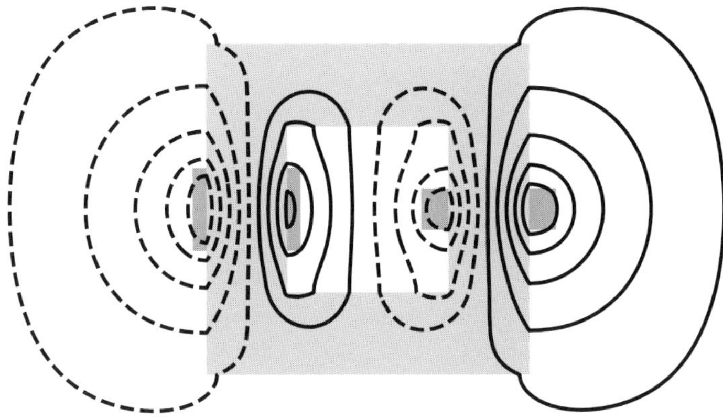

Figure 3.15 Single-phase transformer leakage flux lines ($i_1 = 1$ A, $i_2 = -2$ A, hence total MMF = $N_1 i_1 + N_2 i_2 = 0$).

In contrast, when $i_1 + i_2' \neq 0$, a magnetizing flux component appears. This component of flux travels around the core, and adds on top of the leakage fluxes. Its presence is indicated by flux lines that pass through both coils. In the equivalent circuit, magnetizing flux is associated with the vertical branch inductance L_m. Note that the leakage flux lines travel through the air, which implies that $L_{l1}, L_{l2}' \ll L_m$. Due to this fact, even small deviations of the net MMF from zero cause the picture to change drastically, as illustrated in Figure 3.16. For comparison purposes, the flux lines in all plots are equally spaced apart (by plotting MVP contours spaced by $\Delta A = 4$ μWb/m). It should be emphasized that the flux lines represent the total flux in the device, and should not be interpreted individually as being of the leakage or magnetizing kind. Numerical values of leakage and magnetizing fluxes are provided in Table 3.5. Small deviations from the original balanced MMF condition lead to a rapid increase of the magnetizing flux, whereas the leakage fluxes are not influenced as much.

Table 3.5 Leakage and magnetizing flux values (in μWb).

	$N_1 i_1$	$N_2 i_2$	Φ_{l1}	Φ_{l2}	Φ_m
Figure 3.15	20.00	−20.00	0.9542	−1.0341	0
Figure 3.16(a)	20.02	−20.00	0.9552	−1.0341	0.0366
Figure 3.16(b)	20.06	−20.00	0.9571	−1.0341	0.1097
Figure 3.16(c)	20.18	−20.00	0.9628	−1.0341	0.3291
Figure 3.16(d)	20.38	−20.00	0.9724	−1.0341	0.6948
Figure 3.16(e)	20.66	−20.00	0.9857	−1.0341	1.2067
Figure 3.16(f)	21.02	−20.00	1.0029	−1.0341	1.8650

Formulation in terms of fluxes, permeances, and MMFs. The flux equations can be expressed in terms of actual fluxes (in Wb) instead of flux linkages through a division by the number of turns. The leakage flux of coil 1 is

$$\Phi_{l1} = \frac{L_{l1} i_1}{N_1}. \tag{3.135}$$

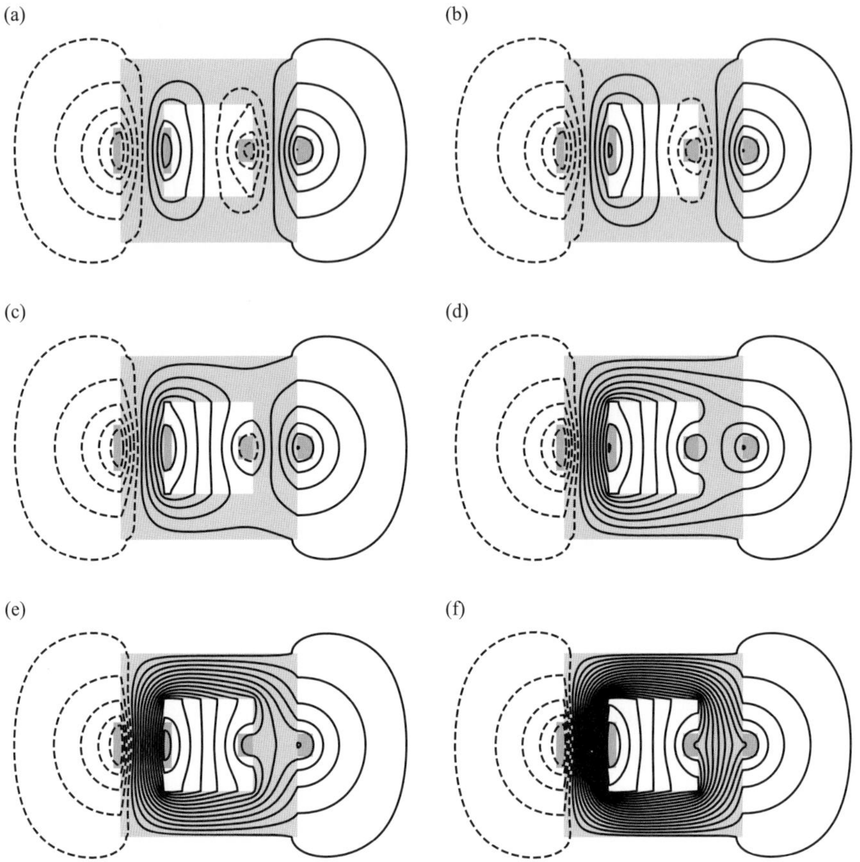

Figure 3.16 Single-phase transformer flux lines for nonzero total MMF: $i_1 = 1.001 + 0.002k^2$ A, $k = 0, 1, \ldots, 5$, and $i_2 = -2$ A.

The leakage flux of coil 2 is

$$\Phi_{l2} = \frac{L'_{l2} i'_2}{N_1} = \frac{L_{l2} i_2}{N_2}. \tag{3.136}$$

The magnetizing flux is

$$\Phi_m = \frac{L_m(i_1 + i'_2)}{N_1} = \frac{L_m}{N_1^2}(N_1 i_1 + N_2 i_2). \tag{3.137}$$

Therefore, we can express the flux linkage vs. current equations as an equivalent set of flux vs. MMF equations, as follows:

$$\Phi_1 = \Phi_{l1} + \Phi_m = \frac{L_{l1}}{N_1^2}(N_1 i_1) + \frac{L_m}{N_1^2}(N_1 i_1 + N_2 i_2), \tag{3.138a}$$

$$\Phi_2 = \Phi_{l2} + \Phi_m = \frac{L_{l2}}{N_2^2}(N_2 i_2) + \frac{L_m}{N_1^2}(N_1 i_1 + N_2 i_2). \tag{3.138b}$$

The L/N^2 terms are called **permeances** $\mathcal{P}$, and they can be measured in Wb/At. These coefficients are independent of the number of turns of each winding, and they encapsulate the effects of device geometry and the materials that are present. Note that the permeances are the inverse reluctances that were introduced in Example 2.15 on page 165. Here, we have expressions with **leakage** and **magnetizing permeances**. For our particular case study, these permeances are

$$\mathcal{P}_{l1} = \frac{L_{l1}}{N_1^2} = 47.7 \text{ nWb/At}, \; \mathcal{P}_{l2} = \frac{L_{l2}}{N_2^2} = \frac{L'_{l2}}{N_1^2} = 51.7 \text{ nWb/At}, \tag{3.139a}$$

$$\mathcal{P}_m = \frac{L_m}{N_1^2} = 1.83 \text{ µWb/At}. \tag{3.139b}$$

Observe that the magnetizing permeance is approximately 35–40 times larger than the leakage permeances. This is consistent with the results of Table 3.5. It signifies that the magnetizing flux is 35–40 times more sensitive to an increase in magnetizing MMF, compared to the sensitivity of the leakage flux of each coil with respect to an increase of its own MMF. Alternatively, we could write this set of equations as

$$\Phi_1 = \mathcal{P}_{11}(N_1 i_1) + \mathcal{P}_{12}(N_2 i_2), \tag{3.140a}$$

$$\Phi_2 = \mathcal{P}_{21}(N_1 i_1) + \mathcal{P}_{22}(N_2 i_2), \tag{3.140b}$$

with $\mathcal{P}_{21} = \mathcal{P}_{12}$, which resemble the flux linkage equations (3.111).

Modeling core loss. The role of the resistance R_c in the equivalent circuit is to represent **hysteresis** and **eddy current** loss in the core, see §2.3.5. These non-conservative effects are not captured by FEA, where the magnetic field is assumed to be lossless. Hence, this element is added to the circuit after the main magnetic coupling between the coils has been determined by an FEA. Its value is typically determined by post-processing the results of the field analysis, or through experimental measurements on a physical device.

The presence of R_c in parallel with the magnetizing inductance in the equivalent circuit can be justified as follows. Suppose that the transformer primary side is connected to an a.c. source of frequency f Hz. Then, this voltage minus a small drop over R_1 and L_{l1} will appear across the magnetizing branch. Even if the material is nonlinear and the currents are not perfectly sinusoidal, the magnetizing branch voltage will be close to sinusoidal. The voltage developed across L_m will be equal to the rate of change of the magnetizing flux linkage:

$$v_m = N_1 \frac{d\Phi_m}{dt}. \tag{3.141}$$

If Φ_m oscillates between $\pm\Phi_{m0}$, the equivalent circuit suggests that the average power dissipated in R_c will be proportional to $f^2\Phi_{m0}^2$. As we have seen in §2.3.5, eddy current loss is proportional to $f^2 B_{\max}^2$, whereas hysteresis loss is proportional to $fB_{\max}^{1.6}$ (per Steinmetz) or $f^a B_{\max}^b$ (more generally). However, the flux density is not uniform within the core, and a more precise calculation would account for the spatial variation of flux density. Since the magnetizing flux dominates, this model of core loss is a reasonable and widely accepted approximation for a circuit-level device model.

Example 3.7 *FEA of an air-gap region.*

The circular air-gap region was the main subject of an extensive study in §2.3.2 on page 122. Nonzero Dirichlet boundary conditions are imposed on the inner and outer boundaries, so an objective of this example is to develop the code for enforcing these conditions, as outlined in §3.2.4. (See Problem 3.15.) More importantly, this example will help us draw certain qualitative conclusions regarding the FEA solution.

Solution

The problem has the following parameters. The inner and outer radii are $r_i = 50$ mm and $r_o = 55$ mm, respectively. A magnetic field with $p = 2$ pole-pairs is created by setting the following inner and outer boundary conditions:

$$A(r_i, \phi) = A_{pi} \cos(p\phi - \phi_{pi}) \,, \quad A(r_o, \phi) = A_{po} \cos p\phi \,, \tag{3.142}$$

with $A_{pi} = A_{po} = 0.02$ Wb/m and $\phi_{pi} = 10°$. (A 10° angular offset may be unrealistic because it leads to enormous torque; from Equation (2.361b), we calculate −3621.9 Nm per unit depth! However, this value is chosen here for visualization purposes.)

The air-gap region is meshed by placing $N_b = 180$ nodes on each boundary, so the angle between consecutive nodes is $\delta\phi = 2\pi/N_b$ rad (2°). All triangles are made approximately equal by imposing a maximum element size $0.5(r_i\delta\phi)^2$. The mesh is not structured, so triangular elements are created in an arbitrary fashion. We thus obtain a mesh with roughly 950 vertices and 1,550 elements. We enforce boundary conditions by fixing the potential of all nodes on the inner and outer boundaries to the values given by Equation (3.142), for $\phi = j\,\delta\phi$, $j = 0, 1, \ldots, N_b-1$. The solution is shown in Figure 3.17. Note how the peaks of the MVP waveform that are located at 0°, ±90°, and 180° (on the outer boundary) coincide with zero flux points. Recall that the radial component of the B-field leads the MVP by 90° electrical, which implies that north poles are located at roughly −45° and 135°, and south poles at 45° and −135°.

Since we have an analytical expression for the solution of this problem (Equation (2.115) on page 132), we can also examine the error between the FEA and the true field

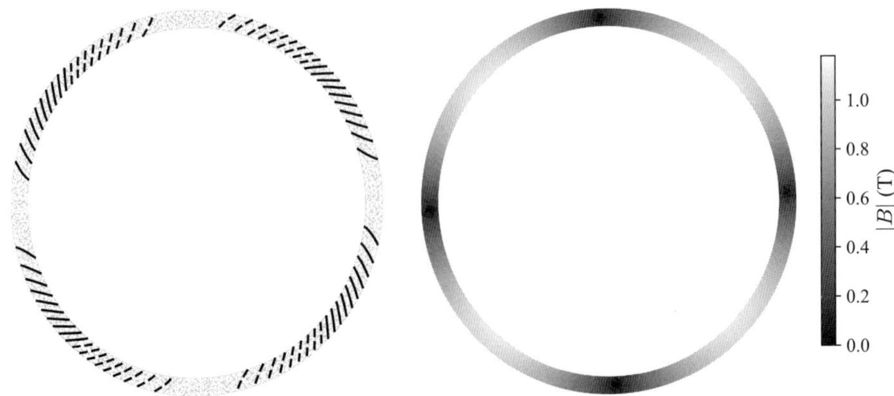

Figure 3.17 Flux lines and B-field magnitude obtained by FEA in an air-gap region.

solution. The MVP error does not appear to have any clearly discernible spatial pattern. Nevertheless, it is interesting to look at the statistical distribution of the MVP error. Figure 3.18(a) shows the error calculated at all internal vertices, whereas Figure 3.18(b) shows the error calculated at all element centroids. (Dirichlet nodes have been excluded so as not to distort the results. Centroids were chosen for no particular reason other than being geometrically well-defined internal element points.) In this case, it is interesting to note that the errors follow more or less a zero-mean Gaussian distribution. It is also interesting to note that the centroid errors are approximately twice as large as the errors at the element vertices (based on the standard deviations σ of the bell curves). The average value of the MVP error magnitude (which follows a half-normal distribution) is $\sigma\sqrt{2/\pi}$, which is approximately 1.7–3.4 μWb/m, or roughly 0.0085–0.017% of the peak value of the MVP at the boundary. This is already a very accurate solution, and we would expect the error to reduce further if we increase the number of elements in the mesh. We have thus reached an important conclusion, which is singled out because it may not have been immediately obvious from the previous discussion; namely, *the FEA solution does not represent an interpolation of the true solution*. The error stems from three sources: (i) from the discretization of the domain; (ii) from the use of piecewise linear functions; and (iii) from the discretization of the boundary, which is no longer a perfect circle but a polygon.

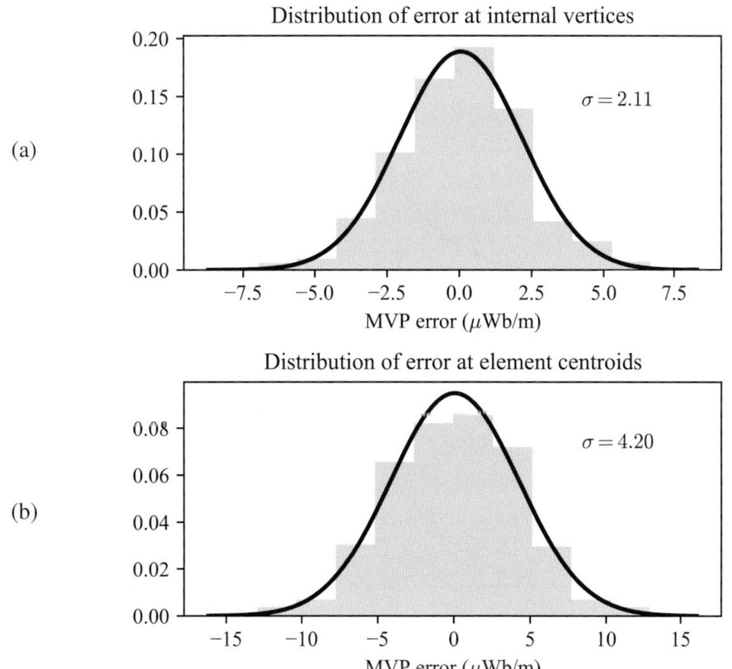

Figure 3.18 Statistics (histograms and probability density functions) of MVP error between FEA and true solution.

Next we examine the error in the B-field. Again, we select the centroids of all elements as observation points, where we decompose the B-field in its radial and tangential components. We also calculate the error in estimating the magnitude of the B-field. The results are given in Figure 3.19. The bell curves are best fits to the data; in this case, the average error is close to zero, but it is apparent that the statistical distributions are not Gaussian, since they exhibit a slight bimodality. Regardless of this, it is clear that the B-field errors are much more significant compared to the MVP errors. This is because we observe errors that amount to roughly 1% of the maximum value of B in the air gap, or two orders of magnitude more than the MVP error. Although we cannot generalize these results to other situations, we should be cognizant of this tendency, as it may affect the calculation of electromagnetic torque in electric machines.

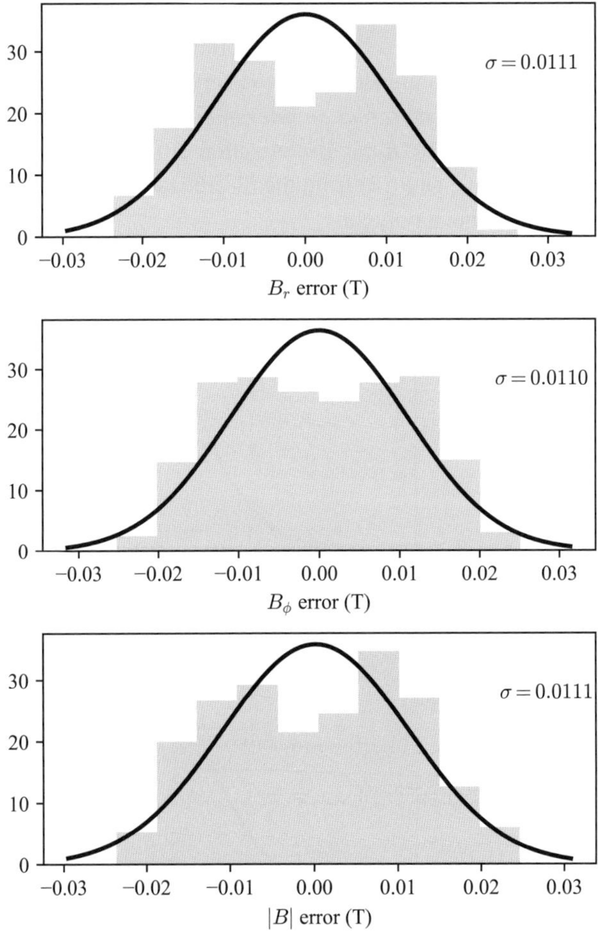

Figure 3.19 Statistics of B-field error between FEA and true solution.

3.5 Solution of Nonlinear Poisson's Equation Using the FEM

In the previous sections, we explained how to analyze magnetically linear devices using the FEM, that is, how to perform FEA. Now we are ready to proceed with the FEA of devices built with isotropic *nonlinear* ferromagnetic materials (devices with permanent magnets will be treated in a subsequent section). Recall that the solution of a 2-D nonlinear magnetostatic problem satisfies the nonlinear Poisson equation (3.48):

$$\frac{\partial}{\partial x}\left(\nu\frac{\partial A}{\partial x}\right)+\frac{\partial}{\partial y}\left(\nu\frac{\partial A}{\partial y}\right)=-J\,. \tag{3.143}$$

We have argued that the solution may be obtained by making the following energy-related functional stationary:

$$I(A)=\iint_{\mathcal{D}} m(x,y,A)\,dx\,dy-\iint_{\mathcal{D}} JA\,dx\,dy\,, \tag{3.144}$$

where

$$m(x,y,A)=\frac{1}{2}\int\limits_{0}^{B^2(A)}\nu(x,y,b^2)\,db^2 \tag{3.145}$$

is the energy density (per unit depth) at position (x, y), which in turn depends on the reluctivity ν of the material occupying that position.

The domain is always meshed so that triangles do not straddle material boundaries. The contribution to the functional from triangle k is

$$I_k(A_k)=\iint_{\Delta_k} m_k(A_k)\,dx\,dy-\iint_{\Delta_k} J_kA_k\,dx\,dy\,, \tag{3.146}$$

where

$$m_k(A_k)=\frac{1}{2}\int\limits_{0}^{B_k^2(A_k)}\nu_k(b^2)\,db^2\,. \tag{3.147}$$

Like before, we assume a linear variation of the solution A_k in each triangle:

$$A_k(x,y,\mathbf{A}^k)=\sum_{i=1}^{3}A_i^k\alpha_i^k(x,y)\,. \tag{3.148}$$

Now Equation (3.96) applies, allowing us to express the upper limit of integration for m_k as a quadratic function of the three local potentials in the triangle and the element stiffness matrix:

$$B_k^2=B_k^2(\mathbf{A}^k)=\frac{1}{\Delta_k}(\mathbf{A}^k)^\top\mathbf{S}^k\mathbf{A}^k\,. \tag{3.149}$$

After performing the integration, the energy density obtains a constant value in each triangle, given by the composite function $m_k(B_k^2(\mathbf{A}^k))$. Therefore, we can express the contribution of the kth element as a function of $\mathbf{A}^k$:

$$I_k(\mathbf{A}^k)=m_k(B_k^2(\mathbf{A}^k))\,\Delta_k-\frac{J_k\Delta_k}{3}\sum_{i=1}^{3}A_i^k\,. \tag{3.150}$$

Stationarity of the functional requires that the partial derivative with respect to each free potential vanishes. For example, with respect to the nth vertex potential (which should not be a Dirichlet node), we have

$$\frac{\partial I}{\partial A_n} = \sum_{k=1}^{N_\Delta} \frac{\partial I_k}{\partial A_n} = 0\,. \tag{3.151}$$

The terms in the sum will be nonzero only for those triangles that contain node n as a vertex. If triangle k is one of these triangles, we calculate its contribution to the partial derivative with respect to the ith vertex potential, $i \in \{1, 2, 3\}$ in the local (disjoint) numbering scheme, corresponding to global node n:

$$\frac{\partial I_k}{\partial A_n} = \frac{\partial I_k}{\partial A_i^k} = \frac{dm_k}{d(B_k^2)} \frac{\partial (B_k^2)}{\partial A_i^k} \Delta_k - \frac{J_k \Delta_k}{3}\,. \tag{3.152}$$

Here, we used the chain rule of differentiation. This leads to

$$\frac{\partial I_k}{\partial A_i^k} = \nu_k(B_k^2(\mathbf{A}^k)) \cdot (\mathbf{S}^k \mathbf{A}^k)_i - \frac{J_k \Delta_k}{3}\,, \tag{3.153}$$

through the fundamental theorem of calculus applied to Equation (3.147), and by expanding and differentiating the terms of the quadratic expression (3.149). The notation $(\mathbf{S}^k \mathbf{A}^k)_i$ means the ith element of a 3×1 vector; this value is obtained by multiplying the ith row of the element stiffness matrix with the element potentials. More concisely, **the contribution to the gradient from the kth triangle** can be expressed as the 3×1 vector

$$\boxed{\mathbf{g}^k = \nu_k \mathbf{S}^k \mathbf{A}^k - \frac{J_k \Delta_k}{3} \begin{bmatrix} 1 & 1 & 1 \end{bmatrix}^\mathsf{T}.} \tag{3.154}$$

Pausing the algebra for a moment, we consider the outcome of adding the contributions from all triangles to the sum in Equation (3.151), and then setting this partial derivative to zero for all nodes. It turns out we can follow the same procedure as we described for the linear magnetics case. We iterate over all triangles in the domain, accounting as we move along for $\nu_k \mathbf{S}^k$ contributions to a sparse $\mathbf{S}$-matrix, and for $J_k \Delta_k / 3$ contributions to a $\mathbf{b}$-vector. The previous Python functions should work without major modification, if any. For instance, the function `build_S_matrix` is ready to go, as it accepts an external vector of reluctivities for each triangular element, which we can update accordingly before calling it. We then apply boundary conditions as explained in §3.2.4, e.g., see Equation (3.89) on page 237. Eventually, we will obtain a system of equations of the form

$$\mathbf{S}(\mathbf{A}) \cdot \mathbf{A} = \mathbf{b}\,. \tag{3.155}$$

The nth row corresponds to $\partial I / \partial A_n = 0$, unless this is a fixed node, in which case it states trivially that $A_n = A_{nf}$. The number of equations will be equal to the number of unknowns, so in principle, we should be able to solve this system.

Nevertheless, $\mathbf{S}(\mathbf{A})$ is a matrix whose coefficients depend on the potentials because of the nonlinear reluctivities. Even though this looks similar to the equation for the linear case (3.78), it is a nonlinear system of equations because $\mathbf{S}$ is no longer constant.

Therefore, it can only be solved using iterative algorithms. Two such algorithms will be outlined next, namely, the **fixed-point method** and the **Newton–Raphson method**. The fixed-point method is easy to implement but suffers from two main disadvantages: convergence to a solution is not guaranteed, and convergence is typically slow. The Newton–Raphson[64] method is more difficult to implement but has rapid convergence.

3.5.1 Fixed-Point Method

In mathematics, a **fixed point** of a function f is a point x^* that is mapped to itself, i.e., a point that satisfies $x^* = f(x^*)$. The fixed-point method attempts to compute such a point recursively, by generating a sequence x, $f(x)$, $f(f(x))$, $f(f(f(x)))$, ... The sequence may converge or not to a fixed point, depending on the function.

From Equation (3.155), we observe that the FEA solution is indeed a fixed point, because it satisfies

$$\mathbf{A} = f(\mathbf{A}) = \mathbf{S}(\mathbf{A})^{-1}\mathbf{b}\,. \tag{3.156}$$

The fixed-point method yields a sequence of solutions, which will be denoted by a superscript within parentheses, $\mathbf{A}^{(1)}$, $\mathbf{A}^{(2)}$, $\mathbf{A}^{(3)}$, ... Within each iteration, the reluctivities are all constant; for elements in magnetically nonlinear regions, the reluctivities depend upon the B-field, which in turn depends upon the latest vertex potentials. Hence, the reluctivity of the kth triangle in the ith iteration is

$$\nu_k^{(i)} = \nu_k(B_k^2(\mathbf{A}^{(i)}))\,. \tag{3.157}$$

The main steps of the fixed-point algorithm are outlined in Box 3.3.

Box 3.3 Programming the FEM: The fixed-point algorithm

1. Set iteration index $i := 1$.
2. Set initial guess $\mathbf{A}^{(1)}$.
3. Calculate $B_k^2(\mathbf{A}^{(i)})$ in each element, and look up the reluctivity $\nu_k^{(i)}$.
4. Build the stiffness matrix $\mathbf{S}^{(i)} = \mathbf{S}(\mathbf{A}^{(i)})$.
5. Calculate a residual r, see Equation (3.158).
6. **If** r is small enough (i.e., less than $\epsilon > 0$), **quit**; **else**:
7. Solve $\mathbf{S}^{(i)}\mathbf{x} = \mathbf{b}$.
8. Update $\mathbf{A}^{(i+1)} := (1 - \xi)\mathbf{A}^{(i)} + \xi\mathbf{x}$.
9. Increment iteration counter, $i := i + 1$.
10. **If** $i > N_{\text{iter}}$, **stop**; **else repeat** from Step 3.

In Step 2, the algorithm is initialized with a guess $\mathbf{A}^{(1)}$. For example, the simplest choice would be to set all potentials to zero. Alternatively, we could establish the solution of a magnetically linear system using fixed values for the reluctivities of nonlinear materials.

The stopping criterion of Step 6 is based on a normalized residual

$$r = \frac{||\mathbf{S}^{(i)}\mathbf{A}^{(i)} - \mathbf{b}||}{||\mathbf{b}||}. \tag{3.158}$$

This quantity determines how far we are from satisfying Equation (3.155). The user defines a stopping criterion $\epsilon > 0$. The normalization with respect to the vector norm, $||\mathbf{b}||$, makes this criterion insensitive to the operating point (current level) in the device.

In Step 8 we have introduced a **relaxation factor** $\xi \in (0, 1]$. The new iterate is a convex combination of the previous iterate and the vector $\mathbf{x}$ suggested by the pure fixed-point iteration. The relaxation factor determines the relative weights of the two components. If ξ is close to zero, then we are being cautious as we are not updating the solution by much each time. Setting ξ closer to 1 means we are being more aggressive. Depending on the problem at hand, the value of ξ could have a significant impact on whether the algorithm converges or not, as well as on the rate of convergence. A proper value can be determined by trial and error.

Lastly, the stopping criterion of Step 10 ensures that the algorithm terminates after a given maximum number of iterations, N_{iter}.

3.5.2 Newton–Raphson Method

The Newton–Rapshon method, also known simply as **Newton's method**, is a *root-finding algorithm*. Here it calculates the solution of the functional stationarity conditions (3.155), written as

$$\mathbf{g}(\mathbf{A}) = \mathbf{S}(\mathbf{A}) \cdot \mathbf{A} - \mathbf{b} = \mathbf{0}, \tag{3.159}$$

where $\mathbf{g} : \mathbb{R}^N \to \mathbb{R}^N$ is a vector-valued function of the MVP solution vector.

Recall that $\mathbf{g}$ contains first derivatives of the functional, $\partial I/\partial A_n$, so it is the **gradient** of $I(\mathbf{A})$. More precisely, $\mathbf{g}$ is a *modified* gradient because it contains the derivatives only with respect to the free nodes. For any fixed node j, the corresponding equation is replaced by $A_j - A_{jf} = 0$. For instance, consider a contrived example with three nodes, where the second one is fixed at potential A_{2f}. We need to solve

$$\mathbf{g}(\mathbf{A}) = \begin{bmatrix} S_{11} & S_{12} & S_{13} \\ 0 & 1 & 0 \\ S_{31} & S_{32} & S_{33} \end{bmatrix} \begin{bmatrix} A_1 \\ A_2 \\ A_3 \end{bmatrix} - \begin{bmatrix} b_1 \\ A_{2f} \\ b_3 \end{bmatrix} = \begin{bmatrix} 0 \\ 0 \\ 0 \end{bmatrix}, \tag{3.160}$$

where the remaining elements of the stiffness matrix, e.g., S_{11}, S_{12}, etc, are nonlinear functions of $\mathbf{A} = (A_1, A_2, A_3)$.

The norm of the gradient can serve as a stopping criterion for the algorithm. The normalized residual

$$r = \frac{||\mathbf{g}(\mathbf{A}^{(i)})||}{||\mathbf{b}||} \tag{3.161}$$

is a measure of how close we are to satisfying the stationarity of the functional. It is normalized to reduce dependence on operating point (i.e., current magnitude). It was written in this form, rather than as in Equation (3.158), because we will not be assembling explicitly a stiffness matrix $\mathbf{S}^{(i)}$. Instead, we will establish the gradient by adding the contributions of Equation (3.154) from each triangle (and then applying boundary conditions).

In the Newton–Raphson method, we proceed with a first-order N-dimensional Taylor expansion of the modified gradient $\mathbf{g}$ around an arbitrary MVP vector $\mathbf{A}$, which we set equal to zero:

$$\mathbf{g}(\mathbf{A} + \mathbf{x}) \approx \mathbf{g}(\mathbf{A}) + \mathbf{H}(\mathbf{A})\,\mathbf{x} = \mathbf{0}\,. \tag{3.162}$$

Solving (3.162) for $\mathbf{x}$ leads to a new estimate of the solution. In other words, the vector

$$\mathbf{x} = -\,[\mathbf{H}(\mathbf{A})]^{-1}\,\mathbf{g}(\mathbf{A}) \tag{3.163}$$

is the approximate step we should take from $\mathbf{A}$ to reach the root. The method converges very rapidly when point $\mathbf{A}$ is close to a root, where we expect $\mathbf{g}$ to be nearly linear. A 1-D illustration of Newton's method as a root-finding algorithm is provided in Figure 3.20.

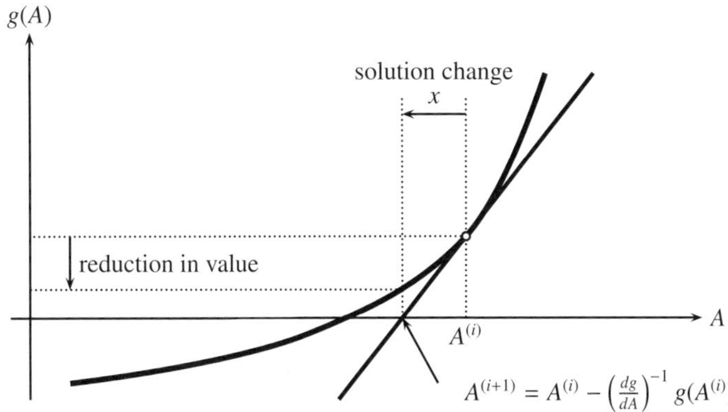

Figure 3.20 Illustration of Newton's method as a root-finding algorithm on a scalar function $g(A)$.

In N dimensions:

$$\mathbf{H}(\mathbf{A}) = \begin{bmatrix} (\nabla g_1)^\top \\ \vdots \\ (\nabla g_N)^\top \end{bmatrix} = \begin{bmatrix} \frac{\partial^2 I}{\partial A_1^2} & \frac{\partial^2 I}{\partial A_1 \partial A_2} & \cdots & \frac{\partial^2 I}{\partial A_1 \partial A_N} \\ \vdots & \vdots & & \vdots \\ \frac{\partial^2 I}{\partial A_N \partial A_1} & \frac{\partial^2 I}{\partial A_N \partial A_2} & \cdots & \frac{\partial^2 I}{\partial A_N^2} \end{bmatrix} \tag{3.164}$$

is a square, symmetric $N \times N$ matrix that contains the first derivatives of the (modified) gradient (i.e., the second derivatives of the functional, except at those rows that correspond to fixed potential values), evaluated at $\mathbf{A}$. This is called a **Hessian matrix**.[65] With a slight abuse of notation we are using the same symbol for the Hessian as for the magnetic field, but it should be clear that this is something different. The Hessian is also the Jacobian matrix of the gradient (see Example 1.24).

The elements of the Hessian can be calculated by summing individual triangle contributions, which in turn are obtained by differentiating Equation (3.153). For the kth triangle, we obtain

$$\frac{\partial^2 I_k}{\partial A_i^k \partial A_j^k} = \frac{\partial}{\partial A_j^k}\left(\nu_k(B_k^2(\mathbf{A}^k))\right) \cdot (\mathbf{S}^k\mathbf{A}^k)_i + \nu_k(B_k^2(\mathbf{A}^k)) \cdot S_{ij}^k. \tag{3.165}$$

Applying the chain rule yields

$$\frac{\partial}{\partial A_j^k}\left(\nu_k(B_k^2(\mathbf{A}^k))\right) = \frac{d\nu_k}{d(B_k^2)} \cdot \frac{\partial(B_k^2(\mathbf{A}^k))}{\partial A_j^k} = \frac{d\nu_k}{d(B_k^2)} \cdot \frac{2}{\Delta_k}(\mathbf{S}^k\mathbf{A}^k)_j. \tag{3.166}$$

Hence, the kth triangle contributes nine terms to the Hessian, for $i, j \in \{1, 2, 3\}$, namely

$$\frac{\partial^2 I_k}{\partial A_i^k \partial A_j^k} = \frac{d\nu_k}{d(B_k^2)} \cdot \frac{2}{\Delta_k}(\mathbf{S}^k\mathbf{A}^k)_i(\mathbf{S}^k\mathbf{A}^k)_j + \nu_k(B_k^2(\mathbf{A}^k)) \cdot S_{ij}^k. \tag{3.167}$$

The contribution from the kth triangle to the Hessian is thus a 3×3 matrix

$$\boxed{\mathbf{H}^k = \frac{2}{\Delta_k}\frac{d\nu_k}{d(B_k^2)}(\mathbf{S}^k\mathbf{A}^k)(\mathbf{S}^k\mathbf{A}^k)^\top + \nu_k\mathbf{S}^k.} \tag{3.168}$$

Enforcing Boundary Conditions

We need to ensure that our algorithm properly accounts for the presence of boundary nodes with fixed potential. For instance, based on the contrived example that was introduced earlier, where the (modified) gradient is

$$\mathbf{g}(\mathbf{A}) = \begin{bmatrix} S_{11} & S_{12} & S_{13} \\ 0 & 1 & 0 \\ S_{31} & S_{32} & S_{33} \end{bmatrix} \begin{bmatrix} A_1 \\ A_2 \\ A_3 \end{bmatrix} - \begin{bmatrix} b_1 \\ A_{2f} \\ b_3 \end{bmatrix}, \tag{3.169}$$

the Hessian would have the following structure:

$$\mathbf{H} = \begin{bmatrix} \times & \times & \times \\ 0 & 1 & 0 \\ \times & \times & \times \end{bmatrix}, \tag{3.170}$$

where "×" denotes elements of arbitrary value. In particular, the second row is obtained as

$$(\nabla g_2)^\top = \begin{bmatrix} \frac{\partial g_2}{\partial A_1} & \frac{\partial g_2}{\partial A_2} & \frac{\partial g_2}{\partial A_3} \end{bmatrix} = \begin{bmatrix} 0 & 1 & 0 \end{bmatrix}, \tag{3.171}$$

since $g_2(\mathbf{A}) = A_2 - A_{2f}$.

Suppose that at the start of the algorithm we initialize the MVP of the second node at its prescribed value, i.e., setting $A_2^{(1)} = A_{2f}$. At this point, the second element of the gradient, $g_2^{(1)} = A_2^{(1)} - A_{2f}$, will evaluate to zero. Hence, the constraint $x_2 = -g_2^{(1)} = 0$ will be imposed from the second equation of $\mathbf{Hx} = -\mathbf{g}$. At the next iteration, the potential of the fixed node will remain the same since $A_2^{(2)} = A_2^{(1)} + x_2 = A_2^{(1)} = A_{2f}$; thus the second element of the gradient, $g_2^{(2)} = A_2^{(2)} - A_{2f}$, will again be zero. Therefore, the algorithm will not change this potential moving forward, keeping it always at its fixed value. In other words, the vector $\mathbf{x}$, which bears the meaning of a change in the solution, will always have a zero as a second element. This implies that we can set to zero all elements of the second column of $\mathbf{H}$ because they are irrelevant, while keeping a value of 1 on the diagonal. The validity of $\mathbf{Hx} = -\mathbf{g}$ is maintained with this manipulation, where $\mathbf{H}$ is the *modified* Hessian obtained after zeroing out the columns related to the Dirichlet nodes.[e] The Hessian is thus made symmetric, thereby allowing for a more efficient numerical solution of the linear system of equations for $\mathbf{x}$.

In summary, enforcing boundary conditions in Newton's method consists of: (i) initializing Dirichlet nodes with their given MVP values; and (ii) modifying the Hessian and gradient, which will have the structure

$$\mathbf{H} = \begin{bmatrix} \times & 0 & \times \\ 0 & 1 & 0 \\ \times & 0 & \times \end{bmatrix} \text{ and } \mathbf{g} = \begin{bmatrix} \times \\ 0 \\ \times \end{bmatrix}. \tag{3.172}$$

Of course, this logic can be extended to an arbitrary number of Dirichlet nodes.

Interpretation as a Minimizing Algorithm

An alternative interpretation of the Newton–Raphson method is that of a *minimizing algorithm*. We approximate locally the functional by a second-order Taylor expansion:

$$I(\mathbf{A} + \mathbf{x}) \approx I(\mathbf{A}) + \mathbf{g}^{\top}\mathbf{x} + \frac{1}{2}\mathbf{x}^{\top}\mathbf{Hx}, \tag{3.173}$$

where the (modified) gradient and Hessian are evaluated at $\mathbf{A}$, and $\mathbf{x}$ represents a change in the MVP that respects boundary conditions. If the Hessian is **positive definite**, i.e., $\mathbf{x}^{\top}\mathbf{Hx} > 0$ for arbitrary $\mathbf{x} \in \mathbb{R}^N, \mathbf{x} \neq \mathbf{0}$, the quadratic approximation is **convex**. Hence, the unique minimizer of the quadratic form is found by setting its gradient to zero, leading to Equation (3.162). In other words, if I was a quadratic function, the vector $\mathbf{x}$ that satisfies $\mathbf{Hx} = -\mathbf{g}$ would be the minimizer of I. This is exactly what happens in the case of linear magnetics. A 1-D illustration is shown in Figure 3.21.

[e] Strictly speaking, we should denote the modified Hessian by a different symbol, e.g., $\mathbf{H}'$. However, we will maintain the original notation $\mathbf{H}$ for the sake of simplicity.

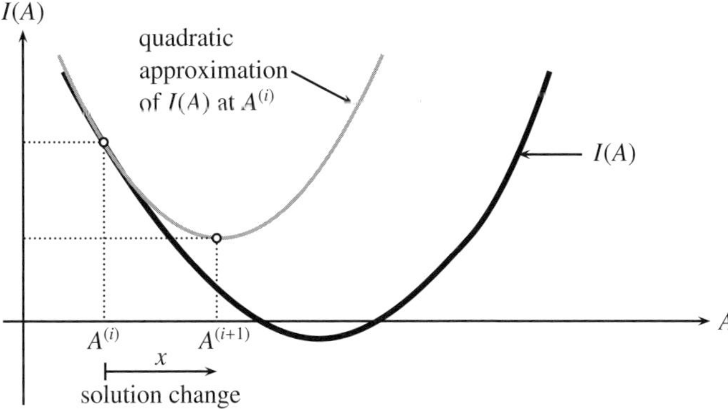

Figure 3.21 Illustration of Newton's method as a minimization algorithm on a convex, scalar function $I(A)$. Here, $x = -(d^2I/dA^2)^{-1} \cdot (dI/dA)$ is the full Newton step that minimizes the quadratic approximation of I at $A^{(i)}$.

In nonlinear problems, the functional is not quadratic, so it cannot be minimized in a single step. Therefore, in the context of a minimizing algorithm, the vector $\mathbf{x}$ should be interpreted as a **descent direction**. This means that if we take a small step in the direction of $\mathbf{x}$, say $t\mathbf{x}$, with $0 < t \ll 1$,[f] we will obtain values of I that are lower than the current value, $I(\mathbf{A})$. Using a first-order Taylor expansion, we see that the functional obtains the value

$$I(\mathbf{A} + t\mathbf{x}) \approx I(\mathbf{A}) + t\mathbf{g}^\top \mathbf{x} = I(\mathbf{A}) - t\mathbf{x}^\top \mathbf{H}\mathbf{x}, \tag{3.174}$$

which is less than $I(\mathbf{A})$ if $\mathbf{H}$ is positive definite, which in turn occurs if the function is locally convex. The algorithm repeats by re-calculating the gradient and Hessian at each point, and then updates the solution so that the functional value is reduced each time. Newton's algorithm belongs to the family of **descent methods** in optimization. The scaling parameter t is often called the **step size** or **step length**. The main steps of the Newton–Raphson algorithm are outlined in Box 3.4.

Box 3.4 Programming the FEM: The Newton–Raphson algorithm

1. Set iteration index $i := 1$.
2. Set initial guess $\mathbf{A}^{(1)}$.
3. Calculate $B_k^2(\mathbf{A}^{(i)})$ in each element, and look up $\nu_k^{(i)}$ and $d\nu_k^{(i)}/d(B_k^2)$.
4. Build the gradient $\mathbf{g}^{(i)}$ and the Hessian $\mathbf{H}^{(i)}$; enforce boundary conditions.
5. Calculate a residual r, see Equation (3.161).
6. **If** r is small enough (i.e., less than $\epsilon > 0$), **quit**; **else**:
7. Solve $\mathbf{H}^{(i)}\mathbf{x} = -\mathbf{g}^{(i)}$.
8. Update $\mathbf{A}^{(i+1)} := \mathbf{A}^{(i)} + t^{(i)}\mathbf{x}$.
9. Increment iteration counter, $i := i + 1$.
10. **If** $i > N_{\text{iter}}$, **stop**; **else repeat** from Step 3.

In Step 2, the algorithm is initialized with a guess $\mathbf{A}^{(1)}$. Similar to the fixed-point method, the simplest choice would be to set all potentials to zero, or we could use the solution of a magnetically linear system. Yet another option is to use a previous nonlinear solution from a nearby operating point.

In Step 4, we calculate both the gradient and the Hessian. Strictly speaking, the convergence decision that is made in Step 6 does not require the Hessian, so this calculation may be performed separately (i.e., we could calculate the Hessian only if the residual is found to be larger than the threshold), to gain a small computational advantage.

In Step 8, we introduced a variable step length $t^{(i)}$. For simplicity, one could use a constant t, $0 < t \leq 1$ (same for all i). Smaller values mean that we are not updating the solution very aggressively; this could help stabilize the algorithm, but convergence will be slow. However, choosing a constant t that universally yields good performance could require trial and error. A good choice may be elusive because of dependence on

[f] Here, the parameter t should not be confused with the symbol for time.

operating point. Alternatively, we can devise algorithms for dynamically selecting a $t^{(i)}$ at each iteration. For example, we could search for the step length that minimizes the function along the descent direction. In optimization theory, this is called an **exact line search**. The main drawback of this approach is the large number of iterations it would take to ensure that we have indeed found the smallest function value, which can be computationally inefficient (since each function evaluation requires the solution of a nonlinear FEA, which is computationally costly).

Another approach is to perform an **inexact line search** for a step length $t^{(i)}$ that reduces the function value sufficiently. One such algorithm is the **backtracking line search method** [1]. In this algorithm, we first take a full step in the descent direction suggested by Newton's method. If the reduction in the function value is acceptable, we stop, otherwise we keep reducing the step length (i.e., we backtrack) until we are satisfied. This process is guaranteed to terminate once the step becomes small enough, since we are searching along a descent direction (assuming convexity). The backtracking line search method requires two parameters, namely, $\alpha \in (0, 0.5)$ and $\beta \in (0, 1)$. The parameter α controls the level of function-value reduction that we consider as acceptable, whereas β controls the backtracking. The algorithm is outlined in Box 3.5.

Box 3.5 Programming the FEM: The backtracking line search algorithm

1. Set $t := 1$.
2. **While** $I(\mathbf{A} + t\mathbf{x}) > I(\mathbf{A}) + \alpha t\, \mathbf{g}^\top \mathbf{x}$, $t := \beta t$.

Impact of *B–H* Curve on Convergence

In general, the convergence of the Newton–Raphson algorithm depends on the function that is being minimized and the initial solution guess. For instance, if a function has multiple local minima, the method may converge to one of these, thus failing to locate the global minimum. Nevertheless, even if a function has a single minimum, Newton's algorithm may still fail to converge; and this is occasionally the case when conducting FEA of electric machines.[g] If we encounter such an issue in practice, we can resolve it by adjusting slightly the *B–H* curve of the materials in the device in order to ensure convergence.

There is a special class of functions where convergence of Newton's algorithm is guaranteed, namely, **strictly convex** functions. A strictly convex function $f : \mathbb{R}^N \to \mathbb{R}$ has the property that

$$f(\theta \mathbf{x}_1 + (1 - \theta)\mathbf{x}_2) < \theta f(\mathbf{x}_1) + (1 - \theta) f(\mathbf{x}_2), \tag{3.175}$$

for any $\theta \in [0, 1]$, and for arbitrary, distinct $\mathbf{x}_1, \mathbf{x}_2 \in \mathbb{R}^N$. A **convex** function (i.e., not strictly convex), satisfies the above equation with a less-than-or-equal-to inequality. In words, a convex function lies below the linear segment connecting any two points in its

[g] Whether the nonlinear magnetostatic functional could actually have multiple stationary points is a question that we will not attempt to answer in this text.

domain. Simply speaking, a convex function has the shape of a bowl, and this is why it has a single minimum point. Furthermore, if a function is twice differentiable, then it is strictly convex if and only if its Hessian is positive definite. Newton's method is guaranteed to converge to the unique minimum of strictly convex functions.

As far as our problem is concerned, we claim that: *if the device consists of materials whose B–H curves satisfy the condition $d\nu/d(B^2) \geq 0$, then convergence of the Newton–Raphson algorithm is guaranteed.*

Proof A first proof of algorithm convergence is based on the fundamental definition of convexity in Equation (3.175). For notational simplicity, let us define a function $g_k : \mathbb{R}^N \to \mathbb{R}$ as $g_k(\mathbf{A}) = B_k^2(\mathbf{A}^k)$. (This function depends only on three elements out of the entire MVP vector.) From Equation (3.149), it is clear that the element stiffness matrix $\mathbf{S}^k$ is **positive semi-definite** since $B_k^2 \geq 0$. This means that g_k is convex, or that

$$g_k(\theta\mathbf{A}_1 + (1-\theta)\mathbf{A}_2) \leq \theta g_k(\mathbf{A}_1) + (1-\theta)g_k(\mathbf{A}_2)\,, \tag{3.176}$$

for any $\theta \in [0,1]$, and for arbitrary $\mathbf{A}_1, \mathbf{A}_2 \in \mathbb{R}^N$.

Now, let us check the convexity of the energy density function $m_k = m_k(B_k^2(\mathbf{A}^k))$. This function may be thought of as a **composition** $m_k = f_k \circ g_k : \mathbb{R}^N \to \mathbb{R}$, where we introduced a function $f_k : \mathbb{R} \to \mathbb{R}$ defined by $f_k(x) = 0.5 \int_0^x \nu_k(\xi)\,d\xi$. In other words, f_k takes $x = B_k^2$ as its argument, whereas m_k takes the MVP $\mathbf{A}^k$ as its argument; however, they both represent the same quantity, energy density. From physical considerations, the function f_k is a **strictly increasing** function. Hence, based on the convexity of g_k, Equation (3.176), we have that

$$f_k(g_k(\theta\mathbf{A}_1 + (1-\theta)\mathbf{A}_2)) \leq f_k(\theta g_k(\mathbf{A}_1) + (1-\theta)g_k(\mathbf{A}_2))\,. \tag{3.177}$$

To proceed, we check the convexity of f_k. To this end, take the first derivative, $f_k'(x) = 0.5\,\nu_k(x)$; then, take the second derivative, $f_k''(x) = 0.5\,d\nu_k/dx = 0.5\,d\nu_k/d(B_k^2)$. When our assumption regarding the B–H curve holds, then $f_k'' \geq 0$, and f_k is convex! This implies that

$$f_k(\theta g_k(\mathbf{A}_1) + (1-\theta)g_k(\mathbf{A}_2)) \leq \theta f_k(g_k(\mathbf{A}_1)) + (1-\theta)f_k(g_k(\mathbf{A}_2))\,. \tag{3.178}$$

Combining Equations (3.177) and (3.178), we obtain

$$f_k(g_k(\theta\mathbf{A}_1 + (1-\theta)\mathbf{A}_2)) \leq \theta f_k(g_k(\mathbf{A}_1)) + (1-\theta)f_k(g_k(\mathbf{A}_2))\,. \tag{3.179}$$

This means that the composite function $m_k = f_k \circ g_k$ is also convex.

Therefore, the $I_k(\mathbf{A})$ contributions to the functional, given by Equation (3.150):

$$I_k(\mathbf{A}^k) = m_k(B_k^2(\mathbf{A}^k))\Delta_k - \frac{J_k\Delta_k}{3}\sum_{i=1}^{3} A_i^k\,, \tag{3.180}$$

are all convex functions of $\mathbf{A}$. (This is because the linear terms do not affect convexity.) The functional $I(\mathbf{A})$ consists of a sum of such terms, so it is also a convex function of $\mathbf{A}$. □

Proof A second proof of algorithm convergence is based on the positive definiteness of the Hessian. Recall that contributions to the Hessian from each element are given by Equation (3.168), written here as

$$\mathbf{H}^k = \frac{2}{\Delta_k}\frac{d\nu_k}{d(B_k^2)}\mathbf{y}\mathbf{y}^\top + \nu_k\mathbf{S}^k\,, \tag{3.181}$$

where $\mathbf{y} = \mathbf{S}^k\mathbf{A}^k$. The positive definiteness of $\mathbf{H}^k$ can be checked by evaluating the sign of $\mathbf{w}^\top\mathbf{H}^k\mathbf{w}$ for arbitrary $\mathbf{w} \in \mathbb{R}^3$. We have

$$\mathbf{w}^\top\mathbf{H}^k\mathbf{w} = \frac{2}{\Delta_k}\frac{d\nu_k}{d(B_k^2)}(\mathbf{w}^\top\mathbf{y})^2 + \nu_k\mathbf{w}^\top\mathbf{S}^k\mathbf{w}\,. \tag{3.182}$$

Clearly, since $\mathbf{S}^k$ is positive semi-definite, the condition $d\nu_k/d(B_k^2) \geq 0$ suffices for $\mathbf{H}^k$ to be positive semi-definite. Therefore, the global Hessian matrix, $\mathbf{H}$, will also be positive semi-definite. In particular, when boundary conditions are applied, the modified Hessian matrix should become positive definite, in which case the functional $I(\mathbf{A})$ would be strictly convex. □

3.6 Building Algorithms for Nonlinear Magnetics

In this section, we will set forth details needed to implement an FEA program with nonlinear ferromagnetic materials. We begin by explaining how nonlinear *B*–*H* curves can be modeled, and then proceed with building the gradient vector and Hessian matrix.

3.6.1 Modeling Nonlinear Materials

A device may contain one or more nonlinear materials, e.g., different steel types. One of the things that need to take place at the start of an FEA program is loading the anhysteretic *B*–*H* characteristic of each material in the memory. Typically, these curves will be provided as discrete data points; for instance, as (H_1, B_1), (H_2, B_2), ... From these, we then need to generate the reluctivity function(s), $\nu(B^2)$. The following function, `load_BH_curve_pchip`, serves this purpose: it loads *B*–*H* data points from a text file, and then creates a reluctivity function that *interpolates* (i.e., passes exactly through) the data.

```
import numpy as np
import scipy.interpolate as spi

def load_BH_curve_pchip(fname='hb.txt'):
    hb = np.loadtxt(fname)
    Hdat = hb[:,0]
    Bdat = hb[:,1]
    nudat = np.divide(Hdat,Bdat)

    Hmax = Hdat[-1]
    Bmax = Bdat[-1]
    B1 = Bdat[0]
    H1 = Hdat[0]
    m1 = 0.5*B1*H1 # field energy density at H1(B1)
```

```
15      assert B1 > 0, \
            'First B point of BH curve is not strictly positive'

        Bsqdat = Bdat**2

20      nuB2_intrpl = spi.PchipInterpolator(Bsqdat,nudat)
        dnudB2_intrpl = nuB2_intrpl.derivative()
        int_nu_intrpl = nuB2_intrpl.antiderivative()

        return nuB2_intrpl, dnudB2_intrpl, int_nu_intrpl,\
25          Bmax, B1, m1
```

The input file should contain two columns, where the first column lists H-field, and the second column lists B-field, in ascending order. The first data point in the file should be a strictly positive $B_1 > 0$; otherwise, the program throws an exception in line 15. It is assumed that the B–H curve is linear for $0 \le B \le B_1$. The final data point is $(H_{\max}, B_{\max})$. In the absence of further information, what happens after this value is up to the user. There are numerous ways we could extrapolate to the complete saturation of the material.

The function returns a **piecewise cubic Hermite**[66] **interpolating polynomial** (pchip) for $\nu(B^2)$. It also outputs the derivative, $d\nu/d(B^2)$, and antiderivative, $\int \nu(b^2)\, db^2$, of the interpolant. These three functions are based on the SciPy PchipInterpolator class. A piecewise cubic interpolant is a function that consists of a family of cubic polynomials, i.e., functions of the form

$$p_j(x) = c_{0j} + c_{1j}(x - x_j) + c_{2j}(x - x_j)^2 + c_{3j}(x - x_j)^3 , \tag{3.183}$$

one for each segment j of the curve, defined by consecutive data points, $[x_j, x_{j+1}]$, $j = 1, 2, \ldots, J$. The pchip library internally determines the coefficients of the polynomials so that the following conditions are met: (i) the curve interpolates the given points; (ii) the curve is smooth (in particular, at the nodes between two consecutive segments), so the first derivative is continuous (however, the second derivative may jump between segments); (iii) the shape of the data (their monotonicity) is preserved.

A different approach for modeling a magnetization curve would be to *fit* a given function to B–H data. Of course, curve fitting is different than interpolating, as it does not necessarily guarantee that the curve would pass exactly through the data, but it would find the function that is "closest" to the data. This would be advantageous if the data is noisy. However, coming up with a functional expression of the nonlinear magnetization curve may be tricky.

Visualizing a Magnetization Curve

The visualization of various aspects of the B–H curve is an important initial step in the process of developing a nonlinear FEA program. Obtaining a satisfactory functional representation of $\nu(B^2)$ can be challenging due to the nonlinearity of the curve and the wide range of the reluctivity. Here we describe how one could inspect the various magnetization functions for validity. The B–H data are loaded using `load_BH_curve_pchip`. All curves are plotted based on pchip representations of $\nu(B^2)$, its derivative, and its antiderivative.

We are provided with the following B–H curve data of a 0.025-in thick non-oriented silicon steel lamination, where the first column is the H-field in A/m, and the second column is the B-field in T.

```
41.4   0.2
56.5   0.4
79.8   0.7
119    1
175    1.2
234    1.3
370    1.4
777    1.5
1283   1.55
2108   1.6
3256   1.65
4727   1.7
8721   1.8
14888 1.9
26023 2.0
65521 2.1
```

To evaluate the pchip functions, we use the following syntax, e.g., for the derivative of the reluctivity:

```
x = np.linspace(Bsqdat[0],Bsqdat[-1],100)
y = dnudB2_intrpl.__call__(x)
```

In particular, the magnetic energy density is calculated as follows:

```
m = m1 + 0.5*int_nu_intrpl.__call__(x)
```

Here, `m1` is the energy density from the linear part of the magnetization curve, calculated in line 14 of `load_BH_curve_pchip`.

The B–H data points and the corresponding $B(H)$ interpolating curve are shown as the first subplot of Figure 3.22. This is usually plotted on a logarithmic scale for the H-field to highlight the behavior at low field values, which is not visible when plotted on a linear scale. The pchip interpolant of the reluctivity $\nu = H/B$ is plotted vs. B^2 in the second subplot of Figure 3.22.

The absolute value of the derivative of the reluctivity, $|\nu'| = |d\nu/d(B^2)|$, which consists of a family of second-order polynomials, is shown in the third subplot of Figure 3.22. In this case, ν' initially obtains negative values. It crosses zero at the third data point ($B_3^2 = 0.7^2 = 0.49$) that corresponds to minimum reluctivity, after which point it remains positive. Due to the logarithmic scale, the zero value corresponds to where the curve dips. The curve is continuous, but discontinuities in its derivative (i.e., the second derivative of ν, which is not guaranteed to be continuous with a pchip interpolation) can be observed. As we have explained previously, the negativity of ν' may cause convergence issues. Manufacturer-provided B–H curve data commonly exhibit this sort of behavior at weak magnetic fields, which may be attributed to the law of Rayleigh. To guarantee convergence, we can "convexify" the functional by modifying the curve close to the origin so that $\nu' \geq 0$. In doing so, the energy density function remains approximately the same, but is now convex. For many electric machine applications, where the bulk of the material is permeated by higher fields, this slight adjustment should not

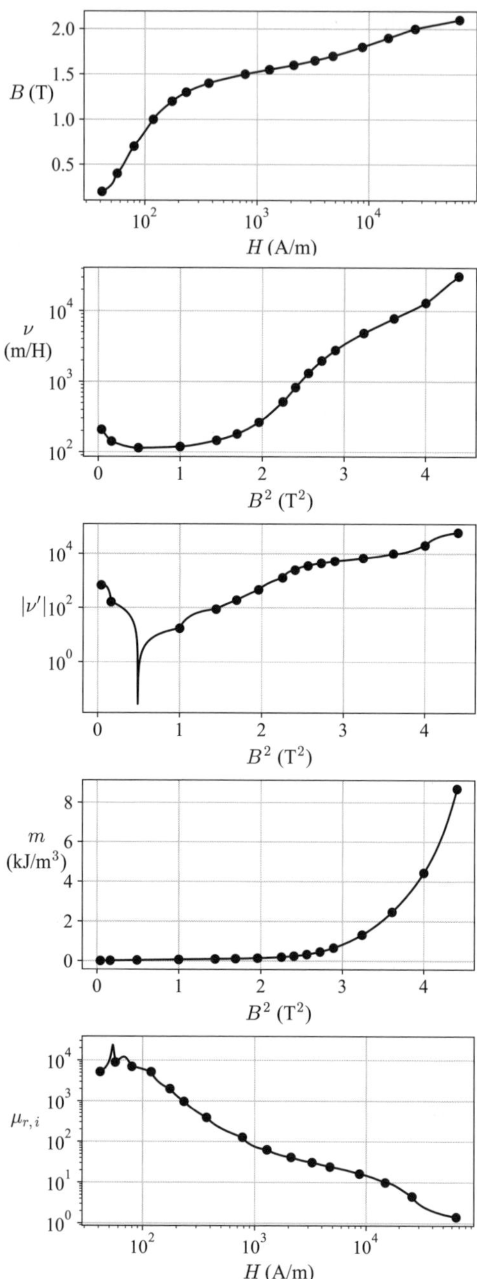

Figure 3.22 Visualization of different aspects of a B–H curve: data points and curve fit.

affect the solution in any noticeable manner. The energy density function is shown in the fourth subplot of Figure 3.22. The function is indeed concave close to the origin, although this is hard to see at this scale.

Another quantity of interest is the **incremental relative permeability**

$$\mu_{r,i} = \frac{1}{\mu_0}\frac{dB}{dH}, \tag{3.184}$$

which indicates the slope of the B–H curve. This can be calculated as

$$\mu_{r,i} = \frac{1}{\mu_0}\left(\frac{dH}{dB}\right)^{-1} \tag{3.185a}$$

$$= \frac{1}{\mu_0}\left[\frac{dH}{d(B^2)}\frac{d(B^2)}{dB}\right]^{-1} \tag{3.185b}$$

$$= \frac{1}{2\mu_0 B}\left[\frac{d(\nu B)}{d(B^2)}\right]^{-1} \tag{3.185c}$$

$$= \frac{1}{2\mu_0 B}\left[\frac{d\nu}{d(B^2)}B + \nu\frac{dB}{d(B^2)}\right]^{-1} \tag{3.185d}$$

$$= \frac{1}{2\mu_0 B}\left[\frac{d\nu}{d(B^2)}B + \nu\frac{1}{2B}\right]^{-1} \Rightarrow \tag{3.185e}$$

$$\mu_{r,i} = \frac{1}{2\mu_0}\left[\frac{d\nu}{d(B^2)}B^2 + \frac{1}{2}\nu\right]^{-1}. \tag{3.185f}$$

The formula employs the derivative of the pchip interpolant, $d\nu/d(B^2)$. The function is plotted vs. $H = \nu(B^2)B$ in the last subplot of Figure 3.22. A correct B–H curve fit should initially have $\mu_{r,i}$ on the order of 1000–10,000, which then should gradually decrease to a value slightly higher than 1.0, depending on the highest value of the H-field in the measurements. Typically, B–H data from material manufacturers are below full saturation. For higher values of H-field, the incremental relative permeability should asymptotically approach unity. After a certain point, the material behaves magnetically like free space, and its magnetization has peaked. Any further increase in the magnetic field creates a change in flux density equal to $\Delta B = \mu_0 \Delta H$. If $\mu_{r,i}$ crosses below 1, then the material exhibits (incrementally) less permeability than free space, which is physically unlikely, if not impossible, to occur in a ferromagnetic material.

Extrapolating a *B–H* Curve

An FEA program will likely need to evaluate points outside the given B–H data range. Prior to convergence, Newton–Raphson iterations may require function evaluations at arbitrary MVP values; and even after convergence, the actual operating point may involve highly saturated regions in the device. Therefore, it is necessary to define explicitly how the B–H curve is extrapolated. The pchip interpolation function comes with a default extrapolation capability, but we should not rely on this.

For low B-field values, i.e., for $0 \le B \le B_1$, we can assume that the B–H curve is linear, connecting the origin to the first data point (H_1, B_1). In this region, we set constant reluctivity, leading to

$$\nu = \nu_1 = \frac{H_1}{B_1} = \frac{2m_1}{B_1^2}, \quad \frac{d\nu}{d(B^2)} = 0, \quad m = \frac{1}{2}\nu_1 B^2. \tag{3.186}$$

The behavior for high B-field values, i.e., for $B \geq B_{\max}$, is actually unknown in the absence of data. However, we are not completely in the dark. We know that the B–H curve slope should asymptotically approach μ_0. Several extrapolation approaches have been proposed in the literature to model this phenomenon. Here, we will describe a simple method where the curve is extrapolated linearly past the last data point with a slope of $\mu_0 = \nu_0^{-1}$. This leads to a reluctivity

$$\nu = \frac{H}{B} = \frac{H_{\max} + \nu_0(B - B_{\max})}{B}, \tag{3.187}$$

a reluctivity derivative

$$\frac{d\nu}{d(B^2)} = \frac{1}{2B^3}(\nu_0 B_{\max} - H_{\max}), \tag{3.188}$$

and an energy density

$$m = m_{\max} + H_{\max}(B - B_{\max}) + \frac{\nu_0}{2}(B - B_{\max})^2, \tag{3.189}$$

where $m_{\max}$ is the energy density at the last data point. The derivation of these formulas is left as an exercise for the reader (see Problem 3.17).

Various aspects of the extrapolated B–H curve are shown in Figure 3.23. We use the same data as previously, but we have eliminated the first two data points that were associated with a negative $d\nu/d(B^2)$. Now, $m(B^2)$ is a convex function.

3.6.2 Building the Gradient and Hessian Matrix

The gradient and Hessian matrix can be assembled using the following Python function:

```python
import numpy as np
import scipy.sparse as sps

def build_grad_Hess(n_tri, n_nodes, triangles, Sk, A, J, \
    delta, nu, dnu):
    """
    Builds gradient and Hessian matrix of the functional.
    """

    g = np.zeros(n_nodes, dtype = float)
    row = np.zeros(9*n_tri, dtype = int)
    col = np.zeros(9*n_tri, dtype = int)
    h = np.zeros(9*n_tri, dtype = float)
    count = 0

    for k in range(n_tri):

        s11 = Sk[k,0]
        s12 = Sk[k,1]
        s13 = Sk[k,2]
        s22 = Sk[k,3]
        s23 = Sk[k,4]
        s33 = Sk[k,5]

        l = triangles[k,1]-1
```

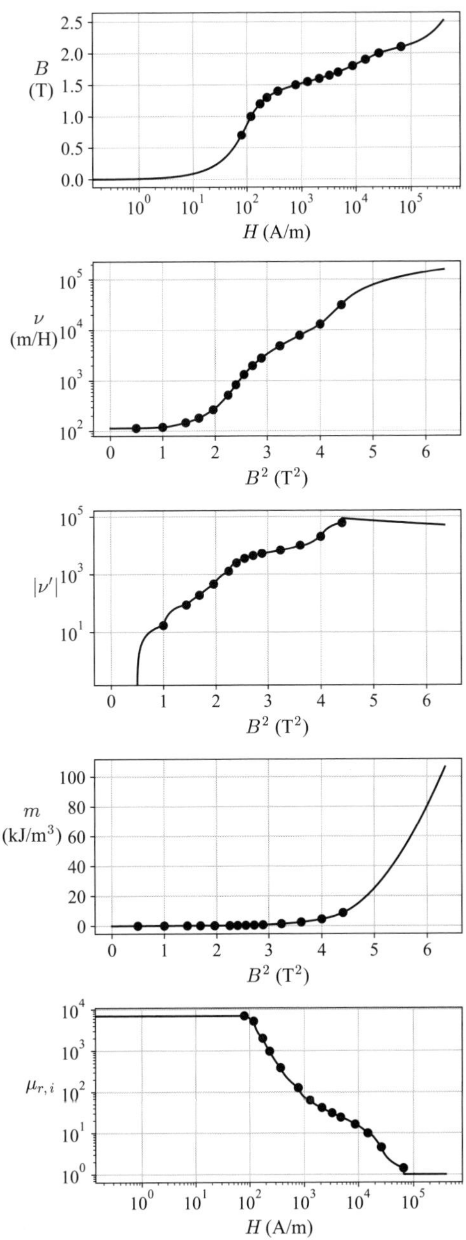

Figure 3.23 Visualization of different aspects of a convexified and extrapolated *B*–*H* curve: data points and curve fit.

```
    m = triangles[k,2]-1
    n = triangles[k,3]-1

    A1 = A[1]
```

```
30          A2 = A[m]
            A3 = A[n]

            dlt = delta[k]
            t = J[k]*dlt/3
35          nuk = nu[k]
            dnuk = dnu[k]

            # elements of Sk*Ak
            SA1 = s11*A1 + s12*A2 + s13*A3
40          SA2 = s12*A1 + s22*A2 + s23*A3
            SA3 = s13*A1 + s23*A2 + s33*A3

            # gradient
            g[l] += nuk*SA1 - t
45          g[m] += nuk*SA2 - t
            g[n] += nuk*SA3 - t

            # Hessian
            gamma = 2/dlt*dnuk
50
            row[count] = l
            col[count] = l
            h[count] = nuk*s11 + gamma * SA1*SA1
            count += 1
55
            row[count] = l
            col[count] = m
            tmp = nuk*s12 + gamma * SA1*SA2
            h[count] = tmp
60          count += 1
            row[count] = m
            col[count] = l
            h[count] = tmp
            count += 1
65
            # similar commands for other elements
            ...

        return g, sps.coo_matrix((h, (row,col)), \
70          shape=(n_nodes,n_nodes)).tocsc()
```

The `build_grad_Hess` function requires information about the mesh, e.g., the number of nodes and triangles, and the element stiffness matrices. It also needs the latest MVP estimate, the current density, and the reluctivity and its derivative for all triangles in the domain. It then iterates over all triangles with the **for** loop starting in line 16, thus collecting contributions to the gradient and Hessian using Equations (3.154) and (3.168), respectively. The Hessian is a sparse matrix (like the global stiffness matrix in the linear case).

Note that boundary conditions need to be enforced outside `build_grad_Hess`. The previous code snippet for modifying the stiffness matrix (see page 246) applies directly for the Hessian. The elements of the gradient vector corresponding to Dirichlet nodes should be set to zero as well.

3.6.3 FEA Program Implementation with Nonlinear Magnetics

The previous routines constitute the main building blocks of a magnetostatic FEA program. The main program structure for a nonlinear solver is outlined in Box 3.6.

Box 3.6 Programming the FEM: Main program structure (nonlinear magnetics)

1. Define and mesh the geometry (call `Triangle`).
2. Load nonlinear *B–H* curve data (call `load_BH_curve_pchip`).
3. Read node and triangle data files (call `load_mesh`).
4. Build element stiffness matrices (call `calculate_pqrDeltaSk`).
5. Assign material properties and current density to triangles.
6. Run the Newton–Raphson algorithm (within each iteration, call `build_grad_Hess` and enforce boundary conditions).
7. Calculate quantities of interest and generate plots by post-processing $\mathbf{A}$.

Depending on the application, the analysis may differ. Nevertheless, the reader should now be able to glue the pieces together to analyze a wide variety of electric machines. Several examples will be provided subsequently in this text. In the example that follows, we explain how to take into account the impact of lamination stacking.

Example 3.8 *Laminated square-core nonlinear inductor.*

We revisit the square-core inductor Example 3.4 on page 248. The only modification is that we substitute the core material by a nonlinear ferromagnetic material with the *B–H* curve of Figure 3.23. The purpose of this example is to help readers write and debug their first nonlinear FEA program. The main outcome is the nonlinear λ vs. i characteristic of the inductor, which is shown in Figure 3.24(a).

Let us assume that the core is constructed as a stack of thin laminations for reducing eddy current loss. The laminations are separated by thin insulation layers. Also, the manufacturing process may not guarantee that they are perfectly planar or of exactly the same width across their cross-section. Therefore, the lamination stack is not a solid piece of ferromagnetic material. To capture this, the concept of a **stacking factor**, $k_{st} < 1$, is introduced. Here, $k_{st} = 0.95$. How should we incorporate this in the analysis?

Solution

One simple way to take this effect into account is as follows. Suppose that the actual geometrical length of the lamination stack is $\ell_{st} = 10$ cm. The length ℓ that we should use for calculating quantities such as flux linkage (see Equation (3.108)) and field energy by a multiplicative scaling of the 2-D FEM quantities is

$$\ell = k_{st}\ell_{st}\,. \tag{3.190}$$

Physically, this approach implies that there is no magnetic flux between laminations.

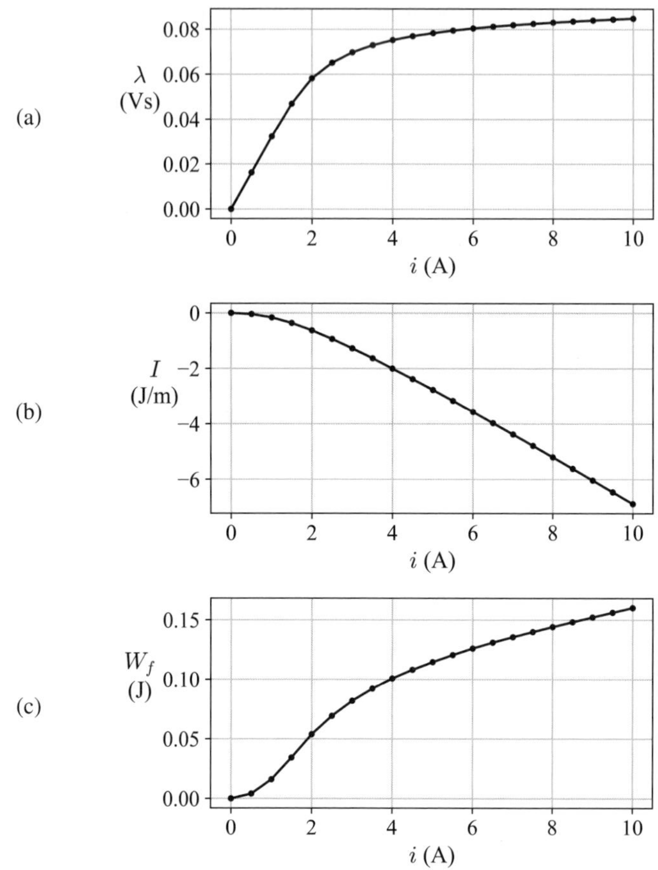

Figure 3.24 (a) Flux linkage vs. current, (b) functional vs. current, and (c) coupling field energy vs. current for the nonlinear square-core inductor.

A second simple method to account for imperfect stacking is by adjusting the *B*–*H* characteristic of the ferromagnetic material, while keeping the depth equal to the physical lamination stack length, $\ell = \ell_{st}$. The idea is to create a modified, effective *B*–*H* curve, as if steel is diluted with a non-magnetic material. In contrast to the previous approach, here we are accounting for the feeble magnetic flux density in the inter-lamination space. To this end, the magnetization M' of the modified material is only $k_{st} \cdot 100\%$ that of steel, where we use primes to denote the modified *B*–*H* characteristic. The new B'–H curve can be obtained from the manufacturer (H, B) data as follows:

$$B' = \mu_0(H + M') \tag{3.191a}$$

$$= \mu_0(H + k_{st}M) \tag{3.191b}$$

$$= \mu_0 H + k_{st}(B - \mu_0 H) \tag{3.191c}$$

$$= (1 - k_{st})\mu_0 H + k_{st}B\,. \tag{3.191d}$$

The two modeling approaches do not provide identical results. However, in this example, the differences are insignificant. The functional is plotted vs. current in Figure 3.24(b). Note that negative values are obtained, which is justified since the functional bears a physical meaning of negative coenergy. The coupling field energy is plotted in Figure 3.24(c), calculated based on the functional I as

$$W_f = \lambda i - W_c = \lambda i + \ell I \,. \tag{3.192}$$

3.7 Devices with Permanent Magnets

The effect of permanent magnets can be incorporated in the FEM in a straightforward manner. Suppose for simplicity that our device uses a single PM material with: initial energy density m_0 (an unknown constant), reluctivity ν_{pm}, and remanence $\vec{\mathbf{B}}_r^k = (B_{rx}^k, B_{ry}^k)$. Within each PM triangle k, we are setting a constant $\vec{\mathbf{B}}_r^k$ value. Nevertheless, the remanence can be a function of position, thus allowing for a non-uniform magnetization.

We have shown that the energy density in PMs is given by Equation (2.82b) on page 119. It consists of three terms:

$$m_k = m_0 + \underbrace{\frac{\nu_{pm}}{2}(B_k^2 + B_{rk}^2)}_{m_{1,k}} \underbrace{- \nu_{pm} \vec{\mathbf{B}}^k \cdot \vec{\mathbf{B}}_r^k}_{m_{2,k}} \,. \tag{3.193}$$

Hence, the contribution to the functional from the kth element is

$$I_k = (m_0 + m_{1,k} + m_{2,k})\Delta_k \,, \tag{3.194}$$

where we are assuming that the PM cannot carry free currents. (In cases where eddy currents flow in PMs, we need to subtract the corresponding J_k term from I_k.)

The contribution of the PM element to the gradient is calculated by taking the partial derivatives with respect to the ith vertex MVP (i.e., the nth node in a global numbering scheme). This yields

$$\frac{\partial I_k}{\partial A_n} = \frac{\partial I_k}{\partial A_i^k} = \frac{\partial m_{1,k}}{\partial A_i^k}\Delta_k + \frac{\partial m_{2,k}}{\partial A_i^k}\Delta_k \,. \tag{3.195}$$

We have already calculated the derivative of the $m_{1,k}$ term when dealing with linear materials. For the derivative of the $m_{2,k}$ term, we recall Equation (3.94) on page 255, which suggests that

$$\frac{\partial m_{2,k}}{\partial A_i^k}\Delta_k = -\frac{\nu_{pm}}{2}(B_{rx}^k r_i^k - B_{ry}^k q_i^k) \,. \tag{3.196}$$

Therefore, **the contribution to the gradient from the kth triangle** is (cf. Equation (3.154))

$$\boxed{\mathbf{g}^k = \nu_{pm}\mathbf{S}^k\mathbf{A}^k - \frac{\nu_{pm}}{2}(B_{rx}^k \mathbf{r}^k - B_{ry}^k \mathbf{q}^k) \,.} \tag{3.197}$$

This implies that PMs can be treated as linear materials. **The contribution to the Hessian from PM-occupied triangles** is simply

$$\boxed{\mathbf{H}^k = \nu_{pm}\mathbf{S}^k\,.} \tag{3.198}$$

Implementing this in the FEA program is straightforward since it requires only minor modifications to the `build_grad_Hess` function.

3.8 FEA of Axisymmetric Problems

Axisymmetric problems were introduced in §1.7.6. In Example 2.16 on page 167, we analyzed a plunger using simplifying assumptions about the field. Now, we will explain how to implement an axisymmetric magnetostatic FEA. We will use the fundamental theoretical results of Box 1.4 on page 80 throughout the analysis.

3.8.1 Magnetically Linear Axisymmetric Problems

Let us first assume magnetic linearity for simplicity. In cylindrical coordinates, the differential volume is $dv = r\,d\phi\,dr\,dz$. Due to symmetry, the functional becomes

$$I(A_\phi) = \frac{1}{2}\iint_{\mathcal{D}} \nu(z,r)\,\|B\|^2\, r\,dz\,dr \;-\; \iint_{\mathcal{D}} JA_\phi\, r\,dz\,dr\,, \tag{3.199}$$

where we have integrated out the angle ϕ. Strictly speaking, this functional needs to be multiplied by 2π radians to obtain a value in joules. Introducing the function

$$\tilde{A} = rA_\phi\,, \tag{3.200}$$

we obtain a functional in terms of $\tilde{A}$:

$$I(\tilde{A}) = \frac{1}{2}\iint_{\mathcal{D}} \nu(z,r)\frac{\|\nabla\tilde{A}\|^2}{r}\,dz\,dr \;-\; \iint_{\mathcal{D}} J\tilde{A}\,dz\,dr\,, \tag{3.201}$$

where the gradient operator is in cylindrical coordinates. Hence, *we minimize this functional over the set of piecewise linear functions* $\tilde{A}$.

The FEM formulation is very similar to what we have presented earlier in this chapter. Now, the cylindrical coordinates z and r play the role of x and y, respectively. The main difference of the axisymmetric functional (3.201) from that of Equation (3.52) is that the integrand of the field energy is divided by r. The algebraic manipulations are otherwise identical, so we do not need to repeat them in a lot of detail.

Therefore, by adapting Equation (3.53), the FEM solution in the kth triangle is a linear function of the spatial coordinates and the values of $\tilde{A}$ at the vertices:

$$\tilde{A}_k(z,r,\tilde{\mathbf{A}}^k) = \sum_{i=1}^{3} \tilde{A}_i^k\,\alpha_i^k(z,r)\,. \tag{3.202}$$

The gradient of $\tilde{A}$ is[h]

$$\nabla \tilde{A}_k(z, r, \tilde{\mathbf{A}}^k) = \begin{bmatrix} (\nabla \tilde{A}_k)_z \\ (\nabla \tilde{A}_k)_r \end{bmatrix} = \frac{1}{2\Delta_k} \begin{bmatrix} (\mathbf{q}^k)^\top \\ (\mathbf{r}^k)^\top \end{bmatrix} \tilde{\mathbf{A}}^k . \tag{3.203}$$

The functional integral is a sum of element contributions as in Equation (3.54a):

$$I(\tilde{A}) = \sum_{k=1}^{N_\Delta} I_k(\tilde{A}_k) = \sum_{k=1}^{N_\Delta} \frac{1}{2} \iint_{\Delta_k} \nu_k \frac{\|\nabla \tilde{A}_k\|^2}{r} \, dz \, dr - \sum_{k=1}^{N_\Delta} \iint_{\Delta_k} J \tilde{A}_k \, dz \, dr . \tag{3.204}$$

To simplify the algebra, we approximate the first integral by setting $r = r_k$, representing the centroid of triangle-k:

$$\frac{1}{2} \iint_{\Delta_k} \nu_k \frac{\|\nabla \tilde{A}_k\|^2}{r} \, dz \, dr \approx \frac{\nu_k}{2r_k} \|\nabla \tilde{A}_k\|^2 \Delta_k = \frac{\nu_k}{2r_k} (\tilde{\mathbf{A}}^k)^\top \mathbf{S}^k \, \tilde{\mathbf{A}}^k . \tag{3.205}$$

This should be a reasonable approximation as long as the triangles are small so that r does not vary by much within each one. The element stiffness matrices are identical to the ones we derived earlier, see Equation (3.28) on page 225, so they can be assembled using the function `calculate_pqrDeltaSk` (in §3.3.2). However, the routine for assembling the global stiffness matrix, `build_S_matrix`, requires a minor modification to reflect the division by r_k in each triangle. Finally, we obtain the solution by imposing a Dirichlet boundary condition on the z-axis with $\tilde{A} = 0$, and solving the linear equation

$$\mathbf{S}\tilde{\mathbf{A}} = \mathbf{b} . \tag{3.206}$$

The B-field satisfies

$$\vec{\mathbf{B}}^k = \begin{bmatrix} B_z^k \\ B_r^k \end{bmatrix} = \frac{1}{2r\Delta_k} \begin{bmatrix} (\mathbf{r}^k)^\top \\ -(\mathbf{q}^k)^\top \end{bmatrix} \tilde{\mathbf{A}}^k \approx \frac{1}{2r_k\Delta_k} \begin{bmatrix} (\mathbf{r}^k)^\top \\ -(\mathbf{q}^k)^\top \end{bmatrix} \tilde{\mathbf{A}}^k . \tag{3.207}$$

Hence, the approximation $r \approx r_k$ leads to a constant B-field in each element. Furthermore, its magnitude satisfies

$$B_k^2 = \frac{\|\nabla \tilde{A}_k\|^2}{r^2} \approx \frac{\|\nabla \tilde{A}_k\|^2}{r_k^2} = \frac{1}{(2r_k\Delta_k)^2} (\tilde{\mathbf{A}}^k)^\top \begin{bmatrix} \mathbf{q}^k & \mathbf{r}^k \end{bmatrix} \begin{bmatrix} (\mathbf{q}^k)^\top \\ (\mathbf{r}^k)^\top \end{bmatrix} \tilde{\mathbf{A}}^k , \tag{3.208}$$

or, in view of Equation (3.28):

$$B_k^2 = \frac{1}{r_k^2 \Delta_k} (\tilde{\mathbf{A}}^k)^\top \mathbf{S}^k \, \tilde{\mathbf{A}}^k . \tag{3.209}$$

These small differences from the 2-D case should be taken into account by modifying our post-processing (e.g., plotting) routines accordingly.

[h] The notation $\mathbf{r}^k$ in the basis function definition carries over from the earlier definition on the x–y plane, but this variable should not be confused with the radial coordinate r. Coincidentally, here, both symbols appear simultaneously, so the reader is cautioned to pay attention to context. For example, the first element of the basis function parameter $\mathbf{r}^k$ is $r_1^k = z_3^k - z_2^k$, whereas that of $\mathbf{q}^k$ is $q_1^k = r_2^k - r_3^k$, where the right-hand-side rs are radial coordinates.

3.8.2 Nonlinear Axisymmetric Problems

The previous ideas can be readily extended to cover the case of nonlinear magnetics, following the steps of §3.5. A suitable functional (in joules per radian) is

$$I(A_\phi) = \iint_{\mathcal{D}} m(z, r, A_\phi)\, r\, dz\, dr - \iint_{\mathcal{D}} J A_\phi\, r\, dz\, dr\,. \tag{3.210}$$

Introducing the function $\tilde{A}$, we obtain the modified functional

$$I(\tilde{A}) = \iint_{\mathcal{D}} m(z, r, \tilde{A})\, r\, dz\, dr - \iint_{\mathcal{D}} J\tilde{A}\, dz\, dr\,, \tag{3.211}$$

where the energy density function is

$$m(z, r, \tilde{A}) = \frac{1}{2} \int_0^{B^2(\tilde{A})} \nu(z, r, b^2)\, db^2\,. \tag{3.212}$$

The contribution from triangle-k is approximated as

$$I_k(\tilde{\mathbf{A}}^k) = m_k(B_k^2(\tilde{\mathbf{A}}^k))\, r_k \Delta_k - \frac{J_k \Delta_k}{3} \sum_{i=1}^{3} \tilde{A}_i^k\,, \tag{3.213}$$

where we have assumed a constant B_k in the triangle. Therefore, the contribution to the partial derivative of the functional with respect to the function value of the nth global node or the ith local node is

$$\frac{\partial I_k}{\partial \tilde{A}_n} = \frac{\partial I_k}{\partial \tilde{A}_i^k} = \frac{dm_k}{d(B_k^2)} \frac{\partial (B_k^2)}{\partial \tilde{A}_i^k} r_k \Delta_k - \frac{J_k \Delta_k}{3} = \frac{\nu_k(B_k^2(\tilde{\mathbf{A}}^k))}{r_k} \cdot (\mathbf{S}^k \tilde{\mathbf{A}}^k)_i - \frac{J_k \Delta_k}{3}\,. \tag{3.214}$$

More concisely, the contribution to the gradient from the kth triangle can be expressed as the 3×1 vector

$$\mathbf{g}^k = \frac{\nu_k}{r_k} \mathbf{S}^k \tilde{\mathbf{A}}^k - \frac{J_k \Delta_k}{3} \begin{bmatrix} 1 & 1 & 1 \end{bmatrix}^\top\,. \tag{3.215}$$

Furthermore, the contribution to the Hessian from triangle-k is

$$\mathbf{H}^k = \frac{2}{r_k^3 \Delta_k} \frac{d\nu_k}{d(B_k^2)} (\mathbf{S}^k \tilde{\mathbf{A}}^k)(\mathbf{S}^k \tilde{\mathbf{A}}^k)^\top + \frac{\nu_k}{r_k} \mathbf{S}^k\,. \tag{3.216}$$

The divisions by r_k in the formulas can be accommodated by a minor modification of the `build_grad_Hess` function.

3.8.3 Flux Linkage Calculation

Following the guidelines of §3.4.1, we can obtain the following formula for calculating the flux linkage of a cylindrical coil:

$$\lambda = \frac{2\pi N}{A_c} \iint_{\mathbf{r} \in S} A_\phi(\mathbf{r})\, r\, dz\, dr = \frac{2\pi N}{A_c} \iint_{\mathbf{r} \in S} \tilde{A}(\mathbf{r})\, dz\, dr\,, \tag{3.217}$$

where N is the number of turns, A_c is the coil area, and S represents an integration surface over positively oriented conductors. If the coil area is meshed with K triangles, the flux linkage can be evaluated from an FEA solution using

$$\lambda = \frac{2\pi N}{A_c} \sum_{k=1}^{K} \frac{\tilde{A}_1^k + \tilde{A}_2^k + \tilde{A}_3^k}{3} \Delta_k . \tag{3.218}$$

3.9 Galerkin's Method

The FEM that we have described in the previous sections is known as **Galerkin's method**, but it is really based on the groundbreaking earlier work of Ritz[67] [2]. All key ingredients and mathematical steps have been already set forth. Here, we present a different viewpoint of the same computational process, summarizing the main ideas. This material is not really necessary for implementation of the FEM; however, it is provided as an introduction to the theoretical foundations of this subject.

3.9.1 Distinguishing Features of Galerkin's Method

The main concept that has emerged thus far is that the true solution to the problem and the approximate solution that we are finding numerically belong in two different **function spaces** or sets of functions. In particular, the true solution w^* needs to be in the space of **square-integrable** functions, denoted by $\mathcal{L}^2$. A function $w : \mathcal{D} \to \mathbb{R}$, $\mathcal{D} \subset \mathbb{R}^2$, is said to belong to $\mathcal{L}^2$ when the integral of its square is finite, i.e., $\iint_{\mathcal{D}} w^2 \, da < \infty$.

We use the canonical Poisson problem $\nabla^2 w = -h$ on $\mathcal{D} \subset \mathbb{R}^2$, with w known (Dirichlet condition) on $\partial\mathcal{D}_1$, and w free on $\partial\mathcal{D}_2$, as an example (see Figure 2.3). If w^* is a solution to this Poisson problem, the following integrals should hold as an identity:

$$\iint_{\mathcal{D}} \eta \cdot (\nabla^2 w^* + h) \, da = \iint_{\mathcal{D}} \eta \cdot 0 \, da = 0 \tag{3.219}$$

and

$$\oint_{\partial\mathcal{D}} \eta \cdot (-w_y^* \, dx + w_x^* \, dy) = \int_{\partial\mathcal{D}_1} 0 \cdot (-w_y^* \, dx + w_x^* \, dy) + \int_{\partial\mathcal{D}_2} \eta \cdot 0 = 0 , \tag{3.220}$$

for every possible η that is clamped to zero on $\partial\mathcal{D}_1$, and that does not cause awkward things to happen (e.g., if η is not bounded then the integrals may not be zero). The function η is called a **test function**, and it belongs to the **test space**. Note that this is what Step 4 of the variational process asserts (see §2.1.2), where the Euler–Lagrange equation becomes the Poisson equation for an appropriate choice of functional (see Equation (2.50)), and η bears the meaning of a variation. Going backwards through an integration by parts, we reach the weak form of Step 2, cf. Equation (3.71):

$$\iint_{\mathcal{D}} (\nabla\eta \cdot \nabla w^* - \eta h) \, da = 0 . \tag{3.221}$$

This leads us to the first salient feature of the Galerkin method.

Galerkin's method feature #1: *The original problem is transformed to its weak form.* By reducing the differentiation order (through an integration by parts), we are removing constraints imposed on candidate solutions. Clearly, the solution that we find may not be suitable for the original Poisson equation, where two derivatives are needed. On the other hand, an important advantage is that we can search for an approximate solution in the space of piecewise linear functions. The gradient of these functions is constant within each triangular element, but it is discontinuous across elements. This is not an issue, since we can break up the integral in Equation (3.221) into an element-wise sum. However, had we attempted to obtain a solution to the original Poisson equation in the space of piecewise linear functions, we would encounter the obstacle that the Laplacian of a piecewise linear function is zero within elements, and is not well-defined across element boundaries.

Hence, to ensure that the weak form is well-defined, we introduce the space $\mathcal{H}^1$ of functions that (i) are square integrable (i.e., $\mathcal{H}^1 \subset \mathcal{L}^2$); and (ii) have *square-integrable first derivatives* (this is what the superscript "1" denotes). In particular, we ask that a candidate solution function $w \in \mathcal{H}^1$ is such that (i) it attains its prescribed value on $\partial\mathcal{D}_1$; and (ii) $\iint_{\mathcal{D}} \nabla w \cdot \nabla w \, da < \infty$. In functional analysis, this is called a **Sobolev**[68] **space**. It is also a **Hilbert**[69] **space**.

Without going into too much detail about the properties of these spaces, essentially this means that we can define an **inner product** between two functions $u, w \in \mathcal{H}^1$ as

$$(u, w) = \iint_{\mathcal{D}} uw \, da \,, \tag{3.222}$$

and the **norm** of a function as

$$\|w\| = \sqrt{(w, w)} \,. \tag{3.223}$$

The geometric concept of orthogonality applies in the sense that two functions are said to be **orthogonal** if $(u, w) = 0$. We also define the **energetic inner product** as

$$(u, w)_E = \iint_{\mathcal{D}} \nabla u \cdot \nabla w \, da \,, \tag{3.224}$$

and the **energetic norm**

$$\|w\|_E = \sqrt{(w, w)_E} \,. \tag{3.225}$$

Using these definitions, the weak form (3.221) may be written as

$$(\eta, w^*)_E = (\eta, h) \,, \tag{3.226}$$

where $w^* \in \mathcal{H}^1$ and $\eta \in \mathcal{H}^1_0$. The test functions η belong in a different Hilbert space, with the requirement that $\eta = 0$ on the boundary $\partial\mathcal{D}_1$ (this is what the subscript "0" denotes).

The set of piecewise linear functions is a subset of $\mathcal{H}^1$, so these functions are appropriate for the Galerkin method. They are the simplest functions that one could use. Of course, $\mathcal{H}^1$ contains many other classes of functions, such as polynomials of higher order, which are also used in the FEM.

Galerkin's method feature #2: *The approximate solution lives in a finite-dimensional function space $\mathcal{X}^N \subset \mathcal{H}^1$.*
From an infinite-dimensional $\mathcal{H}^1$ space, we obtain a finite-dimensional space via a discretization of the domain (e.g., using triangular elements). Here, the superscript N denotes the finite number of basis functions in the solution space $\mathcal{X}^N$, which is called the **trial** or **ansatz space**. The number N equals the number of vertices in an FEA mesh with linear elements because the solution can be reconstructed if we know its value at the vertices:

$$w(x, y) = \sum_{n=1}^{N} W_n \, \phi_n(x, y) \,, \tag{3.227}$$

where W_n are coefficients that multiply N **global basis functions** ϕ_n. In general, even if the basis functions are nonlinear, they are defined so as to have compact support (i.e., they are zero everywhere except in the neighborhood of a vertex), and they must satisfy $\phi_i(x_j, y_j) = \delta_{ij}$.

Note that the boundary of the original domain $\partial\mathcal{D}$ can have an arbitrary shape (e.g., consisting of several smooth pieces that form corners at connection points). In this case, the boundary of the FEM domain will be an approximation of $\partial\mathcal{D}$ by a number of linear segments, as in Figure 3.1. Hence, $\mathcal{H}^1$ and $\mathcal{X}^N$ consist of functions that are defined over domains that are similar but may not be identical. We expect that the discretization of the boundary is such that it does not affect the solution significantly. This source of error will be ignored in what follows, and the two domains will be considered equal. To this end, let us suppose from now on that we are solving a slightly modified problem, where the boundary is adjusted to conform with its piecewise linear approximation.

Furthermore, functions from $\mathcal{H}^1$ are assigned a problem-specific Dirichlet boundary condition on $\partial\mathcal{D}_1$, which could be arbitrary. However, functions from $\mathcal{X}^N$ are limited because they cannot capture arbitrary function variations. For instance, with linear elements, we can only describe a piecewise linear variation of boundary conditions (over the approximated boundary). This second source of error is also ignored in what follows, and we are content with an approximation of the boundary condition. Hence, let us also suppose from now on that we are solving a slightly modified problem by changing the original boundary condition to a piecewise linear approximation.

When the two problem modifications regarding the boundary shape and condition are made, we can claim that *any function from the trial space also belongs in the original Sobolev space*, or that $\mathcal{X}^N \subset \mathcal{H}^1$. We will use this result later on.

The global basis functions ϕ_n are, of course, related to the element basis functions α_i, which were presented in §3.1.2 for linear elements. The ϕ_n are commonly called **hat functions**, and were introduced by Courant.[70] Recall that the α_i resemble pyramids that are defined over a single element (see Figure 3.3). A global basis function ϕ_n also resembles a pyramid that takes the value 1 over the node n, and then drops to zero over all the surrounding elements that contain n as one of their vertices. In other words, ϕ_n is a pyramid that is formed by joining the α_i pyramids of the triangles surrounding node n. Note that the basis functions are "almost" orthogonal to each other since $(\phi_j, \phi_k) = 0$, for all $j \neq k$, except for adjacent nodes where the functions are both nonzero.

Galerkin's method feature #3: *The test space is (essentially) the same as the trial space.*
We have explained that the FEM approximation $w \in \mathcal{X}^N$ minimizes the functional $I(w)$, and equivalently satisfies the weak form (3.221). To ensure this, in the Galerkin method we test against all functions $\eta \in \mathcal{X}_0^N$ using (almost) identical trial and test spaces. The spaces are the same because they contain all piecewise linear functions over the chosen discretization of the domain, but the difference is that functions in $\mathcal{X}_0^N$ have zero coefficients for those hat functions that lie on the Dirichlet boundary. (The boundary hat functions are chopped since they cannot extend past the boundary.) In other words, the test space has fewer degrees of freedom, $N_0 < N$, than the trial space.

Using the inner product definitions, we ask that (cf. Equation (3.226))

$$(\eta, w)_E = (\eta, h) \quad \text{for any } \eta \in \mathcal{X}_0^N . \tag{3.228}$$

Since η can be expressed as a linear combination of $N_0 < N$ basis functions ϕ_j, it suffices that

$$(\phi_j, w)_E = (\phi_j, h) \quad \text{for } j = 1, 2, \ldots, N_0 . \tag{3.229}$$

Using Equation (3.227), we expand this to

$$\sum_{n=1}^{N} (\phi_j, \phi_n)_E \, W_n = (\phi_j, h) \quad \text{for } j = 1, 2, \ldots, N_0 , \tag{3.230}$$

or

$$\sum_{n=1}^{N_0} (\phi_j, \phi_n)_E \, W_n = (\phi_j, h) - \sum_{n=N_0+1}^{N} (\phi_j, \phi_n)_E \, W_n \quad \text{for } j = 1, 2, \ldots, N_0 . \tag{3.231}$$

We have thus formed a linear system of N_0 equations with N_0 unknowns, which can be solved for determining the coefficients W_n, $n = 1, \ldots, N_0$, of the solution to our problem. (We are assuming that the coefficients W_n with $N_0 + 1, \ldots, N$, correspond to Dirichlet nodes, so they are known.)[i]

This linear system corresponds to what we would obtain by minimizing the quadratic functional (3.68) for linear elements. The process that we have described in the previous sections, where the stiffness matrix was assembled on an element-by-element basis, is a computationally efficient method of evaluating all energetic inner products $(\phi_j, \phi_n)_E$ over the entire domain.

3.9.2 Energetic Interpretation of Galerkin's Method

The Galerkin solution has an important property that is highlighted in Box 3.7.

Proof We can visualize the trial space $\mathcal{X}^N \subset \mathcal{H}^1$ (e.g., the set of piecewise linear functions) as a plane. The true solution, which probably does not belong in this space (because it is unlikely to be piecewise linear), can be represented by a point $w^* \in \mathcal{H}^1$, as

[i] The source term h can be an arbitrary function in $\mathcal{H}^1$. The numerical calculation of (ϕ_j, h) may be performed by approximating h by a function $h' \in \mathcal{X}^N$.

Box 3.7 FEM fundamentals: Galerkin orthogonality property

The Galerkin solution w represents the closest we can get within the trial space $\mathcal{X}^N$ to the true solution w^*, with distances being measured by the energetic norm. This means that

$$||w - w^*||_E \leq ||u - w^*||_E \quad \text{for any } u \in \mathcal{X}^N. \tag{3.232}$$

Also, the error $e = w - w^*$ is orthogonal to the space:

$$(e, \eta)_E = 0 \quad \text{for any } \eta \in \mathcal{X}_0^N. \tag{3.233}$$

In other words, the Galerkin solution is a projection of the true solution to the trial space.

shown in Figure 3.25. The error e is a vector from this point to the Galerkin solution w. We can measure the magnitude of the error using either the regular norm or the energy norm; we choose the latter for reasons that will become clear later on. Hence

$$||e||_E^2 = ||w - w^*||_E^2 = \iint_{\mathcal{D}} \nabla(w - w^*) \cdot \nabla(w - w^*)\, da \tag{3.234a}$$

$$= \iint_{\mathcal{D}} \nabla(w - w^*) \cdot \nabla(w - u)\, da + \iint_{\mathcal{D}} \nabla(w - w^*) \cdot \nabla(u - w^*)\, da\,, \tag{3.234b}$$

where we substituted $w - w^* = (w - u) + (u - w^*)$ in the second gradient by introducing an arbitrary $u \in \mathcal{X}^N$. Now, we note that $\eta = w - u \in \mathcal{X}_0^N$ can be interpreted as a test function. (Subtracting the two functions makes the Dirichlet conditions cancel at the boundary, so their difference is in $\mathcal{X}_0^N$.) Hence, the first integral vanishes because both w and w^* satisfy the weak form (3.221). In particular, w^* satisfies the weak form for any $\eta \in \mathcal{H}_0^1$, whereas the Galerkin solution w satisfies it for any $\eta \in \mathcal{X}_0^N$. Since $\eta \in \mathcal{X}_0^N \subset \mathcal{H}_0^1$, we

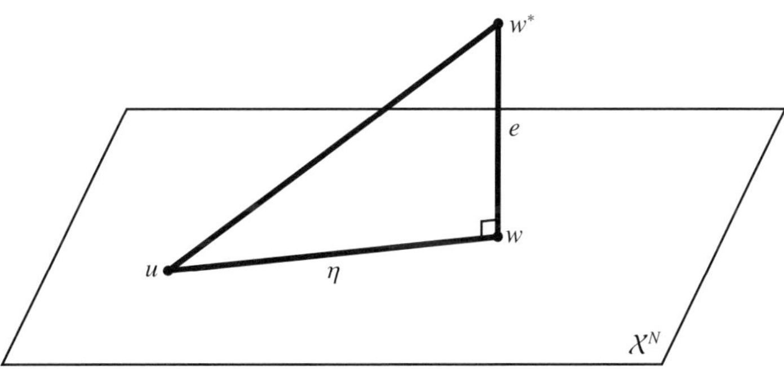

Figure 3.25 Illustration of the Galerkin orthogonality property.

obtain[j]

$$\iint_{\mathcal{D}} \nabla(w - w^*) \cdot \nabla(w - u)\, da = \iint_{\mathcal{D}} \nabla w \cdot \nabla \eta\, da - \iint_{\mathcal{D}} \nabla w^* \cdot \nabla \eta\, da \tag{3.235a}$$

$$= \iint_{\mathcal{D}} \eta h\, da - \iint_{\mathcal{D}} \eta h\, da = 0. \tag{3.235b}$$

We have thus shown orthogonality, i.e., that

$$(e, \eta)_E = 0 \quad \text{for any } \eta \in \mathcal{X}_0^N. \tag{3.236}$$

Furthermore, it follows from Equation (3.234b) that

$$\|e\|_E^2 = (e, u - w^*)_E. \tag{3.237}$$

To proceed, we invoke the Cauchy[71]–Schwarz[72] inequality, which states that the inner product of two functions cannot exceed the product of their norms. For the energetic inner product, we can express the Cauchy–Schwarz inequality as

$$(u, v)_E \leq \|u\|_E\, \|v\|_E. \tag{3.238}$$

Therefore, it follows that

$$\|e\|_E^2 = (e, u - w^*)_E \leq \|e\|_E\, \|u - w^*\|_E, \tag{3.239}$$

and equivalently

$$\|e\|_E \leq \|u - w^*\|_E, \tag{3.240}$$

which signifies that the Galerkin solution has the smallest distance from the true solution in terms of the energetic norm. □

3.9.3 Galerkin's Method for the Nonlinear Poisson Equation in Magnetic Devices

We have introduced Galerkin's method based on the linear Poisson equation in an abstract, mathematical manner. However, electric machine applications typically require solving a nonlinear Poisson equation (see, e.g., Equation (3.48)), or equivalently, computing the stationary point of a nonlinear magnetostatic functional. Our objective here is to identify the main features of the Galerkin method for this particular case of interest.

Weak Form of Nonlinear Poisson Equation

Based on our earlier work, we know that the 2-D magnetostatic functional (in devices without permanent magnets) is

$$I(A) = \frac{1}{2} \iint_{\mathcal{D}} \int_0^{B^2} \nu(x, y, b^2)\, d(b^2)\, dx\, dy - \iint_{\mathcal{D}} AJ\, dx\, dy, \tag{3.241}$$

where $\nu(x, y, B^2)$ is the potentially nonlinear reluctivity function. An explicit dependence of ν on position (x, y) was used due to the presence of different materials in space.

[j] An implicit assumption here is that the FEM does not introduce any error in the source-term integral (η, h).

Adapting the notation of §2.1.2, where A_x and A_y are the partial derivatives of $A(x, y)$, we have

$$F(x, y, A, A_x, A_y) = \frac{1}{2} \int_0^{A_x^2+A_y^2} \nu(x, y, b^2)\, d(b^2) \; - \; JA\,, \tag{3.242}$$

since

$$B^2(A) = \|\nabla A\|^2 = \left(\frac{\partial A}{\partial x}\right)^2 + \left(\frac{\partial A}{\partial y}\right)^2 = A_x^2 + A_y^2\,. \tag{3.243}$$

In this context, it should be noted that $B = \|\nabla A\|$ does not signify the norm of a function, but instead a norm of the vector $\nabla A(x, y)$ at some point in space.

From Step 2 of the variational process, stationarity of the functional is equivalent to

$$\iint_{\mathcal{D}} \left[\eta \frac{\partial F}{\partial A} + \frac{\partial \eta}{\partial x}\frac{\partial F}{\partial A_x} + \frac{\partial \eta}{\partial y}\frac{\partial F}{\partial A_y}\right] dx\, dy = 0\,, \tag{3.244}$$

or

$$\iint_{\mathcal{D}} \left[-\eta J + \eta_x \nu(x, y, B^2) A_x + \eta_y \nu(x, y, B^2) A_y\right] dx\, dy = 0\,, \tag{3.245}$$

for any test function η. This formula was obtained by employing the fundamental theorem of calculus and the chain rule, e.g.

$$\frac{\partial F}{\partial A_x} = \frac{\partial F}{\partial (B^2)}\frac{\partial (B^2)}{\partial A_x} = \frac{1}{2}\nu(x, y, B^2)\frac{\partial (B^2)}{\partial A_x} \tag{3.246}$$

and

$$\frac{\partial (B^2)}{\partial A_x} = 2A_x = 2\frac{\partial A}{\partial x}\,. \tag{3.247}$$

Therefore, stationarity of the functional (3.241) is guaranteed by a function $A^* \in \mathcal{H}^1$ that satisfies the weak form:

$$\iint_{\mathcal{D}} \nu(x, y, B^{*2})\, \nabla\eta \cdot \nabla A^*\, dx\, dy = \iint_{\mathcal{D}} \eta J\, dx\, dy\,, \tag{3.248}$$

for any test function $\eta \in \mathcal{H}_0^1$, where $B^* = \|\nabla A^*\|$. This resembles the weak form (3.71), where $\nu = \nu(x, y)$ was constant within entire sub-regions since materials were assumed to be magnetically linear.

Energetic Inner Product and its Physical Interpretation

The energetic inner product between two functions $A_1, A_2 \in \mathcal{H}^1$ can be defined as

$$(A_1, A_2)_E = \iint_{\mathcal{D}} \nu_E(x, y)\, \nabla A_1 \cdot \nabla A_2\, dx\, dy\,. \tag{3.249}$$

Here, $\nu_E(x, y) > 0$ is an arbitrary function of position. Physically, if the functions A_1 and A_2 are representative of MVP, then ν_E is representative of a reluctivity distribution over the entire space. We do not allow ν_E to depend on either A_1 or A_2. This suffices for a symmetric inner product, $(A_1, A_2)_E = (A_2, A_1)_E$, which any inner product should satisfy. The corresponding energetic norm of a function A is

$$||A||_E = \sqrt{(A, A)_E}\,. \tag{3.250}$$

If we set $\nu_E(x, y) = \nu(x, y, B^{*2})$, thereby equating ν_E with the actual reluctivity of the materials at the solution, then the weak form (3.248) can be written more concisely as (cf. Equation (3.226))

$$(\eta, A^*)_E = (\eta, J) \quad \text{for any } \eta \in \mathcal{H}_0^1\,. \tag{3.251}$$

As the name implies, the energetic norm is related to the coupling field energy; however, we should be cautious when ascribing such a physical interpretation. In a domain that consists entirely of magnetically linear materials, we can set $\nu_E(x, y) = \nu(x, y)$, so

$$||A||_E^2 = \iint_{\mathcal{D}} \nu(x, y)\nabla A \cdot \nabla A \, dx\, dy = \iint_{\mathcal{D}} \nu(x, y) B^2 \, dx\, dy\,. \tag{3.252}$$

Therefore, for magnetically linear devices:

$$||A||_E^2 = 2W_f\,, \tag{3.253}$$

where W_f is the coupling field energy (per unit depth) corresponding to the magnetic potential A. In contrast, it is clear that for magnetically nonlinear devices:

$$||A||_E^2 \neq 2W_f\,, \tag{3.254}$$

even if we set $\nu_E(x, y) = \nu(x, y, B^2)$!

Furthermore, we can relate the energetic inner product to the difference in energy between two magnetization states. First, let us consider a device consisting entirely of magnetically linear materials. Take B_1 and $B_2 = B_1 + \delta B$, associated with a change in the magnetic potential from A_1 to $A_2 = A_1 + \delta A$. We have

$$B_2^2 = (B_1 + \delta B)^2 = ||\nabla(A_1 + \delta A)||^2 = ||\nabla A_1||^2 + 2\nabla A_1 \cdot \nabla(\delta A) + ||\nabla(\delta A)||^2\,, \tag{3.255}$$

where norms are taken point-wise, so

$$B_2^2 - B_1^2 = 2\nabla A_1 \cdot \nabla(\delta A) + ||\nabla(\delta A)||^2 = \nabla(\delta A) \cdot [2\nabla A_1 + \nabla(\delta A)]\,. \tag{3.256}$$

The corresponding difference in field energy (per unit depth) is

$$\delta W_f = W_f(A_2) - W_f(A_1) = \frac{1}{2}\iint_{\mathcal{D}} \nu(x, y)(B_2^2 - B_1^2)\, dx\, dy\,, \tag{3.257}$$

or

$$\delta W_f = (\delta A, A_1)_E + \frac{1}{2}||\delta A||_E^2\,, \tag{3.258}$$

where the energetic inner product was defined by setting $\nu_E = \nu$. For example, suppose $A_1 = A^*$ and $A_2 = A$. Then, the difference in field energy between the FEA solution and the true solution is

$$\delta W_f = (e, A^*)_E + \frac{1}{2}||e||_E^2\,, \tag{3.259}$$

with

$$e = A - A^* . \tag{3.260}$$

When nonlinear materials are involved, the difference in energy is

$$\delta W_f = \frac{1}{2} \iint_{\mathcal{D}} \int_{B_1^2}^{B_2^2} \nu(x, y, b^2)\, d(b^2)\, dx\, dy . \tag{3.261}$$

Using the mean value theorem, the energy density integral equals

$$\frac{1}{2} \int_{B_1^2}^{B_2^2} \nu(x, y, b^2)\, d(b^2) = \frac{1}{2} \nu(x, y, B_3^2)(B_2^2 - B_1^2) , \tag{3.262}$$

for some B_3 between B_1 and B_2. For small field variations, Equation (3.258) is again a valid approximation. In this case, since $B_1 \approx B_3 \approx B_2$, we could set $\nu_E(x, y)$ equal to the reluctivity for either B_1 or B_2.

Orthogonality Property

The proof of the Galerkin orthogonality property (see §3.9.2) hinges on the fact that both the true solution $A^* \in \mathcal{H}^1$ and the FEA solution $A \in \mathcal{X}^N$ satisfy the weak form. We have shown that A^* satisfies the weak form (3.251) with $\nu_E(x, y) = \nu(x, y, B^{*2})$. On the other hand, A will satisfy a *different* weak form with $\nu_E(x, y) = \nu(x, y, B^2)$, where $B = ||\nabla A||$. The difference is due to the fact that linear elements yield a piecewise constant reluctivity since B itself is constant in each element. A small discrepancy could thus be present in the energetic inner products. (This discrepancy will exist even for higher-order elements because it is unlikely that the reluctivity distribution of the approximate solution will match that of the exact solution.)

To proceed, we need to decide which ν_E to use. For instance, setting ν_E to the FEM reluctivity distribution, we obtain

$$(\eta, A)_E = (\eta, J) , \tag{3.263a}$$

$$(\eta, A^*)_E \neq (\eta, J) \quad \text{(but approximately equal)} , \tag{3.263b}$$

for arbitrary $\eta \in \mathcal{X}_0^N$. If the domain mesh consists of very small elements, and if we have a good approximation of the solution so that $B \approx B^*$ within each element, then it is probably safe to assume that the discrepancy is small. It is also helpful that the reluctivity ν within ferromagnetic materials is orders of magnitude smaller than ν_0 (the reluctivity of air and copper). Hence, subtracting the two equations, we obtain an approximate orthogonality property:

$$(\eta, e)_E \approx 0 . \tag{3.264}$$

The energetic norm squared of the error between the FEA solution and the true solution, $e = A - A^*$, is

$$||e||_E^2 = (A - A^*, A - A^*)_E \tag{3.265a}$$

$$\begin{aligned} &= (A - A^*, A - \tilde{A})_E + (A - A^*, \tilde{A} - A^*)_E && (3.265b)\\ &= (\ e\ ,\ \eta\ )_E + (\ e\ , \tilde{A} - A^*)_E && (3.265c)\\ &\approx \quad 0 \quad + (\ e\ , \tilde{A} - A^*)_E\,, && (3.265d)\end{aligned}$$

where $\tilde{A} \in \mathcal{X}^N$ is some other function in the trial space, and $\eta = A - \tilde{A} \in \mathcal{X}_0^N$ is a test function. Invoking Cauchy–Schwarz, we obtain the approximate inequality

$$||e||_E = ||A - A^*||_E \lessapprox ||\tilde{A} - A^*||_E \ \text{ for any } \tilde{A}\,. \tag{3.266}$$

This result indicates that the Galerkin solution is approximately the closest we can get to the true solution in terms of the energetic norm.

Does this mathematical result have a physical significance? It turns out that the answer lies with the functional. We will show that the energetic norm of the error is related to the deviation of the functional from its value at the true solution, $I(A^*)$, which is the minimum over all functions in the Hilbert space $\mathcal{H}^1$. In particular, *the FEA solution is the closest to the true solution in terms of coenergy, among all functions in the trial space.*

Proof First, we take the difference of the functional from the true solution A^* to the FEA solution A, which is

$$I(A) - I(A^*) = W_f(A) - W_f(A^*) - (e, J)\,. \tag{3.267}$$

Using Equation (3.259), and ignoring approximations originating from nonlinearities, we have

$$I(A) - I(A^*) = (e, A^*)_E + \frac{1}{2}||e||_E^2 - (e, J)\,. \tag{3.268}$$

Since $e \in \mathcal{H}_0^1$, in view of Equation (3.251), we obtain

$$I(A) - I(A^*) = \frac{1}{2}||e||_E^2\,. \tag{3.269}$$

Similarly, for any other function $\tilde{A}$ in the space of trial functions $\mathcal{X}^N$, we have

$$I(\tilde{A}) - I(A^*) = \frac{1}{2}||\tilde{A} - A^*||_E^2\,. \tag{3.270}$$

(These hold as approximate equalities in the case of nonlinear materials, for relatively small deviations from A^*.) Now, Equation (3.266) comes in handy, as it suggests that

$$I(A^*) \le I(A) \le I(\tilde{A})\,. \tag{3.271}$$

□

3.10 Summary

The main objective of this chapter was to introduce the FEM. We started with abstract Laplace and Poisson problems, and then gradually adapted the method to the idiosyncrasies of electric machines, including magnetic nonlinearities and the presence of permanent magnet materials. Our approach was centered around the minimization of a functional using linear triangular elements. We also presented implementation details regarding the programming of the method using Python. Finally, we discussed the main features of Galerkin's method, thus establishing a basic theoretical background for the FEM.

Readers now have at their disposal all the basic building blocks for computing and visualizing the solution of a 2-D linear or nonlinear quasi-magnetostatic problem. In the next chapter, we will perform FEA on several types of electrical machines, and we will calculate quantities of interest such as torque and flux linkage.

3.11 Further Reading

In developing this material, we have stood on the shoulders of giants who pioneered the FEM. Notable books that have influenced us are [3–14]. These references contain a wealth of additional information that we could not include in this work, such as different types of elements for 2-D and 3-D FEA, and the numerical properties of the method. A few early papers documenting the adaptation of the FEM (which was originally used for civil and mechanical engineering problems) for the analysis of nonlinear magnetic systems and rotating electrical machinery are [15–29].

Problems

3.1 Provide an argument that $\mathbf{S}^k$ is a positive semi-definite matrix, i.e., show that

$$(\mathbf{W}^k)^\top \mathbf{S}^k \mathbf{W}^k \geq 0 . \tag{3.272}$$

Which choice of $\mathbf{W}^k$ yields zero?

3.2 Derive Equation (3.37) for the two-element example.

3.3 Prove that[k]

$$\iint_{\Delta_k} \alpha_i^k(x, y)\, dx\, dy = \frac{\Delta_k}{3} , \tag{3.273}$$

[k] *Hint.* In problems 3.3 and 3.5, the following result from multivariate calculus will simplify the effort. Let

$$F = \iint_{\mathcal{D}} \phi(x, y)\, dx\, dy ,$$

and suppose $x = f(u, v)$ and $y = g(u, v)$. $\mathcal{D}^*$ is the region of the u–v plane obtained by mapping the original region $\mathcal{D}$ of the x–y plane through this transformation. Then

$$F = \iint_{\mathcal{D}^*} \phi[f(u, v), g(u, v)] \frac{\partial(x, y)}{\partial(u, v)}\, du\, dv ,$$

where the symbol $\partial(x, y)/\partial(u, v)$ denotes the Jacobian of the transformation, defined by the determinant

$$\frac{\partial(x, y)}{\partial(u, v)} = \begin{vmatrix} \frac{\partial x}{\partial u} & \frac{\partial x}{\partial v} \\ \frac{\partial y}{\partial u} & \frac{\partial y}{\partial v} \end{vmatrix} .$$

Establish a transformation that maps (x_1, y_1), (x_2, y_2), and (x_3, y_3) to $(0, 0)$, $(1, 0)$, and $(0, 1)$, respectively. Integrations are much more easily carried out in the u–v plane.

for all i. What does this imply for the average value of the solution over an element?

3.4 Prove that the stiffness matrix is invariant to coordinate axis translation and/or rotation. Provide a physical meaning for this property.

3.5 Suppose that the current density inside a triangular element, $J_k(x, y, \mathbf{J}^k)$, is a linear interpolate of the vertex values $\mathbf{J}^k$, as in Equation (3.53) for the MVP. Prove that

$$\iint_{\Delta_k} A_k(x, y, \mathbf{A}^k)\, J_k(x, y, \mathbf{J}^k)\, dx\, dy = (\mathbf{A}^k)^\top \mathbf{T}^k \mathbf{J}^k , \tag{3.274}$$

where

$$\mathbf{T}^k = \frac{\Delta_k}{12} \begin{bmatrix} 2 & 1 & 1 \\ 1 & 2 & 1 \\ 1 & 1 & 2 \end{bmatrix} . \tag{3.275}$$

3.6 Derive the complete stiffness matrix for Example 3.1.

3.7 Explain why the inductance calculation based on the average MVP method will be independent of source current (in a magnetically linear device).

3.8 Consider the square-core inductor of Example 3.4. Write an FEA program that:

1. Uses the Triangle program to generate a mesh.
2. Determines the magnetic potential at each node.
3. Plots the equipotential contours.
4. Plots the flux density.
5. Plots the A-field along a horizontal axis of symmetry.

In other words, replicate the results of this example. Also, provide a discussion of the results.

3.9 Consider the square-core inductor of Example 3.4. Calculate the inductance from (i) the field energy obtained from an FEA solution; and (ii) the flux linkage. Compare this with an analytically calculated value, obtained using an elementary magnetic equivalent circuit. For these calculations, assume that the device depth is 30 mm.

3.10 Repeat Problem 3.8 by exploiting symmetry, thus meshing only the top half of the device.

3.11 Recall the concepts of Box 2.3 on page 158. Show that the flux linkage expression (3.108) is such that $\boldsymbol{\lambda} \cdot \mathbf{i} = \ell \mathbf{A}^\top \mathbf{b}$, where $\boldsymbol{\lambda}$ is a vector of coil flux linkages, $\mathbf{i}$ contains coil currents, ℓ is the axial length of a 2-D device, and $\mathbf{A}^\top \mathbf{b}$ is the second term of the magnetostatic functional.

3.12 Write a function that calculates the flux linkage of a coil. Use this to replicate the results of Example 3.5. (It is highly unlikely that your triangulation will match exactly what was used in this example, so your answer will probably be slightly different.)

3.13 Replicate the results of Example 3.6.

3.14 Explain why permeances are independent of the number of turns.

3.15 Replicate the results of Example 3.7. Adjust various geometric and mesh-specific problem parameters, and describe what you observe.

3.16 Show that the element Hessian can be expressed in terms of the B-field components as

$$\mathbf{H}^k = \frac{1}{4\Delta_k}\begin{bmatrix}\mathbf{r}^k & -\mathbf{q}^k\end{bmatrix}\left\{2\frac{d\nu_k}{d(B_k^2)}\begin{bmatrix}B_x^k\\ B_y^k\end{bmatrix}\begin{bmatrix}B_x^k & B_y^k\end{bmatrix} + \nu_k\mathbb{I}_2\right\}\begin{bmatrix}(\mathbf{r}^k)^\top\\ (-\mathbf{q}^k)^\top\end{bmatrix}. \tag{3.276}$$

Furthermore, show that the term in the curly brackets is an incremental reluctivity tensor relating $\delta\vec{\mathbf{H}}^k$ and $\delta\vec{\mathbf{B}}^k$.

3.17 Derive the B–H curve extrapolation formulas and replicate the plots of §3.6.1.

3.18 Replicate the results of Example 3.8. You should experiment with both fixed-point and Newton–Raphson algorithms (the latter should converge much faster). To visualize the rate of convergence, you may plot the variation of the normalized residual (on a logarithmic y-axis) and the field energy or coenergy as a function of iteration number. Also, plot the flux lines and the B-field magnitude for various operating points. What conclusions can you draw?

3.19 Consider a PM with uniform magnetization $\vec{\mathbf{M}}_0$. Show that the contribution to the gradient from any internal vertex due to the permanent magnetization is zero. Also, explain what happens for nodes at the boundary of the PM.

References

[1] S. Boyd and L. Vandenberghe, *Convex Optimization.* Cambridge: Cambridge University Press, 2004.

[2] M. J. Gander and G. Wanner, "From Euler, Ritz, and Galerkin to modern computing," *SIAM Review*, vol. 54, no. 4, pp. 627–666, 2012.

[3] J. P. A. Bastos and N. Sadowski, *Electromagnetic Modeling by Finite Element Methods.* New York: Marcel Dekker, 2003.

[4] N. Bianchi, *Electrical Machine Analysis Using Finite Elements.* Boca Raton, FL: CRC Taylor & Francis, 2005.

[5] K. Eriksson, D. Estep, P. Hansbo, and C. Johnson, *Computational Differential Equations.* Lund, Sweden: Studentlitteratur, 1996.

[6] K. Hameyer and R. Belmans, *Numerical Modeling and Design of Electrical Machines and Devices.* Southampton, UK: WIT Press, 1999.

[7] S. R. H. Hoole, Ed., *Finite Elements, Electromagnetics and Design.* Amsterdam: Elsevier Science, 1995.

[8] J.-M. Jin, *The Finite Element Method in Electromagnetics*, 3rd ed. Hoboken, NJ: Wiley-IEEE Press, 2014.

[9] C. Johnson, *Numerical Solution of Partial Differential Equations by the Finite Element Method.* Cambridge: Cambridge University Press, 1987.

[10] J. N. Reddy, *Introduction to the Finite Element Method*, 4th ed. New York: McGraw-Hill, 2019.

[11] S. J. Salon, *Finite Element Analysis of Electrical Machines.* Boston: Springer, 1995.

[12] P. P. Silvester and R. L. Ferrari, *Finite Elements for Electrical Engineers*, 3rd ed. Cambridge: Cambridge University Press, 1996.

[13] G. Strang and G. J. Fix, *An Analysis of the Finite Element Method*, 2nd ed. Wellesley, MA: Wellesley-Cambridge Press, 2008.

[14] O. C. Zienkiewicz, R. L. Taylor, and J. Z. Zhu, *The Finite Element Method: Its Basis and Fundamentals*, 7th ed. Amsterdam: Butterworth-Heinemann, 2013.

[15] P. Silvester and M. V. K. Chari, "Finite element solution of saturable magnetic field problems," *IEEE Trans. Power Apparatus & Systems*, vol. 89, no. 7, pp. 1642–1651, Sep./Oct. 1970.

[16] M. V. K. Chari and P. Silvester, "Analysis of turboalternator magnetic fields by finite elements," *IEEE Trans. Power Apparatus & Systems*, vol. 90, no. 2, pp. 454–464, Mar./Apr. 1971.

[17] P. Silvester, H. S. Cabayan, and B. T. Browne, "Efficient techniques for finite element analysis of electric machines," *IEEE Trans. Power Apparatus & Systems*, vol. 92, no. 4, pp. 1274–1281, Jul. 1973.

[18] M. V. K. Chari, "Finite-element solution of the eddy-current problem in magnetic structures," *IEEE Trans. Power Apparatus & Systems*, vol. 93, no. 1, pp. 62–72, Jan. 1974.

[19] A. Foggia, J. C. Sabonnadiere, and P. Silvester, "Finite element solution of saturated travelling magnetic field problems," *IEEE Trans. Power Apparatus & Systems*, vol. 94, no. 3, pp. 866–871, May 1975.

[20] A. Y. Hannalla and D. C. MacDonald, "Numerical analysis of transient field problems in electrical machines," *Proc. IEE*, vol. 123, no. 9, pp. 893–898, Sep. 1976.

[21] N. A. Demerdash and N. K. Lau, "Flux penetration and losses in solid nonlinear ferromagnetics using state space techniques applied to electrical machines," *IEEE Trans. Magnetics*, vol. 12, no. 6, pp. 1039–1041, Nov. 1976.

[22] N. A. Demerdash and T. W. Nehl, "Use of numerical analysis of nonlinear eddy current problems by finite elements in the determination of parameters of electrical machines with solid iron rotors," *IEEE Trans. Magnetics*, vol. 15, no. 6, pp. 1482–1484, Nov. 1979.

[23] N. A. Demerdash, T. W. Nehl, and F. A. Fouad, "Finite element formulation and analysis of three dimensional magnetic field problems," *IEEE Trans. Magnetics*, vol. 16, no. 5, pp. 1092–1094, Sep. 1980.

[24] M. V. K. Chari, S. H. Minnich, Z. J. Csendes, J. Berkery, and S. C. Tandon, "Load characteristics of synchronous generators by the finite-element method," *IEEE Trans. Power Apparatus & Systems*, vol. 100, no. 1, pp. 1–13, Jan. 1981.

[25] J. W. Dougherty and S. H. Minnich, "Finite element modeling of large turbine generators; calculations versus load test data," *IEEE Trans. Power Apparatus & Systems*, vol. 100, no. 8, pp. 3921–3929, Aug. 1981.

[26] F. A. Fouad, T. W. Nehl, and N. A. Demerdash, "Magnetic field modeling of permanent magnet type electronically operated synchronous machines using finite elements," *IEEE Trans. Power Apparatus & Systems*, vol. 100, no. 9, pp. 4125–4135, Sep. 1981.

[27] S. H. Minnich, M. V. K. Chari, and J. F. Berkery, "Operational inductances of turbine-generators by the finite-element method," *IEEE Trans. Power Apparatus & Systems*, vol. 102, no. 1, pp. 20–27, Jan. 1983.

[28] S. C. Tandon, A. F. Armor, and M. V. K. Chari, "Nonlinear transient finite element field computation for electrical machines and devices," *IEEE Trans. Power Apparatus & Systems*, vol. 102, no. 5, pp. 1089–1096, May 1983.

[29] M. P. Krefta and O. Wasynczuk, "A finite element based state model of solid rotor synchronous machines," *IEEE Trans. Energy Conv.*, vol. 2, no. 1, pp. 21–30, Mar. 1987.

4 Electric Machine FEA Implementation Guidelines

Πολυμαθίη νόον ἔχειν οὐ διδάσκει

"Learning many things does not teach one to have an intelligent mind"

Heraclitus[73]

Whereas the previous chapter introduced the main ideas of the finite element method (FEM), this chapter presents practical finite element analysis (FEA) implementation guidelines specifically for electric machines. First, in §4.1 we will take a look at various methods for calculating force and torque as a post-processing step based on an FEA solution. In §4.2, we will explain how to interface the turning rotor with the stator, without actually re-meshing the part that moves.

Then, we will begin the analysis of various electric machines in earnest, based on our knowledge of magnetostatic FEA. We will define the geometric details of the mesh for each topology, and will then turn our attention to the interpretation of the FEA results and what these imply for the operation of these machines. In §4.3, we will discuss the FEA of two-phase and three-phase round-rotor and salient-pole synchronous machines that have a field winding on the rotor. This includes details regarding the analysis in the rotor reference frame using Park's transformation, and the derivation of *qd* equivalent-circuit parameters from FEA studies. §4.4 is about surface-mounted and interior permanent-magnet synchronous machines, and includes a discussion of power electronic converter control strategies. §4.5 focuses on the switched reluctance machine. Finally, in §4.6, we will explain how the multi-pole periodicity of the magnetic field can be exploited to reduce computational burden.

4.1 Force and Torque Calculation

A theoretical treatment of the development of electromagnetic forces can be found in Chapter 2. We have seen that forces can be obtained either as a derivative of the coenergy (a global scalar quantity) or as integrals of the magnetic field distribution in space. These results need to be adapted to the discretized nature of the FEM.

We will present various methods that have been proposed for computing force or torque in electromechanical devices from an FEA solution. We will not attempt to set forth all such methods that have been devised, which the reader can obtain from the technical literature on this subject. The approaches that will be discussed are based on conventional triangular elements. The underlying algorithms are general purpose, meaning that the calculations are not calibrated with some sort of *a priori* knowledge about the solution. We will refrain from making claims regarding accuracy, or whether one method should be preferred over another. We will, however, be paying attention to the derivation and interpretation of each formula.

4.1.1 Coenergy Method

The most direct and perhaps easiest to code approach is a coenergy method, based on Equation (2.234):

$$F_e = \left.\frac{\partial W_c(\mathbf{i}, \theta)}{\partial \theta}\right|_{\mathbf{i}\ \text{constant}}. \tag{4.1}$$

Here, F_e can be either force or torque, depending on whether the physical meaning of θ is a linear or rotational displacement, respectively.

To implement this method, first we obtain the numerical value of the coenergy (i.e., the negative of the functional $I(A)$, adjusted for device axial length ℓ) for a given winding current vector $\mathbf{i}$ and position θ. Computing the coenergy is relatively straightforward after the Newton–Raphson functional minimization algorithm has converged. We then make a small but finite perturbation in position $\delta\theta$, keeping currents constant. We thus obtain a new MVP solution and the associated coenergy. The force is found as the numerical derivative

$$F_e \approx \frac{W_c(\mathbf{i}, \theta + \delta\theta) - W_c(\mathbf{i}, \theta)}{\delta\theta} = \ell\,\frac{I(A; \mathbf{i}, \theta) - I(A; \mathbf{i}, \theta + \delta\theta)}{\delta\theta}. \tag{4.2}$$

The result is approximate for two reasons: (i) because W_c is not the real coenergy of the system (as we discussed in the previous chapter, $W_c = -\ell I(A)$ is as close as we can get to the actual coenergy value, $-\ell I(A^*)$, with a function A from the trial space $\mathcal{X}^N$), and (ii) because the derivative is computed numerically.

For small displacement, we could save a bit of time by avoiding remeshing the domain, simply stretching the elements in an air-gap layer separating the stationary and movable members. The creation of a distortable air-gap layer is also sensible from a numerical standpoint because then we avoid numerical error associated with working with two potentially different triangulations of the domain. However, this approach requires additional effort to code properly.

The value of $\delta\theta$ that ensures good accuracy in the numerical derivative calculation should be selected carefully. We should not make $\delta\theta$ so small that the difference between $W_c(\mathbf{i}, \theta + \delta\theta)$ and $W_c(\mathbf{i}, \theta)$ approaches the finite precision of the computer; otherwise, round-off errors will affect the answer. On the other hand, we should not make $\delta\theta$ too large either; otherwise, the second- and higher-order terms in the Taylor series expansion of W_c about θ will become significant. It is reasonable to assume that a compromise can be achieved if W_c is "well behaved" for small changes in θ.

A disadvantage of the coenergy method is that it requires two separate FEA solutions. Nevertheless, the Newton–Raphson algorithm at $\theta + \delta\theta$ may converge quickly because it can be initialized from the solution at θ. In contrast, the other methods that will be presented in this chapter can provide the answer from a single FEA solution.

An alternative interpretation of the coenergy method can be obtained if we recall the mean value theorem from calculus. Suppose that our analysis entails the evaluation of the solution at a sequence of positions θ_k, $k = 1, 2, \ldots,$ under constant current. Then, under the mild assumption that the field energy function is differentiable with respect to position θ, the mean value theorem tells us that there exists a point θ' in the interval $[\theta_k, \theta_{k+1}]$ such that

$$F_e(\theta') = \frac{W_c(\mathbf{i}, \theta_{k+1}) - W_c(\mathbf{i}, \theta_k)}{\theta_{k+1} - \theta_k}. \tag{4.3}$$

Thus, we can estimate the force or torque in a device without the need to conduct additional FEA studies, but at the expense of an unknown "phase shift" in the waveform. This method is used in the example that follows.

Example 4.1 *FEA of a cylindrical electromagnetic plunger.*

We will conduct FEA of a cylindrical electromagnetic plunger, similar to the one of Example 2.16 on page 167. The implementation will be based on the material of §3.8 on page 294. The FEA unknown is a modified MVP, $\tilde{A} = rA_\phi$. Our primary interest is to calculate the force using the coenergy method. The ideas presented earlier in this section still apply, but minor modifications are needed since we are not dealing with the same kind of 2-D problem (e.g., we should not be multiplying the functional by ℓ but rather by 2π). Our secondary objective is to calculate the inductance of the solenoid.

Solution

The plunger dimensions are defined in Figure 4.1. Let us suppose that:

- Plunger radius $r_1 = 0.75$ mm.
- Plunger width $w_p = 7.5$ mm.
- Core width $w = 10$ mm.
- Core inner radius $r_2 = 0.85$ mm.
- Core outer radius $r_3 = 3$ mm.
- Core yoke widths $w_1 = 0.2$ mm, $w_2 = w_4 = 0.5$ mm.
- Stator pole radius r_1 (equal to plunger radius).
- Stator pole width $w_3 = 2$ mm.
- The coil is separated by a 0.1-mm gap from the core on all sides. So, the coil width along the radial direction is $w_{cr} = r_3 - r_2 - w_1 - 0.1 = 1.85$ mm, and the axial coil width is $w_{cz} = w - w_2 - w_4 - 0.2 = 8.8$ mm.

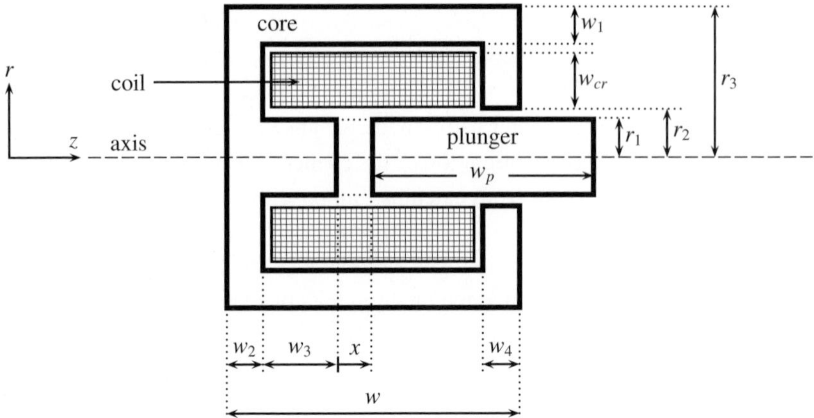

Figure 4.1 Geometric parameter definitions for FEA of an axisymmetric electromagnetic plunger.

- The solenoid has $N = 20$ turns of copper AWG-20 magnet wire, so the slot has packing factor $k_{pf} \approx 0.7$. The coil current is $i = 2$ A.
- The core and plunger are solid steel pieces, with the *B–H* curve that we used previously to plot Figure 3.23 on page 289.

The problem is solved over a rectangular domain, and the boundary condition $\tilde{A} = 0$ is imposed on all four sides. One of the domain sides coincides with the axis of symmetry, $r = 0$. The other three sides are 0.25 mm away from the core and plunger steel. The mesh has approximately 2,700 nodes and 5,200 elements. (These values change depending on the plunger position.)

Illustrative FEA results for three plunger positions, $x = 0$, $x = 1$ mm, and $x = 2$ mm, are depicted in Figure 4.2. We can observe in Figure 4.2(a) that the *B*-field is strongest when the plunger is contracted, reaching values close to 1.5–2 T within the plunger. The flux density reduces radially outwards because the flux spreads over a wider area. At $x = 1$ mm in Figure 4.2(b), the field has weakened by an order of magnitude. The field keeps dropping as we increase x further, as in Figure 4.2(c). We can also observe a significant component of leakage flux through the slot, which increases as the plunger is moved outwards.

For force and inductance calculation, we use a range of closely spaced plunger positions $x \in [0, 0.5]$ mm. At each point, we compute the final, converged value of the nonlinear functional (which equals the negative coenergy) by collecting contributions from all triangles using Equation (3.213) multiplied by a factor of 2π. Then we apply the coenergy method to calculate the force using a numerical derivative as in Equation (4.3). Also, we compute the solenoid inductance as $L(x) = \lambda(x)/i$, where the flux linkage is obtained using Equation (3.218). The results are shown in Figure 4.3. The maximum attractive force is found to be roughly 0.8 N. The force then drops rapidly for increasing values of x. The oscillation of the force waveform close to $x = 0$ could be attributed to nonlinear effects related to the saturation of the core material. Clearly, this

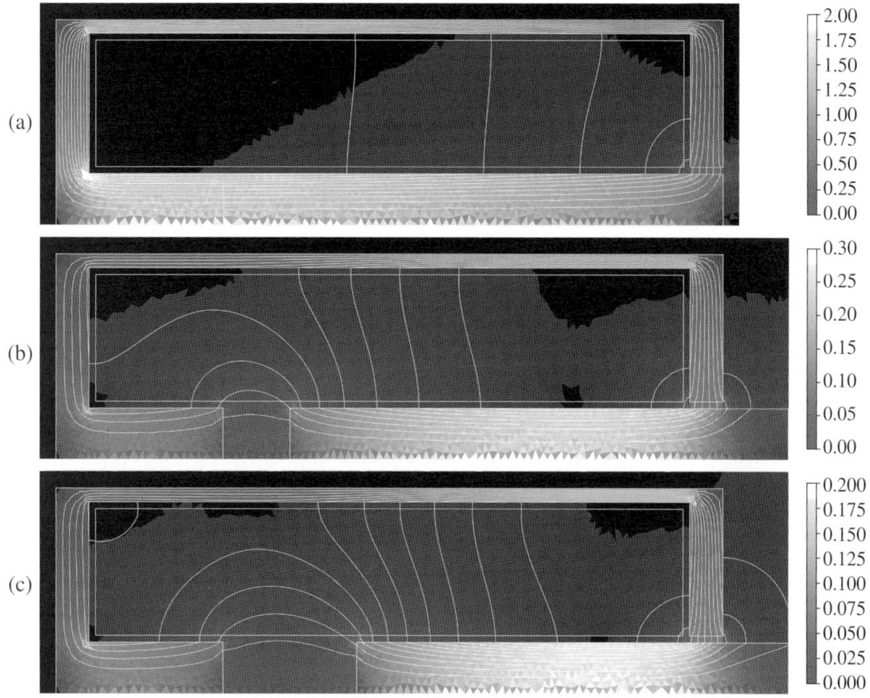

Figure 4.2 Contours of $\tilde{A}$ superimposed on B-field magnitude color plots for different plunger positions. (The figure is cropped on the right, so the plunger is not shown in its entirety in (c).)

plunger design is not very effective for values of x larger than 0.2 mm, where the force is negligible. The inductance is on the order of a few μH, and appears to be more or less proportional to the coenergy, which is representative of the response of a magnetically linear system (although this system is nonlinear). As a sanity check, we are superimposing the results of an elementary magnetic circuit-based analysis using the formulas of Example 2.16. The results exhibit similar trends; however, the magnetic circuit derivation involve several simplifying assumptions, such as the absence of slot leakage flux and uniform flux density over the pole surfaces, affecting the accuracy of the answer.

4.1.2 Maxwell Stress Tensor Methods

The Maxwell stress tensor (MST) method can be used for the calculation of force or torque on objects of arbitrary shape. Hence, it is a generic, powerful method that is widely used. Recall that the MST method for resultant force calculation in a quasi-magnetostatic problem is based on a surface integral of the stress $\vec{\mathbf{t}} \approx \vec{\mathbf{t}}_m$[a], as in Equation (2.328) on page 187:

$$\vec{\mathbf{F}}_e = \oint\!\!\oint_{\partial V} \vec{\mathbf{t}}\, da = \oint\!\!\oint_{\partial V} \nu_0 \left[(\vec{\mathbf{B}} \cdot \hat{\mathbf{n}})\vec{\mathbf{B}} - \tfrac{1}{2}B^2\, \hat{\mathbf{n}}\right] da\,. \tag{4.4}$$

[a] The approximation is because we are dropping the insignificant electromagnetic field momentum term and the feeble contribution from the electric field tensor.

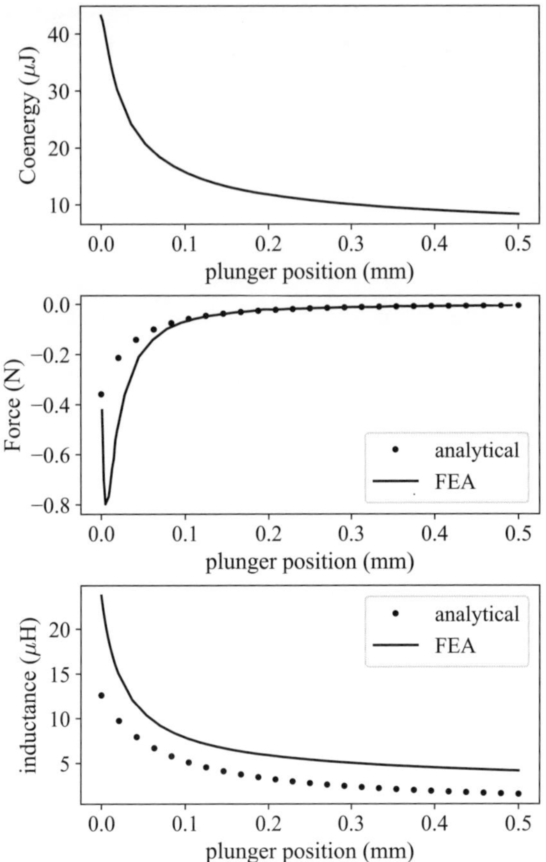

Figure 4.3 Electromagnetic plunger coenergy, force, and inductance as a function of position.

The integration surface should contain the entire moving body. In a 2-D problem, we obtain the resultant force per unit depth as a line integral:

$$\tilde{\mathbf{F}}_e = \vec{\mathbf{F}}_e/\ell = \oint \nu_0 \left[(\vec{\mathbf{B}} \cdot \hat{\mathbf{n}})\vec{\mathbf{B}} - \tfrac{1}{2}B^2\,\hat{\mathbf{n}}\right] dr\,. \tag{4.5}$$

This integration should be interpreted as a component-wise integration of two scalar fields (e.g., $\tilde{F}_x$ and $\tilde{F}_y$).

For example, let us refer to the geometry of the linear-motion electromagnet of Problem 4.12 shown in Figure 4.78 on page 422, where a rectangular path can be readily defined surrounding the moving body. From Equations (2.334) and (2.335), we can express normal and tangential force components (with respect to the integration path) as

$$d\tilde{F}_n = \nu_0\left(B_n^2 - \tfrac{1}{2}B^2\right) dr = \tfrac{\nu_0}{2}(B_n^2 - B_s^2)\, dr \tag{4.6}$$

and

$$d\tilde{F}_s = \nu_0 B_n B_s\, dr\,, \tag{4.7}$$

respectively. Here, B_n and B_s are normal and tangential B-field components along the integration path, taking the values of $\pm B_x$ or $\pm B_y$. In this case, the problem geometry is conducive to the alignment of a rectangular integration path with the coordinate axes, which facilitates a direct evaluation of $\tilde{F}_x$ and $\tilde{F}_y$.

Similarly, electromagnetic torque is given by Equation (2.351):

$$\vec{\tau}_e = \oiint_{\partial V} (\vec{\mathbf{r}} \times \vec{\mathbf{t}})\, da\,. \tag{4.8}$$

In 2-D electric machine geometries, we may integrate over a cylindrical surface of radius R, thus obtaining the z-axis torque per unit depth as the integral of Equation (2.357) on page 193:

$$\tilde{\tau}_{ez} = \tau_{ez}/\ell = \nu_0 R^2 \int_0^{2\pi} B_\phi\, B_r\, d\phi\,, \tag{4.9}$$

where B_ϕ and B_r are tangential and radial components along the integration path (assuming a polar coordinate system positioned at the center of rotation).

In an FEA computer program, the MST integrals can be computed as finite sums. It should be noted that the calculation is approximate for two reasons. (i) The MST force and torque calculation formulas have been derived by vector calculus manipulations on the true (twice-differentiable) field solution. Therefore, an error is introduced whenever we substitute the approximate FEA solution $\vec{\mathbf{B}}$ in these formulas because $\vec{\mathbf{B}}$ deviates from the true solution $\vec{\mathbf{B}}^*$ by an unknown (hopefully small) amount.[b] (ii) The numerical estimation of an integral as a sum involving only a finite number of function values (a procedure also known as numerical quadrature) also introduces an error.

These two sources of error are illustrated in Figure 4.4. The solid line is the function we wish to integrate, e.g., it could be representative of $y = B_\phi\, B_r$ vs. $x = \phi$ from the torque formula (4.9). To evaluate the integral as a sum in the computer, we sample the FEA-based function value at certain points, shown by the white dots. We expect that the sampled values fall within a certain range from the true values, where the range should decrease by using a finer mesh. The figure shows an elementary integration as a sum of rectangular areas. If we find that the function varies rapidly between points, we could reduce the mesh size further to improve the accuracy of the quadrature. Without some *a priori* knowledge about the behavior of the solution, this would lead to an iterative procedure. Also, a mesh size reduction could lead to a more computationally demanding FEA. Instead, a more advanced quadrature method, such as the trapezoidal rule or Simpson's[74] rule, could be used.[c]

[b] In Example 3.7, we saw that the errors between $\vec{\mathbf{B}}$ and $\vec{\mathbf{B}}^*$ tend to be distributed evenly around the true solution. If this is the case, then we would hope that they would integrate to something insignificant, if an adequate number of sample points is used. Of course, this expectation is difficult to generalize, and perhaps such considerations are outside the scope of this text.

[c] Simpson's rule is based on a quadratic approximation of the integrand. If x_0, x_1, and x_2 are three equidistant points, with spacing $h = x_1 - x_0 = x_2 - x_1$, where we have function values f_0, f_1, and f_2, respectively, then it states that

$$\int_{x_0}^{x_2} f(x)\, dx = \int_{x_0}^{x_0+2h} f(x)\, dx \approx \frac{h}{3}(f_0 + 4f_1 + f_2)\,.$$

This formula can be readily extended to a larger interval with many sample points.

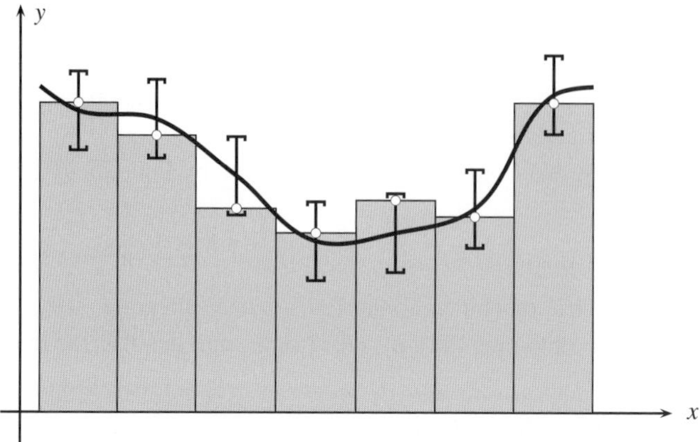

Figure 4.4 The FEA-based calculation of force/torque using the MST method can be interpreted as the numerical integration (quadrature) of a function where the integrand is uncertain.

MST Torque Calculation Implementation

Let us focus on the calculation of torque in an electric machine. The MST method requires an integration path. In theory, if we knew the true solution that lives in the Hilbert space $\mathcal{H}^1$, the choice of path should not matter, as long as it surrounds the entire moving body. However, the numerical quadrature of an FEA solution that lives in $\mathcal{X}^N$ is path dependent. When linear elements are used, $\vec{\mathbf{B}}$ is constant in each triangle. In other words, we can sample a unique field value at an arbitrary point within each triangle. These sample points should lie on the presumed path of integration.

It is customary and practical to integrate on a path that crosses through a thin layer of elements within the air gap, as shown in Figure 4.5. When defining the mesh, we make sure that such a layer is created to enable the calculation of torque. This layer is bounded by two circles that are concentric with the rotor. Suppose the middle of the air-gap layer is a circle of radius R. The inner circle has radius $r_i = R - w/2$, whereas the outer circle has radius $r_o = R + w/2$, where w is the layer width. The layer can be created by adding an adequate number of triangle vertices on the two circles. For instance, we may add $N = 720$ nodes with coordinates

$$x_n = r_o \cos(n-1)\frac{2\pi}{N}, \quad y_n = r_o \sin(n-1)\frac{2\pi}{N}, \tag{4.10}$$

with $n = 1, 2, \ldots, N$. This leads to an outer polygon with 0.5° angular spacing between nodes. We then add another N nodes at r_i. Note that the angular position of the inner nodes can be arbitrary; for instance, we do not have to place them exactly opposite the outer nodes. (This capability will become important when accounting for the rotation of the rotor.) We should choose a layer width such that

$$r_o - r_i = w \approx \frac{2\pi}{N}R \quad \text{(where “}\approx\text{” is not enforced strictly)}, \tag{4.11}$$

so that the triangles are somewhat close to being equilateral.

One possible choice is a *polygonal path of integration*. One way to form a polygonal

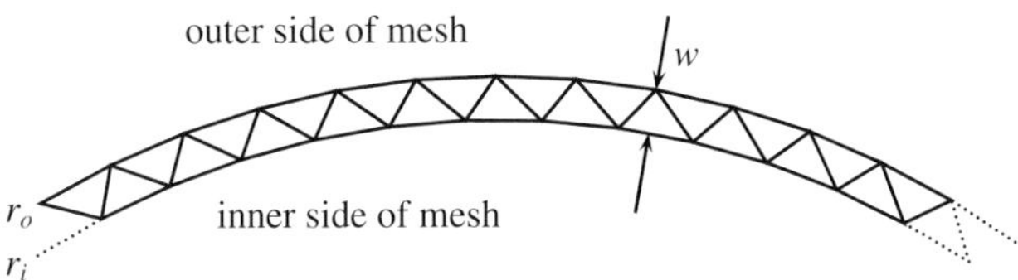

Figure 4.5 Triangulation of a thin layer in the air-gap region. (The surrounding mesh is not shown for clarity.)

path is by connecting triangle sides at a given ratio γ, $0 < \gamma < 1$, with γ increasing outwards (as defined in Figure 4.6). A common choice is to use a path that connects the midpoints of triangle sides ($\gamma = 1/2$) so that the integration path is divided almost equally between triangles. We are thereby exploiting information contained in the *B*-field solution from all air-gap layer triangles to the same extent. In triangles like *BCD* or *ADE*, the length of the line segment that connects the triangle sides is $\xi_k = (1 - \gamma)\ell_k$, where ℓ_k is the length of the triangle side parallel to the line segment. In triangles like *ABD*, the line segment length is $\xi_k = \gamma\ell_k$.

The torque (per unit depth) is computed by evaluating Equation (4.8) as a sum over all triangles in the air-gap layer:

$$\tilde{\tau}_{ez} = \oint (\vec{\mathbf{r}} \times \vec{\mathbf{t}})_z \, dr = \sum_{k \in \mathcal{E}_{ag}} \int_{\vec{\mathbf{r}}_1^k}^{\vec{\mathbf{r}}_2^k} (\vec{\mathbf{r}}^k \times \vec{\mathbf{t}}^k)_z \, dr \,, \tag{4.12}$$

where both $\vec{\mathbf{r}}^k$ and $\vec{\mathbf{t}}^k$ are functions of position in each triangle k from the set of air-gap layer elements $\mathcal{E}_{ag}$, as we are integrating from $\vec{\mathbf{r}}_1^k$ to $\vec{\mathbf{r}}_2^k$. (At this point, we are still using the true solution, as if we knew it.) Because the polygonal path of integration consists of linear segments, the normal vector $\hat{\mathbf{n}}^k$ is constant over each segment. Now, to proceed with the numerical quadrature, we need to specify the location within each triangle where we will sample the integrand, which will determine the value of $\vec{\mathbf{r}}^k$. In the absence of any prior information about the solution, e.g., how rapidly it is changing, we will sample from the midpoint of the polygon edge in triangle k, denoted by $\vec{\mathbf{r}}_0^k =$

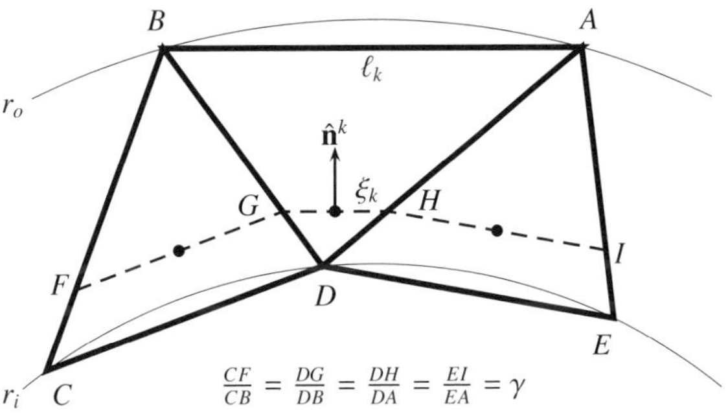

Figure 4.6 Polygonal path of integration for $\gamma = 0.25$ (exaggerated scale).

(x_0^k, y_0^k). Hence, we have obtained the **polygonal path quadrature**

$$\boxed{\tilde{\tau}_{ez} \approx \sum_{k \in \mathcal{E}_{ag}} (\vec{\mathbf{r}}_0^{\,k} \times \vec{\mathbf{t}}_0^{\,k})_z \, \xi_k = \sum_{k \in \mathcal{E}_{ag}} (x_0^k t_y^k - y_0^k t_x^k) \, \xi_k \, ,} \tag{4.13}$$

where $\vec{\mathbf{t}}_0^{\,k} = (t_x^k, t_y^k)$ is the linear-element FEA stress, which is constant in each triangle k. For example, referring to Figure 4.6, $\vec{\mathbf{r}}_0^{\,k}$ would point to the middles of IH, HG, GF, and so on; these points are shown as dots. This vector is simple to find based on the coordinates of the triangle vertices, so this method is relatively easy to code.

Alternatively, we can compute the torque based on a *circular integration path*, as in Equation (4.9). In this case, we draw a circle of radius r (so that $r_i < r < r_o$) within the air-gap layer, as shown in Figure 4.7. A common choice is to pick $r = (r_i + r_o)/2$ so that the integration path is divided almost equally among triangles. To break up the integral into a sum over triangles, we first need to identify the intersections of the circle with the element sides (shown as white dots in the figure). This leads to a set of quadratic equations, so this approach is somewhat more computationally intense and difficult to implement than the polygonal path method.

Once the intersection points have been identified, we can proceed with the calculation of the torque per unit depth:

$$\tilde{\tau}_{ez} = \oint (r\,\hat{\mathbf{n}} \times \vec{\mathbf{t}})_z \, r \, d\phi = r^2 \sum_{k \in \mathcal{E}_{ag}} \int_{\phi_1^k}^{\phi_2^k} (\hat{\mathbf{n}}^k \times \vec{\mathbf{t}}^{\,k})_z \, d\phi \, , \tag{4.14}$$

where ϕ_1^k and $\phi_2^k = \phi_1^k + \delta\phi_k$, $\delta\phi_k > 0$, are the angles of the intersection points with the sides of triangle k. Note that both $\hat{\mathbf{n}}^k$ and $\vec{\mathbf{t}}^{\,k}$ are still functions of ϕ, as if we are using the actual solution. In the absence of other prior information, we shall sample the integrand at the midpoint angle $\phi_0^k = (\phi_1^k + \phi_2^k)/2$. Using an elementary rectangular quadrature rule, this yields the **circular path quadrature**

$$\boxed{\tilde{\tau}_{ez} \approx r^2 \sum_{k \in \mathcal{E}_{ag}} (\hat{\mathbf{n}}_0^k \times \vec{\mathbf{t}}_0^{\,k})_z \, \delta\phi_k = \nu_0 r^2 \sum_{k \in \mathcal{E}_{ag}} B_\phi^k B_r^k \delta\phi_k \, .} \tag{4.15}$$

Here, the angle ϕ_0^k plays a key role: it defines the direction of the unit normal vector $\hat{\mathbf{n}}_0^k$

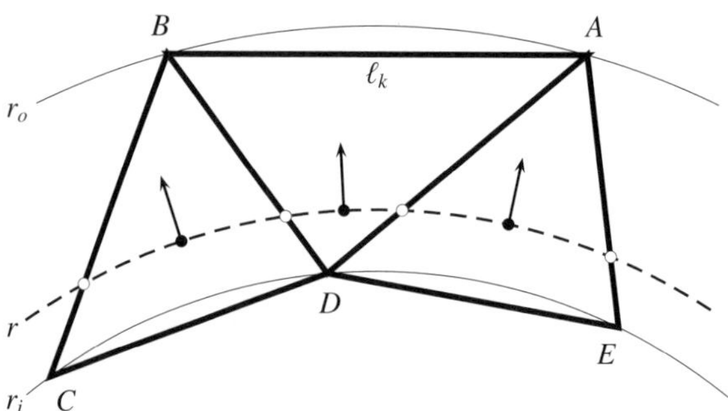

Figure 4.7 Circular path of integration for $r = r_i + 0.25(r_o - r_i)$ (exaggerated scale).

(shown by the arrows in the figure), which in turn enters into the formula for $\vec{\mathbf{t}}_0{}^k$, and also defines the axes of decomposition of the B-field in tangential and normal components, B^k_ϕ and B^k_r, respectively. The resemblance of Equation (4.15) with (4.9) is clear. The two quadrature formulas that we have derived, (4.13) and (4.15), yield similar but not identical answers.

A variant of the MST torque calculation method is commonly known as *Arkkio's method* [1]. This method hinges on the observation that Equation (4.9) yields the same answer regardless of radius. Hence, integrating both sides from r_i to r_o:

$$\tilde{\tau}_{ez} = \frac{\nu_0}{r_o - r_i} \int_{r_i}^{r_o} \int_0^{2\pi} r^2 B_\phi B_r \, d\phi \, dr \tag{4.16a}$$

$$= \frac{\nu_0}{r_o - r_i} \iint_{S_{ag}} r B_\phi B_r \, da \,, \tag{4.16b}$$

where S_{ag} denotes the surface of the air-gap annulus between r_i and r_o. Ideally, the air gap is bounded by two circles; however, in our conventional FEA implementation with triangular elements, the integration surface is formed by the inner and outer polygonal boundaries. A simple quadrature rule is to employ a sum that involves a single point in each triangle, which leads to **Arkkio's method quadrature**

$$\boxed{\tilde{\tau}_{ez} \approx \frac{\nu_0}{r_o - r_i} \sum_{k \in \mathcal{E}_{ag}} r_k B^k_\phi B^k_r \Delta_k \,.} \tag{4.17}$$

As before, we are free to choose where to evaluate the integrand. For instance, we may choose to use the element centroids to determine r_k and the decomposition of the B-field in tangential and radial components. We note that this formula does not require a thin layer of elements in the air gap; therefore, we could apply this to thicker layers and/or layers without any particular mesh pattern. From an implementation perspective, Arkkio's method is arguably the simplest method to code.

4.1.3 Virtual Displacement Method

The virtual displacement method (VDM) is based on Equation (2.229):

$$F_e = -\left.\frac{\partial W_f}{\partial \theta}\right|_{\vec{\mathbf{A}} \text{ constant over conductors}} . \tag{4.18}$$

We have seen an application of this method for torque calculation in §2.4.2, where we performed a virtual distortion of the air-gap region with an analytical expression of the air-gap field and energy in hand.

We will follow the same approach here. The simplest method is by surrounding the moving member with a *single layer* of elements. The method works best if there is a clear separation between the moving body and its surroundings, to allow for this air-gap layer to be defined. The position of the layer is arbitrary; it can be adjacent to the body or further out. This layer acts like a sponge that absorbs movement. It is distorted by keeping its outer side fixed in space, while the inner side is allowed to move in unison with the moving body. While this takes place, the MVP stays constant throughout space.

Hence, the field energy stays everywhere constant except in the air-gap layer, where we must calculate its change due to the distortion. The distortion is virtual in the sense that it does not actually occur, since the calculation is based on a partial derivative calculation.

The distorted layer is entirely in air, so it constitutes a magnetically linear region, which stores an amount of magnetic energy equal to

$$W_{ag} = \frac{\nu_0 \ell}{2} \sum_{k \in \mathcal{E}_{ag}} B_k^2 \Delta_k = \frac{\nu_0 \ell}{2} \sum_{k \in \mathcal{E}_{ag}} (\mathbf{A}^k)^\top \mathbf{S}^k \mathbf{A}^k , \tag{4.19}$$

where $\mathcal{E}_{ag}$ is a set of all element indices in the air-gap layer, ℓ is the device axial length, and $\mathbf{S}^k$ are element stiffness matrices. Therefore, force or torque (per unit depth) can be found by application of Equation (4.18):

$$\tilde{F}_e = \frac{F_e}{\ell} = -\frac{\nu_0}{2} \sum_{k \in \mathcal{E}_{ag}} (\mathbf{A}^k)^\top \frac{\partial \mathbf{S}^k}{\partial \theta} \mathbf{A}^k . \tag{4.20}$$

The main idea of the VDM is to calculate the derivatives of $\mathbf{S}^k$ analytically.

Force Calculation Using the VDM

An example of a virtual distortion for calculation of force is illustrated in Figure 4.8. The drawing represents a virtual displacement along the x-axis, which will yield the corresponding component of force, F_{ex}. Note that the mesh does not need to be aligned with the axes in any way. The black dots at the top and bottom signify that this layer extends upwards and downwards and eventually wraps around the moving part, which in this figure is on the right. The shape of the layer can be arbitrary. The depicted elements are stretched horizontally, whereas elements on the opposite side of the layer are compressed.

We identify two types of triangles in the layer, namely, those with only one vertex that moves, and those with two vertices that move. We will refer to the former as Type-1 triangles, and to the latter as Type-2 triangles. We will also introduce a local node numbering scheme, where the moving node(s) will be node #1 for Type-1 triangles, and nodes #1 and #2 for Type-2 triangles. Note that this local numbering may be different from the local numbering that the meshing program (e.g., Triangle) determines.

Let us recall the pertinent expressions for the kth element stiffness matrix:

$$S_{ij}^k = \frac{1}{4\Delta_k}(q_i^k q_j^k + r_i^k r_j^k) , \tag{4.21}$$

with

$$q_1^k = y_2^k - y_3^k , \qquad q_2^k = y_3^k - y_1^k , \qquad q_3^k = y_1^k - y_2^k , \tag{4.22a}$$

$$r_1^k = x_3^k - x_2^k , \qquad r_2^k = x_1^k - x_3^k , \qquad r_3^k = x_2^k - x_1^k , \tag{4.22b}$$

and

$$\Delta_k = \frac{1}{2}[(x_2^k - x_1^k)(y_3^k - y_1^k) - (y_2^k - y_1^k)(x_3^k - x_1^k)] = \frac{1}{2}(r_3^k q_2^k - q_3^k r_2^k) . \tag{4.23}$$

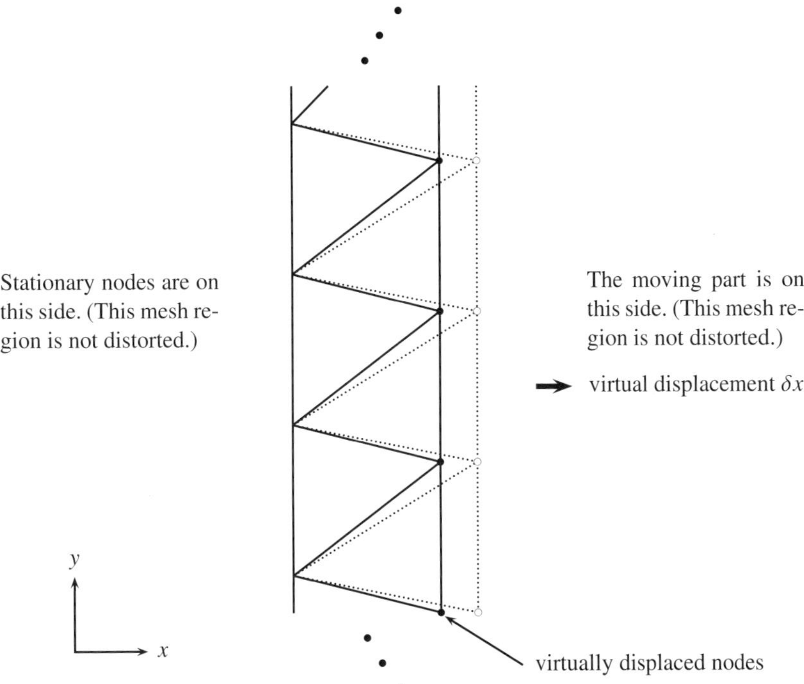

Figure 4.8 Virtual distortion of a single layer of elements surrounding a moving part for force calculation.

Nodes in each element are numbered in a counterclockwise sense so that $\Delta_k > 0$. For a Type-1 triangle (only node #1 moves), it turns out that

$$\frac{\partial q_1^k}{\partial y} = \frac{\partial(y_2^k - y_3^k)}{\partial y} = 0\,, \quad \frac{\partial q_2^k}{\partial y} = \frac{\partial(y_3^k - y_1^k)}{\partial y} = -1\,, \quad \frac{\partial q_3^k}{\partial y} = \frac{\partial(y_1^k - y_2^k)}{\partial y} = 1\,, \quad (4.24a)$$

$$\frac{\partial r_1^k}{\partial x} = \frac{\partial(x_3^k - x_2^k)}{\partial x} = 0\,, \quad \frac{\partial r_2^k}{\partial x} = \frac{\partial(x_1^k - x_3^k)}{\partial x} = 1\,, \quad \frac{\partial r_3^k}{\partial x} = \frac{\partial(x_2^k - x_1^k)}{\partial x} = -1\,, \quad (4.24b)$$

or more concisely

$$\frac{\partial \mathbf{q}^k}{\partial y} = \begin{bmatrix} 0 \\ -1 \\ 1 \end{bmatrix} \quad \text{and} \quad \frac{\partial \mathbf{r}^k}{\partial x} = \begin{bmatrix} 0 \\ 1 \\ -1 \end{bmatrix}. \quad (4.24c)$$

For a Type-2 triangle (only nodes #1 and #2 move), we have

$$\frac{\partial q_1^k}{\partial y} = \frac{\partial(y_2^k - y_3^k)}{\partial y} = 1\,, \quad \frac{\partial q_2^k}{\partial y} = \frac{\partial(y_3^k - y_1^k)}{\partial y} = -1\,, \quad \frac{\partial q_3^k}{\partial y} = \frac{\partial(y_1^k - y_2^k)}{\partial y} = 0\,, \quad (4.25a)$$

$$\frac{\partial r_1^k}{\partial x} = \frac{\partial(x_3^k - x_2^k)}{\partial x} = -1\,, \quad \frac{\partial r_2^k}{\partial x} = \frac{\partial(x_1^k - x_3^k)}{\partial x} = 1\,, \quad \frac{\partial r_3^k}{\partial x} = \frac{\partial(x_2^k - x_1^k)}{\partial x} = 0\,, \quad (4.25b)$$

or

$$\frac{\partial \mathbf{q}^k}{\partial y} = \begin{bmatrix} 1 \\ -1 \\ 0 \end{bmatrix} \quad \text{and} \quad \frac{\partial \mathbf{r}^k}{\partial x} = \begin{bmatrix} -1 \\ 1 \\ 0 \end{bmatrix}. \tag{4.25c}$$

Now, let us evaluate the derivative

$$\frac{\partial S_{ij}^k}{\partial x} = \frac{1}{4\Delta_k}\left(\frac{\partial q_i^k}{\partial x}q_j^k + q_i^k\frac{\partial q_j^k}{\partial x} + \frac{\partial r_i^k}{\partial x}r_j^k + r_i^k\frac{\partial r_j^k}{\partial x}\right) + \frac{1}{4}(q_i^k q_j^k + r_i^k r_j^k)\frac{\partial(\Delta_k^{-1})}{\partial x}, \tag{4.26}$$

where

$$\frac{\partial(\Delta_k^{-1})}{\partial x} = -\frac{1}{2\Delta_k^2}\left(\frac{\partial r_3^k}{\partial x}q_2^k + r_3^k\frac{\partial q_2^k}{\partial x} - \frac{\partial q_3^k}{\partial x}r_2^k - q_3^k\frac{\partial r_2^k}{\partial x}\right). \tag{4.27}$$

Note that all $\partial q/\partial x$ terms vanish. Hence

$$\frac{\partial S_{ij}^k}{\partial x} = \frac{1}{4\Delta_k}\left(\frac{\partial r_i^k}{\partial x}r_j^k + r_i^k\frac{\partial r_j^k}{\partial x}\right) - \frac{1}{8\Delta_k^2}(q_i^k q_j^k + r_i^k r_j^k)\left(\frac{\partial r_3^k}{\partial x}q_2^k - q_3^k\frac{\partial r_2^k}{\partial x}\right), \tag{4.28}$$

or

$$\frac{\partial S_{ij}^k}{\partial x} = \frac{1}{4\Delta_k}\left(\frac{\partial r_i^k}{\partial x}r_j^k + r_i^k\frac{\partial r_j^k}{\partial x}\right) - \frac{S_{ij}^k}{2\Delta_k}\left(\frac{\partial r_3^k}{\partial x}q_2^k - q_3^k\frac{\partial r_2^k}{\partial x}\right). \tag{4.29}$$

As we did for the stiffness matrix in Equation (3.28), we may express this derivative for the entire matrix as follows:

$$\frac{\partial \mathbf{S}^k}{\partial x} = \frac{1}{4\Delta_k}\left[\frac{\partial \mathbf{r}^k}{\partial x}(\mathbf{r}^k)^\top + \mathbf{r}^k\frac{\partial (\mathbf{r}^k)^\top}{\partial x}\right] - \frac{1}{2\Delta_k}\left(\frac{\partial r_3^k}{\partial x}q_2^k - q_3^k\frac{\partial r_2^k}{\partial x}\right)\mathbf{S}^k. \tag{4.30}$$

Hence

$$\frac{\partial \mathbf{S}^k}{\partial x} = \frac{1}{4\Delta_k}\left[\frac{\partial \mathbf{r}^k}{\partial x}(\mathbf{r}^k)^\top + \mathbf{r}^k\frac{\partial (\mathbf{r}^k)^\top}{\partial x}\right] - \frac{q_1^k}{2\Delta_k}\mathbf{S}^k \quad \text{for Type-1 triangles} \tag{4.31}$$

and

$$\frac{\partial \mathbf{S}^k}{\partial x} = \frac{1}{4\Delta_k}\left[\frac{\partial \mathbf{r}^k}{\partial x}(\mathbf{r}^k)^\top + \mathbf{r}^k\frac{\partial (\mathbf{r}^k)^\top}{\partial x}\right] + \frac{q_3^k}{2\Delta_k}\mathbf{S}^k \quad \text{for Type-2 triangles.} \tag{4.32}$$

Similar expressions can be derived for a virtual displacement along the y-axis, that is

$$\frac{\partial \mathbf{S}^k}{\partial y} = \frac{1}{4\Delta_k}\left[\frac{\partial \mathbf{q}^k}{\partial y}(\mathbf{q}^k)^\top + \mathbf{q}^k\frac{\partial (\mathbf{q}^k)^\top}{\partial y}\right] - \frac{1}{2\Delta_k}\left(r_3^k\frac{\partial q_2^k}{\partial y} - \frac{\partial q_3^k}{\partial y}r_2^k\right)\mathbf{S}^k, \tag{4.33}$$

leading to

$$\frac{\partial \mathbf{S}^k}{\partial y} = \frac{1}{4\Delta_k}\left[\frac{\partial \mathbf{q}^k}{\partial y}(\mathbf{q}^k)^\top + \mathbf{q}^k\frac{\partial (\mathbf{q}^k)^\top}{\partial y}\right] - \frac{r_1^k}{2\Delta_k}\mathbf{S}^k \quad \text{for Type-1 triangles} \tag{4.34}$$

and

$$\frac{\partial \mathbf{S}^k}{\partial y} = \frac{1}{4\Delta_k}\left[\frac{\partial \mathbf{q}^k}{\partial y}(\mathbf{q}^k)^\top + \mathbf{q}^k\frac{\partial (\mathbf{q}^k)^\top}{\partial y}\right] + \frac{r_3^k}{2\Delta_k}\mathbf{S}^k \quad \text{for Type-2 triangles.} \tag{4.35}$$

Next, we calculate the contribution to the x-axis force component (per unit depth) from the virtual distortion of element k, by substituting Equations (4.31) and (4.32) in

$$\tilde{F}_{ex}^{k} = -\frac{\nu_0}{2}(\mathbf{A}^k)^\top \frac{\partial \mathbf{S}^k}{\partial x}\mathbf{A}^k . \tag{4.36}$$

For Type-1 triangles, the substitution yields

$$\tilde{F}_{ex}^{k} = -\frac{\nu_0}{2}\left[\frac{1}{2\Delta_k}(\mathbf{A}^k)^\top \frac{\partial \mathbf{r}^k}{\partial x}(\mathbf{r}^k)^\top \mathbf{A}^k - \frac{q_1^k}{2\Delta_k}(\mathbf{A}^k)^\top \mathbf{S}^k \mathbf{A}^k\right], \tag{4.37}$$

whereas for Type-2 triangles, it yields

$$\tilde{F}_{ex}^{k} = -\frac{\nu_0}{2}\left[\frac{1}{2\Delta_k}(\mathbf{A}^k)^\top \frac{\partial \mathbf{r}^k}{\partial x}(\mathbf{r}^k)^\top \mathbf{A}^k + \frac{q_3^k}{2\Delta_k}(\mathbf{A}^k)^\top \mathbf{S}^k \mathbf{A}^k\right]. \tag{4.38}$$

We could actually stop the derivation here and implement the VDM using these formulas. However, let us manipulate the expression further, to see what happens.

It is convenient to multiply the four-matrix term in the above equations by invoking the associative property of matrix multiplication in the following fashion:

$$(\mathbf{A}^k)^\top \frac{\partial \mathbf{r}^k}{\partial x}(\mathbf{r}^k)^\top \mathbf{A}^k = \left[(\mathbf{A}^k)^\top \frac{\partial \mathbf{r}^k}{\partial x}\right]\cdot\left[(\mathbf{r}^k)^\top \mathbf{A}^k\right]. \tag{4.39}$$

Recall Equations (3.94) and (3.96), repeated here for convenience:

$$\vec{\mathbf{B}}^k = \begin{bmatrix} B_x^k \\ B_y^k \end{bmatrix} = \frac{1}{2\Delta_k}\begin{bmatrix} (\mathbf{r}^k)^\top \\ -(\mathbf{q}^k)^\top \end{bmatrix}\mathbf{A}^k, \quad B_k^2 = \frac{1}{\Delta_k}(\mathbf{A}^k)^\top \mathbf{S}^k \mathbf{A}^k . \tag{4.40}$$

Using these formulas and the previously derived expressions for $\partial \mathbf{r}^k/\partial x$, we obtain

$$\tilde{F}_{ex}^{k} = -\frac{\nu_0}{2}\left[(A_2^k - A_3^k)B_x^k - \frac{q_1^k}{2}B_k^2\right] \quad \text{for Type-1 triangles} \tag{4.41}$$

and

$$\tilde{F}_{ex}^{k} = -\frac{\nu_0}{2}\left[(A_2^k - A_1^k)B_x^k + \frac{q_3^k}{2}B_k^2\right] \quad \text{for Type-2 triangles.} \tag{4.42}$$

We now take a closer look at the geometry of the two types of triangular elements, shown in Figure 4.9. We define unit normal and tangential vectors, $\hat{\mathbf{n}}^k$ and $\hat{\mathbf{s}}^k$, respectively, for the two triangles. The unit normal points outwards, away from the moving member, and is normal to triangle side 2–3 for Type-1 elements and side 1–2 for Type-2 elements. The tangential direction is decided as if we are moving in the counterclockwise sense along a path that surrounds the moving member. For the Type-1 triangle, we can express

$$q_1^k = y_2^k - y_3^k = -\ell_{23}^k\, \hat{\mathbf{s}}^k \cdot \hat{\mathbf{j}}, \tag{4.43}$$

that is, as a function of the triangle side length. Also, it can be readily verified using a flux-based argument that

$$A_2^k - A_3^k = -B_n^k \ell_{23}^k , \tag{4.44}$$

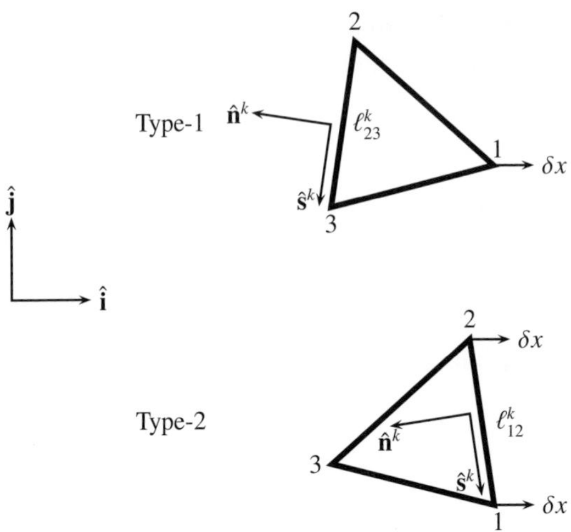

Figure 4.9 Geometry of Type-1 and Type-2 triangular elements.

where $B_n^k = \vec{\mathbf{B}}^k \cdot \hat{\mathbf{n}}^k$. Similarly, for the Type-2 triangle, we have

$$q_3^k = y_1^k - y_2^k = \ell_{12}^k \hat{\mathbf{s}}^k \cdot \hat{\mathbf{j}} \tag{4.45}$$

and

$$A_2^k - A_1^k = -B_n^k \ell_{12}^k . \tag{4.46}$$

It is also evident from the geometry that

$$\hat{\mathbf{s}}^k \cdot \hat{\mathbf{j}} = \hat{\mathbf{n}}^k \cdot \hat{\mathbf{i}} = n_x^k . \tag{4.47}$$

Summarizing, we have shown that, regardless of triangle type:

$$\tilde{F}_{ex}^k = \frac{\nu_0 \ell_k}{2} \left(B_n^k B_x^k - \frac{1}{2} B_k^2 n_x^k \right), \tag{4.48}$$

where $\ell_k = \ell_{23}^k$ or $\ell_k = \ell_{12}^k$ is the appropriate side length. In an identical fashion, it can be shown that the y-component of force is

$$\tilde{F}_{ey}^k = \frac{\nu_0 \ell_k}{2} \left(B_n^k B_y^k - \frac{1}{2} B_k^2 n_y^k \right). \tag{4.49}$$

Combining the two formulas in a vector, we obtain the **VDM force**

$$\boxed{\tilde{\mathbf{F}}_e^k = \nu_0 \frac{\ell_k}{2} \left[\left(\vec{\mathbf{B}}^k \cdot \hat{\mathbf{n}}^k \right) \vec{\mathbf{B}}^k - \frac{1}{2} B_k^2 \hat{\mathbf{n}}^k \right].} \tag{4.50}$$

This expression resembles the MST force formula (4.5). The factor $\ell_k/2$ in the equation is the length of the segment that connects the midpoints of the two triangle sides, which is parallel to side 2–3 in a Type-1 triangle or to side 1–2 in a Type-2 triangle. Therefore, Equation (4.50) is representative of a simple quadrature rule for an MST force integral over a polygonal path that passes through the triangle side midpoints. We have thus shown that *the VDM is numerically equivalent to an MST-based force calculation!*

Torque Calculation Using the VDM

The calculation of torque involves a virtual displacement of nodes in a rotational sense. Let us first consider the case of Type-2 triangles, like the one shown in Figure 4.10, having one stationary vertex on the outer boundary and two moving vertices on the inner boundary. The outer and inner radii are denoted by r_o and r_i, respectively. The triangle is distorted by rotating the two inner points equally by an angle $\delta\theta$. The rotor is assumed to be perfectly aligned with the center of rotation. Note that due to the curvature of the inner circle, the two points are not displaced by the same vector. This is in contrast to what happens when distorting a mesh for force calculation, where the displacement is the same for both points. This discrepancy will reflect on the torque formula.

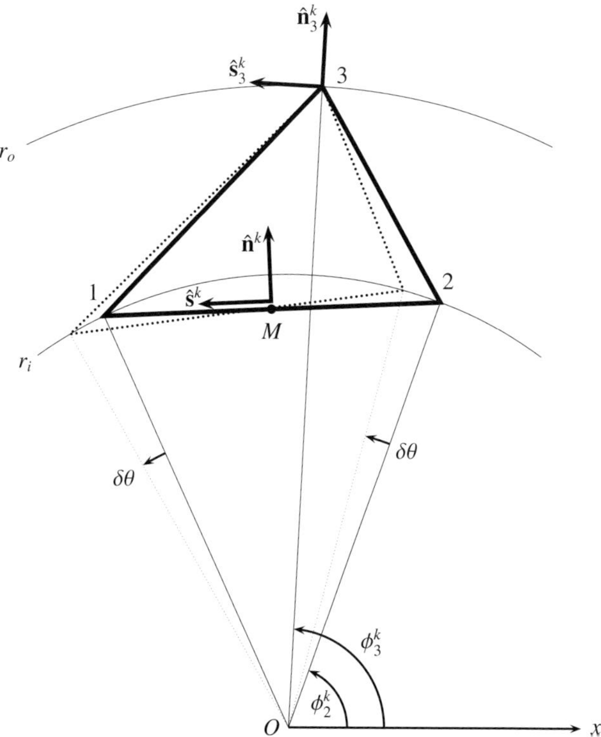

Figure 4.10 Distortion of a Type-2 triangle (exaggerated scale).

When a Type-2 triangle is distorted, we obtain

$$\frac{\partial \mathbf{q}^k}{\partial \theta} = \begin{bmatrix} x_2^k \\ -x_1^k \\ x_1^k - x_2^k \end{bmatrix} \quad \text{and} \quad \frac{\partial \mathbf{r}^k}{\partial \theta} = \begin{bmatrix} y_2^k \\ -y_1^k \\ y_1^k - y_2^k \end{bmatrix}. \tag{4.51}$$

These equations are obtained from Equations (4.22a) and (4.22b) by expressing node coordinates in polar form, and considering that vertex #3 is fixed. For instance, since $x_2^k = r_i \cos\phi_2^k$, as shown in Figure 4.10, we obtain $\partial x_2^k/\partial\theta = -r_i \sin\phi_2^k = -y_2^k$, because

$\partial\phi_2^k/\partial\theta = 1$. Therefore, the derivative of the stiffness matrix is

$$\frac{\partial \mathbf{S}^k}{\partial\theta} = \frac{1}{4\Delta_k}\left[\frac{\partial \mathbf{q}^k}{\partial\theta}(\mathbf{q}^k)^\top + \mathbf{q}^k\frac{\partial(\mathbf{q}^k)^\top}{\partial\theta} + \frac{\partial \mathbf{r}^k}{\partial\theta}(\mathbf{r}^k)^\top + \mathbf{r}^k\frac{\partial(\mathbf{r}^k)^\top}{\partial\theta}\right] - \frac{1}{\Delta_k}\frac{\partial\Delta_k}{\partial\theta}\mathbf{S}^k, \tag{4.52}$$

with

$$\frac{\partial\Delta_k}{\partial\theta} = \frac{1}{2}\left(\frac{\partial r_3^k}{\partial\theta}q_2^k + r_3^k\frac{\partial q_2^k}{\partial\theta} - \frac{\partial r_2^k}{\partial\theta}q_3^k - r_2^k\frac{\partial q_3^k}{\partial\theta}\right) \tag{4.53a}$$

$$= \frac{1}{2}[(y_1^k - y_2^k)(y_3^k - y_1^k) - (x_2^k - x_1^k)x_1^k + y_1^k(y_1^k - y_2^k) - (x_1^k - x_3^k)(x_1^k - x_2^k)] \tag{4.53b}$$

$$= \frac{1}{2}[(y_1^k - y_2^k)y_3^k + (x_1^k - x_2^k)x_3^k], \tag{4.53c}$$

which can be expressed as the dot product

$$\frac{\partial\Delta_k}{\partial\theta} = \frac{1}{2}(\ell_{12}^k\,\hat{\mathbf{s}}^k)\cdot(r_o\,\hat{\mathbf{n}}_3^k). \tag{4.54}$$

Interestingly, the formula contains a dot product between the tangent unit vector at the triangle base (between vertices #1 and #2) and the radially outwards unit normal vector at vertex #3. This term vanishes if the triangle is isosceles ($\ell_{13}^k = \ell_{23}^k$); however, in general it will have a nonzero but relatively small magnitude. The sign of $\hat{\mathbf{s}}^k \cdot \hat{\mathbf{n}}_3^k$ will depend on the relative position of vertex #3 with respect to the midpoint M.

To proceed, we observe that Equation (4.51) can be written as

$$\frac{\partial \mathbf{q}^k}{\partial\theta} = -\mathbf{r}^k + \begin{bmatrix} x_3^k \\ -x_3^k \\ 0 \end{bmatrix} \quad\text{and}\quad \frac{\partial \mathbf{r}^k}{\partial\theta} = \mathbf{q}^k + \begin{bmatrix} y_3^k \\ -y_3^k \\ 0 \end{bmatrix}. \tag{4.55}$$

This implies that (cf. Equation (4.39))

$$(\mathbf{A}^k)^\top\frac{\partial \mathbf{q}^k}{\partial\theta} = -2\Delta_k B_x^k + x_3^k(A_1^k - A_2^k) = -2\Delta_k B_x^k + x_3^k\ell_{12}^k B_n^k, \tag{4.56a}$$

$$(\mathbf{A}^k)^\top\frac{\partial \mathbf{r}^k}{\partial\theta} = -2\Delta_k B_y^k + y_3^k(A_1^k - A_2^k) = -2\Delta_k B_y^k + y_3^k\ell_{12}^k B_n^k, \tag{4.56b}$$

where $B_n^k = \vec{\mathbf{B}}^k \cdot \hat{\mathbf{n}}$ is a normal component of the magnetic field with respect to the triangle side 1–2, which has length ℓ_{12}^k.

It remains to substitute the previous expressions in

$$\tilde{\tau}_{ez}^k = -\frac{\nu_0}{2}(\mathbf{A}^k)^\top\frac{\partial \mathbf{S}^k}{\partial\theta}\mathbf{A}^k, \tag{4.57}$$

which yields, after a few algebraic manipulations, the **VDM contribution to torque from a Type-2 triangle**

$$\boxed{\tilde{\tau}_{ez}^k = \frac{\nu_0\ell_{12}^k r_o}{2}\left[(\vec{\mathbf{B}}^k\cdot\hat{\mathbf{s}}_3^k)(\vec{\mathbf{B}}^k\cdot\hat{\mathbf{n}}^k) + \frac{1}{2}\hat{\mathbf{s}}^k\cdot\hat{\mathbf{n}}_3^k B_k^2\right].} \tag{4.58}$$

The analysis follows identical steps for Type-1 triangles, which have one vertex on the inner boundary and two vertices on the outer boundary. The distortion of a Type-1 triangle is illustrated in Figure 4.11. It can be shown that the **VDM contribution to torque from a Type-1 triangle** is

$$\tilde{\tau}_{ez}^{k} = \frac{\nu_0 \ell_{23}^{k} r_i}{2}\left[(\vec{\mathbf{B}}^{k} \cdot \hat{\mathbf{s}}_1^{k})(\vec{\mathbf{B}}^{k} \cdot \hat{\mathbf{n}}^{k}) + \frac{1}{2}\hat{\mathbf{s}}^{k} \cdot \hat{\mathbf{n}}_1^{k} B_k^2\right]. \tag{4.59}$$

Examining these two equations, we observe the following. (i) Both include an $\ell^k/2$ term, which is the length of the segment connecting the triangle side midpoints; however, the formulas use a different radius, namely the radius of the circle where the single node resides. (ii) They employ a decomposition of the B-field in tangential and normal components at two *different* points within the triangle; these four terms are then mixed together! (iii) There is a small but finite term that involves B_k^2. From these observations, and in contrast to the case of *force* calculation, we conclude that the VDM torque calculation is *not* numerically equivalent to the MST-based approaches that we have presented.

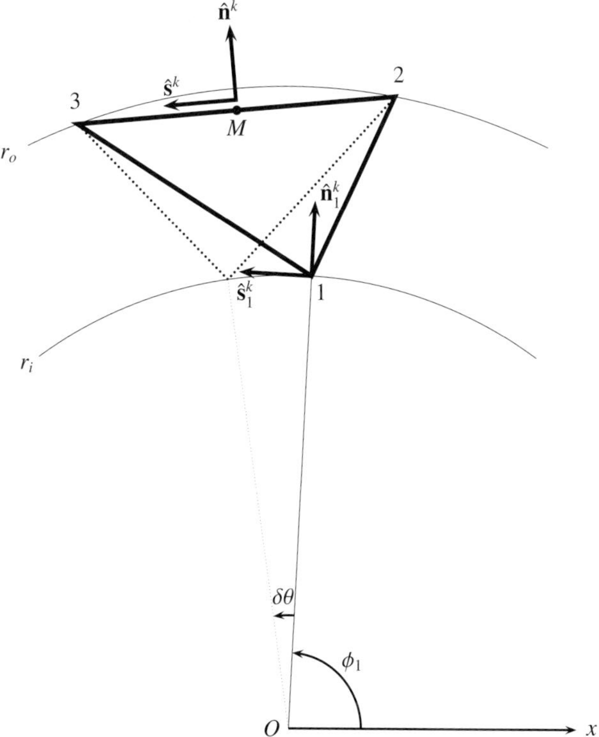

Figure 4.11 Distortion of a Type-1 triangle (exaggerated scale).

4.1.4 An Air-Gap MVP-Based Method

When the air gap is circular, as in most rotating machines, we may use directly the MVP solution in the calculation of torque. The calculation is based on Equation (2.292) on page 182 (derived from a virtual distortion of the air gap), or Equation (2.361b) on page 195 (derived from the MST), which are identical analytical expressions. They are repeated here for convenience:

$$\tau_e = \sum_{p=1}^{\infty} \tau_{ep} = \sum_{p=1}^{\infty} \frac{2\pi\nu_0 \ell p^2}{\zeta^{-p} - \zeta^p} A_{pi} A_{po} \sin(\phi_{po} - \phi_{pi}) \,. \tag{4.60}$$

Recall that this result is based on a decomposition of the MVP at the inner and outer boundaries of an air-gap region as a sum of p-pole-pair fields, that is, as a Fourier series. Here, A_{pi} and A_{po} denote amplitudes of the MVP (a continuous function of position) at the inner and outer boundaries, respectively. For instance, at the inner boundary, the p-pole-pair MVP component is

$$A_p(r = r_i, \phi) = A_{pi} \cos(p\phi - \phi_{pi}) \,, \tag{4.61}$$

where ϕ denotes circumferential position in the air gap.

The FEA solution may be interpreted as a noisy sampling of the true solution at the element vertices. Hence, the idea is to *reconstruct* two continuous functions as Fourier series with terms like those of Equation (4.61) from the samples so that we can then apply Equation (4.60) to calculate torque. In Python, the built-in discrete Fourier transform (DFT) functions of NumPy can be used to obtain the MVP amplitude and phase. In particular, since our signals are real (i.e., not complex), we can use `numpy.fft.rfft`.

Suppose we have a vector of MVP values, $\mathbf{a} = (a_0, a_1, \ldots, a_{N-1})$, over N consecutive, equally spaced positions on either the inner or outer boundary of the air-gap layer spanning 360°. (Note that here we use lowercase for the MVP values because uppercase is reserved for the amplitudes.) If the first point is at an angle ϕ_0 within the air gap, then the $m + 1$ point is at $\phi_m = \phi_0 + 2\pi m/N$, for $m = 0, \ldots, N - 1$.

In NumPy, the DFT is defined as a vector $\bar{\mathbf{A}} = (\bar{A}_0, \bar{A}_1, \ldots, \bar{A}_{N-1})$ of complex numbers:

$$\bar{A}_p = \sum_{m=0}^{N-1} a_m \exp\left\{-2\pi j \frac{mp}{N}\right\} \quad \text{for } p = 0, \ldots, N-1 \,. \tag{4.62}$$

The first term, $\bar{A}_0$, contains the sum of the signal (i.e., the sum of the MVP values), which is a real number. This term does not affect the torque and can thus be ignored.[d] In theory, the maximum component we may estimate is related to the *Nyquist frequency*.[75] Since our sampling rate is $N/2\pi$ points per radian, the Nyquist frequency is half of that, or $N/4\pi$ cycles per radian, which means that the maximum pole-pair that we can estimate is $p_{\max} = N/2$ (assuming N is an even number). The coefficients for $p = N/2+1, \ldots, N-1$ are complex conjugates of the first $N/2-1$ terms for $p = 1, \ldots, N/2-1$, so they do not contain additional information. In practice, because of the FEA sampling

[d] Usually, in an electric machine operating under normal conditions, we would expect $\bar{A}_0 = 0$ due to physical considerations. See the discussion on the analytical solution of the air-gap MVP on page 125.

error, we should define many more points N than the maximum number of poles that we would like to capture. Otherwise, the error will distort our results for the higher pole-pair fields. The fact that these are multiplied by p^2 in the torque formula is not in our favor either. So, we will typically use only the first few components of the DFT, way below the Nyquist limit; let us denote this number as $P < N/2$.

In NumPy, the inverse DFT is defined as

$$a_m = \frac{1}{N}\sum_{p=0}^{N-1} \bar{A}_p \exp\left\{2\pi j \frac{mp}{N}\right\} \quad \text{for } m = 0, \ldots, N-1\,. \tag{4.63}$$

Let us write the complex DFT coefficients in polar form using Euler's formula:

$$\bar{A}_p = \left|\bar{A}_p\right| \exp\{j\phi_{Ap}\}\,. \tag{4.64}$$

Substituting this in Equation (4.63), and using the fact that $2\pi m/N = \phi_m - \phi_0$, we can reconstruct an MVP function of position (at either the inner or outer boundary) that interpolates the samples:

$$A(\phi) = \frac{2}{N}\sum_{p=1}^{P} \left|\bar{A}_p\right| \cos(p(\phi - \phi_0) + \phi_{Ap})\,. \tag{4.65}$$

The interpolation is approximate, $A(\phi_m) \approx a_m$, because we are dropping high-order terms, which we expect to be insignificant.

Suppose that both boundaries have N points. It follows that the torque can be computed by the **MVP-DFT method** as

$$\boxed{\tau_e = 2\pi\nu_0\ell \sum_{p=1}^{P} \frac{p^2}{\zeta^{-p} - \zeta^{p}} \frac{4}{N^2} \left|\bar{A}_{pi}\right| \left|\bar{A}_{po}\right| \sin(\phi_{po} - \phi_{pi})\,,} \tag{4.66}$$

where

$$\phi_{po} - \phi_{pi} = p(\phi_{o0} - \phi_{i0}) - (\phi_{Apo} - \phi_{Api})\,. \tag{4.67}$$

Here, ϕ_{o0} and ϕ_{i0} denote the angular offsets of the first nodes on the outer and inner layers, respectively. Also, $\zeta = r_1/r_2$, $r_1 < r_2$, is the ratio of inner and outer radii *of the boundaries involved in the DFT calculation* (i.e., these radii are not necessarily the same as the air-gap inner or outer boundary).

Example 4.2 *Numerical comparison of torque calculation methods.*

We revisit Example 3.7 on page 270 that involved the FEA of an annular air-gap region with given Dirichlet boundary conditions imposed on the inner and outer boundaries. The objective is to apply the various torque calculation methods that we have presented, and compare the numerical results.

Solution

The annulus geometry is identical to the previous case (r_i = 50 mm, r_o = 55 mm). We impose the following boundary conditions:

$$A(r_i, \phi) = A_{pi}\cos(p\phi - \phi_{pi})\,, \quad A(r_o, \phi) = A_{po}\cos p\phi\,, \tag{4.68}$$

with $p = 2$, $A_{pi} = A_{po} = 0.02$ Wb/m, and $\phi_{pi} = 0.2°$. To apply the methods, we create three layers of triangles. The middle layer is bounded by two concentric regular polygons, whose circumcircles have radii $r_1 = r_i + (r_o - r_i)/3$ and $r_2 = r_i + 2(r_o - r_i)/3$. All four polygons (two outer and two inner) have $N = 180$ nodes each, which are spaced 2° apart. A detail of the mesh is shown in Figure 4.12.

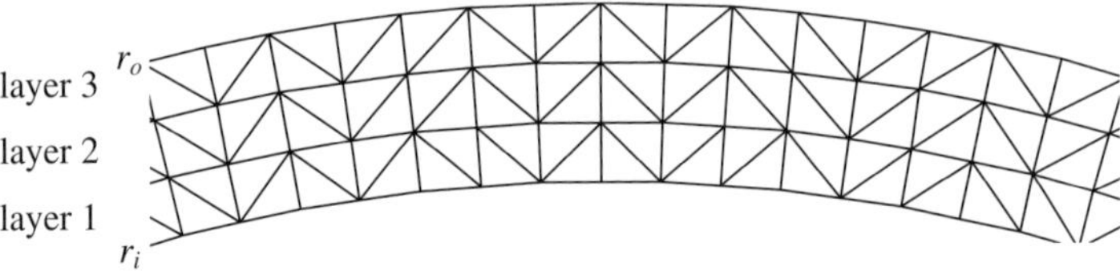

Figure 4.12 Detail of an air-gap mesh.

The various torque calculation methods yield the numerical results listed in Table 4.1. The relative error between the FEA-based results and the analytical torque expression (4.60) is listed as well. We apply the methods to each individual layer, setting $\gamma = 1/2$ to define the polygonal and circular quadrature paths for the MST method. We note that calculating the torque as an average value over all layers seems to yield more accurate answers; however, this is not always necessarily true. Although it may be tempting to reach conclusions regarding the accuracy of the various methods, we will

Table 4.1 Comparison of various torque calculation methods

Method	Torque (Nm)	Error
analytical	−72.807	N/A
polygonal path MST (layer 1)	−72.820	0.018%
polygonal path MST (layer 2)	−72.758	−0.067%
polygonal path MST (layer 3)	−72.839	0.045%
polygonal path MST (average)	−72.806	−0.001%
circular path MST (layer 1)	−72.813	0.008%
circular path MST (layer 2)	−72.886	0.109%
circular path MST (layer 3)	−72.767	−0.054%
circular path MST (average)	−72.822	0.021%
Arkkio's method (layer 1)	−72.803	−0.004%
Arkkio's method (layer 2)	−72.803	−0.005%
Arkkio's method (layer 3)	−72.795	−0.016%
Arkkio's method (entire air gap)	−72.801	−0.008%
VDM (layer 1)	−72.836	0.040%
VDM (layer 2)	−72.713	−0.129%
VDM (layer 3)	−72.884	0.106%
VDM (average)	−72.811	0.006%
MVP-DFT (layer 1)	−72.803	−0.004%
MVP-DFT (layer 2)	−72.807	0.001%
MVP-DFT (layer 3)	−72.810	0.005%
MVP-DFT (average)	−72.807	0.001%

refrain from doing so. The errors can be sensitive to the mesh (e.g., the positioning of the nodes) and other problem parameters. Relative errors on the order of 1% have been observed in perturbations of this problem. Nevertheless, any one of these methods could be acceptable for all practical purposes.

4.2 Rotation without Remeshing

The analysis of rotating electric machines invariably requires the computation of the magnetic field at various rotor positions. It is, therefore, important to discuss the implementation of rotation in an FEA program. Arguably, the simplest way to achieve rotation is by rotating the rotor nodes to the desired angular position (i.e., by updating their coordinates), and then remeshing the full geometric domain. However, this approach is computationally inefficient.

For higher efficiency, we will demonstrate an alternative approach where we mesh the stator and rotor domains only once (at the start of the FEA program). The rotor mesh rotates as a whole, thus we do not need to recompute its element stiffness matrices (see Problem 3.4). Furthermore, to allow rotation to arbitrary angles, we will use a thin air-gap region, resembling those presented in the preceding torque calculation sections, serving as a means to separate stator and rotor. For simplicity, we consider an air gap consisting of a single layer of elements, as illustrated in Figure 4.13. The shape of the stator and rotor meshes containing the bulk of the elements is held fixed; therefore, remeshing the air-gap layer when the rotor is repositioned incurs only a light computational burden.

Suppose the outer node potentials are contained in a vector $\mathbf{A}_o$ of dimension N_o, and the inner node potentials are in a vector $\mathbf{A}_i$ of dimension N_i. Note that these vectors include the potentials at the air-gap boundary. The single-layer air-gap region has a total of N_{ag} nodes, but it does not contain additional nodes internally. Therefore, the air-gap node potentials can be mapped to those of the other nodes by

$$\mathbf{A}_{ag} = \begin{bmatrix} \mathbf{C}_{ao} & \mathbf{C}_{ai} \end{bmatrix} \begin{bmatrix} \mathbf{A}_o \\ \mathbf{A}_i \end{bmatrix} = \mathbf{C}\mathbf{A}\,, \tag{4.69}$$

where $\mathbf{C}_{ao}$ and $\mathbf{C}_{ai}$ are matrices of dimension $N_{ag} \times N_o$ and $N_{ag} \times N_i$, respectively. The role of these matrices is similar to that of the connection matrix encounted in §3.1.4 (but not exactly the same). These sparse matrices contain mostly zeros, and their nonzero elements are equal to 1. The mth row of $\mathbf{C} = \begin{bmatrix} \mathbf{C}_{ao} & \mathbf{C}_{ai} \end{bmatrix}$ should have a single nonzero element. For example, if $\mathbf{C}_{ao,mn} = 1$, then the nth outer mesh node coincides with the mth air-gap node.

The FEM functional can be expressed as

$$I(\mathbf{A}) = I_o(\mathbf{A}_o) + I_{ag}(\mathbf{A}_{ag}) + I_i(\mathbf{A}_i)\,, \tag{4.70}$$

where we separated the contributions from the three regions. The contribution to the

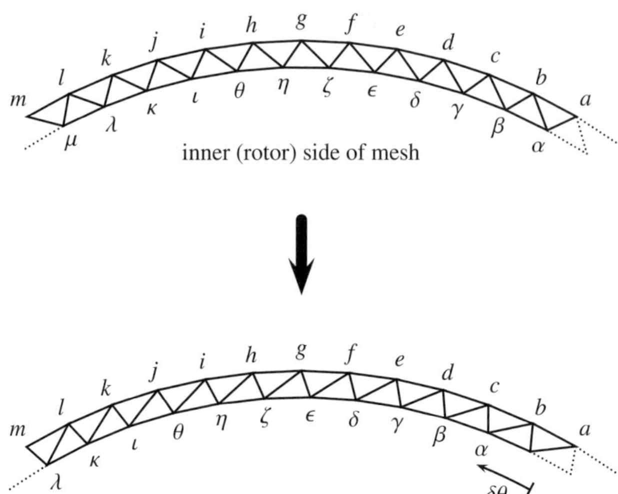

Figure 4.13 An air-gap layer interface between the outer and inner mesh that gets remeshed every time the rotor moves. Here, the rotor has turned by $\delta\theta = 7°$ counterclockwise.

functional from the (magnetically linear, current-free) air-gap region is

$$I_{ag}(\mathbf{A}_{ag}) = \frac{1}{2}\mathbf{A}_{ag}^{\top}\mathbf{S}_{ag}\mathbf{A}_{ag}\,. \tag{4.71}$$

Therefore, combining this with Equations (4.69) and (4.70) yields

$$I(\mathbf{A}) = I_o(\mathbf{A}_o) + \frac{1}{2}\mathbf{A}^{\top}\mathbf{C}^{\top}\mathbf{S}_{ag}\mathbf{C}\mathbf{A} + I_i(\mathbf{A}_i)\,. \tag{4.72}$$

The gradient of the functional can thus be written as

$$\mathbf{g}(\mathbf{A}) = \frac{\partial I}{\partial \mathbf{A}} = \begin{bmatrix} \frac{\partial I}{\partial \mathbf{A}_o} \\ \frac{\partial I}{\partial \mathbf{A}_i} \end{bmatrix} = \begin{bmatrix} \frac{\partial I_o}{\partial \mathbf{A}_o} \\ \frac{\partial I_i}{\partial \mathbf{A}_i} \end{bmatrix} + \mathbf{C}^{\top}\mathbf{S}_{ag}\mathbf{C}\mathbf{A} = \begin{bmatrix} \mathbf{g}_o(\mathbf{A}_o) \\ \mathbf{g}_i(\mathbf{A}_i) \end{bmatrix} + \mathbf{C}^{\top}\mathbf{g}_{ag}(\mathbf{A}_{ag})\,, \tag{4.73}$$

and the Hessian is

$$\mathbf{H}(\mathbf{A}) = \begin{bmatrix} \mathbf{H}_o(\mathbf{A}_o) & \mathbf{0} \\ \mathbf{0} & \mathbf{H}_i(\mathbf{A}_i) \end{bmatrix} + \mathbf{C}^{\top}\mathbf{S}_{ag}\mathbf{C}\,. \tag{4.74}$$

The preceding analysis suggests the algorithm of Box 4.1 for implementing rotation. This is a modified main program structure, similar to the one that was presented in Box 3.6 on page 291, where now solutions are obtained at a number of distinct rotor positions. Implementation details are left as an exercise for the reader. Without loss of generality, we are assuming that the stator is on the outer side of the device.

When meshing the stator and the air-gap regions in Step 1, the meshing software (Triangle) should be instructed that there is a hole in the middle (see §3.3.3 for syntax).

Box 4.1 Programming the FEM: Main program structure (nonlinear magnetics and rotation)

1. Define and mesh the geometry (call `Triangle`) separately for the stator and rotor regions (with the rotor at an arbitrary initial position).
2. Load nonlinear *B–H* curve data (call `load_BH_curve_pchip`).
3. Read stator and rotor node and triangle data files (call `load_mesh`).
4. Build stator and rotor element stiffness matrices (call `calculate_pqrDeltaSk`).
5. Assign material properties to triangles.
6. Identify the nodes at the stator inner and rotor outer boundaries, which are then combined to form the air gap; establish the linking matrix $\mathbf{C}$ (which remains constant throughout the analysis).
7. For each rotor position:
   a. Update current density in triangles (as it may change with time).
   b. Update rotor node coordinates.
   c. Remesh the air-gap region (call `Triangle`).
   d. Run the Newton–Raphson algorithm (within each iteration, call `build_grad_Hess` separately for the stator, rotor, and air gap, and then combine to assemble the global gradient and Hessian matrix; then, enforce boundary conditions).
   e. Calculate quantities of interest, and generate plots by post-processing $\mathbf{A}$.

Also, note that no new nodes should be added to the air-gap layer in order to conform with the surrounding meshes. To enforce this requirement, Triangle can be called with the switch `-YY` that prohibits the insertion of new points on the boundary (as well as in internal segments).

We are now ready to perform FEA on rotating electric machines! The case studies that follow involve various types of machines (namely, wound-rotor synchronous machines, permanent-magnet synchronous machines, and switched reluctance machines), which can all be analyzed using magnetostatic FEA. Each example is described in sufficient detail for FEA implementation, relying on the various algorithms that we have presented thus far; writing these programs is left as an exercise for the reader. The machines that we will study are not all representative of optimally designed devices (except for the Toyota Prius and switched reluctance motors), and some design features have been omitted for simplicity. So, we will still refer to these machines as "elementary" even though they are fully functional, in contrast to the ones we encountered earlier (see, e.g., Example 2.17 on page 169). We will first provide some guidance on how to mesh the stator and rotor. Then, we will perform case studies and interpret the results, focusing on understanding the main principles of operation.

4.3 Wound-Rotor Synchronous Machines

The first type of machine that we analyze is the **wound-rotor synchronous machine (WRSM)**, whose main features were introduced in Example 2.17 on page 169. The WRSM is the most common type of electric generator found in coal-fired, natural gas, nuclear, or hydro power plants. We will first examine a simple two-phase machine with a cylindrical rotor. Then we will analyze a three-phase machine with a salient-pole rotor.

But what exactly are we trying to determine? A primary objective of an electric machine FEA (regardless of machine type) is to determine expressions for the flux linkages of the coils as functions of the currents and the rotor angle. For example, in a two-phase WRSM, we would determine

$$\lambda_a = \lambda_a(i_a, i_b, i_f, \theta_r)\,, \tag{4.75a}$$

$$\lambda_b = \lambda_b(i_a, i_b, i_f, \theta_r)\,, \tag{4.75b}$$

$$\lambda_f = \lambda_f(i_a, i_b, i_f, \theta_r)\,. \tag{4.75c}$$

In addition, we would obtain the torque as

$$\tau_e = \tau_e(i_a, i_b, i_f, \theta_r)\,. \tag{4.76}$$

Here, θ_r denotes the **electrical rotor angle**, which is related to the **mechanical rotor angle** θ_{rm} in a $2p$-pole machine by

$$\theta_r = p\theta_{rm}\,. \tag{4.77}$$

In general, due to magnetic nonlinearity, this process requires running multiple FEA studies over a range of possible currents and rotor angles. This task can be computationally demanding, as the number of calculations grows with the number of windings and the level of accuracy that we desire.

In turn, the flux linkages are related to the winding voltages by

$$v_a = R_s i_a + \frac{d}{dt}\lambda_a(i_a, i_b, i_f, \theta_r)\,, \tag{4.78a}$$

$$v_b = R_s i_b + \frac{d}{dt}\lambda_b(i_a, i_b, i_f, \theta_r)\,, \tag{4.78b}$$

$$v_f = R_f i_f + \frac{d}{dt}\lambda_f(i_a, i_b, i_f, \theta_r)\,. \tag{4.78c}$$

The **mechanical rotor speed** ω_{rm} is related to the **electrical rotor speed** ω_r by

$$\omega_r = p\omega_{rm}\,. \tag{4.79}$$

The speed is governed by the equation of motion. If the machine shaft is assumed to be a rigid mass (which may not always be a valid assumption), then we have

$$\frac{J}{p}\frac{d}{dt}\omega_r = \tau_e(i_a, i_b, i_f, \theta_r) - \tau_m - \tau_f(\omega_r)\,, \tag{4.80}$$

where τ_m is an externally applied mechanical torque on the shaft, τ_f is a friction function, and J is the combined moment of inertia of the machine and its prime mover or load. For a complete system of differential equations, we must also consider that

$$\frac{d}{dt}\theta_r = \omega_r \,. \tag{4.81}$$

For instance, under steady-state conditions, the current waveforms and rotor speed may be assumed to be known functions of time. Then, the time derivatives of the flux linkages may be evaluated via numerical means, leading to the calculation of the winding voltages. Alternatively, the voltages and the mechanical torque could be given functions of time, in which case we would need to solve the previous system of differential equations for the winding currents, the speed, and the rotor angle. Note that these calculations do not account for the effects of core loss that is not captured in a conservative coupling-field-based FEA, which would reflect on the coil flux linkages.

4.3.1 Elementary Two-Phase Round-Rotor Synchronous Machine: Fundamentals of Distributed Windings

Our first example involves an elementary two-phase, two-pole (i.e., $p = 1$) round-rotor WRSM. The stator has two windings a and b, which are positioned orthogonally in space (in terms of the magnetic fields that they produce). The rotor has a single field winding f. Our focus will be on calculating the mapping from currents to flux linkages and torque. This will help us understand the basic principles of operation of the synchronous machine.

Meshing the Stator

We begin by defining the geometry of the stator cross-section, including the shape of the stator slot. In practice, the slot shape can be intricate; however, for simplicity we will use a simple shape that is defined by just a few points and the interconnecting linear segments, as shown in Figure 4.14.

The outer and inner stator radii are denoted by r_{so} and r_{si}, respectively. The stator/air-gap interface has radius r_o (keeping notation consistent with earlier sections). In the FEA mesh, circular arcs are approximated as polygonal curves, formed by a number of points connected by small linear segments. The programmer must define the number of points under each tooth and the number of points on the stator outer boundary. Note that the spacing of points under the teeth at r_{si} is an important parameter that should be chosen carefully because it should be similar to the discretization of the air-gap layer between stator and rotor, which in turn may affect the calculation of torque. Also, a fine discretization of the tooth base may be important for capturing the variation of the B-field in those regions more accurately, which is significant for the estimation of core loss. On the other hand, the discretization of the outer boundary r_{so} is perhaps of lesser significance; however, it may affect the meshing within the stator yoke region, so this parameter should be chosen carefully as well. Since most of the flux is expected to be contained within the stator, the FEA "universe" could end at the stator outer boundary by imposing an $A = 0$ boundary condition. Alternatively, a larger circle may be drawn a small distance away from the stator at radius r_{ob}.

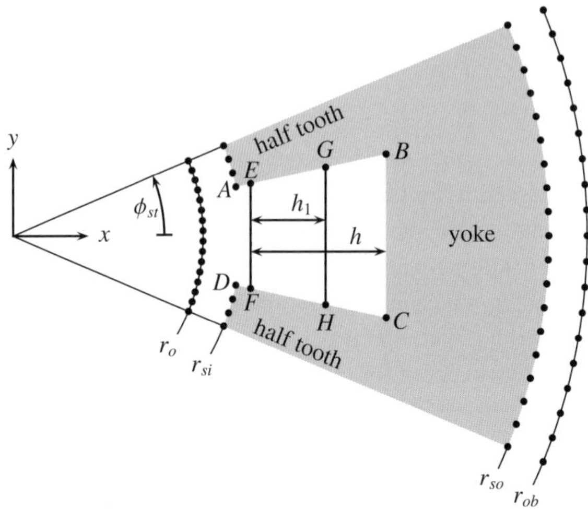

Figure 4.14 Geometry of an elementary stator slot with a two-layer winding (dimensions are somewhat exaggerated for clarity). Dots signify vertices that should be included in the planar straight-line graph.

The stator has n_{ss} evenly spaced slots, which are separated by pieces of ferromagnetic material called **teeth**. Figure 4.14 depicts the prototypical slot, which is defined by the four points $ABCD$. The slot is symmetric with respect to the horizontal, thus forming an isosceles trapezoid. It suffices to specify the coordinates of points A and B. Of course, point A lies on the inner surface at a distance $r_A = r_{si}$ from the origin. However, its angular position ϕ_A is a design parameter that determines the slot opening. The exact position of point B (in terms of both radius r_B and angle ϕ_B) is another design parameter.

A common choice is to have teeth whose side walls are parallel to each other (but note that this is not precisely the case in Figure 4.14, where the teeth are expanding slightly as we move radially outwards). Let us consider the tooth having an axis of symmetry at angle $\phi_{st} = \pi/n_{ss}$ and a constant width w_{st} (this is the first tooth that we encounter in the counterclockwise direction starting from the horizontal in Figure 4.14). This tooth can be defined by points A and B such that

$$r_A \sin(\phi_{st} - \phi_A) = r_B \sin(\phi_{st} - \phi_B) = \frac{w_{st}}{2}\,. \tag{4.82}$$

Also, note that the slope of the tooth wall AB, which will run parallel to the axis of the tooth, will be $\tan\phi_{st}$.

The winding is placed within the trapezoidal region $EBCF$, with $x_E > r_A$. For simplicity, and to avoid the creation of very small triangular elements in the mesh, we assume that the winding extends all the way to the steel; hence, we are ignoring the presence of the slot liner and/or other geometric details regarding the actual winding placement, the effect of which may be incorporated in the slot packing factor parameter k_{pf} (see §3.4.3 on page 262). The vertical line segment GH partitions the winding region in two layers, to allow for slots containing conductors of two different phases. Typically, the two layers have equal cross-sectional area in order to accommodate the same number of turns.

For equal areas, it can be readily shown that the winding region should be subdivided at a ratio h_1/h satisfying the quadratic equation

$$(BC - EF)\left(\frac{h_1}{h}\right)^2 + 2EF\,\frac{h_1}{h} - \frac{BC + EF}{2} = 0\,, \tag{4.83}$$

where $h = x_B - x_E$ is the winding height, $h_1 = x_G - x_E$, and BC, EF denote the segment lengths (assuming $BC > EF$).

The stator mesh can thus be created using Triangle, as explained in Example 3.4 on page 248. We can define the planar straight-line graph (PSLG) as follows. First, we add the vertices on the stator outer boundary, the stator/air-gap layer interface, and the outer universe boundary, if applicable. Then, the inner surface can be defined using a `for` loop iterating over the number of slots, by rotating the vertices of the prototypical slot shown as dots in Figure 4.14. For example, we may add the points by moving along the path in a counterclockwise sense. To avoid defining points twice, the last point on the inner surface shape should be omitted, as it would coincide with the first point of the next slot. Second, we define line segments between points, including the vertical segments separating the two winding layers. Third, we add a hole by defining a point at the center so that this entire inner region is not meshed. Lastly, we define regional attributes, namely, those of steel, air, and the coil layers.

Figure 4.15 shows the stator cross-section of the elementary two-phase, two-pole WRSM that we will analyze. This machine has $n_{ss} = 8$ stator slots. The following dimensions are given: $r_{so} = 110$ mm, $r_{si} = 55$ mm, $w_{st} = 25$ mm, $r_B = 80$ mm, $r_E = 57$ mm, $\ell = 100$ mm (axial stack length). The outer boundary of the mesh is positioned at $r_{ob} = 115$ mm. Also, the air-gap width is $g = 1$ mm, so we set $r_o = r_{si} - g/3 = 54.67$ mm.

Figure 4.15 also defines the winding distribution within the slots, such that a two-pole magnetic field is obtained. Positive polarity (e.g., b) indicates that positive current is flowing out of the page. The magnetic axes of the two windings indicate the directionality of the flux created by positive currents (see also Figure 2.28 on page 171). A counterclockwise rotating magnetic field is obtained by a two-phase alternating current set leading to an $a \to b \to -a \to -b$ magnetic field orientation. This current can be expressed as the balanced set

$$i_a(t) = \sqrt{2}\,I_s\cos(\omega t + \phi_i)\,, \tag{4.84a}$$

$$i_b(t) = \sqrt{2}\,I_s\sin(\omega t + \phi_i)\,, \tag{4.84b}$$

where I_s is the root mean square (r.m.s.) value of the stator current, and ϕ_i is a phase angle. Of course, the actual current in an electric machine may have additional harmonic components; however, we will ignore these here for simplicity.

Since we are given the slot dimensions, we can calculate the area of each half-slot, which is 307.8 mm^2. Assuming that the winding packing factor is $k_{pf} \approx 0.5$ implies that each half-slot has roughly 150 mm^2 of useful conductor surface area. Hence, we place 25 turns of 6-mm^2 (AWG 10) copper conductors (magnet wire) in each half-slot, for a total of $N_s = 100$ series-connected turns in each phase. The particular order of

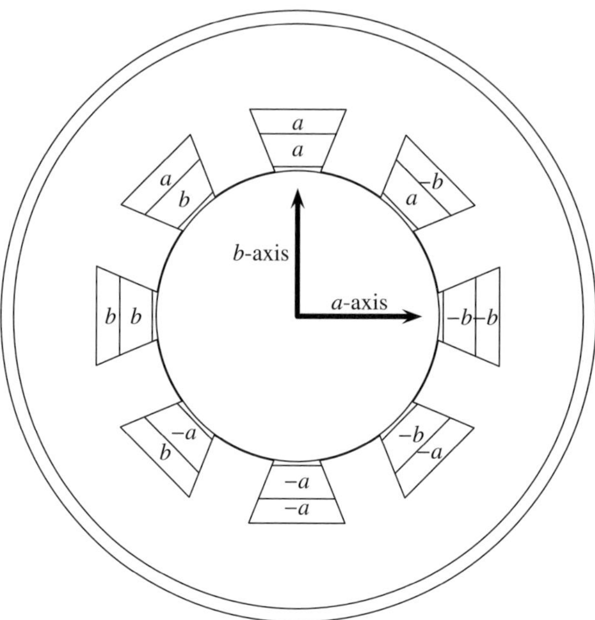

Figure 4.15 Cross-section of elementary two-phase, two-pole WRSM stator PSLG. The innermost "circle" has radius $r_o = r_{si} - g/3$. The outermost "circle" has radius r_{ob}. Quotation marks are used because, strictly speaking, these are not circles; they are polygons inscribed in circles of given radii.

interconnection between the turns, which depends on the winding pattern at the two end regions, does not affect the main operating characteristics of the motor. The stator steel has the B–H curve that we used previously to plot Figure 3.23. The lamination stacking factor is $k_{st} = 0.97$.

Meshing the Rotor

The elementary WRSM has a cylindrical rotor. The field winding is distributed within slots at the surface; so strictly speaking, the shape of the rotor lamination is not perfectly cylindrical. The rotor outer radius that defines the cylinder surface is denoted by r_{ro}, whereas its inner radius is r_{ri}, as shown in Figure 4.16. The center region $r < r_{ri}$ is occupied by the machine shaft, which is of a different material than the outer part of the rotor. A thin layer of elements is present between the outer rotor boundary and the inner air-gap radius r_i.

Figure 4.16 shows the geometry of a prototypical rectangular slot shape $ABCD$. (With a slight abuse of notation, we are reusing the same letters that we used for the stator slot.) The rotor slots are identical, and their axes are mutually displaced by an angle $2\phi_{rt}$; however, we do not cover the entire rotor periphery with slots. Therefore, we will define an angular span for the slot region ϕ_{rs} as the angle between the first and last slot (i.e., between their axes), which has to be a multiple of $2\phi_{rt}$. If the field winding is distributed among n_{rs} slots in total (n_{rs} even), we have $\phi_{rs} = (n_{rs} - 2)\,\phi_{rt}$.

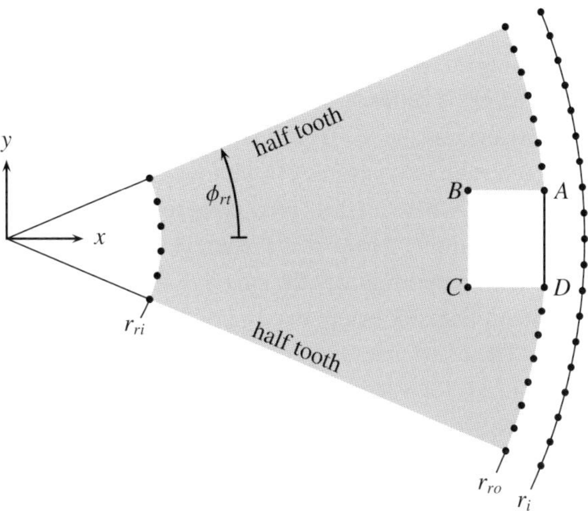

Figure 4.16 Geometry of an elementary cylindrical rotor slot (dimensions are somewhat exaggerated for clarity). Dots signify vertices that should be included in the PSLG.

Figure 4.17 shows the rotor cross-section of our elementary WRSM. This rotor has $n_{rs} = 10$ slots for the field winding; there are five slots on each side spanning $\phi_{rs} = 60°$. The remaining rotor dimensions are $r_{ro} = r_{si} - g = 54$ mm, $r_{ri} = 10$ mm, $\phi_A = \phi_{rt}/2 = 3.75°$, and $r_B = 49$ mm. Recall that the air-gap width is $g = 1$ mm, so we set $r_i = r_{si} - 2g/3 = 54.33$ mm. The rotor slot area is 35.4 mm^2. We place four turns of 6-mm^2 (AWG 10) copper wire in each slot, for a total of $N_f = 20$ turns. The rotor laminations are made from the same material as the stator, and have equal stacking factor (0.97). The shaft material is assumed to be magnetically linear, with a relative permeability of 200.

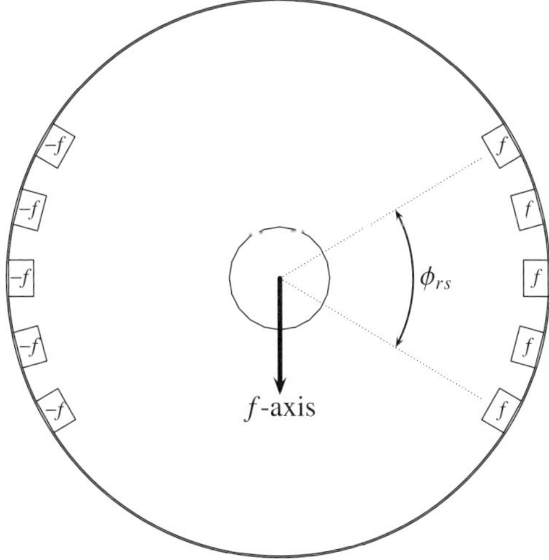

Figure 4.17 Cross-section of elementary two-pole cylindrical WRSM rotor PSLG. The innermost "circle" has radius r_{ri}. The outermost "circle" has radius $r_i = r_{ro} + g/3$.

MMF of Distributed Windings

The magnetizing effect of current in the stator and rotor windings, which are distributed over several slots, is different from that of a concentrated winding (cf. Figure 2.27 on page 170). To analyze the MMF of distributed windings, it is customary to define **MMF vectors**, which we will denote as $\mathcal{F}$ (without an arrow). These vectors are defined on an "electrical plane," where angles correspond to electrical (i.e., not physical, mechanical) angles. In our example, where $p = 1$, there is no difference between the two. An MMF vector has length proportional to the *effective* ampere-turns (that will be defined shortly), and its orientation shows where maximum "pressure" exists in the air gap for magnetic flux to flow. The MMF vector is a useful construct for thinking about the operation of electric machines, for visualization purposes, and for performing simple calculations. However, since we can now solve directly for the magnetic field distribution in space using FEA, the significance of the MMF vector is relegated to a secondary, conceptual role.

Consider a single turn spanning 180° with the positive conductor at $\phi + \pi/2$ and the negative conductor at $\phi - \pi/2$, carrying current i. As discussed in Example 2.17, the air-gap MMF of this turn is a rectangular waveform oscillating between $\pm 0.5i$, and its axis is oriented at angle ϕ. The MMF vector of this single turn is associated with the fundamental component of its MMF waveform, which is $(2/\pi)i\cos\phi$. We can use the complex number

$$\mathcal{F}_{1\text{ turn}} = \frac{2}{\pi} i \exp j\phi \tag{4.85}$$

to represent the single-turn MMF vector in space (the real axis is the x-axis).

The MMF vector of the entire distributed winding is defined as a superposition of the MMF vectors of all turns. For instance, the MMF vector of the a-phase in our case study is

$$\mathcal{F}_a = \frac{1}{\pi} i_a \sum_{k=1}^{n_{ss}} N_{a,k} \exp j(\phi_{ss,k} - \pi/2)\,, \tag{4.86}$$

denoting by $N_{a,k}$ the total number of a-turns in the kth slot (can be positive or negative depending on the directionality), which is positioned at angle $\phi_{ss,k}$. Here, the a-phase conductor distribution in the slots is the vector

$$\mathbf{N}_a = \begin{bmatrix} 0 & 1 & 2 & 1 & 0 & -1 & -2 & -1 \end{bmatrix} \cdot 25\,, \tag{4.87}$$

where we define slot 1 as the one centered on the x-axis in Figure 4.15. The calculation results in a real number, which represents a horizontal vector pointing to the right (as expected):

$$\mathcal{F}_a = \frac{2}{\pi}(2 + \sqrt{2})(25) i_a = \frac{2}{\pi} \underbrace{(0.854)}_{k_{ws}} N_s i_a\,. \tag{4.88}$$

The factor $k_{ws} < 1$ is a **stator winding distribution factor**, indicative of the difference in terms of fundamental components between this distributed winding and a concentrated winding having the same number of turns. Hence, the reason why we used the term "effective" to describe the ampere-turns is because we are accounting for: (i) only the fundamental component of the field in the air gap; and (ii) the distribution of the winding in slots. The combined MMF vector of the stator can be expressed as the vector addition of the two coil MMFs, which are oriented along their respective axes:

$$\mathcal{F}_s = \frac{2}{\pi} k_{ws} N_s (i_a\, \hat{\mathbf{x}} + i_b\, \hat{\mathbf{y}}) \tag{4.89a}$$

$$= \frac{2\sqrt{2}}{\pi} k_{ws} N_s I_s \left[\cos(\omega t + \phi_i)\, \hat{\mathbf{x}} + \sin(\omega t + \phi_i)\, \hat{\mathbf{y}}\right] . \tag{4.89b}$$

In a similar manner, we may calculate the rotor MMF vector using Equations (4.86) and (4.88) for the field winding distribution. The **field winding distribution factor** is

$$k_{wf} = \frac{1 + 2\cos 15^\circ + 2\cos 30^\circ}{5} = 0.933 . \tag{4.90}$$

When the rotor is turned at an arbitrary angle θ_r, the rotor MMF vector is

$$\mathcal{F}_r = \frac{2}{\pi} k_{wf} N_f i_f (\sin\theta_r\, \hat{\mathbf{x}} - \cos\theta_r\, \hat{\mathbf{y}}) \tag{4.91a}$$

$$= \frac{2}{\pi} k_{wf} N_f i_f \left[\cos(\theta_r - 90^\circ)\, \hat{\mathbf{x}} + \sin(\theta_r - 90^\circ)\, \hat{\mathbf{y}}\right] . \tag{4.91b}$$

The total MMF vector is defined as the sum of the stator and rotor vectors:

$$\mathcal{F} = \mathcal{F}_s + \mathcal{F}_r . \tag{4.92}$$

Admittedly, our treatment of the MMF vector has not been very thorough. We anticipate this vector to be indicative of the principal directionality and magnitude of the magnetic field inside the machine. We will confirm this using FEA.

Flux Linkage and Torque: Open-Circuit Conditions

A meshed cross-section of the complete WRSM is shown in Figure 4.18. In this case study, the stator, rotor, and air-gap meshes have 8,063, 6,255, and 2,062 triangular elements, respectively. The programmer should adjust the spacing of the PSLG vertices so that the mesh is acceptable. One way to check whether the mesh is dense enough is to evaluate the convergence of a quantity of interest, e.g., the torque, with increasing number of triangular elements. More advanced FEA programs have algorithms for increasing the mesh resolution adaptively; however, this is outside the scope of this text.

Here, the rotor has turned counterclockwise by $\theta_r = 30^\circ$. The field winding axis is commonly called the **direct axis** or ***d*-axis** of the machine. The direction that lies 90° ahead of the d-axis is called the **quadrature axis** or ***q*-axis**. When the d-axis is aligned with a stator winding axis, this signifies maximum coupling between the two coils. For example, when the rotor angle is 90°, we expect maximum coupling between the field winding and the a-phase (see Figure 4.15).

To confirm if this is really the case, we will excite the field winding with a constant current i_f, and we will monitor the flux linkage of the a-phase as the rotor turns; the stator windings will not be conducting current. At this point, let us not worry about how the field current will be maintained at a perfectly constant level. As a matter of fact, this may be impossible with a simple d.c. source and may require a power electronic

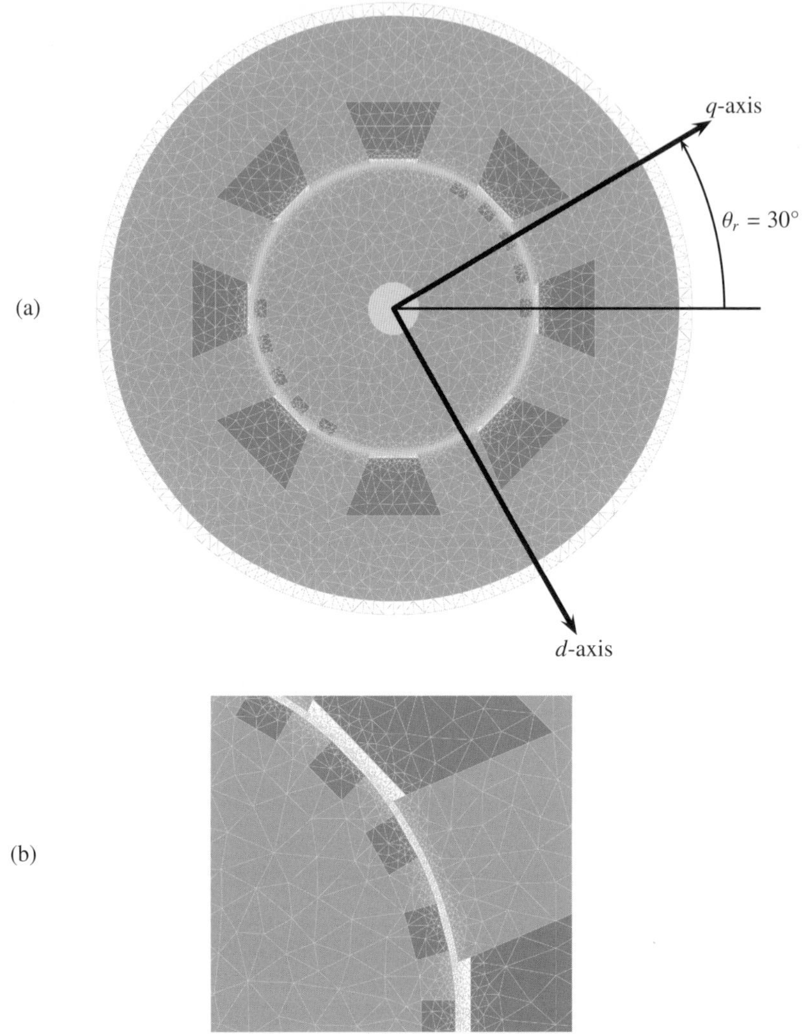

Figure 4.18 (a) Meshed cross-section of elementary two-phase, two-pole round-rotor WRSM. (b) Mesh detail in a region close to the air gap and the q-axis.

converter because a nonzero $d\lambda_f/dt$ will induce oscillations in the current. Nevertheless, this study should not be interpreted as representative of a real operating condition. Rather, we are just calculating the mapping between a given field current and the stator flux linkage, namely, $\lambda_a(i_a = 0, i_b = 0, i_f, \theta_r)$. We will also calculate the field winding flux linkage, $\lambda_f(i_a = 0, i_b = 0, i_f, \theta_r)$. To this end, we perform a sequence of FEA studies for $\theta_r = 0, 3, 6, \ldots, 180°$, and for several field winding currents $i_f = 10, 20, \ldots, 60$ A. Flux linkage is calculated using the average MVP method (see the analysis of §3.4.1, and the main result of Equation (3.108) on page 260).

Note that the upper limit of 60 A corresponds to current density $J_f = 10$ A/mm^2 in the field winding conductors. In practice, this limit is determined by the cooling system design. The total field winding ohmic loss at this current level is

$$P_{f,\text{loss}} = 2\frac{J_f^2}{\sigma}N_f A_f \ell = 2\frac{\left(10^7\right)^2}{5.19 \cdot 10^7}(20)(6 \cdot 10^{-6})(0.1) = 46.2\ \text{W}\,, \tag{4.93}$$

where the conductivity of copper was adjusted to reflect a 50°C operating temperature. The factor 2 accounts for both sides of the winding; we are thus ignoring the losses occurring at the end turns. It should be emphasized that this loss calculation does not affect the magnetostatic FEA; it serves only as a sanity check on the magnitude of the current.

The open-circuit flux linkages λ_a and λ_f are plotted in Figure 4.19. First, we observe that the flux increases almost linearly with field current, which implies that we are operating in a magnetically linear regime. For example, see the flux plots of Figure 4.20 that correspond to a 60-A excitation; in particular, the flux density plots on the right confirm that B remains below 1.5 T throughout the machine. The second observation is that the flux waveforms are not perfectly sinusoidal for the stator winding or constant for the field winding, as in an ideal machine. The harmonic components are due to the effects of the slots. For instance, at $\theta_r = 90°$, when we would have expected maximum coupling for the a-phase, the stator flux exhibits a small dip! This may be attributed to the presence of the stator slots at 0 and 180°, that is, directly along the d-axis. This effectively increases the magnetic reluctance and thus reduces the flux, as shown in Figure 4.20(b). Clearly, this elementary machine design has room for improvement because we would like the open-circuit flux to exhibit much less harmonic distortion. In turn, this would lead to a more sinusoidal **back-emf**, which is the derivative of the open-circuit flux linkage with respect to time. Note that harmonics in the open-circuit flux linkage become amplified in the back-emf waveform.

Now we turn our attention to the electromagnetic torque. The torque waveform obtained from the open-circuit test is shown in Figure 4.21. Torque is calculated as the average of five methods described in §4.1, namely the polygonal path MST method, the circular path MST method, Arkkio's method, the virtual displacement method, and the MVP-DFT method. In an ideal (slotless) machine with a perfectly uniform air gap, we would have expected this waveform to be zero: torque should not be developed by a round-rotor machine with only field excitation. However, the slots are responsible for the creation of a relatively small oscillatory torque component. This represents a reluctance torque that tends to turn the rotor in a direction such that the magnetic reluctance is minimized (akin to the cogging torque in PM machines). In this case, this happens when the d-axis is aligned with the stator teeth, at $\theta_r = 22.5, 67.5, 112.5$, or $157.5°$, which represent stable equilibrium points. These are points where the torque crosses zero and has a negative slope (see Example 2.18 on page 173). In contrast, the points where torque crosses zero with a positive derivative, e.g., $\theta_r = 45°$, are unstable equilibria.

Flux Linkage and Torque: Loaded Conditions

We have argued earlier (see Example 2.17 on page 169) that the "useful" component of torque is obtained by interaction between a stator MMF and a rotor MMF. Like a

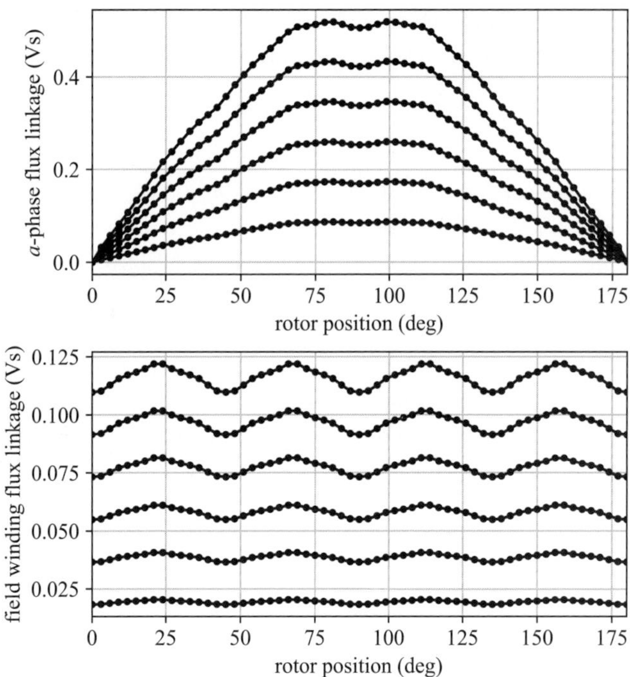

Figure 4.19 Elementary two-phase round-rotor WRSM open-circuit flux linkage waveforms. The curves correspond to six levels of field winding current, $i_f = 10, 20, \ldots, 60$ A.

magnetic dipole, the developed electromagnetic torque attempts to rotate the rotor so that its magnetic axis (i.e., the d-axis) aligns with that of the stator MMF. In that earlier example, where the winding was concentrated in two slots, the magnitude of the torque did not depend on the rotor angle, as in Equation (2.267). However, when the winding is distributed in slots and the air-gap MMF has a more sinusoidal nature, the torque magnitude will vary with rotor angle. We will now validate this principle with FEA.

To interpret the results, it is useful to recall the concept of a rotating magnetic field created by the stator coils (see page 131). We have assumed that the stator currents are given by the balanced set (4.84a) and (4.84b). This means that at time $t_1 = -\phi_i/\omega$, when only the a-phase carries current, the magnetic field created by the stator has a principal orientation through the air gap and the rotor in the horizontal direction. At this time instant, the fundamental pole-pair field is created solely by the action of the a-phase ampere-turns, which obtain their peak value, $(2\sqrt{2}/\pi)k_{ws}N_sI_s$. As time evolves, the a-phase MMF reduces in magnitude, and the b-phase MMF increases. At time $t_2 = t_1 + \pi/(2\omega)$, the field is created solely by the b-phase MMF. At this time instant, the magnetic field has a vertically upwards principal orientation. Between times t_1 and t_2, the field is created by the combined effect of both MMFs, and is rotating counterclockwise inside the machine. Since this is a single pole-pair device ($p = 1$), the angular velocity of the rotating magnetic field equals the electrical frequency, that is, $\omega_p = \omega$ rad/s.

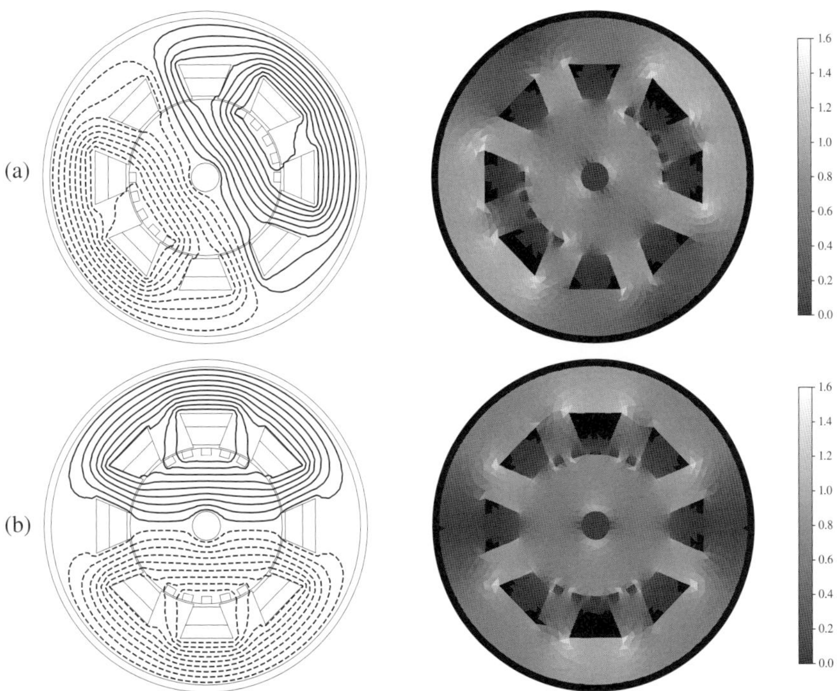

Figure 4.20 Elementary two-phase round-rotor WRSM open-circuit flux lines (left) and flux density (right): (a) $\theta_r = 30°$ and $i_f = 60$ A; (b) $\theta_r = 90°$ and $i_f = 60$ A.

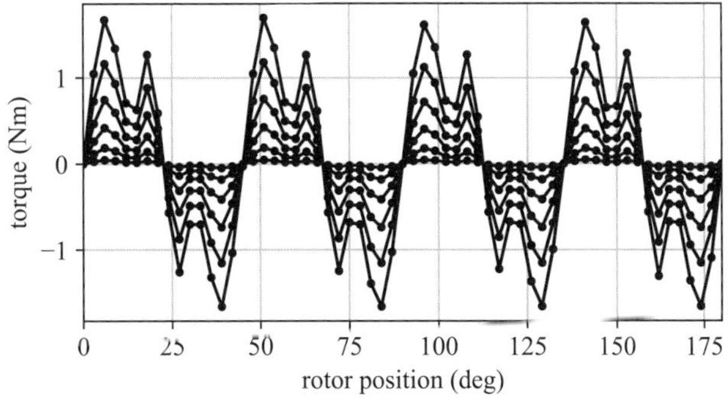

Figure 4.21 Elementary two-phase round-rotor WRSM open-circuit torque. The curves correspond to six levels of field winding current, $i_f = 10, 20, \ldots, 60$ A.

Steady-state conditions are obtained when the rotor is moving at constant synchronous speed ω so that its angle changes as $\theta_r = \omega t$. At time $t = 0$, $\mathcal{F}_r$ is pointing vertically downwards. For example, if $\phi_i = 0°$, then $\mathcal{F}_s$ is continually leading $\mathcal{F}_r$ by 90°. This is representative of an operating point where the device is acting as a motor, where torque is positive. Alternatively, by setting $\phi_i < -90°$, we can make $\mathcal{F}_s$ lag $\mathcal{F}_r$, which represents a generating mode of operation with negative torque. To summarize, given how we have defined axes and angles in this example, $\mathcal{F}_s$ leads $\mathcal{F}_r$ by $\phi_i + 90°$. We can decompose $\mathcal{F}_s$ in components along the q- and d-axes (tied on the rotor). In the steady state, these are

$$\mathcal{F}_{qs} = \|\mathcal{F}_s\| \cos\phi_i\,, \tag{4.94a}$$

$$\mathcal{F}_{ds} = -\|\mathcal{F}_s\| \sin\phi_i\,, \tag{4.94b}$$

where $\|\mathcal{F}_s\| = (2\sqrt{2}/\pi)k_{ws}N_sI_s$. Since the MMF of the field winding is (by definition) on the d-axis, we can express the qd components of the total MMF as

$$\mathcal{F}_q = \mathcal{F}_{qs} = \|\mathcal{F}_s\| \cos\phi_i\,, \tag{4.95a}$$

$$\mathcal{F}_d = \mathcal{F}_{ds} + \mathcal{F}_{dr} = -\|\mathcal{F}_s\| \sin\phi_i + \|\mathcal{F}_r\|\,, \tag{4.95b}$$

where $\|\mathcal{F}_r\| = (2/\pi)k_{wf}N_fi_f$.

As a case study, we will set $i_f = 60$ A (so that $\|F_r\| = 712.6$ At, and we will examine the FEA solution for two stator current magnitudes, namely, (i) $\sqrt{2}I_s = 13.11$ A (leading to $\|F_s\| \approx \|F_r\|$) and (ii) $\sqrt{2}I_s = 26.23$ A (leading to $\|F_s\| \approx 2\|F_r\|$). We will conduct FEA over a broad range of $\mathcal{F}_s$ positions $\phi_i \in [-90, 90]°$, while the rotor is at $\theta_r = 0°$. Performing these FEA studies leads to the flux line and flux density plots shown in Figures 4.22–4.24 for case (i), and in Figures 4.25–4.27 for case (ii). We have superimposed the stator, rotor, and total MMF vectors on the flux line plots, which serve as visual aids for interpreting the results. Note that we keep the number of flux lines and the colormap limits the same in all plots, regardless of B-field magnitude.

We can make the following interesting observations:

- The B-field magnitude plots are like X-rays showing the intensity of magnetic flux throughout the machine. In the previous case study, where the stator was open-circuited, the B-field was generally lower and the machine was only mildly saturated. Now, the presence of the stator MMF increases the B-field throughout the cross-section, and we expect that saturation will play a more prominent role.
- We can confirm that the magnetic field is indeed principally directed as we would expect by the total MMF $\mathcal{F}$. However, the figures are worth a thousand words. The FEA-based plots are much richer and more insightful.
- The directionality of the field inside the rotor agrees with the direction of the total MMF vector $\mathcal{F}$; however, the presence of the stator slots distorts the field as we get closer to the air gap. The presence of the rotor slots also distorts the field, but this is somewhat harder to see. We can observe that the magnetic flux tends to align with the stator teeth in fixed directions in space, even though the MMF vector may be pointing at slightly different angles.
- As the stator MMF turns counterclockwise and the angle formed between $\mathcal{F}_s$ and $\mathcal{F}_r$ increases, the magnitude of the total MMF vector reduces, and the machine becomes demagnetized. In the extreme case where $\mathcal{F}_s$ directly opposes $\mathcal{F}_r$, only leakage flux lines can be seen in the third case of Figure 4.24, and the X-ray is almost entirely black, indicating that the leakage flux density is very low.

It is also interesting to observe the variation of flux linkage and torque as a function of the stator MMF position. These waveforms are plotted in Figure 4.28. In regard to λ_a, we see that its magnitude is generally dictated by the orientation of $\mathcal{F}_s$, as we would

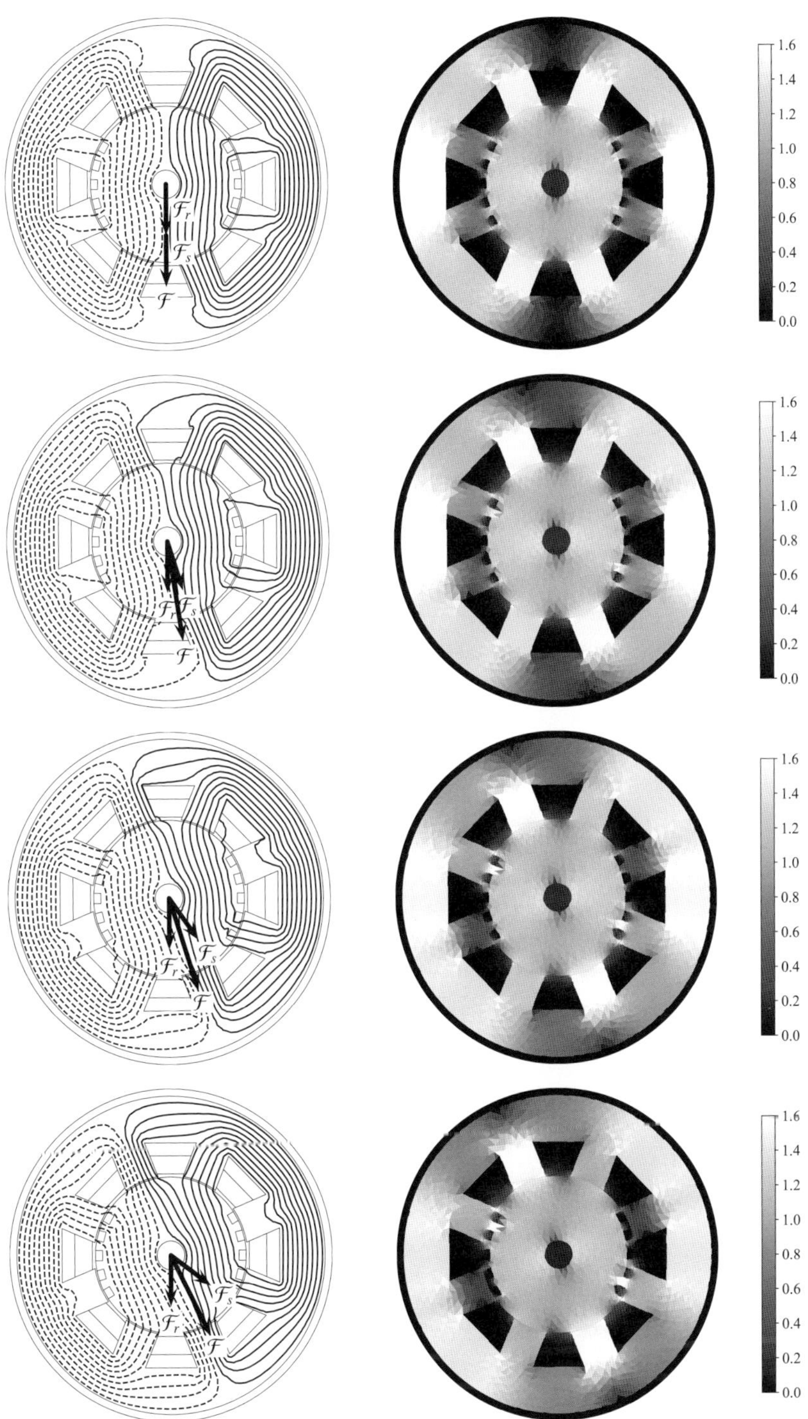

Figure 4.22 Elementary two-phase round-rotor WRSM flux line and flux density plots (A): stator current magnitude $\sqrt{2}I_s = 13.11$ A, stator MMF angle $\phi_i \in \{-90, -72, -54, -36\}°$.

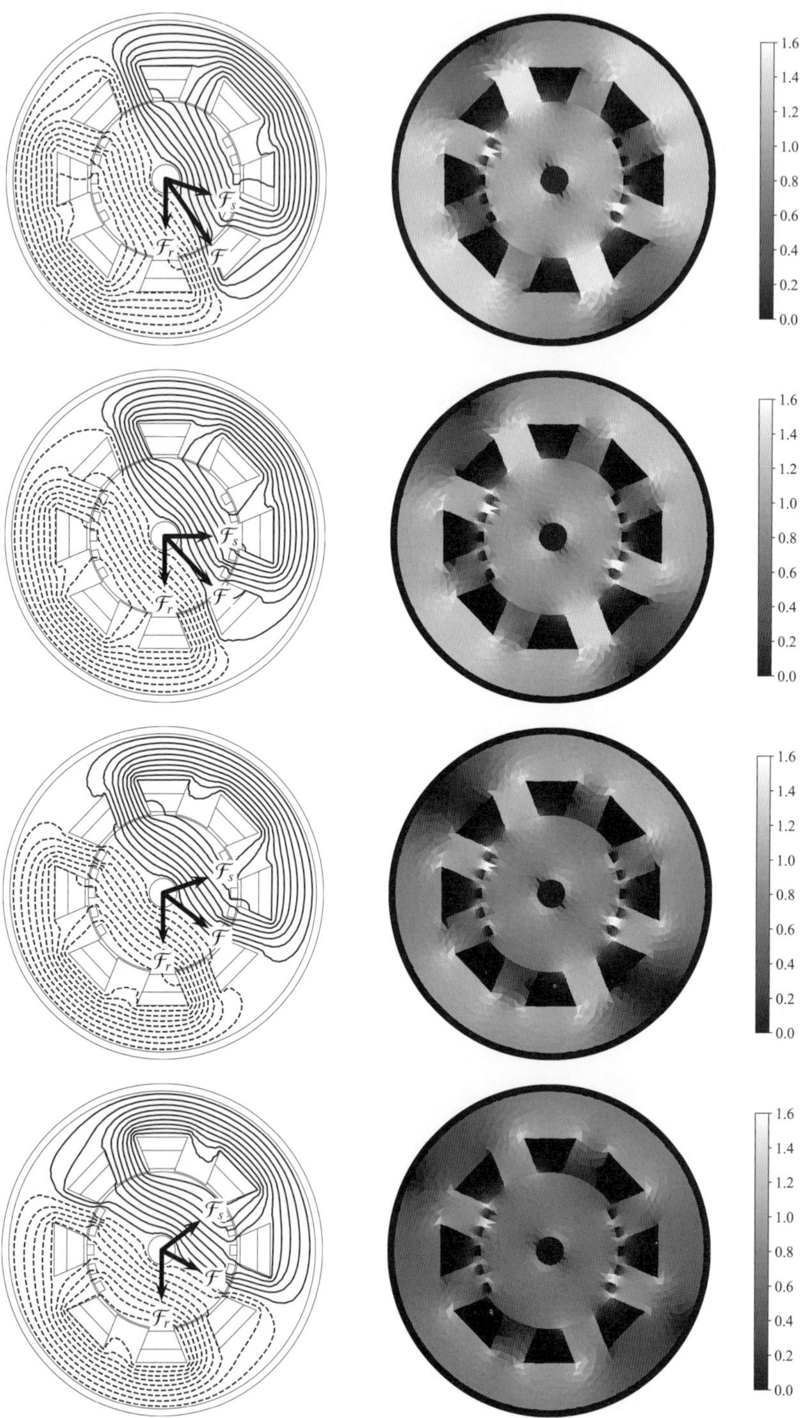

Figure 4.23 Elementary two-phase round-rotor WRSM flux line and flux density plots (B): stator current magnitude $\sqrt{2}I_s = 13.11$ A, stator MMF angle $\phi_i \in \{-18, 0, 18, 36\}°$.

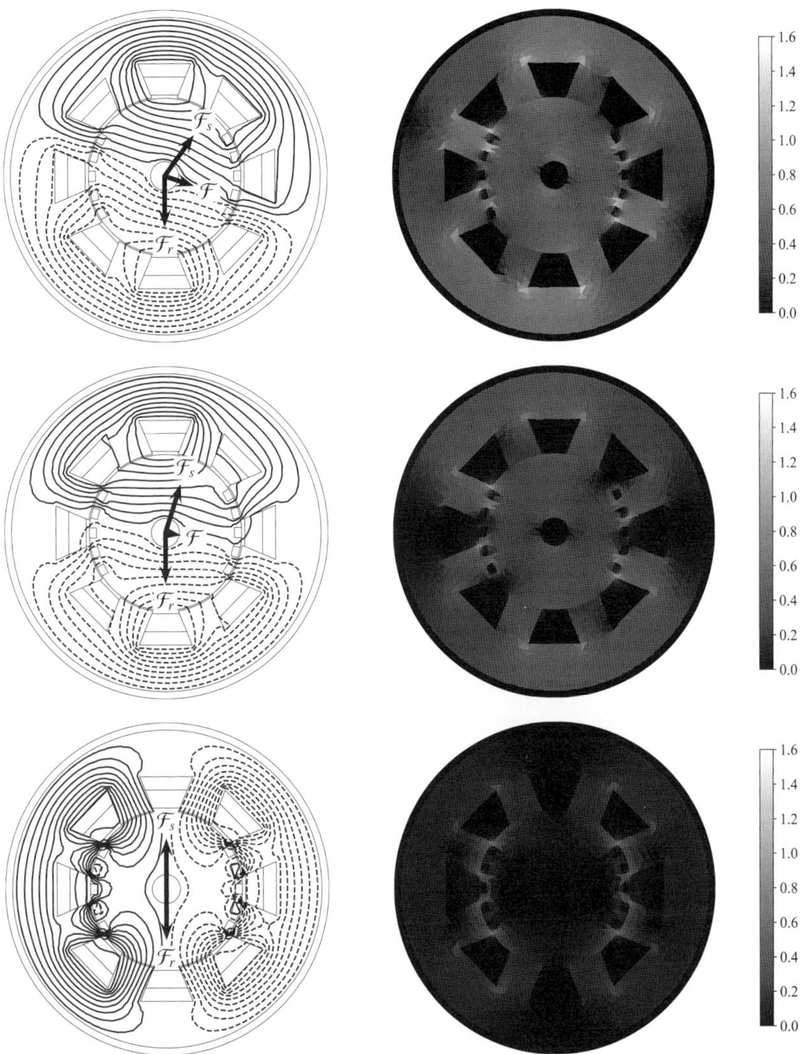

Figure 4.24 Elementary two-phase round rotor WRSM flux line and flux density plots (C): stator current magnitude $\sqrt{2}I_s = 13.11$ A, stator MMF angle $\phi_i \in \{54, 72, 90\}°$.

expect. The a-axis is collinear with the q-axis at this rotor position, so the a-phase flux linkage is driven by $\mathcal{F}_{qs} = \|\mathcal{F}_s\| \cos \phi_i$. The variation is more or less sinusoidal, such that when $\mathcal{F}_s$ is horizontal and aligned with the a-axis, the value of λ_a is maximum. The effect of saturation can be observed by comparing the two peaks. In a linear device, the two peaks would have been proportional to the magnitude of the stator MMF. Here, even though $\mathcal{F}_s$ doubled in magnitude, the flux linkage increased at a smaller rate. Moreover, the shape of λ_a is distorted for the highly saturated case; for example, the curve is flat between 0 and 18°. This can be attributed to the different magnetic field conditions at

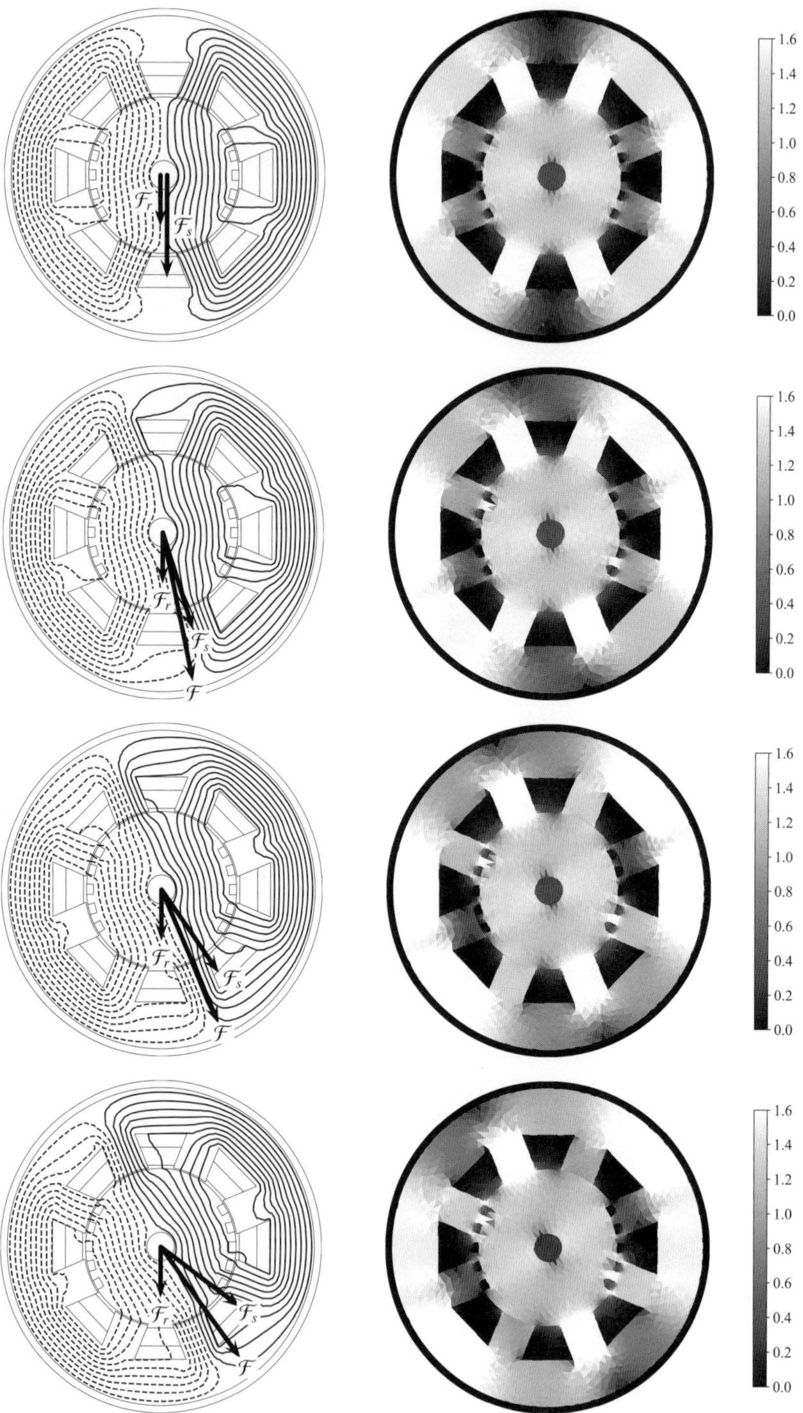

Figure 4.25 Elementary two-phase round-rotor WRSM flux line and flux density plots (D): stator current magnitude $\sqrt{2}I_s = 26.23$ A, stator MMF angle $\phi_i = \{-90, -72, -54, -36\}°$.

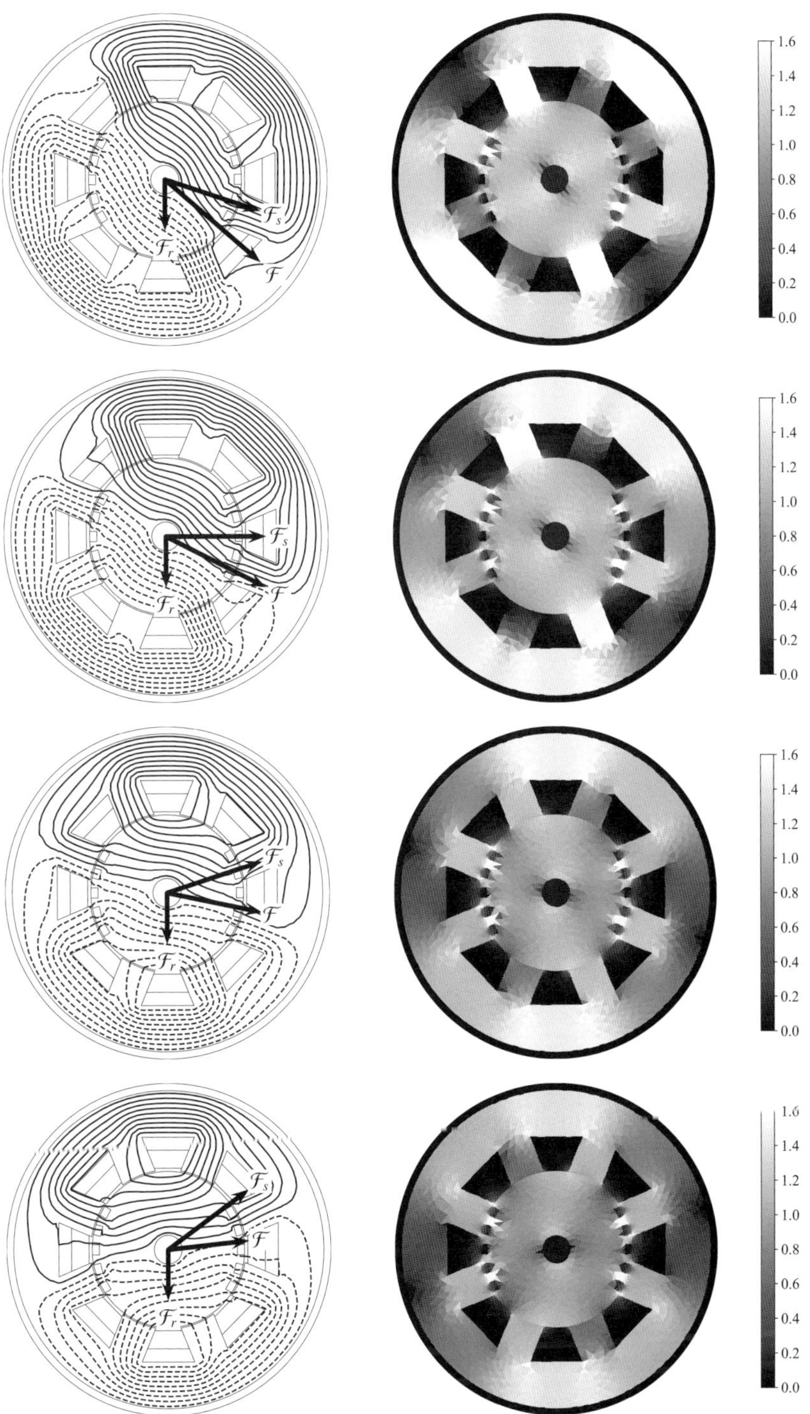

Figure 4.26 Elementary two-phase round-rotor WRSM flux line and flux density plots (E): stator current magnitude $\sqrt{2}I_s = 26.23$ A, stator MMF angle $\phi_i \in \{-18, 0, 18, 36\}°$.

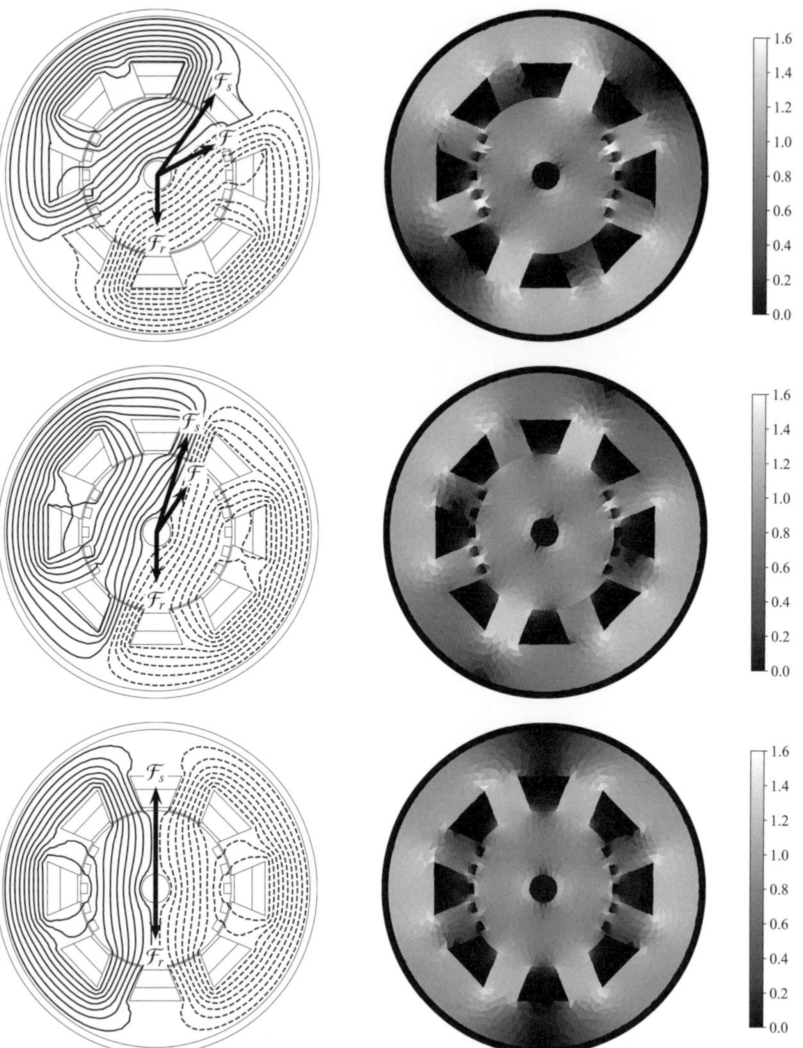

Figure 4.27 Elementary two-phase round-rotor WRSM flux line and flux density plots (F): stator current magnitude $\sqrt{2}I_s = 26.23$ A, stator MMF angle $\phi_i \in \{54, 72, 90\}°$.

$\phi_i \in \{-18, 0, 18\}°$, which can be observed in Figure 4.26. Clearly, the saturation level throughout the machine drops due to the demagnetizing action of $\mathcal{F}_{ds}$. Hence, the same q-axis MMF at $\phi_i = \pm 18°$ yields different flux linkage values. This is illustrative of **cross-coupling due to saturation** between the two axes.

The torque variation is sinusoidal as well. We have previously argued that torque is produced through the interaction of the stator and rotor MMFs (see Example 2.17 on page 169). This qualitative conclusion was reached through a simplified analysis. The FEA results indicate that torque is largely proportional to $\mathcal{F}_q$ or λ_a (i.e., the q-

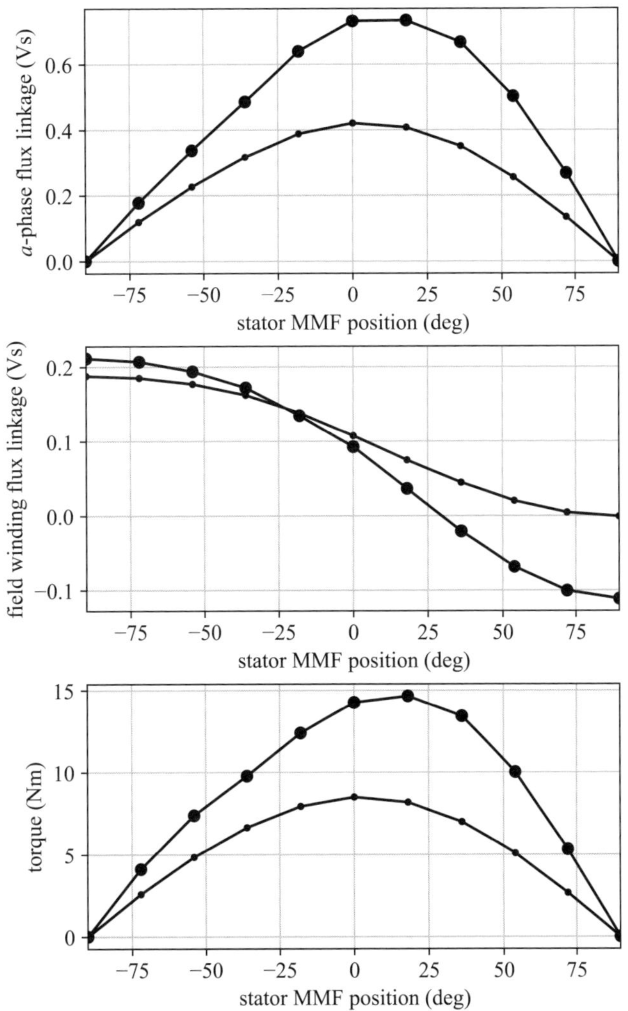

Figure 4.28 Elementary two-phase round-rotor WRSM under load: flux linkage and torque waveforms for $\sqrt{2}I_s = 13.11$ A (smaller line markers) and $\sqrt{2}I_s = 26.23$ A (larger line markers).

axis stator flux). This is an important conclusion: *in round-rotor synchronous machines with distributed windings, the q-axis current is responsible for the creation of torque*. Therefore, we may argue that the torque is roughly proportional to the product of the two MMFs:

$$\tau_e \sim \mathcal{F}_{dr} \cdot \mathcal{F}_{qs} = \|\mathcal{F}_r\| \cdot \|\mathcal{F}_s\| \sin\phi\,, \tag{4.96}$$

where $\phi = \phi_i + \pi/2$ is the angle between $\mathcal{F}_s$ and $\mathcal{F}_r$. The harmonics in the waveform are due to saturation and slotting effects, and do not readily yield to simple analytical arguments.

The variation of the field winding flux linkage λ_f is again illustrative of the effects of saturation. In the first case, $\mathcal{F}_d$ drops from 1425 to 0 At, whereas in the second case, $\mathcal{F}_d$ drops from 2138 to −713 At. The corresponding drop in the flux linkage, however, is disproportional to the MMF reduction: in the first case, it drops by roughly 0.19 Vs, whereas in the second case, it drops by roughly 0.32 Vs. A relevant observation is that, at $\phi_i = -90°$, an increase of $\mathcal{F}_d$ by 50% leads to an increase of only ~12% in λ_f, suggesting that the machine is quite saturated at this point (see Figure 4.25). Interestingly, the curves do not intersect at $\phi_i = 0°$, where in both cases $F_d = F_{dr}$. The heavier saturation of the second case from a higher q-axis MMF leads to a slight reduction of the flux linkage; this is again a manifestation of the cross-coupling effect. (Recall that $\lambda_f \approx 0.11$ Vs at no load, as seen from Figure 4.19.)

4.3.2 Elementary Three-Phase Salient-Rotor Synchronous Machine: Basics of Two-Reaction Theory

Compared to the previous machine, the machine that we study in this section has two main differences: (i) it has a three-phase winding on the stator; and (ii) it has a salient-pole rotor. We shall keep the major dimensions identical, for comparison purposes. Our analysis will proceed along similar steps. However, due to its saliency, this type of machine is conveniently analyzed using a mathematical change of variables in what is called the *qd* reference frame. These details will be presented later on in this section.

In a three-phase machine, a counterclockwise rotating magnetic field is obtained by alternating current forming an *a–b–c* sequence, with the orientation of the winding axes as in Figure 4.29. Such a sequence can be expressed as the balanced set

$$i_a(t) = \sqrt{2}\, I_s \cos(\omega t + \phi_i)\,, \tag{4.97a}$$

$$i_b(t) = \sqrt{2}\, I_s \cos(\omega t + \phi_i - 2\pi/3)\,, \tag{4.97b}$$

$$i_c(t) = \sqrt{2}\, I_s \cos(\omega t + \phi_i + 2\pi/3)\,. \tag{4.97c}$$

Under steady-state conditions, the electrical frequency equals the electrical rotor speed, $\omega = \omega_r$.

Meshing the Stator

The machine has the elementary stator slot design shown in Figure 4.14. The stator dimensions are $r_{so} = 110$ mm, $r_{si} = 55$ mm, $r_B = 80$ mm, $r_E = 57$ mm, $\ell = 100$ mm, $r_{ob} = 115$ mm, $g = 1$ mm, and $r_o = r_{si} - g/3 = 54.67$ mm. We increase the number of slots to $n_{ss} = 24$, which leads to a better air-gap MMF (i.e., one with reduced harmonic content). The stator tooth width is set to $w_{st} = 10$ mm. The number of pole-pairs remains the same ($p = 1$). We thus have a winding that has four slots/pole/phase (whereas the previous two-phase machine had only two slots/pole/phase), which will be distributed in two layers. The winding pattern and stator cross-section are depicted in Figure 4.29. The stator steel has the *B–H* curve that we used previously to plot Figure 3.23. The lamination stacking factor is $k_{st} = 0.97$.

For the given slot dimensions, we can calculate the area of each half-slot, which is

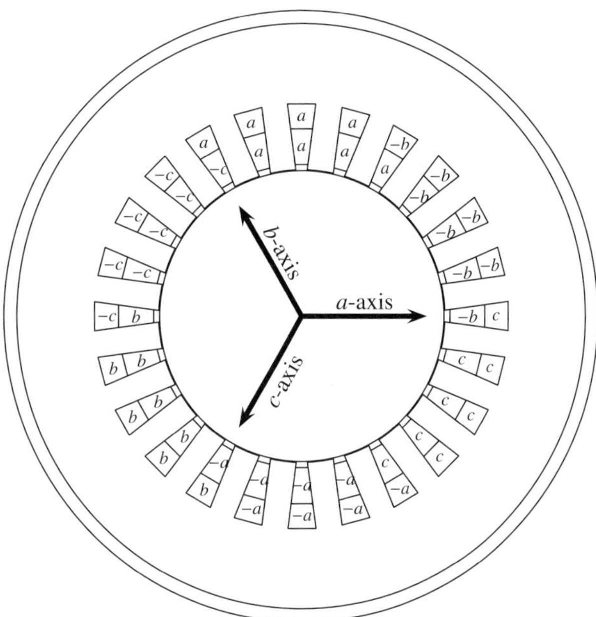

Figure 4.29 Cross-section of elementary three-phase, two-pole WRSM stator PSLG. The innermost "circle" has radius $r_o = r_{si} - g/3$. The outermost "circle" has radius r_{ob}.

90.5 mm^2. Assuming that the winding packing factor is $k_{pf} \approx 0.5$ implies that each half-slot has roughly 45 mm^2 of useful conductor surface area. Hence, we place eight turns of 6-mm^2 (AWG 10) copper conductors (magnet wire) in each half-slot, for a total of $N_s = 64$ series-connected turns in each phase. The per-phase stator winding resistance (ignoring end turns) at 50°C is

$$R_s = 2\frac{N_s \ell}{\sigma A_s} = 41.1 \text{ m}\Omega . \tag{4.98}$$

Meshing the Rotor

Figure 4.30 depicts the cross-section of a salient rotor pole spanning π/p rad (here, $p = 2$). For simplicity, the pole faces are circular arcs that are concentric with the shaft, although in practice they may be shaped differently to achieve a better back-emf. We are ignoring the presence of damper windings, which are typically embedded close to the pole faces. The coordinates of points A–E that define the rotor shape, and the coordinates of points F–I that define the field winding area, are sufficient to generate the geometry. To avoid the creation of small triangles, the coil is adjacent to the steel (i.e., $y_F = y_G = y_B$). The shape is symmetric with respect to the x-axis.

The dimensions used in this case study are $r_i = r_{si} - 2g/3 = 54.33$ mm, $r_{ro} = r_{si} - g = 54$ mm, $r_{ri} = 10$ mm, $r_A = 20$ mm, $\phi_A = \pi/2$ rad, $\phi_B = 0.6\phi_A$, $x_C = x_B + 25 = r_A \cos\phi_B + 25 = 36.76$ mm, $r_D = 49$ mm, $x_G = 35.76$ mm, $x_F = x_G - 15$ mm, and $y_H = y_G + 7 = r_A \sin\phi_B + 7 = 23.18$ mm. The rotor cross-section is shown in Figure 4.31.

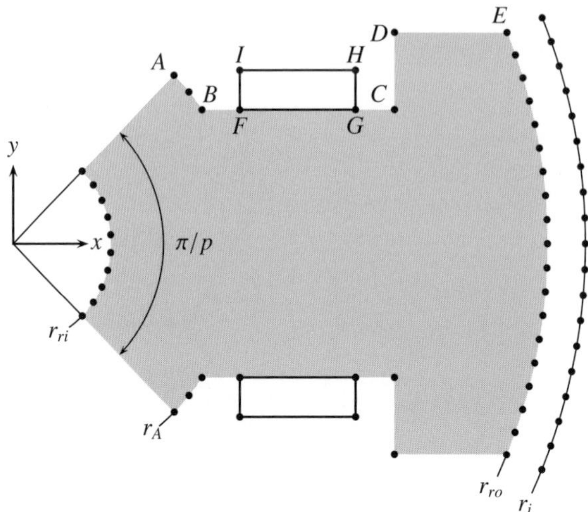

Figure 4.30 Geometry of an elementary salient rotor pole (dimensions are somewhat exaggerated for clarity). Dots signify vertices that should be included in the PSLG.

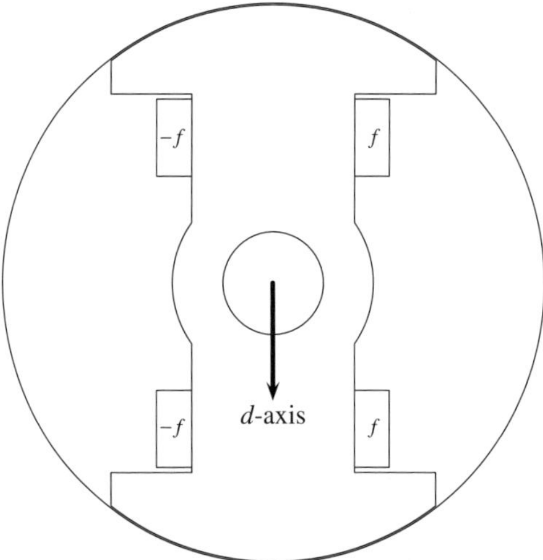

Figure 4.31 Cross-section of elementary two-pole salient-rotor WRSM PSLG. The innermost "circle" has radius r_{ri}. The outermost "circle" has radius $r_i = r_{ro} + g/3$.

The field winding area is $15 \times 7 = 105$ mm^2 on each pole side. We place 10 turns of 6-mm^2 (AWG 10) copper wire in each winding region, for a total of $N_f = 20$ turns. The field winding resistance (ignoring end turns) at 50°C is

$$R_f = 2\frac{N_f \ell}{\sigma A_f} = 12.8 \text{ m}\Omega\,. \tag{4.99}$$

The rotor laminations are made from the same material as the stator, and have the same stacking factor. The shaft material is assumed to be magnetically linear with a relative permeability of 200.

Flux Linkage and Torque: Open-Circuit Conditions

The complete machine is meshed as shown in Figure 4.32. The mesh detail illustrates the small size of the triangular elements in the vicinity of the air gap. In this case study, the stator, rotor, and air-gap meshes have 8,549, 6,191, and 2,062 triangular elements, respectively.

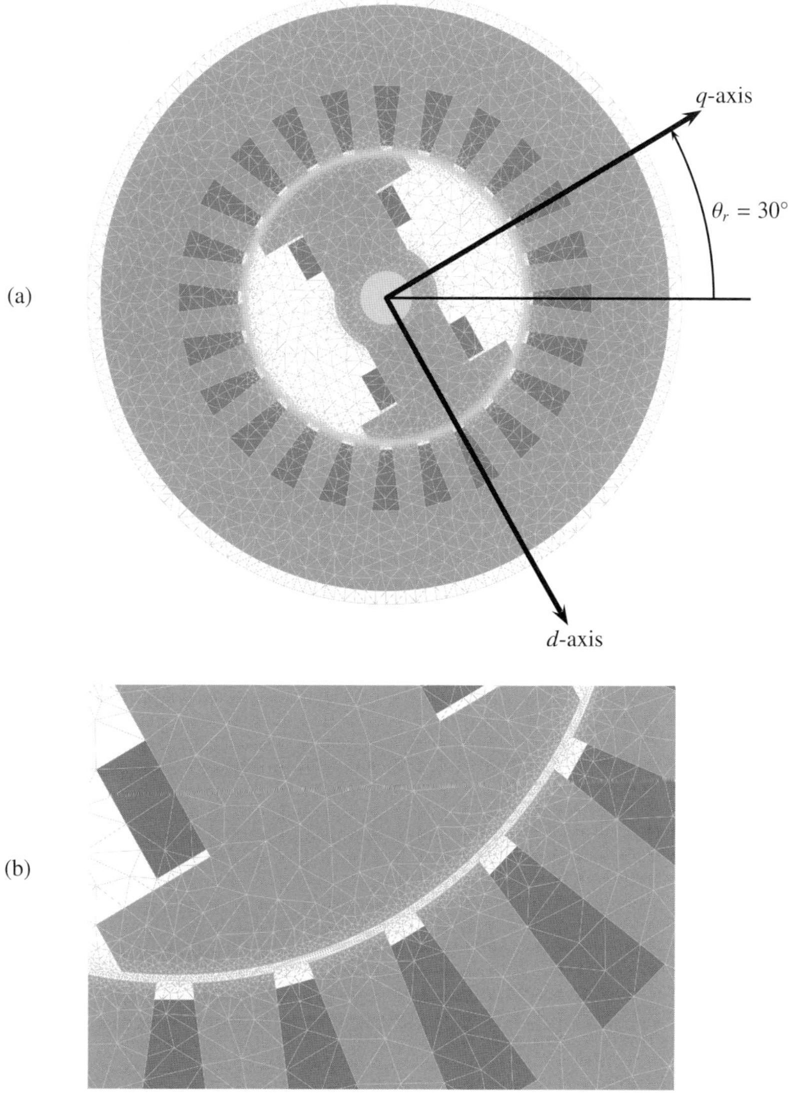

Figure 4.32 (a) Meshed cross-section of elementary three-phase, two-pole wound-rotor WRSM. (b) Mesh detail in a region close to the air gap and the d-axis.

A series of FEA studies is conducted over a grid of field winding currents $i_f = 10, 20, \ldots, 60$ A and rotor positions $\theta_r = 0, 2.5, 5, \ldots, 180°$, while keeping the armature winding open. A snapshot for $i_f = 60$ A and $\theta_r = 30°$ is shown in Figure 4.33. Note that the rotor is heavily saturated at this current level. We can observe the variation of the flux linkages of the stator and field windings in Figure 4.34. Compared to the previous machine that we analyzed, the effect of slotting on the flux linkages has been dramatically reduced. The impact of saturation is also apparent since the flux linkages do not increase proportionally with the MMF. It is also noteworthy that the reluctance torque (not plotted) has been reduced by an order of magnitude.

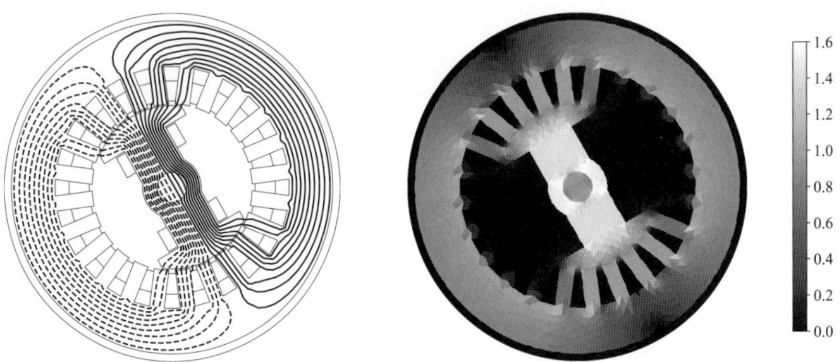

Figure 4.33 Elementary three-phase salient-rotor WRSM open-circuit flux lines (left) and flux density (right) at $\theta_r = 30°$ and $i_f = 60$ A.

Flux Linkage and Torque: Loaded Conditions

Salient-pole WRSMs are typically analyzed in the rotor reference frame. This type of analysis is referred to as **Blondel's two-reaction theory**,[76] first proposed in the late nineteenth century. The approach is based on two magnetic equivalent circuits representing flux along the d- and q-axes, respectively, which exhibit unequal magnetic reluctance, $\mathcal{R}_q > \mathcal{R}_d$. The magnetic circuits are excited by the rotor and stator MMFs. Therefore, one can visualize two magnetizing flux components being present, namely

$$\Phi_{mq} = \frac{\mathcal{F}_q}{\mathcal{R}_q} = \frac{\mathcal{F}_{qs}}{\mathcal{R}_q}, \tag{4.100a}$$

$$\Phi_{md} = \frac{\mathcal{F}_d}{\mathcal{R}_d} = \frac{\mathcal{F}_{dr} + \mathcal{F}_{ds}}{\mathcal{R}_d}, \tag{4.100b}$$

representing the fundamental pole-pair fields. This modeling approach is the subject of circuit-based treatments of synchronous machines. Due to the distributed nature of the windings and the unknown distribution of flux density within the machine, several simplifying assumptions are required. These details are outside the scope of this text because we can calculate the quantities of interest (flux linkages and torque) directly from the FEA. Nevertheless, such understanding is conceptually useful for interpreting the FEA results.

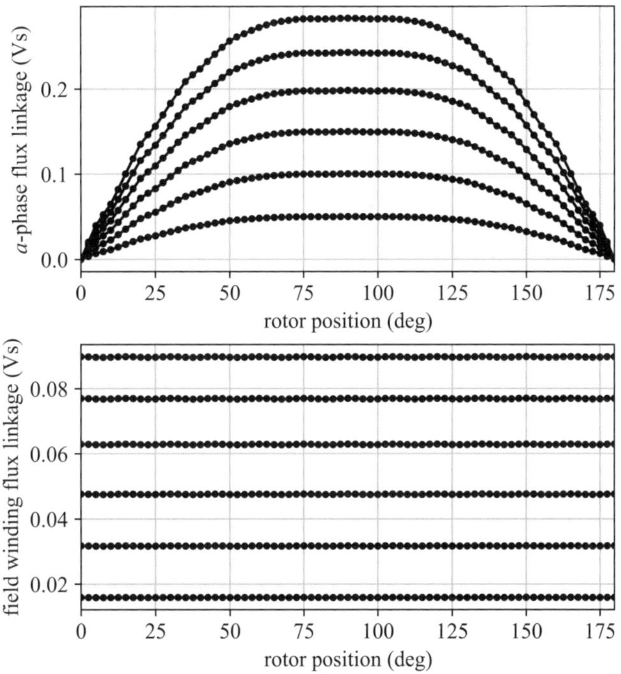

Figure 4.34 Elementary three-phase salient-rotor WRSM open-circuit flux linkage waveforms. The curves correspond to six levels of field winding current, $i_f = 10, 20, \ldots, 60$ A.

By inspection of the winding diagram in Figure 4.29, we calculate the stator winding distribution factor as $k_{ws} \approx 0.95$. The stator MMF vector is

$$\mathcal{F}_s = \frac{2}{\pi} k_{ws} N_s (i_a\,\hat{\mathbf{a}} + i_b\,\hat{\mathbf{b}} + i_c\,\hat{\mathbf{c}})\,, \tag{4.101}$$

where we defined unit vectors along the three axes of the stator windings:

$$\hat{\mathbf{a}} = \hat{\mathbf{x}}\,,\;\; \hat{\mathbf{b}} = -\frac{1}{2}\hat{\mathbf{x}} + \frac{\sqrt{3}}{2}\hat{\mathbf{y}}\,,\;\; \hat{\mathbf{c}} = -\frac{1}{2}\hat{\mathbf{x}} - \frac{\sqrt{3}}{2}\hat{\mathbf{y}}\,. \tag{4.102}$$

Therefore

$$\mathcal{F}_s = \frac{2}{\pi} k_{ws} N_s \left[\left(i_a - \frac{1}{2} i_b - \frac{1}{2} i_c \right) \hat{\mathbf{x}} + \frac{\sqrt{3}}{2} (i_b - i_c)\,\hat{\mathbf{y}} \right]. \tag{4.103}$$

These components are then decomposed along the direct and quadrature axes, depending on the rotor position. Under steady-state conditions, where the machine is excited with a balanced three-phase set of currents (4.97a)–(4.97c) and $\theta_r = \omega t$, the qd stator MMF vectors obtain constant values:

$$\mathcal{F}_{qs} = \frac{3\sqrt{2}}{\pi} k_{ws} N_s I_s \cos\phi_i\,, \tag{4.104a}$$

$$\mathcal{F}_{ds} = -\frac{3\sqrt{2}}{\pi} k_{ws} N_s I_s \sin\phi_i\,. \tag{4.104b}$$

On the other hand, establishing the effective MMF of the field winding in this case requires additional assumptions because it is not as easy to visualize the flux lines. The magnetizing flux lines originating from the pole faces (between the opposing D-points on each pole) are crossing the air gap more or less radially, and link all turns of the field winding. Hence, the air-gap MMF oscillates between $\pm N_f i_f/2$. Flux lines can also originate from the ABC sides of the inner pole surface. We can simplistically imagine these as leaving the rotor perpendicularly to the surface, and traveling more or less on a straight line until they reach the stator surface. Flux lines between points G and C are driven by the entire MMF of the coil. However, the MMF drops linearly between points G and F, after which it is zero. We thus define an angle

$$\alpha = \arccos\left(\frac{x_F + x_G}{2r_{si}}\right), \tag{4.105}$$

with respect to the d-axis. We assume that the rotor MMF is zero between α and $\pi - \alpha$, which define two symmetric points with respect to the q-axis. The MMF is also zero on the opposite side of the rotor. A Fourier analysis of this three-level rotor MMF waveform leads to the fundamental component

$$\mathcal{F}_{dr} = \frac{2}{\pi} \sin\alpha \, N_f i_f \, . \tag{4.106}$$

For the rotor dimensions of this case study, $\sin\alpha \approx 0.86$. It should be noted that this calculation may break down for higher pole-count machines, where other assumptions may be necessary. The calculation of the MMFs is only meant as a conceptual and visual aid in interpreting the FEA results.

We shall imitate the conditions of the previous case study for the two-phase WRSM. We will set $i_f = 60$ A (so that $||\mathcal{F}_r|| = 655.4$ At) and two levels of stator current: (i) $\sqrt{2}I_s = 11.295$ A (so that $||\mathcal{F}_s|| \approx ||\mathcal{F}_r||$) and (ii) $\sqrt{2}I_s = 22.59$ A (so that $||\mathcal{F}_s|| \approx 2||\mathcal{F}_r||$). We will then increase the position of $\mathcal{F}_s$ from $\phi_i = -90°$ to $\phi_i = 90°$, while keeping the rotor locked at $\theta_r = 0°$. The FEA results are plotted in Figures 4.35–4.37 for case (i) and Figures 4.38–4.40 for case (ii).

We highlight the main differences from the round-rotor case:

- The flux lines have a tendency to flow mostly through the pole faces and through the rotor along the d-axis. Very few lines cross the large air gap along the q-axis. This is indicative of a higher q-axis reluctance.
- Another consequence of $\mathcal{R}_q > \mathcal{R}_d$ is that $\mathcal{F}$ is no longer aligned with the principal directionality of the magnetic field. This can be observed in all plots. For instance, in the second plot of Figure 4.36, we have a situation where $\mathcal{F}_d = \mathcal{F}_q$; however, a visual inspection indicates that $\Phi_{md} > \Phi_{mq}$ in this case (even though we have not explained how these fluxes are calculated).
- Interestingly, we observe that q-axis flux is traveling through the poles in the transverse direction, but not directly through the central path of the q-axis. This can be observed clearly in the last two plots of Figure 4.39, which depict a situation where $\mathcal{F}_d$ is negligible compared to $\mathcal{F}_q$.

- In the last plot of Figure 4.37, we have created a situation where $\mathcal{F}_r$ and $\mathcal{F}_s$ are directly opposing each other. The flux lines represent stator and rotor coil leakage paths. The magnetizing flux is practically zero, and the flux density X-ray is at its darkest.

We observe the variation of the flux linkages and torque as a function of the stator MMF position ϕ_i in Figure 4.41. In regard to λ_a and λ_f, the situation is similar as before, although the numerical values are different. We can observe in these waveforms the effects of saturation and cross-saturation between the two axes. Nevertheless, the torque waveform is very different and merits further discussion.

First, we see that torque is always positive for case (i), but goes negative for case (ii) after roughly 36°. Second, we observe that the torque is not sinusoidal and does not follow the previous relationship (4.96), especially in case (ii), although the distortion is also visible but less prominent in case (i). If one employs an analytical approach based on Blondel's two-reaction theory and Park's transformation (which will be introduced shortly in §4.3.3), it can be shown that the torque can be expressed as

$$\tau_e = A\,\mathcal{F}_{dr}\mathcal{F}_{qs} + B\,\mathcal{F}_{qs}\mathcal{F}_{ds} = A\,\|\mathcal{F}_r\| \cdot \|\mathcal{F}_s\| \sin\phi + \frac{B}{2}\|\mathcal{F}_s\|^2 \sin 2\phi\,, \tag{4.107}$$

where A and B are non-negative constants, and $\phi = \phi_i + \pi/2$ is the angle by which $\mathcal{F}_s$ leads the d-axis. In particular, B is a constant that originates from the saliency of the rotor, which is positive when $\mathcal{R}_q > \mathcal{R}_d$ as in this case. The first term is identical to the "magnetic dipole" torque generated by a round-rotor machine. The second term of the torque expression is recognized as a **reluctance torque**. Of course, this formula is only as accurate as the underlying analytical assumptions can ever be, and does not capture all the intricacies that FEA does. This is the same result that was obtained for an elementary reluctance machine using a coupling field energy-based argument in Example 2.18 on page 173, which was expressed as Equation (2.269). (In Example 2.18, which involved a single winding on the stator, we moved the rotor with respect to the a-axis, so the torque expression was in terms of the rotor angle, $\tau_e = i_s^2 L_B \sin 2\theta_r$, with θ_r representing the angle between the q-axis and the a-axis. The angle formed between the d-axis and the stator MMF is $\phi = \pi/2 - \theta_r$, so $\sin 2\theta_r = \sin 2\phi$.) In case (ii), where the stator MMF has doubled, the reluctance term dominates over the weaker first term at the higher range of $\phi > 120°$, leading to negative values for torque.

4.3.3 Analysis in qd Variables Using Park's Transformation

The equations of synchronous machines are commonly manipulated with a transformation of variables called Park's[77] transformation, which is inspired from Blondel's two-reaction theory. Park's transformation conceptually projects a three-phase set of stator variables $\mathbf{f}_{abc}$ to the orthogonal qd-axes of a reference frame that is fixed on the rotor. A third component, which we call the zero component, is just the mean value of the three quantities. Here, $\mathbf{f}$ is representative of current, voltage, or flux linkage. We will provide a concise presentation of the main underlying ideas, which will be validated using FEA.

The forward transformation is defined by

$$\mathbf{f}_{qd0} = \mathbf{K}(\theta_r)\,\mathbf{f}_{abc}\,, \tag{4.108}$$

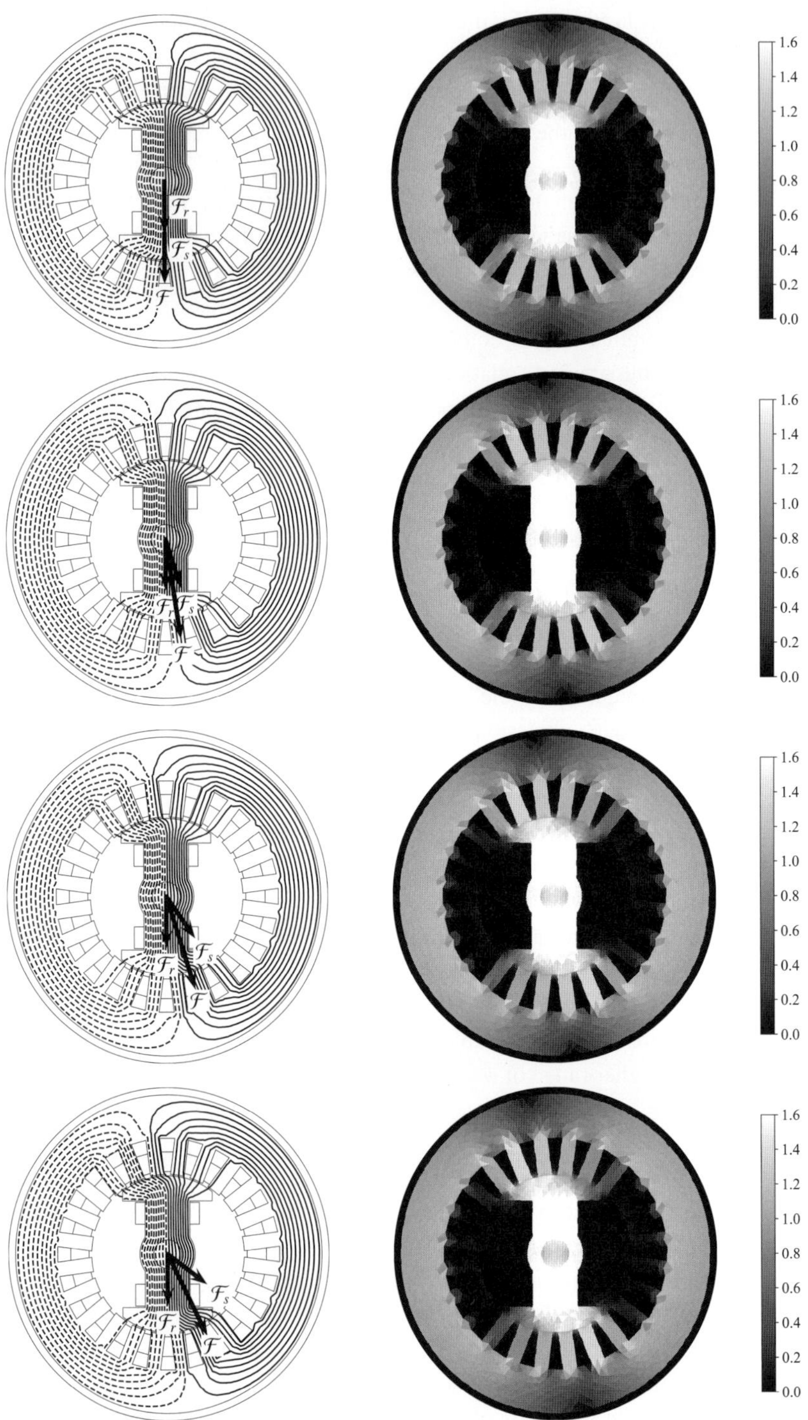

Figure 4.35 Elementary three-phase salient-pole WRSM flux line and flux density plots (A): stator current magnitude $\sqrt{2}I_s = 11.295$ A, stator MMF angle $\phi_i \in \{-90, -72, -54, -36\}°$.

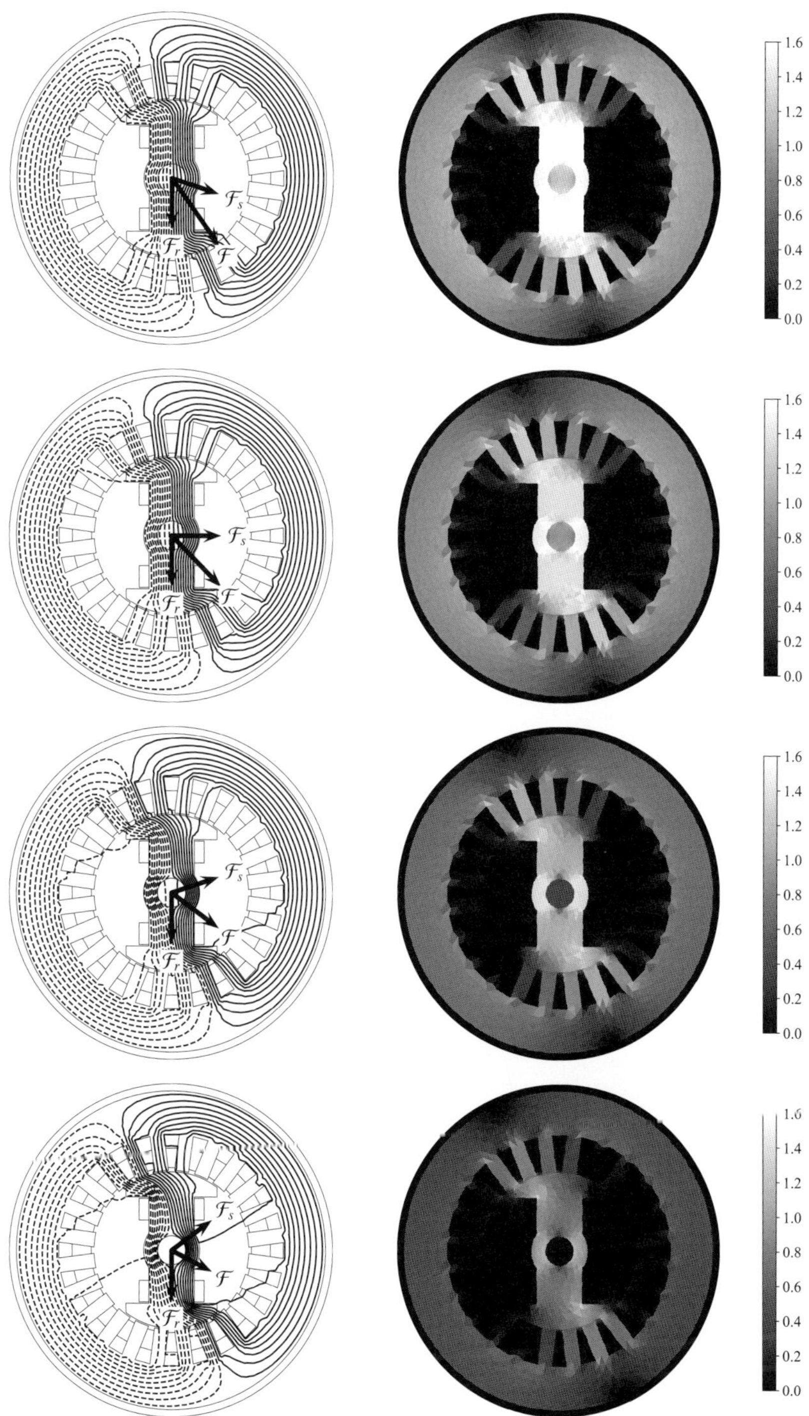

Figure 4.36 Elementary three-phase salient-pole WRSM flux line and flux density plots (B): stator current magnitude $\sqrt{2}I_s = 11.295$ A, stator MMF angle $\phi_i \in \{-18, 0, 18, 36\}°$.

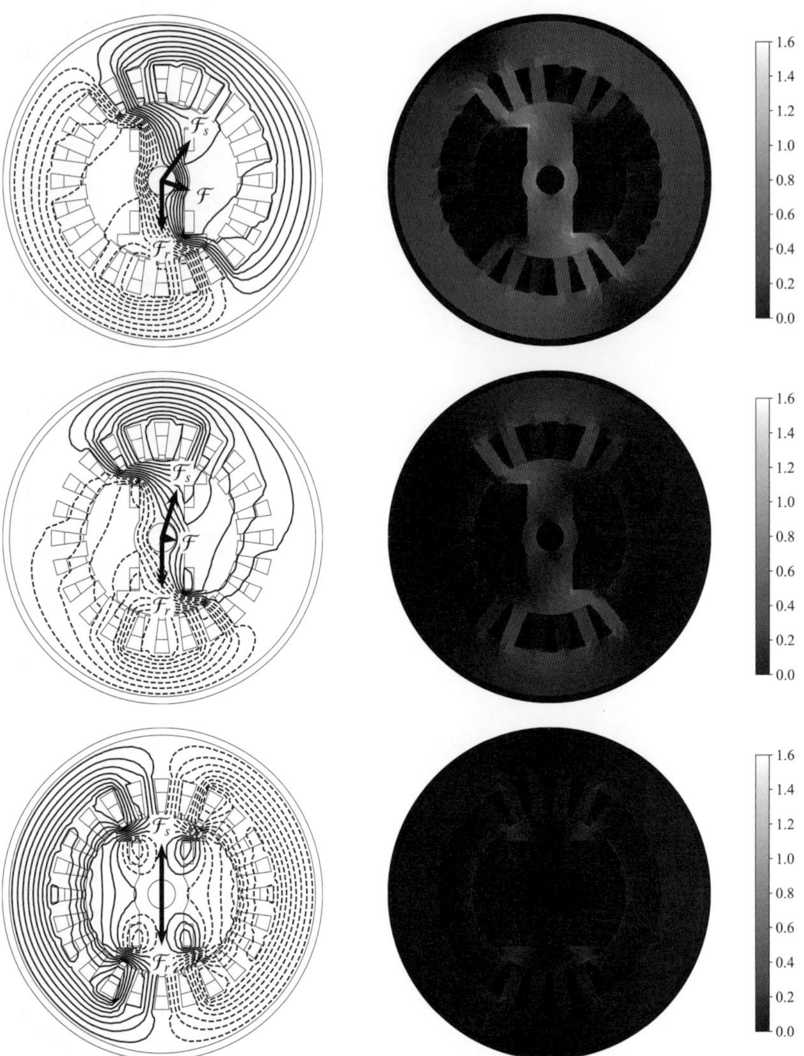

Figure 4.37 Elementary three-phase salient-pole WRSM flux line and flux density plots (C): stator current magnitude $\sqrt{2}I_s = 11.295$ A, stator MMF angle $\phi_i \in \{54, 72, 90\}°$.

where the **f**s are 3 × 1 column vectors. The 3 × 3 transformation matrix depends on the electrical rotor angle. Its exact form depends on how one defines the positions of the axes and the angles. We have defined the q-axis as leading the d-axis, and the angle θ_r is between the q- and a-axes. Hence

$$\mathbf{K}(\theta_r) = \frac{2}{3}\begin{bmatrix} \cos\theta_r & \cos(\theta_r - 2\pi/3) & \cos(\theta_r + 2\pi/3) \\ \sin\theta_r & \sin(\theta_r - 2\pi/3) & \sin(\theta_r + 2\pi/3) \\ 1/2 & 1/2 & 1/2 \end{bmatrix}. \tag{4.109}$$

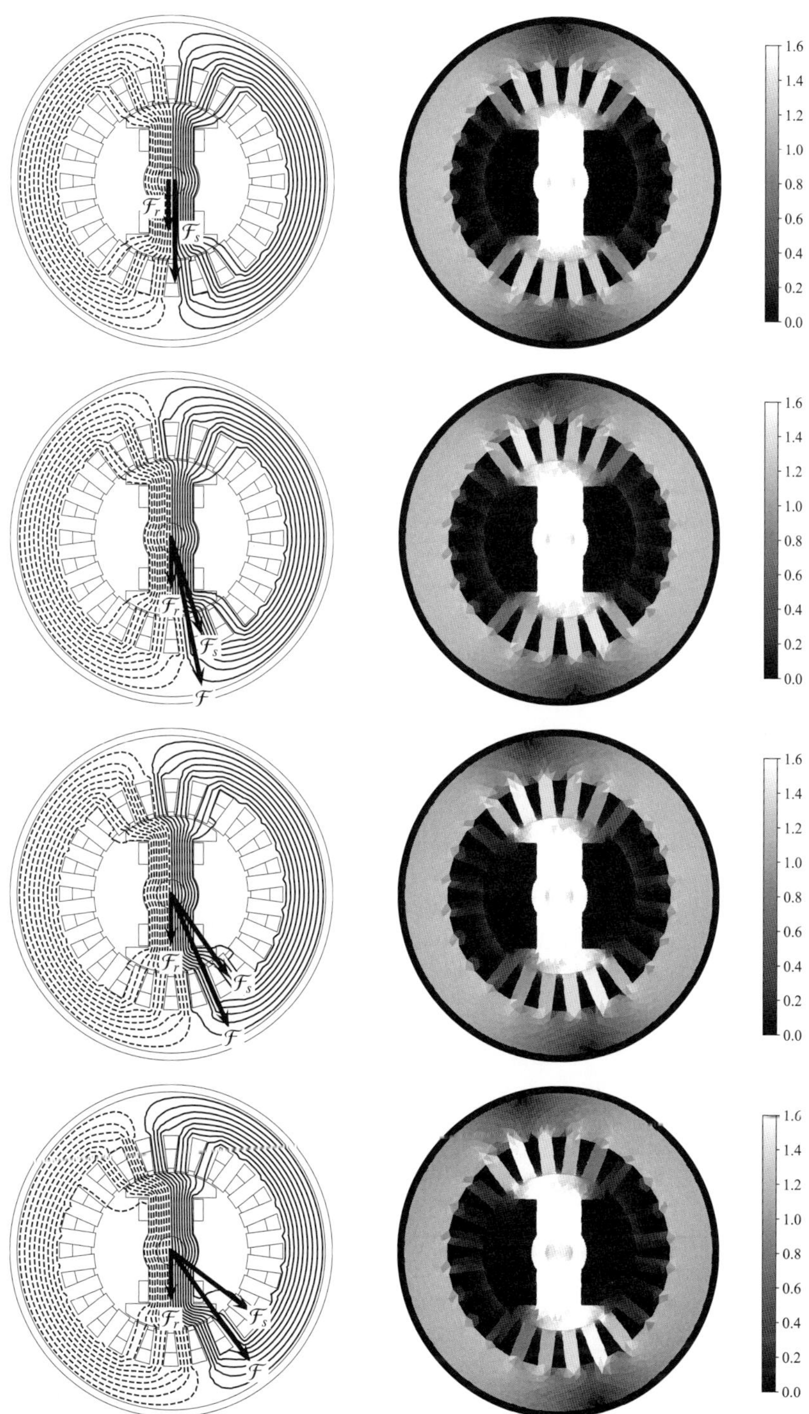

Figure 4.38 Elementary three-phase salient-pole WRSM flux line and flux density plots (D): stator current magnitude $\sqrt{2}I_s = 22.59$ A, stator MMF angle $\phi_i \in \{-90, -72, -54, -36\}°$.

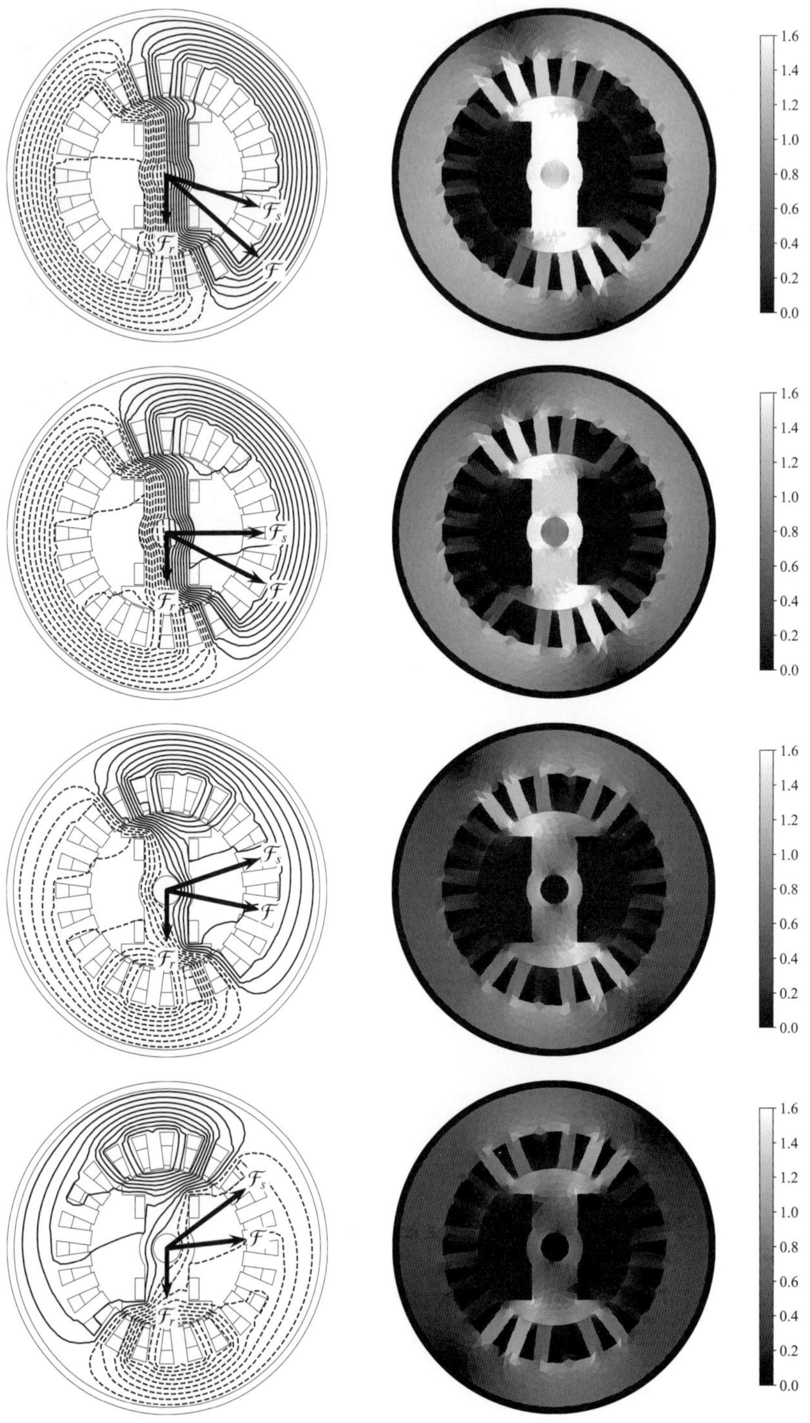

Figure 4.39 Elementary three-phase salient-pole WRSM flux line and flux density plots (E): stator current magnitude $\sqrt{2}I_s = 22.59$ A, stator MMF angle $\phi_i \in \{-18, 0, 18, 36\}°$.

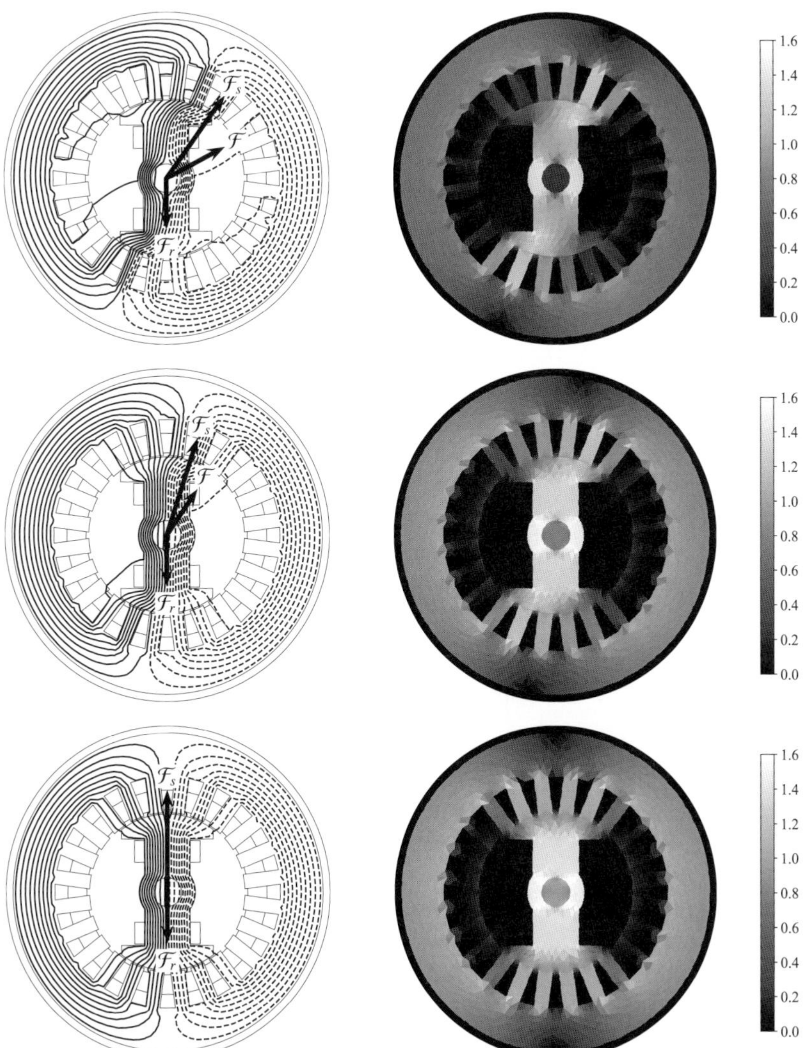

Figure 4.40 Elementary three-phase salient-pole WRSM flux line and flux density plots (T): stator current magnitude $\sqrt{2}I_s = 22.59$ A, stator MMF angle $\phi_i \in \{54, 72, 90\}°$.

Here, $\mathbf{f}_{qd0} = [f_q \;\; f_d \;\; f_0]^\top$ is a vector of transformed variables (quadrature, direct, and zero, respectively). The inverse transformation is

$$\mathbf{f}_{abc} = \mathbf{K}(\theta_r)^{-1}\,\mathbf{f}_{qd0}\,, \tag{4.110}$$

where

$$\mathbf{K}(\theta_r)^{-1} = \begin{bmatrix} \cos\theta_r & \sin\theta_r & 1 \\ \cos(\theta_r - 2\pi/3) & \sin(\theta_r - 2\pi/3) & 1 \\ \cos(\theta_r + 2\pi/3) & \sin(\theta_r + 2\pi/3) & 1 \end{bmatrix}. \tag{4.111}$$

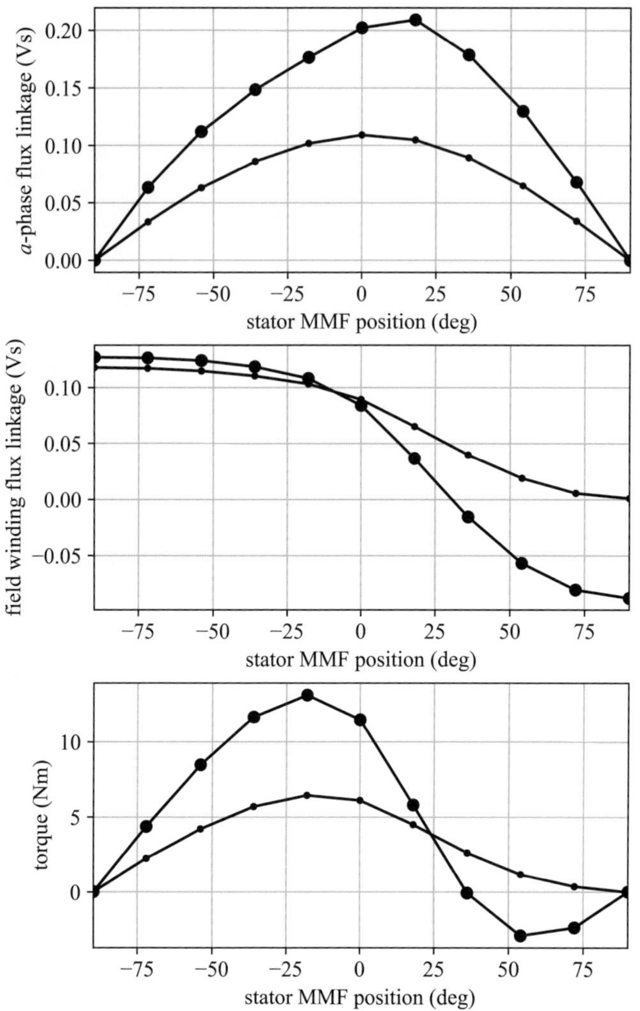

Figure 4.41 Elementary three-phase salient-pole WRSM under load: flux linkage and torque waveforms for $\sqrt{2}I_s = 11.295$ A (smaller line markers) and $\sqrt{2}I_s = 22.59$ A (larger line markers).

The transformation can be applied to any vector $\mathbf{f}_{abc}$ on an instantaneous basis. In particular, under steady-state conditions where the electrical frequency of the currents ω equals the rotor electrical speed $\omega_r = d\theta_r/dt$, the transformation of the balanced set of stator currents (4.97a)–(4.97c) to the rotor reference frame yields (these trigonometric manipulations are left as an exercise for the reader)

$$i_q = \sqrt{2}\, I_s \cos\phi_i\,, \tag{4.112a}$$

$$i_d = -\sqrt{2}\, I_s \sin\phi_i\,, \tag{4.112b}$$

$$i_0 = 0\,. \tag{4.112c}$$

Since we can choose the time $t = 0$ arbitrarily, we have selected it so that the rotor is initially vertical, i.e.

$$\theta_r = \omega t\,. \tag{4.113}$$

The zero component of the current typically vanishes under normal operating conditions, so it can be ignored. The reduced vector is denoted as $\mathbf{f}_{qd}$. From Equations (4.104a) and (4.104b), we have

$$\mathcal{F}_{qs} = \frac{3}{\pi} k_{ws} N_s i_q\,, \tag{4.114a}$$

$$\mathcal{F}_{ds} = -\frac{3}{\pi} k_{ws} N_s i_d\,, \tag{4.114b}$$

which shows the direct relationship between qd currents and stator MMF components.

Similarly, the $qd0$ flux linkages are obtained by transforming the physical three-phase winding flux linkages using $\boldsymbol{\lambda}_{qd0} = \mathbf{K}(\theta_r)\,\boldsymbol{\lambda}_{abc}$. In a conservative coupling field, the *abc* flux linkages are functions of the winding currents and the rotor position:

$$\lambda_a = \lambda_a(i_a, i_b, i_c, i_f, \theta_r)\,, \tag{4.115a}$$

$$\lambda_b = \lambda_b(i_a, i_b, i_c, i_f, \theta_r)\,, \tag{4.115b}$$

$$\lambda_c = \lambda_b(i_a, i_b, i_c, i_f, \theta_r)\,. \tag{4.115c}$$

Recall that we have observed the significant dependency of $\boldsymbol{\lambda}_{abc}$ on the rotor angle under open-circuit conditions. *The main advantage of Park's transformation is that it eliminates (for the most part) the dependence on θ_r so that the elements of $\boldsymbol{\lambda}_{qd0}$ appear as constants for given $\mathbf{i}_{qd}$ (plus a typically negligible ripple component).* Hence, we can express the transformed flux linkages as functions of transformed currents:

$$\lambda_q \approx \lambda_q(i_q, i_d, i_f)\,, \tag{4.116a}$$

$$\lambda_d \approx \lambda_d(i_q, i_d, i_f)\,, \tag{4.116b}$$

$$\lambda_0 \approx 0\,. \tag{4.116c}$$

A similar assumption is used for the field winding:

$$\lambda_f \approx \lambda_f(i_q, i_d, i_f)\,, \tag{4.117}$$

which we have observed (under open-circuit conditions) in Figure 4.34. This simplification is sufficient for many types of machine analysis, including control system design and power system simulation, where the higher-order harmonics are not that important.

We can also transform the phase voltages by applying Park's transformation to the ohmic and inductive voltage drops of the coils, $\mathbf{v}_{qd0} = \mathbf{K}(\theta_r)\,\mathbf{v}_{abc}$. This operation yields

$$v_q = R_s i_q + \omega_r \lambda_d + \frac{d}{dt}\lambda_q\,, \tag{4.118a}$$

$$v_d = R_s i_d - \omega_r \lambda_q + \frac{d}{dt}\lambda_d\,, \tag{4.118b}$$

$$v_0 = R_s i_0 + \frac{d}{dt}\lambda_0\,. \tag{4.118c}$$

A steady-state operating point can be specified by constant $\mathbf{i}_{qd}$ and i_f, corresponding to a balanced three-phase set of constant-amplitude stator currents and fixed excitation on the field, respectively. If Equations (4.116a)–(4.116c) are valid, the time derivatives of the qd flux linkages vanish. We obtain the steady-state voltage equations

$$V_q = \sqrt{2}V_s \cos\phi_v = R_s I_q + \omega_r \Lambda_d \,, \tag{4.119a}$$

$$V_d = -\sqrt{2}V_s \sin\phi_v = R_s I_d - \omega_r \Lambda_q \,, \tag{4.119b}$$

where uppercase symbols are used to denote steady-state conditions. The corresponding phase voltages are

$$v_a(t) = \sqrt{2}\,V_s \cos(\omega t + \phi_v)\,, \tag{4.120a}$$

$$v_b(t) = \sqrt{2}\,V_s \cos(\omega t + \phi_v - 2\pi/3)\,, \tag{4.120b}$$

$$v_c(t) = \sqrt{2}\,V_s \cos(\omega t + \phi_v + 2\pi/3)\,. \tag{4.120c}$$

Torque in *qd* Variables

The expression for electromagnetic torque in the analytical framework of Park's transformation is particularly simple and insightful. To derive this, we can use either an energy or a coenergy approach. Since we have expressed the qd flux linkages as functions of the qd currents in Equations (4.116a) and (4.116b), it is convenient to work with the coenergy.

The coupling field coenergy in a three-phase WRSM is the line integral

$$W_c(\mathbf{i}_{abc}, i_f, \theta_{rm}) = \int_{(\mathbf{0},0,\theta_{rm0})}^{(\mathbf{i}_{abc},i_f,\theta_{rm})} \boldsymbol{\lambda}_{abc}^{\top}\, d\tilde{\mathbf{i}}_{abc} + \lambda_f\, d\tilde{i}_f + \tau_e\, d\tilde{\theta}_{rm}\,. \tag{4.121}$$

The path of integration starts from an initial de-energized state (where all stator and rotor currents are zero) at some arbitrary mechanical rotor angle θ_{rm0}, and ends at a final state of interest $(\mathbf{i}_{abc}, i_f, \theta_{rm})$. In a conservative coupling field, the integral will only depend on the two endpoints. Note that we are allowing the angle to change simultaneously with the other variables, as this is a general expression and not just a convenient choice of integration path for a numerical calculation. In the general case of a $2p$-pole machine, the mechanical rotor angle is related to the electrical rotor angle by $\theta_{rm} = \theta_r/p$.

Now let us transform the stator variables to the rotor reference frame in the integral:

$$W_c(\mathbf{i}_{qd0}, i_f, \theta_{rm}) = \int_{(\mathbf{0},0,\theta_{rm0})}^{(\mathbf{i}_{qd0},i_f,\theta_{rm})} (\mathbf{K}^{-1}\boldsymbol{\lambda}_{qd0})^{\top}\, d(\mathbf{K}^{-1}\,\tilde{\mathbf{i}}_{qd0}) + \lambda_f\, d\tilde{i}_f + \tau_e\, d\tilde{\theta}_{rm}\,. \tag{4.122}$$

The first term of the integrand is

$$(\mathbf{K}^{-1}\boldsymbol{\lambda}_{qd0})^{\top}\, d(\mathbf{K}^{-1}\,\tilde{\mathbf{i}}_{qd0}) = \boldsymbol{\lambda}_{qd0}^{\top}(\mathbf{K}^{-1})^{\top}(d\mathbf{K}^{-1})\tilde{\mathbf{i}}_{qd0} + \boldsymbol{\lambda}_{qd0}^{\top}(\mathbf{K}^{-1})^{\top}\mathbf{K}^{-1} d\tilde{\mathbf{i}}_{qd0}\,. \tag{4.123}$$

After trigonometric manipulations, we can show that

$$(\mathbf{K}^{-1})^{\top}(d\mathbf{K}^{-1}) = d\tilde{\theta}_r \cdot \begin{bmatrix} 0 & 3/2 & 0 \\ -3/2 & 0 & 0 \\ 0 & 0 & 0 \end{bmatrix} \tag{4.124}$$

and

$$(\mathbf{K}^{-1})^{\mathsf{T}}\mathbf{K}^{-1} = \begin{bmatrix} 3/2 & 0 & 0 \\ 0 & 3/2 & 0 \\ 0 & 0 & 3 \end{bmatrix}. \tag{4.125}$$

Therefore

$$W_c(\mathbf{i}_{qd0}, i_f, \theta_{rm}) = \int_{(\mathbf{0},0,\theta_{rm0})}^{(\mathbf{i}_{qd0},i_f,\theta_{rm})} \frac{3}{2}\lambda_q\, d\tilde{i}_q + \frac{3}{2}\lambda_d\, d\tilde{i}_d + 3\lambda_0\, d\tilde{i}_0 + \lambda_f\, d\tilde{i}_f + \left(\frac{3p}{2}\lambda_q i_d - \frac{3p}{2}\lambda_d i_q + \tau_e\right) d\tilde{\theta}_{rm}. \tag{4.126}$$

A property of a conservative field is that the integrand is a gradient of some scalar field. For example, this implies that $(3/2)\lambda_q = \partial W_c/\partial i_q$. More importantly for this proof, it means that

$$\frac{3p}{2}(\lambda_q i_d - \lambda_d i_q) + \tau_e = \frac{\partial W_c}{\partial \theta_{rm}}. \tag{4.127}$$

We can also evaluate the coenergy using a convenient path, moving the rotor first from θ_{rm0} to θ_{rm} while keeping currents to zero, and then proceeding to energize the coils keeping the rotor fixed. Hence

$$W_c(\mathbf{i}_{qd0}, i_f, \theta_{rm}) = \int_{(\mathbf{0},0,\theta_{rm})}^{(\mathbf{i}_{qd0},i_f,\theta_{rm})} \frac{3}{2}\lambda_q\, d\tilde{i}_q + \frac{3}{2}\lambda_d\, d\tilde{i}_d + 3\lambda_0\, d\tilde{i}_0 + \lambda_f\, d\tilde{i}_f. \tag{4.128}$$

We invoke the basic property/assumption of Park's transformation, that is, the *qd* flux linkages are independent of rotor position, as in Equations (4.116a)–(4.116c) and (4.117). This implies that the integrand of Equation (4.128) is independent of rotor position:[e]

$$\frac{\partial W_c(\mathbf{i}_{qd0}, i_f, \theta_{rm})}{\partial \theta_{rm}} = 0. \tag{4.129}$$

Therefore, from Equation (4.127) we obtain **the approximate yet widely used torque expression**

$$\boxed{\tau_e = \frac{3p}{2}(\lambda_d i_q - \lambda_q i_d).} \tag{4.130}$$

Torque is positive when it tends to accelerate the rotor in the counterclockwise sense. Note that this equation holds perfectly well for magnetically nonlinear devices, as long as the coupling field can be assumed to be conservative. It represents the main component of electromagnetic torque generated from the action of the fundamental pole-pair field, and incorporates both the "magnetic dipole" and reluctance components of the torque. However, since the proof hinges on ignoring the position-dependent ripple that is present in the *qd* flux linkages, torque harmonics due to non-idealities such as slotting effects are not captured.

[e] At first glance, this expression is awkward since we have learned that taking the partial of the coenergy with respect to the position yields the torque! However, this is only true if the expression for coenergy is a function of the actual physical *abc* currents. Here, the coenergy is in terms of the transformed *qd* currents.

We can define a local x–y coordinate system tied on the rotor so that the x-axis coincides with the rotor d-axis, and the y-axis is the q-axis. In this coordinate system, we can define stator flux linkage and current vectors as $\vec{\lambda}_s = (\lambda_d, \lambda_q)$ and $\vec{\mathbf{i}}_s = (i_d, i_q)$, respectively. Hence, using Equation (4.130), **the torque vector is the cross product**

$$\boxed{\vec{\tau}_e = \frac{3p}{2}(\vec{\lambda}_s \times \vec{\mathbf{i}}_s)\,.} \tag{4.131}$$

The electromagnetic torque is positive when $\vec{\mathbf{i}}_s$ leads $\vec{\lambda}_s$ and vice versa. Since the cross product is invariant to the choice of coordinate system, this formula can be used to determine the directionality of torque by inspection of the flux lines (that are indicative of the direction of $\vec{\lambda}_s$) and the stator MMF vector $\mathcal{F}_s$, which is collinear with $\vec{\mathbf{i}}_s$. These vectors can be directly or indirectly observed in Figures 4.35–4.40, and the result is confirmed numerically by Figure 4.41.

Observing $qd0$ Variables from FEA Results

It is interesting to observe the $qd0$ variables in the rotor reference frame (i.e., Park's transformation), obtained by transforming the physical variables from the FEA results. This will confirm whether our assumption regarding the negligible dependence of the qd flux linkages on the rotor angle is valid. We use the same currents as before, namely, $i_f = 60$ A, and a balanced three-phase set for the stator coils with two levels: (i) $\sqrt{2}I_s = 11.295$ A and (ii) $\sqrt{2}I_s = 22.59$ A. The stator MMF vector is positioned at fixed angle $\phi_i = 20°$ with respect to the q-axis so that $(i_q, i_d) \in \{(10.61, -3.86), (21.23, -7.73)\}$ A. We note that this is not necessarily representative of a real operating condition, but should be interpreted as the calculation of the current-to-flux-linkage mapping. The rotor is rotated by 360° with a separate FEA every 1°. The abc flux linkages λ_{abc} are transformed to the rotor reference frame in a post-processing stage using Equation (4.108).

The winding flux linkage waveforms are shown in Figure 4.42. In an ideal machine, the a-phase flux linkage should vary sinusoidally as the rotor turns. However, λ_a exhibits a significant amount of third-harmonic distortion, which becomes more pronounced with the increased current of case (ii). The field winding flux linkages have dominant d.c. components plus relatively small ripple. Since $i_d < 0$ in this case, its demagnetizing effect is reflected in the reduction of λ_f from case (i) to case (ii). The ripple of λ_f has 24 oscillations per cycle, equal to the number of stator teeth, clearly due to stator slotting.

The fundamental frequency waveforms shown in gray color in Figure 4.42 are obtained from the analysis of λ_{qd0}, calculated by applying Park's transformation to λ_{abc}, and plotted in Figure 4.43. In particular, the fundamental waveforms represent the inverse transformation of the average values of λ_{qd} over one electrical cycle:

$$\lambda_{a,1} = \overline{\lambda_q}\cos\theta_r + \overline{\lambda_d}\sin\theta_r\,, \tag{4.132}$$

where $(\overline{\lambda_q}, \overline{\lambda_d}) \in \{(74.4, 212.1), (148.7, 96.1)\}$ mVs. Alternatively, we could write

$$\lambda_{a,1} = \Lambda_{s,1}\cos(\theta_r + \phi_\lambda)\,, \tag{4.133}$$

where $(\Lambda_{s,1}, \phi_\lambda) \in \{(224.8, -70.7), (177.0, -32.9)\}$ (mVs,°). We note that the waveforms of λ_{qd} are periodic every 60°. (A proof of this fact is left as an exercise for the reader.)

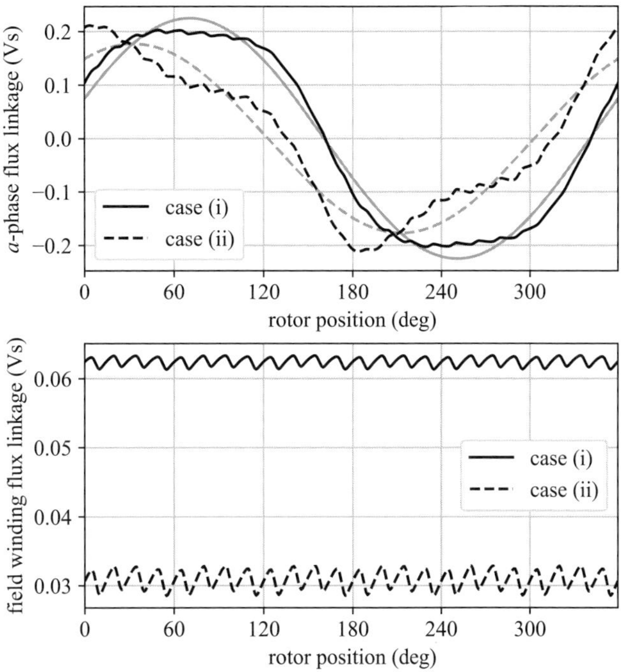

Figure 4.42 Elementary three-phase salient-rotor WRSM: variation of a-phase and field winding flux linkages with rotor position for two cases of stator current. The black (solid and dashed) lines are FEA results. The gray lines in the top plot are fundamental frequency components.

Interestingly, the waveform of $\lambda_0 \neq 0$, even though $i_0 = 0$; this flux has zero mean and a main third-harmonic component of small magnitude.

Figure 4.44(a) shows the torque calculated from the FEA solution, whereas Figure 4.44(b) shows the torque calculated from Equation (4.130). Due to the periodicity of λ_{qd}, it is observed that the torque is also periodic every 60°. The waveforms appear to have the same mean value; however, the harmonic content of the FEA torque is higher, with a very significant 24th harmonic, which is even more pronounced in case (ii). The torque ripple is ~2–3 Nm peak-to-peak. This is not surprising, since we explained that the analytical formula neglects the effect of non-idealities.

As a matter of fact, it could be argued that the waveforms of Figure 4.44(b) are contrived because Equation (4.130) is *only* valid when the qd flux linkages are independent of rotor position, since this was the main assumption of its proof. One possible way to work around this inconsistency and yet employ the formula is by using position-averaged values of the flux linkages. Substituting the mean values $\overline{\lambda_q}$ and $\overline{\lambda_d}$ in Equation (4.130), we obtain $\overline{\tau_e} \in \{3.808, 4.783\}$ Nm. A calculation of the mean value using the FEA torque data yields $\overline{\tau_e} \in \{3.806, 4.780\}$ Nm. The difference is in the third decimal digit. We thus conclude that *the equation* $\tau_e = (3p/2)(\overline{\lambda_d}\, i_q - \overline{\lambda_q}\, i_d)$ *is an acceptable approximation for the calculation of average torque.*[f]

[f] Recall that here we have assumed balanced sinusoidal three-phase currents, in which cases i_q and i_d are constant. However, if both currents and flux linkages have harmonic content, the averaging is trickier since, in general, $\overline{ab} \neq \overline{a}\,\overline{b}$.

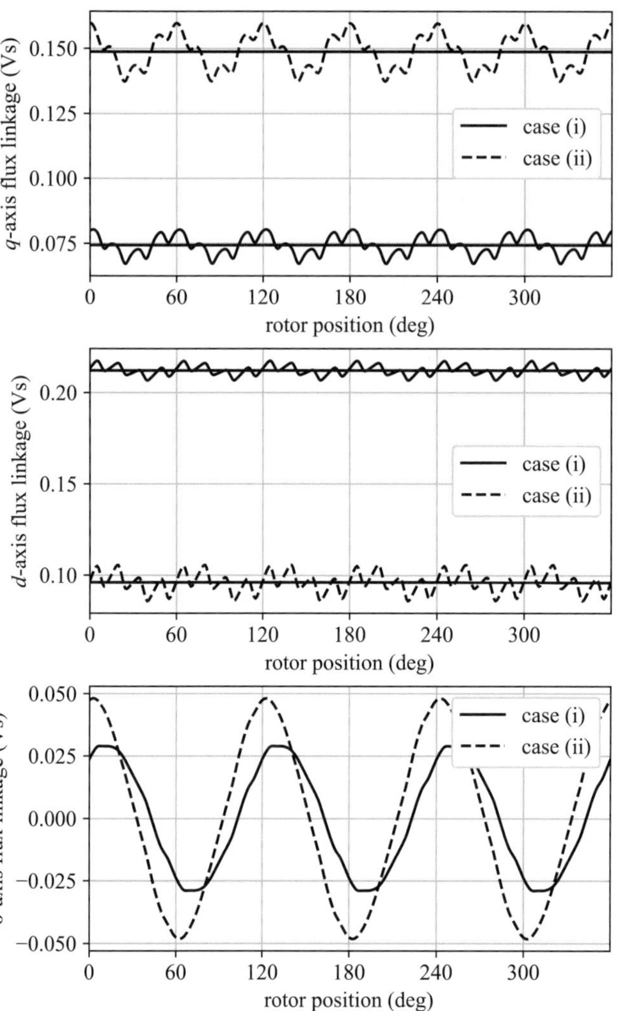

Figure 4.43 Elementary three-phase salient-rotor WRSM: variation of $qd0$ flux linkages with rotor position.

4.3.4 Derivation of a WRSM qd Equivalent Circuit from FEA

The equivalent circuit of a three-phase WRSM resembles that of a transformer (see Example 3.6 on page 264). Now we have two separate circuits for the qd-axes of the machine as per Blondel's two-reaction theory. We will demonstrate a way to derive the parameters of these circuits using FEA. Even though the machine is magnetically non-linear, we will derive linear circuits with constant inductances (independent of currents and rotor position), hoping that they can represent the operation over a broad range. Furthermore, we shall decouple the two axes, even though we are aware of the cross-coupling effect due to saturation. Finally, we concede to capturing only the fundamental pole-pair field in the machine.

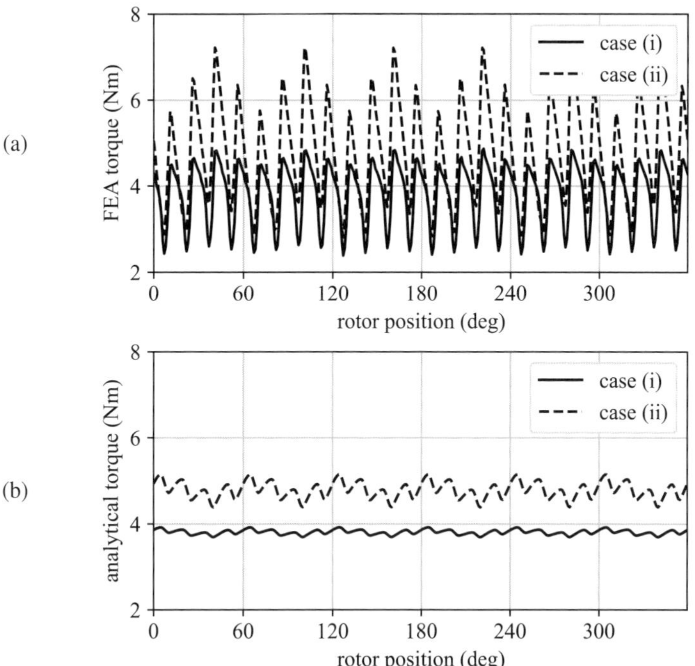

Figure 4.44 Elementary three-phase salient-rotor WRSM: variation of torque with rotor position. (a) FEA result. (b) Calculation based on *qd* components.

The analysis begins by writing down the flux linkage equations of the coils:

$$\begin{bmatrix} \lambda_{abc} \\ \lambda_f \end{bmatrix} = \begin{bmatrix} \mathbf{L}_s & \mathbf{L}_{sf} \\ \mathbf{L}_{sf}^\top & L_f \end{bmatrix} \begin{bmatrix} \mathbf{i}_{abc} \\ i_f \end{bmatrix}. \tag{4.134}$$

Magnetic linearity implies that the inductance matrices are independent of currents, but they still depend on θ_r. Note that $\mathbf{L}_s$ is a 3×3 symmetric matrix, and $\mathbf{L}_{sf}$ is a 3×1 column vector. Applying Park's transformation leads to

$$\begin{bmatrix} \lambda_{qd0} \\ \lambda_f \end{bmatrix} = \begin{bmatrix} \mathbf{K}\mathbf{L}_s\mathbf{K}^{-1} & \mathbf{K}\mathbf{L}_{sf} \\ \mathbf{L}_{sf}^\top\mathbf{K}^{-1} & L_f \end{bmatrix} \begin{bmatrix} \mathbf{i}_{qd0} \\ i_f \end{bmatrix}. \tag{4.135}$$

Then, imposing the empirical conditions that we expect the transformed inductances to satisfy, we obtain the following machine model:

$$\begin{bmatrix} \lambda_q \\ \lambda_d \\ \lambda_0 \\ \lambda_f \end{bmatrix} = \begin{bmatrix} L_q & 0 & 0 & 0 \\ 0 & L_d & 0 & L_{df} \\ 0 & 0 & L_0 & 0 \\ 0 & \frac{3}{2}L_{df} & 0 & L_f \end{bmatrix} \begin{bmatrix} i_q \\ i_d \\ i_0 \\ i_f \end{bmatrix}, \tag{4.136}$$

where all inductance values are constant. Let us ignore the λ_0 equation for simplicity.

We introduce a reflected field winding current

$$i'_f = \frac{2}{3}\frac{k_{wf}N_f}{k_{ws}N_s}i_f\,, \tag{4.137}$$

where the field winding distribution factor is $k_{wf} = \sin\alpha$ as per Equation (4.106). (In a round-rotor WRSM, we would use a different expression for the distribution factor.) Conceptually, this implies that every ampere of i'_f generates the same MMF as an ampere of i_d, as if we modified the rotor winding to be identical to that of the stator. Hence, i'_f and i_d contribute equally to the d-axis magnetizing flux Φ_{md}. For the particular machine under study, we have $i'_f = 0.188\,i_f$.

We then separate the flux linkages into leakage plus magnetizing components by defining the corresponding inductances:

$$L_d = L_{ls} + L_{md}\,, \tag{4.138a}$$

$$L_f = L_{lf} + L_{mf}\,. \tag{4.138b}$$

In our elementary machine, this separation is really necessary only for the d-axis, where we have two coupled coils. In real machines, it would be required to separate the q-axis inductance as in $L_q = L_{ls} + L_{mq}$ because of the presence of damper windings on the q-axis. Therefore

$$\lambda_d = L_{ls}i_d + L_{md}i_d + \frac{3}{2}\frac{k_{ws}N_s}{k_{wf}N_f}L_{df}i'_f\,. \tag{4.139}$$

In view of our assumption regarding the equivalent magnetizing effect of i_d and i'_f, we deduce that

$$L_{md} = \frac{3}{2}\frac{k_{ws}N_s}{k_{wf}N_f}L_{df}\,. \tag{4.140}$$

Substituting this in the equation for λ_f and introducing i'_f:

$$\lambda_f = \frac{k_{wf}N_f}{k_{ws}N_s}L_{md}i_d + \frac{3}{2}\frac{k_{ws}N_s}{k_{wf}N_f}L_{mf}i'_f + \frac{3}{2}\frac{k_{ws}N_s}{k_{wf}N_f}L_{lf}i'_f\,. \tag{4.141}$$

We define a reflected field winding flux linkage as

$$\lambda'_f = \frac{k_{ws}N_s}{k_{wf}N_f}\lambda_f\,. \tag{4.142}$$

For the machine under study, $\lambda'_f = 3.541\,\lambda_f$. Therefore

$$\lambda'_f = L_{md}i_d + \frac{3}{2}\left(\frac{k_{ws}N_s}{k_{wf}N_f}\right)^2 L_{mf}i'_f + \frac{3}{2}\left(\frac{k_{ws}N_s}{k_{wf}N_f}\right)^2 L_{lf}i'_f\,. \tag{4.143}$$

In view again of our assumption regarding the magnetizing equivalence of i'_f and i_d, we deduce that

$$L_{md} = \frac{3}{2}\left(\frac{k_{ws}N_s}{k_{wf}N_f}\right)^2 L_{mf}\,. \tag{4.144}$$

We also define

$$L'_{lf} = \frac{3}{2}\left(\frac{k_{ws}N_s}{k_{wf}N_f}\right)^2 L_{lf}\,. \tag{4.145}$$

The final form of the flux linkage equations in qd variables is thus

$$\lambda_q = L_q i_q, \tag{4.146a}$$

$$\lambda_d = L_{ls} i_d + L_{md}(i_d + i'_f), \tag{4.146b}$$

$$\lambda'_f = L'_{lf} i'_f + L_{md}(i_d + i'_f). \tag{4.146c}$$

Combining the $qd0$ voltage equations (4.118a)–(4.118c) with the above flux linkage equations leads to the equivalent circuits of the elementary WRSM shown in Figure 4.45. We have introduced the reflected field winding voltage and resistance:

$$v'_f = \frac{k_{ws} N_s}{k_{wf} N_f} v_f \tag{4.147}$$

and

$$r'_f = \frac{3}{2}\left(\frac{k_{ws} N_s}{k_{wf} N_f}\right)^2 r_f, \tag{4.148}$$

respectively. In real machines, these circuits are typically augmented by additional branches on the rotor side to represent the effects of the damper windings and/or eddy currents flowing in the rotor steel.

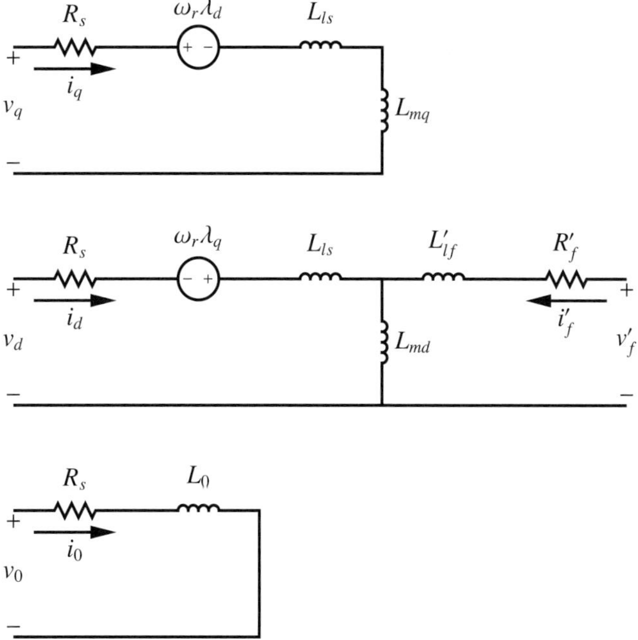

Figure 4.45 Equivalent $qd0$ circuits of elementary WRSM.

The model derivation is complete, and now the objective is to estimate the values of L_q, L_{md}, L_{ls}, and L'_{lf} using FEA. We have three equations and four unknowns, so clearly this cannot be achieved with a single study. Furthermore, even if we could rely on a single study to obtain these values, it would not be appropriate because these parameters vary (hopefully not by much if our model is reasonable) over the operating range of the machine. We will conduct a number K of FEA studies at various current levels and rotor angles, which will lead to an over-determined linear system of equations that will be solved via least squares. Let us place the unknown inductances in a column vector $\mathbf{x} = [L_q \ L_{md} \ L_{ls} \ L'_{lf}]^\top$. The kth FEA study, $k \in \{1, \ldots, K\}$, will be conducted using i_q^k, i_d^k, $i_f'^k$, and θ_r^k as inputs. The kth FEA observation will be $\boldsymbol{\lambda}^k = [\lambda_q^k \ \lambda_d^k \ \lambda_f'^k]^\top$. Each time we run a study, we append the following equations to the system:

$$\begin{bmatrix} i_q^k & 0 & 0 & 0 \\ 0 & i_d^k + i_f'^k & i_d^k & 0 \\ 0 & i_d^k + i_f'^k & 0 & i_f'^k \end{bmatrix} \mathbf{x} = \boldsymbol{\lambda}^k . \tag{4.149}$$

The final system is of the form $\mathbf{A}\mathbf{x} = \boldsymbol{\lambda}$, with $\mathbf{A}$ a $3K \times 4$ matrix and $\boldsymbol{\lambda}$ a $3K \times 1$ column vector. The least-squares solution is

$$\hat{\mathbf{x}} = (\mathbf{A}^\top \mathbf{A})^{-1} \mathbf{A}^\top \boldsymbol{\lambda} . \tag{4.150}$$

In the absence of information about a particular operating region of interest, we will sample currents uniformly (and independently) from the following ranges: $i'_f \in [2, 12]$ A, $i_q \in [1, 24]$ A, $i_d \in [-24, 0]$ A. Rotor angles will be sampled uniformly in the range $\theta_r \in [0, 60]°$, taking advantage of the periodicity of λ_{qd} and λ_f.[g] The evolution of the least-squares estimate with iteration number is shown in Figure 4.46. The estimate varies widely at the beginning, and then eventually stabilizes after enough data points are collected. The final estimate after 200 samples is $L_q = 6.9$ mH, $L_{md} = 25.6$ mH, $L_{ls} = 1.1$ mH, and $L'_{lf} = 0.5$ mH.

The predictive accuracy of the linear flux linkage model is checked against the previous results of Figures 4.42–4.44. We have studied two cases, where the stator currents were $(i_q, i_d) \in \{(10.61, -3.86), (21.23, -7.73)\}$ A and the referred value of field current was $i'_f = 0.188 \cdot 60 = 11.28$ A (same in both). Using these values, the model predicts

$$\lambda_q = L_q i_q \in \{73.2, 146.5\} \text{ mVs} , \tag{4.151a}$$

$$\lambda_d = L_{ls} i_d + L_{md} (i_d + i'_f) \in \{185.7, 82.4\} \text{ mVs} , \tag{4.151b}$$

$$\lambda'_f = L'_{lf} i'_f + L_{md} (i_d + i'_f) \in \{195.6, 96.5\} \text{ mVs} , \tag{4.151c}$$

$$\lambda_f = \lambda'_f / 3.541 \in \{55.2, 27.2\} \text{ mVs} , \tag{4.151d}$$

$$\tau_e = (3p/2)(\lambda_d i_q - \lambda_q i_d) \in \{3.38, 4.32\} \text{ Nm} . \tag{4.151e}$$

Overall, the predictions are reasonable, but the relative error compared to the FEA results is significant. Clearly, for higher accuracy the inductive parameters of the model should be adjusted according to the current levels in the machine due to saturation, thus leading to a nonlinear model. In general, *the validity of such linear machine models deteriorates the further we move away from a nominal operating point*, so they should be used with caution.

[g] Arguably, we could use, e.g., Latin hypercube sampling over the distribution of currents and rotor angle for better performance.

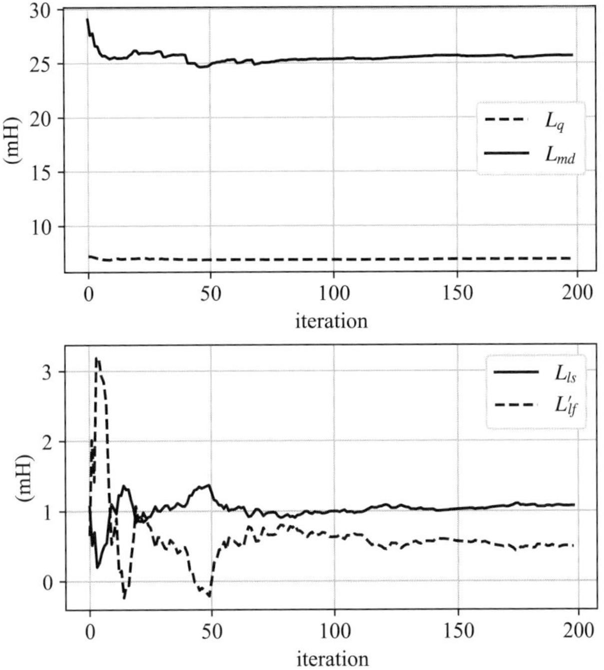

Figure 4.46 Least-squares parameter identification of elementary WRSM.

4.4 Permanent-Magnet Synchronous Machines

The second type of machine that we analyze is the **permanent-magnet synchronous machine (PMSM)**, which we introduced in Example 2.19 on page 175. The incorporation of PMs in FEA was discussed in §3.7 on page 293. PMSMs come in a variety of sizes and configurations. For instance, they are commonly used as **brushless d.c. motors** found in tools and industrial motion control systems. The PMSMs that we will study have a three-phase distributed winding on the stator, similar to a wound-rotor synchronous machine. The main difference lies in the rotor design, which does not have a field winding. Instead, magnetic excitation comes from permanent magnets that are placed on the rotor. Hence, the PMSM is typically more efficient than a similar wound-rotor machine. We will conduct two case studies, first on a surface-mounted PMSM and second on an interior PMSM (the motor of the 2004 Toyota Prius).

4.4.1 Elementary Surface-Mounted PMSM: Principles of Operation

We analyze an elementary three-phase, four-pole ($p = 2$) surface-mounted PMSM. In this simple design, the magnets are glued on the rotor surface and they project outwards.

Meshing the Stator

We use the stator slot definition that was introduced earlier for the elementary WRSM, shown in Figure 4.14 on page 338. The main stator dimensions are not modified: r_{so} =

110 mm, r_{si} = 55 mm, r_B = 80 mm, r_E = 57 mm, ℓ = 100 mm, and r_{ob} = 115 mm. The air gap (i.e., the distance between the inner stator and outer rotor surfaces) is increased to g = 11 mm in order to accommodate the PMs that have depth d_{pm} = 10 mm. Let us denote the width of the air-gap interface layer as $w_{agl} = r_o - r_i$. In the previous case studies, this was equal to $g/3$, but this is no longer possible in a surface-mounted PMSM. This parameter is now determined by r_{si} and the PM outer surface radius, r_{pm} = 54 mm. Hence, we set $w_{agl} = (r_{si} - r_{pm})/3 = 0.33$ mm and $r_o = r_{si} - w_{agl} = 54.67$ mm. Also, we increase the number of stator slots to $n_{ss} = 36$ (three slots/pole/phase) in order to obtain a decent four-pole air-gap MMF. The stator tooth width is set to w_{st} = 7 mm. The stator steel has the *B–H* curve that we used previously to plot Figure 3.23 on page 289. The lamination stacking factor is $k_{st} = 0.97$.

The winding is distributed in two layers and follows the pattern listed in Table 4.2, where the first slot is centered horizontally. This nine-slot winding pattern spans one pole. Over the next pole pitch, that is, in slots 10–18, we repeat the same pattern by reversing the conductor polarity. Then the pattern repeats again over the next two poles, that is, in slots 19–36. The stator cross-section is plotted in Figure 4.47. Since this machine has two pole-pairs, there exist two sets of magnetic axes.

The half-slot area is 56.8 mm^2. Assuming that the winding packing factor is $k_{pf} \approx$ 0.5 implies that each half-slot has roughly 28.4 mm^2 of useful conductor surface area. Hence, we place five turns of 6-mm^2 (AWG 10) copper conductors (magnet wire) in

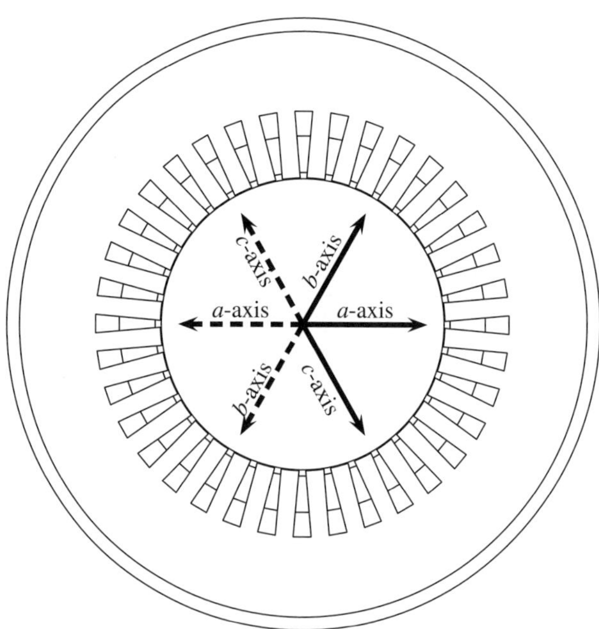

Figure 4.47 Cross-section of elementary four-pole PMSM stator PSLG. The innermost "circle" has radius $r_o = r_{si} - w_{agl}$. The outermost "circle" has radius r_{ob}.

Table 4.2 Winding pattern in 36-slot surface-mounted PMSM (first nine slots)

slot number	1	2	3	4	5	6	7	8	9
inner layer	$-b$	$-b$	$-b$	a	a	a	$-c$	$-c$	$-c$
outer layer	c	$-b$	$-b$	$-b$	a	a	a	$-c$	$-c$

each half-slot. We can connect the winding of each pole-pair in parallel or in series. Let us connect in series for a total of $N_s = 60$ turns in each phase.

Meshing the Rotor

In practice, the PMs can be shaped in various ways. Here, for simplicity, we define a rotor geometry where the PMs are arc-shaped with a cross-section that is defined by two circular arcs and line segments for the sides. Each magnet occupies an angular span $\theta_{pm} < \pi/p$ on the rotor surface. A single pole is shown in Figure 4.48. The magnet depth in the radial direction is $d_{pm} = r_{pm} - r_{ro}$. We assume that the permanent magnetization is uniform and is in the radial direction. To implement this, we assign a remanence vector $\vec{\mathbf{B}}_r^k = B_r\hat{\mathbf{r}}^k$ in each PM triangle, where $\hat{\mathbf{r}}^k$ points from the shaft center (that coincides with the origin of the coordinate system) to the centroid of triangle k.

The numerical values of the rotor dimensions are set to $r_i = r_{si} - 2w_{agl} = 54.33$ mm, $r_{pm} = r_{si} - 3w_{agl} = 54$ mm, $d_{pm} = 10$ mm, $r_{ro} = r_{pm} - d_{pm} = 44$ mm, $r_{ri} = 20$ mm, and $\theta_{pm} = (2/3)\cdot(\pi/p) = \pi/3 = 60°$. The rotor laminations are made from the same material as the stator, and have identical stacking factor. The PMs are ferrite magnets with the following typical parameters: remanence $B_r = 0.35$ T (at the operating temperature of the motor) and relative permeability $\mu_{r,pm} = 1.06$. The shaft material is magnetically

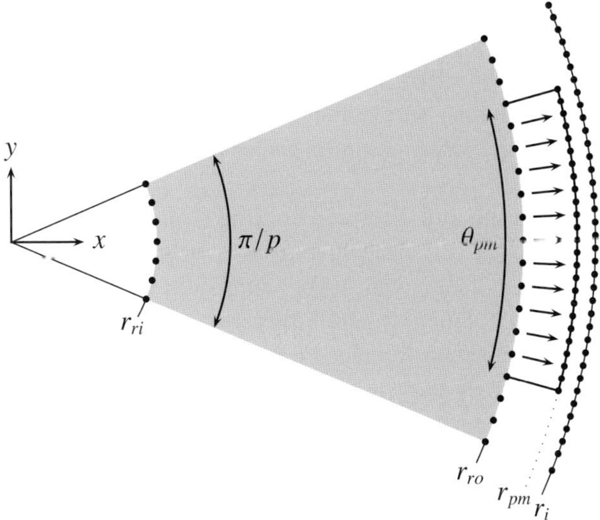

Figure 4.48 Geometry of an elementary surface-mounted PMSM north rotor pole with radial magnetization. South rotor poles are defined identically to the north poles, but their magnetization points inwards. Dots signify vertices that should be included in the PSLG.

linear with $\mu_{r,\text{sh}} = 200$. Additional PSLG points could be added to the PM side walls and/or their interior, if needed. Note that the numbers of PSLG vertices on the PM inner and outer surfaces do not need to be equal.

Flux Linkage and Torque: Open-Circuit Conditions

The meshed PMSM is shown in Figure 4.49. In this case study, the stator, rotor, and air-gap meshes have 9,423, 6,485, and 2,062 triangular elements, respectively. Since this is a four-pole machine, there are two sets of (non-orthogonal) qd-axes that are tied on the rotor. The d-axes are centered on the north rotor poles, where PM flux is oriented from the rotor to the stator, similarly to a WRSM. The q-axes are centered between PMs, leading their respective d-axes by 90° electrical. The figure shows the mechanical rotor angle θ_{rm}, which is related to the electrical rotor angle by $\theta_r = p\theta_{rm}$. By convention, we measure the rotor angle as the angle between one of the a-axes (at $\phi = 0$) and one of the q-axes. In Figure 4.49, the rotor is turned by $\theta_{rm} = 75°$ counterclockwise.

First, we conduct a series of FEA studies with the stator winding open over a range of rotor positions $\theta_{rm} = 0°, 0.625°, \ldots, 90°$, that is, over 180° electrical. The angular spacing is selected to obtain sufficient resolution for capturing slotting effects ($0.625° = (1/16) \cdot (360°/n_{ss})$). A snapshot for $\theta_{rm} = 0°$ is shown in Figure 4.50(a). At this angle, the two north rotor poles are located at −45° and 135°. Note that, in this machine, the magnetizing effect of the PMs alone is not strong enough to saturate the device. We can observe the variation of the a-phase flux linkage in Figure 4.51(a). Interestingly, the effect of slotting on the flux linkage is negligible.

The open-circuit torque in PM motors is commonly referred to as the **cogging torque**. In an ideal machine, the cogging torque is zero. However, a non-uniform air gap causes the appearance of a small reluctance torque component. This waveform is obtained from the FEA and plotted in Figure 4.52(a), where slotting effects are clearly visible, causing a ±0.15-Nm oscillation. It should be noted that some of the harmonics present could be attributed to the effects of remeshing the air-gap interface layer as the rotor turns, which impacts the torque calculation somewhat. However, such numerical issues become less significant when the machine is loaded and the torque increases, as we shall see next.

Flux Linkage and Torque: Loaded Conditions

We will now add balanced three-phase current to the stator windings. It is convenient to work in qd variables, which were introduced via Park's transformation in the analysis of the WRSM. Since we have been consistent in our definitions of axes and angles, the reference frame transformation translates identically to the case of the PMSM.

As a first case study, we will set $i_q = 30$ A and $i_d = 0$ A. This means that we are imposing a stator MMF along the q-axis, which rotates in synchronism with the rotor. The FEA solution at $\theta_r = 0°$ is shown in Figure 4.50(b). Compared with the open-circuit conditions right above, it may be observed that the magnetic field has rotated counter-clockwise, and that its magnitude has intensified. It is also interesting to observe the flux linkage and torque waveforms as the rotor turns, which are shown in Figures 4.51(b) and 4.52(b), respectively.

Adding i_q causes a phase shift and a magnitude increase in the flux linkage, which

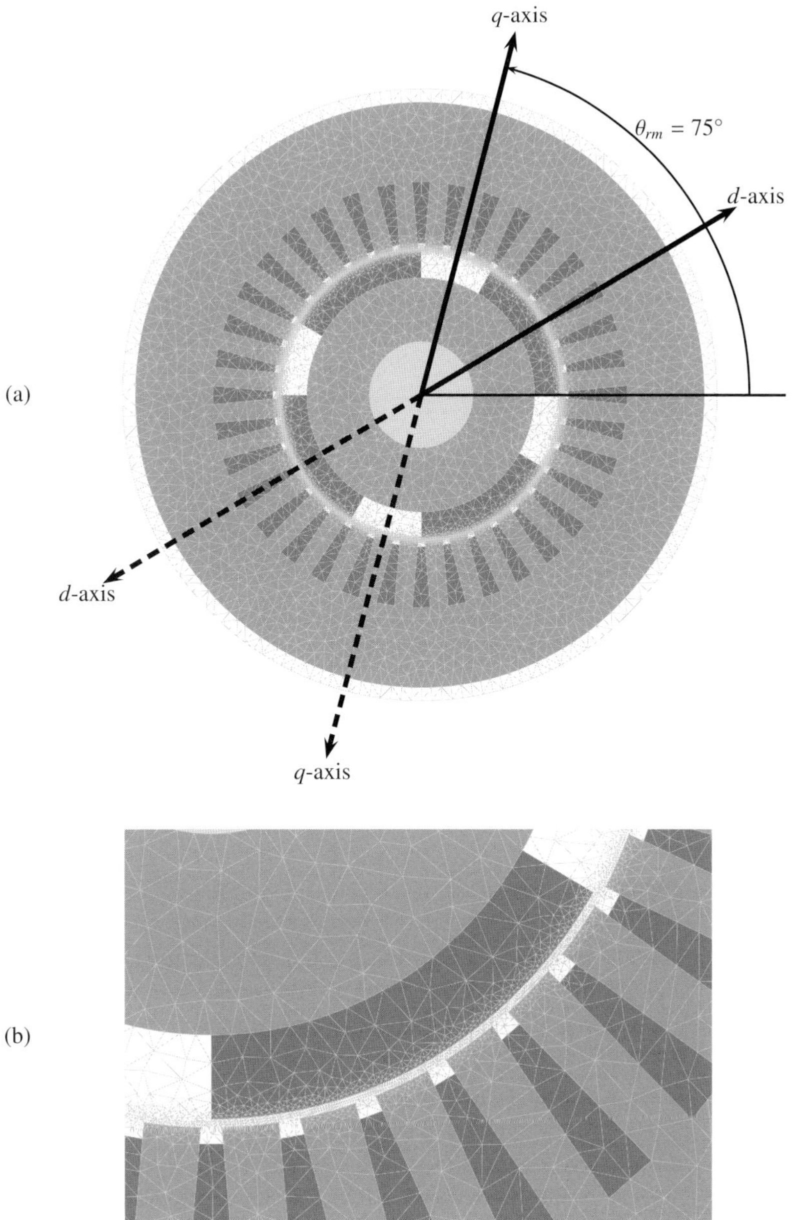

Figure 4.49 (a) Meshed cross-section of elementary three-phase, four-pole surface-mounted PMSM. (b) Mesh detail in a region close to the air gap and a south rotor pole.

otherwise remains quite smooth. The qd flux linkages, shown in Figure 4.53, are again periodic every 60°. They have a d.c. component plus a relatively minor ripple, with the sixth harmonic being dominant. Note how i_q has led to a nonzero λ_q, and that $\lambda_d \approx 0.096$ Vs corresponds to the peak value of the open-circuit λ_a, which means that λ_d

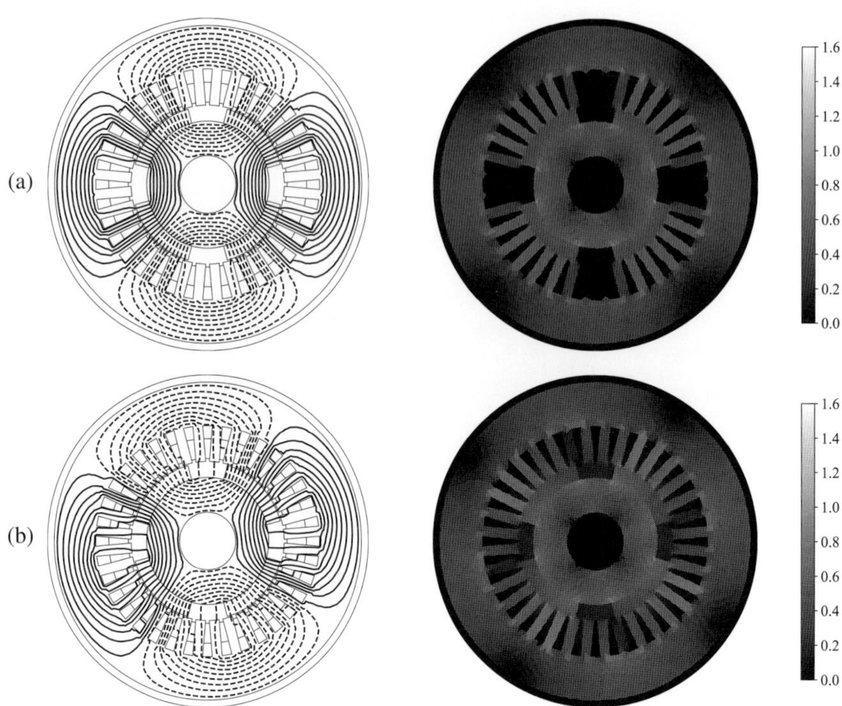

Figure 4.50 Elementary surface-mounted PMSM flux lines (left) and flux density (right) at $\theta_{rm} = 0°$: (a) open-circuit conditions; (b) operation under load with $i_q = 30$ A, $i_d = 0$ A.

(generated by the PMs) is approximately the same in both cases. On the other hand, λ_0 exhibits a small third-harmonic oscillation around zero.

The flow of positive i_q is responsible for the generation of positive electromagnetic torque, as dictated by Equation (4.130). The average torque value is 8.6 Nm. The cogging torque component is still present, but a stronger sixth-harmonic component has also developed, leading to a ±0.4 Nm oscillation around the average value. This **torque ripple** is responsible for increased noise and vibration from the motor. Note that the ripple arises even with (almost) perfectly sinusoidal flux linkage and winding current. If the current has additional harmonics, e.g., due to pulse-width modulation of the terminal voltage, then these would create additional (switching-frequency) torque harmonics. The presence of the sixth-harmonic components (and multiples thereof) in the *qd* flux linkages and the torque are attributed to the distribution of the winding in slots in three-phase machines.

It is instructive to generate a **torque contour plot** on the i_q–i_d plane. The process to obtain this is straightforward: we first define a grid of operating points on the *qd*-current plane; then for each point, we conduct a series of FEA studies over 60° electrical (exploiting the periodicity of the torque waveform) and calculate the average torque. Here, we set $i_q > 0$ so that the PMSM operates as a motor. The plot is generated with the Matplotlib Pyplot `contour` function (see Example 1.13 on page 19). The result

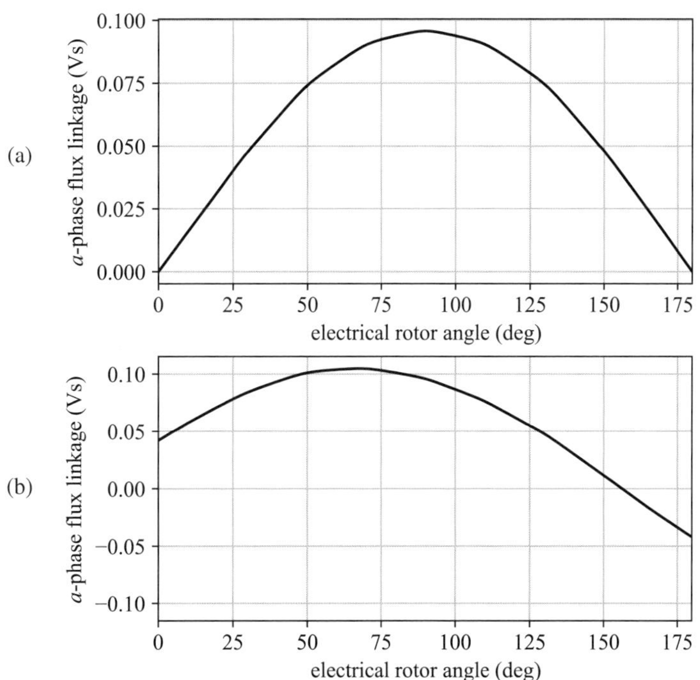

Figure 4.51 Elementary surface-mounted PMSM a-phase flux linkage waveforms: (a) open-circuit conditions; (b) operation under load with $i_q = 30$ A, $i_d = 0$ A.

is shown in Figure 4.54. For this machine, we observe that the torque is essentially proportional to i_q due to magnetic linearity over the given range, and that i_d has a minor influence (adding positive i_d increases the torque slightly). Hence, to minimize ohmic loss, we should operate the machine with $i_d = 0$, which can be readily achieved with a PMSM power electronic drive. In conclusion, a surface-mounted PMSM is, for all practical purposes, a round-rotor machine.

4.4.2 Derivation of a PMSM *qd* Equivalent Circuit from FEA

The *qd* equivalent circuit of a three-phase PMSM is a simplified variant of the WRSM circuit (see §4.3.4 on page 376) since the PMSM does not have rotor windings. The magnetizing effect of the PMs can be represented by a constant term. We will repeat the exercise of deriving the circuit parameters based on FEA.

Assuming magnetic linearity, Park's transformation leads to the following flux linkage equations:

$$\begin{bmatrix} \lambda_q \\ \lambda_d \\ \lambda_0 \end{bmatrix} = \begin{bmatrix} L_q & 0 & 0 \\ 0 & L_d & 0 \\ 0 & 0 & L_0 \end{bmatrix} \begin{bmatrix} i_q \\ i_d \\ i_0 \end{bmatrix} + \begin{bmatrix} 0 \\ \lambda_m \\ 0 \end{bmatrix}, \tag{4.152}$$

where all inductance values are constant, and where $\lambda_m > 0$ represents the d-axis flux

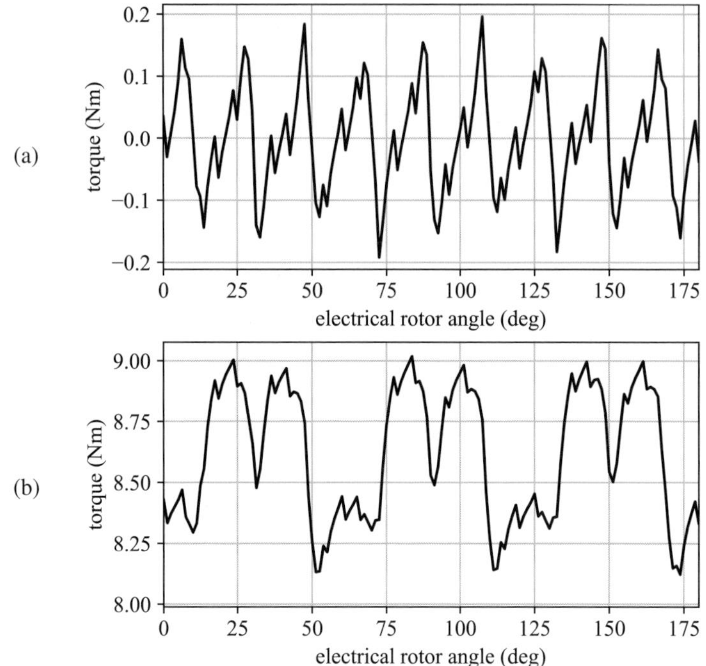

Figure 4.52 Elementary surface-mounted PMSM torque waveforms: (a) open-circuit (cogging) torque; (b) operation under load with $i_q = 30$ A, $i_d = 0$ A.

linkage due to the PMs. We shall ignore the λ_0 equation for simplicity. In surface-mounted PMSMs, we expect that $L_d \approx L_q$. However, this model is valid also for interior PM machines, which exhibit significant saliency ($L_q > L_d$).

The objective is to estimate the values of L_q, L_d, and λ_m using FEA. We will follow the same procedure as before: we will conduct a number K of FEA studies at various current levels and rotor angles, which will lead to an over-determined linear system of equations that will be solved via least squares. Let us place the unknown parameters in a column vector $\mathbf{x} = [L_q \; L_d \; \lambda_m]^\top$. The kth FEA study, $k \in \{1, \ldots, K\}$, will be conducted using i_q^k, i_d^k, and θ_r^k as inputs. The kth FEA observation will be $\boldsymbol{\lambda}^k = [\lambda_q^k \; \lambda_d^k]^\top$. Each time we run a study, we append the following equations to the system:

$$\begin{bmatrix} i_q^k & 0 & 0 \\ 0 & i_d^k & 1 \end{bmatrix} \mathbf{x} = \boldsymbol{\lambda}^k . \tag{4.153}$$

The final system is of the form $\mathbf{A}\mathbf{x} = \boldsymbol{\lambda}$, with $\mathbf{A}$ a $2K \times 3$ matrix and $\boldsymbol{\lambda}$ a $2K \times 1$ column vector. The least-squares solution is $\hat{\mathbf{x}} = (\mathbf{A}^\top\mathbf{A})^{-1}\mathbf{A}^\top\boldsymbol{\lambda}$.

Let us sample currents uniformly (and independently) from the following ranges: $i_q \in [0, 50]$ A, $i_d \in [-25, 25]$ A. Electrical rotor angles will be sampled uniformly in the range $\theta_r \in [0, 60]^\circ$, taking advantage of the periodicity of $\boldsymbol{\lambda}_{qd}$. The evolution of the least-squares estimate with iteration number is shown in Figure 4.55. In this case, it does not take long for the estimate to converge, presumably because this machine is

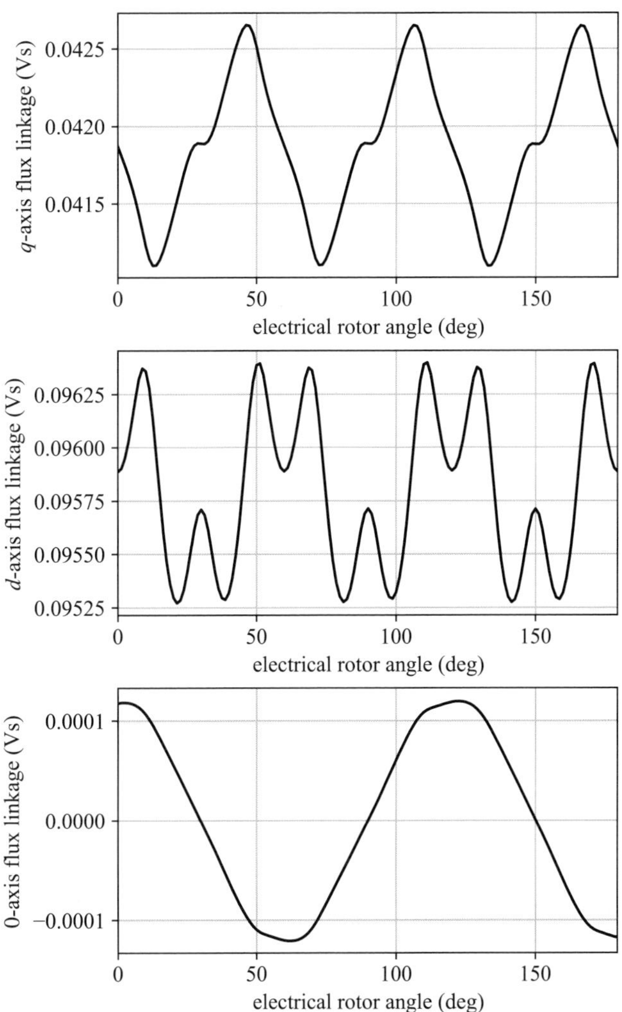

Figure 4.53 Elementary surface-mounted PMSM: variation of $qd0$ flux linkages with rotor position.

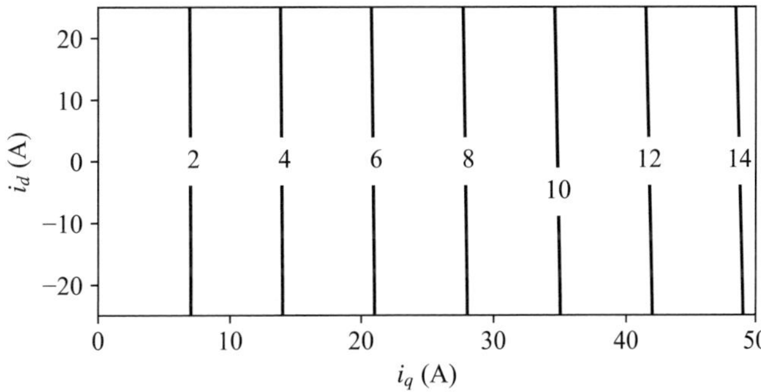

Figure 4.54 Elementary surface-mounted PMSM torque contour map. Torque contour units are in Nm.

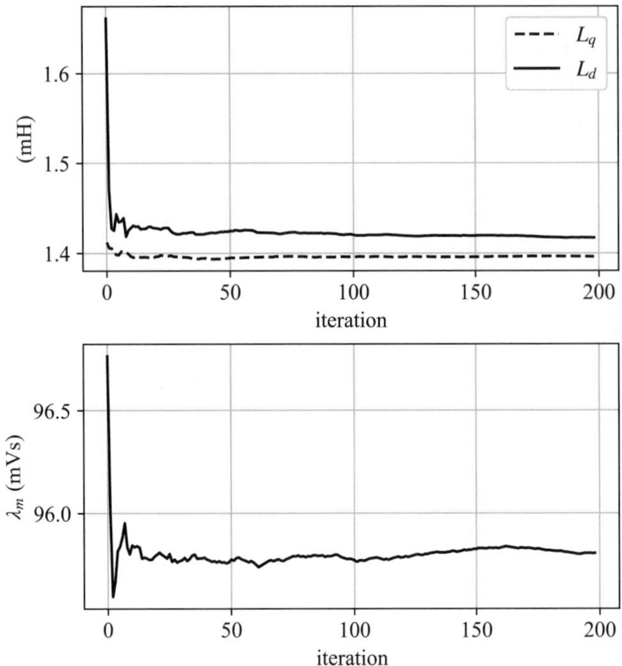

Figure 4.55 Least-squares parameter identification of elementary surface-mounted PMSM.

not strongly saturated, thus the linear flux linkage model is a good representation. Here, adding samples has primarily the effect of "averaging out" the (small) dependence of the qd flux linkages on rotor position. The final estimate after 200 samples is L_q = 1.40 mH, L_d = 1.42 mH, and λ_m = 95.8 mVs. The fact that $L_d > L_q$ is because the PMs (positioned along the d-axis) have relative permeability $\mu_{r,pm} = 1.06 > 1$.

The predictive accuracy of the linear flux linkage model is checked against the previous results of Figures 4.53 and 4.54. For i_q = 30 A and i_d = 0 A, the model predicts

$$\lambda_q = L_q i_q = 41.9 \text{ mVs}, \tag{4.154a}$$

$$\lambda_d = L_d i_d + \lambda_m = 95.8 \text{ mVs}, \tag{4.154b}$$

$$\tau_e = (3p/2)(\lambda_d i_q - \lambda_q i_d) = (3p/2)[\lambda_m i_q + (L_d - L_q) i_d i_q] = 8.6 \text{ Nm}. \tag{4.154c}$$

The model predictions are excellent, compared to the FEA results.

4.4.3 Interior PMSM: Fundamentals of qd-Current Control

As a second case study, we will analyze the interior PMSM found in the powertrain of the 2004 Toyota Prius hybrid electric vehicle, which is representative of a real-world design. This is an eight-pole (i.e., p = 4) motor that is rated 50 kW (at the base speed of 1300 rpm) and 400 Nm, which has been studied extensively in the technical literature [2–7]. The PMs in this machine are made of NdFeB (neodymium iron boron), and they are placed in pockets inside the rotor in a V-shaped configuration.

Meshing the Stator

The main stator dimensions are outer radius r_{so} = 134.5 mm, inner radius r_{si} = 80.95 mm, axial length ℓ = 83.6 mm, mesh outer boundary radius r_{ob} = 140 mm, and air-gap length g = 0.73025 mm (the number of decimal digits in this parameter suggests that it has been measured very precisely due to its importance). The number of stator slots is n_{ss} = 48 (two slots/pole/phase). The machine is constructed with 0.014-in thick laminations, and the stacking factor is assumed to be k_{st} = 0.95. The *B*–*H* curve of the lamination steel is defined by the following data points (first column lists *H*-values in A/m, second column lists *B*-values in T):

```
500     1.39
1000    1.47
2500    1.57
5000    1.67
10000   1.79
17500   1.89
23100   1.95
30000   1.99
100000  2.1
200000  2.245
300000  2.38
```

The stator slot has a more complicated design than the ones we have used in the elementary machines thus far. Figure 4.56 shows a prototypical slot *ABCDEFGHI* that is centered horizontally. The slot is symmetric with respect to the *x*-axis. Close to the air gap, the ends of the teeth extrude both ways, thus reducing the slot opening width. The slot opening is $y_A - y_I$ = 1.93 mm wide at $r = r_{si}$. The tooth-end side *AB* is horizontal and has a thickness of $x_B - x_A$ = 1.016 mm. Therefore, x_B = 81.96 mm and $y_B = y_A$ = 0.965 mm. The coil area is to the right of the vertical segment *BH*. Point *C* is at x_C = 82.45 mm, y_C = 1.56 mm, and point *D* is at x_D = 111.02 mm, y_D = 3.44 mm. These coordinates are such that the slope of the upper slot wall *CD* is

$$\frac{y_D - y_C}{x_D - x_C} = \tan \phi_{st} = \tan \frac{\pi}{n_{ss}}, \tag{4.155}$$

where ϕ_{st} denotes the angle of the subsequent stator tooth axis. This condition is necessary to achieve a constant tooth width. The tooth is capped by a circular arc. Point *E* is at x_E = 114.45 mm, y_E = 0 mm. The slot depth is $x_E - r_{si}$ = 33.5 mm.

The winding is distributed in a single layer and follows the pattern listed in Table 4.3. This six-slot winding pattern spans one pole. Over the next pole pitch, that is, in slots

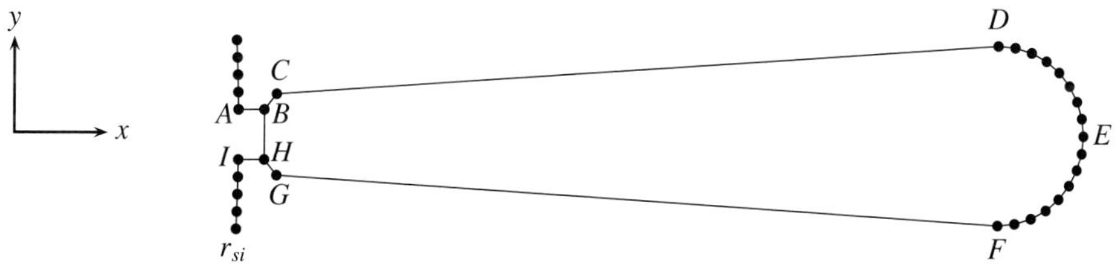

Figure 4.56 Interior PMSM stator slot geometry. Dots are PSLG points.

7–12, we repeat the same pattern by reversing the conductor polarity. Then the pattern repeats again over the next six poles, that is, in slots 13–48. For one of the (four) a-axes to be horizontal, which is our usual convention, the first slot needs to be centered at $\phi = 18.75°$. The stator cross-section is illustrated in Figure 4.57 over 90° mechanical (360° electrical).

Table 4.3 Winding pattern in 48-slot interior PMSM (first six slots)

slot number	1	2	3	4	5	6
phase	a	a	$-c$	$-c$	b	b

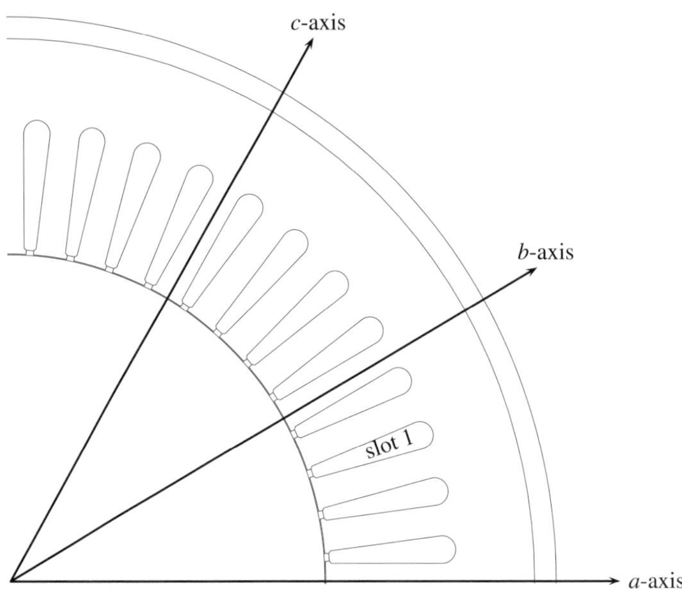

Figure 4.57 Detail of IPMSM stator PSLG. The innermost "circle" has radius $r_o = r_{si} - g/3$. The outermost "circle" has radius r_{ob}.

The slot area is 162.4 mm^2. The stator wiring is made of a bundle of 13 × 0.65-mm^2 (AWG 19) copper conductors, for a total of A_s = 8.5 mm^2 of copper area in the cable. Each slot contains nine turns, so the slot packing factor is calculated as $k_{pf} \approx 0.5$. The winding is connected in series over the eight poles for a total of $N_s = 72$ turns per phase. The per-phase stator winding resistance (ignoring end turns) at 21°C is calculated as

$$R_s = 2\frac{N_s \ell}{\sigma A_s} = 24.6 \text{ m}\Omega . \qquad (4.156)$$

Note that the experimentally measured value of the stator winding resistance (at the terminals, including end turns) is reported to be R_s = 69 mΩ. Of course, this is the value that should be used in voltage drop and loss calculations. In this motor, the end turns extend approximately 4 cm past the lamination stack on each side (recall ℓ = 8.4 cm), and this increases the resistance substantially. At a rated current level of 250 A, r.m.s.

and an operating temperature of 60°C, the ohmic loss is 15 kW. If the power output at the shaft is assumed to be 50 kW, the efficiency is approximately 77% (ignoring all other sources of loss). However, this is not the most efficient operating point of this motor, which is designed for maximum efficiency at torque levels around 100 Nm and speeds around 2000 rpm (power output around 20 kW), where the combined motor/inverter efficiency reportedly exceeds 93%.

Meshing the Rotor

The main rotor dimensions are air-gap layer inner radius $r_i = r_{si} - 2g/3 = r_o - g/3 = 80.46317$ mm, outer radius $r_{ro} = 80.21975$ mm, and shaft radius $r_{ri} = 55.5$ mm. The laminations are from the same material as the stator. The shaft is modeled as a linear material with relative permeability $\mu_{r,\mathrm{sh}} = 200$. The rotor cross-section is shown in Figure 4.58 over 90°. Interestingly, the motor designers have added small dimples on the rotor surface at ±45° electrical from the rotor pole axes in a bid to reduce the torque ripple and the iron loss. These are assumed to be 0.5 mm deep and 4° wide.

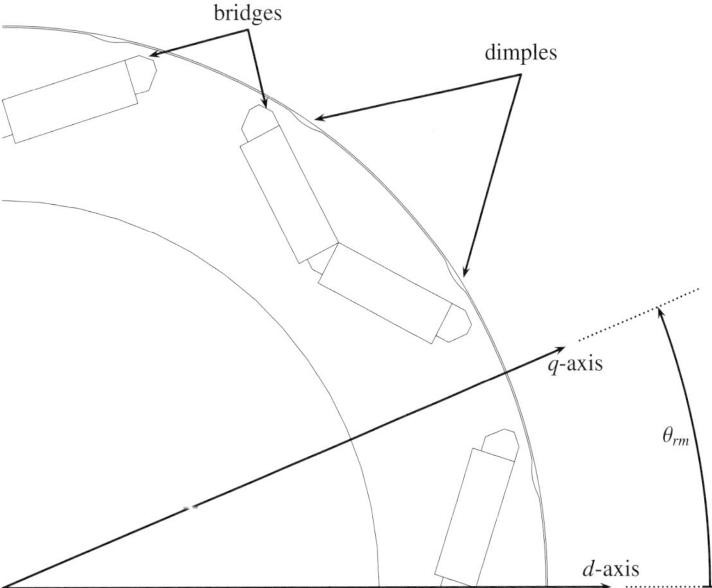

Figure 4.58 Detail of IPMSM rotor PSLG (plotted for $\theta_{rm} = 22.5°$, where the rotor d-axis coincides with the stator a-axis). The innermost "circle" has radius r_{ri}. The outermost "circle" has radius $r_i = r_{ro} + g/3$.

The PMs are placed within pockets in a V-shaped configuration forming an angle of 145°. The prototypical pocket geometry is illustrated in Figure 4.59, which depicts the upper part of a rotated V (looking like a "less than" < sign). The figure also lists the coordinates of the points defining the pocket. Both magnets in the V have the same polarity, and combined they tend to create magnetic flux oriented along the KA-axis of

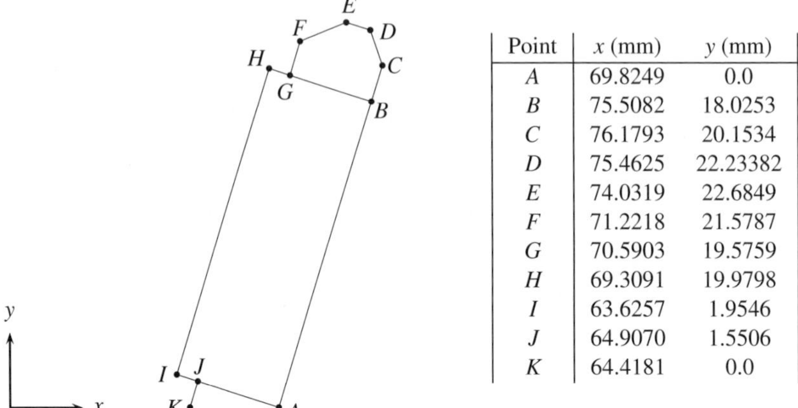

Point	x (mm)	y (mm)
A	69.8249	0.0
B	75.5082	18.0253
C	76.1793	20.1534
D	75.4625	22.23382
E	74.0319	22.6849
F	71.2218	21.5787
G	70.5903	19.5759
H	69.3091	19.9798
I	63.6257	1.9546
J	64.9070	1.5506
K	64.4181	0.0

Figure 4.59 Interior PMSM rotor pocket geometry. Dots are PSLG points. The pocket is symmetric with respect to a horizontal axis through KA.

the V. The PMs are rectangular and are embedded tightly within the rectangle $ABHI$, which has dimensions 6.5 × 18.9 mm. The FEA mesh does not model the small gap between the PMs and the surrounding steel. The remaining regions $BCDEFG$ and AJK are air pockets. Note that the CD side of the pocket lies very close to the rotor outer surface, forming a narrow bridge. Here, these pockets are modeled with a few linear segments, but in practice they may be shaped more intricately. Their purpose is to act as **flux barriers** along the d-axis. Effectively, they reduce the d-axis inductance in order to obtain a large negative $L_d - L_q$ value. In view of the torque equation

$$\tau_e = \frac{3p}{2}(\lambda_d i_q - \lambda_q i_d) = \frac{3p}{2}[\lambda_m i_q + (L_d - L_q) i_d i_q], \tag{4.157}$$

this saliency can be exploited to generate an additional reluctance torque component from the motor through the injection of negative i_d. The NdFeB magnets have remanence $B_r = 1.2$ T (adjusted for operating temperature) and relative permeability $\mu_{r,pm} = 1.06$. Their magnetization is assumed to be uniform and parallel to the short pocket sides AI and BH.

Flux Linkage and Torque: Open-Circuit Conditions

The meshed PMSM is shown in Figure 4.60. The stator, rotor, and air-gap layer meshes have 23,093, 15,819, and 4,162 triangular elements, respectively. First, we conduct a series of FEA studies with the stator winding open over 180° electrical, with 16 rotor positions per slot. A snapshot for $\theta_{rm} = 0°$ is shown in Figure 4.61(a). At this angle, the four north rotor poles are located at −112.5°, −22.5°, 67.5°, and 157.5°.

In this machine, something very interesting takes place. Under open-circuit conditions, a sizeable fraction of the PM flux (roughly 10%) flows around the magnets without crossing the air gap, representing a leakage flux that does not link the stator windings. This flux heavily saturates the thin bridges between the rotor surface and the pock-

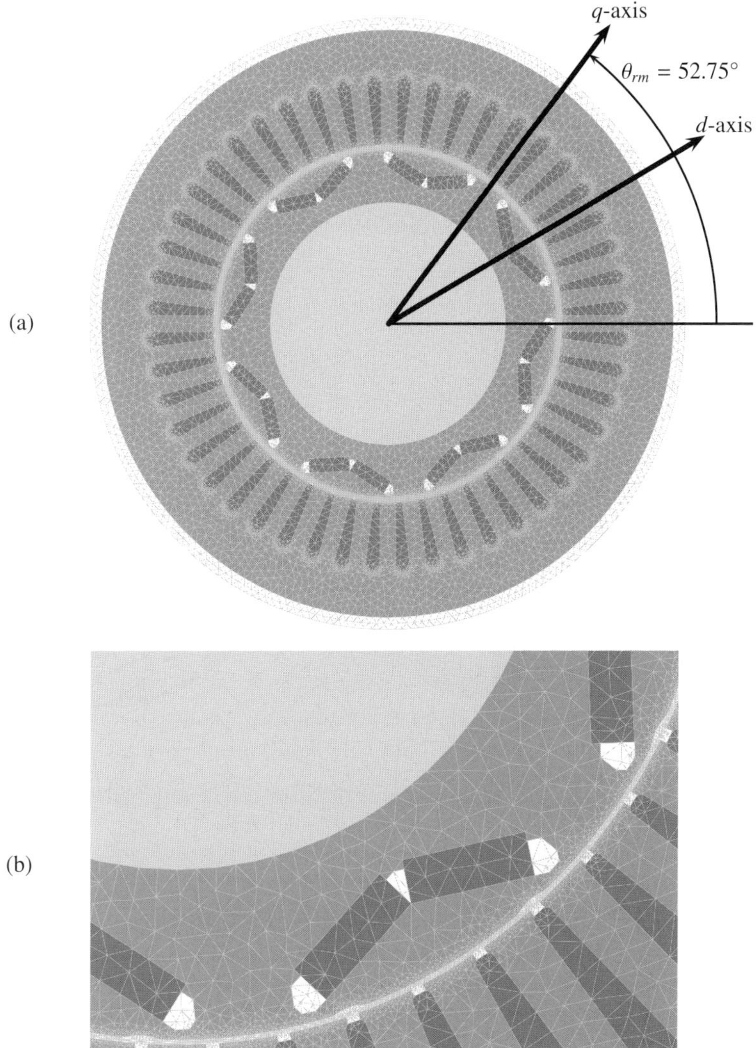

Figure 4.60 (a) Meshed cross-section of interior PMSM. (b) Mesh detail in a region close to the air gap and a north rotor pole.

ets, where the B-field is around 2 T! The saturation of the bridges is critical for the operation of the motor because it does not allow even more flux to circulate around the rotor, and forces the remaining PM flux to cross the air gap. From a motor designer's perspective, the shape of the bridges is an important consideration, which also affects the structural properties of the rotor.

We can observe the variation of the a-phase flux linkage in Figure 4.62(a), where again the effect of slotting is negligible. However, the flux linkage waveform has a flat peak of approximately 0.2 Vs. The cogging torque is plotted in Figure 4.63(a), where a ±3-Nm oscillation can be observed.

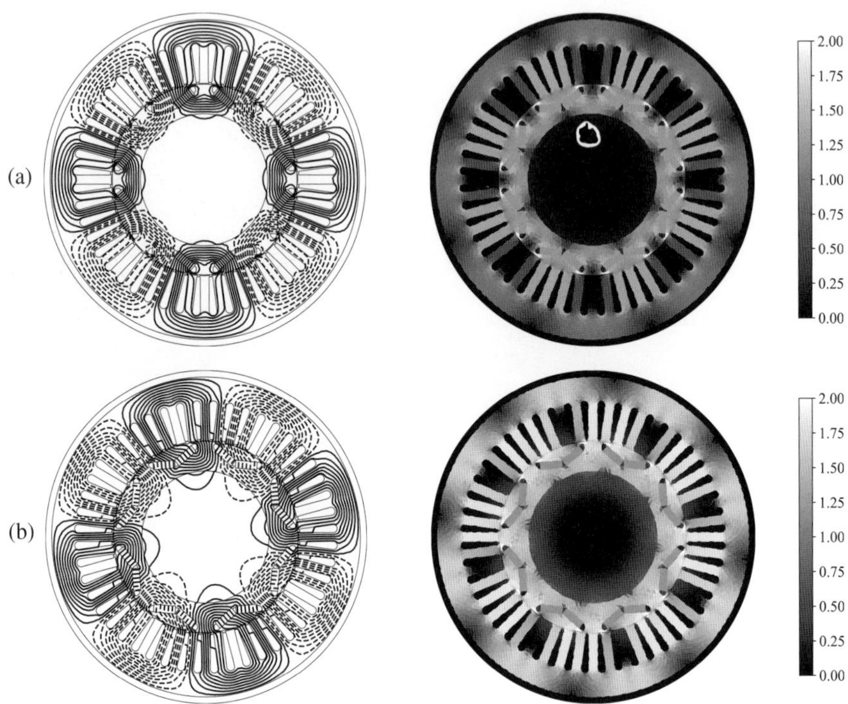

Figure 4.61 Interior PMSM flux lines (left) and flux density (right) at $\theta_{rm} = 0°$: (a) open-circuit conditions; (b) operation under load with $i_q = 100$ A, $i_d = 0$ A.

Flux Linkage and Torque: Loaded Conditions

To study the effects of loading, we add balanced three-phase current to the stator windings, thus creating a synchronously rotating MMF. As a first case study, we set i_q = 100 A and $i_d = 0$ A. The FEA solution at $\theta_r = 0°$ is shown in Figure 4.61(b). Compared with the open-circuit conditions, it may be observed that the magnetic field has rotated counterclockwise, and that its magnitude has intensified throughout the motor.

The flux linkage and torque waveforms are shown in Figures 4.62(b) and 4.63(b), respectively. Adding i_q causes a phase shift (roughly by 60°) and a doubling of the flux linkage magnitude. This can be verified by the decomposition of the flux linkages in *qd* coordinates, shown in Figure 4.64. As expected, these have a d.c. component plus a relatively minor ripple, with the sixth harmonic being dominant. The average value $\overline{\lambda_d} \approx 0.2$ Vs is similar to the peak value of the open-circuit λ_a fundamental component, which appears to be slightly higher than 0.2 Vs. Note that λ_0 exhibits a third-harmonic oscillation around zero, which is two orders of magnitude higher than in the surface-mounted PMSM that we studied earlier. The average torque value is 119.7 Nm. The torque ripple is significant: harmonics that are multiples of the sixth are present, leading to a ±8-Nm oscillation around the average value.

As a second case study, we generate contour plots of the average torque and the *qd*

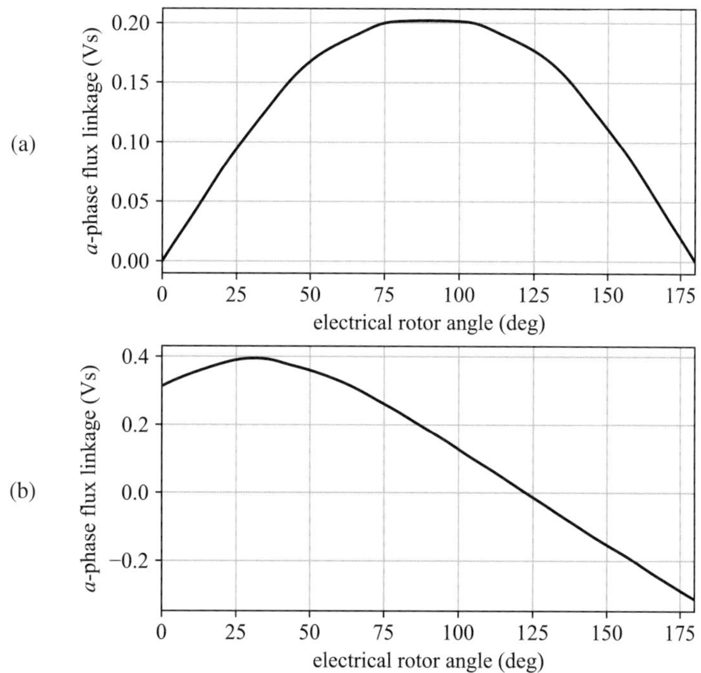

Figure 4.62 Interior PMSM a-phase flux linkage waveforms: (a) open-circuit conditions; (b) operation under load with $i_q = 100$ A, $i_d = 0$ A.

flux linkages. These are obtained for a grid of qd currents, $\mathbf{i}_{qd} \in [0, 360] \times [-360, 100]$ A with a spacing of 20×20 A. The averaging is performed using 10 rotor positions over 60° electrical. It should be noted that this requires $19 \cdot 24 \cdot 10 = 4{,}560$ FEA studies. At the time of writing this text, each (i_q, i_d, θ_r) point requires roughly 2–3 seconds on a personal computer, mainly due to the number of iterations required for Newton's method (8–10, depending on the current level); so, overall this study takes several hours to complete. However, it contains a significant amount of information about the performance of the motor over all conceivable operating conditions.

The FEA-based torque contours are shown in Figure 4.65. These are quite different from the contours of the surface-mounted PMSM in Figure 4.54, which were mostly vertical. In the interior PMSM machine, the rotor saliency is the primary reason for the shape of the contours. To explain this conceptually, let us recall the torque equation (4.157). This formula suggests that, in a magnetically linear motor, the contours should be rectangular hyperbolas. Solving for i_q, we have

$$i_q = \frac{2}{3p} \frac{\tau_e}{\lambda_m + (L_d - L_q) i_d}, \tag{4.158}$$

which is an equation of the form $y = A/(x - x_0)$. This implies that the asymptotes are the lines

$$i_d = -\frac{\lambda_m}{L_d - L_q} > 0 \text{ and } i_q = 0. \tag{4.159}$$

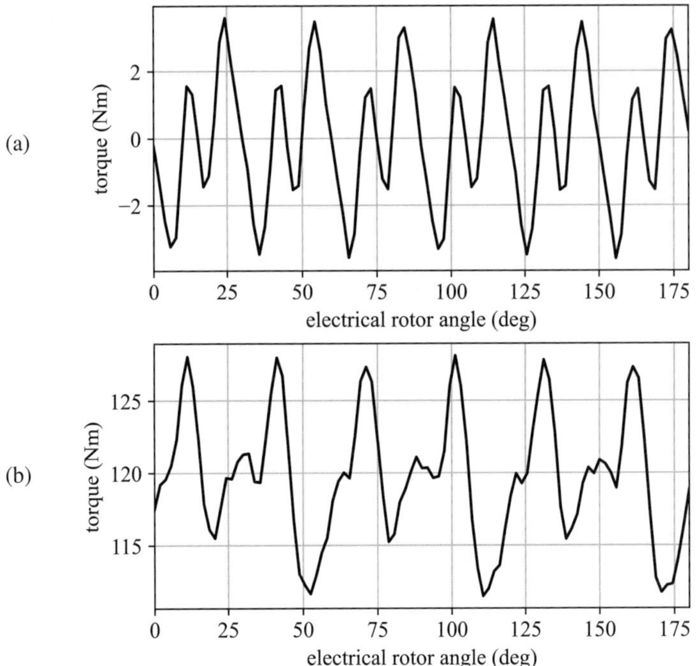

Figure 4.63 Interior PMSM torque waveforms: (a) open-circuit (cogging) torque; (b) operation under load with $i_q = 100$ A, $i_d = 0$ A.

Since the coefficient $L_d - L_q < 0$, the hyperbola is "flipped." However, this is only an analytical result, which should not be used for numerical purposes due to the magnetic nonlinearity of this device.

The torque contour plot is useful because it can inform the control strategy of a PMSM drive that aims to operate the motor in the most efficient manner. We can answer the question: *How can a desired amount of torque be generated with the minimum amount of current so that the ohmic loss is minimum?* Alternatively: *What is the maximum amount of torque that can be generated with a given r.m.s. current?* Using Equations (4.112a) and (4.112b), we find that the r.m.s. value of the stator phase current, I_s, is related to the qd currents by

$$i_q^2 + i_d^2 = (\sqrt{2}\, I_s)^2 \,. \tag{4.160}$$

This equation represents a circle on the i_q–i_d plane that is centered on the origin and has radius $\sqrt{2}\, I_s$. For instance, let us assume that the rated current of this motor is 250 A, r.m.s. Then, the **current limit** of the motor is represented with the black circle of radius 353.6 A in Figure 4.65. At least in theory, the motor could operate at any (i_q, i_d) point inside the circle, and should be capable of generating electromagnetic torque well in excess of 400 Nm. The largest gray circle has a radius of 267 A, and is chosen to be tangent to the 400-Nm torque contour. The point of tangency is the answer to the questions that we posed earlier: it represents the minimum current for getting 400 Nm out

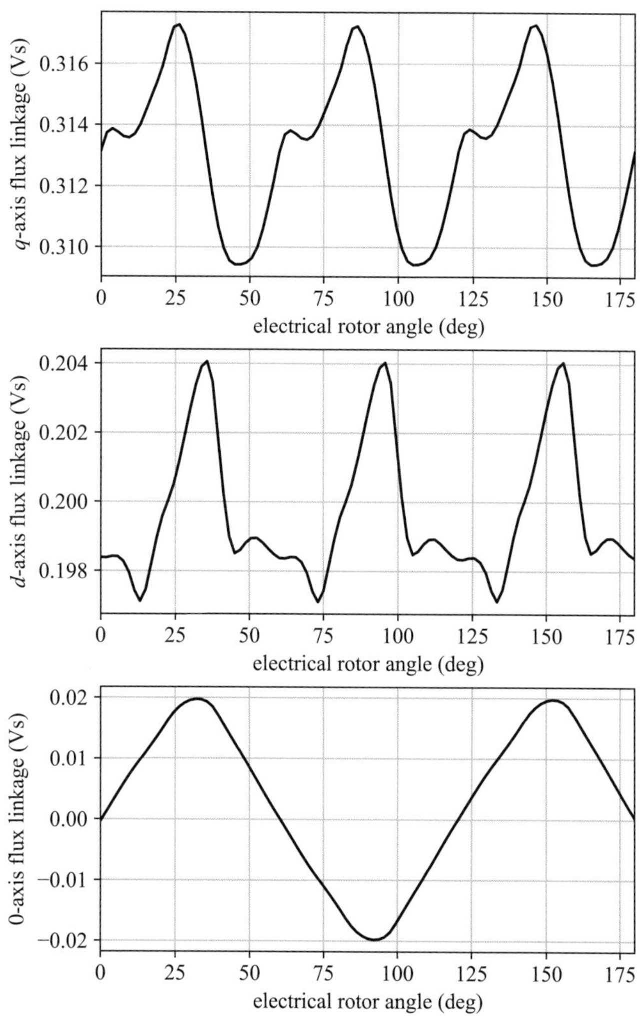

Figure 4.64 Interior PMSM: variation of $qd0$ flux linkages with rotor position.

of the motor, and the maximum torque that the machine can produce with 267 A, peak. This is called **maximum torque per ampere** (MTPA) operation. The smaller gray circles are tangent to the remaining torque contours. Joining all MTPA points (shown as black dots), we obtain the **MTPA curve**. We can conclude that, in interior PMSMs, it is beneficial from a torque perspective to add negative i_d, or equivalently to advance the stator current phase angle $\phi_i = \arctan(-i_d, i_q)$.

The flux linkage contour plots are shown in Figure 4.66. In a magnetically linear machine, these contours would have formed a perfect rectangular grid. However, due to magnetic nonlinearity, this is no longer the case. These curves are interesting in their own right, but they are also useful as intermediate quantities for calculating the voltage.

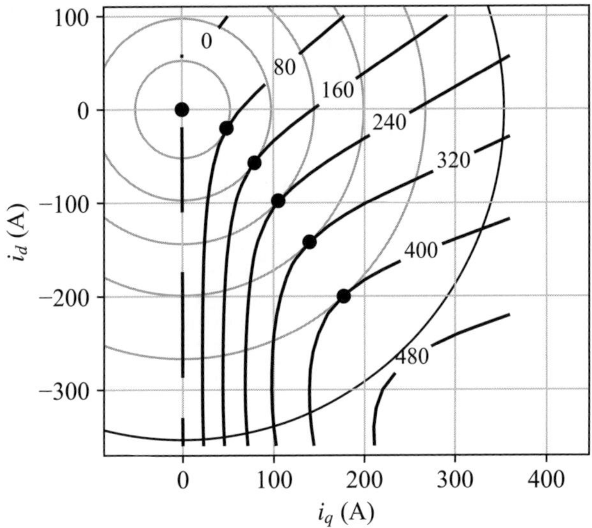

Figure 4.65 Interior PMSM torque contour plot. Torque contour units are in Nm.

Recall that the steady-state qd voltages are

$$V_q = R_s I_q + \omega_r \Lambda_d \,, \tag{4.161a}$$

$$V_d = R_s I_d - \omega_r \Lambda_q \,. \tag{4.161b}$$

Therefore, we can find $V_q(I_q, I_d; \omega_r)$ and $V_d(I_q, I_d; \omega_r)$, and then we can calculate the peak line-to-neutral voltage

$$V_{\text{peak}}(I_q, I_d; \omega_r) = \sqrt{V_q^2 + V_d^2} \,, \tag{4.162}$$

over the i_q–i_d domain, for any given value of rotor speed as a parameter. This calculation assumes that the voltages are purely sinusoidal, which is not entirely true since the flux linkage waveform contains harmonics. It is certainly possible to do a more precise calculation of the peak voltage by numerically differentiating λ_{qd} with respect to time; however, this approximation suffices for our purpose, which is to understand how the voltage limitation of the power electronic inverter affects the operation of the motor.

For proper operation of a three-phase inverter (e.g., one that employs a space-vector modulation scheme), the d.c.-link voltage should be greater than the line-to-line peak value. Therefore, the voltages should satisfy

$$\sqrt{3}\, k_{dc} V_{\text{peak}} \leq V_{dc} \,, \tag{4.163}$$

where the $\sqrt{3}$ factor converts the line-to-neutral to line-to-line voltage, and $k_{dc} > 1$ is a safety factor that accounts for non-idealities that we have neglected, such as harmonics in the voltage waveform, core loss, or the possibility that the d.c.-link voltage may fluctuate around its nominal value. For the 2004 Toyota Prius, the nominal d.c.-link voltage is $V_{dc} = 500$ V, and we set $k_{dc} = 1.1$. For this analysis, we will also assume that rated current is 267 A, peak. Also, we adjust the winding resistance value to an operating

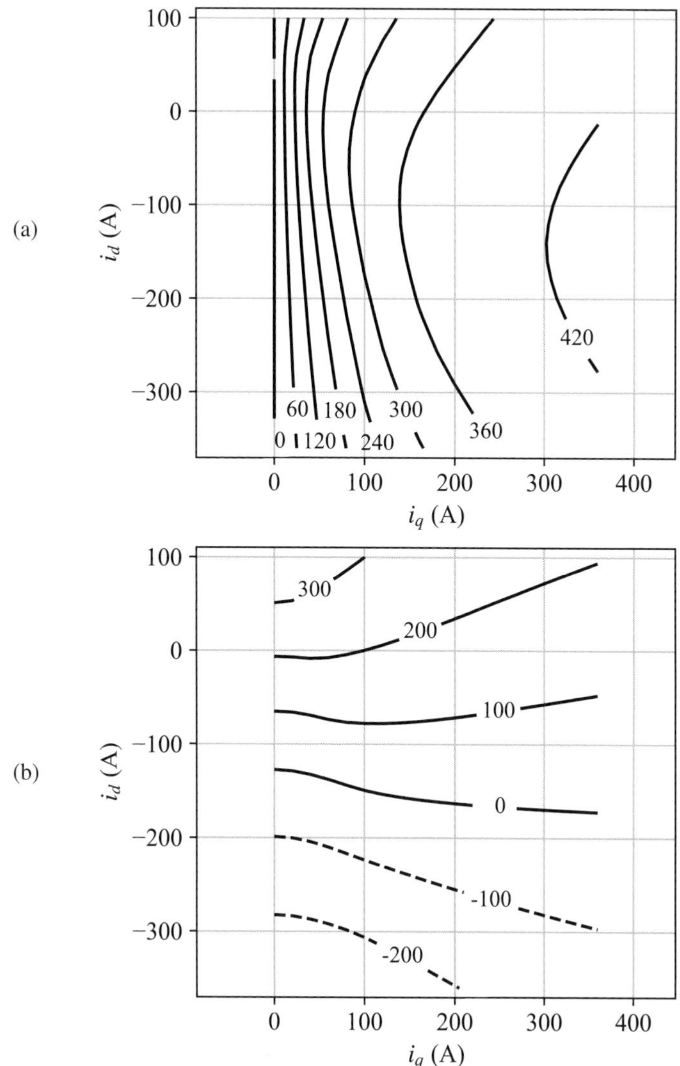

Figure 4.66 Interior PMSM flux linkage contour plots: (a) q-axis; (b) d-axis. The flux linkage unit is 1 mVs.

temperature of 100°C: $R_s = 90.4$ mΩ. The limiting voltage contours are plotted in black color in Figure 4.67 for various rotor speeds. These are quasi-elliptical in shape, and they shrink with increasing rotor speed. The gray curves are the earlier torque contours.

It is useful to think of the motor operation in terms of the maximum torque that can be produced at any given speed. We usually define three ranges of operation, as follows:

1. At low speeds, from standstill to about 1550 rpm, the motor can operate on any point along the MTPA curve up to rated torque (i.e., 400 Nm). This is a **constant torque** operation limit.

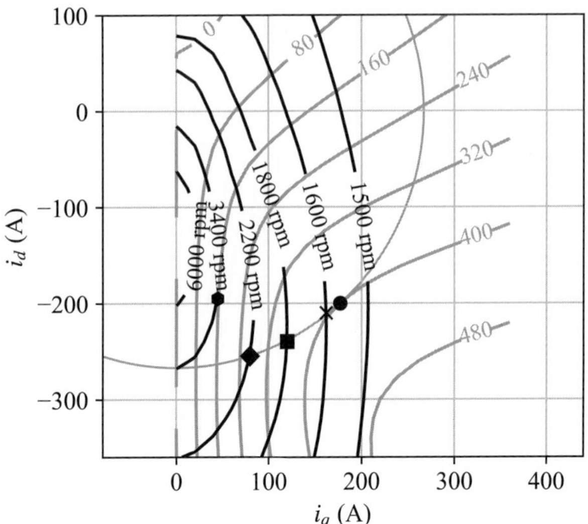

Figure 4.67 Interior PMSM voltage limit plot.

2. At higher speeds, e.g., at 1600 rpm, we observe that the drive is not capable of operating on the 400-Nm contour because of the voltage limit. To maximize torque, we must simultaneously decrease i_q and i_d (shifting to the $\times$ point), where the motor is producing less than 400 Nm. In other words, for speeds higher than 1550 rpm, we are at a regime where both current and voltage limits are active, commonly called the **constant power** operation limit. At this limit, the product of current and voltage r.m.s. values is constant, which in the language of a.c. circuits translates to constant *apparent* power. Other examples of points in this regime are the square for 1800 rpm, and the diamond for 2200 rpm, where the torque is roughly 340 and 270 Nm, respectively.
3. At speeds higher than approximately 2200 rpm, the voltage limit becomes the only active constraint because the maximum torque is obtained within the current limit circle. For example, see the hexagon for the 3400-rpm curve. This is called the **constant voltage** operation limit. In this regime, as the speed increases from 2200 to 6000 rpm (rated speed for this motor), the maximum torque drops from 270 to 80 Nm.

The maximum torque vs. speed curve of this interior PMSM is conceptually illustrated in Figure 4.68. Of course, a similar curve can be plotted for other motor designs based on the same principles. The modification of the currents in this way, which leads to a gradual reduction of the flux inside the motor, is commonly referred to as **flux weakening**. This reduction of the flux at high speeds is necessary for not exceeding the inverter maximum voltage limit. A simple argument to justify this can be readily made from the voltage equations, where after neglecting the stator resistance, we may conclude that $V_{\text{peak}} \approx \omega_r \Lambda_{\text{peak}}$. In general, we need to gradually reduce the q-axis current towards zero, and we must inject just the right amount of negative d-axis current to counteract the PM flux (from Figure 4.66, this value is $i_d \approx -125$ A).

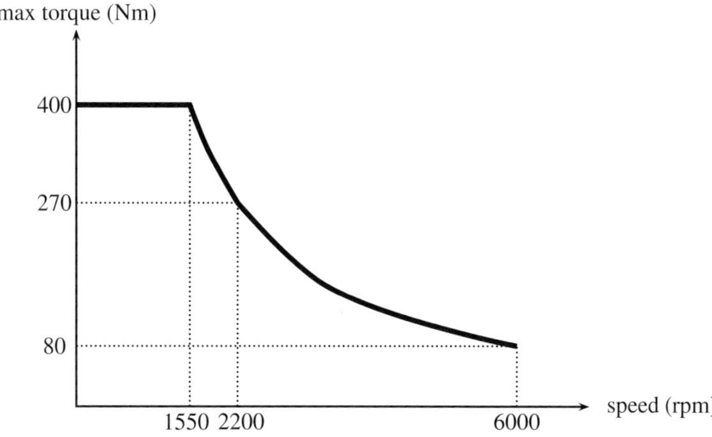

Figure 4.68 Maximum torque vs. speed characteristic of interior PMSM.

4.5 Switched Reluctance Machines

The third type of machine that we analyze is the **switched reluctance machine (SRM)**. The basic principle of operation of reluctance machines was introduced in Example 2.18 on page 173. We have explained through a coupling field energy-based approach that the developed torque tends to align the rotor so that the magnetic reluctance is minimized (and the inductance is maximized). An SRM does not have rotor windings. Nevertheless, both stator and rotor have a salient structure leading to the production of a significant reluctance torque component. This is of sufficient magnitude to make the SRM the motor of choice in many vehicular applications, especially those that require a rugged and reliable motor.

The SRM name also indicates that the excitation has a switched nature, as opposed to the sinusoidal nature of the voltages and currents in synchronous and induction machines with distributed windings. In contrast, an SRM has *concentrated* windings, and the currents are far from sinusoidal. Simply speaking, the phases are switched on/off through the action of a power electronic converter alternating the polarity of a d.c. voltage across the windings. In common SRM implementations, a positive phase current is established by applying a $+V_{dc}$ voltage across the terminals, and subsequently it is extinguished by switching to $-V_{dc}$.

Meshing the Stator

SRMs are defined by the number of teeth on the stator and rotor. The SRM that we will analyze has $n_{st} = 8$ stator teeth and $n_{rt} = 6$ rotor teeth. Its stator is depicted in Figure 4.69. This design is commonly referred to as an 8/6 SRM, although other combinations are possible, such as 6/4 and 12/8 SRMs. The number of stator and rotor teeth always differs to avoid rotor positions where torque cannot be produced due to the simultaneous alignment of all teeth. The teeth are also commonly referred to as *poles*; however, this may be somewhat confusing. As we have learned, in synchronous and

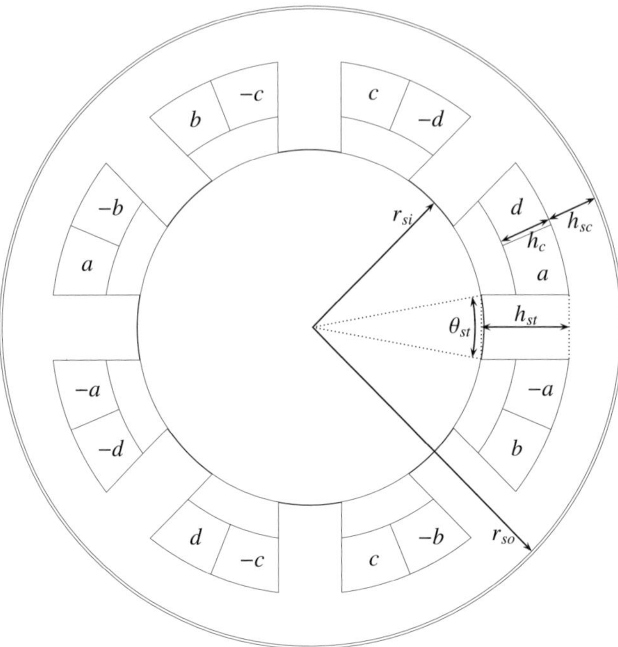

Figure 4.69 SRM stator PSLG. The innermost "circle" has radius $r_o = r_{si} - g/3$. The outermost "circle" has radius r_{ob}.

induction machines, the (dominant) number of poles determines the variation of the air-gap flux, which varies as $\cos p\phi$ in the air gap. In the case of the SRM, the magnetic flux travels through a subset of opposing stator and rotor teeth according to which phase is excited, so the number of teeth exceeds the number of *active* magnetic poles at any given instant. For instance, when the a-phase is excited in this SRM, flux has a tendency to flow horizontally from left to right, and only two poles are formed on the stator surface, i.e., a north pole on the left and a south pole on the right. Also, the flux variation in the air gap is not sinusoidal.

This 8/6 SRM design is a vehicular motor obtained from [8]. The main stator dimensions are outer radius r_{so} = 173.5 mm, inner radius r_{si} = 96.95 mm, stator core thickness h_{sc} = 28.1 mm (i.e., the thickness of the back-iron or stator yoke annulus), axial length ℓ = 190.9 mm, mesh outer boundary radius r_{ob} = 175 mm, and air-gap length g = 0.955 mm. The teeth have parallel sides, and their shape is defined by[h] (i) the tooth height h_{st} = 49 mm, which is the distance from the innermost surface (at r_{si}) to the stator yoke along the tooth side, and (ii) the tooth angle θ_{st} = 21°, which is the angle formed by the tooth arc (at r_{si}). Note that these dimensions are not all independent; they should satisfy the following geometric constraint for consistency:

$$\sqrt{(r_{si}\cos(\theta_{st}/2) + h_{st})^2 + (r_{si}\sin(\theta_{st}/2))^2} + h_{sc} = r_{so}\,. \tag{4.164}$$

[h] In the machines that we meshed previously, we chose to describe the geometry of the slots. Of course, describing the teeth is an equivalent way of parametrizing the geometry.

The stacking factor is assumed to be $k_{st} = 0.97$. The B–H curve of the lamination steel is defined by the following data points (first column lists H-values in A/m, second column lists B-values in T):

```
500     1.413
1000    1.492
2500    1.593
5000    1.684
10000   1.802
15000   1.875
20000   1.923
25000   1.957
30000   1.983
50000   2.050
75000   2.103
100000  2.146
200000  2.289
330000  2.459
```

In SRMs, concentrated windings are formed by wrapping conductors around teeth, so each stator slot contains conductors from two phases. For simplicity in the definition of the PSLG, the slot area is divided into two equal parts as shown in Figure 4.69. (An alternative and perhaps more realistic PSLG would have separated coils, leaving some space between them.) A coil height parameter is thus introduced to measure how much of the slot is occupied by the coil. In this machine, the coil height is $h_c = 29.4$ mm in the radial direction. In other words, the coil region is bounded by two circular arcs and two line segments on the sides: the innermost arc is at radius $r_{so} - h_{sc} - h_c$, and the outermost arc is at radius $r_{so} - h_{sc}$. The conductor area is found to be 988 mm^2 per slot per phase.

Assuming that the winding packing factor is $k_{pf} \approx 0.4$ implies that each half-slot has roughly 395 mm^2 of useful conductor surface area. We place 11 turns of $A_s = 33$-mm^2 copper magnet wire (AWG 3) around each pole, so the total number of series-connected turns per phase is $N_s = 22$. Let us suppose that the maximum (instantaneous) current density is $J_{\max} = 10$ A/mm^2, representing a thermal constraint in an air-cooled motor. (This limit accounts for the fact that the SRM current in general follows a trapezoidal waveform, and that each phase is turned on for roughly 50% of the time.) Hence, the phase current cannot exceed a peak value of $I_{\max} = J_{\max} A_s = 330$ A. The per-phase stator winding resistance (ignoring end turns) at 50°C is

$$R_s = 2\frac{N_s \ell}{\sigma A_s} = 4.9 \text{ m}\Omega\,. \tag{4.165}$$

As a sanity check, we calculate the instantaneous power loss at maximum current, $I_{\max}^2 R_s = 534$ W. Note that this value is an order of magnitude less than the ohmic loss (roughly 15 kW) of the Prius PMSM at maximum current, implying that this SRM design may be better suited for continuous operation at rated output. (This is merely a comparison between two particular motor designs; this claim should not be generalized to all PMSMs and SRMs.)

Meshing the Rotor

The SRM rotor geometry can be described by just a few parameters. We parametrize the rotor using the number of teeth $n_{rt} = 6$, the outer radius $r_{ro} = r_{si} - g = 95.99$ mm, the inner shaft radius $r_{ri} = 30$ mm, the tooth arc angle $\theta_{rt} = 23°$ (at r_{ro}), and the tooth height $h_{rt} = 33.4$ mm (the teeth have parallel sides). These dimensions are illustrated in Figure 4.70. The lamination material and stacking factor are assumed to be identical to those of the stator. The shaft is modeled as a linear material with relative permeability $\mu_{r,\text{sh}} = 200$.

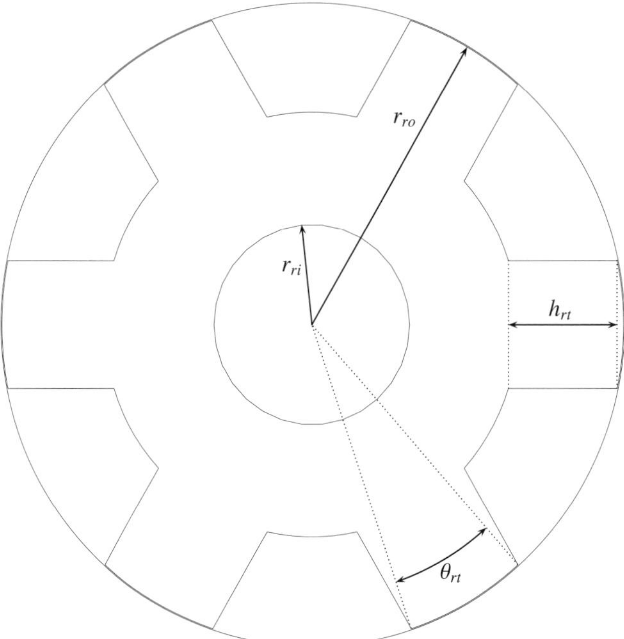

Figure 4.70 SRM rotor PSLG. The innermost "circle" has radius r_{ri}. The outermost "circle" has radius $r_i = r_{ro} + g/3$.

Operating Principle

To understand the main operating principle of an SRM, let us look at its cross-section at a sequence of rotor positions, as shown in Figure 4.71. The stator coils (see Figure 4.69) need to be turned on and off at just the right times so that the developed reluctance torque always tends to pull in the counterclockwise sense. At each one of these positions, one needs to ask the question: *which coil(s) should we energize for obtaining the maximum reluctance torque in the desired direction?*

- Let's begin at position (A). Here, if we excite either the horizontal or vertical magnetic axes (phase a or c, respectively), the torque will be zero. If we excite the axis oriented at $-45°/135°$ (phase b), we obtain torque in the clockwise sense because these stator poles will attract the closest rotor poles at $-30°/150°$. So, it is clear that our only feasible option is to excite the axis oriented at $45°/-135°$ (phase d) to attract the rotor poles at $30°/-150°$.

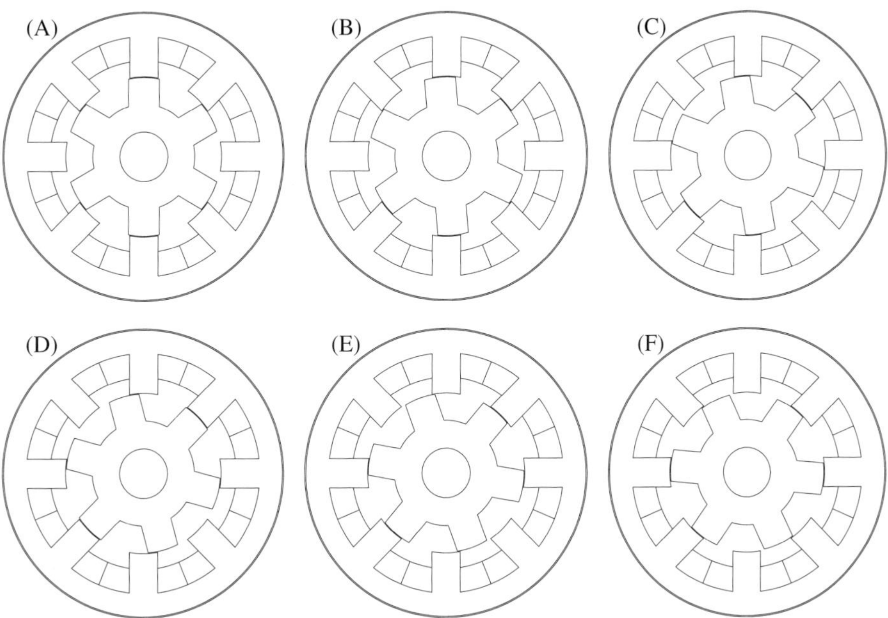

Figure 4.71 SRM at a sequence of rotor positions (relative rotation by 5°).

- At position (B), the rotor has turned counterclockwise by 5°. We must ensure that phases *c* and *b* remain turned off because they would produce a torque in the opposite direction. Hence, we keep phase *d* energized. However, since the build-up of current takes some amount of time, we may also start the energization of phase *a*. The exact point where we would turn on this phase cannot be determined precisely by this kind of analysis, but this could be as early as right after position (A), where the rotor teeth are slightly less than 30° away from the leading stator pole.
- At position (C), the teeth that are pulled by phase *d* are now at 40°/−140°. It is probably prudent to de-energize phase *d* (if this hasn't started already) since current cannot be extinguished instantaneously The current in phase *a* should have reached close to its peak value at this point, so phase *a* should be now playing the dominant torque-producing role, although phase *d* still contributes to the torque but to a lesser extent. We have learned that torque is proportional to the rate of change of the inductance with respect to position (assuming magnetic linearity, see Example 2.14 on page 164). At this position, the self-inductance of phase *d* is almost at its maximum value, so its rate of change is small. On the other hand, the self-inductance of phase *a* is increasing much more rapidly. Hence, phase-*a* current is much more significant for the development of torque than phase-*d* current at this position.
- At position (D), we must make certain that phase *d* is totally switched off because if it still carries current it would lead to a torque component in the opposite direction once the rotor moves ever slightly counterclockwise. Phase *a* should be the only phase that is conducting. Moreover, this is the earliest point that we can begin the turn-on process for phase *b*. (The situation is similar to position (A) regarding the transition of current from phase *d* to phase *a*.)

- Position (E) is similar to position (B), in the sense that this represents an intermediate point in the transition of current from phase *a* to phase *b*. Here, phases *c* and *d* must remain de-energized.
- Position (F) is similar to position (C). Here, the current in phase *a* should have started its decline, whereas the current in phase *b* should have risen to a large value.

In summary, an 8/6 SRM has four stator phases *abcd*, which are energized sequentially from a four-phase power electronic converter, allowing only positive current to flow in the coils. The timing of the switching events is critical for the correct operation of the motor drive, so accurate position feedback is necessary. During the transition of current from one phase to the next, two phases are conducting simultaneously; otherwise, a single phase is conducting. It is noteworthy that the *abcd* sequence that leads to counterclockwise rotation causes the stator field to rotate *clockwise*. From an angle of 0° when the *a*-phase is energized, the field axis rotates to −45°, −90°, and finally to −135° when phase *d* is switched on. Interestingly, when the *d*-to-*a* transition takes place, the field assumes its original 0° orientation, so a full rotation of the stator field is never completed. However, this is of little consequence since the reluctance torque is not affected by the directionality of the field.

To describe the switching of the phases mathematically, we must first define the mechanical rotor angle more precisely. Suppose that position (A) shown in Figure 4.71 is defined as $\theta_{rm} = 0°$. This means, for instance, that position (C) corresponds to $\theta_{rm} = 10°$, and position (F) corresponds to $\theta_{rm} = 25°$. We define an **electrical angle with respect to phase** x as

$$\theta_{rx} = n_{rt} \cdot \left[\theta_{rm} + \frac{2\pi}{n_{st}}(x-1)\right], \tag{4.166}$$

where $x \in \{1, 2, 3, 4\}$ represents phase *a*, *b*, *c*, or *d*, respectively. This formula implies that the electrical angle of each phase will complete a cycle every $360°/n_{rt} = 60°$ mechanical; or equivalently, that each phase will be activated $n_{rt} = 6$ times over every single mechanical rotation. Within these 60°, there are four phases that are activated consecutively, so each phase is activated 15° mechanical or $n_{rt} \cdot 15° = 90°$ electrical after the previous one.

Hence, as far as phase *d* is concerned: position (A) corresponds to $\theta_{r4} = 6 \cdot 45° \cdot 3 = 810°$ electrical, which is equivalent to 90° electrical; and position (D) corresponds to $\theta_{r4} = 6 \cdot (15° + 45° \cdot 3) = 900°$ electrical, which is equivalent to 180° electrical. Or as far as phase *a* is concerned: position (A) corresponds to $\theta_{r1} = 0°$; and position (D) corresponds to $\theta_{r1} = 6 \cdot 15° = 90°$ electrical. In other words, this definition leads to minimum self-inductance for phase x at $\theta_{rx} = 0°$, and maximum self-inductance at $\theta_{rx} = 180°$. Also, based on our discussion, we now understand that the earliest we may turn on a given phase x is at $\theta_{rx} = 0°$, and that the current must be extinguished completely by the time the angle has reached $\theta_{rx} = 180°$ electrical, after which point phase x should remain open-circuited until $\theta_{rx} = 360°$ electrical.

4.5.1 Magnetostatic Analysis

The complete meshed SRM is shown in Figure 4.72. The stator, rotor, and air-gap layer meshes have 19,794, 13,048, and 3,808 triangular elements, respectively.

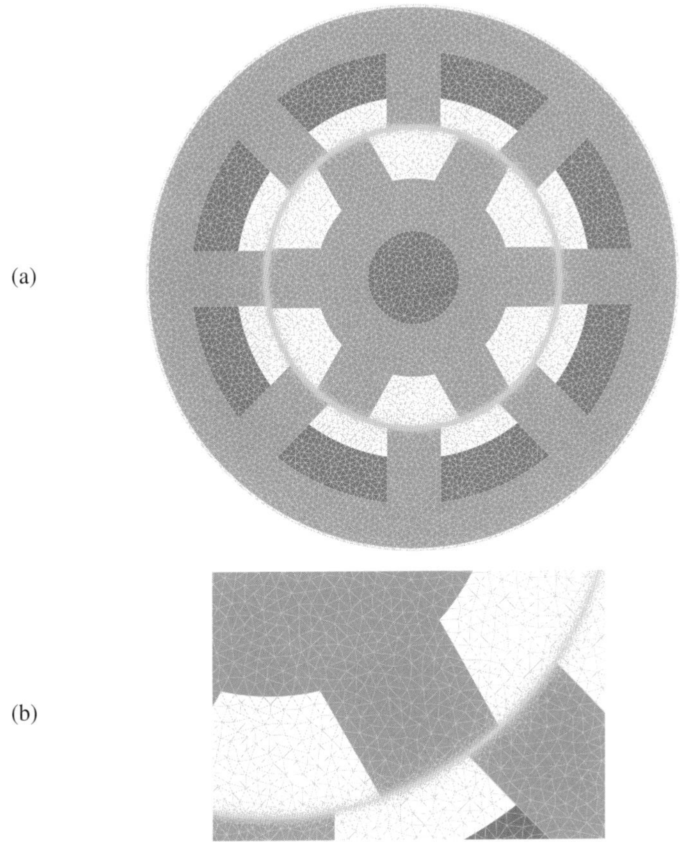

Figure 4.72 (a) Meshed cross-section of SRM. (b) Mesh detail in a region close to the air gap.

Current Transitions

The first SRM case study consists of a sequence of magnetostatic FEAs to observe the changing fields inside the machine during the transition of current from one phase to another. In Figure 4.73, we have the transition from phase d to phase a. The four subplots correspond to positions (A)–(D) of Figure 4.71, where $\theta_{r1} \in \{0, 30, 60, 90\}°$. We assume (arbitrarily) that, at these four snapshots, the currents are $(i_a, i_b, i_c, i_d) \in \{(0, 0, 0, 300), (100, 0, 0, 200), (200, 0, 0, 100), (300, 0, 0, 0)\}$ A. In Figure 4.74, we have the subsequent transition from phase a to phase b. The four subplots are positions (D)–(F) of Figure 4.71 plus a new position (G), corresponding to $\theta_{r1} \in \{90, 120, 150, 180\}°$. At these four snapshots, the currents are $(i_a, i_b, i_c, i_d) \in \{(300, 0, 0, 0), (200, 100, 0, 0), (100, 200, 0, 0), (0, 300, 0, 0)\}$ A.

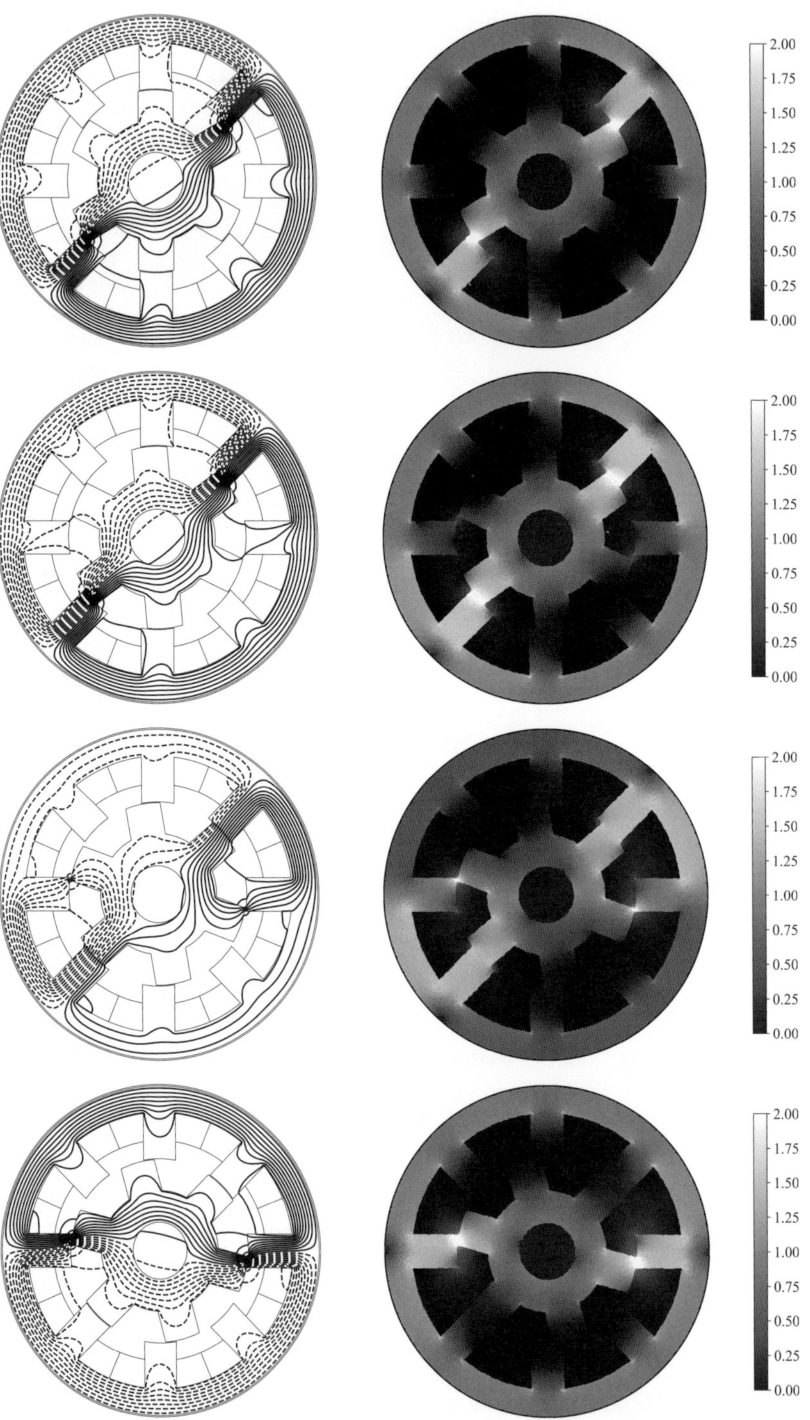

Figure 4.73 SRM flux lines (left) and flux density (right) during a transition of current from phase *d* to phase *a*.

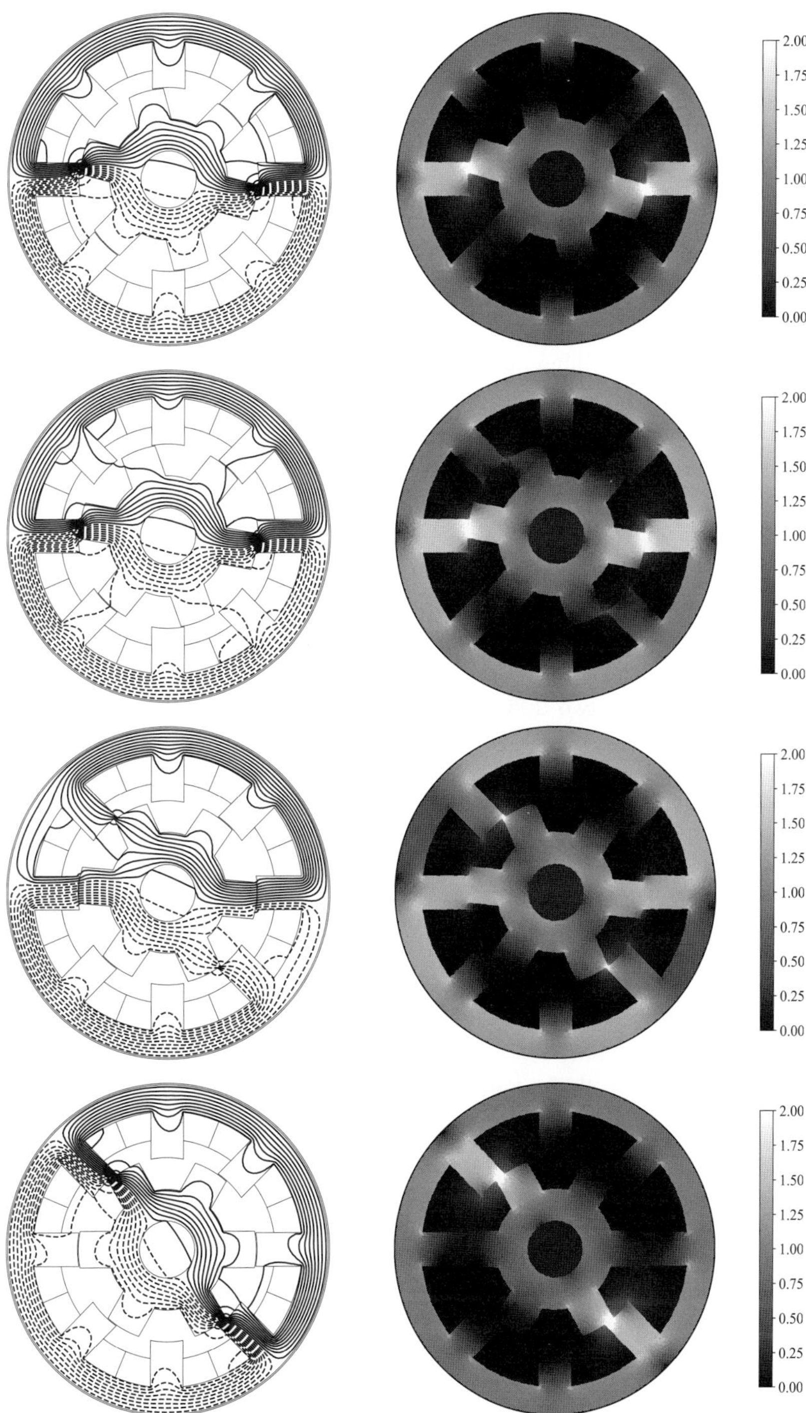

Figure 4.74 SRM flux lines (left) and flux density (right) during a transition of current from phase *a* to phase *b*.

Note that in terms of the fields, the two transitions are not identical. This is because in the first case, the field directionality changes from $-135°$ to $0°$, whereas in the second case, the field rotates from $0°$ to $-45°$. The next two transitions from phase b to phase c and from phase c to phase d are identical to this latter one. In general, we observe that the B-field tends to be quite elevated at opposing corners of stator and rotor poles, especially when these are not perfectly aligned, since flux tends to flow through these regions of low reluctance.

Torque and Flux Linkage vs. Angle with Single-Phase Excitation

The second SRM case study involves rotation with phase a carrying constant current and the remaining three phases open, to observe the variation of reluctance torque and flux linkage. We use seven current levels, $i_a \in \{50, 100, \ldots, 350\}$ A, and a set of rotor angles $\theta_{rm} \in \{0, 2, 4, \ldots, 30\}°$. As we have explained, this is not representative of a real operating condition. In general, a time-domain analysis of SRM operation requires knowledge of the torque and flux linkage functions over a 3-D domain, defined by (positive) currents in any two consecutive phases and the rotor angle over 180° electrical.

The result is shown in Figure 4.75 for a range of 30° mechanical, which corresponds to the 180° electrical that we would activate phase a for positive torque generation. Note that the torque in Figure 4.75(a) is an odd function around the origin, whereas the flux linkage in Figure 4.75(b) is an even function of rotor position. Figures 4.75(b) and (c) are two different ways of visualizing in two dimensions the 3-D surface of $\lambda_a(i_a, \theta_{rm}; i_b = 0, i_c = 0, i_d = 0)$, so they contain exactly the same data. The flux linkage in Figure 4.75(b) is parametrized by current value, whereas in Figure 4.75(c) it is parametrized by rotor angle.

The behavior resembles that of the elementary reluctance machine of Example 2.18. In that case, we had obtained a sinusoidal waveform for the torque; however, the SRM torque exhibits a more complex behavior. We observe that at $\theta_{rm} = 0°$, where the stator poles are directly in between two consecutive rotor poles (position (A) of Figure 4.71), the reluctance torque starts out at zero. This is an unstable equilibrium point since any small deviation from this position in either direction will tend to accelerate the rotor away from this angle. The torque increases rapidly and peaks somewhere in the range $[10, 15]°$ (between positions (C) and (D), respectively), where the waveform flattens. Then it drops gradually to reach the second equilibrium point at $\theta_{rm} = 30°$, where two rotor poles are directly opposite from the a-phase poles. This is a stable equilibrium point because small deviations from this position in either direction will tend to pull the rotor back.

As expected, the torque increases with current level. Interestingly, the increase in torque appears to be roughly proportional to the current. This is counterintuitive because in a magnetically linear device, the torque would be $\tau_e = (1/2)\,(dL_a/d\theta_{rm})\,i_a^2$. Hence, torque would increase with the square of the current. Of course, the SRM is nonlinear, so this formula does not apply. The flux also increases with current level. Initially, at low rotor angles where the air-gap reluctance dominates, the increase is roughly linear with respect to current. However, at higher rotor angles (e.g., larger than 10°), saturation of the stator and rotor poles causes the flux to become a nonlinear function of current.

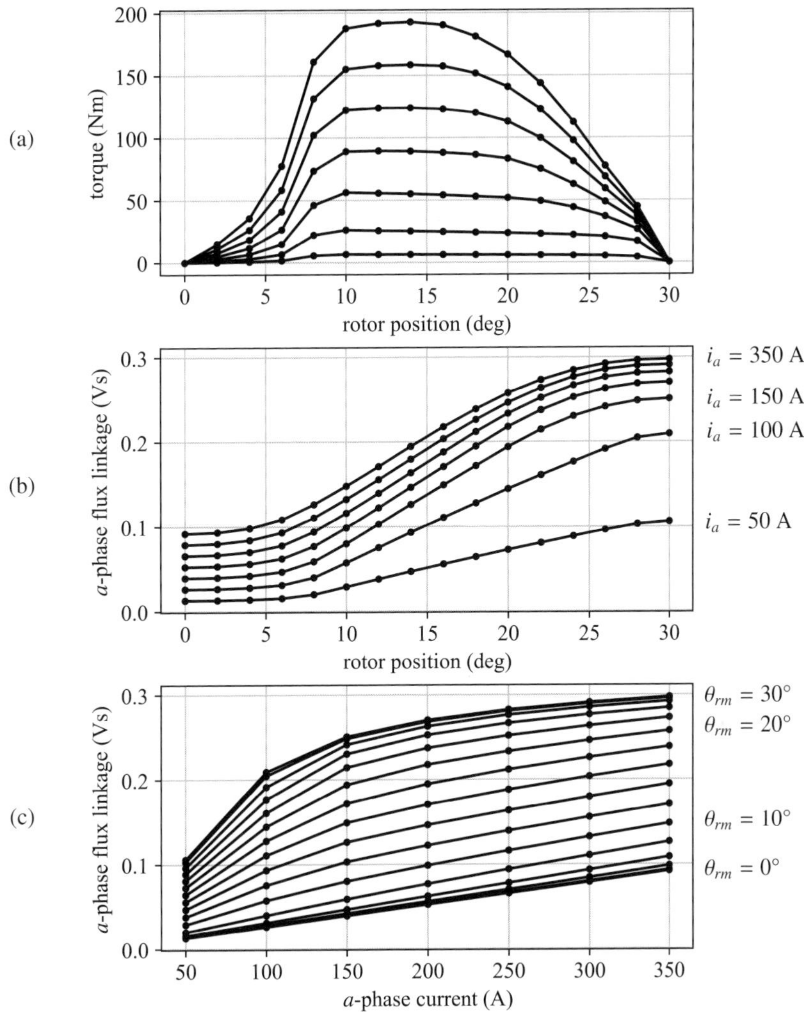

Figure 4.75 SRM torque and flux linkage variation with various levels of a-phase excitation. Subplots (a) and (b) are for $i_a \in \{50, 100, \ldots, 350\}$ A. Subplot (c) is for $\theta_{rm} \in \{0, 2, 4, \ldots, 30\}°$

4.6 Exploiting Periodicity

Electric machines typically have multiple pole-pairs. In order to reduce the computational burden, we can perform FEA on a single pole-pair or even a single pole by taking advantage of the periodicity of the magnetic field. This is an outcome of periodicity in both the geometric design and the current sources. We can exploit periodicity by meshing only a minimal fraction of the geometry, and virtually replicating the mesh a number of times to cover the entire 360°. Therefore, the number of unknowns of the FEA problem is reduced, leading to significant computational gains from the underlying matrix operations.

4.6.1 Analysis of a Single Pole-Pair (Periodic Conditions)

Figure 4.76 depicts the geometry of a single pole-pair spanning $2\pi/p$ radians. This is the domain that gets meshed in the FEA. The rotor is assumed to be internal, and the depicted "pizza slice" (circular sector) $\alpha\beta\gamma$ is rotated to signify that its position is arbitrary. Note that β forms part of the inner air-gap boundary at $r = r_i$, and η forms part of the outer air-gap boundary at $r = r_o$.

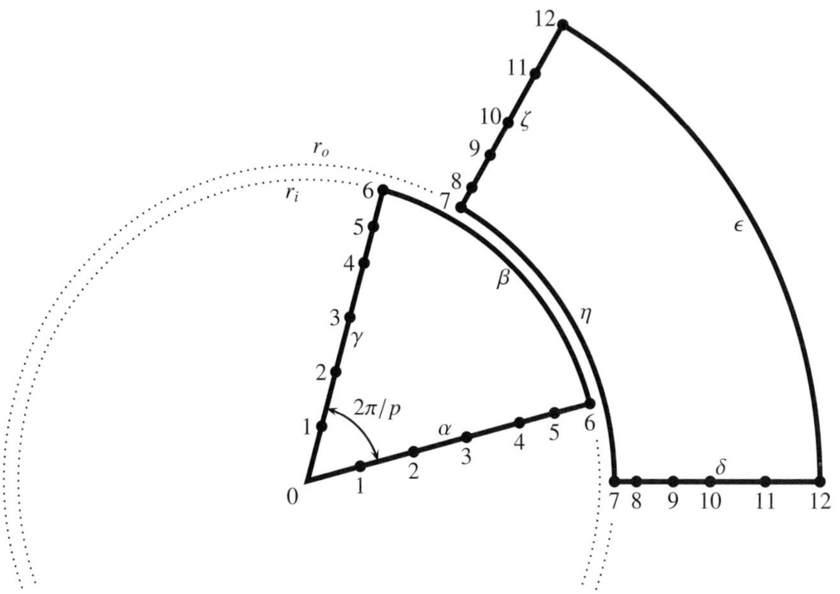

Figure 4.76 Definitions for the analysis of a single pole-pair.

In a machine that is periodic over every pole-pair, the fields satisfy

$$A(r, \phi + \tfrac{2\pi}{p}) = A(r, \phi)\,, \tag{4.167a}$$

$$B_r(r, \phi + \tfrac{2\pi}{p}) = B_r(r, \phi)\,, \tag{4.167b}$$

$$B_\phi(r, \phi + \tfrac{2\pi}{p}) = B_\phi(r, \phi)\,, \tag{4.167c}$$

for arbitrary r, ϕ. In particular, the MVP is identical at the radial boundaries, namely α/γ and δ/ζ. To model this situation in an FEA, we must ensure that the two boundaries are segmented identically so that vertices are found at exactly the same radial positions. For instance, these would be the nodes designated 1–6 on the rotor and 7–12 on the stator, which appear on either side.[i]

Suppose that the vector of rotor MVPs is arranged as

$$\mathbf{A}_r = \begin{bmatrix} \mathbf{A}_i^\top & \mathbf{A}_\alpha^\top & \mathbf{A}_\gamma^\top \end{bmatrix}^\top , \tag{4.168}$$

[i] It is certainly possible to use arbitrary shapes for the α/γ and/or δ/ζ boundaries, as long as the shapes are identical on each side. In other words, it is not necessary to use linear segments, which are used here for simplicity and without loss of generality. This could be a convenient choice for delineating the shape of geometrical features.

where $\mathbf{A}_i$ contains all rotor (inner) MVPs including the center node 0 and the nodes on the air-gap boundary β, but excluding those of the boundaries α/γ. The contribution of the entire (360° mechanical) rotor domain to the functional is denoted as I_r, which equals p times the sum of contributions from elements in the circular sector $\alpha\beta\gamma$.

First, we form the gradient and Hessian for the $\alpha\beta\gamma$ domain with respect to $\mathbf{A}_r$, using the algorithms of §3.6; however, we must further adjust these to account for the periodic constraint

$$\mathbf{A}_\gamma = \mathbf{A}_\alpha \, . \tag{4.169}$$

The original (unconstrained) gradient contains the partial derivatives of the functional with respect to all nodes of the sector:

$$\mathbf{g}_r = \begin{bmatrix} \partial I_r/\partial \mathbf{A}_i \\ \partial I_r/\partial \mathbf{A}_\alpha \\ \partial I_r/\partial \mathbf{A}_\gamma \end{bmatrix} = \begin{bmatrix} \mathbf{g}_i \\ \mathbf{g}_\alpha \\ \mathbf{g}_\gamma \end{bmatrix} . \tag{4.170}$$

The MVPs of the γ-nodes are no longer independent variables. In the calculations, these nodes are essentially to be treated as α-boundary nodes. The set of independent variables is now

$$\mathbf{A}'_r = \begin{bmatrix} \mathbf{A}_i^\top & \mathbf{A}_\alpha^\top \end{bmatrix}^\top . \tag{4.171}$$

So, the modified gradient should include the partial derivatives with respect to $\mathbf{A}_i$ and $\mathbf{A}_\alpha$ only. Note that

$$\mathbf{A}_r = \mathbf{C}_r \mathbf{A}'_r \, , \tag{4.172}$$

where we have introduced a rectangular node association matrix

$$\mathbf{C}_r = \begin{bmatrix} \mathbb{I}_{N_i} & \mathbf{0} \\ \mathbf{0} & \mathbb{I}_{N_\alpha} \\ \mathbf{0} & \mathbb{I}_{N_\alpha} \end{bmatrix} , \tag{4.173}$$

which includes identity and zero matrices of the appropriate size.[j] The subscripts N_i and N_α denote the number of elements in $\mathbf{A}_i$ and $\mathbf{A}_\alpha$, respectively. (Clearly, in a realistic FEA, most probably we will have $N_\alpha \gg 6$.) Hence, we obtain the modified (reduced) gradient

$$\mathbf{g}'_r = p \begin{bmatrix} \mathbf{g}_i \\ \mathbf{g}_\alpha + \mathbf{g}_\gamma \end{bmatrix} = p \mathbf{C}_r^\top \mathbf{g}_r \, . \tag{4.174}$$

Similarly, from the unmodified Hessian

$$\mathbf{H}_r = \begin{bmatrix} \mathbf{H}_{ii} & \mathbf{H}_{i\alpha} & \mathbf{H}_{i\gamma} \\ \mathbf{H}_{\alpha i} & \mathbf{H}_{\alpha\alpha} & \mathbf{H}_{\alpha\gamma} \\ \mathbf{H}_{\gamma i} & \mathbf{H}_{\gamma\alpha} & \mathbf{H}_{\gamma\gamma} \end{bmatrix} , \tag{4.175}$$

[j] Here, for the sake of simplicity, we have grouped the nodes in a particularly convenient way. However, in an FEA implementation, the nodes can be listed in arbitrary order because it is really not necessary to have them ordered in this same manner. In this case, the $\mathbf{C}_r$ matrix will still contain only zeros and ones, but these may be dispersed.

we obtain the modified (reduced) Hessian

$$\mathbf{H}_r' = p\begin{bmatrix} \mathbf{H}_{ii} & \mathbf{H}_{i\alpha} + \mathbf{H}_{i\gamma} \\ \mathbf{H}_{\alpha i} + \mathbf{H}_{\gamma i} & \mathbf{H}_{\alpha\alpha} + \mathbf{H}_{\alpha\gamma} + \mathbf{H}_{\gamma\alpha} + \mathbf{H}_{\gamma\gamma} \end{bmatrix} = p\mathbf{C}_r^\top \mathbf{H}_r \mathbf{C}_r \,. \tag{4.176}$$

On the stator side, we arrange the stator MVPs as

$$\mathbf{A}_s = \begin{bmatrix} \mathbf{A}_o^\top & \mathbf{A}_\delta^\top & \mathbf{A}_\zeta^\top \end{bmatrix}^\top , \tag{4.177}$$

where $\mathbf{A}_o$ contains all stator (outer) MVPs except those of the δ/ζ boundary nodes, but including those along the ϵ/η boundaries. The set of independent stator variables is

$$\mathbf{A}_s' = \begin{bmatrix} \mathbf{A}_o^\top & \mathbf{A}_\delta^\top \end{bmatrix}^\top , \tag{4.178}$$

due to the periodic constraint

$$\mathbf{A}_\zeta = \mathbf{A}_\delta \,. \tag{4.179}$$

Following an identical procedure, we obtain the modified gradient and Hessian of the stator-side functional I_s with respect to $\mathbf{A}_s'$:

$$\mathbf{g}_s' = p\mathbf{C}_s^\top \mathbf{g}_s, \ \mathbf{H}_s' = p\mathbf{C}_s^\top \mathbf{H}_s \mathbf{C}_s \,, \tag{4.180}$$

where

$$\mathbf{C}_s = \begin{bmatrix} \mathbb{I}_{N_o} & \mathbf{0} \\ \mathbf{0} & \mathbb{I}_{N_\delta} \\ \mathbf{0} & \mathbb{I}_{N_\delta} \end{bmatrix} . \tag{4.181}$$

These terms reflect the fact that I_s equals p times the functional contributions from the $\delta\epsilon\zeta\eta$ domain.

Finally, an air-gap region is created by repeating the nodes on the β/η boundaries (including the corner points α_6/δ_7) p times in order to form two circles (or strictly speaking, two polygons), as shown in Figure 4.76. This air-gap region is then meshed in its entirety, and the air-gap stiffness matrix $\mathbf{S}_{ag}$ is calculated. The air-gap nodes are associated with the rotor and stator nodes with a linking matrix $\mathbf{C}$, following the ideas of §4.2. If the air-gap region consists of a single layer of elements, then the number of air-gap nodes is $N_{ag} = p(N_\beta + N_\eta + 2)$. The air-gap node potentials are mapped to those of the stator and rotor nodes by

$$\mathbf{A}_{ag} = \begin{bmatrix} \mathbf{C}_{as} & \mathbf{C}_{ar} \end{bmatrix} \begin{bmatrix} \mathbf{A}_s' \\ \mathbf{A}_r' \end{bmatrix} = \mathbf{C}\mathbf{A}' , \tag{4.182}$$

where $\mathbf{A}'$ contains all independent MVPs, of which there are $N_s = N_o + N_\delta$ from the stator side plus $N_r = N_i + N_\alpha$ from the rotor side. Hence, $\mathbf{C}_{as}$ and $\mathbf{C}_{ar}$ have dimension $N_{ag} \times N_s$ and $N_{ag} \times N_r$, respectively. The linking matrix $\mathbf{C}$ is sparse with all nonzero elements equal to 1. If the air-gap region forms a single layer, then each row of $\mathbf{C}$ has a single "1." Each nonzero column of $\mathbf{C}$ has "1" p times.

The device-level functional is

$$I(\mathbf{A}') = I_s(\mathbf{A}_s') + \frac{1}{2}\mathbf{A}'^\top \mathbf{C}^\top \mathbf{S}_{ag} \mathbf{C}\mathbf{A}' + I_r(\mathbf{A}_r') \,. \tag{4.183}$$

The gradient and Hessian of the functional with respect to the free nodes are

$$\mathbf{g} = \frac{\partial I}{\partial \mathbf{A}'} = \begin{bmatrix} \frac{\partial I}{\partial \mathbf{A}'_s} \\ \frac{\partial I}{\partial \mathbf{A}'_r} \end{bmatrix} = \begin{bmatrix} \frac{\partial I_s}{\partial \mathbf{A}'_s} \\ \frac{\partial I_r}{\partial \mathbf{A}'_r} \end{bmatrix} + \mathbf{C}^\top \mathbf{S}_{ag} \mathbf{C} \mathbf{A}' = \begin{bmatrix} \mathbf{g}'_s \\ \mathbf{g}'_r \end{bmatrix} + \mathbf{C}^\top \mathbf{g}_{ag} \tag{4.184}$$

and

$$\mathbf{H} = \begin{bmatrix} \mathbf{H}'_s & \mathbf{0} \\ \mathbf{0} & \mathbf{H}'_r \end{bmatrix} + \mathbf{C}^\top \mathbf{S}_{ag} \mathbf{C}\,, \tag{4.185}$$

respectively. Dirichlet boundary conditions can be enforced by modifying these matrices as described in earlier sections. This would typically entail eliminating all ϵ boundary nodes as well as the corner δ node (e.g., δ_{12}) from the equations. A symmetric Hessian matrix is thus obtained. Implementation details are left as an exercise for the reader.

4.6.2 Analysis of a Single Pole (Antiperiodic Conditions)

Even more reduction can be obtained if the machine exhibits antiperiodicity over each pole. The illustration of Figure 4.76 can be reused for this case; the main difference being that now the angular span is a single pole pitch, that is, π/p radians. The field should satisfy the antiperiodic conditions

$$A(r, \phi + \tfrac{\pi}{p}) = -A(r, \phi)\,, \tag{4.186a}$$

$$B_r(r, \phi + \tfrac{\pi}{p}) = -B_r(r, \phi)\,, \tag{4.186b}$$

$$B_\phi(r, \phi + \tfrac{\pi}{p}) = -B_\phi(r, \phi)\,, \tag{4.186c}$$

for arbitrary r, ϕ. The radial and tangential components of the B-field change sign in consecutive domains. However, this does not impact the coupling field energy (i.e., the first term of the magnetostatic functional) whose value depends on B^2. Furthermore, in terms of the current sources, symmetry implies that

$$J(r, \phi + \tfrac{\pi}{p}) = -J(r, \phi)\,, \tag{4.187}$$

so the integral $\iint JA\,da$ (i.e., the second term of the functional) is identical over consecutive domains. Therefore, we conclude that each single-pole domain contributes the same amount to the overall functional, which thus equals $2p$ times the individual contributions.

On the rotor side, the boundary constraint that the FEA solution must satisfy is

$$\mathbf{A}_\gamma = -\mathbf{A}_\alpha\,. \tag{4.188}$$

Using the same set of independent variables as before ($\mathbf{A}_i$ and $\mathbf{A}_\alpha$), we define a rectangular matrix

$$\mathbf{C}_r = \begin{bmatrix} \mathbb{I}_{N_i} & \mathbf{0} \\ \mathbf{0} & \mathbb{I}_{N_\alpha} \\ \mathbf{0} & -\mathbb{I}_{N_\alpha} \end{bmatrix}, \tag{4.189}$$

which allows us to express the reduced gradient and Hessian concisely as

$$\mathbf{g}_r' = 2p \begin{bmatrix} \mathbf{g}_i \\ \mathbf{g}_\alpha - \mathbf{g}_\gamma \end{bmatrix} = 2p\mathbf{C}_r^\top \mathbf{g}_r \,, \tag{4.190}$$

$$\mathbf{H}_r' = 2p \begin{bmatrix} \mathbf{H}_{ii} & \mathbf{H}_{i\alpha} - \mathbf{H}_{i\gamma} \\ \mathbf{H}_{\alpha i} - \mathbf{H}_{\gamma i} & \mathbf{H}_{\alpha\alpha} - \mathbf{H}_{\alpha\gamma} - \mathbf{H}_{\gamma\alpha} + \mathbf{H}_{\gamma\gamma} \end{bmatrix} = 2p\mathbf{C}_r^\top \mathbf{H}_r \mathbf{C}_r \,. \tag{4.191}$$

Following an identical procedure for the stator side, where we must satisfy the constraint

$$\mathbf{A}_\zeta = -\mathbf{A}_\delta \,, \tag{4.192}$$

we obtain the modified gradient and Hessian:

$$\mathbf{g}_s' = 2p\mathbf{C}_s^\top \mathbf{g}_s, \; \mathbf{H}_s' = 2p\mathbf{C}_s^\top \mathbf{H}_s \mathbf{C}_s \,, \tag{4.193}$$

where

$$\mathbf{C}_s = \begin{bmatrix} \mathbb{I}_{N_o} & \mathbf{0} \\ \mathbf{0} & \mathbb{I}_{N_\delta} \\ \mathbf{0} & -\mathbb{I}_{N_\delta} \end{bmatrix} . \tag{4.194}$$

The air-gap region is created by repeating the nodes on the β/η boundaries (including the corner points α_6/δ_7) $2p$ times. The mapping of air-gap to stator/rotor MVPs is again given by an equation like (4.182). Now, the nonzero elements of the linking matrix $\mathbf{C}$ equal $+1$ or -1, depending on the angular distance of the node from the corresponding "base" domain (i.e., based on the polarity of its domain). If the air-gap region does not have internal nodes, then each row of $\mathbf{C}$ has a single "±1." Each nonzero column of $\mathbf{C}$ should have a "$+1$" p times and a "-1" p times.

Example 4.3 *Single-pole FEA of the Toyota Prius IPMSM.*

We illustrate these concepts on the Toyota Prius IPMSM, which was analyzed in §4.4.3. This eight-pole machine is geometrically identical over each pole (i.e., over 45°). Also, the current sources (conductors and PMs) are antiperiodic over each pole, that is, they obey Equation (4.187), according to the six slot/pole winding pattern of Table 4.3. Hence, the conditions that need to hold for conducting FEA over a single-pole domain are satisfied.

Our single-pole triangulation is similar to the previous one in terms of triangle size. Our mesh has 2,886 elements in the stator and 1,948 elements in the rotor, so a reduction roughly by a factor of 8 has been obtained. A similar reduction is obtained in terms of the number of nodes. The air gap has $520 \cdot 8 = 4{,}160$ elements, which is approximately the same number as before, as we are meshing the entire circle with identical angular resolution; however, since none of these nodes is independent, there is no negative computational impact. Figure 4.77, which should be compared with Figure 4.61, is representative of a single-pole FEA solution. Due to the reduction in the number of variables, the Python code has accelerated by approximately one order of magnitude!

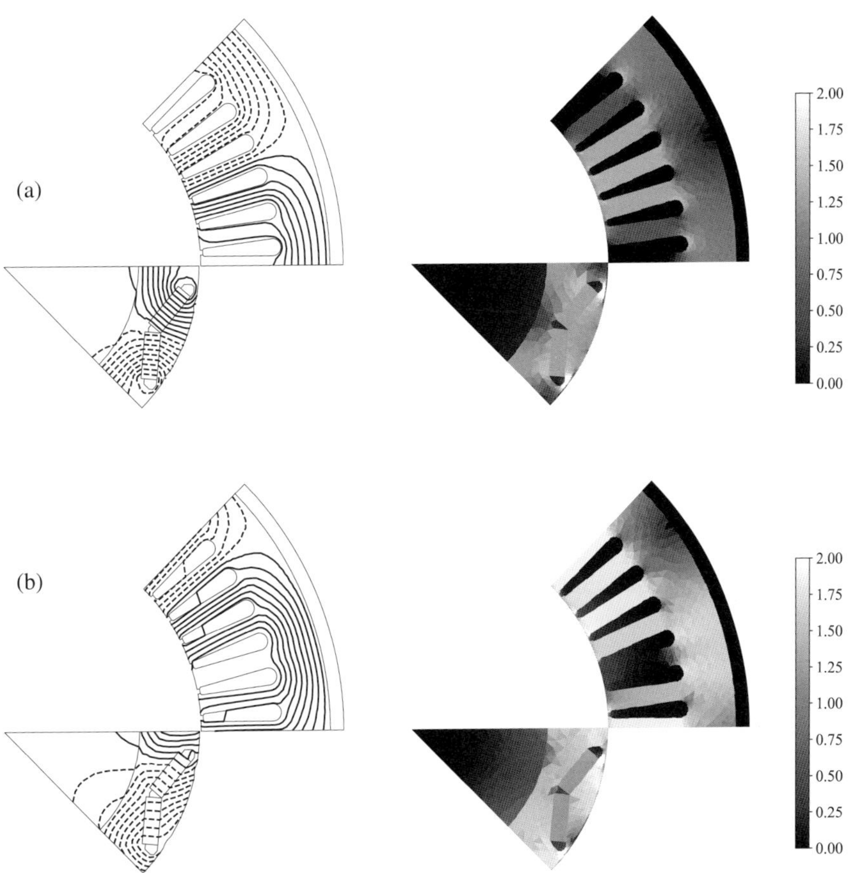

Figure 4.77 Interior PMSM flux lines (left) and flux density (right) at $\theta_{rm} = 0°$: (a) open-circuit conditions; (b) operation under load with $i_q = 100$ A, $i_d = 0$ A.

4.7 Summary

In this chapter, we worked through the specifics of electric machine FEA. We took a deep dive into various methods for calculating electromagnetic force and torque from the FEA solution, which is an important aspect of electric machine analysis. We also discussed how to implement rotation efficiently, avoiding a remeshing of the entire rotor. Then, we studied the operation of various common types of electric machines, including wound-rotor synchronous machines, PM synchronous machines, and switched reluctance machines. Our understanding was enhanced through FEA, which allowed us to observe the fields within the devices. Finally, we explained how to take advantage of the periodicity in the solution for maximum computational speed.

In the next chapter, we will learn how to implement the FEM to tackle problems in the time domain.

4.8 Further Reading

The calculation of electromagnetic force is a fascinating topic. Due to the presence of magnetized matter, the underlying physics can be quite involved. Furthermore, the numerical aspects of accurate force/torque calculation within the FEM, including the complications arising from movement, can be daunting. For further reading on methods that have been developed for force/torque calculation, the reader may refer to [9–27].

Problems

4.1 Replicate the results of Example 4.1.

4.2 Prove that the MST-torque polygonal path quadrature formula (4.13) is exact for a (hypothetical) $\vec{\mathbf{B}}$ that is constant within each element. In other words, show that the integration over a linear segment of $\vec{\mathbf{r}} \times \vec{\mathbf{t}}$, where $\vec{\mathbf{t}}$ is constant, yields $(\vec{\mathbf{r}}_0 \times \vec{\mathbf{t}})\xi$, where $\vec{\mathbf{r}}_0$ is the midpoint of the segment, and ξ is the segment length.

4.3 Suppose that we have a (hypothetical) $\vec{\mathbf{B}}$ that is constant within each element. Then, show that an exact calculation would yield

$$\tilde{\tau}_{ez} = r^2 \sum_{k \in \mathcal{E}_{ag}} \sin \delta\phi_k (\hat{\mathbf{n}}_0^k \times \vec{\mathbf{t}}_0^k)_z = \nu_0 r^2 \sum_{k \in \mathcal{E}_{ag}} \sin \delta\phi_k \, B_\phi^k B_r^k \,, \tag{4.195}$$

instead of the MST-torque circular path quadrature formula (4.15).

4.4 Show, by means of a contrived example, that the MST-torque formulas (4.13) and (4.15) yield different answers.

4.5 Write Python functions that implement the MST-based torque calculations given by Equations (4.13), (4.15), and (4.17).

4.6 Prove the VDM equations (4.33), (4.34), (4.35), and (4.49).

4.7 Write a Python function that implements the VDM for force calculation.

4.8 Write a Python function that implements the VDM for torque calculation.

4.9 Perform the algebraic manipulations needed to derive the VDM torque formulas (4.58) and (4.59).

4.10 Derive the following alternative form of the VDM torque formula (4.58), valid for a Type-2 triangle:

$$\tilde{\tau}_{ez}^k = \frac{\nu_0 \ell_{12}^k r_o}{2} (\vec{\mathbf{B}}^k \cdot \hat{\mathbf{s}}_b^k)(\vec{\mathbf{B}}^k \cdot \hat{\mathbf{n}}_b^k) \,. \tag{4.196}$$

This formula involves the decomposition of the B-field along two new orthogonal directions. The subscript "b" corresponds to a direction defined by the bisector of the angle formed by the two lines that connect the rotation center O to the midpoint M of side 1–2 and vertex #3, respectively.

Also, derive the analogous formula for a Type-1 triangle:

$$\tilde{\tau}_{ez}^k = \frac{\nu_0 \ell_{23}^k r_i}{2} (\vec{\mathbf{B}}^k \cdot \hat{\mathbf{s}}_b^k)(\vec{\mathbf{B}}^k \cdot \hat{\mathbf{n}}_b^k) \,, \tag{4.197}$$

where the bisector refers to the angle formed between the lines OM and $O1$.

Finally, for the usual case where both boundaries have the same number of nodes, N, which are equally separated by the angle $\delta\phi = 2\pi/N$, prove that the two previous expressions yield (regardless of triangle type)

$$\tilde{\tau}_{ez}^{k} = \nu_0 r_i r_o \sin\left(\frac{\delta\phi}{2}\right)(\vec{\mathbf{B}}^k \cdot \hat{\mathbf{s}}_b^k)(\vec{\mathbf{B}}^k \cdot \hat{\mathbf{n}}_b^k). \tag{4.198}$$

4.11 Write a Python function that implements the MVP-DFT torque calculation given by Equation (4.66).

4.12 Consider the rectangular U–I electromagnet shown in Figure 4.78. Suppose the device axial length is $\ell = 30$ mm. The ferromagnetic core material has the nonlinear B–H characteristic shown in Figure 3.23 on page 289. The moving member has one degree of freedom, denoted by the variable x.

Create a mesh for this geometry so that the triangulation looks similar to the one depicted in Figure 4.79 for $x = 5$ mm (upper half of geometry). Note the presence of layers of elements surrounding the moving body, which we have created on purpose for facilitating the application of the MST method and the VDM.

For coil current $i = 10$ A, calculate the electromagnetic force using:

1. The coenergy method. Justify your choice of δx.
2. The MST method. Compare the results of the following paths:
   a. A vertical path going down the left column of air-gap elements, right next to the middle of the air gap.
   b. A vertical path going down the right column of air-gap elements, right next to the middle of the air gap.
   c. A vertical path going down the right column of air-gap elements through the midpoints of edges.
   d. A path partially constructed from path (c) above, but totally enclosing the moving member by connecting the midpoints of edges immediately surrounding the body.
3. The VDM by application of Equations (4.37) and (4.38) over the layer of triangles immediately surrounding the moving member.
4. Analytic expressions derived from a simplified magnetic equivalent circuit (see Example 2.15, page 165).

Clarifying remarks:

- Paths 2a–2c are not completely surrounding the moving body, hence they are not entirely valid choices. However, since they reach the outer boundary of the problem, where the B-field is weak, the result should be approximately the same as the one obtained by path 2d.
- Paths 2a and 2b are not great choices. Geometrically, they represent essentially the same line in space, but they yield two different answers due to the discontinuity of the B-field in adjacent triangles. We should be better served by paths 2c and 2d, which pass through midpoints of edges.

4.13 Replicate (to the extent possible) the results of Example 4.2. Experiment with the mesh and/or other problem parameters, and discuss what you observe.

4.14 Replicate the results of §4.3.1. Repeat the studies for a range of rotor positions.

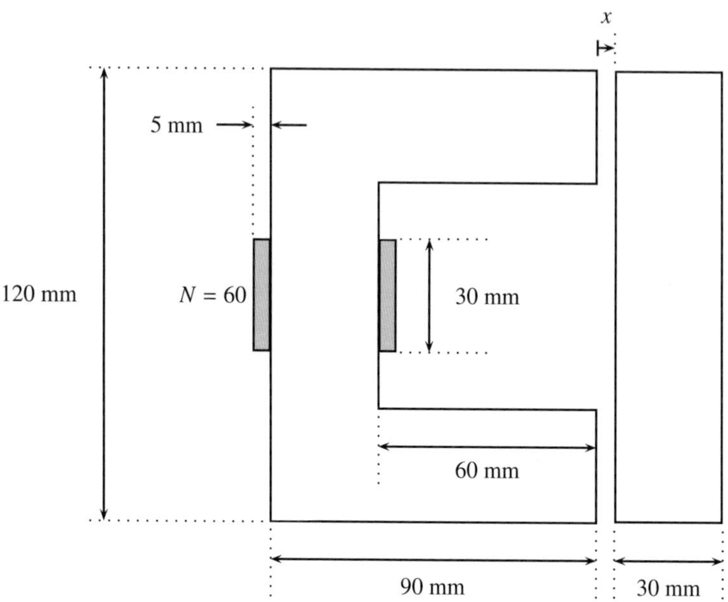

Figure 4.78 Geometry of a simple electromagnet.

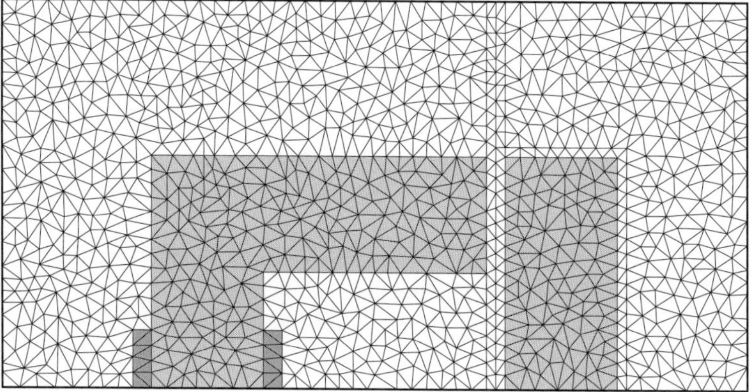

Figure 4.79 FEA mesh example.

4.15 Replicate the results of §4.3.2. Repeat the studies for a range of rotor positions.

4.16 Replicate the results of §4.3.3. Calculate the three-phase voltage waveforms for this operating condition (including all harmonics).

4.17 Consider a synchronous machine operating under conditions like those described in §4.3.3. Prove that λ_{qd} are periodic every 60° electrical.

4.18 Replicate the results of §4.3.4. Experiment with different current ranges.

4.19 Replicate the results of §4.4.1.

4.20 Replicate the results of §4.4.3. Calculate and plot the line-to-line back-emf waveform (under open-circuit conditions).

4.21 Replicate the results of §4.5.

4.22 Replicate the results of §4.4.3 using a single-pole FEA.

References

[1] A. Arkkio, "Analysis of induction motors based on the numerical solution of the magnetic field and circuit equations," *Acta Polytechnica Scandinavica, Electrical Engineering Series*, no. 59, 1987.

[2] K. Muta, M. Yamazaki, and J. Tokieda, "Development of new-generation hybrid system THS II – Drastic improvement of power performance and fuel economy," in *SAE World Congress & Exhibition*, Detroit, MI, Mar. 2004.

[3] R. H. Staunton, P. J. Otaduy, J. M. KcKeever, S. C. Nelson, J. M. Bailey, S. Das, and R. L. Smith, "PM motor parametric design analyses for a hybrid electric vehicle traction drive application — interim report," Oak Ridge National Laboratory, Tech. Rep. ORNL/TM-2004/120, Jul. 2004.

[4] M. Kamiya, "Development of traction drive motors for the Toyota hybrid system," *IEEJ Trans. Industry Applications*, vol. 126, no. 4, pp. 473–479, 2006.

[5] R. H. Staunton, C. W. Ayers, L. D. Marlino, J. N. Chiasson, and T. A. Burress, "Evaluation of 2004 Toyota Prius hybrid electric drive system," Oak Ridge National Laboratory, Tech. Rep. ORNL/TM-2006/423, May 2006.

[6] J. S. Hsu, C. W. Ayers, C. L. Coomer, R. H. Wiles, T. A. Burress, S. L. Campbell, K. T. Lowe, and R. T. Michelhaugh, "Report on Toyota/Prius motor torque capability, torque property, no-load back emf, and mechanical losses," Oak Ridge National Laboratory, Tech. Rep. ORNL/TM-2004/185, May 2007.

[7] T. A. Burress, C. L. Coomer, S. L. Campbell, A. A. Wereszczak, J. P. Cunningham, L. D. Marlino, L. E. Seiber, and H. T. Lin, "Evaluation of the 2008 Lexus LS 600H hybrid synergy drive system," Oak Ridge National Laboratory, Tech. Rep. ORNL/TM-2008/185, Jan. 2009.

[8] K. M. Rahman, B. Fahimi, G. Suresh, A. V. Rajarathnam, and M. Ehsani, "Advantages of switched reluctance motor applications to EV and HEV: Design and control issues," *IEEE Trans. Ind. Appl.*, vol. 36, no. 1, pp. 111–121, Jan./Feb. 2000.

[9] C. J. Carpenter, "Surface-integral methods of calculating forces on magnetized iron parts," *Proc. IEE-Part C: Monographs*, vol. 107, no. 11, pp. 19–28, 1960.

[10] E. M. H. Kamerbeek, *On the theoretical and experimental determination of the electromagnetic torque in electrical machines.* Technische Hogeschool Eindhoven, 1970.

[11] K. Reichert, H. Freundl, and W. Vogt, "The calculation of forces and torques within numerical magnetic field calculation methods," in *Proc. Conf. on the Computation of Magnetic Fields (COMPUMAG)*, Oxford, 1976, pp. 64–73.

[12] A. A. Abdel-Razek, J. L. Coulomb, M. Feliachi, and J. C. Sabonnadiere, "The calculation of electromagnetic torque in saturated electric machines within combined numerical and analytical solutions of the field equations," *IEEE Trans. Magnetics*, vol. 17, no. 6, pp. 3250–3252, 1981.

[13] J. L. Coulomb, "A methodology for the determination of global electromechanical quantities from a finite element analysis and its application to the evaluation of magnetic forces, torques and stiffness," *IEEE Trans. Magnetics*, vol. 19, no. 6, pp. 2514–2519, Nov. 1983.

[14] J. L. Coulomb and G. Meunier, "Finite element implementation of virtual work principle for magnetic or electric force and torque computation," *IEEE Trans. Magnetics*, vol. 20, no. 5, pp. 1894–1896, Sep. 1984.

[15] B. Davat, Z. Ren, and M. Lajoie-Mazenc, "The movement in field modeling," *IEEE Trans. Magnetics*, vol. 21, no. 6, pp. 2296–2298, Nov. 1985.

[16] J. Penman and M. D. Grieve, "Efficient calculation of force in electromagnetic devices," *IEE Proceedings*, vol. 133B, no. 4, pp. 212–216, Jul. 1986.

[17] S. McFee, J. P. Webb, and D. A. Lowther, "A tunable volume integration formulation for force calculation in finite-element based computational magnetostatics," *IEEE Trans. Magnetics*, vol. 24, no. 1, pp. 439–442, Jan. 1988.

[18] Z. Ren, "Comparison of different force calculation methods in 3D finite element modelling," *IEEE Trans. Magnetics*, vol. 30, no. 5, pp. 3471–3474, Sep. 1994.

[19] W. Cai, P. Pillay, and K. Reichert, "Accurate computation of electromagnetic forces in switched reluctance motors," in *Proc. 5th Intern. Conf. on Electrical Machines and Systems (ICEMS)*, vol. 2, Shenyang, China, Aug. 2001, pp. 1065–1071.

[20] O. J. Antunes, J. P. A. Bastos, and N. Sadowski, "Using high-order finite elements in problems with movement," *IEEE Trans. Magnetics*, vol. 40, no. 2, pp. 529–532, 2004.

[21] F. Henrotte, G. Deliége, and K. Hameyer, "The eggshell approach for the computation of electromagnetic forces in 2D and 3D," *COMPEL - The international journal for computation and mathematics in electrical and electronic engineering*, vol. 23, no. 4, pp. 996–1005, 2004.

[22] R. Sanchez-Grandia, R. Vives-Fos, and V. Aucejo-Galindo, "Magnetostatic Maxwell's tensors in magnetic media applying virtual works method from either energy or co-energy," *Eur. Phys. J. Appl. Phys.*, vol. 35, pp. 61–68, 2006.

[23] O. Barre, P. Brochet, and M. Hecquet, "Experimental validation of magnetic and electric local force formulations associated to energy principle," *IEEE Trans. Magnetics*, vol. 42, no. 4, pp. 1475–1478, Apr. 2006.

[24] A. Bossavit, "On forces in magnetized matter," *IEEE Trans. Magnetics*, vol. 50, no. 2, pp. 229–232, Feb. 2014.

[25] B. Silwal, P. Rasilo, L. Perkkiö, M. Oksman, A. Hannukainen, T. Eirola, and A. Arkkio, "Computation of torque of an electrical machine with different types of finite element mesh in the air gap," *IEEE Trans. Magnetics*, vol. 50, no. 12, pp. 1–9, Dec. 2014.

[26] K. T. McDonald, "Methods of calculating forces on rigid, linear magnetic media," 2017. [Online]. Available: https://physics.princeton.edu/~mcdonald/examples/

[27] A. Bermúdez, A. Rodríguez, and I. Villar, "Extended formulas to compute resultant and contact electromagnetic force and torque from Maxwell stress tensors," *IEEE Trans. Magnetics*, vol. 53, no. 4, pp. 1–9, Apr. 2017.

5 Problems in the Time Domain

Χρόνου φείδου

"Use time sparingly"

Delphic maxim attributed to Chilon[78]

Our analysis thus far has been entirely magnetostatic. The current sources were assumed to be imposed externally, and the finite element analysis (FEA) solution was representative of mutually independent snapshots in time. In this chapter, we will explain how to perform studies of electromechanical devices in the time domain. The main difference from before is that we will relax the constraint of known, uniform current within conductors. Thus, we will be solving for the distribution of the current as a function of time given an external electromagnetic excitation.

We begin with an overview of the relevant physics of induced currents in §5.1. Then, in §5.2, we explain how the 2-D finite element method (FEM) is formulated in the time domain. We proceed with a discussion of a few elementary problems that admit analytical solutions in §5.3, such as rectangular and circular conductors. This will help us gain valuable insights regarding eddy currents. The topic of §5.4 is FEA under sinusoidal steady-state conditions using phasors. §5.5 highlights some important differences between our previous winding modeling approach in the low-frequency regime and the higher frequencies of interest in this chapter. Finally, in §5.6, we set forth the methodology for analysis of arbitrary electrical transients in the time domain, including the effects of motion.

5.1 Physics of Induced Currents

Our first goal is to recall the physics that govern how current is induced by a changing magnetic field. Thus far, we have considered the presence of externally applied currents, which were uniformly distributed over the cross-sections of relatively thin winding conductors. Essentially, we have developed what could be called a "current-in, flux linkage-out" model. Now we shall develop a reverse, "voltage-in, current-out" model. Modeling the induced currents will allow us to study devices with relatively thick conductors, where the current density is not necessarily uniform.

Recall that the electrons inside a moving conductor experience an electric field

$$\vec{\mathbf{E}} = -\nabla\varphi - \frac{\partial \vec{\mathbf{A}}}{\partial t} - \vec{\mathbf{v}} \times \vec{\mathbf{B}}\,, \tag{5.1}$$

where φ is an electric potential, the magnetic field is $\vec{\mathbf{B}} = \nabla \times \vec{\mathbf{A}}$, and $\vec{\mathbf{v}}$ is the velocity with respect to a stationary reference frame (i.e., the frame of $\vec{\mathbf{B}}$). If the conductivity of the material is σ, the current density is related to the electric field by

$$\vec{\mathbf{J}} = \sigma\vec{\mathbf{E}} = \sigma\left(-\nabla\varphi - \frac{\partial \vec{\mathbf{A}}}{\partial t} - \vec{\mathbf{v}} \times \vec{\mathbf{B}}\right). \tag{5.2}$$

This equation connects the free current with the electromagnetic field, and forms the basis for the analysis that follows.

5.1.1 Electric Field in Two-Dimensional Problems

We shall focus on 2-D problems, where current is aligned with the z-axis, and motion is constrained on the x–y plane. Hence, the electric potential inside conductors satisfies

$$\frac{\partial\varphi}{\partial x} = \frac{\partial\varphi}{\partial y} = 0\,, \tag{5.3}$$

since an electric field cannot be present on the x–y plane (otherwise, current would flow in this direction). This implies that *every point on a conductor cross-section has the same electric potential.*

Furthermore, we can determine how the potential varies along the z-axis. Since $\nabla\cdot\vec{\mathbf{J}} = 0$, it follows that $\nabla\cdot\vec{\mathbf{E}} = 0$ inside conductors. Setting the divergence of Equation (5.1) to zero, we obtain

$$\nabla^2\varphi + \frac{\partial(\nabla\cdot\vec{\mathbf{A}})}{\partial t} + \nabla\cdot(\vec{\mathbf{v}} \times \vec{\mathbf{B}}) = 0\,. \tag{5.4}$$

In 2-D problems, both $\nabla\cdot\vec{\mathbf{A}} = 0$ and $\nabla\cdot(\vec{\mathbf{v}} \times \vec{\mathbf{B}}) = 0$ because $\vec{\mathbf{A}}$ and $\vec{\mathbf{v}} \times \vec{\mathbf{B}}$ only have a z-component for which $\partial/\partial z = 0$. Therefore

$$\nabla^2\varphi = \frac{\partial^2\varphi}{\partial z^2} = 0\,. \tag{5.5}$$

Hence, the electric potential inside conductors varies linearly as

$$\varphi = c_0 + c_1 z\,, \tag{5.6}$$

with constant c_0, c_1. In other words, the electrostatic component of the field in the conductor is constant:

$$-\nabla\varphi = -c_1\hat{\mathbf{k}}\,. \tag{5.7}$$

This relationship is useful for interconnecting conductors to form coils, and for interfacing these with external circuits. For example, if the axial length is ℓ, the potential difference between the two ends of a conductor is $c_1\ell$.

5.1.2 Motional EMF in Two-Dimensional Problems

The motional electromotive force (EMF) is a voltage that is associated with the component of the electric field due to motion inside a magnetic field. In 2-D problems, this electric field component can be expressed as

$$\vec{\mathbf{v}} \times \vec{\mathbf{B}} = (v_x B_y - v_y B_x)\,\hat{\mathbf{k}} = -\left(v_x \frac{\partial A_z}{\partial x} + v_y \frac{\partial A_z}{\partial y}\right)\hat{\mathbf{k}}\,, \tag{5.8}$$

which implies (see Equation (1.170)) that[a]

$$\vec{\mathbf{v}} \times \vec{\mathbf{B}} = -(\vec{\mathbf{v}} \cdot \nabla)\vec{\mathbf{A}}\,. \tag{5.9}$$

Therefore, an alternative way of writing Equation (5.1) is

$$\vec{\mathbf{E}} = -\nabla\varphi - \frac{d\vec{\mathbf{A}}}{dt}\,, \tag{5.10}$$

where we used the total (material or convective) derivative of the magnetic vector potential (MVP):

$$\frac{d\vec{\mathbf{A}}}{dt} = \frac{\partial \vec{\mathbf{A}}}{\partial t} + (\vec{\mathbf{v}} \cdot \nabla)\vec{\mathbf{A}}\,. \tag{5.11}$$

Equation (5.10) is important because it suggests that *the electric field inside a moving conductor depends on the rate of change of the magnetic field observed in the reference frame of the conductor*. Hence, as we did previously for the rotors of the electric machines that we analyzed, we can mesh moving domains only once, and we can calculate the electric field based on a numerical approximation of the time derivative of nodal MVPs.

5.2 Formulation of Two-Dimensional FEA in the Time Domain

Even though this chapter is dealing with time-domain problems, the requirement for stationarity of the *magnetostatic* functional still applies. Recall that this property leads to $\nabla\times\vec{\mathbf{H}} = \vec{\mathbf{J}}$, which is representative of quasi-magnetostatic conditions. This is still the differential equation that the magnetic field must satisfy at each instant, even though the current density now depends on the derivative of the MVP. Therefore, we can still solve the problem by minimizing the magnetostatic functional on an instantaneous basis.

We recall Equation (3.146), which describes the contribution to the 2-D functional from triangle k in a ferromagnetic material with nonlinear B–H curve:

$$I_k(A_k) = \frac{1}{2}\iint\limits_{\Delta_k} \int\limits_{0}^{B_k^2(A_k)} \nu_k(b^2)\, db^2\, dx\, dy - \iint\limits_{\Delta_k} J_k A_k\, dx\, dy\,. \tag{5.12}$$

[a] Generally, in 3-D problems, by applying the identity (1.175) we can show that

$$\vec{\mathbf{v}} \times \vec{\mathbf{B}} = -(\vec{\mathbf{v}} \cdot \nabla)\vec{\mathbf{A}} + \nabla(\vec{\mathbf{v}} \cdot \vec{\mathbf{A}})\,.$$

We still assume a linear variation of the solution A_k in each triangle:

$$A_k(x, y, \mathbf{A}^k) = \sum_{i=1}^{3} A_i^k \alpha_i^k(x, y). \tag{5.13}$$

The main difference from the previous formulation is in the second integral of Equation (5.12), which needs to be re-evaluated because the current density is no longer constant within each element. Since

$$J_k(x, y) = \sigma_k E_k(x, y) = \sigma_k \frac{V_k}{\ell} - \sigma_k \frac{dA_k(x, y, \mathbf{A}^k)}{dt} = J_{1,k} + J_{2,k}(x, y, \dot{\mathbf{A}}^k), \tag{5.14}$$

the current has two components. The first component

$$J_{1,k} = \sigma_k \frac{V_k}{\ell} \tag{5.15}$$

is an ohmic term that is proportional to the potential difference

$$V_k = \varphi_k(-\ell/2) - \varphi_k(\ell/2) \tag{5.16}$$

at the two ends of the conductor, and it is constant over the triangle (but it could be a function of time). However, the second component depends on the time derivatives of the MVP at the vertices, $\dot{\mathbf{A}}^k$, leading to a linear variation within the triangle:

$$J_{2,k}(x, y, \dot{\mathbf{A}}^k) = -\sigma_k \sum_{i=1}^{3} \dot{A}_i^k \alpha_i^k(x, y). \tag{5.17}$$

This term is representative of induced currents due to a changing magnetic field. Hence, we have (see Problem 3.5 on page 308)

$$\iint_{\Delta_k} J_{2,k}(x, y, \dot{\mathbf{A}}^k)\, A_k(x, y, \mathbf{A}^k)\, dx\, dy = (\mathbf{J}_2^k)^\top \mathbf{T}^k \mathbf{A}^k, \tag{5.18}$$

where

$$\mathbf{J}_2^k = -\sigma_k \begin{bmatrix} \dot{A}_1^k & \dot{A}_2^k & \dot{A}_3^k \end{bmatrix}^\top \tag{5.19}$$

and

$$\mathbf{T}^k = \frac{\Delta_k}{12} \begin{bmatrix} 2 & 1 & 1 \\ 1 & 2 & 1 \\ 1 & 1 & 2 \end{bmatrix}. \tag{5.20}$$

We thus conclude that the contribution to the functional from the kth element (in a ferromagnetic material) is

$$I_k(\mathbf{A}^k) = m_k(B_k^2(\mathbf{A}^k))\,\Delta_k - \frac{J_{1,k}\Delta_k}{3} \begin{bmatrix} 1 & 1 & 1 \end{bmatrix} \mathbf{A}^k - (\mathbf{J}_2^k)^\top \mathbf{T}^k \mathbf{A}^k. \tag{5.21}$$

It may be important in a given application to model the presence of eddy currents in permanent magnets (PMs) (see §3.7 on page 293 for general PM modeling considerations). In this case, the energy density function m_k is given by Equation (3.193), and $J_{1,k} = 0$, but the $J_{2,k}$ term of Equation (5.21) remains the same.

Therefore, a general expression for **the contribution to the gradient from the kth triangle** is (cf. Equations (3.154) on page 274 and (3.197) on page 293):

$$\boxed{\mathbf{g}^k = \nu_k \mathbf{S}^k \mathbf{A}^k - \frac{\nu_k}{2}(B^k_{rx}\mathbf{r}^k - B^k_{ry}\mathbf{q}^k) - \frac{J_{1,k}\Delta_k}{3}\begin{bmatrix}1 & 1 & 1\end{bmatrix}^{\mathrm{T}} + \sigma_k \mathbf{T}^k \dot{\mathbf{A}}^k\,.} \tag{5.22}$$

The contribution to the Hessian is identical to Equation (3.168) on page 278. After the full gradient vector is assembled, we have to solve an equation of the form

$$\boxed{\mathbf{g} = \mathbf{T}\dot{\mathbf{A}} + \mathbf{S}\mathbf{A} - \mathbf{b} = \mathbf{0}\,.} \tag{5.23}$$

The matrix $\mathbf{T}$ collects element-wise contributions from $\sigma_k \mathbf{T}^k$ matrices, similarly to how the global stiffness matrix $\mathbf{S}$ collects contributions of $\nu_k \mathbf{S}^k$ terms. The vector $\mathbf{b}$ collects contributions from exogenous currents and PMs. This is a **differential-algebraic system of equations (DAE)**. In a DAE, in contrast to a full system of ordinary differential equations (ODEs), we do not have equations for all the derivatives. Here, we only have MVP derivatives for the vertices of *conductor* triangles. In other words, the sparse matrix $\mathbf{T}$ will have a number of all-zero rows. Mathematically, we would say that $\mathbf{T}$ has low rank. Since $\mathbf{S} = \mathbf{S}(\mathbf{A})$ if we are modeling magnetic nonlinearity, we may have to solve a nonlinear DAE.

In what follows, we will explore two ways to solve this equation. Our first approach will be in the sinusoidal steady state, which would be a valid approach for linear magnetics, where the DAE can be solved using phasors. Then, we will explain how the DAE can be solved by time stepping, which would be the most general approach that can handle nonlinear materials and time-domain transients.

5.3 Eddy Current Problems with Analytical Solutions

Before proceeding with FEA studies, it is instructive to study various eddy current problems that are known to have an analytical solution. The examples that we will present will involve relatively simple 1-D and 2-D geometries, and should be helpful for acquiring a deeper conceptual understanding of induced currents. For a more in-depth treatment of such problems, the reader may refer to [1–3].

5.3.1 Eddy Currents in a Semi-Infinite Plate

We consider a semi-infinite conductive plate under the influence of an external time-varying magnetic field. This is essentially a 1-D geometry, where the plate is centered on the x–y plane and variables only change along the z-axis, as shown in Figure 5.1 (identical to Figure 2.21). If the plate is relatively thin, we are looking at a lamination of an electric machine; however, the analysis that follows applies in general regardless of thickness. For consistency with the previous analysis, we keep the same orientation for the coordinate system, as if laminations are stacked along the z-axis (as in a 2-D FEA).

Our model assumes that $\partial/\partial x = 0$ and $\partial/\partial y = 0$, as if the plate extends infinitely

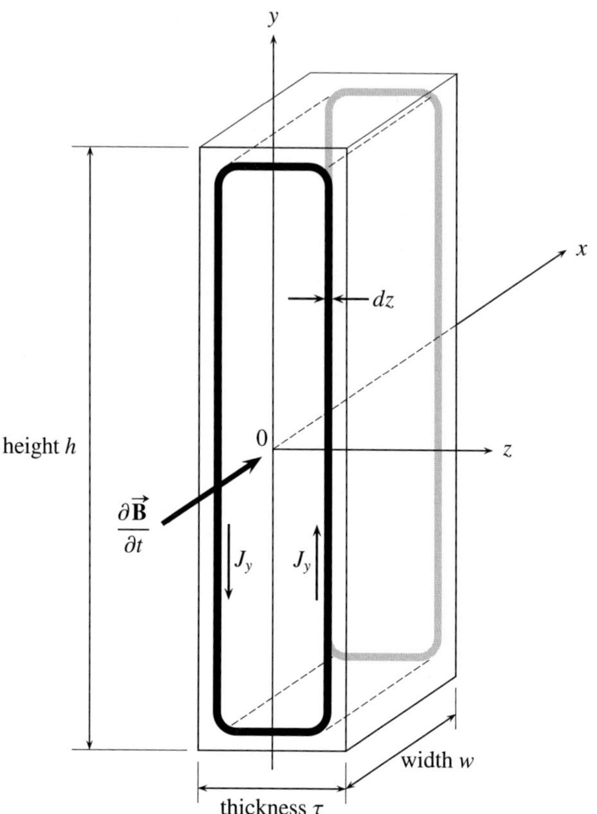

Figure 5.1 3-D view of a conductive plate for eddy current analysis under the influence of an external magnetic field.

along the x- and y-axes. In our earlier analysis of eddy currents (in the context of a thin lamination for the calculation of eddy current loss in §2.3.5), we had ignored the impact of the magnetic field created by the induced currents themselves. We will now relax this simplifying assumption, and will seek a more accurate solution of the magnetic field. Hence, we will allow the magnetic field to vary along the z-direction over the plate cross-section, $z \in [-\tau/2, \tau/2]$.

Combining $\nabla \times \vec{\mathbf{E}} = -\partial \vec{\mathbf{B}}/\partial t$ with the relationships $\vec{\mathbf{B}} = \mu_0 \mu_r \vec{\mathbf{H}}$ and $\vec{\mathbf{J}} = \sigma \vec{\mathbf{E}}$, we obtain (cf. Equation (2.170))

$$\frac{\partial J_y}{\partial z} = \sigma \mu_0 \mu_r \frac{\partial H_x}{\partial t}, \tag{5.24}$$

where σ and μ_r are the material conductivity and relative permeability, respectively. This equation relates the rate of change of the magnetic field (with respect to time) to the rate of change of the induced current density within the plate (with respect to position). In our previous approach to this problem, we had assumed that the magnetic field (and its dependence on time) is known within the plate, so we were able to solve this first-order equation directly for the variable on the left-hand side. However, since now we are accounting for the magnetic field produced by the induced currents, this equation all by itself is not sufficient. So, we also invoke Ampère's law, $\nabla \times \vec{\mathbf{H}} = \vec{\mathbf{J}}$, which was not strictly enforced in our prior approach. Hence

$$\frac{\partial H_x}{\partial z} = J_y \,, \tag{5.25}$$

thereby relating the magnetic field inside the plate to the induced current. Differentiating with respect to z, we obtain

$$\frac{\partial^2 H_x}{\partial z^2} = \frac{\partial J_y}{\partial z} \,. \tag{5.26}$$

Combining this equation with (5.24) yields the partial differential equation (PDE)

$$\frac{\partial^2 H_x}{\partial z^2} = \sigma \mu_0 \mu_r \frac{\partial H_x}{\partial t} \,, \tag{5.27}$$

which is called a **diffusion equation**. We shall impose the boundary conditions

$$H_x(-\tau/2, t) = H_x(\tau/2, t) = H_{x,\mathrm{ext}}(t) \,, \tag{5.28}$$

based on a given, externally imposed magnetic field.

Solution for Sinusoidal Excitation Using Phasors

Suppose that the external excitation is sinusoidal with respect to time:

$$H_{x,\mathrm{ext}}(t) = H_{\max} \cos \omega t \,. \tag{5.29}$$

We seek a steady-state solution of the form

$$H_x(z,t) = H_{x0}(z) \cos(\omega t + \phi(z)) \,, \tag{5.30}$$

where the amplitude H_{x0} and phase ϕ are functions of position. To proceed, it is convenient to introduce **phasors**. A phasor appears if we write the solution as the real part of a complex expression as follows:

$$H_x(z,t) = \mathrm{Re}\left\{H_{x0}(z)\, e^{j\phi(z)}\, e^{j\omega t}\right\} = \mathrm{Re}\left\{\tilde{H}_x(z)\, e^{j\omega t}\right\} , \tag{5.31}$$

where we denoted the phasor using a tilde:

$$\tilde{H}_x(z) = H_{x0}(z)\, e^{j\phi(z)} \,. \tag{5.32}$$

Note that in this definition of a phasor, the magnitude is associated with the peak value of the waveform. (Alternatively, a phasor can be defined based on the r.m.s. value of the signal.) The explicit dependence on time can thus be eliminated, and the PDE (5.27) becomes a second-order complex ODE in terms of the phasor:

$$\frac{d^2 \tilde{H}_x}{dz^2} = j\omega\sigma\mu_0\mu_r \tilde{H}_x = a^2 \tilde{H}_x \,, \tag{5.33}$$

where we defined a constant

$$a = \sqrt{j\omega\sigma\mu_0\mu_r} = \sqrt{j} \cdot \sqrt{\omega\sigma\mu_0\mu_r} = \frac{1+j}{\sqrt{2}} \sqrt{\omega\sigma\mu_0\mu_r} \,. \tag{5.34}$$

The general solution of Equation (5.33) is

$$\tilde{H}_x(z) = C_1 e^{az} + C_2 e^{-az} . \tag{5.35}$$

The complex parameter a that appears in the exponents of the PDE solution is customarily expressed in terms of a real-valued parameter

$$\delta = \frac{1+j}{a} = \frac{\sqrt{2}e^{j\pi/4}}{a} , \tag{5.36}$$

or

$$\boxed{\delta = \sqrt{\frac{2}{\omega\sigma\mu_0\mu_r}} .} \tag{5.37}$$

This is called the **skin depth** (in meters). Hence

$$e^{\pm az} = e^{\pm(1+j)z/\delta} . \tag{5.38}$$

Therefore, the skin depth has the physical significance of a distance (in the positive or negative direction, accordingly) that leads to an attenuation of each component of the solution by a factor $1/e \approx 0.368$, as well as a phase change of $-90°$. This variation of current magnitude with distance leads to the so-called **skin effect**. Simply speaking, as we shall see from the solution that we will obtain immediately afterwards, current tends to accumulate closer to the conductor surface as the frequency increases.

The skin depth depends on the material properties and the frequency of the applied excitation. Typical values for a few commonly used materials in electric machines are illustrated in Figure 5.2. Note that the skin depth of steel, due to its high permeability, is lower than that of copper and aluminum. For an operating frequency of 50–60 Hz, electrical steel has $\delta \approx 1.5$ mm, which justifies the use of laminations (an order of magnitude thinner) to form magnetic cores. On the other hand, copper and aluminum have $\delta \approx 10$ mm, which signifies that we should avoid using solid conductors of larger dimensions since we would be underutilizing the material.

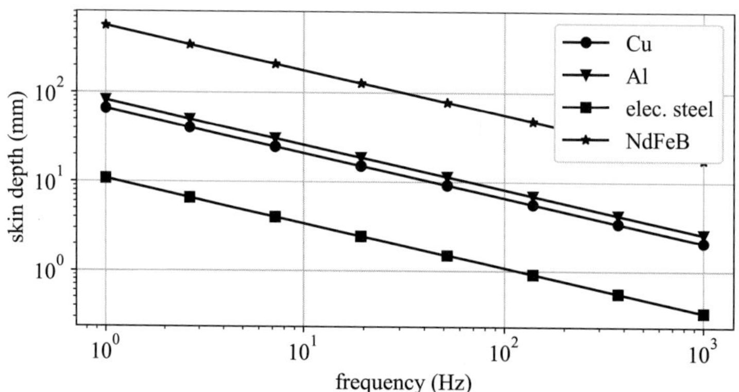

Figure 5.2 Skin depth for various materials as a function of frequency.

Now let us continue with the solution. Applying boundary conditions for $z = \pm\tau/2$:

$$C_1 e^{-a\tau/2} + C_2 e^{a\tau/2} = C_1 e^{a\tau/2} + C_2 e^{-a\tau/2} = H_{\max}\,, \tag{5.39}$$

leading to

$$(C_1 + C_2)(e^{a\tau/2} + e^{-a\tau/2}) = 2H_{\max}\,, \tag{5.40a}$$

$$(C_1 - C_2)(e^{a\tau/2} - e^{-a\tau/2}) = 0\,. \tag{5.40b}$$

Equivalently, using a hyperbolic cosine for notational expediency:

$$C_1 = C_2 = \frac{H_{\max}}{2\cosh(a\tau/2)}\,. \tag{5.41}$$

Therefore, the magnetic field solution is

$$\tilde{H}_x(z) = H_{\max}\frac{\cosh az}{\cosh(a\tau/2)}\,. \tag{5.42}$$

The dependence on a hyperbolic cosine (of a complex number) implies that the magnetic field is attenuated inside the plate, reaching its smallest magnitude at the middle. We obtain the current density by differentiating the magnetic field as per Equation (5.25) written in phasor form:

$$\tilde{J}_y(z) = aH_{\max}\frac{\sinh az}{\cosh(a\tau/2)}\,. \tag{5.43}$$

Note that $\tilde{J}_y(-z) = -\tilde{J}_y(z)$, which means that the current circulates within the plate, as shown in Figure 5.1, attaining its maximum value at the surfaces.

Demagnetizing Impacts of Eddy Currents

If the plate has height h along the y-axis, we can calculate the flux phasor by integrating Equation (5.42):

$$\tilde{\Phi} = h\int_{-\tau/2}^{\tau/2} \tilde{B}_x(z)\,dz = \tau h B_{\max}\frac{\tanh(a\tau/2)}{a\tau/2}\,, \tag{5.44}$$

where $B_{\max} = \mu_0\mu_r H_{\max}$. If $\tau \ll \delta$, this equation implies that

$$\tilde{\Phi} \approx \tau h B_{\max}\,, \tag{5.45}$$

indicating that the flux created by the induced currents is negligible. However, if δ is comparable to τ (or smaller), then the flux gets limited by the action of the induced currents, which increasingly resist the magnetic field (Lenz's law). We can define an **equivalent complex permeability** for the overall behavior of the plate as

$$\mu_e = \frac{\tilde{B}_e}{H_{\max}} = \frac{\tilde{\Phi}}{\tau h H_{\max}} = \mu_0\mu_r\frac{\tanh(a\tau/2)}{a\tau/2}\,, \tag{5.46}$$

as if there is an equivalent uniform flux density $\tilde{B}_e = \tilde{\Phi}/\tau h$. The magnitude and phase of the normalized quantity $\mu_e/\mu_0\mu_r$ are plotted in Figure 5.3. It can be seen that for plates as thick as roughly $\tau \approx 0.2\delta$, the impact of eddy currents is negligible. For increased thickness beyond $\tau \approx \delta$, the permeability magnitude (and, hence, the flux through the plate) drops rapidly, and the ability of the plate to carry flux quickly deteriorates, due to the action of the eddy currents that oppose the external magnetic field.

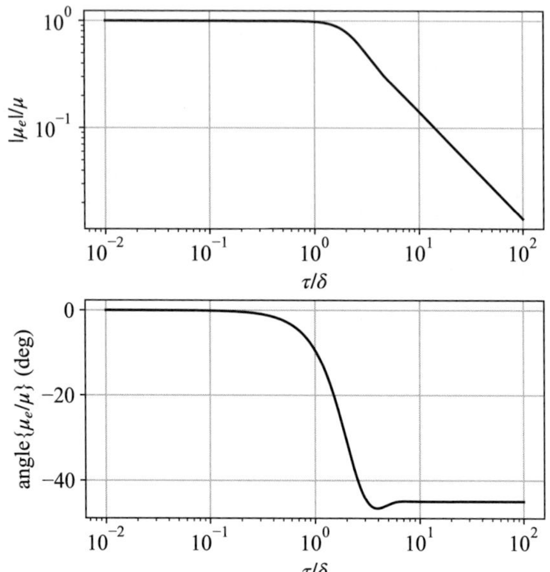

Figure 5.3 Semi-infinite plate complex permeability magnitude and phase.

Loss Calculations

The calculation of loss within the semi-infinite plate is based on integrating the Joule loss density, J_y^2/σ, over time and space. This was undertaken earlier in §2.3.5 based on a simplified expression for the induced current density, whereas now we have an accurate solution based on the diffusion equation.

At each point within the plate, the time-average of the loss density is $|\tilde{J}_y(z)|^2/2\sigma$. We can show using algebraic manipulations of Equation (5.43)[b] that the square of the current density peak value (i.e., the square of the phasor magnitude) is

$$|\tilde{J}_y(z)|^2 = 2\left(\frac{H_{\max}}{\delta}\right)^2 \frac{\cosh(2z/\delta) - \cos(2z/\delta)}{\cosh(\tau/\delta) + \cos(\tau/\delta)}\,. \tag{5.47}$$

To calculate the overall eddy current loss, we integrate this expression analytically over the spatial coordinate. This leads to the loss per unit surface area of the plate (wh). Integrating:

$$\overline{p_{\text{eddy}}} = \frac{1}{2\sigma}\int_{-\tau/2}^{\tau/2} |\tilde{J}_y(z)|^2\,dz = \frac{H_{\max}^2}{\sigma\delta}\,\frac{\sinh(\tau/\delta) - \sin(\tau/\delta)}{\cosh(\tau/\delta) + \cos(\tau/\delta)}\quad (\text{W/m}^2)\,. \tag{5.48}$$

In laminations where $\tau \ll \delta$, the Taylor expansion of the trigonometric terms yields

$$\overline{p_{\text{eddy}}} \approx \frac{H_{\max}^2}{\sigma\delta}\,\frac{(\tau/\delta)^3}{6} = \frac{\sigma\pi^2 f^2 \tau^3 B_{\max}^2}{6}\quad (\text{W/m}^2)\,, \tag{5.49}$$

[b] These manipulations are based on trigonometric and hyperbolic function identities, such as

$$\sinh(x + jy) = \sinh x \cos y + j\cosh x \sin y\,,$$
$$2\sinh^2 x = \cosh 2x - 1\,.$$

which is essentially identical to Equation (2.179). In the opposite case, where $\tau \gg \delta$, the loss attains the value

$$\overline{p_{\text{eddy}}} \approx \frac{H_{\text{max}}^2}{\sigma\delta} \quad (\text{W/m}^2)\,. \tag{5.50}$$

An alternate method for calculating loss is based on Poynting's theorem (see §2.3.1, page 121). The instantaneous power absorbed by the plate can be found as the surface integral of the Poynting vector $-\vec{\mathbf{S}} = -\vec{\mathbf{E}} \times \vec{\mathbf{H}}$ over its two sides. On the plate surface, the electric field is $\vec{\mathbf{E}}(\pm\tau/2) = \vec{\mathbf{J}}(\pm\tau/2)/\sigma$. The outwards unit normal is $\hat{\mathbf{n}} = -\hat{\mathbf{k}}$ and $\hat{\mathbf{n}} = \hat{\mathbf{k}}$, at $z = -\tau/2$ and $z = \tau/2$, respectively. Therefore

$$-\vec{\mathbf{S}}(\pm\tau/2)\cdot\hat{\mathbf{n}}(\pm\tau/2) = -\frac{1}{\sigma}[J_y(\pm\tau/2)\,\hat{\mathbf{j}} \times H_x(\pm\tau/2)\,\hat{\mathbf{i}}]\cdot(\pm\hat{\mathbf{k}}) \tag{5.51a}$$

$$= \pm\frac{1}{\sigma}J_y(\pm\tau/2)H_x(\pm\tau/2)\,. \tag{5.51b}$$

By virtue of the odd symmetry of the current and the identical boundary conditions in this case, the total (i.e., accounting for both sides) instantaneous power absorption per unit surface area is

$$p_{\text{eddy}} = \frac{2}{\sigma}J_y(\tau/2)H_x(\tau/2) = 2E_y(\tau/2)H_x(\tau/2)\,. \tag{5.52}$$

This is the product of two sinusoidal quantities. The phase of the surface magnetic field is zero, serving as the reference in our solution. However, the phase of the surface current density (and thus the electric field) is determined by Equation (5.43):

$$\tilde{E}_y(\tau/2) = E_0\angle\phi_E\,. \tag{5.53}$$

Therefore, the time-average eddy current loss (per unit surface area) is

$$\overline{p_{\text{eddy}}} = 2\cdot\left(\frac{1}{2}E_0H_{\text{max}}\cos\phi_E\right) \quad (\text{W/m}^2)\,. \tag{5.54}$$

This expression may be rewritten in terms of the surface phasors as

$$\overline{p_{\text{eddy}}} = 2\cdot\text{Real}\left\{\frac{1}{2}\tilde{E}_y\tilde{H}_x^*\right\}_{z=\tau/2} \quad (\text{W/m}^2)\,, \tag{5.55}$$

where the asterisk denotes complex conjugate. This is reminiscent of a similar expression involving $\tilde{V}\tilde{I}^*$ in a.c. circuit analysis. It can be shown that this yields the same result as Equation (5.48).

5.3.2 Elliptical Model of Magnetic Hysteresis

As we explained in §2.3.5 on page 152, hysteresis literally means in Greek that something is lagging behind. In magnetics, it is the flux density B that lags behind the magnetic field intensity H. Under sinusoidal steady-state conditions, this physical effect can be modeled in a relatively simple yet approximate way by assuming that the phasors (oriented along the x-axis in our case study) are related as

$$\tilde{B}_x = \mu_0\mu_r e^{-j\theta_h}\tilde{H}_x = \mu_h\tilde{H}_x\,. \tag{5.56}$$

Here, θ_h is a **hysteresis angle**, meaning that the B-field lags the H-field by θ_h/ω seconds in the time domain. We are thus ignoring harmonics, assuming that both fields are sinusoidal. The **hysteretic permeability** μ_h is a complex number.

This model leads to an elliptical B–H curve, which is an approximation of the real hysteresis loop, like the one depicted in Figure 2.22 on page 153. To derive the equation of the ellipse, it is convenient to find the locus of points on an H vs. B/μ plane, where we have scaled the ordinate (vertical) axis by $\mu = \mu_0\mu_r$. On this modified coordinate system, it can be readily shown that the locus forms a rotated ellipse, whose major axis is directed along the 45° line, and minor axis is along the 135° line. This is shown in Figure 5.4(a). Increasing θ_h makes the ellipse larger; however, the direction of the axes remains fixed. This ellipse has a width of $2\sqrt{2}H_{\max}\cos(\theta_h/2)$ along the major axis, and a height of $2\sqrt{2}H_{\max}\sin(\theta_h/2)$ along the minor axis. Figure 5.4(b) shows what the ellipse looks like on the original B–H plane. (It still looks like an ellipse because this is a linear transformation.) As we know, the ellipse area bears the physical meaning of hysteresis loss per unit volume. Hence, this model suggests that the time-average hysteresis loss is

$$\overline{p_{\text{hyst}}} = \frac{\pi \sin\theta_h}{\mu_0\mu_r} f B_{\max}^2 \quad (\text{W/m}^3)\,. \tag{5.57}$$

This model yields a quadratic dependence on the maximum flux density $B_{\max} = \mu H_{\max}$, whereas the original Steinmetz equation (2.187) (which was established experimentally) uses an exponent of 1.6. Note that since the magnetic field is not uniform inside the plate, Equation (5.57) represents the pointwise loss density, and cannot be used directly by a multiplication with the volume (unless the effect of eddy currents is negligible). Instead, we must rely on the field solution.

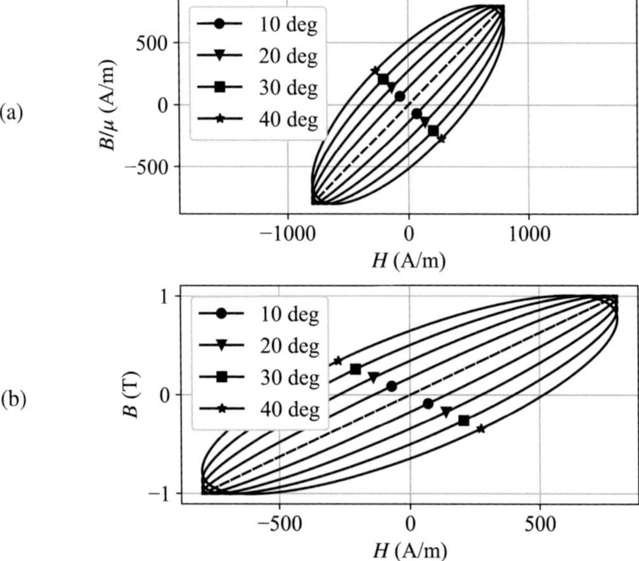

Figure 5.4 Elliptical hysteresis curve approximations, parametrized by the hysteresis angle θ_h.

We may repeat the analysis of §5.3.1 using μ_h in lieu of μ. Equation (5.33) becomes

$$\frac{d^2\tilde{H}_x}{dz^2} = j\omega\sigma\mu_h\tilde{H}_x = b^2\tilde{H}_x\,, \tag{5.58}$$

with a new constant

$$b = ae^{-j\theta_h/2} = \frac{\sqrt{2}}{\delta}e^{j(\pi/4-\theta_h/2)} = b_{\text{re}} + jb_{\text{im}}\,. \tag{5.59}$$

The current density is thus (cf. Equation (5.43))

$$\tilde{J}_y(z) = bH_{\text{max}}\frac{\sinh bz}{\cosh(b\tau/2)}\,. \tag{5.60}$$

It can be readily shown that (cf. Equation (5.47))

$$|\tilde{J}_y(z)|^2 = 2\left(\frac{H_{\text{max}}}{\delta}\right)^2\frac{\cosh(2b_{\text{re}}z) - \cos(2b_{\text{im}}z)}{\cosh(b_{\text{re}}\tau) + \cos(b_{\text{im}}\tau)}\,, \tag{5.61}$$

which can be integrated for an analytical calculation of the eddy current loss.

The total loss in the plate (eddy current plus hysteresis loss) can be obtained by application of Poynting's theorem, using a formula like (5.55). For reference, these loss components (per unit surface area) are

$$\overline{p_{\text{eddy}}} = \frac{H_{\text{max}}^2}{\sqrt{2}\sigma\delta}\,\frac{\sinh(b_{\text{re}}\tau)/\cos\phi_b - \sin(b_{\text{im}}\tau)/\sin\phi_b}{\cosh(b_{\text{re}}\tau) + \cos(b_{\text{im}}\tau)} \quad (\text{W/m}^2) \tag{5.62}$$

and

$$\overline{p_{\text{total}}} = \frac{\sqrt{2}H_{\text{max}}^2}{\sigma\delta}\,\frac{\sinh(b_{\text{re}}\tau)\cos\phi_b - \sin(b_{\text{im}}\tau)\sin\phi_b}{\cosh(b_{\text{re}}\tau) + \cos(b_{\text{im}}\tau)} \quad (\text{W/m}^2)\,, \tag{5.63}$$

where $\phi_b = \text{angle}\{b\} = \pi/4 - \theta_h/2$.

The magnetic field is given by

$$\tilde{H}_x(z) = H_{\text{max}}\frac{\cosh bz}{\cosh(b\tau/2)}\,. \tag{5.64}$$

Hence, we can calculate the flux phasor by integrating Equation (5.64):

$$\tilde{\Phi} = h\int_{-\tau/2}^{\tau/2}\tilde{B}_x(z)\,dz = \tau hB_{\text{max}}e^{-j\theta_h}\frac{\tanh(b\tau/2)}{b\tau/2}\,. \tag{5.65}$$

The equivalent permeability is

$$\mu_e = \frac{\tilde{B}_e}{H_{\text{max}}} = \frac{\tilde{\Phi}}{\tau hH_{\text{max}}} = \mu_h\frac{\tanh(b\tau/2)}{b\tau/2}\,. \tag{5.66}$$

5.3.3 The Complex Poynting Vector

In the preceding case study, we calculated loss based on the Poynting vector $\vec{\mathbf{S}}$. First we found the instantaneous power flow into the plate, which was then averaged for the time-average loss (i.e., eddy current plus hysteresis loss, if present). When a piece of equipment is operating in the sinusoidal steady-state, it is convenient to introduce the complex Poynting vector.

We define **complex vector fields** based on phasors, for example

$$\tilde{\mathbf{E}} = \tilde{E}_x \hat{\mathbf{i}} + \tilde{E}_y \hat{\mathbf{j}} + \tilde{E}_z \hat{\mathbf{k}} \quad \text{and} \quad \tilde{\mathbf{H}} = \tilde{H}_x \hat{\mathbf{i}} + \tilde{H}_y \hat{\mathbf{j}} + \tilde{H}_z \hat{\mathbf{k}} . \tag{5.67}$$

The **complex Poynting vector** is defined as

$$\mathbf{S} = \tilde{\mathbf{E}} \times \tilde{\mathbf{H}}^* . \tag{5.68}$$

Note that the formula involves the complex conjugate of the magnetic field. The complex Poynting vector is a complex number but not a phasor because the product of the electric and magnetic field phasors is not representative of a sinusoidally varying physical quantity. So, we do not use a tilde over **S**.

In the quasi-magnetostatic case, Poynting's theorem (see Equation (2.89) on page 122) is expressed as

$$-\oiint_{\partial\mathcal{D}} \vec{\mathbf{S}} \cdot d\mathbf{a} = \iiint_{\mathcal{D}} \frac{1}{\sigma} \vec{\mathbf{J}} \cdot \vec{\mathbf{J}}\, dv + \iiint_{\mathcal{D}} \vec{\mathbf{H}} \cdot \frac{\partial \vec{\mathbf{B}}}{\partial t}\, dv . \tag{5.69}$$

Recall that this result was obtained by elementary vector calculus manipulations starting from $\nabla \cdot \vec{\mathbf{S}}$. The surface integral on the left-hand side, which is obtained by application of Gauss' law over the volume $\mathcal{D}$, is positive in the outward direction; but due to the presence of the minus sign, it represents the flow of power inward, which becomes Joule loss plus power absorbed by the magnetic field.

Identical operations can be repeated using the complex vectors instead of the physical fields starting from $\nabla \cdot \mathbf{S}$. This process begins by applying the vector calculus identity

$$\nabla \cdot \mathbf{S} = \nabla \cdot (\tilde{\mathbf{E}} \times \tilde{\mathbf{H}}^*) = \tilde{\mathbf{H}}^* \cdot (\nabla \times \tilde{\mathbf{E}}) - \tilde{\mathbf{E}} \cdot (\nabla \times \tilde{\mathbf{H}}^*) . \tag{5.70}$$

Substituting the curls by Maxwell's equations in phasor form (ignoring displacement current):

$$\nabla \times \tilde{\mathbf{E}} = -j\omega \tilde{\mathbf{B}} \quad \text{and} \quad \nabla \times \tilde{\mathbf{H}}^* = \tilde{\mathbf{J}}^* , \tag{5.71}$$

and using constitutive laws for an isotropic material:

$$\tilde{\mathbf{B}} = \mu_h \tilde{\mathbf{H}} \quad \text{and} \quad \tilde{\mathbf{J}} = \sigma \tilde{\mathbf{E}} , \tag{5.72}$$

we have

$$\nabla \cdot \mathbf{S} = -j\omega\mu_h \tilde{\mathbf{H}} \cdot \tilde{\mathbf{H}}^* - \frac{1}{\sigma} \tilde{\mathbf{J}} \cdot \tilde{\mathbf{J}}^* . \tag{5.73}$$

The permeability

$$\mu_h = \mu_0 \mu_r e^{-j\theta_h} = \mu_{\mathrm{re}} + j\mu_{\mathrm{im}} , \tag{5.74}$$

with $\mu_{\mathrm{re}} > 0$ and $\mu_{\mathrm{im}} \leq 0$, could be complex if we incorporate hysteretic effects using an elliptical model. Note that the material is essentially magnetically linear (i.e., it does not saturate) since μ_r is constant. Applying the divergence theorem and dividing by 2:

$$-\frac{1}{2} \oiint_{\partial\mathcal{D}} \mathbf{S} \cdot d\mathbf{a} = \iiint_{\mathcal{D}} \frac{j\omega\mu_h}{2} \tilde{\mathbf{H}} \cdot \tilde{\mathbf{H}}^*\, dv + \iiint_{\mathcal{D}} \frac{1}{2\sigma} \tilde{\mathbf{J}} \cdot \tilde{\mathbf{J}}^*\, dv . \tag{5.75}$$

Separating real and imaginary parts:

$$-\frac{1}{2}\text{Real}\left\{\oiint_{\partial\mathcal{D}} \mathbf{S}\cdot d\mathbf{a}\right\} = \iiint_{\mathcal{D}} \frac{1}{2\sigma}\tilde{\mathbf{J}}\cdot\tilde{\mathbf{J}}^*\,dv + \iiint_{\mathcal{D}} \frac{\omega\cdot(-\mu_{\text{im}})}{2}\tilde{\mathbf{H}}\cdot\tilde{\mathbf{H}}^*\,dv \tag{5.76a}$$

and

$$-\frac{1}{2}\text{Imag}\left\{\oiint_{\partial\mathcal{D}} \mathbf{S}\cdot d\mathbf{a}\right\} = \iiint_{\mathcal{D}} \frac{\omega\mu_{\text{re}}}{2}\tilde{\mathbf{H}}\cdot\tilde{\mathbf{H}}^*\,dv\,. \tag{5.76b}$$

Hence, Equation (5.76a) implies that *integrating one-half the real part of the complex Poynting vector over the boundary of $\mathcal{D}$ yields the time-average Joule plus hysteresis loss inside $\mathcal{D}$*. The hysteresis loss is determined by the imaginary part of the complex permeability (cf. Equation (5.57)).

But what meaning can we ascribe to Equation (5.76b) for the imaginary part? To this end, we evaluate the time-average of $BH/2$ over a period $T = 2\pi/\omega$, which is

$$\overline{m} = \frac{1}{T}\int_{t_0}^{t_0+T} \frac{1}{2}\vec{\mathbf{B}}(t)\cdot\vec{\mathbf{H}}(t)\,dt = \frac{1}{4}\mu_{\text{re}}\,\tilde{\mathbf{H}}\cdot\tilde{\mathbf{H}}^*\,, \tag{5.77}$$

for arbitrary t_0. We call this quantity a **time-average stored magnetic energy density**, even though the presence of hysteresis implies that we may not be able to recover it fully from the field. Integrating over the volume, we obtain the **time-average stored magnetic energy**

$$\overline{W_f} = \frac{1}{4}\iiint_{\mathcal{D}} \mu_{\text{re}}\tilde{\mathbf{H}}\cdot\tilde{\mathbf{H}}^*\,dv\,. \tag{5.78}$$

Because of this relationship, the parameter μ_{re} is also called the **effective permeability**. Therefore, we conclude that *integrating one-half the imaginary part of the complex Poynting vector over the boundary of $\mathcal{D}$ equals 2ω times the average energy stored in the magnetic field within $\mathcal{D}$.*

The Relationship of Real Power, Reactive Power, and Impedance with the Complex Poynting Vector

The real and imaginary parts of the complex Poynting vector are analogous to the real and reactive power in a.c. electric circuits. In fact, the reader may recall from elementary electromagnetism that these powers are derived from application of Poynting's theorem over a volume surrounding an entire piece of equipment. Let us take a closer look.

In the case of a single-phase magnetically linear device operating under sinusoidal steady-state conditions, we define voltage and current phasors at the terminals as

$$\tilde{V} = V\underline{/\phi_v} \quad\text{and}\quad \tilde{I} = I\underline{/\phi_i}\,, \tag{5.79}$$

respectively (based on the peak values of the waveforms). Then, we define the **real and reactive power** absorbed by the circuit as the real and imaginary parts of the voltage and conjugate current phasor product:

$$P + jQ = \frac{1}{2}\tilde{V}\tilde{I}^*\,. \tag{5.80}$$

Therefore

$$P = \frac{1}{2} VI \cos(\phi_v - \phi_i)\,, \tag{5.81a}$$

$$Q = \frac{1}{2} VI \sin(\phi_v - \phi_i)\,. \tag{5.81b}$$

The angle

$$\phi = \phi_v - \phi_i \tag{5.82}$$

is called the **power factor angle** of the circuit. It can be readily shown that the real power P equals the time-average power consumed by the circuit:

$$P = \overline{p(t)} = \frac{1}{T} \int_{t_0}^{t_0+T} v(t)\, i(t)\, dt\,. \tag{5.83}$$

The physical meaning of the reactive power Q is somewhat more elusive, but it will be revealed based on the Poynting vector, as will be explained shortly.

The circuit **impedance** is defined as

$$Z = \frac{\tilde{V}}{\tilde{I}} = R + jX = R + j\omega L\,. \tag{5.84}$$

The **resistance** R and **reactance** X in this equation are the real and imaginary parts of Z. Here, the **inductance** is defined as $L = X/\omega$ since we are assuming that our device can only store magnetic energy (under quasi-magnetostatic conditions); otherwise, the reactance also includes a capacitive term. Both R and L are functions of frequency. To differentiate R from its d.c. value, we may also call it the **a.c. resistance** and denote it by R_{ac}. Therefore, substituting $\tilde{V} = Z\tilde{I}$ in Equation (5.80):

$$P + jQ = \frac{1}{2} Z\tilde{I}\tilde{I}^* = \frac{1}{2} Z\,|\tilde{I}|^2\,, \tag{5.85}$$

or

$$P = \frac{1}{2} R\,|\tilde{I}|^2\,, \tag{5.86a}$$

$$Q = \frac{1}{2} X\,|\tilde{I}|^2\,. \tag{5.86b}$$

These circuit variables are related to the complex Poynting vector. Let us surround the device by a closed surface, as shown in Figure 5.5, which is a simplified 2-D representation of the 3-D physical reality. The surface is far away from the device, where fields are negligible. Furthermore, in the frequency regime of interest, there is no electromagnetic radiation of power, so the integration of Poynting's vector on the bottom, right, and top surfaces amounts to zero. In particular, the surface on the left is taken to be perpendicular to the terminal conductors. Hence, the MVP due to current flowing in the terminal conductors is also normal to this surface (ignoring external fields).

We can express the Poynting vector as

$$\mathbf{S} = \tilde{\mathbf{E}} \times \tilde{\mathbf{H}}^* = -(\nabla\tilde{\varphi} + j\omega\tilde{\mathbf{A}}) \times \tilde{\mathbf{H}}^*\,, \tag{5.87}$$

introducing the phasors of the electric potential and the MVP. Since $\tilde{\mathbf{A}} \times \tilde{\mathbf{H}}^*$ is a vector

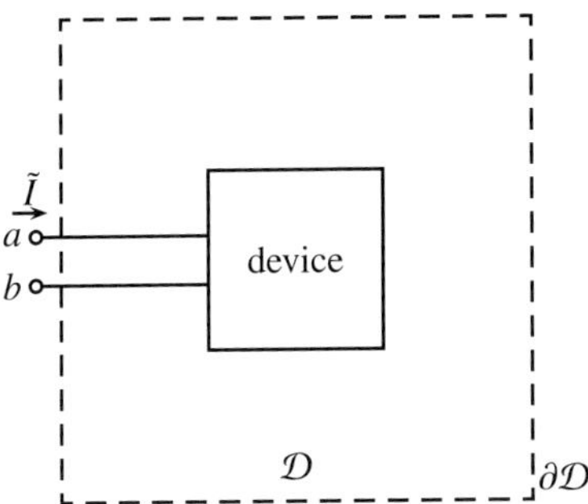

Figure 5.5 Closed surface surrounding a single-phase magnetic device. The surface is penetrated by the terminal conductors for application of Poynting's theorem.

that is tangent to the left surface (by virtue of the perpendicularity of the MVP to this plane), its integral over the left surface is zero. Even if this is not exactly the case, the frequency is sufficiently low so that $j\omega\tilde{\mathbf{A}}$ is negligible. So, for all practical purposes, on the left surface we can approximate

$$\mathbf{S} = \tilde{\mathbf{E}} \times \tilde{\mathbf{H}}^* \approx -\nabla\tilde{\varphi} \times \tilde{\mathbf{H}}^* . \tag{5.88}$$

Applying the vector calculus identity (1.160), we have

$$\mathbf{S} = \tilde{\varphi}\, \nabla \times \tilde{\mathbf{H}}^* - \nabla \times (\tilde{\varphi}\tilde{\mathbf{H}}^*) , \tag{5.89}$$

or equivalently

$$\mathbf{S} = \tilde{\varphi}\, \tilde{\mathbf{J}}^* - \nabla \times (\tilde{\varphi}\tilde{\mathbf{H}}^*) . \tag{5.90}$$

The surface integral of the second term can be found using Stokes' theorem: it becomes a line integral of $\tilde{\varphi}\tilde{\mathbf{H}}^*$ on the boundary of the left surface, which evaluates to zero by virtue of the attenuation of the fields at that distance from the device and the terminals. Therefore, **the complex Poynting vector is related to the real and reactive power by**

$$\boxed{-\frac{1}{2}\iint_{\partial\mathcal{D}} \mathbf{S}\cdot d\mathbf{a} = \frac{1}{2}(\tilde{\varphi}_a - \tilde{\varphi}_b)\,\tilde{I}^* = \frac{1}{2}\tilde{V}\tilde{I}^* = P + jQ .} \tag{5.91}$$

As expected, the real power P is related to the integral of the real part of $\mathbf{S}$ over $\partial\mathcal{D}$ due to its physical significance. More interestingly, we may conclude that **the reactive power is related to the time-average magnetic field energy by**

$$\boxed{Q = \frac{1}{2}\omega L|\tilde{I}|^2 = 2\omega\overline{W_f} .} \tag{5.92}$$

Hence, the inductance is related to the time-average magnetic field energy by

$$\overline{W_f} = \frac{1}{4}L\,|\tilde{I}|^2 . \tag{5.93}$$

The circuit impedance can be obtained from a 2-D analytical or FEA solution, based on the surface integrals of Joule loss and field energy density, respectively. If the domain $\mathcal{D}$ extends to "infinity," that is, if we know the field everywhere in space, then we are accounting for the entire magnetic flux and energy. In this case, the calculation yields the total impedance of the device. In the examples that we have presented, $\mathcal{D}$ tightly encloses the device since we only know the field inside the plate and up to its surface. In such circumstances, the above calculation yields what is known as the **internal impedance** of the component because we are not accounting for the external magnetic flux and energy. For a more rigorous view of how electrical circuits can be developed from electromagnetic fields, the reader may consult physics texts, such as the ones listed at the end of Chapter 1.

Example 5.1 *Origins of the term "skin depth."*

Show that the skin depth can be associated with a notional layer (or skin) containing all the induced current (even though this is not what is really happening).

Solution

Suppose the plate is infinitely thick. In this case, the solution of Equation (5.33) has a single component

$$\tilde{H}_x(z) = H_{\text{max}} e^{-az} \,. \tag{5.94}$$

Here, we have moved the $z = 0$ plane to coincide with the plate surface so that the plate is where $z \geq 0$. In the time domain, the field is

$$H_x(z,t) = H_{\text{max}} e^{-z/\delta} \cos(\omega t - z/\delta) \,. \tag{5.95}$$

This equation describes a wave that propagates with velocity $\delta\omega$ m/s inside the material, while attenuating by $1/\delta$ Np/m (nepers[79] per meter).

The current density phasor is obtained by differentiation per Equation (5.25):

$$\tilde{J}_y(z) = -aH_{\text{max}} e^{-az} = \tilde{J}_{y0} e^{-az} \,, \tag{5.96}$$

where $\tilde{J}_{y0}$ is the surface current density. The total current phasor per unit width in the plate (assumed positive in the negative y-axis direction) is

$$\tilde{I}/w = -\int_0^\infty \tilde{J}_y(z)\, dz = -\frac{\tilde{J}_{y0}}{a} = -\frac{\tilde{J}_{y0}\,\delta}{\sqrt{2}} e^{-j\pi/4} = H_{\text{max}} \,. \tag{5.97}$$

Hence, the total induced current is in phase with the applied magnetic field, $\tilde{I} = I\underline{/0^\circ}$. One may interpret this as if there is a uniform distribution of the r.m.s. surface current density, $|\tilde{J}_{y0}|/\sqrt{2}$, over a layer of depth δ.

Moreover, the time-average eddy current loss per surface area (wh) can be calculated as

$$\overline{p_{\text{eddy}}} = \frac{1}{2\sigma}\int_0^\infty |\tilde{J}_y(z)|^2\, dz = \frac{H_{\text{max}}^2}{2\sigma\delta} = \frac{I^2}{2\sigma\delta w^2} = \left(\frac{I}{\sqrt{2}}\right)^2 R\frac{1}{wh} \,, \tag{5.98}$$

where we defined an equivalent resistance

$$R = \frac{h}{\sigma\delta w} \,. \tag{5.99}$$

This result may be interpreted as if all current is uniformly flowing within a skin of depth δ.

Example 5.2 *Frequency response of semi-infinite plate.*
A steel plate has thickness $\tau = 1$ cm, height $h = 1$ m, relative permeability $\mu_r = 1000$ (assumed constant), and conductivity $\sigma = 2 \cdot 10^6$ S/m. The plate forms part of a magnetic circuit that is driven by a coil with $N = 100$ turns. The coil is connected to a sinusoidal current source, capable of supplying variable-frequency current from d.c. to 1 kHz. At d.c. (and very low frequencies), the flux density through the plate is uniform, and the flux density-to-current ratio is $B_{\max}/i = 0.05$ T/A (flux and current are in phase). Hysteresis effects are ignored. Calculate the frequency response of the plate.

Solution
First, we examine the variation of magnetic field and current magnitude within the plate. These functions are plotted in Figure 5.6 for three excitation frequencies (10, 100, and 1000 Hz), assuming that the coil current is $i = 20$ A. The demagnetization of the material with frequency is obvious, even as low as 10 Hz. We can also observe the skin effect, which tends to push the induced current increasingly closer to the plate surfaces.

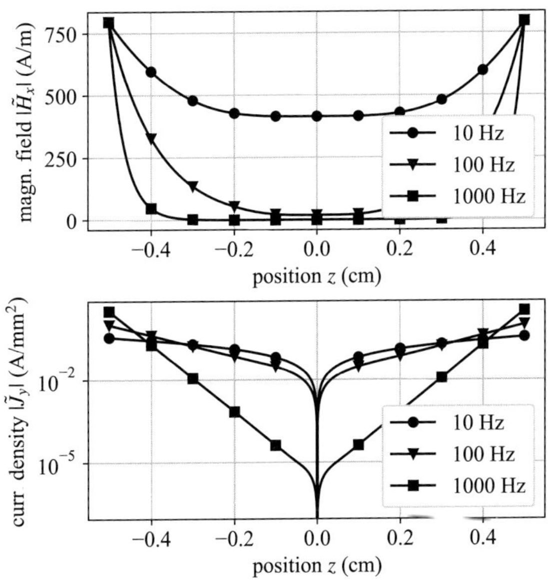

Figure 5.6 Semi-infinite plate magnetic field and current density.

In an equivalent circuit of this hypothetical device, the plate will correspond to a circuit element with impedance

$$Z = \frac{j\omega N\tilde{\Phi}}{\tilde{I}} = j\omega N\tau h \frac{\mu_e}{\mu_0\mu_r}\frac{B_{\max}}{i}, \tag{5.100}$$

using the equivalent complex permeability of Equation (5.46). The impedance is plotted in Figure 5.7. At low frequencies, the plate looks like an ideal inductor (whose response is indicated by the dashed line). At higher frequencies, the rate of increase of impedance magnitude reduces (on a $\log|Z|$ vs. $\log\omega$ plot, the slope drops by one half), and the power factor angle drops from 90° to 45°. This signifies the presence of a resistive component, corresponding to eddy current ohmic loss.

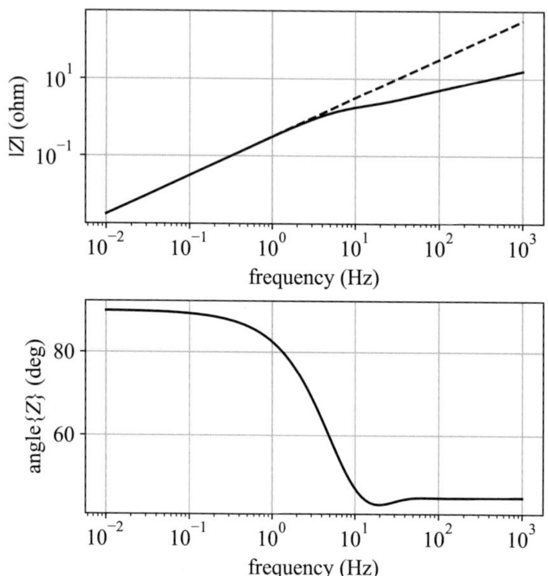

Figure 5.7 Semi-infinite plate impedance.

5.3.4 The Semi-infinite Plate as a Conductor

The semi-infinite plate can also be used as a simple model for a current-carrying rectangular conductor, such as a strip conductor or a busbar in electrical substations (i.e., any conductor with dimensions satisfying $\tau \ll w \ll h$). For electric machines, this model is useful for understanding the behavior of hairpin windings, which are being considered for electric vehicle traction motors [4].

The geometry is shown in Figure 5.8, which is essentially a rotated version of Figure 5.1 depicting the external circuit connection. A sinusoidal current of peak value I is flowing along the y-axis, imposed by an external circuit that connects at $y = \pm h/2$. The phase of the current is taken as the angle reference. The main problem is to determine the current density distribution within the conductor, that is, along the z-axis, neglecting variations along the other two axes. For consistency with the previous case study, we have maintained the relative orientation of the axes with respect to the plate.

The field obeys the same differential equation (5.33). However, the boundary conditions are now different. Applying Ampère's law around a rectangular path that tightly encloses the strip on an arbitrary x–z plane, and assuming there is no other current/magnetic field source nearby (i.e., ignoring the effect of the current flowing in the external

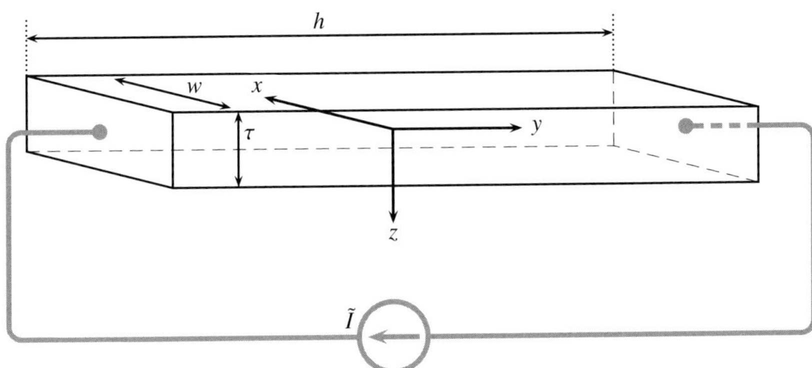

Figure 5.8 3-D view of a rectangular conductor connected to an external circuit.

circuit), we obtain

$$\tilde{H}_x(\tau/2) = -\tilde{H}_x(-\tau/2) = \frac{I}{2w} . \tag{5.101}$$

Therefore, the magnetic field solution is

$$\tilde{H}_x(z) = \frac{I}{2w} \frac{\sinh az}{\sinh(a\tau/2)} , \tag{5.102}$$

and the current density is (by differentation)

$$\tilde{J}_y(z) = \frac{aI}{2w} \frac{\cosh az}{\sinh(a\tau/2)} . \tag{5.103}$$

Since the d.c. current density is $J_{dc} = I/\tau w$, we have

$$\tilde{J}_y(z) = J_{dc} \frac{a\tau}{2} \frac{\cosh az}{\sinh(a\tau/2)} . \tag{5.104}$$

The magnitude of the current density (normalized with respect to J_{dc}) is plotted in Figure 5.9(a) for three different values of the ratio τ/δ. Note that $\tilde{J}_y(-z) = \tilde{J}_y(z)$. The current remains essentially uniformly distributed for $\tau/\delta < 1$. However, for higher ratios, the skin effect is responsible for an uneven distribution of the current in the conductor.

The Busbar Loss and a.c. Resistance

It can be shown by integrating $|\tilde{J}_y|^2$ (using similar manipulations as before) that the loss per unit length (h) of this conductor is

$$\overline{p_{\text{eddy}}} = \frac{I^2}{4w\sigma\delta} \frac{\sinh(\tau/\delta) + \sin(\tau/\delta)}{\cosh(\tau/\delta) - \cos(\tau/\delta)} \ \text{(W/m)} . \tag{5.105}$$

This power loss can also be expressed in terms of the a.c. resistance:

$$\overline{p_{\text{eddy}}} = \frac{1}{2} I^2 R_{ac} , \tag{5.106}$$

where the a.c. resistance per unit length is

$$R_{ac} = \frac{1}{2w\sigma\delta} \frac{\sinh(\tau/\delta) + \sin(\tau/\delta)}{\cosh(\tau/\delta) - \cos(\tau/\delta)} . \tag{5.107}$$

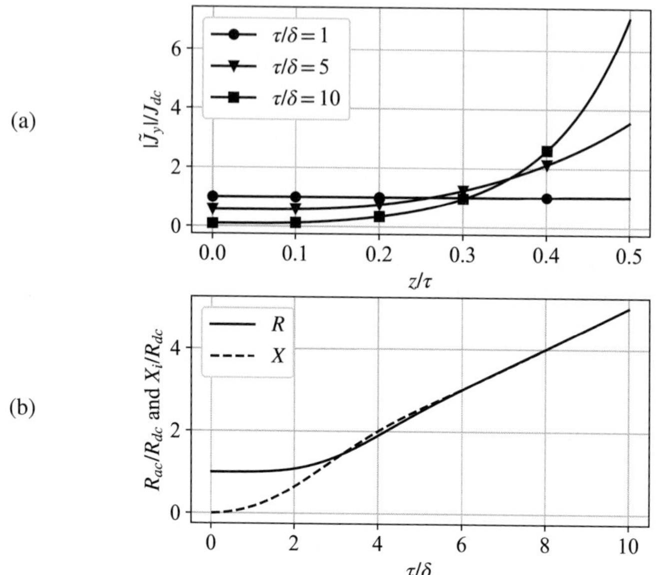

Figure 5.9 Busbar: (a) current density and (b) a.c. resistance/reactance.

Since the d.c. resistance of the busbar per unit length is

$$R_{dc} = \frac{1}{\sigma \tau w}, \tag{5.108}$$

we have

$$R_{ac} = R_{dc} \frac{\tau}{2\delta} \frac{\sinh(\tau/\delta) + \sin(\tau/\delta)}{\cosh(\tau/\delta) - \cos(\tau/\delta)}. \tag{5.109}$$

The normalized a.c. resistance is plotted in Figure 5.9(b). The skin effect leads to an increase of the resistance and higher loss for the same current.

The Busbar Impedance

A complex impedance $Z_i = R_{ac} + jX_i$ can be calculated for the busbar based on the complex Poynting vector, using Equations (5.85) and (5.91). The subscript "i" reminds us that this is only the internal impedance. On the two sides of the busbar, we have

$$-\mathbf{S} = \frac{1}{\sigma} \tilde{J}_y \tilde{H}_x^* \hat{\mathbf{k}}\Big|_{z=\pm\tau/2} = \pm \frac{aI^2}{4\sigma w^2} \frac{\cosh(a\tau/2)}{\sinh(a\tau/2)} \hat{\mathbf{k}}. \tag{5.110}$$

Close to the terminals on the planes $y = \pm h/2$, the Poynting vector is perpendicular to the surface normal vector, so the contribution to the surface integral is zero. Integrating $-\mathbf{S} \cdot \hat{\mathbf{n}}$ over both sides of the conductor, we obtain the impedance per unit length

$$Z_i = R_{ac} + jX_i = \frac{a}{2\sigma w} \frac{\cosh(a\tau/2)}{\sinh(a\tau/2)}, \tag{5.111a}$$

and after separating real and imaginary parts:

$$Z_i = R_{dc} \frac{\tau}{2\delta} \left[\frac{\sinh(\tau/\delta) + \sin(\tau/\delta)}{\cosh(\tau/\delta) - \cos(\tau/\delta)} + j \frac{\sinh(\tau/\delta) - \sin(\tau/\delta)}{\cosh(\tau/\delta) - \cos(\tau/\delta)} \right]. \tag{5.111b}$$

Note that R_{ac} is identical to the previous calculation of Equation (5.109). The normalized reactance X_i/R_{dc} is plotted together with the a.c. resistance in Figure 5.9(b).

5.3.5 Semi-infinite Plate with Twin Return Conductor

In the previous section, we analyzed an isolated plate-like conductor. Now, we will calculate the solution when an identical conductor carrying the return current is placed parallel to this one. Suppose that the second conductor is positioned somewhere in the $z > \tau/2$ space. The exact position does not matter, since our model assumes an infinite length. The return conductor carries current in the opposite direction than its twin (e.g., we could imagine that the current source of Figure 5.8 is replaced by a second strip). We will calculate the field and the current inside the first conductor.

Under the presence of the return conductor, a new set of boundary conditions is obtained. Ampère's law yields

$$\tilde{H}_x(\tau/2) = \frac{I}{w} \quad \text{and} \quad \tilde{H}_x(-\tau/2) = 0\,. \tag{5.112}$$

Therefore, the magnetic field is concentrated in the space between the two conductors. Solving the ODE with these boundary conditions, we find the following magnetic field:

$$\tilde{H}_x(z) = \frac{I}{w}\frac{\sinh(a(z+\tau/2))}{\sinh(a\tau)} \tag{5.113}$$

and the current density

$$\tilde{J}_y(z) = \frac{aI}{w}\frac{\cosh(a(z+\tau/2))}{\sinh(a\tau)} = J_{dc}a\tau\frac{\cosh(a(z+\tau/2))}{\sinh(a\tau)}\,. \tag{5.114}$$

The current distribution is no longer symmetric inside the conductor, as current tends to accumulate towards the inner surface, as shown in Figure 5.10(a). Compared to the isolated conductor case, the surface current density magnitude is higher in this case.

The complex Poynting vector on the surface of the first conductor where the magnetic field is nonzero is

$$-\mathbf{S} = \frac{1}{\sigma}\,\tilde{J}_y\tilde{H}_x^*\,\hat{\mathbf{k}}\Big|_{z=\tau/2} = \frac{aI^2}{\sigma w^2}\frac{\cosh(a\tau)}{\sinh(a\tau)}\hat{\mathbf{k}}\,. \tag{5.115}$$

Therefore, we obtain the following internal impedance, which accounts for the energy in only one of the twin conductors, and is expressed per unit length (h):

$$Z_i = R_{ac} + jX_i = R_{dc}\frac{\tau}{\delta}\left[\frac{\sinh(2\tau/\delta)+\sin(2\tau/\delta)}{\cosh(2\tau/\delta)-\cos(2\tau/\delta)} + j\frac{\sinh(2\tau/\delta)-\sin(2\tau/\delta)}{\cosh(2\tau/\delta)-\cos(2\tau/\delta)}\right]. \tag{5.116}$$

The resistive and inductive components are plotted in Figure 5.10(b). Compared to the isolated busbar case, the values have roughly doubled. This increase in the ohmic loss, which is due to the presence of the second conductor, is called the **proximity loss**.

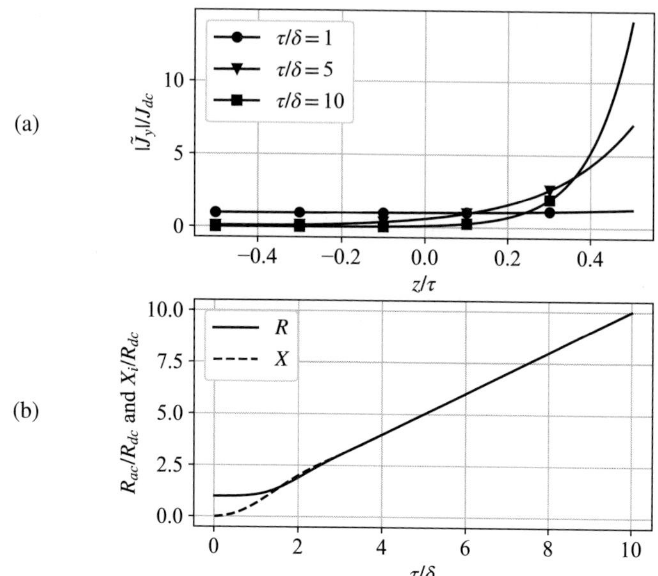

Figure 5.10 Twin busbar: (a) current density and (b) a.c. resistance/reactance.

5.3.6 Circular Conductor

In Example 1.22 on page 36, we studied the magnetic field of an infinitely long circular wire with uniform current distribution. Now, we will revisit this problem, but we will account for the uneven distribution of current in the conductor due to the skin effect. The conductor radius is R.

We use cylindrical coordinates and phasors for the field quantities (assuming that we are in a sinusoidal steady state). A current phasor $\tilde{I}$ is directed along the z-axis. We have

$$\nabla \times \tilde{\mathbf{H}} = \tilde{\mathbf{J}} = \sigma \tilde{\mathbf{E}} . \tag{5.117}$$

Applying the curl once more, substituting the curl of the electric field with the negative of the time derivative of the B-field, and using the curl-of-the-curl vector calculus identity (1.163), we arrive at the diffusion equation

$$\nabla^2 \tilde{\mathbf{H}} = j\omega\mu\sigma \tilde{\mathbf{H}} = a^2 \tilde{\mathbf{H}} . \tag{5.118}$$

The problem exhibits cylindrical symmetry, so $\partial/\partial\phi = 0$. Also, $\partial/\partial z = 0$ due to the infinite length of the wire. The magnetic field only has a ϕ-component. Expanding the vector Laplacian in cylindrical coordinates yields the second-order ODE

$$r^2 \frac{d^2 \tilde{H}_\phi}{dr^2} + r \frac{d\tilde{H}_\phi}{dr} - (a^2 r^2 + 1)\tilde{H}_\phi = 0 . \tag{5.119}$$

The general solution of this ODE employs modified Bessel[80] functions of the first and second kind, of order 1:[c]

$$\tilde{H}_\phi(r) = CI_1(ar) + DK_1(ar)\,. \tag{5.120}$$

However, K_1 is unbounded at the origin, so $D = 0$ for physical reasons. The solution must satisfy the boundary condition (Ampère's law)

$$\tilde{H}_\phi(R) = \frac{I}{2\pi R}\,. \tag{5.121}$$

Therefore

$$\tilde{H}_\phi(r) = \frac{I}{2\pi R}\frac{I_1(ar)}{I_1(aR)} \tag{5.122}$$

and

$$\tilde{J}_z(r) = \frac{1}{r}\frac{d(r\tilde{H}_\phi)}{dr} = \frac{I}{2\pi R}\frac{1}{I_1(aR)}\left(\frac{I_1(ar)}{r} + \frac{dI_1(ar)}{dr}\right). \tag{5.123}$$

Using the following property of Bessel functions, which relates the derivative of I_1 to the zero-order function I_0:

$$I_1'(x) = I_0(x) - \frac{1}{x}I_1(x)\,, \tag{5.124}$$

we obtain

$$\tilde{J}_z(r) = \frac{aI}{2\pi R}\frac{I_0(ar)}{I_1(aR)} = J_{dc}\frac{aR}{2}\frac{I_0(ar)}{I_1(aR)}\,. \tag{5.125}$$

The normalized current distribution magnitude is plotted in Figure 5.11(a) for different values of the parameter R/δ. Using the complex Poynting vector, we can also readily find an expression for the internal impedance per unit length:

$$Z_i = R_{dc}\frac{aR}{2}\frac{I_0(aR)}{I_1(aR)}\,, \tag{5.126}$$

where the d.c. resistance per unit length is $R_{dc} = 1/(\sigma\pi R^2)$. This is computed numerically and illustrated in Figure 5.11(b). The solution has similarities with that of the rectangular conductor. The skin effect is again significant for $R/\delta > 1$. A notable difference is seen in the normalized ac resistance, which is higher by 1/4 from the internal reactance at higher values of R/δ (whereas previously, these two were asymptotically equal).

5.4 Phasor-Based FEA

When the objective of the analysis is a magnetically linear device that operates in a sinusoidal steady state, it is convenient to use a phasor-based FEA. In this case, every field variable is varying sinusoidally with a given frequency ω. Hence, all field quantities can be described by complex vector fields (phasors), for example, as in Equation (5.67). In a 2-D FEA implementation, we seek a solution where the MVP belongs to the space

[c] By convention, these are denoted as I_1 and K_1. Please note that I_1 should not be confused with the current I.

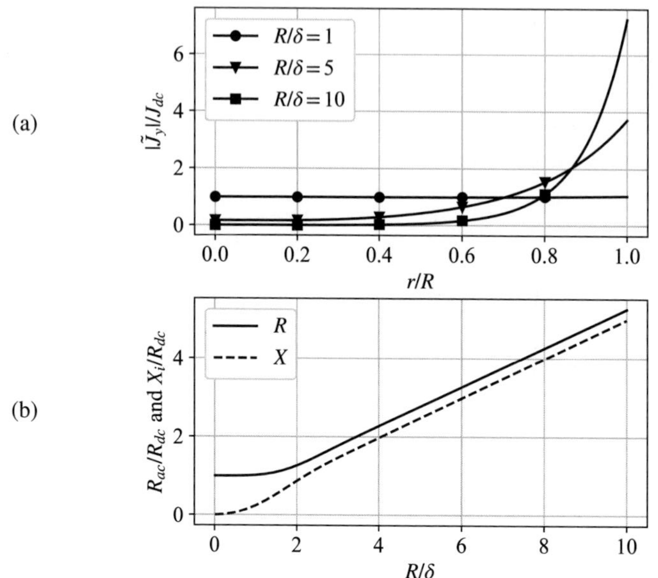

Figure 5.11 Circular conductor: (a) current density and (b) a.c. resistance/reactance.

of piecewise linear phasors over the given mesh of the domain. Inside each triangle, the MVP depends on the vertex MVP phasors:

$$\tilde{A}_k(x, y, \tilde{\mathbf{A}}^k) = \sum_{i=1}^{3} \tilde{A}_i^k \alpha_i^k(x, y) . \tag{5.127}$$

In particular, we must ensure that the mesh has adequate resolution close to the conductor surfaces in order to capture the variation of the field due to the skin effect with sufficient accuracy.

Transforming Equation (5.23) to the phasor domain, we obtain a linear system of equations with complex coefficients:

$$(\mathbf{S} + j\omega\mathbf{T})\tilde{\mathbf{A}} = \tilde{\mathbf{b}} . \tag{5.128}$$

The right-hand-side vector $\tilde{\mathbf{b}}$ incorporates the impact of sinusoidal excitation by a.c. voltage sources ($\tilde{V}$) acting at the terminals of conductors. This vector assembles contributions from individual elements based on Equations (5.15) and (5.22), where

$$\tilde{J}_{1,k} = -\sigma_k \frac{\tilde{V}_k}{\ell} . \tag{5.129}$$

The right-hand-side vector may also reflect the impact of a sinusoidal boundary condition (i.e., the effect of an external field), as explained in §3.2.4 and Box 3.1 on page 238. Note that PMs are equivalent to a constant excitation, so their impact should not be considered in this formulation. If the device contains stationary PMs, then we should superimpose the solution of a separate magnetostatic problem to the solution of this problem.

The implementation is very similar to that of a magnetostatic problem. A main difference is that we must account for the extra $j\omega\mathbf{T}$ terms when assembling the linear system. The sparse linear solve routines from SciPy work seamlessly with complex matrices. Complex numbers can be defined in Python, e.g., as follows:

```
>>> a = 1 + 2j
>>> b = 3 - 2j
>>> a + b
(4+0j)
```

Hence, creating the matrices can be performed as shown by the below code snippet:

```
S,b = build_S_matrix(n_nodes, triangles, Sk, nu, J_ext, delta)
T   = build_T_matrix(n_nodes, triangles, ind_cond, sigma, delta)
ST  = S + 1j*2*pi*freq*T
```

We can use the same `build_S_matrix` function (see §3.3.2 on page 242). The argument `J_ext` implies that we are feeding the function with only the externally imposed current (if any) for use in the $\tilde{\mathbf{b}}$ vector. We also need to code a new `build_T_matrix` function that will assemble the $\mathbf{T}$ matrix by iterating only over conductor triangles and collecting $\sigma_k\mathbf{T}^k$ contributions based on Equation (5.20). The conductor elements are indexed within the global `triangles` data structure by the array `ind_cond`. This will be very similar to the stiffness matrix assembly function, so it is left as an exercise for the reader. Boundary conditions can be enforced exactly as we have done previously, by modifying elements of the ST sparse matrix. Finally, the MVP phasor array $\tilde{\mathbf{A}}$ can be obtained by a sparse linear system solve as follows:

```
import scipy.sparse.linalg as spsl
A = spsl.spsolve(ST,b)
```

5.4.1 Phasor-Based FEA in Lossy Materials

In §2.2.3, where we discussed the variational approach for solving the magnetostatic problem, we explained that the functional is representative of the system coenergy under the fundamental assumption of a conservative coupling field. However, in the preceding sections of this chapter, we encountered materials that were inherently lossy due to eddy currents and/or hysteresis, for which the conservative field assumption no longer applies. These materials can be modeled with the complex-valued equivalent permeability of Equation (5.66), having a constitutive law of the form $\tilde{\mathbf{B}} = \mu_e\tilde{\mathbf{H}}$ with $\mu_e = |\mu_e|e^{-j\theta_e}$, $\theta_e \geq 0$ so that B lags behind H. Nevertheless, it is indeed possible to conduct phasor-based FEA of devices with lossy magnetic materials.

To see this, imagine that we are solving a sequence of magnetostatic problems in time, as we are cycling through the sinusoidal steady state. You may think of the reluctivity in each triangle ν_k as a function of time such that $\vec{\mathbf{B}}^k$ and $\vec{\mathbf{H}}^k$ are both sinusoidal and exhibit the appropriate phase shift θ_e. Now, take the contribution to the gradient from the kth triangle at an arbitrary time instant, given by Equation (5.22):

$$\mathbf{g}^k = \nu_k\mathbf{S}^k\mathbf{A}^k - \frac{J_{1,k}\Delta_k}{3}\begin{bmatrix}1 & 1 & 1\end{bmatrix}^{\mathrm{T}} + \sigma_k\mathbf{T}^k\dot{\mathbf{A}}^k. \tag{5.130}$$

(We have ignored the presence of PMs.) Since the element stiffness matrix is

$$\mathbf{S}^k = \frac{1}{4\Delta_k}\left[\mathbf{q}^k(\mathbf{q}^k)^\top + \mathbf{r}^k(\mathbf{r}^k)^\top\right], \tag{5.131}$$

and the H-field is related to the MVP by (see Equation (3.94) on page 255)

$$\begin{bmatrix} H_x^k \\ H_y^k \end{bmatrix} = \nu_k \begin{bmatrix} B_x^k \\ B_y^k \end{bmatrix} = \frac{\nu_k}{2\Delta_k}\begin{bmatrix} (\mathbf{r}^k)^\top \\ -(\mathbf{q}^k)^\top \end{bmatrix}\mathbf{A}^k, \tag{5.132}$$

the elemental gradient contribution can be written as

$$\mathbf{g}^k = \frac{1}{2}\begin{bmatrix}\mathbf{r}^k & -\mathbf{q}^k\end{bmatrix}\begin{bmatrix} H_x^k \\ H_y^k \end{bmatrix} - \frac{J_{1,k}\Delta_k}{3}\begin{bmatrix}1 & 1 & 1\end{bmatrix}^\top + \sigma_k \mathbf{T}^k\dot{\mathbf{A}}^k. \tag{5.133}$$

By means of this algebraic trick, we have obtained an expression where all time-varying quantities are sinusoidal. We can thus convert this to phasor form, which yields

$$\tilde{\mathbf{g}}^k = (\nu_e \mathbf{S}^k + j\omega\sigma_k \mathbf{T}^k)\,\tilde{\mathbf{A}}^k - \frac{\tilde{J}_{1,k}\Delta_k}{3}\begin{bmatrix}1 & 1 & 1\end{bmatrix}^\top. \tag{5.134}$$

Hence, we conclude that we can use the complex equivalent reluctivity $\nu_e = 1/\mu_e$ while assembling the stiffness matrix for a phasor-based FEA.

5.4.2 Calculating the B-Field Phasors and the Core Loss

After obtaining the MVP phasors from a 2-D phasor-based FEA, it may be desirable to calculate the flux density phasors, e.g., for visualization purposes. In particular, we have a phasor for each component of $\vec{\mathbf{B}}$. Using Equation (3.94) on page 255, the magnetic flux density phasors in the kth triangle can be expressed as

$$\begin{bmatrix} \tilde{B}_x^k \\ \tilde{B}_y^k \end{bmatrix} = \frac{1}{2\Delta_k}\begin{bmatrix} (\mathbf{r}^k)^\top \\ -(\mathbf{q}^k)^\top \end{bmatrix}\tilde{\mathbf{A}}^k. \tag{5.135}$$

These phasors can be used to visualize the variation of the flux density over time at any point of interest in the device. In general, a phase shift will be present between the two components, which causes the $\vec{\mathbf{B}}^k(t)$ vector to describe an ellipsoidal trajectory.

To see this, let the coordinates of the locus be

$$X = B_x^k \cos\omega t' \quad \text{and} \quad Y = B_y^k \cos(\omega t' - \phi), \tag{5.136}$$

where t' represents a time origin such that $\omega t' = \omega t + \text{angle}\{\tilde{B}_x^k\}$, and $\phi = \text{angle}\{\tilde{B}_x^k\} - \text{angle}\{\tilde{B}_y^k\}$ is the phase difference between the two phasors. We have

$$\frac{Y}{B_y^k} = \cos\omega t' \cos\phi + \sin\omega t' \sin\phi = \frac{X}{B_x^k}\cos\phi + \sin\omega t' \sin\phi, \tag{5.137}$$

which leads, after elementary algebraic manipulations, to

$$\left(\frac{X}{B_x^k}\right)^2 - 2\frac{XY}{B_x^k B_y^k}\cos\phi + \left(\frac{Y}{B_y^k}\right)^2 - \sin^2\phi = 0. \tag{5.138}$$

(This precludes the case where either of the two components happens to be zero; then

$B_x^k B_y^k = 0$, and we have a trivial case with a purely horizontal or vertical vector pulsating along its axis.) This is a quadratic form that is indeed the equation of an ellipse, provided that $\phi \neq 0$ (in which case, the phasor components are in phase, and thus the phasor locus collapses to a line segment). In general, the ellipse will be rotated from the horizontal. An example is shown in Figure 5.12, where $\tilde{B}_x^k = 1\angle 0°$ and $\tilde{B}_y^k = 0.6\angle 45°$ so that the vector is turning clockwise. The dots are spaced equally in time, therefore the angular velocity is not constant, but it is higher at the co-vertices of the ellipse.

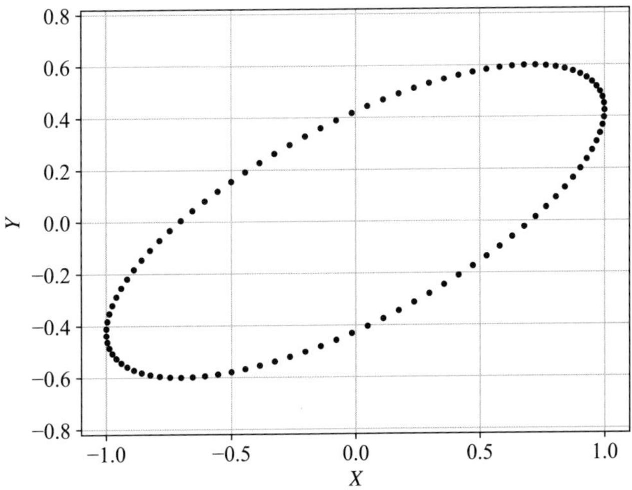

Figure 5.12 Ellipsoidal locus of flux density vector under sinusoidal steady-state conditions.

For calculating core loss, the squared B-field magnitudes can be computed from Equation (5.135) or directly from the MVP as

$$|\tilde{B}_x^k|^2 = \tilde{B}_x^k \cdot (\tilde{B}_x^k)^* = \frac{1}{4\Delta_k^2}(\tilde{\mathbf{A}}^k)^{\mathrm{H}}\mathbf{r}^k(\mathbf{r}^k)^{\mathsf{T}}\tilde{\mathbf{A}}^k, \tag{5.139a}$$

$$|\tilde{B}_y^k|^2 = \tilde{B}_y^k \cdot (\tilde{B}_y^k)^* = \frac{1}{4\Delta_k^2}(\tilde{\mathbf{A}}^k)^{\mathrm{H}}\mathbf{q}^k(\mathbf{q}^k)^{\mathsf{T}}\tilde{\mathbf{A}}^k, \tag{5.139b}$$

where the superscript "H" denotes the conjugate transpose (also known as the Hermitian transpose). Note that we did not calculate a total magnitude of the $\vec{\mathbf{B}}$ vector using the Pythagorean theorem, which in the context of an ellipsoidal variation does not make much physical sense. A simple method to calculate the core loss would be to invoke a Steinmetz formula twice, that is, separately for each component (although this may not be physically entirely justifiable).

Example 5.3 *FEA of rectangular conductor in external time-varying magnetic field.* As a first phasor-based FEA case study, we conduct a 2-D FEA of a long rectangular copper conductor in air under the influence of an external magnetic field. This bears similarity to the analysis of the semi-infinite plate in §5.3.1. A main difference is that the cross-section is finite, as shown in Figure 5.13. In other words, the width w is no longer considered to be much greater than the thickness τ. Here, the aspect ratio is $w/\tau = 2$. Therefore, the solution will vary in both x and y. (Note that we have reverted to the familiar x–y plane for the 2-D FEA. This set of axes is different from the one we used in the previous section, where the solution exhibited a variation along the z-axis.)

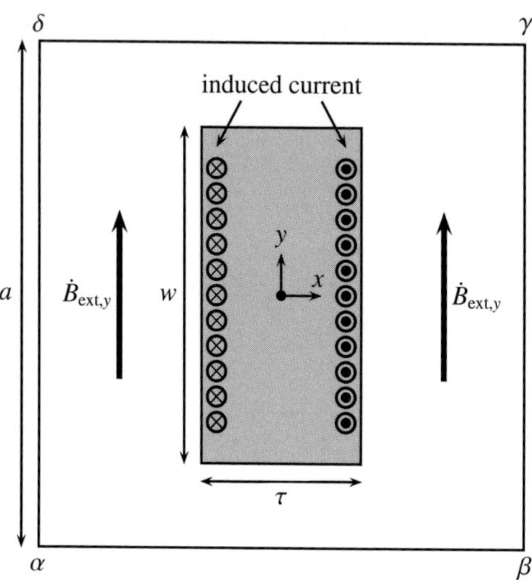

Figure 5.13 Cross-section of rectangular conductor under the influence of time-varying external magnetic field.

Solution

Let us suppose that the external field is oriented along the y-axis. This is captured by imposing a nonzero Dirichlet boundary condition along the α–β and γ–δ sides of the boundary. Suppose the external magnetic field is uniform and varies sinusoidally with time:

$$B_{\text{ext},y} = B_{\max} \cos \omega t \, . \tag{5.140}$$

This is modeled by a linear variation of the MVP at the top and bottom boundaries:

$$A(x, a/2) = A(x, -a/2) = -xB_{\max} \cos \omega t \, , \tag{5.141}$$

where a is the width and height of the square $\alpha\beta\gamma\delta$ boundary. In the FEA program, we define the boundary condition as a phasor that has the same phase angle as the magnetic field (i.e., zero):

$$\tilde{A}(x, -a/2) = \tilde{A}(x, a/2) = -xB_{\max} \, . \tag{5.142}$$

On the left and right boundaries, we set

$$\tilde{A}(-a/2, y) = (a/2)B_{\max} \quad \text{and} \quad \tilde{A}(a/2, y) = -(a/2)B_{\max} \, , \tag{5.143}$$

so that the B-field is tangent to these sides.

Suppose that the conductor dimensions are $\tau = 2$ cm and $w = 4$ cm, and that its conductivity is $\sigma = 58$ MS/m. The external field has a peak value of 1 mT and a frequency of 200 Hz. This leads to a skin depth $\delta = 4.67$ mm, approximately 4.3 times smaller than the thickness. The FEA is conducted over a fine mesh with 26,886 elements (roughly equal in size) and 13,524 vertices. The conductor is centered in a square of dimension $a = 6$ cm. Even with this large number of unknowns, the solution can be obtained within a matter of just a few seconds since the linear equation (5.128) can be solved very efficiently. After computing the MVP phasors $\tilde{\mathbf{A}}$, we obtain the current density phasors by

$$\tilde{\mathbf{J}}_c = -j\omega\sigma\tilde{\mathbf{A}}_c\,, \tag{5.144}$$

where the subscript c denotes that we are referring to vertices of conductor elements.

We can visualize the results by plotting the solution for a few instants in time. Here, we have selected the set of phase angles $\omega t \in \{0, 30, 60, 90, 120\}°$. We compute the time-domain quantities by, e.g.

$$A(t) = \mathrm{Re}\{\tilde{A}e^{j\omega t}\}\,. \tag{5.145}$$

The results are shown in Figure 5.14. The first five subplots are snapshots of the magnetic flux lines and the induced current density in the conductor. First, we observe the demagnetizing effects of the induced currents, which tend to bend the flux lines away from the conductor. (Note that the solid contours indicate positive MVP values, whereas the dashed contours indicate negative MVP.) Second, we are plotting the distribution of current density inside the conductor. The white shades indicate large current in the positive z-axis (coming out of the page), whereas dark shades indicate large current in the negative z-axis (into the page). Light gray implies zero current. We are looking at Lenz's law in action! At time zero, the external magnetic field attains its peak value (hence, its derivative is zero). However, due to diffusion delays, induced currents are still flowing inside the conductor, as if it is still increasing. Afterwards, the external magnetic field starts decreasing. At $\omega t = 90°$, the external field is zero, but its rate of change is the largest (in the negative sense). We can see that the polarity of the induced currents is such that their magnetic field is resisting the change of the external field.

The final plot, which has been created using a Matplotlib Pyplot `tricontourf` command, shows the magnitude of the induced current phasor as filled contour regions. Here, black indicates zero current, whereas white indicates maximum current. The current tends to concentrate close to the conductor surfaces and the corners, in particular. To compute the eddy current loss, we can integrate numerically the r.m.s. value of the current squared over the conductor surface. The loss contributed by element k is (invoking again the results of Problem 3.5 on page 308):

$$\frac{1}{2\sigma}\iint\limits_{\Delta_k} |\tilde{J}_k|^2\,dx\,dy = \frac{1}{2\sigma}\iint\limits_{\Delta_k} \tilde{J}_k\tilde{J}_k^*\,dx\,dy = \frac{1}{2\sigma}(\tilde{\mathbf{J}}^k)^{\mathrm{H}}\mathbf{T}^k\tilde{\mathbf{J}}^k\,. \tag{5.146}$$

The total loss per unit length can be expressed as

$$\overline{p_{\mathrm{eddy}}} = \frac{1}{2\sigma^2}\tilde{\mathbf{J}}^{\mathrm{H}}\mathbf{T}\tilde{\mathbf{J}} = \frac{\omega^2}{2}\tilde{\mathbf{A}}^{\mathrm{H}}\mathbf{T}\tilde{\mathbf{A}}\quad(\mathrm{W/m})\,. \tag{5.147}$$

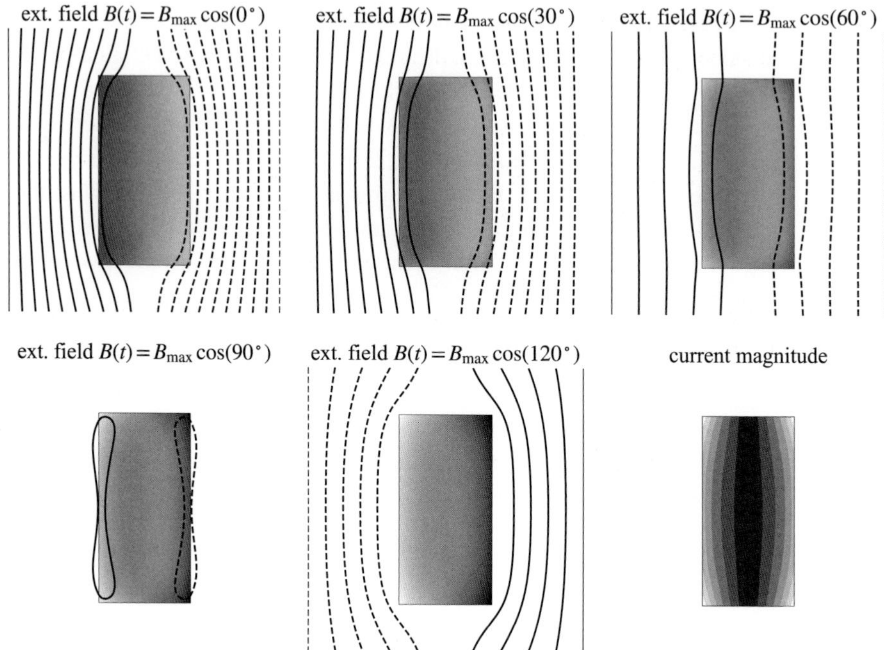

Figure 5.14 Magnetic field and induced currents in conductor under the influence of external magnetic field.

In this case, the FEA-based eddy current loss is 262 mW/m. As a sanity check, we may compare this to the result of Equation (5.48), accounting for the finite width of the conductor by means of a simple multiplication by w. This answer is 97 mW/m, so the elementary 1-D formula significantly underestimates the loss. This could be attributed to the higher intensity of the current density at the corners as well as to the difference in the boundary conditions.

Example 5.4 *FEA of rectangular conductor moving in magnetic field.*
As a second phasor-based FEA case study, we consider the same rectangular copper conductor as in the previous example, but now the conductor moves inside a time-invariant external field. Suppose that the conductor moves horizontally to the right with velocity v_x. The magnetic field has a vertical (y) orientation and is given by the sinusoidal function

$$B_y(x) = B_{\max} \sin 2\pi kx\,, \tag{5.148}$$

where x is the position with respect to a stationary reference frame, and $k > 0$ is a given spatial frequency (i.e., the field is periodic every k^{-1} m). The MVP is, therefore, oriented along the z-axis and given by

$$A(x) = \frac{B_{\max}}{2\pi k} \cos 2\pi kx\,. \tag{5.149}$$

This case study resembles a squirrel-cage induction machine rotor bar slipping by the stator magnetic field, although the arrangement in this elemental example is different.

Solution

We will conduct FEA in the reference frame of the conductor. To model this, we will move the FEA domain of Figure 5.13 inside the external magnetic field. The mesh is identical as in the previous example. If the conductor center position is

$$x(t) = v_x t\,, \tag{5.150}$$

we can set the boundary conditions on an arbitrary point of the FEA boundary displaced by $x_b \in [-a/2, a/2]$ from the conductor center based on Equation (5.149):

$$A(x_b) = \frac{B_{\max}}{2\pi k}\cos(2\pi k(v_x t + x_b))\,. \tag{5.151}$$

Hence, in the conductor reference frame, the magnetic excitation is sinusoidal with frequency $\omega = 2\pi k v_x$, and we can use phasor-based FEA for the analysis. The boundary conditions are defined as phasors:

$$\tilde{A}(x_b) = \frac{B_{\max}}{2\pi k}\angle 2\pi k x_b\,. \tag{5.152}$$

Note that if the spatial periodicity k^{-1} is very large compared to the dimensions of the conductor, this sinusoidal FEA model can serve as an approximation of a conductor moving with constant velocity inside a uniform magnetic field (e.g., by taking a time snapshot when the conductor is moving through the peak of the external field).

What makes this case study even more interesting is the fact that the mean value of dA/dt over the conductor cross-section is nonzero. If left unchecked, this would lead to a nonzero net current flow in the z-direction. (In the previous example, we did not worry about this because symmetry ensured zero current.) Hence, since we are modeling an isolated conductor, we must enforce an additional constraint on the problem so that the net current is zero. In particular, we should constrain the currents of the conductor elements $\tilde{i}_k$ so that their sum satisfies

$$\sum_{k\in\mathcal{E}_c} \tilde{i}_k - \tilde{I} = 0\,, \tag{5.153}$$

where $\mathcal{E}_c$ is the set of conductor element indices. Equivalently, in terms of current densities:

$$\sum_{k\in\mathcal{E}_c}\iint_{\Delta_k} \tilde{J}_k(x,y)\,dx\,dy = \sum_{k\in\mathcal{E}_c}\frac{\tilde{J}_1^k + \tilde{J}_2^k + \tilde{J}_3^k}{3}\Delta_k = \tilde{I} = 0\,. \tag{5.154}$$

Recalling Equation (5.14), this sum becomes

$$\sum_{k\in\mathcal{E}_c}\left(\sigma\frac{\tilde{V}}{\ell}\Delta_k - \sigma j\omega\frac{\tilde{A}_1^k + \tilde{A}_2^k + \tilde{A}_3^k}{3}\Delta_k\right) = \tilde{I} = 0\,, \tag{5.155}$$

or

$$\frac{\tilde{V}}{R_{dc}} - j\omega\sum_{k\in\mathcal{E}_c}\sigma\frac{\tilde{A}_1^k + \tilde{A}_2^k + \tilde{A}_3^k}{3}\Delta_k = \tilde{I} = 0\,, \tag{5.156}$$

where

$$\tilde{V} = \tilde{\varphi}(-\ell/2) - \tilde{\varphi}(\ell/2) \tag{5.157}$$

is the (to be determined) motional EMF phasor between the conductor terminals, and R_{dc} is the d.c. resistance. This is a linear equality constraint of the form

$$\tilde{V}/R_{dc} - j\omega\mathbf{c}^\top\tilde{\mathbf{A}} = \tilde{I} = 0\,, \tag{5.158}$$

where $\mathbf{c}^\top$ collects element-wise contributions from terms like

$$\mathbf{c}^k = \sigma\frac{\Delta_k}{3}\begin{bmatrix}1 & 1 & 1\end{bmatrix}^\top. \tag{5.159}$$

The $\mathbf{c}$ vector can be related to the $\mathbf{b}$ vector of the gradient, which collects contributions

$$\mathbf{b}^k = \frac{\tilde{V}}{\ell}\mathbf{c}^k\,. \tag{5.160}$$

Therefore, we must solve the following augmented system of equations:

$$\begin{bmatrix}\mathbf{S} + j\omega\mathbf{T} & -j\omega\mathbf{c} \\ -j\omega\mathbf{c}^\top & (j\omega\ell)/R_{dc}\end{bmatrix} \cdot \begin{bmatrix}\tilde{\mathbf{A}} \\ \tilde{V}/(j\omega\ell)\end{bmatrix} = \begin{bmatrix}\mathbf{0} \\ \tilde{I}\end{bmatrix}, \tag{5.161}$$

written in this form so that the system matrix is symmetric, and to highlight how the more general case of an external impedance or nonzero imposed current could be represented. For us, $\tilde{I} = 0$, leading to a zero right-hand side; however, the solution is nonzero once boundary conditions are enforced.

From elementary physics, we recall that a moving conductor inside a (uniform) magnetic field develops a motional EMF with magnitude $v_x B_y \ell$ and polarity such that current would tend to flow out of the page. However, in our case study, the magnetic field is sinusoidal, and we deviate from the elementary scenario due to the induced currents. To make for an interesting study, we set the conductor velocity to $v_x = 10$ m/s and the spatial periodicity to $k^{-1} = 0.1$ m, leading to an excitation frequency of 100 Hz. The skin depth is thus $\delta = 6.6$ mm, roughly three times smaller than the conductor thickness. The FEA results are depicted in Figure 5.15. In general, the induced currents have a demagnetizing effect. The motional EMF is found to be $\tilde{V}/\ell = 3.1\angle 95.4^\circ$ mV/m, significantly lower than the low-frequency motional EMF of $10\angle 90^\circ$ mV/m. An eddy current loss of 62.6 mW/m is also calculated. This power must equal the rate of work done by the mechanical force that maintains constant conductor velocity.

Example 5.5 *FEA of rectangular conductor with twin return.*

As a third phasor-based case study, we will conduct an FEA of a rectangular conductor or busbar. A twin return conductor will be nearby, as shown in Figure 5.16. This setup is similar to Example 1.33 on page 72, and the theoretical analysis of §5.3.5 should provide insights regarding the skin and proximity effects. The center-to-center distance between the conductors is d. Solving the FEA yields the current density over the conductor cross-section, which will tend to flow close to the left conductor surface with increasing frequency. Our main goal in this case study is to calculate the line impedance as a function of frequency.

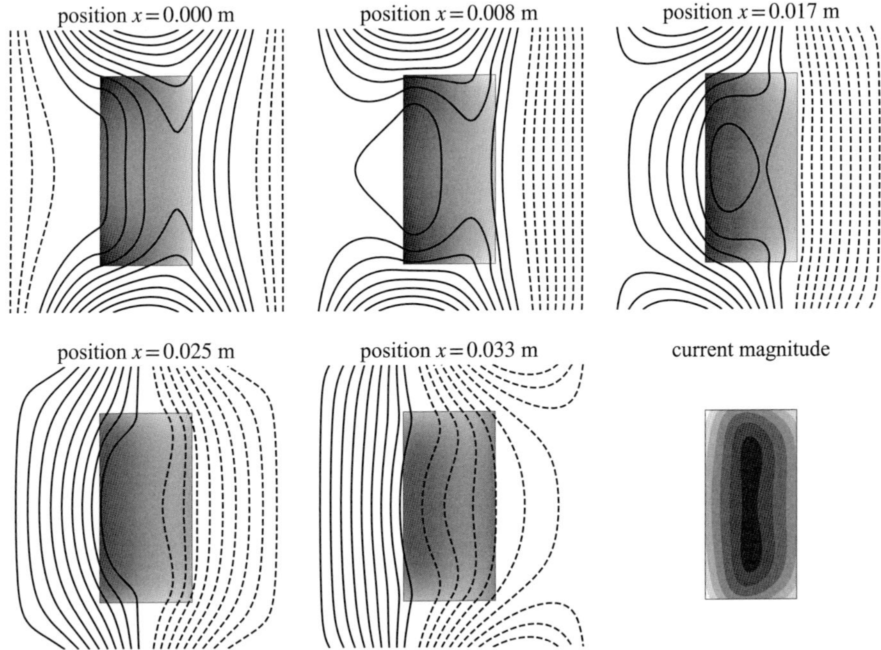

Figure 5.15 Magnetic field and induced currents in moving conductor inside time-invariant, sinusoidal magnetic field.

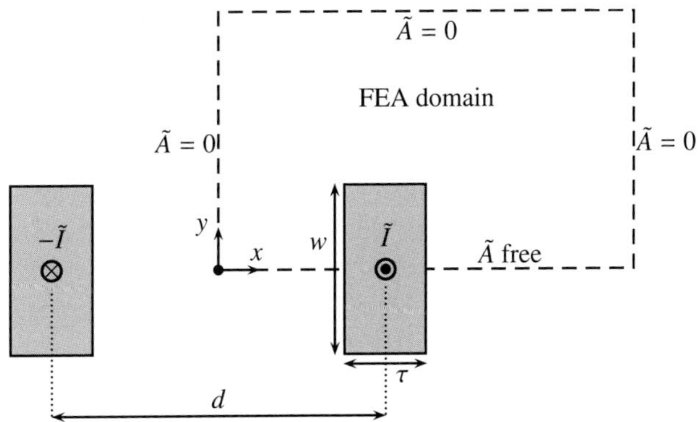

Figure 5.16 Cross-section of rectangular conductor transmission line.

Solution

To reduce computational burden, we exploit the symmetry of this topology, restricting the FEA domain to the first quadrant. A homogeneous Neumann boundary condition is placed on the $y = 0$ axis since we expect the flux lines to cross this boundary perpendicularly. The $\tilde{A} = 0$ condition on the $x = 0$ axis is imposed by virtue of the symmetry. The $\tilde{A} = 0$ conditions on the other two sides indicate that the magnetic field does not escape this rectangular boundary. Of course, this constraint is somewhat artificial since we know that the magnetic field will extend to infinity; however, it suffices for the purposes of this case study.

Let us assume that the conductor dimensions are $\tau = 2$ cm, $w = 4$ cm, and that the axial length is $\ell = 1$ m. The conductor distance is $d = 3$ cm. The material is copper with conductivity $\sigma = 58$ MS/m. The imposed a.c. voltage at the terminals of each conductor of the transmission line is $\tilde{V}$. The voltage has opposite polarity on each side. In this study, we will conduct a frequency sweep where the excitation will vary in the range $f \in [0.1, 1000]$ Hz. At the maximum frequency of 1 kHz, the skin depth is 2.09 mm, which is about 10 times less than the thickness. Therefore, we must ensure that the mesh inside the conductor is adequately refined to capture the variation of current density close to the surface. We may use a different mesh for each frequency; however, for simplicity we shall use an identical mesh consisting of 18,831 triangles and 9,504 nodes, with a much finer resolution inside the conductor. (This has the added advantage of eliminating numerical noise due to a different mesh from one study to the next.)

According to Equation (5.22), the gradient of the functional contributes voltage-dependent $\tilde{J}_{1,k}\Delta_k/3$ terms to the $\tilde{\mathbf{b}}$ right-hand-side vector, which eventually has the form $\tilde{\mathbf{b}} = \mathbf{c}\tilde{V}/\ell$, where $\mathbf{c}$ is a constant vector. Hence, from Equation (5.128), the nodal MVP phasors are proportional to the applied voltage (as expected since this is a linear system). After solving the FEA problem, we can find the total current density in element k as

$$\tilde{J}_k = \tilde{J}_{1,k} + \tilde{J}_{2,k} = \sigma\frac{\tilde{V}}{\ell} - j\omega\sigma\tilde{A}_k(x, y, \tilde{\mathbf{A}}^k)\,. \tag{5.162}$$

The total current is found by adding the currents of the conductor elements $\tilde{i}_k$:

$$\tilde{I} = 2\sum_{k\in\mathcal{E}_c}\tilde{i}_k = 2\sum_{k\in\mathcal{E}_c}\iint_{\Delta_k}\tilde{J}_k(x,y)\,dx\,dy = 2\sum_{k\in\mathcal{E}_c}\frac{\tilde{J}_1^k + \tilde{J}_2^k + \tilde{J}_3^k}{3}\Delta_k\,, \tag{5.163}$$

where $\mathcal{E}_c$ is the set of element indices of the upper-half conductor on the right. Hence, we can calculate the single-conductor impedance $Z_1 = \tilde{V}/\tilde{I}$, which is independent of the applied voltage. Multiplication of Z_1 by a factor of 2 yields the transmission line impedance (i.e., the loop impedance) per unit length:

$$Z = R + j\omega L = 2\frac{\tilde{V}}{\tilde{I}} \quad (\Omega/\text{m})\,. \tag{5.164}$$

Note that since we are accounting for the magnetic field throughout space, this is the total impedance and not just the internal impedance that we calculated earlier in the analytical examples.

Alternatively, we can obtain the resistance and inductance from the equations of the time-average Joule loss and stored magnetic energy, respectively, namely

$$P = \frac{1}{2}R|\tilde{I}|^2 = 4\cdot\frac{1}{2\sigma^2}\tilde{\mathbf{J}}^{\mathrm{H}}\mathbf{T}\tilde{\mathbf{J}} \quad (\text{W/m}) \tag{5.165}$$

and

$$\overline{W_f} = \frac{1}{4}L|\tilde{I}|^2 = 4\cdot\frac{1}{4}\iint_{\mathcal{D}}\nu|\tilde{B}|^2\,dx\,dy = 4\cdot\frac{1}{4}\tilde{\mathbf{A}}^{\mathrm{H}}\mathbf{S}\tilde{\mathbf{A}} \quad (\text{J/m})\,, \tag{5.166}$$

where the multiplication by a factor of 4 accounts for the entire conductor and its twin. Note that we could limit the integration domain to cover just the conductor surface in order to calculate the internal inductance, if needed.

A snapshot of the FEA flux lines at 100 Hz is shown in Figure 5.17. The snapshot is taken when the voltage phase angle is 70° so that the conductor current (which lags the voltage) is close to its peak value. At this frequency, the skin and proximity effects are still relatively weak. Post-processing the FEA results, we obtain the frequency-dependent a.c. line resistance and inductance, which are illustrated in Figure 5.18. Due to the increase of the a.c. resistance, this busbar becomes inefficient for frequencies higher than roughly 100 Hz. Interestingly, the inductance is not constant, but drops with frequency. Decreasing the conductor distance further leads to an increase in the a.c. resistance at higher frequencies due to stronger proximity effect, and to an overall drop in the loop inductance due to the decrease of the loop area.

Figure 5.17 Magnetic flux lines from rectangular busbar.

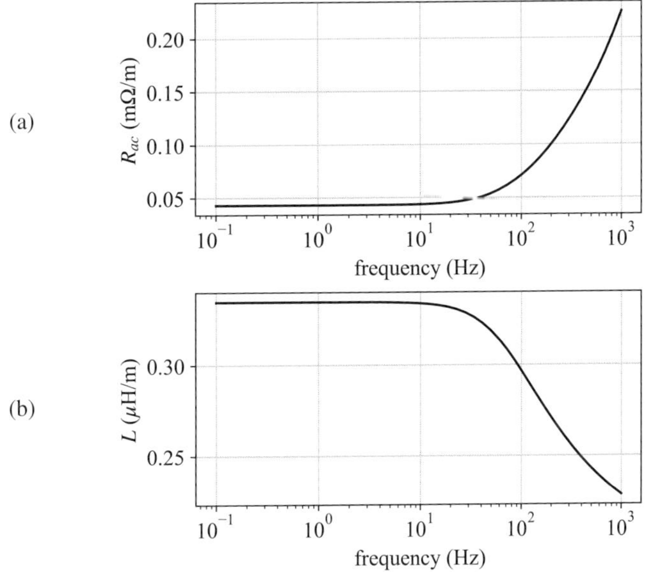

Figure 5.18 Transmission line: (a) a.c. resistance and (b) inductance per unit length.

Example 5.6 *FEA of conductor inside a slot.*

As a fourth phasor-based FEA case study, we will study a rectangular conductor placed in a slot within a ferromagnetic material. The magnetic circuit is completed via a small air gap and a second piece of ferromagnetic material, as shown in Figure 5.19, forming a U–I-shaped core. The return conductor is assumed to be far away, and is not included in the study. This elementary setup resembles a rotor bar of a squirrel-cage induction machine or a stator bar winding.

Solution

The conductor cross-section is 2×4 cm, the axial length is $\ell = 0.1$ m, and the air gap is 3 mm. The piece containing the slot is a square of side 6 cm, and the bottom piece is 6×2 cm. The conductor material is copper with conductivity $\sigma = 58$ MS/m, and the relative permeability of the core is $\mu_r = 1000$. The core is assumed to be laminated. The conductor carries a net sinusoidal current of magnitude 1000 A, peak, and frequency $f = 10$ Hz. The mesh has 7,253 vertices and 14,410 triangles, the majority of those within the slot and the core. (Here, we have used a finer mesh in these regions.) A Dirichlet boundary condition ($\tilde{A} = 0$) is imposed on the outer boundary. The FEA amounts to solving Equation (5.161), setting $\tilde{I} = 1000\underline{/0^\circ}$. The terminal voltage is thus found to be $\tilde{V} = 36.7\underline{/84.1^\circ}$ mV.

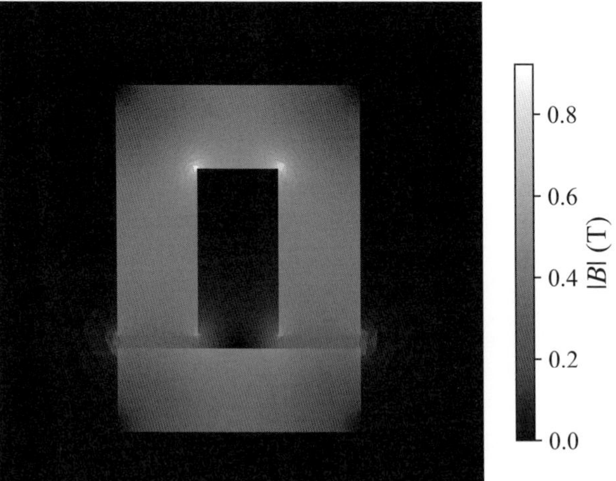

Figure 5.19 Bar conductor carrying a.c. current inside a slot: mesh and magnitude of B-field.

Representative FEA results are shown in Figure 5.20. The d.c. resistance is $R_{dc} = 2.16\ \mu\Omega$, whereas the a.c. resistance is $R_{ac} = 3.76\ \mu\Omega$. This is a significant increase for this frequency level, which would not have occurred had the conductor been lying in free space (see Figure 5.18). Notably, the skin effect is not symmetric. We can observe from the last plot that the current density is higher close to the air gap. Conceptually, this is attributed to the presence of leakage flux, or those flux lines that cross the slot horizontally without making it across the air gap.

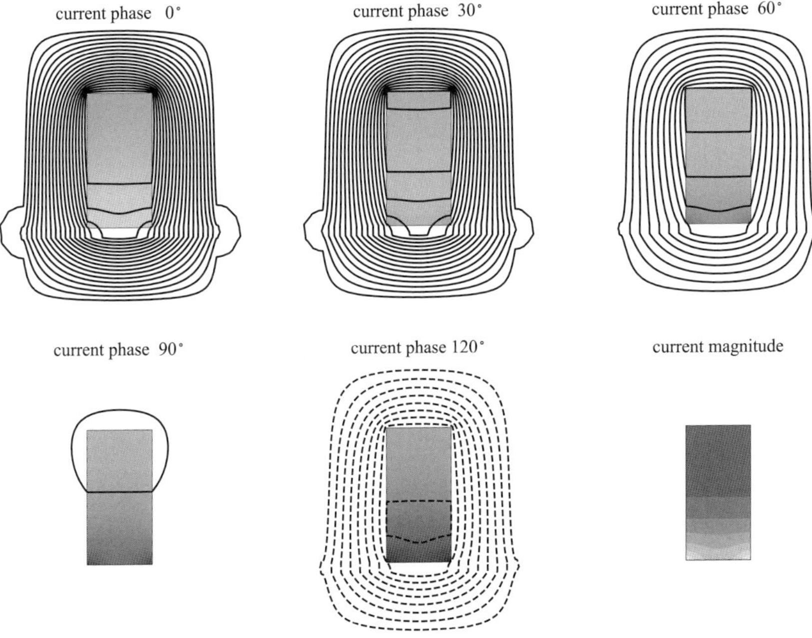

Figure 5.20 Magnetic field and induced currents in bar conductor inside a slot.

5.5 Higher-Frequency Winding Model

We have observed the effects of time-varying magnetic fields on conductors. In an electric machine, these conductors could be part of a winding that is connected to an external circuit. So, our objective here is to extend the winding energy balance theory developed earlier for the low-frequency regime (see §2.3.6 on page 154) to higher frequencies. In the low-frequency regime, the skin effect may be neglected, and we can safely assume that the current is uniformly distributed over the conductor cross-section. This simplification leads to the average MVP method for calculating the flux linkage (see §3.4.1 on page 257); however, these results are invalid at higher frequencies where the skin depth is commensurate with the conductor dimensions. It should be noted that if the frequency is sufficiently high, the presence of parasitic capacitances between the wires as they are closely packed inside the slots, as well as between them and the slot walls, will become increasingly important. Nevertheless, we are still ignoring these effects. Capacitive effects typically become significant at frequencies higher than tens of kHz.

In particular, we pose the following question: *Is it possible to define the resistance and flux linkage of a winding based on field quantities when the current is nonuniformly distributed?* The previous results (e.g., using a complex Poynting vector approach, or as we did in Example 5.5) indicate that this is possible. However, these results were obtained for the special case of sinusoidal steady-state conditions; our objective here is to seek equations that hold more generally, *on an instantaneous basis during arbitrary transients.*

Suppose that the device comprises K circuits (coils). The total current entering conductor k from an external circuit is a surface integral over its positive terminal cross-section S_k:

$$i_k = \iint_{\tilde{\mathbf{r}}\in S_k} \vec{\mathbf{J}}(\tilde{\mathbf{r}}) \cdot d\mathbf{a} = \iint_{\tilde{\mathbf{r}}\in S_k} J(\tilde{\mathbf{r}})\, da\,. \tag{5.167}$$

The variable $\tilde{\mathbf{r}} \in S_k$ is the starting point of individual filaments (the tilde does not denote a phasor in this case). Of course, the flow of current is perpendicular to the conductor terminal surface, but the current density can vary over the surface. We recall Equation (2.196) for the voltage at the terminals of a filament belonging to circuit k and originating at $\tilde{\mathbf{r}}$, which can be written as

$$V_k = J(\tilde{\mathbf{r}})\frac{\ell_k}{\sigma_k} + \frac{d\psi(\tilde{\mathbf{r}})}{dt}\,, \tag{5.168}$$

where ℓ_k is the length of the coil (assumed to be identical for all filaments) and σ_k its conductivity. The variable ψ denotes the flux linking a single filament, which was defined in Equation (2.193) based on a line integral of the MVP from the beginning to the end of the filament:

$$\psi(\tilde{\mathbf{r}}) = \int_{a(\tilde{\mathbf{r}})}^{b(\tilde{\mathbf{r}})} \vec{\mathbf{A}}(\mathbf{r}) \cdot d\mathbf{r}\,. \tag{5.169}$$

The terminal voltage V_k is the same for each filament of circuit k. Our definitions for the resistance and the flux linkage will be based on energy considerations.

To this end, we find the total power absorbed by circuit k by integrating the filament powers $V_k\, di_k$ over the cross-section S_k, leading to

$$V_k i_k = \iint_{\tilde{\mathbf{r}}\in S_k} V_k J(\tilde{\mathbf{r}})\, da = \frac{\ell_k}{\sigma_k}\iint_{\tilde{\mathbf{r}}\in S_k} J^2(\tilde{\mathbf{r}})\, da + \iint_{\tilde{\mathbf{r}}\in S_k} J(\tilde{\mathbf{r}})\frac{d\psi(\tilde{\mathbf{r}})}{dt}\, da\,. \tag{5.170}$$

We are after an equation of the form

$$V_k = R_k i_k + \frac{d\lambda_k}{dt}\,, \tag{5.171}$$

where the physical meaning of the two terms is related to the energy balance: Ri^2 should correspond to Joule loss, and $i \cdot d\lambda/dt$ should yield the power absorbed by the magnetic field. We will thus divide Equation (5.170) by

$$i_k = J_{k,dc} A_k\,, \tag{5.172}$$

where $A_k = |S_k|$ is the cross-sectional area (not to be confused with the MVP), and $J_{k,dc}$ is a uniform current density that we would have obtained at d.c. without the skin effect. Hence, we obtain

$$V_k = \frac{\ell_k}{A_k\sigma_k}\frac{i_k}{A_k}\iint_{\tilde{\mathbf{r}}\in S_k}\left(\frac{J(\tilde{\mathbf{r}})}{J_{k,dc}}\right)^2 da + \frac{1}{A_k}\iint_{\tilde{\mathbf{r}}\in S_k}\frac{J(\tilde{\mathbf{r}})}{J_{k,dc}}\frac{d\psi(\tilde{\mathbf{r}})}{dt}\, da\,. \tag{5.173}$$

This leads to the following **generalized definitions for the coil resistance and rate of change of flux linkage**:

$$\boxed{R_k = \frac{1}{A_k} \iint\limits_{\tilde{\mathbf{r}} \in S_k} \left(\frac{J(\tilde{\mathbf{r}})}{J_{k,dc}} \right)^2 R_{k,dc} \, da} \tag{5.174}$$

and

$$\boxed{\frac{d\lambda_k}{dt} = \frac{1}{A_k} \iint\limits_{\tilde{\mathbf{r}} \in S_k} \frac{J(\tilde{\mathbf{r}})}{J_{k,dc}} \frac{d\psi(\tilde{\mathbf{r}})}{dt} \, da \,,} \tag{5.175}$$

respectively. We may thus also define the winding flux linkage as

$$\lambda_k = \frac{1}{A_k} \iint\limits_{\tilde{\mathbf{r}} \in S_k} \frac{J(\tilde{\mathbf{r}})}{J_{k,dc}} \psi(\tilde{\mathbf{r}}) \, da \,. \tag{5.176}$$

These equations are *weighted spatial averages that depend on a given normalized current density distribution*. At low frequencies, the weighting factor is unity, so we obtain $R_k = R_{k,dc}$ and the average MVP flux linkage. However, in the most general case, the conclusion is that *the instantaneous resistance and inductance are not constant*.

These results are mostly of a theoretical nature. In engineering practice, the use of the a.c. resistance and inductance under steady-state conditions is the most common modeling approach. However, the results of a time-domain FEA contain the low-level information necessary for calculating these quantities under arbitrary transient conditions. Hence, should one wish to compute the instantaneous balance of power entering the terminals of a conductor, it can be done using this approach. In this case, substituting the flux linkage equation (5.169), we may express $d\lambda_k/dt$ as an integral over the winding volume V_k:

$$\frac{d\lambda_k}{dt} = \frac{1}{i_k} \iiint\limits_{V_k} \vec{\mathbf{J}}(\mathbf{r}) \cdot \frac{d\vec{\mathbf{A}}(\mathbf{r})}{dt} \, dv \,. \tag{5.177}$$

This form is more suitable for integration with FEA since it employs the MVP. Of course, this is the same result we obtained earlier as Equation (2.163) on page 147.

5.6 Time-Stepping FEA

Consider an initial-value problem defined by the ODE

$$\dot{y} = f(t, y) \,, \tag{5.178}$$

where $f : [0, \infty) \times \mathbb{R} \to \mathbb{R}$ is a known function, and the initial condition is $y(0) = y_0$. In order to solve this problem numerically, suppose that we discretize time by a sequence of time instants t_n, $n = 0, 1, 2, \ldots, N_f$. We refer to the distance of the time instants as a **time step**, $h_n = t_n - t_{n-1}$. We may use a fixed or variable time step. Solving the ODE means finding estimates of the values y_n at each time instant. Numerous ODE numerical integration algorithms have been proposed. Here, we will use a generalized form of the **trapezoidal rule**.

The generalized trapezoidal rule obtains the solution sequentially by

$$y_{n+1} = y_n + h_{n+1} \cdot \left[\beta f(t_n, y_n) + (1 - \beta) f(t_{n+1}, y_{n+1}) \right] , \tag{5.179}$$

with $0 \leq \beta \leq 1$, that is, by means of a convex combination of the derivative function values at the two consecutive time points. If $\beta = 1$, we obtain the **forward Euler** integration scheme, which is an explicit algorithm where y_{n+1} is obtained directly from the previous value. If $\beta = 0$, we obtain the **backward Euler** scheme, which is an implicit algorithm where y_{n+1} is obtained as the solution of Equation (5.179). The special case $\beta = 0.5$ yields the classical trapezoidal rule, and is often referred to as a **Crank**[81]–**Nicolson**[82] method; this is an implicit method as well.[d] In general, implicit ODE integration schemes exhibit better numerical properties [5]. So, for practical purposes, $\beta \in \{0, 0.5\}$.

5.6.1 Time Stepping Using Newton's Method

We adapt the trapezoidal rule to the case of a quasi-magnetostatic problem without mechanical motion, where we need to solve the DAE (5.23)

$$\mathbf{T}\dot{\mathbf{A}} + \mathbf{S}\mathbf{A} - \mathbf{b} = \mathbf{0} , \tag{5.180}$$

over $t \in [0, t_f]$, for $\mathbf{A}(t) \in \mathbb{R}^N$, with initial conditions $\mathbf{A}(0) = \mathbf{A}_0$ and $\dot{\mathbf{A}}(0) = \mathbf{0}$. This means that the device is initially at rest, so we set the initial MVPs to the solution of $\mathbf{S}\mathbf{A}_0 = \mathbf{b}$. (In the absence of PMs, the MVPs will be zero.) This assumption is introduced because initializing an induced current distribution under arbitrary time-varying conditions can be difficult.

Let us first assume that $\mathbf{b} = \mathbf{b}(t)$ is a known function of time. This would be the case if the device is connected to a current source or if it is excited by an external field through the boundary conditions (as in Example 5.4). The matrix $\mathbf{T}$ is typically of low rank (because conducting regions often are a subset of the cross-section), and represents effects of induced eddy current flow. Furthermore, $\mathbf{T}$ is constant since conductors do not change shape; however, $\mathbf{S}$ could be time-varying due to its dependence on the MVP in nonlinear materials.

Discretizing the DAE using the generalized trapezoidal rule:

$$(1/h_{n+1})(\mathbf{T}\mathbf{A}_{n+1} - \mathbf{T}\mathbf{A}_n) + \beta \mathbf{S}_n \mathbf{A}_n + (1 - \beta)\mathbf{S}_{n+1}\mathbf{A}_{n+1} - \beta \mathbf{b}_n - (1 - \beta)\mathbf{b}_{n+1} = \mathbf{0} , \tag{5.181}$$

and by rearranging terms:

$$[(1 - \beta)\mathbf{S}_{n+1} + (1/h_{n+1})\mathbf{T}]\mathbf{A}_{n+1} + [\beta \mathbf{S}_n - (1/h_{n+1})\mathbf{T}]\mathbf{A}_n - \beta \mathbf{b}_n - (1 - \beta)\mathbf{b}_{n+1} = \mathbf{0} . \tag{5.182}$$

Here, the subscripts n and $n + 1$ denote the time instant, rather than the corresponding

[d] The Crank–Nicolson method appears more often in the literature in the context of solving PDEs using the finite difference method (an alternative to the FEM). However, in our variational formulation we have integrated out the spatial coordinates, and thus we are solving an ordinary differential (and algebraic) equation.

element from the MVP vector. If the device is magnetically linear, the stiffness matrix is constant (i.e., independent of the MVP) and the time-stepping equation is linear. However, in devices containing nonlinear magnetic materials, the time stepping involves solving a nonlinear equation of the form $\mathbf{f}(\mathbf{A}_{n+1}) = \mathbf{0}$ for the MVP at the next time instant, $\mathbf{A}_{n+1}$, given the current MVP, $\mathbf{A}_n$, and the external excitation $\mathbf{b}$ at both instants.

Using the Newton–Raphson algorithm (see §3.5.2 on page 276), the $(i + 1)$th MVP iterate of the $(n + 1)$th time step is obtained by solving

$$\mathbf{f}(\mathbf{A}_{n+1}^{(i)}) + \mathbf{J}(\mathbf{A}_{n+1}^{(i)}) \cdot (\mathbf{A}_{n+1}^{(i+1)} - \mathbf{A}_{n+1}^{(i)}) = \mathbf{0}\,, \tag{5.183}$$

where $\mathbf{J}$ is the Jacobian matrix of $\mathbf{f}$:

$$\mathbf{J}(\mathbf{A}_{n+1}^{(i)}) = \left.\frac{\partial \mathbf{f}}{\partial \mathbf{A}_{n+1}}\right|_{\mathbf{A}_{n+1}^{(i)}} = (1-\beta)\,\mathbf{H}(\mathbf{A}_{n+1}^{(i)}) + (1/h_{n+1})\mathbf{T}\,. \tag{5.184}$$

With a slight abuse of notation, we use the same symbol for the Jacobian as for the current density, but we hope that the context clarifies any confusion. We have thus expressed the Jacobian of the DAE as a function of the Hessian $\mathbf{H}$ of the magnetostatic functional. Hence, this calculation can be implemented in a straightforward manner using previously developed code, i.e., the Python function `build_grad_Hess` that was presented in §3.6.2 on page 288. It should be noted that if $\beta = 1$ (i.e., for a forward Euler numerical integration scheme), then the Jacobian becomes $\mathbf{J} = (1/h_{n+1})\mathbf{T}$, which is a singular matrix; therefore, this approach is only suitable for the trapezoidal rule or the backward Euler scheme (i.e., for $\beta < 1$). Also note that

$$\mathbf{f}(\mathbf{A}_{n+1}^{(i)}) = (1-\beta)\mathbf{g}(\mathbf{A}_{n+1}^{(i)}) + \beta\mathbf{g}(\mathbf{A}_n) + (1/h_{n+1})\mathbf{T}(\mathbf{A}_{n+1}^{(i)} - \mathbf{A}_n)\,, \tag{5.185}$$

so we can use the magnetostatic functional gradient $\mathbf{g}$ that is also returned by the function `build_grad_Hess`. There is no superscript on $\mathbf{A}_n$ because this is the final, converged MVP of the Newton–Raphson algorithm from the previous time step. The algorithm terminates once the residual is smaller than a given threshold. One may define an absolute residual

$$r = \|\mathbf{f}(\mathbf{A}_{n+1}^{(i)})\|\,, \tag{5.186}$$

or a normalized residual (cf. Equation (3.161) on page 276)

$$r = \frac{\|\mathbf{f}(\mathbf{A}_{n+1}^{(i)})\|}{\|(1-\beta)\mathbf{b}_{n+1} + \beta\mathbf{b}_n\|}\,. \tag{5.187}$$

Note that the DAE Jacobian $\mathbf{J}$ is symmetric since both $\mathbf{H}$ and $\mathbf{T}$ are symmetric matrices. Also, if the magnetostatic Hessian $\mathbf{H} > 0$, i.e., if it is positive definite (true under mild constraints on the B–H curve, as discussed on page 281), the DAE Jacobian will also be positive definite (for $\beta < 1$). To prove this, for arbitrary $\mathbf{A} \in \mathbb{R}^N$, $\mathbf{A} \neq \mathbf{0}$, we have

$$\mathbf{A}^\top\mathbf{J}\mathbf{A} = (1-\beta)\underbrace{\mathbf{A}^\top\mathbf{H}\mathbf{A}}_{>0} + (1/h_{n+1})\underbrace{\mathbf{A}^\top\mathbf{T}\mathbf{A}}_{\geq 0} > 0\,, \tag{5.188}$$

since $\mathbf{T} \geq 0$. Hence, $\mathbf{J} > 0$, which bodes well for the convergence of the Newton–Raphson algorithm.

The main steps of an elementary time-stepping algorithm (with fixed time step h)

are outlined in Box 5.1. The algorithm has an outer time-stepping loop that stops after N_f steps are computed, and an inner Newton–Raphson loop that terminates after a maximum of N_{iter} iterations. Note that in Step 3i, the parameter $t^{(i)}$ has nothing to do with time! For consistency, we kept the same notation as earlier for the step size of the Newton–Raphson method.

Box 5.1 Programming the FEM: Time stepping with nonlinear magnetics using the Newton–Raphson algorithm

1. Set time index $n := 0$.
2. Initialize $\mathbf{A}_0$ (e.g., using a magnetostatic FEA) and $\dot{\mathbf{A}}_0$ (e.g., $\dot{\mathbf{A}}_0 = \mathbf{0}$).
3. Solve a nonlinear equation for the MVP at the next time step $(n + 1)$:
   a. Set N–R iteration index $i := 1$.
   b. Set initial guess $\mathbf{A}_{n+1}^{(1)}$ (e.g., $\mathbf{A}_{n+1}^{(1)} := \mathbf{A}_n$).
   c. Calculate $B_k^2(\mathbf{A}_{n+1}^{(i)})$ in each element k, and look up $\nu_k^{(i)}$ and $d\nu_k^{(i)}/d(B_k^2)$.
   d. Build the gradient $\mathbf{g}(\mathbf{A}_{n+1}^{(i)})$ and the Hessian $\mathbf{H}(\mathbf{A}_{n+1}^{(i)})$; enforce boundary conditions.
   e. Calculate the function value $\mathbf{f}(\mathbf{A}_{n+1}^{(i)})$ and the residual r.
   f. **If** r is small enough (i.e., less than $\epsilon > 0$), **quit** (go to Step 4); **else**:
   g. Calculate the Jacobian $\mathbf{J}(\mathbf{A}_{n+1}^{(i)})$.
   h. Solve $\mathbf{J}(\mathbf{A}_{n+1}^{(i)})\,\mathbf{x} = -\mathbf{f}(\mathbf{A}_{n+1}^{(i)})$ for the Newton step $\mathbf{x}$.
   i. Update $\mathbf{A}_{n+1}^{(i+1)} := \mathbf{A}_{n+1}^{(i)} + t^{(i)}\mathbf{x}$, where $t^{(i)}$ is a step size.
   j. Increment N–R iteration counter, $i := i + 1$.
   k. **If** $i > N_{\text{iter}}$, **stop; else repeat** from Step c.
4. Increment time index, $n := n + 1$.
5. **If** $n == N_f$, **stop; else repeat** from Step 3.

5.6.2 Time Stepping with External Circuit (No Eddy Currents, No Motion)

A "simple" time-stepping analysis is conducted when the device under study is stationary and does not contain solid conductors; in other words, when it can be safely assumed that the skin effect is negligible, and the current density within its conductors is uniform. This "low-frequency" assumption leads to a lumped-parameter model (see also page 156), wherein $\mathbf{T} = \mathbf{0}$. For example, this type of analysis could be applicable to single- or multi-phase transformers and inductors.

A common case of interest is the analysis of a system where the device modeled using the FEM is energized by an external circuit with voltage sources. Obviously, there are many conceivable topologies for such circuits. For instance, the armature of a three-phase device can be either Y- or Δ-connected, so appropriate constraints must be applied on the currents and/or the voltages. Furthermore, the external circuit can have arbitrary topology, or it could even be a switched power electronic circuit. For a switched circuit in particular, a variable time step is needed (i.e., h_n should not be constant) so that the integration algorithm can capture the switching instants precisely.

Suppose the device has M coils, so let us introduce the vector of coil currents

$$\mathbf{i} = \begin{bmatrix} i_1 & i_2 & \dots & i_M \end{bmatrix}^\top . \tag{5.189}$$

It is convenient to express the vector $\mathbf{b}$ of the functional gradient as

$$\mathbf{b} = \mathbf{b}_0 + \sum_{m=1}^{M} \mathbf{B}_m i_m = \mathbf{b}_0 + \mathbf{B}\mathbf{i}\,, \tag{5.190}$$

where $\mathbf{b}_0$ is a constant that captures the effect of PMs (if any are present). Hence, the columns $\mathbf{B}_m$ of the $N \times M$ matrix $\mathbf{B}$ (not to be confused with the flux density) represent contributions to $\mathbf{b}$ from each individual coil current. As we have explained in §3.4.3 on page 262, the FEA value of the current density in coil regions with turns density n (turns per square meter) is $J = \pm ni$, where the sign depends on the polarity of the winding. Hence, $\mathbf{B}_m$ is assembled by terms of the form $(n_m \Delta_k/3)[1 \quad 1 \quad 1]^\top$, summed over elements k belonging to coil m. Furthermore, the flux linkages may be obtained via the average MVP method as a linear combination of coil MVPs (see Equation (3.108) on page 260). The coil flux linkages can thus be expressed based on the matrix $\mathbf{B}$:

$$\boldsymbol{\lambda} = \ell \mathbf{B}^\top \mathbf{A}\,. \tag{5.191}$$

For the sake of simplicity, let us consider a rudimentary topology where the FEA coils are galvanically isolated and connected individually to external RL (magnetically linear) branches. Hence, using vector notation, the circuit satisfies the ODE

$$\mathbf{v}_{\text{ext}} = \mathbf{R}_{\text{ext}}\mathbf{i} + \mathbf{L}_{\text{ext}}\frac{d}{dt}\mathbf{i} + \mathbf{v}\,, \tag{5.192}$$

where $\mathbf{v}_{\text{ext}}$ is an $M \times 1$ vector of external voltage sources, $\mathbf{R}_{\text{ext}}$ and $\mathbf{L}_{\text{ext}}$ are $M \times M$ resistance and inductance matrices, and $\mathbf{v}$ is an $M \times 1$ vector of voltages at the FEA coil terminals. These voltages are given by

$$\mathbf{v} = \mathbf{R}\mathbf{i} + \frac{d}{dt}\boldsymbol{\lambda}\,, \tag{5.193}$$

where $\mathbf{R}$ is an $M \times M$ diagonal matrix of coil resistance values. Therefore, combining the previous expressions, we obtain the DAE[e]

$$\mathbf{S}\mathbf{A} - \mathbf{B}\mathbf{i} - \mathbf{b}_0 = \mathbf{0}\,, \tag{5.194a}$$

$$\ell \mathbf{B}^\top \frac{d}{dt}\mathbf{A} + \mathbf{L}_{\text{ext}}\frac{d}{dt}\mathbf{i} + (\mathbf{R}_{\text{ext}} + \mathbf{R})\mathbf{i} - \mathbf{v}_{\text{ext}} = \mathbf{0}\,. \tag{5.194b}$$

For a more concise formulation, we may augment the MVP vector (typically containing thousands of elements) with the currents (only a handful):

$$\mathbf{A}' = \begin{bmatrix} \mathbf{A} \\ \mathbf{i} \end{bmatrix}, \tag{5.195}$$

[e] For more physical "symmetry" between the equations, one may wish to multiply the first equation, which is obtained from the magnetostatic functional expressed on a per-unit length basis, by the axial length ℓ.

so that the joint FEA/circuit DAE can be expressed using modified/augmented vectors and matrices (denoted by primes) as

$$\mathbf{T}'\dot{\mathbf{A}}' + \mathbf{S}'\mathbf{A}' - \mathbf{b}' = \mathbf{0}\,. \tag{5.196}$$

The modified matrices are structured as follows:

$$\mathbf{T}' = \begin{bmatrix} \mathbf{0}_{N\times N} & \mathbf{0}_{N\times M} \\ \ell\mathbf{B}^\top & \mathbf{L}_{\text{ext}} \end{bmatrix}, \quad \mathbf{S}' = \begin{bmatrix} \mathbf{S} & -\mathbf{B} \\ \mathbf{0}_{M\times N} & \mathbf{R}_{\text{ext}} + \mathbf{R} \end{bmatrix}, \quad \mathbf{b}' = \begin{bmatrix} \mathbf{b}_0 \\ \mathbf{v}_{\text{ext}} \end{bmatrix}. \tag{5.197}$$

The resultant time-stepping formula is identical in form, $\mathbf{f}(\mathbf{A}'_{n+1}) = \mathbf{0}$, with Equation (5.182):

$$[(1-\beta)\mathbf{S}'_{n+1} + (1/h_{n+1})\mathbf{T}']\mathbf{A}'_{n+1} + [\beta\mathbf{S}'_n - (1/h_{n+1})\mathbf{T}']\mathbf{A}'_n - \beta\mathbf{b}'_n - (1-\beta)\mathbf{b}'_{n+1} = \mathbf{0}\,. \tag{5.198}$$

In expanded form, this function is

$$\begin{aligned} \mathbf{f}(\mathbf{A}^{(i)}_{n+1}, \mathbf{i}^{(i)}_{n+1}) &= (1-\beta)\begin{bmatrix} \mathbf{S}(\mathbf{A}^{(i)}_{n+1}) & -\mathbf{B} \\ \mathbf{0}_{M\times N} & \mathbf{R}_{\text{ext}} + \mathbf{R} \end{bmatrix}\begin{bmatrix} \mathbf{A}^{(i)}_{n+1} \\ \mathbf{i}^{(i)}_{n+1} \end{bmatrix} + \beta\begin{bmatrix} \mathbf{S}(\mathbf{A}_n) & -\mathbf{B} \\ \mathbf{0}_{M\times N} & \mathbf{R}_{\text{ext}} + \mathbf{R} \end{bmatrix}\begin{bmatrix} \mathbf{A}_n \\ \mathbf{i}_n \end{bmatrix} \\ &+ (1/h_{n+1})\begin{bmatrix} \mathbf{0}_{N\times N} & \mathbf{0}_{N\times M} \\ \ell\mathbf{B}^\top & \mathbf{L}_{\text{ext}} \end{bmatrix}\begin{bmatrix} \mathbf{A}^{(i)}_{n+1} - \mathbf{A}_n \\ \mathbf{i}^{(i)}_{n+1} - \mathbf{i}_n \end{bmatrix} - \begin{bmatrix} \mathbf{b}_0 \\ \beta\mathbf{v}_{\text{ext},n} + (1-\beta)\mathbf{v}_{\text{ext},n+1} \end{bmatrix}. \end{aligned} \tag{5.199}$$

This expression can be simplified if we recall that the study is initialized from a d.c. operating point, which implies that $\mathbf{S}\mathbf{A}_0 = \mathbf{b}_0 + \mathbf{B}\mathbf{i}_0$. Hence, using induction, we can show that the time-stepping process will yield $\mathbf{S}\mathbf{A}_n = \mathbf{b}_0 + \mathbf{B}\mathbf{i}_n$, for all n. Therefore, Equation (5.199) becomes

$$\begin{aligned} \mathbf{f}(\mathbf{A}^{(i)}_{n+1}, \mathbf{i}^{(i)}_{n+1}) &= (1-\beta)\begin{bmatrix} \mathbf{S}(\mathbf{A}^{(i)}_{n+1}) & -\mathbf{B} \\ \mathbf{0}_{M\times N} & \mathbf{R}_{\text{ext}} + \mathbf{R} \end{bmatrix}\begin{bmatrix} \mathbf{A}^{(i)}_{n+1} \\ \mathbf{i}^{(i)}_{n+1} \end{bmatrix} + \beta\begin{bmatrix} \mathbf{0}_{N\times N} & \mathbf{0}_{N\times M} \\ \mathbf{0}_{M\times N} & \mathbf{R}_{\text{ext}} + \mathbf{R} \end{bmatrix}\begin{bmatrix} \mathbf{A}_n \\ \mathbf{i}_n \end{bmatrix} \\ &+ (1/h_{n+1})\begin{bmatrix} \mathbf{0}_{N\times N} & \mathbf{0}_{N\times M} \\ \ell\mathbf{B}^\top & \mathbf{L}_{\text{ext}} \end{bmatrix}\begin{bmatrix} \mathbf{A}^{(i)}_{n+1} - \mathbf{A}_n \\ \mathbf{i}^{(i)}_{n+1} - \mathbf{i}_n \end{bmatrix} - \begin{bmatrix} (1-\beta)\mathbf{b}_0 \\ \beta\mathbf{v}_{\text{ext},n} + (1-\beta)\mathbf{v}_{\text{ext},n+1} \end{bmatrix}. \end{aligned} \tag{5.200}$$

This is what we would have obtained had we used the trapezoidal rule on the differential part of the DAE only, i.e., on (5.194b), while keeping the algebraic part as an equality constraint. The DAE Jacobian for the Newton–Raphson method is independent of $\mathbf{i}^{(i)}_{n+1}$:

$$\mathbf{J}(\mathbf{A}^{(i)}_{n+1}) = (1-\beta)\begin{bmatrix} \mathbf{H}(\mathbf{A}^{(i)}_{n+1}) & -\mathbf{B} \\ \mathbf{0}_{M\times N} & \mathbf{R}_{\text{ext}} + \mathbf{R} \end{bmatrix} + (1/h_{n+1})\begin{bmatrix} \mathbf{0}_{N\times N} & \mathbf{0}_{N\times M} \\ \ell\mathbf{B}^\top & \mathbf{L}_{\text{ext}} \end{bmatrix}. \tag{5.201}$$

It should be noted that the Jacobian in this case is a (sparse) *nonsymmetric* matrix, which is an important consideration while choosing an appropriate solver for the update equation (5.183). However, we can obtain a symmetric system with algebraic manipulations, which could be beneficial computationally. The key idea is to eliminate the currents by noting that the bottom-right block of the Jacobian,

$$\mathbf{J}_{2,2} = (1-\beta)(\mathbf{R}_{\text{ext}} + \mathbf{R}) + (1/h_{n+1})\mathbf{L}_{\text{ext}}\,, \tag{5.202}$$

is invertible by virtue of the positive definiteness of the resistance and inductance matrices. This manipulation is referred to as a Kron[83] reduction. Let us thus write Equation (5.183) as two separate equations:

$$\mathbf{J}_{1,1}\Delta\mathbf{A} + \mathbf{J}_{1,2}\Delta\mathbf{i} = -\mathbf{f}_1\,, \tag{5.203a}$$

$$\mathbf{J}_{2,1}\Delta\mathbf{A} + \mathbf{J}_{2,2}\Delta\mathbf{i} = -\mathbf{f}_2\,. \tag{5.203b}$$

Then, subtracting $\mathbf{J}_{1,2}\mathbf{J}_{2,2}^{-1}$ times the second equation from the first one, we obtain

$$(\mathbf{J}_{1,1} - \mathbf{J}_{1,2}\mathbf{J}_{2,2}^{-1}\mathbf{J}_{2,1})\Delta\mathbf{A} = -\mathbf{f}_1 + \mathbf{J}_{1,2}\mathbf{J}_{2,2}^{-1}\mathbf{f}_2\,. \tag{5.204}$$

Now, the reduced Jacobian matrix is

$$\mathbf{J}_{1,1} - \mathbf{J}_{1,2}\mathbf{J}_{2,2}^{-1}\mathbf{J}_{2,1} = (1-\beta)\cdot[\mathbf{H}(\mathbf{A}_{n+1}^{(i)}) + (\ell/h_{n+1})\mathbf{B}\mathbf{J}_{2,2}^{-1}\mathbf{B}^\top]\,. \tag{5.205}$$

This is a *symmetric* matrix because both $\mathbf{H}$ and $\mathbf{J}_{2,2}$ are symmetric. Furthermore, if $\mathbf{H} > 0$, the reduced Jacobian will be positive definite and thus invertible.[f] A two-step algorithm is thus applicable: first, calculate $\Delta\mathbf{A}$ using Equation (5.204), and second, obtain $\Delta\mathbf{i}$ from

$$\mathbf{J}_{2,2}\Delta\mathbf{i} = -\mathbf{f}_2 - \mathbf{J}_{2,1}\Delta\mathbf{A}\,. \tag{5.206}$$

Example 5.7 *Transformer energization.*

Let us revisit Example 3.5 on page 263, involving an elementary single-phase transformer with a 2:1 turns ratio. All dimensions remain identical. The core, which was magnetically linear previously, is now assembled with a nonlinear ferromagnetic lamination, which has the *B–H* curve of Figure 3.23 on page 289. (Magnetic hysteresis is neglected.) A unity stacking factor is assumed. The present case study is a time-domain energization transient. The primary side is connected to a 10-V (r.m.s.), 100-Hz a.c. voltage source through a switch that closes at time $t = 0$, while the secondary side is connected to a series-connected *RL* load with power factor 0.8. We shall apply the above DAE time stepping formulation to this problem, closely following the steps in Box 5.1.

Solution

Suppose that the primary side has 20 turns of AWG-13 copper wire, and the secondary side has 10 turns of AWG-9 wire. The two coil areas are thus packed approximately at 40%. The circuit has the following parameters:

$$\mathbf{R} = \begin{bmatrix} 17.58 & 0 \\ 0 & 3.48 \end{bmatrix} \text{m}\Omega,\ \mathbf{R}_{\text{ext}} = \begin{bmatrix} 8.79 & 0 \\ 0 & 69.53 \end{bmatrix} \text{m}\Omega,\ \mathbf{L}_{\text{ext}} = \begin{bmatrix} 0.1 & 0 \\ 0 & 0.083 \end{bmatrix} \text{mH}\,, \tag{5.207}$$

and

$$\mathbf{v}_{\text{ext}} = \begin{bmatrix} \sqrt{2}\,10\sin(2\pi 100t) \\ 0 \end{bmatrix} \text{V}\,. \tag{5.208}$$

The time-stepping solver uses $\beta = 0.5$ (Crank–Nicolson scheme), a constant time step

[f] This also implies that the original Jacobian is invertible, by means of a Schur[84] complement argument.

$h = 200$ μs, a constant Newton–Raphson update step size $t = 0.6$, and a stopping criterion $\epsilon = 10^{-8}$ (for the normalized residual).[g] We simulate the device for 1 s or 100 cycles, which is roughly how long it takes to reach the steady state. The quantities of interest are the two coil currents and flux linkages, which are obtained by post-processing the MVP at each time step. Differentiating the flux linkages with respect to time yields the back-emf, which is then used to obtain the terminal voltages after accounting for the ohmic drops in the coils. The magnetizing current

$$i_m = i_1 - \frac{i_2}{2}, \tag{5.209}$$

which is proportional to the magnetizing MMF (via N_1), is also of interest because it is indicative of the saturation level within the core. (The minus sign signifies a convention wherein the two MMFs are opposing each other.)

The simulation results are shown in Figures 5.21 and 5.22 for the initial five cycles and the steady state, respectively. We observe that the saturation in the core during the first few cycles reaches very high levels, based on the harmonic content in the waveforms. The magnetizing-flux mean-path length in this device is 0.36 m, so when the magnetizing current reaches a value of 80 A, Ampère's law implies that the average magnetic field inside the core is $20 \cdot 80/0.36 = 4444.4$ A/m, which is way above the knee of the B–H curve.[h] At steady state, the magnetizing current is more sinusoidal, and reaches a peak of approximately 2 A, which corresponds to a magnetic field of 111.1 A/m, so we are operating below the knee of the B–H curve, where the device is roughly magnetically linear. Hence, at steady state, the voltages and currents are sinusoidal, and they exhibit the relationships that one would expect, that is, $v_1/v_2 \approx N_1/N_2 = 2$ and $i_1/i_2 \approx 1/2$.

We proceed with a discussion of a few common topologies that can be represented with this kind of model. This is not meant to be a comprehensive listing or an abstract methodology for arbitrary networks. Nevertheless, the examples should help the reader formulate the equations for circuits of interest.

End Turns

The preceding formulation can be used to capture the effects of winding end turns. These can become part of the external circuit, if it is safe to assume that the magnetic field of the end turns does not couple the FEA region. In general, the magnetic field of the end turns penetrates into the steel close to the end regions; however, in many cases this is a reasonable simplifying assumption to make for justifying a 2-D FEA, instead of a more complicated and computationally heavy 3-D analysis. The windings may have multiple end connections, but since the resistances and inductances of the end turns

[g] This simulation takes several hours to complete on today's hardware. Arguably, it may be more efficient to first extract a mapping from the currents to the flux linkages using FEA, and then to employ this function inside a variable-step ordinary differential equation solver.

[h] The voltage phase at the instant when the switch is closed is an important parameter affecting this transient.

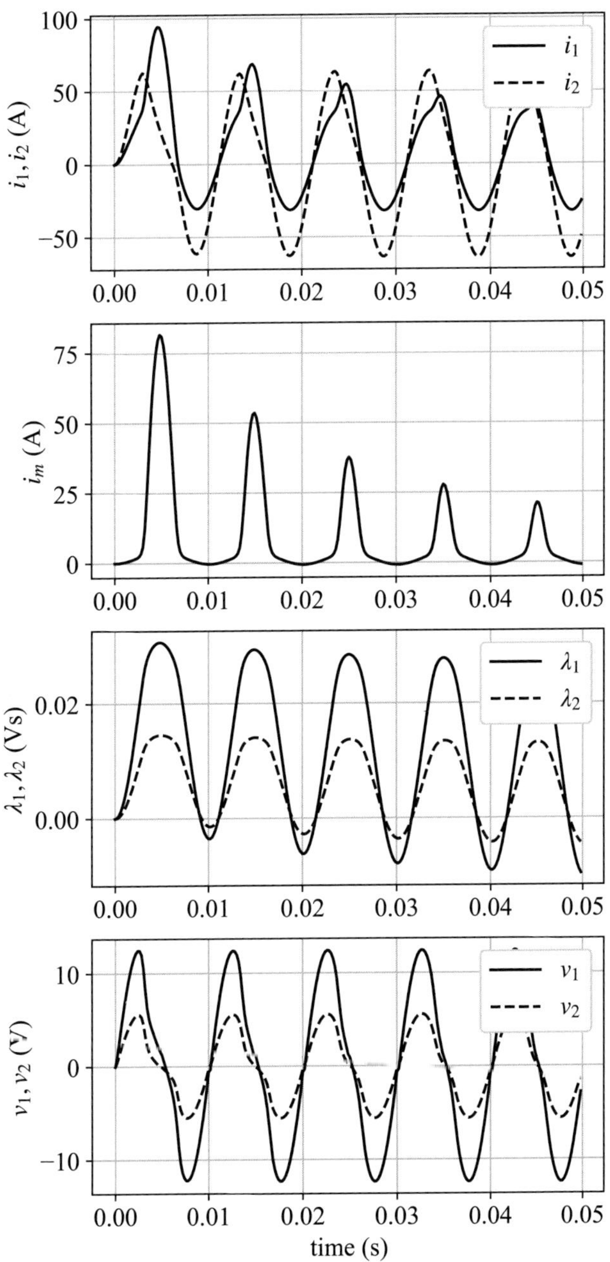

Figure 5.21 Transformer energization transient (first five cycles).

are electrically series-connected, they can be lumped together and incorporated into the external resistance and inductance matrices. Cross-coupling effects between the end turns can be taken into account by adding mutual-inductance terms in $\mathbf{L}_{\text{ext}}$. Analytical expressions for calculating these parameters can be found in the literature, e.g., in [6].

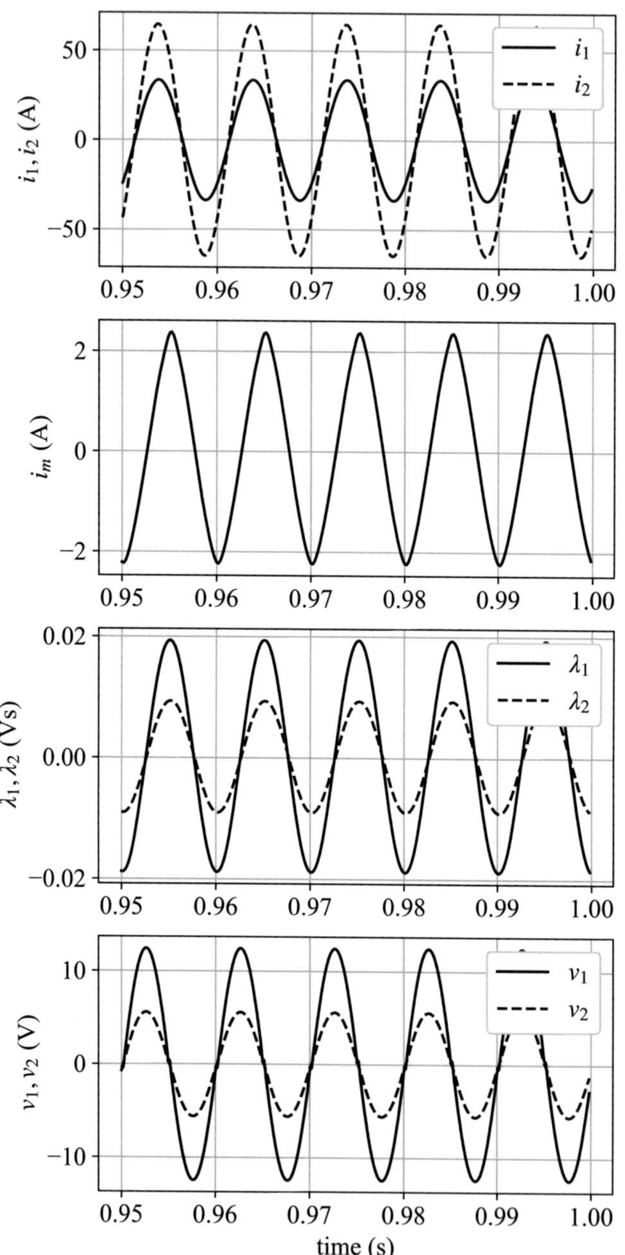

Figure 5.22 Transformer energization transient (steady state).

Three-Phase Y-connection without Neutral Wire

Suppose that the device under study is a three-phase device (i.e., $M = 3$) with Y-connected windings without neutral wire. The windings are series connected to the external circuit, as shown in Figure 5.23. In this topology, the phase currents must sum to

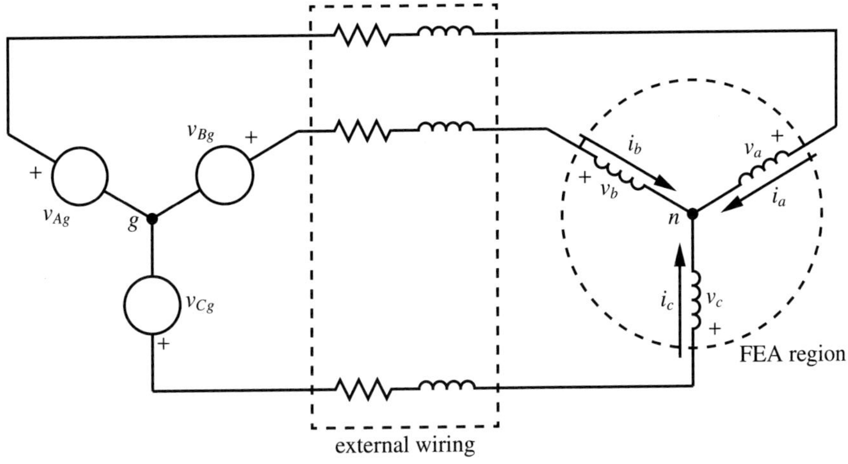

Figure 5.23 Three-phase Y-connected FEA device without neutral wire. (For the sake of illustration, the voltage source is also shown as Y-connected, and its neutral point is the "ground." The d.c. resistance of the FEA region coils is not shown for simplicity.)

zero due to KCL at the neutral point of the star. This constraint can be expressed as

$$i_a + i_b + i_c = \mathbf{1}_3^{\mathsf{T}}\mathbf{i} = 0\,, \tag{5.210}$$

where $\mathbf{1}_3$ is a column vector of ones. Further suppose that we measure the external voltages

$$\mathbf{v}_{\text{ext}} = \begin{bmatrix} v_{Ag} & v_{Bg} & v_{Cg} \end{bmatrix}^{\mathsf{T}}\,, \tag{5.211}$$

with respect to a common point in the external circuit that we call the "ground"; whereas the machine voltages across the coils,

$$\mathbf{v} = \begin{bmatrix} v_a & v_b & v_c \end{bmatrix}^{\mathsf{T}}\,, \tag{5.212}$$

are line-to-neutral. The sources could form a three-phase sinusoidal balanced set; however, this is not necessary. For instance, the voltage sources could be unbalanced, or even pulse-width-modulated waveforms output from an inverter (albeit such a simulation would be computationally expensive).

Application of KVL yields

$$\mathbf{v}_{\text{ext}} = \mathbf{R}_{\text{ext}}\mathbf{i} + \mathbf{L}_{\text{ext}}\frac{d}{dt}\mathbf{i} + \mathbf{v} + v_{ng}\mathbf{1}_3\,, \tag{5.213}$$

where v_{ng} is the potential difference between the neutral point of the star winding and ground. The FEA region terminal voltages are still given by Equation (5.193). We define the augmented MVP vector

$$\mathbf{A}' = \begin{bmatrix} \mathbf{A} \\ \mathbf{i} \\ v_{ng} \end{bmatrix}. \tag{5.214}$$

The joint FEA/circuit DAE is of the form (5.196), with matrices

$$\mathbf{T}' = \begin{bmatrix} \mathbf{0}_{N\times N} & \mathbf{0}_{N\times 3} & \mathbf{0}_{N\times 1} \\ \ell\mathbf{B}^\top & \mathbf{L}_\text{ext} & \mathbf{0}_{3\times 1} \\ \mathbf{0}_{1\times N} & \mathbf{0}_{1\times 3} & 0 \end{bmatrix}, \quad \mathbf{S}' = \begin{bmatrix} \mathbf{S} & -\mathbf{B} & \mathbf{0}_{N\times 1} \\ \mathbf{0}_{3\times N} & \mathbf{R}_\text{ext} + \mathbf{R} & \mathbf{1}_3 \\ \mathbf{0}_{1\times N} & \mathbf{1}_3^\top & 0 \end{bmatrix}, \quad \mathbf{b}' = \begin{bmatrix} \mathbf{b}_0 \\ \mathbf{v}_\text{ext} \\ 0 \end{bmatrix}. \tag{5.215}$$

Essentially, we have added one equation (the current constraint) and one new variable (v_{ng}) to the system of equations.

Three-Phase Δ-Connection

If the device is Δ-connected, as shown in Figure 5.24, we relate the winding currents with those of the external circuit using KCL:

$$\mathbf{i}_\text{ext} = \begin{bmatrix} 1 & -1 & 0 \\ 0 & 1 & -1 \\ -1 & 0 & 1 \end{bmatrix} \begin{bmatrix} i_a \\ i_b \\ i_c \end{bmatrix} = \mathbf{Q}\mathbf{i}. \tag{5.216}$$

Note that the external currents sum to zero; however, the Δ-connected winding currents could contain a circulating component that, by definition, does not affect the line currents. (This is driven by a zero-axis time-varying flux linkage in the coils, λ_0.) Applying KVL around the three loops formed by each side of the Δ and the two external lines it connects to, yields

$$\mathbf{Q}^\top\mathbf{v}_\text{ext} = \mathbf{Q}^\top\mathbf{R}_\text{ext}\mathbf{i}_\text{ext} + \mathbf{Q}^\top\mathbf{L}_\text{ext}\frac{d}{dt}\mathbf{i}_\text{ext} + \mathbf{v}. \tag{5.217}$$

The transpose of the $\mathbf{Q}$-matrix that relates the Δ-currents to the line currents is also used to form the line-to-line voltages, selecting the appropriate external circuit voltage drop terms. Using Equations (5.193) and (5.216), we eliminate the device voltages and the external currents:

$$\mathbf{Q}^\top\mathbf{v}_\text{ext} = (\mathbf{Q}^\top\mathbf{R}_\text{ext}\mathbf{Q} + \mathbf{R})\,\mathbf{i} + (\mathbf{Q}^\top\mathbf{L}_\text{ext}\mathbf{Q})\frac{d}{dt}\mathbf{i} + \ell\mathbf{B}^\top\frac{d}{dt}\mathbf{A}. \tag{5.218}$$

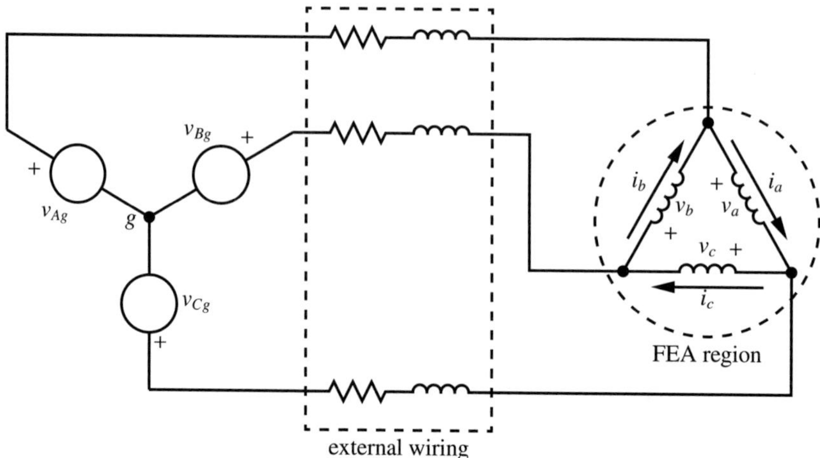

Figure 5.24 Three-phase Δ-connected FEA device.

Observe that by summing the three equations, all $\mathbf{Q}^\top\mathbf{x}$ terms vanish, and we obtain the equation of the circulating current component (cf. (4.118c) on page 371):

$$0 = R\mathbf{1}_3^\top\mathbf{i} + \ell(\mathbf{B}_1^\top + \mathbf{B}_2^\top + \mathbf{B}_3^\top)\frac{d}{dt}\mathbf{A}. \tag{5.219}$$

The augmented MVP vector is defined as

$$\mathbf{A}' = \begin{bmatrix}\mathbf{A}\\ \mathbf{i}\end{bmatrix}. \tag{5.220}$$

The joint FEA/circuit DAE is of the form (5.196), with matrices

$$\mathbf{T}' = \begin{bmatrix}\mathbf{0}_{N\times N} & \mathbf{0}_{N\times 3}\\ \ell\mathbf{B}^\top & \mathbf{Q}^\top\mathbf{L}_{\text{ext}}\mathbf{Q}\end{bmatrix}, \quad \mathbf{S}' = \begin{bmatrix}\mathbf{S} & -\mathbf{B}\\ \mathbf{0}_{3\times N} & \mathbf{Q}^\top\mathbf{R}_{\text{ext}}\mathbf{Q} + \mathbf{R}\end{bmatrix}, \quad \mathbf{b}' = \begin{bmatrix}\mathbf{b}_0\\ \mathbf{Q}^\top\mathbf{v}_{\text{ext}}\end{bmatrix}. \tag{5.221}$$

5.6.3 Time Stepping with External Circuit and Eddy Currents (No Motion)

Now, let us consider devices containing conductive regions allowing the flow of eddy currents. Windings with uniform current distribution can also be present. This case requires a simple modification of the equations that we have presented, namely (i) incorporating a nonzero $\mathbf{T}$ matrix in the DAE and (ii) adding constraints for the net current of the conducting regions. In the absence of external connections, we should impose a zero net current constraint, as we did in Example 5.4. Otherwise, we must model the interconnections of the conducting regions via the end regions.

We add one current constraint for each conducting region, relating the corresponding net current i_c with the terminal voltage v_c. Each of these can be expressed as

$$i_c = \frac{v_c}{R_c} - \mathbf{c}^\top\frac{d}{dt}\mathbf{A}, \tag{5.222}$$

where $\mathbf{c}$ is a column vector assembled from terms like (5.159), i.e., $(\sigma\Delta_k/3)[1\ \ 1\ \ 1]^\top$. Collecting all conductor currents and voltages in the vectors $\mathbf{i}_c$ and $\mathbf{v}_c$, respectively, we have

$$\mathbf{R}_c\mathbf{i}_c = \mathbf{v}_c - \mathbf{R}_c\mathbf{C}^\top\frac{d}{dt}\mathbf{A}, \tag{5.223}$$

introducing a diagonal resistance matrix $\mathbf{R}_c$ and a matrix $\mathbf{C}$ whose columns are the $\mathbf{c}$-vectors. It can be readily shown that

$$\mathbf{R}_c\mathbf{C}^\top = \ell\mathbf{B}_c^\top, \tag{5.224}$$

where the columns of $\mathbf{B}_c$ are assembled similarly to those of $\mathbf{B}$, using a turns density of 1 turn over the area of the conducting region.

Let us assume that the filamentary windings are connected to an elementary external circuit, as in §5.6.2. We have thus arrived at the DAE for $\mathbf{A}$, $\mathbf{i}$, $\mathbf{v}_c$, and $\mathbf{i}_c$:

$$\mathbf{T}\frac{d}{dt}\mathbf{A} + \mathbf{S}\mathbf{A} - \mathbf{B}\mathbf{i} - \mathbf{B}_c\mathbf{R}_c^{-1}\mathbf{v}_c - \mathbf{b}_0 = \mathbf{0}, \tag{5.225a}$$

$$\ell\mathbf{B}_c^\top\frac{d}{dt}\mathbf{A} + \mathbf{R}_c\mathbf{i}_c - \mathbf{v}_c = \mathbf{0}, \tag{5.225b}$$

$$\ell \mathbf{B}^\top \frac{d}{dt}\mathbf{A} + \mathbf{L}_{\text{ext}}\frac{d}{dt}\mathbf{i} + (\mathbf{R}_{\text{ext}} + \mathbf{R})\mathbf{i} - \mathbf{v}_{\text{ext}} = \mathbf{0}\,, \tag{5.225c}$$

$$\text{terminal equations for conducting regions: } \mathbf{g}_c(\cdot) = \mathbf{0}\,. \tag{5.225d}$$

The terminal equations depend on the case study. For instance, if we require that the conductor currents are zero, then $\mathbf{g}_c(\cdot) = \mathbf{i}_c$. In general, the $\mathbf{g}_c$ equations are obtained by applying KCL/KVL to the external connections (similarly to how we establish the external circuit equation for the windings), so these typically are linear equations involving $\mathbf{i}_c$, $\mathbf{v}_c$, and/or their time derivatives.

As a somewhat more concrete example, suppose that the eddy-current regions are mutually galvanically isolated, and that they are connected in series with external *RL* circuits with parameters placed in diagonal matrices $\mathbf{R}_{c,\text{ext}}$, $\mathbf{L}_{c,\text{ext}}$, and voltage-source vector $\mathbf{v}_{c,\text{ext}}$. For instance, this could be the model of an inductor or transformer with bar windings. Also, suppose that we are not interested in having the voltages $\mathbf{v}_c$ as variables in the formulation (e.g., because we may not care to know their values); we can thus eliminate them, as we have done for the winding voltages $\mathbf{v}$. This leads to the following DAE for $\mathbf{A}$, $\mathbf{i}$, and $\mathbf{i}_c$:

$$(\mathbf{T} - \ell \mathbf{B}_c\mathbf{R}_c^{-1}\mathbf{B}_c^\top)\frac{d}{dt}\mathbf{A} + \mathbf{S}\mathbf{A} - \mathbf{B}\mathbf{i} - \mathbf{B}_c\mathbf{i}_c - \mathbf{b}_0 = \mathbf{0}\,, \tag{5.226a}$$

$$\ell \mathbf{B}^\top \frac{d}{dt}\mathbf{A} + \mathbf{L}_{\text{ext}}\frac{d}{dt}\mathbf{i} + (\mathbf{R}_{\text{ext}} + \mathbf{R})\mathbf{i} - \mathbf{v}_{\text{ext}} = \mathbf{0}\,, \tag{5.226b}$$

$$\ell \mathbf{B}_c^\top \frac{d}{dt}\mathbf{A} + \mathbf{L}_{c,\text{ext}}\frac{d}{dt}\mathbf{i}_c + (\mathbf{R}_{c,\text{ext}} + \mathbf{R}_c)\mathbf{i}_c - \mathbf{v}_{c,\text{ext}} = \mathbf{0}\,. \tag{5.226c}$$

5.6.4 Time Stepping With Motion

Time stepping with motion comes in two flavors. The first and more straightforward one is where motion is a prescribed function of time. This can be solved using the previous approach with minor modifications (namely, that we should update the mesh and the stiffness matrix at each time step at the given position). For example, we could time-step a PMSM or a WRSM rotating at constant speed, connected to a sinusoidal three-phase voltage source, thereby observing harmonics in the current waveforms; whereas, in the magnetostatic case studies of the previous chapter, we imposed sinusoidal currents and observed the ripple in the flux linkages (in turn, this reflects on the terminal voltages).

The second flavor, which is of interest here, is when motion is governed by a set of differential equations, which need to be solved for the position and velocity of the moving member. This complicates the assembly of the stiffness matrix because the position of the moving component is not exactly known at the next time instant. We will focus this material on the case of rotational motion; linear motion can be treated in a similar fashion. This model could be used, for instance, to study a squirrel-cage induction motor start-up transient. However, it is arguably the most computationally demanding type of FEA.

Hence, we augment the DAE with the mechanical equations (see (4.80) and (4.81) from page 336)

$$M\frac{d}{dt}\omega_{rm} = \tau_e(\mathbf{A}, \theta_{rm}) - \tau_m(\theta_{rm}, \omega_{rm}, t) - \tau_f(\omega_{rm}), \tag{5.227a}$$

$$\frac{d}{dt}\theta_{rm} = \omega_{rm}, \tag{5.227b}$$

where τ_e is the electromagnetic torque (by convention here, this is positive for motor action), τ_m is an externally applied mechanical load torque on the shaft (depending on the application, this could be a function of position, speed, or time), τ_f is a viscous friction function, and M is the combined moment of inertia of the machine and its prime mover or load (notation changed from J to avoid confusion with the Jacobian). Based on our previous work in §4.1, we know that there are various methods for calculating electromagnetic torque based on the FEA solution. In principle, any one of these methods can be used. Let us calculate the torque using the VDM:

$$\tau_e = -\frac{\ell\nu_0}{2}\sum_{k\in\mathcal{E}_{ag}}(\mathbf{A}^k)^\top\frac{\partial\mathbf{S}^k}{\partial\theta_{rm}}\mathbf{A}^k, \tag{5.228}$$

as it conveniently exposes directly the MVP values (only the air-gap elements $\mathcal{E}_{ag}$ are involved) and the angle; whereas the other torque calculation methods, e.g., the MST or Arkkio's method, have been expressed based on the B-field, so further algebraic manipulations may be needed.

For example, suppose that we have a three-phase Y-connected device without neutral wire. An augmented MVP vector is defined as

$$\mathbf{A}' = \begin{bmatrix}\mathbf{A} & \mathbf{i} & v_{ng} & \omega_{rm} & \theta_{rm}\end{bmatrix}^\top. \tag{5.229}$$

In the absence of eddy currents (without loss of generality, but merely to reduce somewhat the complexity of the equations), the joint FEA/circuit/mechanical DAE is of the form (5.196), with matrices

$$\mathbf{T}' = \begin{bmatrix}\mathbf{0}_{N\times N} & \mathbf{0}_{N\times 3} & \mathbf{0}_{N\times 1} & \mathbf{0}_{N\times 1} & \mathbf{0}_{N\times 1}\\ \ell\mathbf{B}^\top & \mathbf{L}_{\text{ext}} & \mathbf{0}_{3\times 1} & \mathbf{0}_{3\times 1} & \mathbf{0}_{3\times 1}\\ \mathbf{0}_{1\times N} & \mathbf{0}_{1\times 3} & 0 & 0 & 0\\ \mathbf{0}_{1\times N} & \mathbf{0}_{1\times 3} & 0 & M & 0\\ \mathbf{0}_{1\times N} & \mathbf{0}_{1\times 3} & 0 & 0 & 1\end{bmatrix}, \tag{5.230a}$$

$$\mathbf{S}' = \begin{bmatrix}\mathbf{S} & -\mathbf{B} & \mathbf{0}_{N\times 1} & \mathbf{0}_{N\times 1} & \mathbf{0}_{N\times 1}\\ \mathbf{0}_{3\times N} & \mathbf{R}_{\text{ext}}+\mathbf{R} & \mathbf{1}_3 & \mathbf{0}_{3\times 1} & \mathbf{0}_{3\times 1}\\ \mathbf{0}_{1\times N} & \mathbf{1}_3^\top & 0 & 0 & 0\\ \mathbf{0}_{1\times N} & \mathbf{0}_{1\times 3} & 0 & 0 & 0\\ \mathbf{0}_{1\times N} & \mathbf{0}_{1\times 3} & 0 & -1 & 0\end{bmatrix}, \quad \mathbf{b}' = \begin{bmatrix}\mathbf{b}_0\\ \mathbf{v}_{\text{ext}}\\ 0\\ \tau_e-\tau_m-\tau_f\\ 0\end{bmatrix}. \tag{5.230b}$$

The main difference from the previous cases is that the fourth element of the $\mathbf{b}'$-vector (i.e., the net torque τ) is a function of the unknowns. This nonzero gradient affects the Jacobian of the time-stepping formula (5.198), which becomes

$$\mathbf{J}(\mathbf{A}_{n+1}^{(i)}) = (1-\beta)\begin{bmatrix} \mathbf{H}(\mathbf{A}_{n+1}^{(i)}) & -\mathbf{B} & \mathbf{0}_{N\times 1} & \mathbf{0}_{N\times 1} & \mathbf{0}_{N\times 1} \\ \mathbf{0}_{3\times N} & \mathbf{R}_{\text{ext}}+\mathbf{R} & \mathbf{1}_3 & \mathbf{0}_{3\times 1} & \mathbf{0}_{3\times 1} \\ \mathbf{0}_{1\times N} & \mathbf{1}_3^\top & 0 & 0 & 0 \\ -\partial\tau/\partial\mathbf{A} & \mathbf{0}_{1\times 3} & 0 & -\partial\tau/\partial\omega_{rm} & -\partial\tau/\partial\theta_{rm} \\ \mathbf{0}_{1\times N} & \mathbf{0}_{1\times 3} & 0 & -1 & 0 \end{bmatrix}$$
$$+ (1/h_{n+1})\begin{bmatrix} \mathbf{0}_{N\times N} & \mathbf{0}_{N\times 3} & \mathbf{0}_{N\times 1} & \mathbf{0}_{N\times 1} & \mathbf{0}_{N\times 1} \\ \ell\mathbf{B}^\top & \mathbf{L}_{\text{ext}} & \mathbf{0}_{3\times 1} & \mathbf{0}_{3\times 1} & \mathbf{0}_{3\times 1} \\ \mathbf{0}_{1\times N} & \mathbf{0}_{1\times 3} & 0 & 0 & 0 \\ \mathbf{0}_{1\times N} & \mathbf{0}_{1\times 3} & 0 & M & 0 \\ \mathbf{0}_{1\times N} & \mathbf{0}_{1\times 3} & 0 & 0 & 1 \end{bmatrix}, \quad (5.231)$$

where

$$\partial\tau/\partial\mathbf{A} = \partial\tau_e/\partial\mathbf{A}\,, \quad (5.232a)$$
$$\partial\tau/\partial\omega_{rm} = -\partial\tau_m/\partial\omega_{rm} - \partial\tau_f/\partial\omega_{rm}\,, \quad (5.232b)$$
$$\partial\tau/\partial\theta_{rm} = \partial\tau_e/\partial\theta_{rm} - \partial\tau_m/\partial\theta_{rm}\,. \quad (5.232c)$$

The derivatives of τ_m and τ_f can be readily calculated based on the torque functions specified by the problem statement. So, let us focus briefly our attention on the derivatives of the electromagnetic torque.

Differentiating the electromagnetic torque equation (5.228) with respect to the MVPs, which is a quadratic form with symmetric matrix, we obtain

$$\frac{\partial\tau_e}{\partial\mathbf{A}} = -\ell\nu_0 \sum_{k\in\mathcal{E}_{ag}} (\mathbf{A}^k)^\top \frac{\partial\mathbf{S}^k}{\partial\theta_{rm}}\,. \quad (5.233)$$

The derivative of the electromagnetic torque with respect to position may be evaluated by taking the second derivative of the air-gap element stiffness matrices:

$$\frac{\partial\tau_e}{\partial\theta_{rm}} = -\frac{\ell\nu_0}{2} \sum_{k\in\mathcal{E}_{ag}} (\mathbf{A}^k)^\top \frac{\partial^2\mathbf{S}^k}{\partial\theta_{rm}^2}\mathbf{A}^k\,. \quad (5.234)$$

After some elementary algebra, we obtain

$$\frac{\partial^2\mathbf{S}^k}{\partial\theta_{rm}^2} = -\frac{2}{\Delta_k}\frac{\partial\Delta_k}{\partial\theta_{rm}}\frac{\partial\mathbf{S}^k}{\partial\theta_{rm}} - \frac{1}{\Delta_k}\frac{\partial^2\Delta_k}{\partial\theta_{rm}^2}\mathbf{S}^k + \frac{1}{4\Delta_k}\frac{\partial}{\partial\theta_{rm}}\left[\frac{\partial\mathbf{q}^k}{\partial\theta_{rm}}(\mathbf{q}^k)^\top + \mathbf{q}^k\frac{\partial(\mathbf{q}^k)^\top}{\partial\theta_{rm}} + \frac{\partial\mathbf{r}^k}{\partial\theta_{rm}}(\mathbf{r}^k)^\top + \mathbf{r}^k\frac{\partial(\mathbf{r}^k)^\top}{\partial\theta_{rm}}\right]. \quad (5.235)$$

These expressions can be calculated using Equation (4.52) on page 328 and other auxiliary formulae in that section. (These details are left as an exercise for the reader.)

The time-stepping algorithm follows the steps outlined in Box 5.1 on page 468; however, there is an important difference due to the rotating nature of this problem. As we know, an initial guess of the unknowns is required for the Newton–Raphson inner loop (Step 3b). As far as the MVPs and circuit variables are concerned, a simple scheme could be to initialize them using the converged values from the previous time step. The mechanical variables may be initialized using a forward Euler integration from the previous time step, that is

$$\theta^{(1)}_{rm,n+1} = \theta_{rm,n} + h_{n+1}\omega_{rm,n} , \tag{5.236a}$$

$$\omega^{(1)}_{rm,n+1} = \omega_{rm,n} + h_{n+1}\tau_n/M . \tag{5.236b}$$

In each iteration i, before updating the DAE Jacobian, we must first rotate the rotor at the newest position-iterate, $\theta^{(i)}_{rm,n+1}$, remesh the air-gap region, and reassemble the global gradient and Hessian, as explained in §4.2 on page 333.

5.7 Summary

In this final chapter, we focused on how to analyze (and implement in Python) problems involving the spatial distribution of eddy currents as an unknown function of time. After a brief review of physics, we studied how the skin effect impacts the electromagnetic properties of conductors, assisted by a variety of analytically tractable examples. We proceeded with the formulation of phasor-based FEA in magnetically linear (but potentially lossy) devices operating in the sinusoidal steady state. Finally, we presented the main ideas of numerical integration in the time domain, underpinning the analysis of magnetically nonlinear devices driven by external circuits, including the case of mechanical motion.

5.8 Further Reading

This concludes the material of this text; however, it could be the starting point for future explorations. There are many fascinating topics that we have not covered, which could be of interest to the electric machine analyst, such as higher-order elements, 3-D FEM, alternative electromagnetic formulations (e.g., boundary element methods), error analysis, meshing and adaptive meshing algorithms, sparse linear solvers, DAE solvers, various formulations for eddy current problems, advanced hysteresis models, methods for modeling the end-turn regions, thermal modeling, mechanical (noise, vibration, and harshness) analysis, multi-physics modeling, and FEA-based optimization for machine design. The literature on these topics is rich and constantly growing, reflecting the high level of research activity in electric machines. The various references that appear throughout the book could serve as a launch pad for further study.

Problems

5.1 Prove that $\mathbf{T}^k$ is a positive definite matrix.

5.2 Prove Equation (5.47). Then plot $|\tilde{J}_y(z)|$ for various choices of τ/δ.

5.3 Plot the fraction containing the sinusoidal terms in Equation (5.48) as a function of τ/δ. Determine using an analytical calculation the peak value of the loss, and the relationship between thickness and skin depth at this point. When is it safe to neglect the eddy current loss?

5.4 Prove that Equation (5.55) is equivalent to Equation (5.48).

5.5 Prove Equations (5.62) and (5.63). Plot these expressions as a function of τ/δ for various choices of θ_h. (It may be convenient to normalize the result by dividing by $H_{\max}^2/\sigma\delta$.)

5.6 Replicate the results of Example 5.2. Repeat the analysis by considering the effects of magnetic hysteresis. Model the material with an elliptic hysteresis loop defined by $\theta_h = 10°$.

5.7 Prove Equation (5.105).

5.8 Consider the elliptical locus of the B-field given by Equation (5.138). Determine the angle of rotation of the ellipse major axis with respect to the x-axis.

5.9 Replicate the results of Example 5.3.

5.10 Replicate the results of Example 5.4. Repeat the analysis for multiple values of velocity and spatial field periodicity, and discuss the results.

5.11 Replicate the results of Example 5.5. Repeat the analysis for multiple values of the conductor distance d, and discuss the results.

5.12 Replicate the results of Example 5.6.

5.13 Replicate the results of Example 5.7.

5.14 Form the DAE Jacobian corresponding to Equation (5.215). Show how a symmetric Jacobian can be obtained using algebraic manipulations (e.g., by eliminating $\mathbf{i}$ and v_{ng} from the equations). Establish whether the Jacobian is invertible.

5.15 Form the DAE Jacobian corresponding to Equation (5.221). Show how a symmetric Jacobian can be obtained using algebraic manipulations. Establish whether the Jacobian is invertible.

5.16 Prove Equation (5.224).

5.17 The DAE Jacobian for case studies with rotational motion requires the evaluation of Equation (5.235). Derive expressions for $\partial^2\Delta_k/\partial\theta_{rm}^2$, as well as $\partial^2\mathbf{q}^k/\partial\theta_{rm}^2$ and $\partial^2\mathbf{r}^k/\partial\theta_{rm}^2$.

References

[1] J. Lammeraner and M. Štafl, *Eddy Currents*. London: Iliffe, 1966.

[2] R. L. Stoll, *The Analysis of Eddy Currents*. Oxford: Clarendon Press, 1974.

[3] E. E. Kriezis, T. D. Tsiboukis, S. M. Panas, and J. A. Tegopoulos, "Eddy currents: Theory and applications," *Proc. IEEE*, vol. 80, no. 10, pp. 1559–1589, Oct. 1992.

[4] N. Bianchi and G. Berardi, "Analytical approach to design hairpin windings in high performance electric vehicle motors," in *IEEE Energy Conversion Congress and Exposition (ECCE)*, Portland, OR, Sep. 2018, pp. 4398–4405.

[5] W. Gautschi, *Numerical Analysis*, 2nd ed. Basel, Switzerland: Birkhäuser, 2012.

[6] S. D. Sudhoff, *Power Magnetic Devices: A Multi-Objective Design Approach*, 2nd ed. Hoboken, NJ: Wiley-IEEE Press, 2021.

Notes

1 Plato (c. 428–c. 348 BC). Athenian philosopher. Student of Socrates, teacher of Aristotle. Founder of the Academy, the first institution of higher learning in the Western world. This epigraph is from Plato's *Laws*, Book 6, Section 753e.

2 It could be argued that a literal interpretation of the inscription is an oxymoron since one's primary purpose for visiting the Academy would probably be to learn geometry! Perhaps the inscription's real purpose was to convey a message to prospective students, namely, the belief that natural phenomena can be understood using geometrical concepts and a philosophical, inquiring mindset.

3 James Clerk Maxwell (1831–1879). Scottish physicist, famous for unifying the theory of classical electromagnetism.

4 Johann Carl Friedrich Gauss (1777–1855), aka "the greatest mathematician since antiquity." Enough said!

5 Michael Faraday (1791–1867). English scientist and experimentalist who conducted a famous experiment with two coils on Aug. 29, 1831. He was also an accomplished chemist. The SI unit of capacitance is named after him.

6 André-Marie Ampère (1775–1836). French physicist and mathematician, considered one of the founders of electromagnetism. The SI unit of the electric current is named in his honor.

7 Nikola Tesla (1856–1943). Serbian-American electrical engineer, inventor of the induction motor, and proponent of a.c. transmission.

8 The term SI comes from the French *Système International d'unités*, i.e., the International System of Units.

9 René Descartes (1596–1650). French philosopher, mathematician, and scientist, aka "the father of modern western philosophy."

10 Euclid of Alexandria (mid fourth to mid third century BC). Greek mathematician, aka "the father of geometry."

11 Pythagoras of Samos (c. 570–495 BC). Greek philosopher and mathematician, with a rather famous theorem named after him.

12 John Henry Poynting (1852–1914). English physicist, who derived the theorem bearing his name that explains the conservation of energy in an electromagnetic field.

13 Joseph Henry (1797–1878). American scientist, who discovered the phenomenon of self-inductance, and after whom the SI unit of inductance is named.

14 Charles-Augustin de Coulomb (1736–1806). French military engineer and physicist, after whom the SI unit of charge is named.

15 Alessandro Giuseppe Antonio Anastasio Volta (1745–1827). Italian physicist and chemist, who invented the electric battery. The physical quantity that we call voltage is named in his honour.

16 Georg Friedrich Bernhard Riemann (1826–1866). German mathematician, who had great impact on modern mathematics. Gauss was his doctoral advisor. Countless things are named after him.

17 Brook Taylor (1685–1731). English mathematician, best known for the theorem that bears his name.

18 Hans Christian Ørsted (1777–1851). Danish physicist and chemist. He discovered that an electric current creates a magnetic field in 1820, by noticing a compass needle deflect when an electric current was nearby.
19 Carl Gustav Jacob Jacobi (1804–1851). German mathematician, who made numerous fundamental contributions.
20 Pierre-Simon, marquis de Laplace (1749–1827). French mathematician, physicist, and astronomer. The famous transform (that engineers may be familiar with) is named after him, among numerous other things.
21 Siméon Denis Poisson (1781–1840). Prolific French mathematician and physicist. Laplace and Lagrange were his doctoral advisors.
22 Ludvig Valentin Lorenz (1829–1891). Danish physicist and mathematician. Not to be confused with Hendrik Lorentz.
23 Mikhail Vasilyevich Ostrogradsky (1801–1862). Russian mathematician and physicist.
24 Paul Adrien Maurice Dirac (1902–1984). English physicist and Nobel prize winner, who made great contributions to quantum physics.
25 Sir George Gabriel Stokes (1819–1903). British physicist and mathematician. The famous Navier–Stokes equations that describe the motion of fluids are named after him (and Claude-Louis Navier).
26 William Thomson, 1st Baron Kelvin (1824–1907). Scots-Irish physicist and engineer, after whom the SI unit of temperature is named.
27 George Green (1793–1841). British mathematician, who worked on the mathematics underpinning electromagnetic theory. Remarkably, he was self-taught.
28 Peter Gustav Lejeune Dirichlet (1805–1859). German mathematician, who made contributions to number theory and mathematical physics.
29 Carl Neumann (1832–1925). German mathematician, who made contributions to mathematical physics.
30 Hermann Ludwig Ferdinand von Helmholtz (1821–1894). German physician and physicist. Apart from his numerous contributions to physics, he is also known for his pioneering study of the human eye.
31 Wilhelm Eduard Weber (1804–1891). German physicist, collaborator of Gauss. Introduced the symbol c for the speed of light, and invented the first telegraph.
32 Jean-Baptiste Biot (1774–1862). French physicist and astronomer, who made significant contributions to optics and our understanding of polarized light. He also showed that some rocks found on earth are of extraterrestrial origin.
33 Félix Savart (1791–1841). French physicist and mathematician, who also worked in acoustics.
34 Hendrik Antoon Lorentz (1853–1928). Famous Dutch physicist and Nobel prize winner. Worked on the theory of relativity with Einstein.
35 Erik Ivar Fredholm (1866–1927). Swedish mathematician.
36 Boris Grigoryevich Galerkin (1871–1945). Soviet mathematician and civil engineer, whose work set the stage for modern FEA.
37 Archimedes of Syracuse (c. 287–c. 212 BC). Greek mathematician and engineer, perhaps the greatest of all time. This epigraph is quoted by Pappus, *Synagoge*, Book 8, Section 11.
38 Leonhard Euler (1707–1783). Swiss mathematician, scientist, and engineer. One of the greatest mathematicians in history, perhaps the most prolific ever. His name is pronounced "Oiler."
39 Joseph-Louis Lagrange (1736–1813). Italian mathematician. He worked in Berlin, Prussia and later in Paris, France. You may visit his tomb at the Panthéon.
40 Johann Bernoulli (1667–1748). Swiss mathematician. Euler was his student.
41 Sir Isaac Newton (1642–1726). English philosopher, and one of the greatest scientists of all time.

42 Eugenio Beltrami (1835–1900). Italian mathematician.
43 Pierre-Ernest Weiss (1865–1940). French physicist, who developed the theory of magnetism.
44 John William Strutt, 3rd Baron Rayleigh (1842–1916). Famous and highly decorated English physicist and Nobel prize winner.
45 Anders Celsius (1701–1744). Swedish astronomer. The centigrade temperature scale is named in his honor.
46 James Watt (1736–1819). Scottish inventor, after whom the SI unit of power is named. His invention of the modern steam engine was of great importance during the Industrial Revolution.
47 Georg Simon Ohm (1789–1854). German physicist, who discovered the famous law that relates voltage, current, and resistance.
48 Jean-Baptiste Joseph Fourier (1768–1830). French mathematician and physicist. He worked on the theory of heat flow, and discovered the greenhouse effect.
49 James Prescott Joule (1818–1889). English physicist, who studied the relationship between heat and mechanical work.
50 Werner von Siemens (1816–1892). German inventor and founder of the company that still bears his name.
51 Jean Bernard Léon Foucault (1819–1868). French physicist. His pendulum hangs from the dome of the Panthéon in Paris.
52 Charles Proteus Steinmetz (1865–1923). German-American electrical engineer and pioneer of a.c. electricity. Among his numerous contributions are the ones on hysteresis and a.c. circuit analysis using complex numbers (phasors).
53 Heinrich Friedrich Emil Lenz (1804–1865). Russian physicist.
54 Adrien-Marie Legendre (1752–1833). Famous French mathematician.
55 John Hopkinson (1849–1898). British physicist and electrical engineer. He is credited with the invention of three-phase distribution.
56 Leopold Kronecker (1823–1891). German mathematician. Dirichlet was his doctoral supervisor.
57 Luigi Aloisio Galvani (1737–1798). Italian scientist, who is credited with the discovery of bioelectromagnetism.
58 Gottfried Wilhelm (von) Leibniz (1646–1716). Prolific German polymath and philosopher. Among other countless contributions, he is credited with the discovery of calculus (alongside Isaac Newton). He introduced the integral sign $\int$.
59 Pierre de Fermat (1607–1665). French lawyer and mathematician hobbyist. He is famous for his important contributions to calculus and number theory. His "last theorem" is legendary.
60 This epigraph is attributed to Pythagoras, see, e.g., Aristotle, *Metaphysics*, Book 1, Section 987b [10].
61 For more information on Triangle – A Two-Dimensional Quality Mesh Generator and Delaunay Triangulator, see: `www.cs.cmu.edu/%7equake/triangle.html`
62 Boris Nikolaevich Delaunay (1890–1980). Soviet mathematician and master mountain climber.
63 Jakob Steiner (1796–1863). Swiss mathematician, who specialized in geometry.
64 Joseph Raphson (1648–1715). English mathematician, who discovered the root-finding method that bears his name at about the same time as Isaac Newton.
65 Ludwig Otto Hesse (1811–1874). German mathematician. Kirchhoff was his student, among other notable scientists.
66 Charles Hermite (1822–1901). French mathematician. A crater on the Moon is named in his honor.
67 Walther Ritz (1878–1909). Swiss physicist. A lunar crater is named after him.
68 Sergei Lvovich Sobolev (1908–1989). Soviet mathematician.
69 David Hilbert (1862–1943). German mathematician.

70 Richard Courant (1888–1972). German-American mathematician and pioneer of the FEM.
71 Augustin-Louis Cauchy (1789–1857). Famous French mathematician and physicist, who had a profound impact in numerous areas.
72 Karl Hermann Amandus Schwarz (1843–1921). German mathematician.
73 Heraclitus of Ephesus (c. 535–c. 475 BC). Greek philosopher. This epigraph can be found in: H. Ritter et L. Preller, *Historia Philosophiae Graecae*. Editio octava, quam curavit Eduardus Wellmann. Gotha, 1898, Fragment 31.
74 Thomas Simpson (1710–1761). British mathematician. The quadrature rule that bears his name was discovered much earlier.
75 Harry Nyquist (1889–1976). American electronic engineer.
76 André Blondel (1863–1938). French engineer. Inventor of the oscillograph (a precursor of the modern oscilloscope), and pioneer of electrical power engineering.
77 Robert H. Park (1902–1994). American electrical engineer, best known for the transformation that bears his name.
78 Chilon of Sparta (c. 600–c. 520 BC). One of the Seven Sages of Greece.
79 John Napier (1550–1617). Scottish landowner, who discovered logarithms.
80 Friedrich Wilhelm Bessel (1784–1846). German astronomer and mathematician.
81 John Crank (1916–2006). British mathematician and physicist, who worked on partial differential equations.
82 Phyllis Nicolson (1917–1968). British mathematician and physicist, who worked on the heat equation.
83 Gabriel Kron (1901–1968). Hungarian American electrical engineer.
84 Issai Schur (1875–1941). Russian mathematician.

Index

action, 104
air gap, 123
  leakage flux in, 137
  magnetic energy stored in, 131
  magnetic field in, 129
  power flow through, 124, 134–136, 196
  rotating magnetic field in, 132
  virtual distortion of, 181
Ampère, André-Marie, 2
  law for the magnetic field, 2, 69
antiperiodic conditions, 417
Archimedes, 96
armature, 122
asynchronous machine, 196
axisymmetric problem, 78, 167, 313
  linear FEA, 294
  nonlinear FEA, 296

B–H characteristic
  anhysteretic, 117, 154
  elliptical, 436
  extrapolation, 287
  hysteretic, 152, 436
  impact on convergence, 281
  interpolation, 283
  modification of, due to stacking, 291
  nonlinear, 117, 152
  saturation, 117
  visualization, 284
back-emf, *see* electromotive force
BASIC, 11
Beltrami, Eugenio, 101
  identity, 101
Bernoulli, Johann, 100
Bessel, Friedrich, 448
Biot, Jean-Baptiste, 60
Biot–Savart, law of, 60
Blondel, André, 360
bound current, *see* current density
brachistochrone, 100
brushless d.c. motor, 381

calculus of variations, 97
  and Maxwell's equations, 107
  for determining equations of motion, 104
  fundamental lemma of, 99
  principle of stationary action, 104
Cauchy, Augustin-Louis, 302
Celsius, Anders, 119
charge density, 2
  as source of the electrostatic field, 59
  moving, 142
Chilon, 425
coenergy, *see* coupling field
coercivity, 119
cogging torque, *see* electromagnetic torque
coil, *see* winding
collocation method, 85
conductivity, 142
conductor, 140
connection matrix, 226
conservation law
  of momentum, 187
  of angular momentum, 192
  of charge, 2, 141
  of energy
    in electric circuit, 156
    in an a.c. electric machine, 197
    in an electromechanical device, 121
    in the electromagnetic field, 122
constitutive law, 2
  for dielectric polarization, 110
  for ferromagnetic materials, 114, 117
  for permanent magnets, 114, 118
convex function, 279, 281
coordinate system
  Cartesian, 4
  change of basis, 6
  orthogonal transformation, 6
  orthonormality of axes, 39
  polar, 5
  reflection of orthogonal basis, 6
  rotation of orthogonal basis, 6
  spherical, 25
  transformation matrix, 7
core loss, 148, 149, 269, 453
Coulomb, Charles, 16
  gauge, 43
coupling field, 120

coenergy, 158, 159, 312
conservative, 121, 161
energy, 146, 147, 156, 160, 163
in air gap, 131
in magnetically linear system, 161
quadratic form of, 161
through winding, 157
time average of, 439
energy density, 148
time average of, 439
mutual coupling between coils, 263
Courant, Richard, 299
Crank, John, 466
cross product, 8
and triangle area calculation, 10
anti-commutative property of, 8
as a determinant, 8
distributive property of, 9
cross-coupling, *see* magnetic axis
curl, 33
as a determinant, 33
in polar coordinates, 35
invariance to axes selection, 41
current density, 2
as source of the magnetostatic field, 60
as surface density, 38
from moving charge, 142
in bar conductor inside a slot, 462
in busbar, 445, 447
in circular conductor, 449
in conductor element, 428
in conductors, 426
in lamination, 150
in moving conductor, 458
in rectangular conductor, 455
in semi-infinite plate, 433
of bound current, 68
of free current, 37, 68

del, 27
Delaunay, Boris, 252
triangulation, 252
delta function, 48
derivative
convective, 44
material, 44
total, 44
Descartes, René, 4
differential algebraic equation
discretized form, 466
for time stepping with eddy currents, 477
for time stepping with external circuit, 469
for time stepping with motion, 479
general form for time-domain FEA, 429
diffusion equation, 431
Dirac, Paul, 48
direct axis, 343
directional cosine, 39
Dirichlet, Peter, 54
disjoint numbering system, 226
divergence, 30
in Cartesian coordinates, 30
invariance to axes selection, 40
theorem, 47
volume integral of, 47
dot product, 4
algebraic definition of, 4
commutative property of, 5
geometric definition of, 5
invariance to orthogonal transformation, 6
orthogonality property of, 5
doubly fed induction generator, 198
drift velocity, 141

eddy currents, 149
in lamination, 150
in semi-infinite plate, 429
loss due to, 151, 434, 437, 445, 455
resistance in equivalent circuit, 269
specific loss due to, 152
electric displacement field, 2
electric field, 2
in 2-D problems, 426
in conductor reference frame, 427
in moving conductors, 426
irrotational, 43
unit of, 16
electrical angle, 129
electrical frequency, 132
electrical rotor speed, 336
electromagnet
cylindrical, 167, 313
horseshoe, 165, 189
electromagnetic field
angular momentum of, 191, 192
momentum of, 182, 184, 187
stress tensor, 184
force calculation based on, 188
geometric interpretation of, 188
electromagnetic force
based on coupling field coenergy, 164, 312
based on coupling field energy, 163, 321
based on energy balance, 205
based on Maxwell stress tensor, 187, 315
on straight wire, 144
resultant, 180
electromagnetic torque
based on air-gap MVP DFT, 330
based on coupling field coenergy, 164, 312
based on coupling field energy, 163, 327
based on energy balance, 206
based on Maxwell stress tensor, 193, 195, 317
based on virtual distortion of air gap, 182
cogging component of, 164, 384, 395

contour plot of, 386, 397, 401
developed by elementary machine, 173, 174, 178
developed by PMSM, 384, 386, 394–396, 401
developed by SRM, 412
developed by WRSM, 345, 355, 363, 375
equivalence of methods, 195
in *qd* variables, 372
on permanent magnet, 211
reluctance component of, 174, 363
resultant, 180
ripple component of, 386
electromotive force, 142, 154
back-emf, 345
electrostatic field
calculating by variational approach, 110
created by charge distribution, 59
created by point charges, 16
curl of, 59
divergence of, 59
Maxwell's equations for, 3, 109
potential of, 19, 60
elementary electric machine, 169
end turns, 472
energy balance, *see* conservation of energy
equations of motion, 104
Euclid, 4
Euler, Leonhard, 98
backward integration scheme, 466
forward integration scheme, 466
Euler–Lagrange equation
for a 1-D problem, 99
for a 2-D problem, 104
for equation of motion, 105

Faraday, Michael, 2
law for the electromagnetic field, 2
law of induction, 155
Fermat, Pierre, 214
field winding, 123
filament, 141
fixed point, 275
flux, 77
decomposition in *qd* components, 360
in two-reaction theory, 360
leakage component of, 137, 253, 266
in interior PMSM, 394
magnetizing component of, 266
main path of, 253
flux barrier, 394
flux density, 2
as function of element MVP, 255, 295
in air gap, 129, 270
in materials, 68
in PMSM, 384, 394, 396
in SRM, 409
in WRSM, 348, 360, 362
magnitude of, as function of element MVP, 255, 295
of dipole, 66
phasor of, 452
unit of, 4, 16
flux lines
in air gap, 127, 270
in bar conductor inside a slot, 462
in PMSM, 384, 394, 396
in SRM, 409
in transformer, 266
in WRSM, 348, 360, 362
of transmission line, 74, 461
plotting of, 253
refraction of, 71
flux linkage
calculation from axisymmetric FEA, 296
calculation from FEA, 258
calculation of, using the average MVP, 260
contour plot of, 399
equations, general form in machines, 336
in *qd* variables, 374, 385, 396
in PMSM, 384, 395, 396
in SRM, 412
in WRSM, 345, 360, 363, 374
of filament, 145, 155
of winding, 157
flux weakening, 402
force
electromagnetic, *see* electromagnetic force
Lorentz, *see* Lorentz, Hendrik
FORTRAN, 11
Foucault, Léon, 149
currents, *see* eddy currents
Fourier, Joseph, 126
Fredholm, Ivar, 82
frictional loss, 121, 198
function
ansatz, 299
norm of, 298
energetic, 298
orthogonality property of, 298
space of, 297
square integrable, 297
test, 297
trial, 299
functional, 97
for electrostatic problem, 110
for magnetostatic problem, 114
contribution from axisymmetric element to, 295, 296
contribution from element to, 273
in axisymmetric device, 294
relationship to electric circuit variables, 158
for motion of particle, 104
for Poisson's equation, 107

stationarity of, 97, 221, 234

Galerkin, Boris, 85
  method, 297
    orthogonality property of, 300, 305
Galvani, Luigi, 197
Gauss, Carl Friedrich, 2
  law for the electric field, 2
  law for the magnetic field, 2
  pillbox, 32
  surface, 48
  theorem, 47
global basis function, 299
gradient, 27
  and conservative vector fields, 29
  as direction of steepest increase, 28
  as directional derivative, 28
  as necessary condition for optimality, 29
  for phasor-based FEA, 452
  in Cartesian coordinates, 27
  in first-order Taylor expansion, 28
  invariance to axes selection, 40
  of magnetostatic functional, 276
    assembly of, 288, 334
    contribution from element to, 274, 296, 429
    contribution from PM element to, 293
    enforcing boundary conditions, 279
  orthogonality to level sets, 28
  volume integral of, 48
Green, George, 53
  first identity, 53
  second identity, 53
  theorem, 55

Helmholtz, Hermann von, 56
  theorem, 56
Henry, Joseph, 16
Heraclitus, 311
Hermite, Charles, 284
  piecewise cubic polynomial interpolation, 284
Hesse, Ludwig, 277
  Hessian matrix, *see* Hessian matrix
Hessian matrix
  of magnetostatic functional, 277
    assembly of, 288
    contribution from nonlinear element to, 278, 296
    contribution from PM element to, 294
    enforcing boundary conditions, 279
Hilbert, David, 298
Hopkinson, John, 167
  law for magnetic circuits, 167
hysteresis
  angle, 436
  elliptical model of, 435
  in equivalent circuit, 269
  in ferromagnetic materials, 152
  in permanent magnets, 119
  loss due to, 154, 436
  major loop, 153
  minor loop, 153

impedance, 440, 441
  internal, 442
  of busbar, 446
  of circular conductor, 449
  of transmission line, 461
inductance, 160, 440
  calculation from FEA, 261
  leakage component of, 266
  magnetizing component of, 266
  matrix, 160
    positive definiteness of, 161
  reciprocity property of, 161
inner product
  between functions, 97, 298
    energetic, 298, 303
  between vectors, *see* dot product
integration by parts
  in Euler–Lagrange derivation, 99, 103
  in two dimensions, 55
interface conditions
  for normal component of magnetic field, 32, 70
  for tangential component of magnetic field, 37, 71
  modeling as smooth changes, 181
isoline, 19
isotropic materials, 117

Jacobi, Carl, 40
  Jacobian matrix, 40
    of DAE, 467
Joule, James, 142
  law of conductor heating, 142

Kelvin, William Thomson, 52
Kron, Gabriel, 471
Kronecker, Leopold, 184

Lagrange, Joseph-Louis, 98
  Lagrangian, for equations of motion, 104
lamination, 149
  eddy currents in, *see* eddy currents
  stacking factor, 291
Laplace, Pierre-Simon, 41
  equation, 41
    weak form of, 221
  Laplacian operator, 41
  vector Laplacian operator, 43
    on magnetic vector potential, 43
Legendre, Adrien-Marie, 158
Leibniz, Gottfried, 202
Lenz, Emil, 155
level set, 19
line search, 281
  backtracking method for, 281
linear element, 222
  basis function for, 223

contribution to functional, 428
contribution to functional by, 232
Lorentz, Hendrik, 66
force, 66, 144, 183
derived by variational approach, 106
on test charge, 105
Lorenz, Ludvig, 43
gauge, 43
lumped-parameter model, 158

magnetic axis, 171
cross-coupling due to saturation, 354
magnetic charge density, 205
magnetic circuit, 167
magnetic dipole, 66
moment, 66
torque on, 66
magnetic field
energy stored in, *see* coupling field
in ferromagnetic materials, 69
rotating, 122, 131, 134, 346
angular velocity of, 132
magnetic field intensity, 2, 69
unit of, 36
magnetic flux, *see* flux
magnetic flux density, *see* flux density
magnetic hysteresis, *see* hysteresis
magnetic induction, *see* flux density
magnetic linearity, 159
magnetic permeability, *see* permeability
magnetic polarization, *see* magnetization
magnetic poles, 128, 175
magnetic scalar potential, 210
magnetic vector potential, 43, 60
equipotentials as flux lines, 74, 80
gauge fixing, 62
gradient of, 74, 80
in air gap, 126, 270
in linear element, 231, 255
of dipole, 65
of long straight wire, 45
properties in 2-D problems, 74
properties in axisymmetric problems, 80
relationship to flux in 2-D problems, 77
rotating in air gap, 132
magnetization, 67
as function of magnetic field, 69
curve, *see* *B–H* characteristic
in ferromagnetic materials, 68
in permanent magnets, 70
macroscopic relationship to flux density, 69
modeled as equivalent currents, 68
magnetomotive force, 166, 266
drop, 172
of a distributed winding, 342
vector, 342
decomposition in *qd* components, 348
magnetostatic field, 60
curl of, 60
divergence of, 60
Maxwell's equations for, 3, 113
of long straight wire, 16, 36, 45
potential, *see* magnetic vector potential
solving by a variational approach, 114
Matlab, 12
differences with NumPy, 19
similarity with Python Matplotlib Pyplot, 18
Matplotlib
contour function, 19
equal axes in plots, 19
Pyplot library, 18
quiver function, 17, 19
semilogy plotting command, 52
show command, 19
streamplot function, 17, 19
tricontour function, 253
tricontourf function, 455
tripcolor function, 253
matrix transpose, 7
maximum torque per ampere, 399
maximum torque vs. speed curve, 401
Maxwell, James Clerk, 2
equations
differential form, 2
for stationary fields, 3
in magnetized matter, 183
integral form, 2
macroscopic form, 2
stress tensor, *see* electromagnetic field
mechanical angle, 129
mechanical rotor speed, 336
mesh, 220
data structures, 238
plotting, 252
using Triangle software, 240
moments, method of, 81
momentum
angular, 190
linear, 185

nabla, 27
Napier, John, 442
Neumann, Carl, 54
Newton, Isaac, 100
method for root finding, *see* Newton–Raphson algorithm
second law of, 185, 190
Newton–Raphson algorithm
for nonlinear magnetostatic FEA, 276
for nonlinear time stepping FEA, 468
Nicolson, Phyllis, 466
norm
of function, *see* function
of vector, 4

NumPy, 11
  n-dimensional array (ndarray), 11
  cross function, 12
  dot function, 12
  element-wise operations on vectors, 12
  evenly spaced numbers (linspace), 18
  linear algebra (linalg) package, 12
  meshgrid function, 18
  norm function, 12
Nyquist, Harry, 330

Ohm, Georg, 120
  loss, 121, 142, 156
Ørsted, Hans, 36
Ostrogradsky, Mikhail, 47
  theorem, 47

packing factor, 262
parallelepiped volume, based on scalar triple product, 13
Park, Robert, 363
  transformation to the rotor reference frame, 363
pchip, *see* Hermite, Charles
periodic conditions, 414
permanent magnets, 70, 118
  FEM modeling of, 293
permanent-magnet synchronous machine, 197, 381
  elementary, 175
  equivalent circuit, 387
  interior, 390
  surface-mounted, 381
permeability
  complex, 433, 451
  effective, 439
  hysteretic, 436
  incremental, 287
  of free space, 16, 60
  of material, 69
  relative, 69
  single-valued function of, 114
permeance, 269
  leakage component of, 269
  magnetizing component of, 269
permittivity, 16, 59
phasor, 431
planar straight-line graph, 249
Plato, 1
point matching method, 85
Poisson, Siméon Denis, 41
  equation, 41
    for the electrostatic problem, 108
    integration of, 53
    nonlinear form of, 230
    solution by variational calculus, 107
    weak form of, 234, 302
positive definite matrix, 279, 282
power
  electrical, at filament terminals, 144
  exhanged between filament and field, 144
  mechanical, at shaft, 196
  reactive, 439
  real, 439
  rotational motion, 14
power factor, 440
Poynting, John Henry, 8
  complex vector, 438
  theorem, 122, 201, 438
    relationship with real and reactive power, 439
  vector, 121
principle of least action, 104
principle of stationary action, 104
proximity loss, 447
Pythagoras, 4, 219
Python, 11
  accessing array elements, 12
  chained assignments, 18
  comments, 11
  complex numbers, 451
  for loop, 18, 52
  function definition, 51
  importing libraries, 12
  list, 12
  list comprehension, 52
  loading data from file, 239
  long line wrapping, 51
  math symbols in plots, 52
  Matplotlib, *see* Matplotlib
  NumPy, *see* NumPy
  plotting a mesh, 252
  printing to screen, 12
  quiver plot, *see* Matplotlib
  range, 18
  raw strings, 52
  running a program, 11
  scalar field visualization, *see* Matplotlib
  SciPy, *see* SciPy
  streamline plot, *see* Matplotlib
  string concatenation, 52
  syntax, 12
  vector field visualization, *see* Matplotlib
  zero-based indexing, 12

quadrature axis, 343

radial flux, 127
Raphson, Joseph, 275
Rayleigh, John, 117
reactance, 440
reactive power, *see* power
real power, *see* power
relaxation factor, 276
reluctance, 166
reluctance machine, 173, 403
reluctance torque, *see* electromagnetic torque
reluctivity, 117
  complex, 452

remanence, 118
resistance
  a.c., 440, 445
  of busbar, 445
  of filament, 156
  of winding, 157
resistivity, 151
Riemann, Bernhard, 21
  sum, 21
rigid-body motion, 178
Ritz, Walther, 297
rotating magnetic field, *see* magnetic field
rotor, 123
  angle of, 169

Savart, Félix, 60
scalar field, 15
  gradient of, *see* gradient
  line integral of, 20
  surface integral of, 23
  visualization, 19
  volume integral of, 26
Schur, Issai, 471
Schwarz, Hermann, 302
SciPy, 86
  integration library, 86
  interpolation library, 284
  sparse linear algebra library, 247
  sparse matrix, 244
    in COO format, 245
    in CSC format, 247
    in CSR format, 247
Siemens, Werner, 142
Simpson, Thomas, 317
skin depth, 432, 442
skin effect, 432
slip, 196
slip power, 197
slip rings, 197
slot, 170
  fill factor, 262
Sobolev, Sergei, 298
sparse matrix, 229, 242
  stored in compressed sparse column format, 247
  stored in compressed sparse row format, 247
  stored in coordinate format, 245
squirrel-cage induction machine, 197
stability, 174
stator, 123
steady-state operation, 195
Steiner, Jakob, 252
Steinmetz, Charles Proteus, 153
  hysteresis loss formula, 153
step length, 280
stiffness matrix
  assembly of, 228, 242
  enforcing boundary conditions, 237, 246
  global, 227
  of single element, 225
Stokes, George, 52
  theorem, 52
subsynchronous operation, 196
supersynchronous operation, 196
susceptibility, 69
switched reluctance machine, *see* reluctance machine
synchronous machine, 196
  analysis in $qd0$ axes, 360
  cylindrical wound-rotor, 337
  equivalent circuit, 376
  permanent-magnet, 381, 390
  salient-pole wound-rotor, 356

Taylor, Brook, 21
teeth, 338
Tesla, Nikola, 4
test charge, 106
three-phase Δ-connection, 476
three-phase Y-connection, 474
time step, 465
torque, 15
  electromagnetic, *see* electromagnetic torque
torque ripple, *see* electromagnetic torque
trace, 40
transformer
  equivalent circuit of, 264
  single-phase, 263, 471
trapezoidal rule, 466
triangle area
  based on coordinates of vertices, 10, 222
  imposing maximum area constraint, 252
two-reaction theory, 360

variation, 98
  of function in $\mathbb{R}^2$, 102
variational calculus, *see* calculus of variations
vector, 3
  angle between vectors, 5, 8
  cross product, *see* cross product
  dot product, *see* dot product
  field, *see* vector field
  high-dimensional, 4
  in Cartesian coordinates, 4
  inner product, *see* dot product
  length of, 4
  magnitude of, 4
  norm of, 4
  notation, 3, 4
  of unit length, 4
  orthogonality, 5
  physical, 3
  reflection of, 7
  rotation of, 7
  scalar triple product, 13, 14
  vector product, *see* cross product

vector triple product, 13, 15
vector calculus, 1
identities, 42
vector field, 15
circulation of, 20
complex, 438
conservative, 21, 29
divergence of, *see* divergence
incompressible, 30
irrotational, 33
line integral of, 20
solenoidal, 30
surface integral of, 24
interpretation as flux, 24
on closed surface, 24
visualization of, 16
volume integral of, 27
vector product, *see* cross product
virtual displacement method, 321, 479
Volta, Alessandro, 16
voltage
ellipses, contour plot of, 401
induced, 154
of winding, 157
Watt, James, 120
Weber, Wilhelm, 60
Weiss, Pierre, 117
windage loss, 198
winding, 140
concentrated, 260
distributed, 260
distribution factor of, 342
flux linkage of, 157
higher-frequency model of, 463
low-frequency model of, 157
mutual coupling, 263
resistance of, 157
voltage across, 157
wound-rotor induction machine, 197
wound-rotor synchronous machine, 197